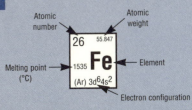

IIIA	IVA	VA	VIA	VIIA	1s²
5 10.81 **B** 2300 (He) 2s²2p	6 12.011 **C** ~3500 (He)2s²2p²	7 14.0067 **N** -209.9 (He) 2s²2p³	8 15.9994 **O** -218.4 (He) 2s²2p⁴	9 18.998403 **F** -219.6 (He) 2s²2p⁵	10 20.179 **Ne** -248.7 (He) 2s²2p⁶

Group headers: IIIA, IVA, VA, VIA, VIIA

B (5, 10.81, 2300, (He) 2s²2p)
C (6, 12.011, ~3500, (He)2s²2p²)
N (7, 14.0067, -209.9, (He) 2s²2p³)
O (8, 15.9994, -218.4, (He) 2s²2p⁴)
F (9, 18.998403, -219.6, (He) 2s²2p⁵)
Ne (10, 20.179, -248.7, (He) 2s²2p⁶)

Al (13, 26.98154, 660, (Ne) 3s²3p)
Si (14, 28.0855, 1415, (Ne)3s²3p²)
P (15, 30.97376, 44.1, (Ne)3s²3p³)
S (16, 32.06, 112.8, (Ne)3s²3p⁴)
Cl (17, 35.453, -100.9, (Ne)3s²3p⁵)
Ar (18, 39.948, -189.2, (Ne)3s²3p⁶)

IB, IIB

Cu (29, 63.546, 1083, (Ar) 3d¹⁰4s)
Zn (30, 65.38, 419.6, (Ar) 3d¹⁰4s²)
Ga (31, 69.72, 29.8, (Ar)3d¹⁰4s²4p)
Ge (32, 72.59, 937, (Ar)3d¹⁰4s²4p²)
As (33, 74.9216, 817.4, (Ar)3d¹⁰4s²4p³)
Se (34, 78.95, 217.4, (Ar)3d¹⁰4s²4p⁴)
Br (35, 79.904, -7.2, (Ar)3d¹⁰4s²4p⁵)
Kr (36, 83.80, -156.6, (Ar)3d¹⁰4s²4p⁶)

Ag (47, 107.868, 962, (Kr) 4d¹⁰5s)
Cd (48, 112.41, 321, (Kr) 4d¹⁰5s²)
In (49, 114.82, 156.6, (Kr)4d¹⁰5s²4p)
Sn (50, 118.69, 232, (Kr)4d¹⁰5s²5p²)
Sb (51, 121.75, 630, (Kr)4d¹⁰5s²5p³)
Te (52, 124.60, 449, (Kr)4d¹⁰5s²5p⁴)
I (53, 126.9045, 113.5, (Kr)4d¹⁰5s²5p⁵)
Xe (54, 131.30, -111.9, (Kr)4d¹⁰5s²5p⁶)

Au (79, 196.9665, 1064, (Xe)4f¹⁴5d¹⁰6s)
Hg (80, 200.59, -38.9, (Xe)4f¹⁴5d¹⁰6s²)
Tl (81, 204.37, 303.5, (Xe)4f¹⁴5d¹⁰6s²6p)
Pb (82, 207.2, 327.5, (Xe)4f¹⁴5d¹⁰6s²6p²)
Bi (83, 208.9804, 271, (Xe)4f¹⁴5d¹⁰6s²6p³)
Po (84, (209), 254, (Xe)4f¹⁴5d¹⁰6s²6p⁴)
At (85, (210), 113.5, (Xe)4f¹⁴5d¹⁰6s²6p⁵)
Rn (86, (222), -71, (Xe)4f¹⁴5d¹⁰6s²6p⁶)

Tb (65, 158.9254, 1356, (Xe)4f⁸5d6s²)
Dy (66, 162.50, 1412, (Xe)4f¹⁰6s²)
Ho (67, 164.9304, 1474, (Xe)4f¹¹6s²)
Er (68, 167.26, 1529, (Xe)4f¹²6s²)
Tm (69, 168.9342, 1545, (Xe)4f¹³6s²)
Yb (70, 173.04, 819, (Xe)4f¹⁴6s²)
Lu (71, 174.967, 1663, (Xe)4f¹⁴5d6s²)

Bk (97, (247), (Rn)5f⁸6d7s²)
Cf (98, (251), (Rn)5f⁹6d7s²)
Es (99, (252))
Fm (100, (257))
Md (101, (258))
No (102, (259))
Lr (103, (260))

Atomic number — Atomic weight — Element — Melting point (°C) — Electron configuration

26   55.847
 Fe
 1535
 (Ar) 3d⁶4s²

ENGINEERING MATERIALS SCIENCE

ENGINEERING MATERIALS SCIENCE

Milton Ohring

Department of Materials Science and Engineering
Stevens Institute of Technology
Hoboken, New Jersey

ACADEMIC PRESS

San Diego   New York   Boston   London   Sydney   Tokyo   Toronto

Front Cover Photograph: Single crystal superalloy turbine blades used in a jet aircraft engine (courtesy of Howmet Corporation).
Back Cover Photograph: Pentium microprocessor integrated circuit chip (courtesy of INTEL Corporation).
Both products rank among the greatest materials science and engineering triumphs of the 20th century.

This book is printed on acid-free paper. ∞

Academic Press, Inc.
A Division of Harcourt Brace & Company
525 B Street, Suite 1900, San Diego, California 92101-4495

United Kingdom Edition published by
Academic Press Limited
24-28 Oval Road, London NW1 7DX

Library of Congress Cataloging-in-Publication Data

Ohring, Milton, date.
    Engineering materials science / by Milton Ohring.
        p.   cm.
    Includes index.
    ISBN 0-12-524995-0
    1. Materials science.   2. Materials.   I. Title.
TA403.037   1995
620.1'1--dc20   94-40228
                                                                 CIP

PRINTED IN THE UNITED STATES OF AMERICA
95   96   97   98   99   00   DO   9   8   7   6   5   4   3   2   1

This book is dedicated to the women in my life;

Ahrona,

   Feigel, Rochelle, and Yaffa,

      and to the memory of my mother, Mollie.

Human existence depends on compassion and knowledge.
Knowledge without compassion is inhuman; compassion
without knowledge is ineffective.

Victor F. Weisskopf

CONTENTS

3 STRUCTURE OF SOLIDS

4 POLYMERS, GLASSES, CERAMICS, AND NONMETALLIC MIXTURES

5 THERMODYNAMICS OF SOLIDS

6 KINETICS OF MASS TRANSPORT AND PHASE TRANSFORMATIONS

7 MECHANICAL BEHAVIOR OF SOLIDS

8 MATERIALS PROCESSING AND FORMING OPERATIONS

9 HOW ENGINEERING MATERIALS ARE STRENGTHENED AND TOUGHENED

10 DEGRADATION AND FAILURE OF STRUCTURAL MATERIALS

11

ELECTRICAL PROPERTIES OF METALS, INSULATORS, AND DIELECTRICS

12

SEMICONDUCTOR MATERIALS AND DEVICES: SCIENCE AND TECHNOLOGY

13

OPTICAL PROPERTIES OF MATERIALS

14 MAGNETIC PROPERTIES OF MATERIALS

15 FAILURE AND RELIABILITY OF ELECTRONIC MATERIALS AND DEVICES

PREFACE

The title *Engineering Materials Science* concisely describes the contents of this book and reflects its point of view. Long before there was a *science* to characterize them, man-made *engineering* materials were used for structural applications, to ease the drudgery of existence, and to enrich life. It was only in the late 19th century that scientific methods allowed a better understanding of engineering materials already in use for millennia. The result was improvement in their properties and more reliable ways of producing them. But science constantly advances, and it was not long before totally new materials like polymers, nuclear materials, semiconductors, and composites—and innovative ways to process and characterize them—were developed. Today, materials engineering and science advances are more closely synchronized. Often, however, science surges ahead, as with the discovery of high-temperature superconductors, making it necessary to establish engineering goals for their exploitation. Conversely, engineering progress frequently raises scientific questions that must be answered to advance the state of the art, for example, in the manufacture of integrated circuits. As reflected in the book's title, these divisions between the science and the engineering are often blurred and sometimes seamless.

This book has three broad objectives and missions. The first is to present a panoramic sweep of the materials field in order to frame it within the borders of the larger scientific and engineering disciplines. By introducing the vocabulary of materials, the way they are characterized, and the concerns involved in producing myriad reliable products from them, I hope to foster a cultural appreciation of the role materials play in technology and society.

The second is to use the subject of materials as the vehicle for connecting the prior freshman/sophomore science courses to the engineering courses that follow. Because the materials field lies at the interface between the science and engineering disciplines, it is admirably suited to serve as the connection in engineering curricula. Therefore, the book aims to expose the practical implications of abstract scientific formulas, to provide realistic limits to ideal models and systems, and to bolster the engineering design process. The transition from theory to practice is evident in the many practical applications presented.

Finally, I have attempted to better prepare those of you who will grapple with day-to-day materials problems in your professional engineering careers. The issues will most likely involve designing the properties needed for a particular component or application, selecting the best material for the job, choosing a manufacturing process to make it, and analyzing why it failed or broke in service. You may then find, as I have, that virtually every engineering problem is ultimately a materials problem. The best preparation for confronting problems that materials pose is a thorough grounding in the scientific fundamentals that underscore their nature. These invariant principles are also the basis for developing totally new materials and addressing the host of future challenges and opportunities they present.

This book is intended to be an introductory text for engineering undergraduates of all disciplines. The only prerequisites assumed are typical preengineering courses: freshman chemistry, physics, and mathematics. The introductory chapter, which I hope will not be glossed over, defines the roles of materials within broad engineering and societal contexts. Chapter 2 provides a modern view of the *electronic* structure of solids and how atoms are bound in them. The *physical* structure or location of atoms in the important engineering solids—metals, polymers, ceramics, and semiconductors—is the primary subject of Chapters 3 and 4. Minimization of energy and the consequences of thermodynamic equilibrium are overriding constraints that govern the stability of material systems. They are dealt with in Chapter 5, while its companion, Chapter 6, is concerned with time-dependent processes involving atom movements that enable structural and chemical modification of materials. The scientific underpinning provided by Chapters 2–6 is part of the core subject matter that will, to a greater or lesser extent, be part of the syllabus of all introductory materials courses.

Chapters 7–10 are devoted to the *mechanically* functional materials—their properties and behavior, how they are processed, how they can be strengthened and toughened, and how they degrade and fail in service. Similarly, Chapters 11–15 focus on the behavior, processing, and reliability of *electrically, optically,* and *magnetically* functional materials and devices. The properties and performance of these electronic materials are a testament to the exploitation of their structures through highly controlled processing. These last nine chapters present two broad avenues of occasionally linked engineering applications.

It is customary for textbooks to include more material than can be covered in a one-semester course, and the same is true here. This should enable instructors to tailor the subject matter to courses with different student audiences. Importantly, it will allow interested readers to explore topics that range beyond the confines of the course syllabus. Regardless of the intended student population, however, the book emphasizes fundamentals. Frequently, the focus is on a single class of materials that display unique properties, but just as often properties of different materials are compared. Therefore, "all you ever wanted to know about ceramics" will not be found in a single chapter but is distributed throughout several relevant chapters. The same is true for metals, polymers, and semiconductors. This comparative approach enables parallels in material behavior to be recognized amid clear differences in their individual natures. Conversely, it allows important distinctions among outwardly similar behaviors to be made. The continual juxtaposition of microscopic and macroscopic points of view sharpens and illuminates these issues. Thus, at times (microscope-like) examination of phemomena at atomic and even subatomic levels is required, but often the unaided eye is fully capable of discerning and analyzing bulk (macroscopic) material behavior. Throughout I have tried to promote the subject matter and above viewpoint in as lively a manner as possible.

More than 80 illustrative examples are fully worked out in the text. In addition there are over 500 problems and questions for student assignments dealing with analysis, materials design, and materials selection. Answers to a selected number of problems are also included.

The copyrighted interactive software that accompanies this book contains modules on different materials topics and is both computationally and graphically oriented. It runs on IBM-compatible computers and, once learned, can help solve many numerical problems presented in this as well as other texts on the subject.

ACKNOWLEDGMENTS

Authors fear that they may not show sufficient awareness of their debt to the people who make their books possible. In this vein I hope that I will be forgiven by the many for citing only the few. Among my colleagues I would like to thank Professor Edward Whittaker for conscientiously reading and reviewing several chapters on electronic properties. I am additionally grateful to Dr. W. Moberly, and Professors W. Carr, D. Smith, and A. Freilich for their critical comments. However, the bulk of the review process was performed by many excellent but anonymous reviewers who are the unsung heroes of this book. They offered encouragement, and much good criticism and advice which I tried to incorporate. I am solely responsible for any residual errors in my understanding of concepts and wording of text.

Some very special people to whom I am indebted sent me the beautiful photographs that elevate and give life to the text. In particular, the contribution of George Vander Voort must be singled out. This exceptional metallographer not only provided many excellent micrographs but taught me a bit of metallurgy as well. I also wish to acknowledge Dr. R. Anderhalt and Mr. Jun Yeh for their respective scanning electron and optical micrographs.

The software was the result of a multi-man-year team effort in which major contributions were made by Dmitry Genken, Tom Harris, Dr. Dan Schwarcz, and Eugene Zaremba. Professors W. Carr and D. Sebastian are owed thanks for evaluating the software.

I am grateful to all of those at Academic Press who had a part in producing this book including Jane Ellis who encouraged me to undertake its writing, Dr. Zvi Ruder for his support of this project, and Deborah Moses who so capably guided and coordinated the complex publishing process. Others who deserve thanks for many favors are Noemia Carvalho, Pat Downes, Dick Widdecombe, and Dale Jacobson.

INTRODUCTION TO MATERIALS SCIENCE AND ENGINEERING

1.1. MATERIALS RESOURCES AND THEIR IMPLICATIONS

1.1.1. A Historical Perspective

The designation of successive historical epochs as the Stone, Copper, Bronze, and Iron Ages reflects the importance of materials to mankind. Human destiny and materials resources have been inextricably intertwined since the dawn of history; however, the association of a given material with the age or era that it defines is not only limited to antiquity. The present nuclear and information ages owe their existences to the exploitation of two remarkable elements, uranium and silicon, respectively. Even though modern materials ages are extremely time compressed relative to the ancient metal ages they share a number of common attributes. For one thing, these ages tended to define sharply the material limits of human existence. Stone, copper, bronze, and iron meant successively higher standards of living through new or improved agricultural tools, food vessels, and weapons. Passage from one age to another was (and is) frequently accompanied by revolutionary, rather than evolutionary, changes in technological endeavors.

It is instructive to appreciate some additional characteristics and implications of these materials ages. For example, imagine that time is frozen at 1500 BC and we focus on the Middle East, perhaps the world's most intensively excavated region with respect to archaeological remains. In Asia Minor (Turkey) the ancient Hittites were already experimenting with iron, while close by to the east in Mesopotamia (Iraq), the Bronze Age was in flower. To the immediate north in Europe, the south in Palestine, and the west in Egypt, peoples were enjoying the benefits of the Copper and Early Bronze Ages. Halfway around the world to the east, the Chinese had already melted iron and demonstrated a remarkable genius for bronze, a copper–tin alloy that is stronger and easier to cast than pure copper. Further to the west on the Iberian Peninsula (Spain and Portugal), the Chalcolithic period, an overlapping Stone and Copper Age held sway, and in North Africa survivals of the Late Stone Age were in evidence. Across the Atlantic Ocean the peoples of the Americas had not yet discovered bronze, but like others around the globe, they fashioned beautiful work in gold, silver, and copper, which were found in nature in the free state (i.e., not combined in oxide, sulfide, or other ores).

Why materials resources and the skills to work them were so inequitably distributed cannot be addressed here. Clearly, very little technological information diffused or was shared among peoples. Actually, it could not have been otherwise because the working of metals (as well as ceramics) was very much an art that was limited not only by availability of resources, but also by cultural forces. It was indeed a tragedy for the Native Americans, still in the Stone Age three millennia later, when the white man arrived from Europe armed with steel (a hard, strong iron--carbon alloy) guns. These were too much of a match for the inferior stone, wood, and copper weapons arrayed against them. Conquest, colonization, and settlement were inevitable. And similar events have occurred elsewhere, at other times, throughout the world. Political expansion, commerce, and wars were frequently driven by the desire to control and exploit materials resources, and these continue unabated to the present day.

When the 20th century dawned the number of different materials controllably exploited had, surprisingly, not grown much beyond what was available 2000 years earlier. A notable exception was steel, which ushered in the Machine Age and revolutionized many facets of life. But then a period ensued in which there was an explosive increase in our understanding of the fundamental nature of materials. The result was the emergence of polymeric (plastic), nuclear, and electronic materials, new roles for metals and ceramics, and the development of reliable ways to process and manufacture useful products from them. Collectively, this modern Age of Materials has permeated the entire world and dwarfed the impact of previous ages.

Only two representative examples of a greater number scattered throughout the book will underscore the magnitude of advances made in materials within a historical context. In Fig. 1-1 the progress made in increasing the strength-to-density (or weight) ratio of materials is charted. Two implications of these advances have been improved aircraft design and energy savings in transportation systems. Less visible but no less significant improvements made in abrasive and cutting tool materials are shown in Fig. 1-2. The 100-fold tool cutting speed increase in this century has resulted in efficient machining and manufacturing

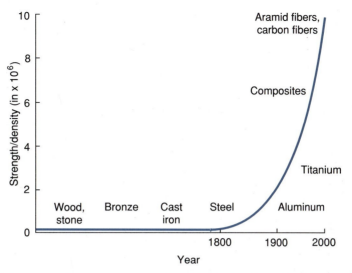

FIGURE 1-1 Chronological advances in the strength-to-density ratio of materials. Optimum safe load-bearing capacities of structures depend on the strength-to-density ratio. The emergence of aluminum and titanium alloys and, importantly, composites is responsible for the dramatic increase in the 20th century. Reprinted with permission from *Materials Science and Engineering for the 1990s*. Copyright 1989 by the National Academy of Sciences. Courtesy of the National Academy Press, Washington, D.C.

processes that enable an abundance of goods to be produced at low cost. Together with the dramatic political and social changes in Asia and Europe and the emergence of interconnected global economies, the prospects are excellent that more people will enjoy the fruits of the earth's materials resources than at any other time in history.

1.1.2. The Materials Cycle

The United States has an enormous investment in the scientific and engineering exploitation of materials resources. A very good way to visualize the scope of this activity is to consider the **total materials cycle** shown in Fig. 1-3. The earth is the source of all resources that are mined, drilled for, grown, harvested, and so on. These virgin raw materials (ores, oil, plants, etc.) must undergo a first level of processing that involves extraction, refining, and purification. Bulk quantities of metals, chemicals, fibers, and other materials are then made available for further processing into so-called engineering materials like alloys, ceramics, and polymers. In converting **bulk** materials into **engineering** materials, we give them extra worth or **added value.** Cheap bulk materials in relatively simple form (e.g., ingots, powder) and large tonnage are thus transformed into more costly products that have more desirable compositions or higher levels of purity or more convenient shapes. Next, further processing, manufacturing, and assembly operations are required to create the finished products that we

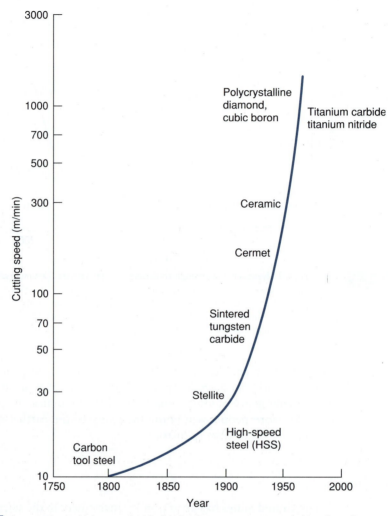

FIGURE I-2 Increase in machining speed with the development over time of the indicated cutting tool materials. Adapted from M. Tesaki and H. Taniguchi, High speed cutting tools: Sintered and coated, *Industrial Materials* **32**, 64 (1984).

use, such as cars, computers, consumer electronics, machine tools, sporting goods, and structures. Value, often considerable in magnitude, is again added to the product. These goods are used until either they cease to function because of degradation or failure (corrosion, mechanical breakage, etc.) or they become obsolete because of the appearance of new, more efficient, and cheaper models. At this point they are discarded as junk. In the past they became part of the world's garbage dumps, completing the cycle. Today it is increasingly more common to recycle these junked articles and reclaim the valuable materials contained within them. Not only is this cost effective, but the environmental imperatives are obvious. Short-circuiting the total materials cycle through recycling is not new. It has been practiced since antiquity for the precious metals, and for centuries in the case of copper alloys and steel.

The right side of Fig. 1-3 defines the special arena of materials science and engineering; aside from a few issues discussed in this chapter the remainder of the book is concerned with this same area of activity.

1.1.3. Materials, Energy, and the Environment

Exploitation of materials resources has an impact on the other resources of the earth. As Fig. 1-4 suggests, **materials, energy,** and the **environment** are the broad natural resources used by humans to fashion their way of life. The complex triangular interactions among these resources, mediated by governments, serve to encompass and limit all human industrial activities. An interesting historical example of this interplay that occurred between 1450 and 1600 in England (and elsewhere in Europe as well) led to the great **Timber Famine.** The confluence of several factors—voyages of discovery with large-scale build-

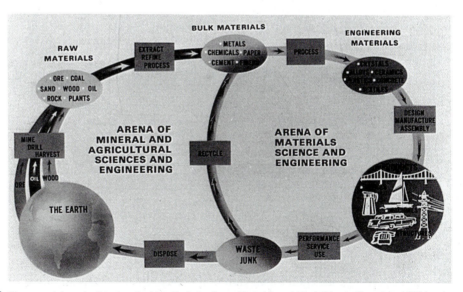

FIGURE I-3 The materials cycle. Reprinted with permission from *Materials and Man's Needs.* Copyright 1975 by the National Academy of Sciences. Courtesy of the National Academy Press, Washington, D.C.

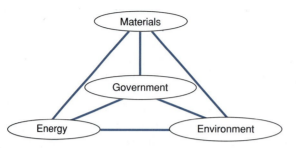

Interactions among materials, energy, and environmental resources.

ing of wooden ships outfitted with iron and bronze armaments, the huge demand for books (and paper) following the invention of printing with movable type, and the general rise in living standard during the Renaissance—created an insatiable demand for both metals and energy. Charcoal, produced from wood, was the major source of this energy. Iron production was grossly inefficient in terms of energy usage: One ton of product required 200 cords of wood! Before long, large forest tracts fell to the ax. Deserts were created and they spread in Spain and North Africa. Salvation for this crisis came only with the use of an alternative source of energy, namely, coal.

A current example of resource interactions involving production of copper is instructive. The concentration of copper in exploitable ores in the United States has fallen to less than 0.6%. To produce $\sim 1.7 \times 10^6$ tons (1.55×10^9 kg) of refined copper (the annual need), about one billion tons of rock must be excavated. If piled 100 ft (30 m) high, this mass would cover some 6.5 square miles (16.8 km^2). In addition, 220×10^{12} BTU (2.32×10^{11} MJ), or about 0.2% of the nation's total energy consumption, is required. Lastly, 3.3×10^6 tons of slag and 3.5×10^6 tons of sulfur dioxide gas are produced, adversely affecting the environment.

Additional important and sobering information on our materials resources is found in Table 1-1, Fig. 1-5, and Table 1-2. These deal with the potential availability of the elements, their geographic distribution, and the energy required to extract them. The most abundant elements in the earth's crust are listed in Table 1-1. An often asked question is: How long will a given resource last? This of course depends on the rate at which it is being consumed. Some materials are presently being consumed at an exponential rate.

The United States' reliance on imported raw materials is depicted together with sources of supply in Fig. 1-5. Each nation has its own list of materials it must import. Included are many critical metals that act like vitamins to strengthen steel and other alloys used in tools and dies, jet and rocket engines, assorted machinery, and processing equipment.

Table 1-2 lists the amounts of energy required to produce a number of materials. The data are most complete for metals produced from either oxide

TABLE I-I	ABUNDANCE (WEIGHT PERCENT) AND DISTRIBUTION OF ELEMENTS ON EARTH[a]

Crust		Seawater		Atmosphere	
Oxygen	47	Oxygen	85.7	Nitrogen	79
Silicon	27	Hydrogen	10.8	Oxygen	19
Aluminum	8	Chlorine	1.9	Argon	2
Iron	5	Sodium	1.05	Carbon dioxide	0.04
Calcium	4	Magnesium	0.127		
Sodium	3	Sulfur	0.089		
Potassium	3	Calcium	0.040		
Magnesium	2	Potassium	0.038		
Titanium	0.4	Bromine	0.0065		
Hydrogen	0.1	Carbon	0.003		
Phosphorus	0.1	Strontium	0.0013		
Manganese	0.1	Boron	0.00046		
Fluorine	0.06				
Barium	0.04				
Strontium	0.04				
Sulfur	0.03				
Carbon	0.02				

[a] The total mass of the crust to a depth of 1 km is 3×10^{21} kg; the mass of the oceans is 10^{20} kg; that of the atmosphere is 5×10^{18} kg.
From M. F. Ashby and D. R. Jones, *Engineering Materials 1: An Introduction to Their Properties and Applications*, Pergamon Press, Oxford (1980).

TABLE I-2	PROCESS ENERGY FOR MATERIALS PRODUCTION

	Energy in 10^6 BTU per ton			
	Process energy[a]	Theoretical energy[b]	Efficiency (%)[c]	From-Scrap energy[d]
Titanium	359	16.0	4.4	
Magnesium ingot	339	20.1	5.9	10
Aluminum ingot	197	25.2	12.8	12
Nickel	89	3.1	3.5	15
Copper	48	1.8	3.7	18
Zinc	48	4.2	8.7	18
Steel slabs	22	5.7	26.0	13
Lead	18	0.8	4.4	10
Silicon	74	25.1	34	
Polyethylene (high density)[e]	23			
Polyethylene (low density)[e]	46			
Glass[e]	15			

[a] Based on data from Battelle–Columbus Laboratories. The process energy is for oxide ore reduction to metal. Mining and ore beneficiation are not included.
[b] The theoretical energy is the **free energy** of oxide formation (see Section 5.2.1.).
[c] Percent efficiency = (theoretical energy/process energy) × 100.
[d] Industrial estimates.
[e] Polymer and glass process energies are based on Stanford Research Institute estimates.
Source: All data except for that in bottom three rows from H. H. Kellogg, *Journal of Metals*, p. 30 (Dec. 1976).

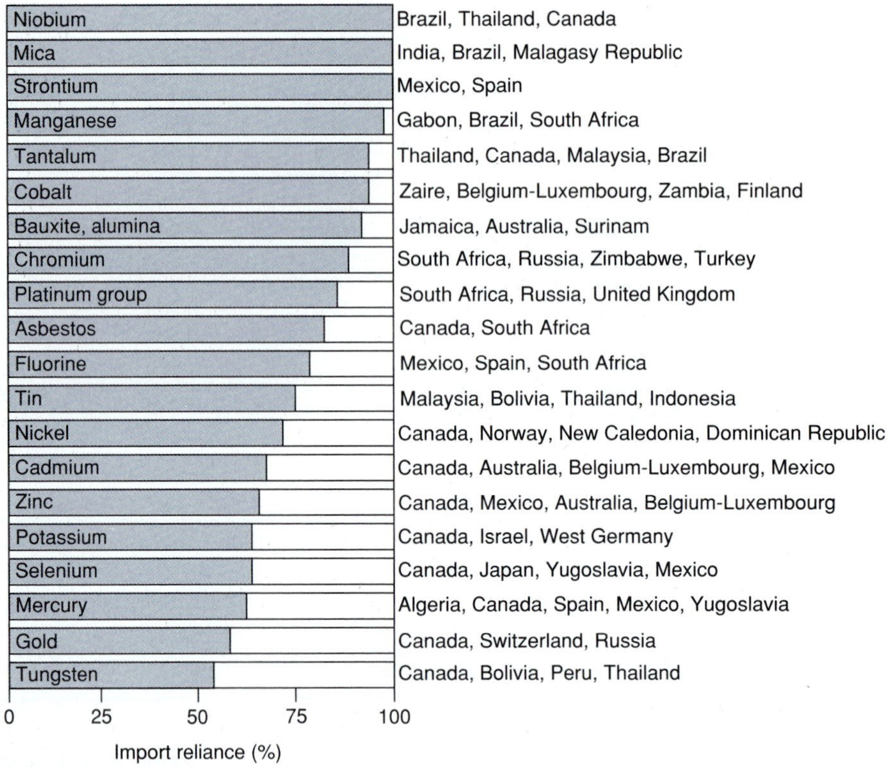

Material	Import reliance (%)	Countries
Niobium		Brazil, Thailand, Canada
Mica		India, Brazil, Malagasy Republic
Strontium		Mexico, Spain
Manganese		Gabon, Brazil, South Africa
Tantalum		Thailand, Canada, Malaysia, Brazil
Cobalt		Zaire, Belgium-Luxembourg, Zambia, Finland
Bauxite, alumina		Jamaica, Australia, Surinam
Chromium		South Africa, Russia, Zimbabwe, Turkey
Platinum group		South Africa, Russia, United Kingdom
Asbestos		Canada, South Africa
Fluorine		Mexico, Spain, South Africa
Tin		Malaysia, Bolivia, Thailand, Indonesia
Nickel		Canada, Norway, New Caledonia, Dominican Republic
Cadmium		Canada, Australia, Belgium-Luxembourg, Mexico
Zinc		Canada, Mexico, Australia, Belgium-Luxembourg
Potassium		Canada, Israel, West Germany
Selenium		Canada, Japan, Yugoslavia, Mexico
Mercury		Algeria, Canada, Spain, Mexico, Yugoslavia
Gold		Canada, Switzerland, Russia
Tungsten		Canada, Bolivia, Peru, Thailand

Import reliance (%): 0   25   50   75   100

FIGURE 1-5   U.S. reliance on imported raw materials and the countries that supply them. Included are metals critical to defense needs and aerospace industries. Data adapted from U.S. Bureau of Mines.

ores or scrap. Actual process energies are compared with theoretical amounts needed assuming 100% efficiency. Note that immense amounts of energy are required to reduce aluminum oxide to aluminum metal, a fact reflected by siting smelters near sources of cheap hydroelectric power. Another interesting example in this regard is silicon, one of the most abundant elements on earth. The energy figure given in Table 1-2 is for the production of metallurgical-grade silicon. For making semiconductor chips, however, electronic-grade silicon (EGS) possessing extraordinary purity is required, and this means additional large expenditures of energy. It has been estimated that 620 kilowatt hours (kWh) of energy is required to produce 1 kg of bulk EGS. As 1 kWh = 3413 BTU (3.60 MJ), on a one-ton basis this translates to 1920×10^6 BTU.

An important lesson learned from Table 1-2 is that substantial energy savings can be realized when metal scrap is remelted. Recycling offers a very satisfying and necessary way to conserve our precious materials and energy resources in an environmentally acceptable manner. This is particularly true for polymers, which are a latecomer to recycling technology. The following example illustrates the potential energy savings resulting from recycling aluminum.

EXAMPLE 1-1

a. Assuming Al has an average heat capacity of 25 J/mol-K (law of Dulong and Petit), latent heat of fusion of 10.8 kJ/mol, and melting point of 660°C, what is the *minimum* energy required to melt 1 ton of recycled aluminum beverage cans?

b. What is the minimum energy required to reduce Al_2O_3 to Al?

ANSWER a. On a 1-mol or 27-g basis, the energy E absorbed in heating Al from 25°C to 660°C is E = 25 J/mol-K × (660°C − 25°C) = 15.9 kJ/mol. To this, 10.8 kJ must be added for melting, yielding a total of 26.7 kJ/mol. One ton is equal to 2000 lb × 454 g/lb = 9.08×10^5 g. Therefore, heating and melting 1 ton of Al require 26.7 kJ/mol × (9.08×10^5 g/27 g/mol) = 8.98×10^5 kJ (8.50×10^5 BTU).

b. Some elementary chemistry texts address such problems. Reference to Fig. 5-2 can provide additional understanding. The chemical (free) energy involved in reducing Al_2O_3 by the reaction

$$\tfrac{2}{3} Al_2O_3 = \tfrac{4}{3} Al + O_2$$

to make $\tfrac{4}{3}$ mol of virgin Al is about 250 kcal when referred to 298 K. This corresponds to 785 kJ/mol or 2.64×10^7 kJ/ton (25×10^6 BTU/ton). Thus, only about (8.98×10^5 kJ/ton)/(2.64×10^7 kJ/ton) × 100 = 3.4% of the energy required to create Al from ore is needed to produce it by melting scrap.

In practice these estimates are somewhat optimistic and more energy is required to produce Al from scrap. Rather than 3.4%, 6% is more realistic, but even this amount represents dramatic savings.

Not only is energy consumed to produce materials, but it takes energy to make convenient sources of energy! The solar cell is a case in point.

EXAMPLE 1-2

The energy required to produce a complete solar cell module is estimated to be 2170 kWh per square meter. How long will it take to recover this energy if an average of 1 kW of solar power falls on each square meter of earth? Assume that the solar panel operates only 5 hours a day, is 12% efficient, and has an area of 1 m².

ANSWER In 1 day the amount of solar energy converted into electrical energy is 1 kW × 5 h × 0.12 = 0.6 kWh. Therefore, the number of days required is

2170 kWh/0.6 kWh/day = 3617 days. The module must operate almost 10 years before it recovers the energy used for its production.

Similar material–energy budgets must be considered for other energy generation strategies.

1.2. MATERIALS AND ENGINEERING

1.2.1. Materials–Design–Processing

Engineers must integrate the triad of terms shown in Fig. 1-6 to create reliable manufactured or fabricated goods. It is important to stress that these factors must be optimally integrated, simultaneously. Good designs of components and systems are defeated if fabrication is impractical or if the wrong materials are selected. Aside from misuse and natural catastrophes, virtually all examples of failure in structures, machine components, consumer goods, and so on are attributable to faults in, or a lack of optimization of, one or a combination of factors related to design, processing, or materials properties. Spectacular examples of failure caused by all three contributing factors occurred on all too many of the 5000 welded steel merchant ships during the Second World War. A dozen ships actually broke in half, like that shown in Fig. 1-7; another 1000 sustained cracks of varying severity. This damage was not the result of enemy attack but rather occurred in cold seawater, sometimes while the ships were moored at dockside. Square bulwark plates that acted as stress concentrators were deemed to constitute poor **design.** Misplaced weld strikes and insufficient weld penetration initiated cracks, exposing **processing** defects. Lastly, and most importantly, the steel **material** selected had insufficient toughness and displayed a fatal tendency to fracture in a brittle manner at slightly cool water temperatures.

Much more is at stake than the embarrassment of engineers who have assured us of the safety of their designs. Loss of life, the legal implications of product liability, and failure to compete in the marketplace are the legacies of

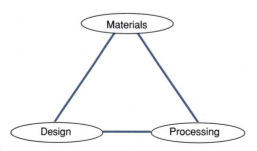

FIGURE I-6   Interactions among materials, design, and processing in manufacturing.

FIGURE 1-7   Tanker that broke in two at dock, viewed from port side. U.S. Coast Guard photograph.

not confronting the subtle issues of materials selection, properties, quality, and reliability.

1.2.2. Processing–Structure–Properties–Performance

It is now popular to subdivide the content of the discipline of materials science and engineering into four major themes: processing, structure, properties, and performance. Indeed, many of the chapters in this book can readily be identified with one or another of these broad topics. Equal in importance to their individual natures, however, is the synergistic manner in which they interact and influence one another. Clearly, the **performance** of a transistor reflects the **processing** or way it was manufactured, which in turn influences the atomic and electronic **structure** of the constituent materials and the **properties** each exhibits. It is instructive to delineate briefly the attributes of these themes.

Structure–properties.   The scientific core of the subject is embodied in structure–property relationships that are intrinsic to the very nature of materials. Humans have always been able to recognize such properties as hardness, mechanical strength, ductility or malleability, reflectivity, and color and roughly distinguish between the properties of different materials. But to understand the nature of materials properties at the level needed to alter them controllably and predictably requires a deep knowledge of structure. Despite our long involvement with materials, this understanding of atomic and crystal structures has emerged only in this century.

Processing.   Casting and mechanical shaping of metals, melting and blowing of glass, and firing and glazing of ceramics are examples of traditional materials processing. Each was an art (as well as the substance of art) for a long time, but now science is in the ascendency. Modern scientific experimental techniques have made precise measurement of time-dependent changes in composition, structure, and properties possible, enabling a better understanding of the stages of change that materials experience during their production. And if change cannot be measured, computers can now be programmed to model and predict materials behavior during processing. Such information can then be used to control processes, enabling greater and more consistent yields of desired products. It is important to reiterate that processing is the only way to add value to materials.

Performance.   The litmus test of engineering is frequently whether a component or system survives the rigors of field use. Provided there are no design flaws, it is up to the materials to meet performance requirements. Processing, of course, influences the latter by fixing the structure and the properties derived from them. Because of the broader implications of performance, aspects of failure and reliability of both mechanical and electronic materials and components are extensively addressed later in this book.

In summary, the links between **structure–properties** and basic science, at one extreme, and between **performance** and societal needs, at the other, are schematically depicted in Fig. 1-8. In this representation **processing** connects scientific ideas and practical understanding to create the vitality necessary to advance the field of materials.

1.2.3. Materials Selection

A key function of materials specialists (e.g., scientists, metallurgists, ceramists, polymer engineers) is to synthesize useful materials and produce them in bulk quantities. On the other hand, civil, mechanical, chemical, and electrical engineers must frequently **select** materials to:

1. Implement designs for new products.
2. Improve existing products. In this case, new materials can cut production costs and rectify design shortcomings.

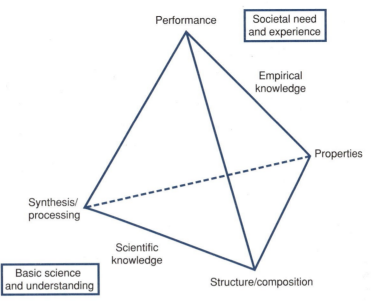

Performance

Societal need and experience

Empirical knowledge

Properties

Synthesis/ processing

Scientific knowledge

Basic science and understanding

Structure/composition

FIGURE 1-8 Scope of materials science and engineering. After M. C. Flemings and D. R. Sadoway, in *Frontiers of Materials Education*, MRS Proc., Vol. 66 (1985).

3. Address emergency situations. These can include failure of the product with loss of consumer confidence, production problems, and shortages in the sources of supply requiring replacement materials.

Thousands of materials are potentially available in assorted sizes and shapes but, not infrequently, the choice usually dwindles to far fewer candidates. Among the issues that must be considered when making the decision to select an engineering material are the following:

1. **Essential property or properties of interest.** This is frequently the most important consideration. In rare cases there may only be one suitable material. If the need is to reach 4 K, only liquid helium will do. Similarly, for catalyst applications there is presently no good alternative to platinum. More commonly, however, there are materials choices. If, for example, mechanical property requirements are uppermost and a certain strength level is essential, many candidates will probably exist. Then other supplementary property requirements should be considered (e.g., density or weight per unit volume, toughness or resistance to fracture, environmental stability, resistance to elevated temperature softening, hardness and wear resistance, appearance) and rated according to importance.

2. **Fabricability.** If materials cannot be readily fabricated or assembled, they cease to be viable choices. Included under fabricability is the ability to be shaped by mechanical processes such as rolling, forging (hammering), and

TABLE 1-3　APPROXIMATE PRICES OF MATERIALS (~1989)

Material	U.S. dollars/ton
Diamonds (industrial high quality)	500,000,000
Platinum	16,500,000
Gold	14,500,000
Silicon	
As extracted	1,300
Single-crystal wafers	10,000,000
Integrated circuits	1,500,000,000
Palladium	4,050,000
Silver	190,000
Carbon fiber-reinforced polymer (CFRP)	170,000
Glass fiber-reinforced polymer (GFRP)	39,000
Glass fibers	1,500
Carbon fibers	45,000
Epoxy resin	6,000
Silicon carbide	
For engineering ceramics	27,000
For abrasives	1,400
For refractory bricks	750
Tungsten	19,500
Cobalt	17,000
Titanium	8,300
Polycarbonate (pellets)	5,300
Brass	
60/40 as sheets	3,750
Ingots	1,650
Aluminum	2,400
Copper	2,600
Steel	
Stainless	2,700
High-speed	1,200
Low alloy	750
Mild—as angles	350
Cast iron (gray)	830
Hardwood	
Teak veneer	1,650
Structural	530
Rubber	
Synthetic	1,400
Natural (commercial)	870
Polystyrene	1,300
Lead (sheet)	1,200
Polyethylene	1,100
Polyvinyl chloride	1,000
Glass	750
Alumina	350
Concrete (reinforced beams)	330
Cement	70

From C. Newey and G. Weaver, *Materials in Action Series: Materials Principles and Practice*, Butterworths, London (1990).

machining, and to be joined by welding, soldering, and other processes. Examples abound of otherwise excellent materials that cannot be drilled or welded. Unlike some materials properties that can be quantified by a numerical value, for example, a melting point of 1083°C or a load-bearing capacity per unit area or tensile strength of 346 MPa (50,000 pounds of force per square inch or psi), fabricability reflects a complex combination of properties. Therefore, it is usually ranked on a qualitative scale, for example, excellent, mediocre, or poor machinability.

3. **Availability.** Optimal materials may not always be available in the desired amount, composition, or shape. Shortages can cause dramatic price rises. An example is cobalt, which is employed in the manufacture of turbine blades and surgical prostheses. In the early 1980s political turmoil in Zaire and Zambia, the major producers, caused the price of the metal to spiral tenfold within days on commodity markets. Materials substitutions are a frequent response to availability problems. The reason is that the user often requires a *property* rather than a *material*. Irrespective of availability, substitutions are part of the natural evolution of materials usage. Numerous examples exist and include use of polyvinyl chloride (PVC) instead of brass and copper tubing in plumbing, and polymer fibers instead of natural materials in clothing.

4. **Cost.** Materials selection is not possible without consideration of costs. Many basic engineering materials are sold as **commodities,** which means that their price is relatively fixed because there are numerous sources of supply. These and other materials are listed together with their approximate prices in Table 1-3. Prices will vary, however, depending on size, shape, and quantity purchased. The increased value of a number of materials due to additional processing should be noted. An outstanding example in this regard is silicon, where the conversion of metallurgical-grade silicon to semiconductor-grade wafers is accompanied by a 7000-fold increase in price. And after processing into integrated circuits, some wafers are worth many times their weight in gold! Engineering materials such as mild (low-carbon) steel, wood, concrete, glass, and certain polymers have found such widespread use because they are relatively cheap. Initial purchase price is not the only cost consideration in materials selection, however. Subsequent labor, production, and assembly costs associated with use of a particular material must also be taken into account.

In summary, a *concurrent* systems approach that synergistically integrates design, manufacturing, and materials selection must be adopted if optimum benefits are to be realized.

1.3. ENGINEERING MATERIALS AND SELECTED APPLICATIONS

In this section a brief survey of important engineering materials and their properties is offered as an introduction to the remainder of the book. By engineering materials we mean those that have survived the ordeal of testing and selection to be incorporated into commercial products and structures. Other classifications of engineering solids exist but our primary concern is with metals, ceramics, polymers, and semiconductors.

1.3.1. Metals

High reflectivity, thermal conductivity (ability to conduct heat) and malleability (ability to be thinned by hammering) combined with low electrical resistance are some of the attributes we associate with metals. The availability of very large numbers of mobile electrons that permeate metals is responsible for facilitating transport of charge and thermal energy. They also establish the electric fields that prevent the penetration of visible light, causing it to be reflected instead. In addition, metals are generally strong (high tensile strength), stiff (large modulus of elasticity), and tough (resistant to fracture). At the same time they are ductile and tend to deform in a plastic manner prior to breaking. Thus metals are readily shaped and formed by mechanical forces. The reason for these desirable mechanical properties is traceable to an atomic bonding that allows atoms to slide readily past one another.

In order of tonnage, irons and steels are the most widely used metals. Next, but far behind, are aluminum and copper together with their numerous alloys. The list of other commercially employed metals is very long and includes nickel, titanium, zinc, tin, lead, manganese, chromium, tungsten, and the precious metals gold, silver, and platinum. Metals, particularly steels and superalloys (containing Ni, Cr, Co, Ti), fill many critical technological functions for which there are no practical substitutes. They are the substances of which machine tools, dies, and energy generation equipment are made. In addition, the reactors and processing equipment used to produce polymers, ceramics, and semiconductors are constructed almost exclusively of metals. Some 99% of a Boeing 747's weight is due to metals (82% aluminum, 13% steel, 4% titanium, 1% fiber glass).

One of the greatest metallurgical achievements in this century has been the development of the modern jet engine. Its severely stressed and thermally exposed components are shown schematically in Fig. 1-9 (see color plate). Actual turbine blades are reproduced on the book's cover; their production and properties are discussed in Chapters 8 and 9. By giving patients a new lease on life, metallic orthopedic implants (Fig. 1-10) that replace damaged bones represent a great advance in modern medical practice. Interestingly, they are made of similar jet engine alloys.

1.3.2. Ceramics

Used by mankind even longer than metals, ceramics are composed of metallic and nonmetallic elements drawn from opposite ends of the Periodic Table. The primary examples include metal oxides (Al_2O_3, TiO_2, ZrO_2, SiO_2, Na_2O,

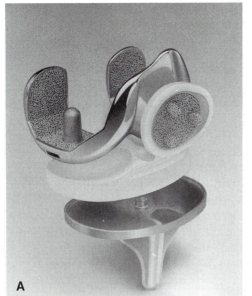

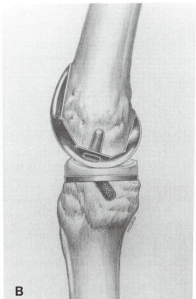

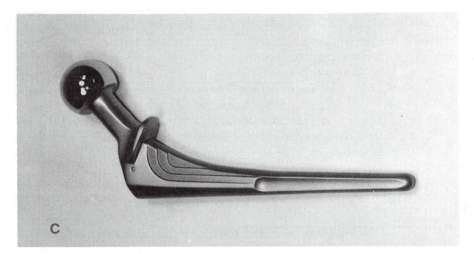

FIGURE 1-10 (A) Porous coated anatomic (PCA) knee prosthesis made of a Co–Cr alloy. The plastic is ultrahigh-molecular-weight polyethylene. Courtesy of C. Lizotte, Howmedica. (B) Method of knee prosthesis fixation. Courtesy of C. Lizotte, Howmedica. (C) Total hip replacement femoral hip stem and head cast made from a Co–Cr–Mo alloy. The prosthesis is implanted with bone cement. Courtesy of J. E. Malayter, Implex.

K_2O, Li_2O, etc.). They are occasionally used singly in products such as fused quartz (SiO_2) tubing and alumina (Al_2O_3) substrates for mounting electronic components. But more frequently, various combinations of these compounds are required to meet the property demands required of various glasses for containers and optical applications, refractory bricks that line industrial furnaces, decorative glazes for pottery, and heat-resistant cookware. Metal carbides (SiC, TiC, WC, BC), nitrides (Si_3N_4, TiN, BN), and borides (TiB_2) are examples of other ceramic materials that have recently assumed critical importance in applications where extreme hardness and/or high-temperature strength are required. This accounts for the widespread use of TiC and TiN coatings on cutting tools to extend their life during machining operations.

Ceramic materials are found in crystalline form, where atoms are geometrically arranged in a repetitive fashion, as well as in the amorphous, noncrystalline state that we associate with glasses. Unlike metals, virtually no electrons are around to conduct either charge or heat, so ceramics serve as good electrical and thermal insulators. Their extremely high melting points signify that their atoms are very strongly bonded to one another. Under applied loading such atoms are not expected to move past each other easily. Thus ceramics and glasses are strong when compressed, but exhibit very little plastic stretching when pulled by tensile forces. Rather, they have a fatal tendency to fracture in a brittle manner under tension. Although strenuous efforts have been made to toughen ceramics, their use as structural materials, aside from concrete, is limited. On the other hand, applications have moved well beyond refractories, bathroom fixtures, and whiteware. Recent decades have witnessed the emergence of many new technical applications for ceramics, including magnets that are electrical insulators, dielectric materials for capacitors and high-frequency circuits, sensors that detect oxygen and water, and assorted devices that convert thermal, stress, and optical stimuli into electrical signals. And of course we must not forget the $YBa_2Cu_3O_7$ high-temperature superconductors that hold so much promise in diverse technologies.

The development of an advanced lightweight silicon nitride gas turbine for automobile and power plant applications illustrates the exciting potential of ceramic materials. Several key components for these purposes are displayed in Fig. 1-11.

1.3.3. Polymers

Close your eyes, reach out, and the chances are good that you have touched a polymer! More polymers are now produced on a volume basis than all of the metals combined, a result of their low average density ($\sim$1.5 Mg/m^3 or 1.5 g/cm^3 compared with $\sim$7.5 Mg/m^3 for typical metals). Their low cost and density, coupled with ease in processing and durability, have been key reasons for the substitution of polymers for metals, glasses, and natural products. The result has been the widespread use of polymers in clothing, fabrics, food and beverage containers, packaging of all kinds, furniture, and sporting equipment. An assortment of polymer products is shown in Fig. 1-12. In one of the more significant trends, the use of plastics in automobiles has increased from $\sim$10% by weight in the late 1970s to about 20% in the early 1990s. Correspondingly,

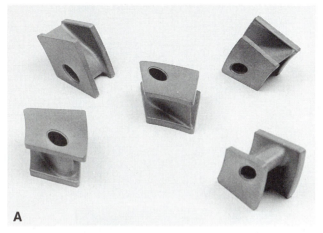

FIGURE 1-11 Silicon nitride turbine components. (A) Gas turbine stator vanes. (B) Turbocharger turbine wheel. (C) Liquid metering valves. Courtesy of AlliedSignal Aerospace.

the weight of U.S. autos has almost halved in the same time from 4000 lb to less than 2500 lb. The impact of this on energy conservation strategies has been dramatic.

Long-chain, very high molecular weight hydrocarbons are the basic building blocks of polymers. The chains either lie together intertwined like a plate of spaghetti strands to create the linear polymers, or are joined at nodes to form three-dimensional networks. Polymers of the latter type (thermosets, e.g., epoxies, phenolics, and silicones) are generally stronger than those of the former type (thermoplastics, e.g., polyethylene, polyvinyl chloride, and nylon). Both tend to have an amorphous structure because the ordering of atoms is not regular. Although there are large variations, they are typically more than 10 times less stiff and strong than metals. In addition, polymers tend to undergo large time-dependent deformations under load, adversely affecting the dimensional stability of components.

To address such shortcomings, polymers have been creatively altered by modifying the chemistry of individual chains and by mixing polymers together. But more importantly, incorporation of thin fibers of glass and carbon into a polymer matrix (typically epoxy and nylon) has resulted in the creation of stiffer and stronger **composites**. These extremely popular and increasingly important, high strength-to-weight ratio engineering materials capitalize on the stiffness of the fiber and the easy formability of an inexpensive polymer matrix. The resulting composite behavior represents a volume average of the beneficial properties of the individual constituents; separately, neither is as useful an engineering material as the composite. The concept of composites has also been extended to metal and ceramic matrices, resulting in an array of high-performance materials that are now widely used in aircraft bodies and components, hulls of small sea craft, automotive parts, and sporting goods (e.g., tennis rackets, skis, fishing rods). A representative collection of applications employing composites, shown in Fig. 1-13, illustrates their pervasive use and the extent to which they have substituted for traditional materials. Composites are extensively treated in Chapter 9.

1.3.4. Semiconductors

Unlike the other materials considered, which have altered our world primarily in evolutionary ways, semiconductors have unleashed a technological revolution in many areas. They have made possible the computers and telecommunica-

---

FIGURE 1-12   Examples of polymer use: (A) Polycarbonate (LEXAN) reusable and recyclable school milk bottles have replaced paperboard cartons. Courtesy of GE Plastics. (B) Fire helmet made of polyetherimide (ULTEM) can withstand a temperature of 260°C (500°F) for 2.5 minutes. Highly light transparent and heat-resistant polyphthalate carbonate (LEXAN) is used for the shield. Courtesy of E. D. Bullard Company and GE Plastics. (C) Passenger-side air bag deployment doors made of Du Pont thermoplastic elastomer (DYM 100) and used on the 1992 Lincoln Town Car. Courtesy of DuPont Automotive.

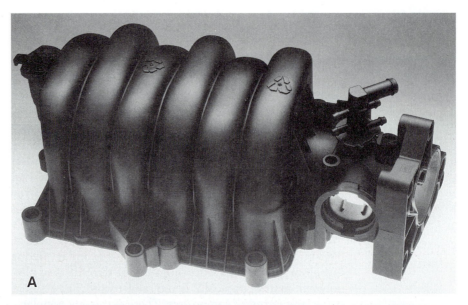

FIGURE 1-13 Examples of composite use: (A) Glass-reinforced nylon (ZYTEL) air intake manifold used on 1993 General Motors V-6 engines. Courtesy of DuPont Automotive. (B) Bicycle whose frame is made of a composite containing carbon fibers in an epoxy resin matrix (TACTIX). Courtesy of The Allsop Company and Dow Plastics (a business group of the Dow Chemical Co.). (C) Voyager aircraft. Most of the structural members and body of this plane were made from graphite–epoxy composites. Courtesy of Hercules Aerospace Products Group and SACMA.

FIGURE I-13 *(continued).*

tions systems that process and transmit huge amounts of information for business, health care, educational, scientific, military, and recreational purposes. Semiconductor devices have added sensitivity, capability, and reliability to electronic sensing, measurement, and control functions, as well as to all consumer electronics equipment.

Elemental silicon is the semiconductor material used for computer logic and memory applications where immense numbers of transistors are grouped together on integrated circuit (IC) chips (Fig. 1-14). In this case each Si wafer undergoes a demanding processing regimen that includes the growth of a very pure glass (**ceramic**) thin film, deposition of **metal** conductors for contacts and interconnections, and use of **polymers** to encapsulate or package the finished IC. Silicon is also used to fabricate discrete semiconductor devices like diodes and transistors; however, for optoelectronic applications requiring lasers, light-emitting diodes, and photodetecting diodes and transistors, compound semiconductors based on gallium arsenide (GaAs) and indium phosphide (InP) are employed.

Electrical conduction in semiconducting materials falls somewhere between ceramic and polymer insulators, on the one hand, and metallic conductors, on

FIGURE 1-14 Integrated circuits at various stages of magnification. For parts A and B, see color plates. (C) Scanning electron microscope image of an integrated circuit. The distance between the arrows is 0.5 μm, a width that is 150 times smaller than the diameter of the human hair. Courtesy of P. Chaudhari, IBM Corporation.

the other. But this distinction is very simplistic because semiconductors, unlike other materials, exhibit interesting electrical phenomena and effects when they join together at junctions with other semiconductors, metals, or insulators; it is these effects that are exploited in devices. Chapters 12 and 13 are devoted to a discussion of the nature of these materials and their operation in devices.

1.4. CONCLUSION

This introductory chapter has attempted to dramatize the role of materials within broader historical and societal contexts, as well as within engineering and manufacturing activities. Several concentric levels of connections conveniently serve to define the impact of materials. Most broadly, **materials–energy–environment** concerns address the wider community, whereas **materials–design–processing** constraints are more narrowly limited to engineering and manufacturing issues. **Processing–structure–properties–performance** interactions are the most focused of all and refer to specific materials.

It is the last tetrad of terms with which the remainder of the book is concerned. At the core of the subject is the **structure–properties** scientific base. Enveloping the core are the engineering concerns of **processing** and **performance.** For millennia, processing and performance of engineering solids were what mattered, and because the former was an art, the latter was very variable in practice. Now structure–property relationships are recognized as the key to understanding the fundamental nature of materials, improving their properties, and creating new materials. For this reason many of the chapters elaborate on dual scientific structure–engineering property themes.

Collectively, Chapters 2 to 6 are seasoned with a scientific flavor that exposes the nature of atomic electrons in solids and how atoms are geometrically positioned and held together in materials. Allied issues of stability and change in materials as a function of composition and temperature are also addressed. Chapters 7 to 15 are more engineering oriented and deal broadly with the mechanical and electrical properties of materials. How to achieve desired properties through processing and avoid degradation of performance are intimately connected issues. In keeping with the book's title these science–engineering distinctions are often blurred and sometimes seamless.

Additional Reading

* M. F. Ashby and D. R. H. Jones, *Engineering Materials 1: An Introduction to Their Properties and Applications,* Pergamon Press, Oxford (1980).
* M. F. Ashby and D. R. H. Jones, *Engineering Materials 2: An Introduction to Microstructures, Processing and Design,* Pergamon Press, Oxford (1986).
* C. Newey and G. Weaver, *Materials in Action Series: Materials Principles and Practice,* Butterworths, London (1990).
* *Scientific American* **217**, No. 3 (1967) and **255**, No. 4 (1986) are entirely devoted to materials.

The following texts broadly treat materials science and engineering at an introductory undergraduate level.

D. Askeland, *The Science and Engineering of Materials,* 3rd ed., PWS–Kent, Boston (1994).
* C. R. Barrett, W. D. Nix, and A. S. Tetelman, *The Principles of Engineering Materials,* Prentice-Hall, Englewood Cliffs, NJ (1973).
W. Callister, *Materials Science and Engineering: An Introduction,* 3rd ed., Wiley, New York (1994).
J. Shackelford, *Introduction to Materials Science for Engineers,* 3rd ed., Macmillan, New York (1992).
W. Smith, *Principles of Materials Science and Engineering,* 2nd ed., McGraw–Hill, New York (1990).
P. A. Thornton and V. J. Colangelo, *Fundamentals of Engineering Materials,* Prentice-Hall, Englewood Cliffs, NJ (1985).
L. Van Vlack, *Elements of Materials Science and Engineering,* 6th ed., Addison–Wesley, Reading, MA (1989).
J. Wulff *et al., The Structure and Properties of Materials,* Vols. I–IV, Wiley, New York (1964).

* Recommended reading.

QUESTIONS AND PROBLEMS

1-1. Select a material or product with which you are familiar, for example, a glass bottle or an aluminum can, and trace a materials cycle for it spanning from the source of raw materials to the point at which it is discarded.

1-2. A titanium orthopedic prosthesis weighs 0.5 kg and sells for 1000 dollars. Insertion of the implant into a patient requires an operation and a 5-day hospital stay. Comment on the approximate ratio of the prosthesis cost to the total cost of this surgical procedure to the patient.

1-3. What is the approximate added value in a contact lens manufactured from a polymer?

1-4. Approximately what percentage of the U.S. annual consumption of ~80 QUADs of energy is used in making some 100 million tons of steel a year? *Note*: 1 QUAD = 10^{15} BTU.

1-5. A QUAD of energy is roughly 1% of U.S. annual consumption. Suppose this need is to be supplied by solar cells that are 15% efficient. How much land area would have to be set aside for light collection assuming the cells could operate 12 hours a day? *Note*: 1 BTU = 1.06×10^3 J.

1-6. Materials play a critical role in generating energy. Mention some ways materials are used in generating energy.

1-7. Why does it take more energy to extract metals from oxide ores that are leaner in metal content?

1-8. Hundred-watt incandescent light bulbs are typically rated to last 1000 hours and cost approximately 1 dollar. Very roughly compare the bulb cost with that of the energy consumed in service.

1-9. Automobile metal scrap from shredded engines and bodies consist of steel (iron base alloys), aluminum, and zinc base alloys in the form of baseball-sized chunks and pieces. These metals are separated from each other and then recycled.
a. Design a method to physically separate steel from the other two metal alloys.
b. Aluminum and zinc pieces are physically parted according to their difference in density, 2.7 g/cm^3 versus 7.1 g/cm^3. Suggest a separation scheme exploiting this property difference.

1-10. a. What is the weight of oxygen in the earth's crust?
b. If 96% of the earth's volume is oxygen, what can you say about the relative sizes of metal and oxygen atoms?

1-11. Suppose iron in the upper kilometer of the earth's crust were predominantly in the form Fe_2O_3 and could be recovered by the reduction reaction

$$2Fe_2O_3 + 3C = 3CO_2 + 4Fe$$

a. If world consumption of iron is 5×10^{11} kg annually, and carbon is used solely for extraction purposes, how long would the iron last?
b. How much carbon would be consumed in extracting it?
c. Mention factors and conditions that would radically alter your answers.

1-12. Why is it more difficult to recycle polymers than metals?

1-13. a. Polymer production is increasing at a rate of about 15% per year. How long will it take to double?

b. If the amount of steel consumed doubles in approximately 20 years, what is the percentage increase in consumption each year?

1-14. The earliest reference to metal working in the *Bible* occurs in *Genesis* 4:22 in which the metalsmith "Tubal-Cain, the forger of every cutting instrument of brass and iron" is introduced. Comment on the historical implications of this verse fragment.

1-15. It has been theorized that metals were first produced from oxide ores by ancient potters glazing and firing clay wares. Suggest a possible scenario for such an occurrence.

1-16. How many products can you name that contain:

a. Metals, semiconductors, ceramics, and polymers?

b. Metals, ceramics, and polymers?

1-17. Sports equipment has undergone and continues to undergo great changes in the choice of materials used. Focus on the equipment used in your favorite sport and specify:

a. What materials change or substitution has occurred.

b. Reasons for the change.

1-18. Materials-designated technologists (i.e., metallurgists, ceramists) have been called engineers' engineers. Why?

1-19. A silicon wafer is processed to make many microprocessor chips measuring 2 cm long $\times$ 2 cm wide $\times$ 0.07 cm thick. If they each sell for 1000 dollars, how many times their weight in gold are they worth? *Note*: The weight of the devices and components on the chip is negligible compared with that of Si.

1-20. "A good measure of the technological advance of a civilization is the temperature which it could attain." Explain what is meant by this statement.

1-21. Speculate on the implications of the following events and trends with respect to the kinds of materials needed and the methods required to produce them:

a. The end of the Cold War

b. The desire to improve the environment and eliminate pollution

c. The shift from a manufacturing to a service economy

d. An increasingly global economy

e. Curbing energy consumption

2

ELECTRONS IN ATOMS AND SOLIDS: BONDING

2.1. INTRODUCTION

It is universally accepted that atoms influence materials properties, but which subatomic portions of atoms (e.g., electrons, nuclei consisting of protons, neutrons) influence which properties is not so obvious. Before addressing this question it is necessary first to review several elementary concepts introduced in basic chemistry courses. Elements are identified by their atomic numbers and atomic weights. Within each atom is a nucleus containing a number of positively charged protons that is equal to the atomic number (Z). Circulating about the nucleus are Z electrons that maintain electrical neutrality in the atom. The nucleus also contains a number of neutrons; these are uncharged.

Atomic weights (M) of atoms are related to the sum of the number of protons and neutrons. But this number physically corresponds to the actual weight of an atom. Experimentally, the weight of Avogadro's number ($N_A = 6.023 \times 10^{23}$) of carbon atoms, each containing six protons and six neutrons, equals 12.00000 g, where 12.00000 is the atomic weight. One also speaks about **atomic mass units** (amu): 1 amu is one-twelfth the mass of the most common isotope of carbon, ^{12}C. On this basis the weight of an electron is 5.4858×10^{-4} amu and protons and neutrons weigh 1.00728 and 1.00867 amu, respectively. Once the atomic weight of carbon is taken as the standard, M values for the other elements are ordered relative to it. A mole of a given element weighs M grams and contains 6.023×10^{23} atoms. Thus, if we had only 10^{23} atoms of copper, by a simple proportionality they would weigh $1/6.023 \times 63.54 = 10.55$ g (0.01055 kg). Note that the atomic weight of Cu, as well as most of the other elements in the Periodic Table including carbon, is not an integer. The reason for this is that elements exist as isotopes (some are radioactive, most are not), with nuclei having different numbers of neutrons. These naturally occurring isotopes are present in the earth's crust in differing abundances, and when a weighted average is taken, nonintegral values of M result. If compounds or molecules (e.g., SiO_2, GaAs, N_2) are considered, the same accounting scheme is adopted except that for atomic quantities we substitute the corresponding molecular ones.

EXAMPLE 2-1

a. What weights of gallium and arsenic should be mixed together for the purpose of compounding 1.000 kg of gallium arsenide (GaAs) semiconductor?

b. If each element has a purity of 99.99999 at.%, how many impurity atoms will be introduced in the GaAs?

Note: $M_{Ga} = 69.72$ g/mol, $M_{As} = 74.92$ g/mol, $M_{GaAs} = 144.64$ g/mol.

ANSWER a. The amount of Ga required is $1000 \times (69.72/144.64) = 482$ g. This corresponds to 482/69.72 or 6.91 mol Ga or, equivalently, to $6.91 \times 6.023 \times 10^{23} = 4.16 \times 10^{24}$ Ga atoms. Similarly, the amount of As needed is also 6.91 mol, or 518 g. The equiatomic stoichiometry of GaAs means that 4.16×10^{24} atoms of As are also required.

b. Impurity atoms introduced by Ga + As atoms number $2 \times (0.00001/100) \times 6.91 \times 6.023 \times 10^{23} = 8.32 \times 10^{17}$. Because the total number of Ga + As atoms is 8.32×10^{24}, the impurity concentration corresponds to 10^{-7}, or 1 part in 10 million.

Returning to the subatomic particles, we note that electrons carry a negative charge of -1.602×10^{-19} coulombs (C); protons carry the same magnitude of charge, but are positive in sign. Furthermore, an electron weighs only 9.108×10^{-28} g, whereas protons and neutrons are about 1840 times heavier. In a typical atom in which $M = 60$, the weight of the electrons is not quite 0.03% of the total weight of the atom. Nevertheless, when atoms form solids, it is basically the electrons that control the nature of the bonds between the atoms, the electrical conduction behavior, the magnetic effects, the optical properties, and the chemical reactions between atoms. In contrast, the subnuclear particles and even nuclei, surprisingly, contribute very little to the story of this book. Radioactivity, the effects of radiation, and the role of high-energy ion beams in semiconductor processing (ion implantation) are exceptions. One reason is that nuclear energies and forces are enormous compared with what

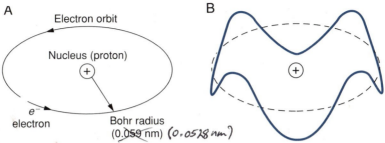

FIGURE 2-1 (A) Model of a hydrogen atom showing an electron executing a circular orbit around a proton. (B) De Broglie standing waves in a hydrogen atom for an electron orbit corresponding to $n = 4$.

atoms experience during normal processing and use of materials. Another reason is that the nucleus is so very small compared with the extent to which electrons range. For example, in hydrogen, the smallest of the atoms (Fig. 2-1A), the single electron circulates around the proton in an orbit whose radius, known as the Bohr radius, is 0.059 nm long [1 nm = 10^{-9} m = 10 Å (angstroms)]. The radius of a proton is 1.3×10^{-6} nm, whereas nuclei, typically ~$M^{1/3}$ times larger, are still very much smaller than the Bohr radius. Before a pair of atomic nuclei move close enough to interact, the outer electrons have long since electrostatically interacted and repelled each other. The preceding considerations make it clear why the next topic addressed is the atomic electrons.

2.2. ATOMIC ELECTRONS IN SINGLE ATOMS

Before the role of electrons in solids can be appreciated it is necessary to understand their behavior in single isolated atoms. The simplest model of an atom assumes it to be a miniature solar system at whose center is a positively charged nucleus (the sun) surrounded by a cloud of orbiting negatively charged atomic electrons (the planets). Charge is distributed so that the atom is electrically neutral. We now know these atomic electrons display a complex dynamical behavior (the larger the Z, the more complex the behavior) governed by the laws of quantum mechanics. The underlying philosophy and mathematical description of quantum theory are quite involved. Nevertheless, the resulting concepts and laws that derive from them can be summarized for our purposes in terms of a few relatively simple equations and rules that are discussed in turn.

2.2.1. Wave/Particle Duality

The term *wave/particle duality* suggests that both particles and waves have a dual nature. For the most part particles (a baseball, a stone, an electron) obey the classical Newtonian laws of mechanics. Occasionally, however, they reveal another, mysterious side of their character and behave like waves. An electron speeding down the column of an electron microscope produces such wavelike diffraction effects upon interaction with materials. Similarly, waves that exhibit standard optical effects like refraction and diffraction can surprisingly dislodge electrons from metals by impinging on them under certain conditions. Our first instinct is to attribute this **photoelectric effect** to a mechanical collision, attesting to the particle-like nature of waves. Wave/particle duality is expressed by the well-known de Broglie relationship

$$\lambda = h/p = h/mv. \tag{2-1}$$

In this important formula λ is the effective wavelength, and p is the momentum of the associated particle whose mass is m and velocity v. Planck's constant h ($h = 6.62 \times 10^{-34}$ J-s) bears witness to the fact that quantum effects are at play here.

The de Broglie relationship provides a way to rationalize the stability of electron orbits in atoms. Classically, circulating electrons ought to emit electromagnetic radiation, lose energy in the process, and spiral inward finally to collapse into the nucleus. But this does not happen and atoms survive in so-called **stationary states.** Why? If the de Broglie waves associated with the electron in hydrogen were **standing waves,** as shown in Fig. 2-1B, they would retain their phase, and orbits would persist intact despite repeated electron revolutions. But, if the waves did not close on themselves, they would increasingly interfere destructively with one another and, with each revolution, move more and more out of phase. Electron orbit disintegration would then be inevitable. This same notion of standing waves is used again later in the chapter (Section 2.4.3.2) to derive the energies of electrons in metals.

2.2.2. Quantized Energies

We know from experience that both particles (objects) and waves possess energy. The kinetic energies of gas molecules and the heating effects of laser light are examples. It outwardly appears that the energies can assume any values whatever. But this is not true. A fundamental law of quantum theory holds that energies of particles and waves, or more appropriately **photons,** can assume only certain fixed or quantized values. For photons, the energy is given by Planck's formula

$$E = h\nu = hc/\lambda. \tag{2-2}$$

Here ν is the frequency of the photon which travels at the speed of light c, where $c = \nu\lambda$ ($c = 2.998 \times 10^8$ m/s). Interestingly, the vibrations of atoms in solids give rise to waves or **phonons,** whose energies are given by a similar formula:

$$E = (n + \tfrac{1}{2})h\nu, \quad n = 1, 2, 3, \ldots. \tag{2-3}$$

In this expression n is a quantum number that assumes integer values. This means that solids absorb and emit thermal energy in discrete quanta of $h\nu$ when their atoms vibrate with frequency ν. In solids, ν is typically 10^{13} Hz.

More relevant to our present needs is the fact that the energies of electrons in atoms are also quantized, meaning that they can assume only certain discrete values. The case of the single electron in the hydrogen atom may be familiar to readers. Here the energy levels (E) are given by the Bohr theory as

$$E = \frac{-2\pi^2 m_e q^4}{h^2 n^2} = \frac{-13.6}{n^2} \text{ eV}, \qquad n = 1, 2, 3, \ldots \qquad (2\text{-}4)$$

The electron mass and charge are m_e and q, respectively. Again, n is an integer known, in this case, as the principal quantum number. The closer they are to the nucleus, the lower the energies of the electrons. And because the electron energy inside the atom is less than that outside it (where the zero-energy-level reference is assumed), a negative sign is conventionally used. Resulting energy levels are enumerated in Fig. 2-2. At any given time only one level can be occupied. In an unexcited hydrogen atom the electron resides in the ground state ($n = 1$) while the other levels or states are vacant. Absorption of 13.6 eV of energy (1 eV/atom = 1.60×10^{-19} J/atom = 96,500 J/mol) will excite the electron sufficiently to eject it and thus ionize the hydrogen atom.

For other atoms a very crude estimate of the electron energies is given by

$$E = -13.6 Z^2/n^2 \text{ eV}, \qquad n = 1, 2, 3, \ldots, \qquad (2\text{-}5)$$

where Z, the atomic number, serves to magnify the nuclear charge. The effect of core electron screening of the nuclear charge is neglected in this formula.

2.2.3. The Pauli Principle

In reality, multielectron atoms and hydrogen as well are more complicated than the simple Bohr model of Fig. 2-1. Electrons orbiting closer to the nucleus shield outer electrons from the pull of the nuclear charge. This complicates their motion sufficiently that different electrons are not at the same energy level; furthermore, electron energies are not easily calculated. In general, more advanced theories indicate that the electron dynamics within all atoms is characterized by four quantum numbers n, l, m, and s. These arise from solutions to the celebrated Schrödinger equation, a cornerstone of the modern quantum description of atoms. Specifically, the three-dimensional motion of electrons is embodied in the quantum numbers n, l, and m.

The **principal quantum number** is still n and it can assume only integer values 1, 2, 3, $\ldots$, as before. Electrons are now organized into shells. When $n = 1$ we speak of the K electron shell, while for the L and M shells, $n = 2$ and $n = 3$, respectively. As will become evident after introduction of the other quantum numbers, there are 2 electrons in the K shell, 8 in the L shell, 18 in the M shell, and so on.

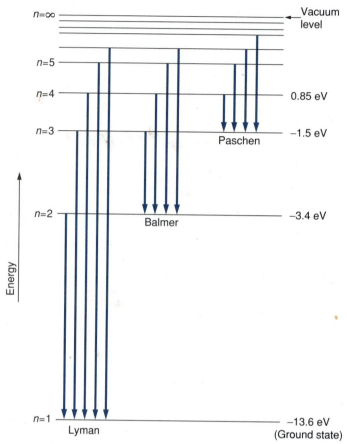

FIGURE 2-2 Electron energy levels in the hydrogen atom. The ground state corresponds to the electron level $n = 1$, with $E = -13.6$ eV. For the $n = \infty$ level, corresponding to $E = 0$, the electron has gained the 13.6 eV energy needed to ionize hydrogen. Electron transitions from upper levels to the $n = 1$ level are known to spectroscopists as the Lyman series; to the $n = 2$ level, the Balmer series; to the $n = 3$ level, the Paschen series.

The angular momentum arising from the rotational motion of orbiting electrons is also quantized or forced to assume specific values which are in the ratio of whole numbers. Recognition of this fact is taken by assigning to the **orbital quantum number** l values of $0, 1, 2, \ldots, n - 1$. The shape of electron orbitals is essentially determined by the l quantum number. When $l = 0$ we speak of s electron states. These electrons have no net angular momentum, and as they move in all directions with equal probability, the charge distribution is spherically symmetrical about the nucleus. For $l = 1, 2, 3, \ldots$, we have corresponding $p, d, f, \ldots$ states.

A third quantum number, m, specifies the orientation of the angular momentum along a specific direction in space. Known as the **magnetic quantum number,** m takes on integer values between $+l$ and $-l$, that is $-l, -l+1, \ldots, +l-1, +l$.

Lastly there is the **spin quantum number,** s, in recognition of the fact that electrons spin as they simultaneously orbit the nucleus. Because there are only two orientations of spin angular momentum, up or down, m assumes $-\frac{1}{2}$ and $+\frac{1}{2}$ values. We return to electron spin in Chapter 14 in connection with ferromagnetism.

The Pauli principle states that no two electrons in an atom can have the same four quantum numbers.

Let us apply the Pauli principle to an atom of sodium. As $Z = 11$ we have to specify a tetrad of quantum numbers (n, l, m, s) for each of the 11 electrons:

$1s$ states $(K$ shell)	$(1, 0, 0, +\frac{1}{2})$ and $(1, 0, 0, -\frac{1}{2})$;
$2s$ states $(L$ shell)	$(2, 0, 0, +\frac{1}{2})$ and $(2, 0, 0, -\frac{1}{2})$;
$2p$ states $(L$ shell)	$(2, 1, 0, +\frac{1}{2})$, $(2, 1, 0, -\frac{1}{2})$, $(2, 1, 1, +\frac{1}{2})$, $(2, 1, 1, -\frac{1}{2})$, $(2, 1, -1, +\frac{1}{2})$, and $(2, 1, -1, -\frac{1}{2})$;
$3s$ state $(M$ shell)	$(3, 0, 0, +\frac{1}{2})$.

Another way to identify the electron distribution in sodium is $1s^2\, 2s^2\, 2p^6\, 3s^1$ and similarly for other elements. In shorthand notation the integers and letters are the principal and orbital quantum numbers, respectively, and the superscript number tells how many electrons have the same n and l values.

2.2.4. Electron Energy Level Transitions

The atomic model that has emerged to this point includes a set of electrons, each having a unique set of numbers that distinguishes it from others of the same atom. Furthermore, the energies of the electrons depend primarily on n and, to a lesser extent, on l. This means that an electron energy level scheme like that for hydrogen (Fig. 2-2) exists for every element. In the case of hydrogen the electron in the ground state $(n = 0)$ can make a transition to an excited state provided (1) the state is vacant, and (2) the electron gains enough energy. The energy needed is simply the difference in energy between the two states.

Likewise, energy is released from a hydrogen atom when it is deexcited, a process that occurs when an electron descends from an occupied excited state to fill an unoccupied lower-energy state. Photons are frequently involved in both types of transitions as schematically indicated in Fig. 2-3. If the energy levels in question are E_1 and E_2, such that $E_2 > E_1$, then

$$E_2 - E_1 = \Delta E = h\nu_{12} = hc/\lambda. \tag{2-6}$$

As a computational aid, if ΔE is expressed in electron volts and λ in micrometers, then

$$\Delta E \text{ (eV)} = 1.24/\lambda \text{ (μm)}. \tag{2-7}$$

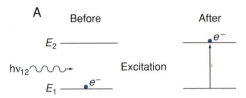

FIGURE 2-3   Schematic representation of electron transitions between two energy levels: (A) Excitation. (B) Deexcitation.

EXAMPLE 2-2

What is the wavelength of the photon emitted from hydrogen during an $n = 4$ to $n = 2$ electron transition?

ANSWER   Using Eq. 2-4, $\Delta E = E_4 - E_2 = -13.6\,(1/4^2 - 1/2^2) = 2.55$ eV. From Eq. 2-7, $\lambda = 1.24/\Delta E = 1.24/2.55 = 0.486\ \mu$m. This wavelength of 486 nm, or 4860 Å, falls in the blue region of the visible spectrum.

2.2.5. Spatial Distribution of Atomic Electrons

It makes a big difference whether electrons move freely in a vacuum as opposed to being confined to orbit atoms. For one thing, in the former case electron motion is usually viewed in classical terms. This means that the electron can have any energy; there is no quantum restriction that limits energies to only certain fixed values as is the case for electrons confined to atoms. We also imagine that the actual location of classical electrons is very precisely limited to coincide within their intrinsic dimensions. According to quantum mechanics and, in particular, the Heisenberg uncertainty principle,

$$\Delta p \cdot \Delta x = h/2\pi, \qquad (2\text{-}8)$$

this view is too simplistic. Rather, Eq. 2-8 suggests that the more precisely the electron momentum p is known, the less we know about its position x, and vice versa. The Δ's in Eq. 2-8 represent the uncertainties in p and x.

A cardinal feature of the wave mechanics approach in quantum theory is the notion that electron waves can have a presence or time-averaged charge density that extends in space, sometimes well beyond atomic dimensions. We speak of **wave functions** that mathematically describe the regions where the electronic charge can most probably be found. If they could be observed with a microscope, the picture would be strange and bewildering. For example, pictorial representations of the geometrically complex charge distributions in hydrogen-like *s*, *p*, and *d* states are shown in Fig. 2-4. It takes some doing to imagine that a single electron can split its existence between two or more distinct locations simultaneously.

2.3. FINGERPRINTING ATOMS

The previously described electronic theory of atoms is not only the substance of textbooks and the culture of science. A number of very important and sophisticated commerical instruments have appeared in recent years that capitalize on the very concepts just introduced. Their function is to perform qualitative as well as quantitative analysis of atoms in very localized regions of solids. Developed largely for the semiconductor industry to characterize the composition of electronic materials and devices, their use has spread to include the analysis of all classes of inorganic and organic, as well as biological, materials. In this section we focus on the emission of both X rays and electrons and learn how to fingerprint or identify the excited atoms from which they originate. **Optical spectroscopy,** a science that has been practiced for a long time, also fingerprints atoms but by measuring the wavelength of visible light derived from low-energy, *outer*-electron-level transitions. In contrast **X-ray spectroscopy,** which we consider next, relies on emission of X rays from the deeper, high-energy **core** electron levels.

2.3.1. X-Ray Spectroscopy

There is hardly a branch of scientific and engineering research and development activity today that does not make use of a **scanning electron microscope** (SEM), shown in Fig. 2-5; many examples of remarkable SEM images are found throughout this book and others have been widely reproduced in the printed and electronic media. The SEM is more fully discussed in Section 3.6.3. Our interest in the SEM here is as a source of finely focused electrons that impinge on a specimen under study. This causes electronic excitations within a volume that typically extends 1000 nm deep and measures about 10^8 nm^3. Typical bacteria and red blood cells are considerably larger. Within the volume probed the chemical composition of impurity particles, structural features, and local regions of the matrix can be analyzed. The incident (~30-keV) electrons are sufficiently energetic to knock out atomic electrons from previously occupied levels as schematically indicated in Fig. 2-6. An electron from a more

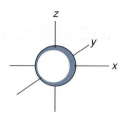

s states

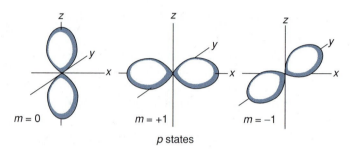

p states

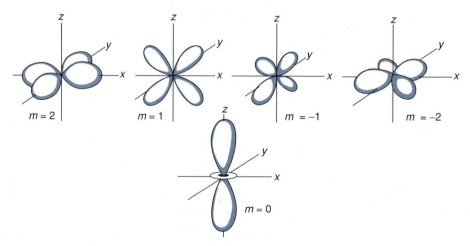

d states

FIGURE 2-4 Pictorial representation of the charge distribution in hydrogen-like *s*, *p*, and *d* wavefunctions. *s* states are spherically symmetrical, whereas *p* states have two charge lobes, or regions of high electron density extending along the axes of a rectangular coordinate system. *d* states typically have four charge lobes.

FIGURE 2-5 Photograph of a modern scanning electron microscope with X-ray energy dispersive analysis detector. The left- and right-hand monitors display the X-ray spectrum and specimen image, respectively. Operation of the SEM is controlled by a personal computer. Courtesy of Philips Electronic Instruments Company, a Division of Philips Electronics North America Corporation.

energetic level can now fall into the vacant level, a process accompanied by the emission of a photon.

To see how this works in practice, consider the electron energy level diagram for titanium metal depicted in Fig. 2-7A, where values for the energies of K, L, and M electrons are quantitatively indicated.

EXAMPLE 2-3

If an electron vacancy is created in the K shell of Ti and an L_3 electron fills it, what are the energy and wavelength of the emitted photon?

ANSWER This problem is identical in spirit to Example 2-2. The energy of the photon is $E_{L_3} - E_K = [-455.5 - (-4966.4)] = 4510.9$ eV. By Eq. 2-7, this corresponds to a wavelength of 2.75×10^{-4} μm, or 0.275 nm. Therefore, the photon is an X ray spectrally identified as K_α. The X-ray emission spectrum diagram for Ti is shown in Fig. 2-7B.

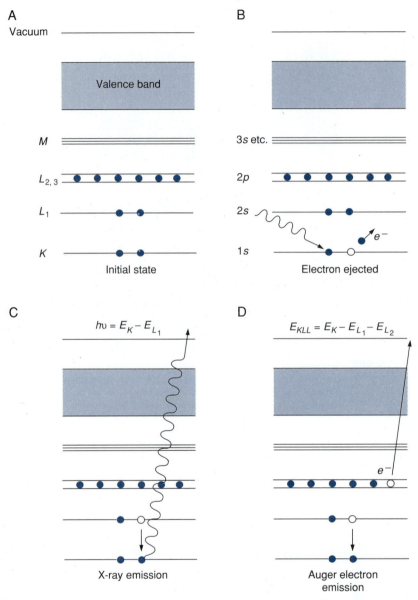

FIGURE 2-6 Schematic representation of electron energy transitions: (A) Initial state. (B) Incident photon or electron ejects K-shell electron. (C) X-ray emission when $2s$ electron fills vacant level. (D) Auger electron emission process.

A

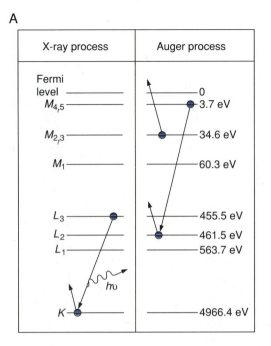

C

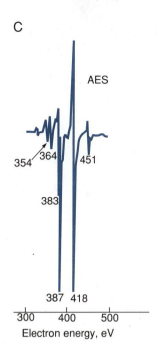

B

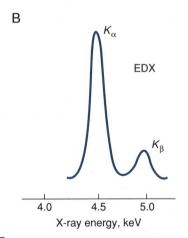

FIGURE 2-7   Electron excitation processes in titanium. (A) Energy level scheme. All electron energies are negative in magnitude. (B) X-ray emission spectrum of Ti. Only the K_α and K_β lines are shown. (C) Auger electron spectral lines for Ti.

During analysis, the electron beam actually generates electron vacancies in all levels. Therefore, many transitions occur simultaneously, and instead of a single X-ray line there is an entire spectrum of lines. Each element has a unique X-ray spectral signature that can be used to identify it unambiguously. Importantly, high-speed pulse processing electronics enables multielement analysis to be carried out simultaneously and in a matter of seconds. X rays emitted from the sample are sensed by a nearby cryogenically cooled semiconductor photon detector attached to the evacuated SEM column and analyzed to determine their energy spectrum. Results are displayed as a spectrum of signal intensity versus X-ray energy, hence the acronym **EDX** (**E**nergy **D**ispersive **X** ray) for the method. For multielectron atoms, Eq. 2-5 suggests that core levels for many useful elements can be expected to have X-ray energies of tens of kilo-electron volts. Extensive tables of the X-ray emission spectra of the elements exist to aid in materials identification.

An allied technique makes use of incident energetic X-ray or gamma ray photons, rather than electrons. They also induce identical electron transitions and X-ray emission, so that from the standpoint of identification of atoms, there is no difference. This so-called **X-ray fluorescence** technique is used when the specimen cannot withstand electron bombardment or when it is not feasible to place it into the vacuum chamber of an SEM. Chemical analyses of pigments in oil paintings and inks in paper currency have been made this way to expose forgeries.

2.3.2. Auger Electron Spectroscopy

Electron transitions within and between outer and core electrons are involved in the technique of Auger electron spectroscopy (AES), as indicated in Fig. 2-6D. But, unlike EDX where X-ray photons are emitted, so called Auger (pronounced oh-zhay) electrons are released in **AES**. A low-energy electron beam (~2 keV) impinges on the specimen and creates the initial vacancy that is filled by an outer electron; however, the photon that is normally created never exits the atom. Instead, it transfers its energy to another electron, the Auger electron. A collection of Auger electrons with varying intensity and unique energies does emerge to establish the wiggly spectrum that is the unambiguous signature of the atom in question. The AES spectrum for Ti is shown in Fig. 2-7C. Because core and not valence (or chemical bonding) electrons are involved in both EDX and AES, it makes little difference in which chemical state the Ti atoms exist: pure Ti, $TiAl_3$, TiO_2, $(C_5H_5)_2 TiI_2$, and so on. Titanium is detected independently of the other elements to which it is bonded.

What makes AES so special is that the low-energy Auger electrons can penetrate only ~1–2 nm (10–20 Å) of material. This means that AES is limited to detection of atoms located in the uppermost 1- to 2-nm-thick surface layers of the specimen. Because it is a surface-sensitive technique, extreme cleanliness and a very high vacuum environment are required during analysis to prevent

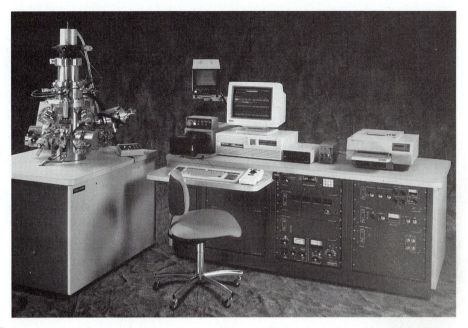

FIGURE 2-8 Photograph of an Auger electron spectrometer. Courtesy of Perkin–Elmer Physical Electronics Division.

further surface contamination. The atomic detection limit of both EDX and AES techniques is about 1%. In the case of AES where the incident electron beam spot size is $\sim$100 nm, the volume sampled for analysis is $[\pi \, (100)^2/4] \times 1 = 7850$ nm^3, which corresponds to about 400,000 atoms. Thus, under optimum conditions it is possible to detect about 4000 impurity atoms in the analysis. This astounding capability does not come cheaply, however. The commercial AES spectrometer shown in Fig. 2-8 costs more than half a million dollars. Auger analysis is used to determine the composition of surface films and contaminants in semiconductors, metals, and ceramics. An example of its use is shown in Fig. 2-9.

Despite the fact that sophisticated analytical methods have been introduced here, both EDX and AES rely on energy transitions involving core electron levels. In closing, our debt to the framers of the quantum theory of electrons in atoms should be appreciated.

2.4. ELECTRONS IN MOLECULES AND SOLIDS

2.4.1. Forming a Hydrogen Molecule

What happens to the atomic electrons when two or more widely separated atoms are brought together to form a molecule? This is an important question

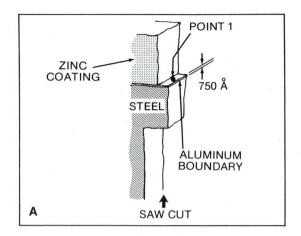

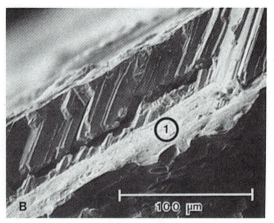

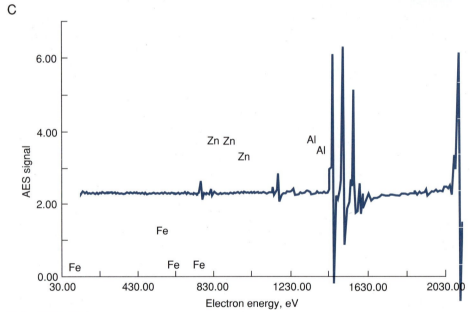

FIGURE 2-9 Case study involving AES. In hot-dip galvanizing, steel acquires a protective zinc coating after being dipped into a molten Zn bath. The iron and zinc react to form a compound that sometimes flakes off during fabrication. To prevent Fe–Zn reaction, 0.10% aluminum is added to the Zn bath. The very thin Al-rich layer that forms prevents the two metals from reacting. (A) The test specimen in this application shown schematically. (B) A magnified image of the Al-rich layer is analyzed at point 1. (C) The resulting AES spectrum detected at point 1 reveals the presence of Al. Courtesy of Perkin–Elmer Physical Electronics Division.

because by extension to many atoms, we can begin to understand condensation to the solid state. Whatever else happens, however, there must be a final overall reduction in the total energy of individual atoms when they interact and form stable molecules or solids. The quantum theories involved are very complex. One approach stresses replacement of **atomic orbitals** or wavefunctions by an equal number of **molecular orbitals.** To see how this works, let us consider two distant atoms of hydrogen (H) that are brought together to form a hydrogen molecule (H_2). Focusing only on the spherically symmetric $1s$ charge distribution, we note little change in each atom as long as they are far apart. But when they move close enough so that the two $1s$ charge clouds overlap, the rules of quantum chemistry suggest two new molecular charge distributions. In the first of these, the electron density is enhanced in the region between the nuclei (Fig. 2-10A). This preponderance of negative charge binds the positively charged nuclei together; an energy reduction occurs in this so-called **bonding orbital.** In Fig. 2-10B the second possibility is shown. Here the electron charge density is enhanced on the side of each nucleus away from the other nucleus. Because little negative charge is left between the positive nuclei, the latter strongly repel each other on approach, and the energy rises. This is the case of the **antibonding orbital.** The overall electron energy versus distance of approach for both orbitals is shown in Fig. 2-10C. Of note are the energy minimum and stable molecular configuration for the bonding orbital. But when the nuclei move too close, their mutual electrostatic repulsion begins to exceed the attraction caused by the electron density between them, and the energy rises.

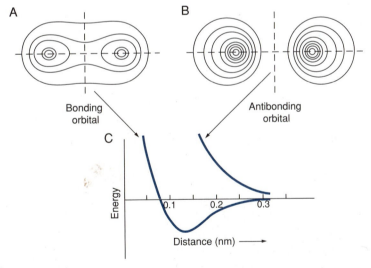

FIGURE 2-10 (A) Representation of electron cloud contours in the two-hydrogen-atom **bonding** orbital. There is a high electron density between the protons. (B) Representation of electron cloud contours in the two-hydrogen-atom **antibonding** orbital. The electron density between the protons is low. (C) Electron energies of the bonding and antibonding orbitals as a function of the internuclear distance.

In summary, recurring themes in the remainder of this chapter are:

1. The splitting of one energy level into two (or more generally, N levels for N electrons)
2. The minimum in the energy at an interatomic distance corresponding to the stable molecule (or solid)
3. The rise in energy when the nuclei approach too closely.

2.4.2. Electron Transfer in Ionic Molecules

At a higher level of complexity consider the formation of a LiF molecule through interaction between initially distant and neutral lithium and fluorine atoms (Fig. 2-11). As we all know, Li is extremely reactive chemically (especially with water) because of the strong tendency to shed its $2s$ valence electron. But this does not happen normally because Li can be stored as metal. An **ionization energy** of 5.4 eV is required to remove the electron and create a Li^+ ion. Similarly, the F atom exhibits a strong **electron affinity,** meaning that incorporation of an electron into its incomplete $2p$ shell is encouraged because it leads to an energy reduction of 3.6 eV. Interestingly, during ionization, the Li atom radius shrinks from 0.157 to 0.078 nm as it becomes a Li^+ ion. The reason is that the higher effective charge on the nucleus in the ion causes it to draw the remaining electrons toward it. Alternately, a lower effective nuclear charge in F^- causes the ion to expand to 0.133 nm from the initially uncharged F atomic radius of 0.071 nm. Information on atomic and ionic radii of the elements is given in Appendix A. Unlike the symmetrical electronic charge distribution in the hydrogen molecule, the electron density in the bonding orbital is large close to F^- and small near Li^+. The reverse is true for the antibonding orbital.

It should be noted that during electron transfer between these atoms as they draw close, there is actually a surprising *increase* in energy of 5.4 eV $-$ 3.6 eV = 1.8 eV! Why do they then react to form molecules? The answer is that the ions are not as close to each other as they could be. Account must be taken of the electrostatic force of attraction between the ions, which falls off with the distance (r) between them as r^{-2}. The potential energy of interaction (U) associated with the force between these point charges is given by

$$U = \frac{-q^2}{4\pi\varepsilon_0 r},\tag{2-9}$$

where ε_0 is a constant (known as the permittivity of vacuum) that has a value of 8.85×10^{-12} C/V-m. As r shrinks, Eq. 2-9 reveals that a point is reached where $U = -1.8$ eV, and the energy is recovered; bringing the ions closer together lowers the energy further, leading to the net energy reduction required for bonding. But, if the Li and F ions are squeezed together even more tightly, the energy rises. This time the negatively charged, core electron clouds begin to repel each other. These repulsive forces have a very short range and hence their contribution to U is assumed to vary with r as $+Br^{-n}$, where both B and

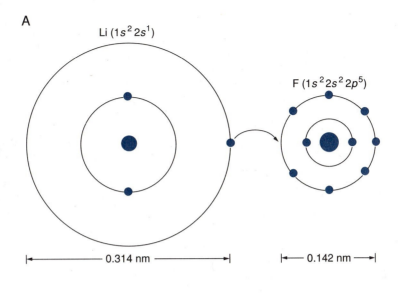

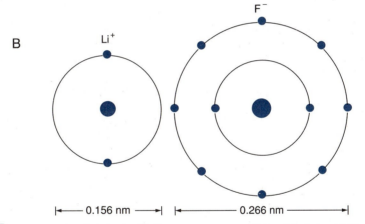

FIGURE 2-11 Formation of a LiF molecule. (A) Electronic structure and size of the isolated lithium and fluorine atoms. (B) Electron transfer from Li to F creating an ionic bond between Li^+ and F^- ions.

n are constants. Because n typically ranges from 5 to 10, as r shrinks r^{-n} rises very rapidly, as seen in Fig. 2-12. Therefore, by adding attractive and repulsive components

$$U = -q^2/4\pi\varepsilon_0 r + Br^{-n}. \tag{2-10}$$

Nature knows how to strike a compromise, however, and the equilibrium distance r_0 prevails in the molecule when the attractive and repulsive forces balance; r_0 can be calculated by minimizing U with respect to r. Simple differen-

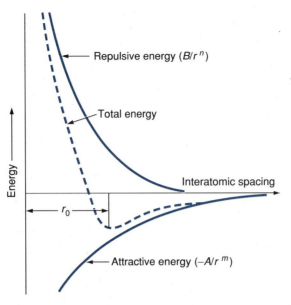

FIGURE 2-12   Potential energy of interaction between two ions or atoms as a function of interatomic spacing. For two monovalent ions of opposite charge, $A = q^2/4\pi\varepsilon_0$, and $m = 1$.

tiation and solving for r yields $r_0 = (4\pi\varepsilon_0 nB/q^2)^{1/(n-1)}$, when $dU/dr = 0$. The quantity $q^2/4\pi\varepsilon_0$ is evaluated to be 2.30×10^{-28} N-m^2, a number that will be useful in Example 2-5.

The model just presented is essentially classical in that quantum theory was not used in any significant way. Nevertheless, a simple extension of these ideas is all that is needed to quantitatively describe the bonding of electropositive and electronegative ions, not in a single molecule, but in alkali halide solids containing huge numbers of atoms. This is done in Section 2-5.

2.4.3. Electrons in Metals

2.4.3.1. Core Electrons

Electrons in metals experience a fate different from that of electrons in ionic materials. Neutral metal atoms have electron complements like those shown for Ti (Fig. 2-7). As the individual metal atoms approach each other they do not transfer electrons, one to the other—alkali halide style—because this is not profitable energetically. Rather, both the core and the outer electrons begin to interact because of electrostatic effects. At any distance of approach the $1s$ electrons of a given atom will interact less with the core $1s$ electrons of surrounding atoms than will the corresponding pairs of $2s$ electrons. The reason, of course, is that the outer $2s$ electrons will be closer to each other. And similarly, $2p$ and $3s$ electrons will interact even more strongly than $2s$ electrons.

Their charge clouds extend further from the nucleus, allowing greater overlap with the electron clouds of the neighboring atoms. As the interatomic separation distance is reduced, the strength of the electronic interactions intensifies in proportion to the extent of overlap.

The resultant alteration of individual electron levels is shown for sodium metal in Fig. 2-13. Broadening of initially discrete levels is the key feature of this figure; it occurs as a consequence of the Pauli principle. Atomic electrons within the interaction volume are forbidden to have the same four quantum numbers n, l, m, and s. Instead, the quantum numbers for each electron assume slightly different values. This changes the overall electron energy in the process; in fact, energies are split into as many levels as there are electrons. Some new levels are higher than the original one, elevating the energy and contributing to antibonding effects; but importantly, new lower energy levels are produced and they are in the majority. As a reuslt, atoms are bound together in the solid because a net energy reduction is achieved.

What emerges are **energy bands,** one for each of the original discrete atomic electron levels, that now contain enormous numbers of very closely spaced energy levels. The highest and lowest of these energy levels define the band-width, which changes as a function of interatomic spacing (Fig. 2-13). As before, very complex compromises are made to offset the attractive and repulsive interactions between electrons and ion cores. In sodium and metals of higher Z, the lower-energy core level bandwidths are narrow and do not overlap. It

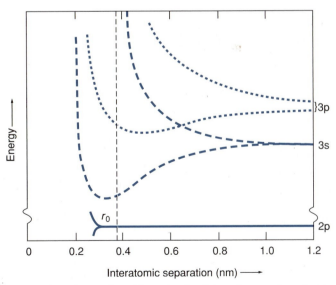

FIGURE 2-13 Calculated overlapping $3s$ and $3p$ energy levels in sodium metal as a function of interatomic separation. After J. C. Slater.

is only the high-energy atomic electron levels that tend to be broadly split and experience overlap at the equilibrium spacing in the solid.

2.4.3.2. Conduction Electrons

The outermost valence electrons of metal atoms undergo similar, but even more extensive splitting than the core electrons. Because the properties of these electrons are so critical to the bonding, chemical, and electrical properties of metals as well as semiconductors, they are variably known as **valence** or **conduction** electrons. They have been the subject of much theoretical and experimental study and a simplified description of their properties is presented here.

One may imagine that a well or box of length L with impenetrable walls, shown in Fig. 2-14A, encloses this dense collection of electrons (e.g., one per atom) that are totally **delocalized** from the atoms. It is the nature of metals not to tolerate voltage differences or gradients across them but to develop a short circuit instead. Therefore, it is reasonable to assume that these electrons only see a constant electric potential (V), whose magnitude we may take to be zero everywhere. But electrons at zero potential are, by definition, free, and the de Broglie relationship implies they ought to have waves associated with them. If an electron moving in the x direction is considered, the longest wavelength (λ) it can have is $2L$. Because the electron changes velocity (v) as it reflects from each wall, all of its possible associated waves have nodes there. The amplitude of these waves must vanish at the walls because electrons can never extend past them. This is a consequence of the essentially infinite value of V outside the well.

In order of decreasing wavelength, the allowable values of λ are

$$\lambda = 2L/1, \quad 2L/2, \quad 2L/3, \quad 2L/4, \quad 2L/5, \ldots, 2L/n_x$$
$$(n_x = \text{integer}), \qquad (2\text{-}11)$$

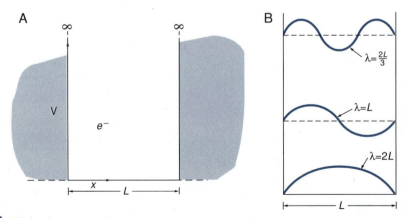

FIGURE 2-14 (A) An electron *particle* in a cubic box that has impenetrable walls. (B) Allowed electron wavefunctions for the particle in the box.

as indicated in Fig. 2-14B. Mathematically they can be represented as sine waves of the form $A \sin 2\pi x/\lambda$, where x is the distance from the left wall and A is a constant. The kinetic energy of these electrons is given by

$$E = \tfrac{1}{2} m_e v^2 = \frac{p^2}{2m_e}. \tag{2-12}$$

As $\lambda = h/p$ (Eq. 2-1), combination of Eqs. 2-1, 2-11, and 2-12 yields

$$E = \frac{h^2 n_x^2}{8m_e L^2}. \tag{2-13}$$

In this formula n_x is an effective quantum number that describes electron motion or wave propagation in the x direction. If account is taken of similar behavior in the y and z directions by including quantum numbers n_y and n_z in Eq. 2-13, the result is simply

$$E = h^2 (n_x^2 + n_y^2 + n_z^2)/8m_e L^2. \tag{2-14}$$

As with atoms, four quantum numbers are still involved but the rules for selecting them are much simpler. Each electron state is defined by a triad of integer quantum numbers (n_x, n_y, n_z) and contains two electrons with antiparallel spins, $+\tfrac{1}{2}$ and $-\tfrac{1}{2}$. Conduction electrons must also obey the Pauli principle; therefore, no two electrons can have the same four quantum numbers.

EXAMPLE 2-4

a. How many conduction electron states are there in a cube of gold 1.0 cm^3 in volume?

b. Identify the 10 lowest quantum states.

c. What is the energy spacing between the two lowest levels?

d. Suppose that one of the states having the highest energy has quantum numbers $n_x = 3.83 \times 10^7$, $n_y = 0$, and $n_z = 0$. What is the highest occupied energy level?

ANSWER   a. Each Au atom contributes one $6s$ conduction electron. Dimensional analysis shows that the number of Au atoms per cubic centimeter is $N_{Au} = N_A$ (atoms/mol) $\times \rho$ (g/cm^3)/M_{Au} (g/mol). For Au, $M_{Au} = 197$ and ρ (the density) $= 19.3$ g/cm^3. After substitution, $N_{Au} = 6.023 \times 10^{23} \times 19.3/(197) = 5.90 \times 10^{22}$ cm^{-3}. This number of electrons can be accommodated in half as many, or 2.95×10^{22} states, because of the two spin orientations.

b. The 10 lowest quantum states are [100], [010], [001], [110], [101], [011], [111], [200], [020], and [002]. Only 20 electrons can occupy these 10 states as each state accommodates two electrons. Equation 2-14 indicates that many of these different states (e.g., the first three) have the same energy. This phenomenon, known as *degeneracy*, is common for large values of E because there are many equivalent combinations of the sums of squares of three numbers. It is obvious that the conduction electron distribution is highly degenerate.

c. The lowest energy is $E_{100} = h^2\{1 + 0 + 0\}/8m_eL^2$. After substitution, $E_{100} = (6.62 \times 10^{-34}$ J-s$)^2 \times 1/[8 \times (9.11 \times 10^{-31}$ kg$) \times (10^{-2}$ m$)^2] = 6.01 \times 10^{-34}$ J $= 3.75 \times 10^{-15}$ eV. The next higher energy level is E_{110} with twice the energy or 7.50×10^{-15} eV. Therefore, the spacing between levels is 3.75×10^{-15} eV.

d. After substitution in Eq. 2-14, $E = \{(6.62 \times 10^{-34})^2 \times [(3.83 \times 10^7)^2 + 0 + 0]\}/[8 \times (9.11 \times 10^{-31}) \times (10^{-2})^2] = 8.82 \times 10^{-19}$ J, or 5.51 eV. The width of the band is the difference in energy between highest and lowest levels, or 5.51 eV.

This illustrative example is instructive for a number of reasons. From it one can appreciate the sort of numbers that describe electrons in metals. In even a tiny specimen there are huge numbers of them and they populate an extremely densely spaced set of discrete states. Because 10^{22} to 10^{23} electrons per cubic centimeter are compressed into a band that has an energy spread of only ~5 eV, the conduction electrons are imagined to populate a continuum of states and described in such terms as free electron gas or sea.

The elementary picture of metals that emerges views the ion cores as being immersed within a conduction electron sea. The latter also serves as a kind of glue that binds the nuclei together in the solid. Electrons move readily in response to very small applied dc voltages or electric fields, and this accounts for the high electrical conductivity of metals. High reflectivity of metals is also a consequence of the free electrons. When visible light shines on a metal the electromagnetic radiation excites these electrons to higher energy levels. But, as they fall back to lower levels, photons are reemitted as reflected light. The role of the free electrons in electrical conduction and other electronic and optical phenomena within metals and semiconductors is addressed again in Chapters 11, 12, and 13.

In real metals, electrons are not so hopelessly confined to the impenetrable well of the above example. It is possible to transfer some electrons from metals into an adjacent vacuum by heating them to a sufficiently high temperature. The operation of vacuum tubes, including the TV picture tube, depends on this thermionic emission of electrons. As noted earlier, incident photons will also remove electrons provided they have sufficient energy. In both cases "sufficient" means the work function energy ($q\Phi$), a kind of electrostatic barrier that prevents conduction electrons from leaving the metal. A frequently assumed model for the electron structure of metals is depicted in Fig. 2-15, where the work function is typically 2–5 eV. Values for the highest occupied electron energy level (Fermi energy) and work function distinguish different metals.

2.4.4. Electrons in Covalent Materials

2.4.4.1. Nonmetals

Electrons in many important engineering materials like diamond, Si, GaAs, polymers, and SiC participate in covalent bonding effects. But substances like

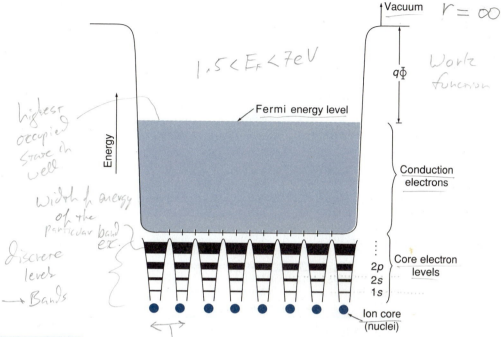

The following are handwritten annotations surrounding the figure:

E_F - fermi energy

$r = \infty$

$1.5 < E_F < 7 eV$

Work function

highest occupied state in well

Width of energy of the particular band ex.

discrete levels → Bands

FIGURE 2-15 Simple electron model of a metal consisting of individual unbroadened and broadened atomic core levels, surmounted by a distribution of conduction electrons.

Lattice spacing in crystal ~5 Å

H_2, which was treated above, F_2, and CH_4, together with hundreds of thousands of organic molecules, also exhibit covalent bonding. This type of bonding relies on neighborly sharing of electrons. Bond energies and lengths have been determined for a great number of covalently bonded atomic pairs and are listed in Table 2-1.

For elements to bond covalently they must have at least a half-filled outer electron shell, for example, half of eight or four electrons in outer s plus p electron states. In the case of a chlorine atom there are seven such electrons (two $3s$ plus five $3p$). Two Cl atoms share one electron between them and thus attain the eight necessary to fill each outer shell and achieve a stable Cl_2 bonding configuration. To generally account for the number of covalent bonds required between elements a rule has been formulated: *If the group number of the Periodic Table in which the element falls is N, then 8 − N covalent bonds are necessary to complete the outer electron shell.* On this basis, Cl ($N = 7$) requires one and C($N = 4$) requires four covalent bonds.

2.4.4.2. Semiconductors

Elements in the fourth column of the Periodic Table are the most important examples of covalent solids from an engineering standpoint. For this reason

| TABLE 2-1 | COVALENT BOND ENERGIES[a] AND LENGTHS |

Bond	Energy (kcal/mol)	Energy (kJ/mol)	Bond length (nm)
C—C	81	340	0.154
C=C	148	620	0.134
C≡C	194	810	0.120
C—H	100	420	0.109
C—N	69	290	0.148
C—O	84	350	0.143
C=O	172	720	0.121
C—F	105	440	0.138
C—Cl	79	330	0.177
O—H	110	460	0.097
O—O	33	140	0.145
O=O	96	400	0.121
O—Si	103	430	0.16
N—O	43	180	0.12
N—H	103	430	0.101
H—H	105	440	0.074
Cl—Cl	57	240	0.199
H—Cl	103	430	0.127

[a] Energy values are approximate as they depend on chemical environment. All values are negative for bond formation (energy is released) and positive for bond breaking (energy is absorbed).

Source: J. Waser, K. N. Trueblood, and C. M. Knobler, *Chem One*, 2nd ed., McGraw–Hill, New York (1980).

let us consider what happens to the electrons when silicon atoms are brought close together to form the solid. At first, electrons stemming from the same energy level (e.g., $3s$–$3s$ and $3p$–$3p$) overlap and split as described previously. But as atoms draw even closer electrons do not solely overlap with electrons of the same orbit. Rather, $3s$ and $3p$ electrons overlap or mix in a significant way to yield the energy band structure shown in Fig. 2-16. This phenomenon is known as **hybridization** and leads to four hybridized electron orbitals that radiate outwardly from each silicon atom in a manner suggested by Fig. 2-17A for diamond (which has the same structure) and explicitly in Fig. 3-6A. Each of these electrons pairs off with another like itself from neighboring silicon atoms to cement strongly directed covalent bonds. Complete electron sharing means that eight electrons will now surround each silicon atom and totally fill all of its four valence band states. There are no more electrons to go around and, therefore, the conduction band is empty. Unlike the situation for Na (Fig. 2-13) at the equilibrium distance (r_0), there is now a **gap** in the electron energy levels. There are no electron states and, therefore, no electrons possessing energies within the span defined by the energy gap, E_g.

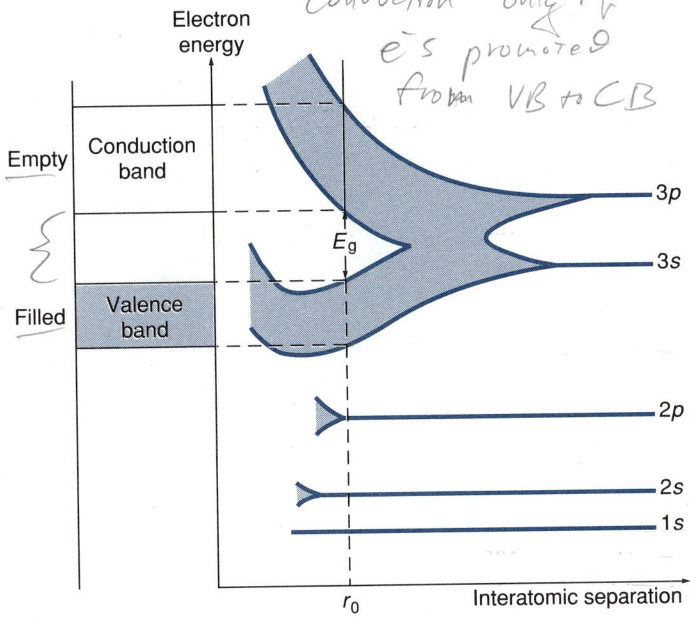

[handwritten notes:] ✱ a metal has no band gap
Si = Semiconductor because band gsp is small (1.1eV)
Conduction only if e⁻'s promoted from VB to CB

[handwritten annotations on figure: "Band Gap"]

FIGURE 2-16 Formation of energy bands in silicon as isolated atoms are brought together. The admixture or hybridization of the 3s and 3p electrons leaves an energy gap in the bands at the equilibrium spacing.

In order of increasing electron energy the picture that emerges at 0 K is:

1. Unbroadened $1s$, $2s$, and $2p$, core levels.
2. A much broadened, filled band of admixed $3s$–$3p$ states commonly known as the **valence band**
3. An **energy gap**
4. A broadened, empty **conduction band** of admixed $3s$–$3p$ states

Electrons in diamond and germanium have similar band structures and display comparable covalent bonding effects. One difference is that $2s$–$2p$ and $4s$–$4p$ hybridized states are involved respectively in C and Ge. And whereas the energy gap (E_g) is 1.12 eV in silicon, E_g = ~6 eV in C and E_g = 0.68 eV in Ge. The low expenditure of energy required to promote electrons into the conduction band facilitates charge motion and makes solid-state semiconductor devices possible. Further discussion of the implications of the electronic structure of semiconductors is deferred until Chapter 12.

2.4.4.3. Carbon

Carbon is a truly remarkable element that exists in a number of polymorphic forms, each possessing strikingly different structures and properties. It has already been noted that carbon atoms in diamond form covalent bonds with one another the way silicon atoms do. The four tetrahedral bonds reflect the

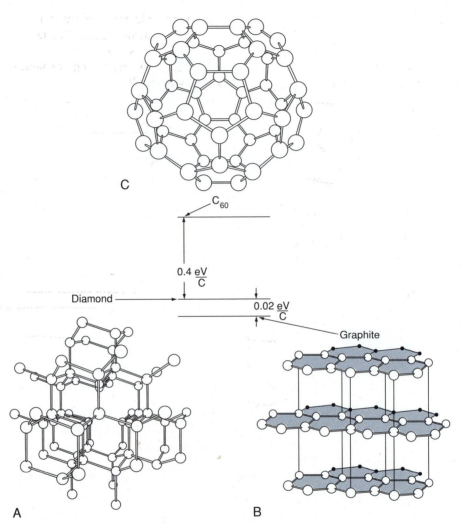

FIGURE 2-17   Polymorphic forms of carbon: (A) Diamond. (B) Graphite. (C) Buckminsterfullerenes. The inset shows that in order of increasing carbon bond stability are the polymorphs C_{60}, diamond, and graphite. After A. H. Hebard, AT&T Bell Laboratories.

geometric configuration; electronically, the bonds consist of four so-called sp^3 hybridized orbitals. The latter confer diamond's fabled hardness and high stiffness. With no conduction electrons, diamond is a poor conductor of electricity. Recent years have witnessed a plethora of new diamond materials that have been prepared in bulk and thin film form. Varied applications include abrasive grit for grinding, hard cutting tools for machining, and wear-resistant coatings on computer hard disk drives.

Graphite, another polymorphic form of carbon, consists of a hexagonal layered structure (Fig. 2-17B). Instead of three-dimensional tetrahedral bonding as in diamond, each carbon atom shares electrons with three other coplanar carbon atoms. In the planes or layers, strong covalent bonds hold atoms together. Weaker, so-called van der Waals forces (see Section 2.5.2) emanating from the fourth electron act normal to these layers. They serve to bond the layers to one another, but not strongly, accounting for the slippery, lubricated feel of graphite. This is the reason for the **anisotropic** behavior of graphite, meaning it exhibits different properties depending on direction. Desirable properties include chemical inertness, high electrical conductivity, stiffness, and temperature stability, as well as a low thermal expansion coefficient and density. These attributes are the basis of graphite use in furnace crucibles, large electrodes, electrical contacts, and engineering composites. Normally graphite is formed by subjecting carbon black particles, tar, pitch, resins, and other materials to the extremely high temperature of ~2500°C, where it is the stable form of carbon.

A last and intriguing polymorph of carbon was discovered in 1985 and has the structure shown in Fig. 2-17C. Years earlier Buckminster Fuller invented the geodesic dome. As this architectural structure bears a strong resemblance to the depicted molecules, the latter are known as *buckminsterfullerenes*; and because they resemble a soccer ball, they are also known as *bucky balls*. The molecule is composed of no fewer than 60 carbon atoms, and denoted by C_{60}. Each carbon atom bonds covalently to three others, forming both hexagonal and pentagonal rings. Presently a scientific curiosity, applications for C_{60} are being explored.

Unlike these polymorphs, carbon also bonds to elements other than carbon (e.g., H, O, N, S) in hydrocarbons and organic compounds. Polymeric materials are the most important engineering examples and they are introduced in Chapter 4.

2.5. BONDING IN SOLIDS

2.5.1. Interaction of Atomic Pairs

Until now a mostly microscopic view of solids has emerged in terms of behavior of individual atomic electrons. Next in the structural hierarchy are individual atoms. Interestingly, there is a way of differentiating solids based on *atomic* interactions that parallels interactions between *electrons*. The similarity lies in the nature of the potential energy of interaction (U) between electrons or atoms as their distance of separation (r) is reduced. Complexities of quantum electron levels, energies, and charge distributions are replaced in favor of the integrated response of a pair of atoms. These are assumed to interact in a classical manner that can be described in mathematically simple ways. We have already considered bonding in an ionic **molecule** using this approach in Section 2.4.2. Additionally, Figs. 2-10, 2-12, 2-13, and 2-16 should be contrasted for their distinctions and compared for their similarities.

The huge collection of atoms that condense to form a solid is very different from an interacting pair of atoms that form a molecule. But there still are attractive and repulsive forces exerted by neighboring atoms. With little loss in generality, what is true for a pair of interacting atoms can be extended to the entire solid. A commonly employed expression for U, applicable to atoms in solids, is given by

$$U = -A/r^m + Br^{-n}, \qquad (2\text{-}15)$$

where A, B, m, and n are constants. As before, the second term represents the repulsive interaction between atoms while the first term is the corresponding attractive contribution to binding. If $m = 1$ we have the case of the alkali halide solid. Although A, B, and n generally have different values in the ionic solid compared with the molecule (Eq. 2-10), a plot of Eq. 2-15 resembles that of Fig. 2-12. In condensed materials with weaker secondary bonds that keep molecules attached to one another (e.g., polymers, ice, liquid gases), m and n are frequently taken as 6 and 12, respectively. A number of physical associations can then be made using Eq. 2-15 and they are considered in turn.

2.5.1.1. Equilibrium Distance of Atomic Separation (r_0)

The equilibrium distance of atomic separation (r_0) was considered previously in Section 2.4.1. Differentiating U with respect to r yields $dU/dr = mA/r^{m+1} - nB/r^{n+1}$. At the equilibrium separation, $dU/dr = 0$, and solving yields for the critical value of r ($=r_0$),

$$r_0 = (nB/mA)^{1/(n-m)}. \qquad (2\text{-}16)$$

In crystalline solids r_0 can be associated with the lattice parameter a, of which more will be said in the next chapter.

2.5.1.2. Binding Energy (U_0)

In forming a solid, N widely spaced atoms in a gas phase can be imagined to condense according to the chemical reaction

$$NG \rightarrow NS + U_0,$$

where G and S refer to the gas and solid atoms. In the process there is an energy reduction of amount U_0. Usually known as the **binding** or **bonding energy,** it is given by the depth of the energy trough. If generalized to a mole of atoms, U_0 would refer to the molar energy to break bonds and decompose the solid. The reason can be seen by considering the reverse chemical reaction where input of (thermal) energy U_0 would cause S to convert to G, and atoms to separate from r_0 to $r = \infty$. A practical measure of the magnitude of U_0 is the heat liberated when elements react to form the solid. The sublimation energy of the solid (energy to transfer atoms from the solid to the vapor) is another widely used measure of U_0. Both measures of U_0 are entered in Table 2-2 for selected solids. Experimentally, U_0 is generally proportional to the melting temperature of the solid. Substitution of r_0 (Eq. 2-16) into Eq. 2-15 yields a theoretical estimate of U_0, provided A and B are known.

2.5.1.3. Forces between Atoms

In mechanical systems the force F is often equivalent to $-dU/dr$. But this quantity was already calculated above and is given by

$$F = -dU/dr = -a/r^M + br^{-N}, \qquad (2-17)$$

where $a = mA$, $b = nB$, $M = m + 1$, and $N = n + 1$. The equations repesenting F and U are thus algebraically similar. So is the plot of F versus r (Fig. 2-18A), but it is shifted to the right of the $U–r$ curve. This curve must cross the axis, or be zero at $r = r_0$, the equilibrium spacing, because the forces of atomic attraction balance those of repulsion at this critical distance. If the $F–r$ curve is flipped over (as in Fig. 2-18B), it is now consistent with easy-to-visualize force displacement behavior; more importantly, it resembles the stress–strain curve, a representation used in Chapter 7.

TABLE 2-2 VALUES OF MELTING POINT (T_M), BINDING ENERGY (U_0), YOUNG'S MODULUS (E_y), AND COEFFICIENT OF THERMAL EXPANSION (α) FOR SELECTED MATERIALS

Material	T_M (°C)	U_0 (kJ/mol)[a]			E_y (GPa)	α (10^{-6} °C^{-1})
NaCl	801	765*	413**	211***	49	40
KCl	776	688*	436**	198***		100
LiF	842	988*	612**	269***	112	
MgO	2800	3891*	603**		293	13.5
SiO$_2$	1710		879**		94	0.5
Al$_2$O$_3$	2980		1674**		379	8.8
Si$_3$N$_4$	1900		749**		310	2.5
WC	2776				720	4.0
Diamond	~4350			713***	1050	1
Si	1415			366***	166	3
Ge	937			384***	129	
Al	660			319***	70	22
Cu	1083			338***	115	17
Fe	1534			398***	205	12
W	3410			~770***	384	4.4
Polyethylene					~0.35	~150
Nylon					~2.8	~100

[a] The three measures of U_0 are physically different: * Energy liberated upon formation from the ions at 298 K. ** Heat liberated upon formation from the elements (atoms) at 298 K. *** Heat of sublimation.

Data were taken from various sources.

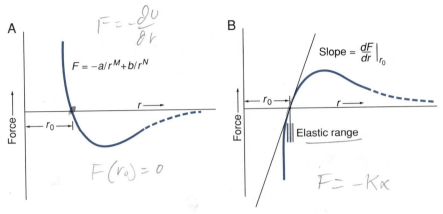

FIGURE 2-18 Force between atoms as a function of the interatomic spacing. (A) Plot of $F = -dU/dr$ versus r. (B) Plot of $F = +dU/dr$ versus r. The range of elastic response is indicated.

Let us focus our attention on the unloaded solid, that is, $F = 0$, $r = r_0$. If the solid is extended by pulling or applying a tensile load, atomic bonds are stretched and $r > r_0$. And similarly, if the solid is compressed, atoms are squeezed together and the bonds contract. If these loads are not too high and then removed, the solid snaps back elastically like a spring. In fact, atomic bonds are often thought of as springs whose dynamical behavior is familiar. For an atomic displacement of $\delta = r - r_0$, the force on the atomic bonds is given by $F = -k_S (r - r_0)$, where k_S is the spring constant. The negative sign means that a positive δ establishes an oppositely directed restoring force. A mathematical definition of the spring constant is $k_S = -(dF/dr)_{r=r_0}$, and this is equivalent to $k_S = (d^2U/dr^2)_{r=r_0}$. For a given δ, atomic forces between atoms are large when k_S is large; this is the case in stiff solids. For example, average atomic bonds in bronze (a copper–tin alloy) are stiffer than those in lead, and this is reflected in the choice of the former metal in bells.

2.5.1.4. Young's Modulus

Solids are placed in a state of **stress** by applying external loads to their surfaces, in which case they **strain**. Just how these loads are transmitted through the solid to exert forces on and stretch individual atomic bonds is not obvious. Nevertheless, extending the spring analogy, we may write a similar relationship between the stress and strain. Although these terms are defined more precisely in Chapter 7, stress is defined here as $-F/r_0^2$ (or bond force per unit *atomic* area). Similarly, the dimensionless strain is assumed to be defined by δ/r_0. **Hooke's law** expresses the fact that stress is linearly proportional to strain, that is, $-F/r_0^2 = E_y \delta/r_0$. The constant of proportionality, E_y, is known as the elastic or **Young's modulus** and is given by $-r_0^{-1} F/\delta$. But $F = -dU/dr$ and $\delta \sim dr$, so that E_y can be approximately expressed as $E_y \sim -r_0^{-1} (dF/dr)_{r_0}$ or

$$E_y \sim r_0^{-1} (d^2U/dr^2)_{r_0}. \tag{2-18}$$

Materials whose U vs r curves are deeply set (large U_0) and highly curved at r_0 tend to be stiff and have high elastic moduli. Physically, strains are small in high-E_y materials because the steep U vs r walls imply a sharp energy increase with r.

Young's moduli of selected materials are listed in Table 2-2. In general the higher the melting point of the solid, the larger the modulus of elasticity.

EXAMPLE 2-5

a. How does Young's modulus depend on r_0 in alkali halide solids?

b. If the equilibrium distance between Na and Cl ions in rocksalt is 0.282 nm (2.82×10^{-10} m), estimate the binding energy of a single NaCl molecule if it is assumed that $n = 9.4$.

ANSWER   a. Young's modulus is given by $E_y \sim -(1/r_0)\,(dF/dr)_{r_0}$. Differentiation of Eq. 2-17 with respect to r_0 yields $(dF/dr)r_0 = m(m + 1)A/r_0^{n+2} - n(n + 1)B/r_0^{m+2}$. Through Eq. 2-16, the constants B and A are related by $B = (mA/n)r_0^{(n-m)}$. Substituting for B yields $(dF/dr)r_0 = m(m - n)A/r_0^{m+2}$, or $E_y = -m(m - n)A/r_0^{m+3}$.

In alkali halides $m = 1$ and, therefore, E_y varies as $\sim 1/r_0^{1+3} = r_0^{-4}$. The predicted dependence of E on the inverse fourth power of r_0 has been observed.

b. In NaCl, $m = 1$ and $A = q^2/4\pi\varepsilon_0 = 2.30 \times 10^{-28}$ N $-$ m^2. Simple substitution of the expression for r_0 (Eq. 2-16) into Eq. 2-15 yields $U_0 = -Ar_0^{-m}\,(1 - m/n)$ after a bit of algebra. Upon evaluation, $U_0 = -2.30 \times 10^{-28}\,(2.82 \times 10^{-10})^{-1}\,(1 - 1/9.4) = 7.29 \times 10^{-19}$ N $-$ m (J). For a mole of Na$^+$–Cl$^-$ ion interactions, $U_0 = -6.02 \times 10^{23} \times 7.29 \times 10^{-19} = -439$ kJ.

A value of U_0 for *solid* NaCl based on experiment is -765 kJ/mol. The cause of the discrepancy is the neglect of an exactly calculable factor known as the *Madelung constant*. This constant, related to the geometry of the electrostatic interaction of all the Na$^+$ and Cl$^-$ ions in the ordered solid state, has a value of 1.748 in the rocksalt structure (See Fig. 3-7). Multiplying by this factor yields $U_0 = -1.748 \times 439 = -767$ kJ/mol, in excellent agreement with experiment. Note that U_0 refers here to the energy reduction when *ions* (not *atoms*) condense to a solid (see Table 2-2).

2.5.1.5. Thermal Expansion

When atoms in a solid vibrate they undergo displacements that are limited by the walls of the U versus r potential well. Imagine a perfectly *symmetrical* or parabolic well; the thermally induced atomic motion in the positive and negative r directions will always average to $r = r_0$. And as the temperature is raised the amplitude of vibration increases but the average value of r is still r_0. Hence there is no thermal expansion.

Consider, however, what happens when the potential well is *asymmetrical*, as shown in Fig. 2-19. At temperature T_1 the average displacement or midpoint between the well walls falls slightly to the right of r_0. And at T_2 ($T_2 > T_1$) the average position lies even further to the right of r_0, and so on. In this case the material thermally expands.

All materials expand to one extent or another when heated and the coefficient of thermal expansion (α) is the material property that measures this tendency. By definition

$$\alpha = \frac{1}{L}\frac{dL}{dT},$$
(2-19)

where L is a unit length of material. Because the attractive portion of the U vs r curve is less steep than the repulsive component, the displacement of the mean equilibrium position relative to r_0 increases with T, and α has a positive value. Thermal expansion coefficients for selected materials are listed in Table 2-2. In general, small coefficients of thermal expansion are observed in high-melting-point materials and vice versa.

EXAMPLE 2-6

Show that Young's modulus decreases with temperature, i.e., $dE_y/dT < 0$.

ANSWER   To demonstrate this we use the formula $E_y = (-m(m - n) A)/r_0^{m+3}$ from illustrative problem Example 2-5. Differentiating, $dE_y/dT = +[m(m - n) A(m + 3)/r_0^{m+4}] (dr_0/dT)$. If r_0 is substituted for L, then by Eq. 2-19, $\alpha = (1/r_0)(dr_0/dT)$. Finally, $dE_y/dT = + [m(m - n)A(m + 3)/r_0^{m+3}] \alpha$. Because $n > m$, and all other terms are positive including α, the sign of dE_y/dT is negative.

The decline in E_y with a rise in T is indeed observed in materials. This subject is addressed again in Section 7.2.4.

2.5.2. Bonding Classifications

In Chapter 1 it was mentioned that metals, ceramics, semiconductors, and polymers constitute our engineering materials. Indeed, the remainder of the book is devoted to their properties. But from a scientific standpoint there are other ways to classify solids. Of the main bonding categories in this new classification scheme—**metallic, covalent, and ionic**—three have already been introduced. But nature only broadly tolerates classification schemes, the less so when so few categories are available to compartmentalize so much phenomena.

Metals like Au, Ag, and Cu, covalent solids like diamond, and ionic materials like NaCl and other alkali halides are reasonably pure exemplars of the three

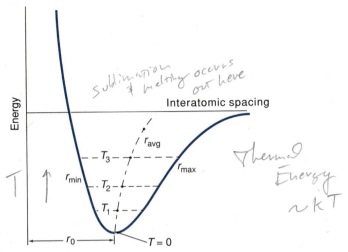

(handwritten annotations on figure:) Sublimation & melting occurs out here — Thermal Energy ~kT

FIGURE 2-19 Schematic representation of atom vibrating in an asymmetrical well defined by the U vs r potential. In this case the average displacement increases with temperature and there is thermal expansion.

distinct bonding mechanisms; however, atoms in most materials are bonded by an admixture of mechanisms. Transition metals with incomplete inner shells (e.g., Fe, Ni, W, etc.) have some covalent character through nearest-atomic-neighbor sharing of core electrons. This gives metallic bonds a directional character that strengthens them, resulting in high melting points. Metals whose core electron complement is complete (e.g., Zn, Mg) have no way to bond covalently and largely melt at lower temperatures. Compound semiconductors like GaAs have a small amount of ionic character, CdTe more so. Conversely, ionic solids like Al_2O_3 and SiO_2 have strong covalent character. Sorting it all out is nightmarish. The thing to remember is that these three bonding mechanisms cement strong **primary** bonds by **transferring** (ionic) and **sharing** (covalent, metallic) electrons. Primary bonds have energies ranging from ~1 to 10 eV/atom (~100 to 1000 kJ/mol).

The remaining bonding mechanisms establish weak **secondary** bonds typically possessing only a tenth (or less) of the energy of primary bonds. Although primary bonds largely originate from direct point charge interactions, secondary bonds stem from second-order coupling of weaker **dipole** charge distributions. A dipole consists of closely spaced positive and negative centers of charge. For example, in the water molecule oxygen is the seat of negative charge while an equal amount of positive charge is covalently shared with the two hydrogen atoms. The molecular geometry of water is V-shaped, with an effective dipole composed of the negative charge at the O apex and positive charge along the two legs set at an angle of 105°. The dangling H atoms or protons (as they have lost electrons to the oxygen) can couple to oxygen atoms of other water molecules in chains. In ice they polymerize into a three-dimensional array that extends throughout the solid as shown in Fig. 2-20. **Hydrogen bonding** is the name given to this bonding mechanism because protons are the bridge that connects the anions of molecules together. This bonding mechanism is also operative between molecules containing **permanent** dipoles composed of hydrogen and strongly electronegative elements, for example, HF and HCl. Hydrogen bonds are appreciably weaker than chemical bonds but stronger than the **van der Waals** bonds considered next.

The most important examples of secondary bonding arise from van der Waals interactions. Dipoles are also involved here but they are not permanent in nature. The classic example of such bonding occurs in inert gases. Outer electron shells are completely filled and the molecule is not only electrically neutral, but the negative charge appears to be symmetrically distributed about the positive nucleus. What then is the mechanism that causes atoms of argon to condense, first as a liquid and then as a solid, as the temperature is successively reduced? Even in the vapor phase slight deviations in the perfect gas law for all kinds of gases were recognized by van der Waals because molecules weakly attracted one another. Apparently, weak dipole moments form and disappear as small electronic charge imbalances are continually created about the nucleus. The orientation of these fluctuating dipoles constantly changes in neighboring atoms or molecules. But on average, weak dipole–dipole interactions exert slight tugs of attachment and nondirectional bonds are produced. Such fluctuating dipole bonds are collectively known as van der Waals bonds. The attractive energy of interaction falls off rapidly with distance, that is, as r^{-6}, in

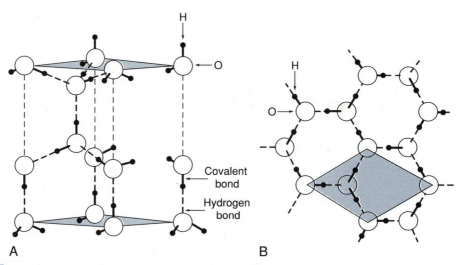

Covalent bond

Hydrogen bond

A   B

FIGURE 2-20 Structure of ice shown in (A) side and (B) top views. (Also see Fig. 3-22D.) Intramolecular covalent bonds exist within the H_2O molecule. Intermolecular hydrogen bonds link H_2O molecules together. From K. M. Ralls, T. H. Courtney, and J. Wulff, *Introduction to Materials Science and Engineering*, Wiley, New York (1976).

contrast to the slower r^{-1} decay in ionic bonding. Bonding energies of ~0.1 eV are typical.

To complicate bonding issues further there are materials that simultaneously possess both primary and secondary bonds. Nitrogen gas is an example. In a single molecule the atoms are covalently bound, but in liquid nitrogen at 77 K van der Waals forces hold the molecules together. Simply pouring the liquid gas on a warmer surface is sufficient to destroy the weak secondary bonding. Long-chain molecules in polymeric solids are very important engineering examples and are treated at length in Chapter 4. So-called **intramolecular** (within the molecule) carbon–carbon and carbon–hydrogen bonds are covalent and strong. Covalent C–C bonds along the spine are admixed with the far weaker **intermolecular** (between molecules) van der Waals bonds that link adjacent linear molecules together. Property admixtures can therefore be expected based on the proportion of intra- and intermolecular bonding.

Not only does van der Waals bonding occur in atoms and molecules of the same type, but it occurs across interfaces of dissimilar materials; indeed it is responsible for adhesion effects in films and coatings. For example, the molecular composition and nature of paint constituents (organic plus ionic) differ from those of the metal surface they coat. Primary chemical bonds across such interfaces do not form and adhesion is a problem that is overcome by weak van der Waals bonds. There are three categories of van der Waals bonds (London, Debye, and Keesom types) dependent on whether neither, one, or both of the paired atoms or molecules possess electric dipoles.

2.5.3. Material Properties

The chapter ends with a comparison among the different bonding mechanisms and the material property trends they foster. This is done in a self-explanatory way in Table 2-3, where assorted properties are compared. Because there are reasonably large property variations within particular bonding groups, comparisons are approximate.

2.6. PERSPECTIVE AND CONCLUSION

This chapter has provided a panoramic view of what materials are composed of and what holds them together. More than any other chapter in the book it enables distinctions to be drawn between materials and their properties. A dichotomy exists between the quantum description of solids in terms of atomic *electrons*, on the one hand, and classical models involving interacting pairs of *atoms* that attract and repel one another, on the other hand. In the former, the approach is to enumerate possible electron states (defined by quantum numbers) of individual atoms, their occupation by electrons, and their fate when atoms are brought together in solids. The latter assumes that whole

TABLE 2-3	BONDING TYPES AND PROPERTY TRENDS

| Property | Bonding type | | | |
	Metallic	Ionic	Covalent	Secondary
Magnitude of U_0	Large (3)[a]	Very large (1)	Very large (2)	Very small (4)
Melting point	High (3)	High (1, 2)	High (1, 2)	Very low (4)
Magnitude of E_y	High (2)	High (1)	High (1)	Very small (4)
Magnitude of α	Small (2)	Small (3, 4)	Small (3, 4)	Large (1)
Hardness (strength)	High (3)	High (2)	High (1)	Very low (4)
Toughness	High (1)	Low (4)	Low (2)	Low (3)
Density	High (l)	Medium (3)	Medium (2)	Low (4)
Electrical conductivity	High (1)	Low (3, 4)	Medium (2)	Low (3, 4)
Thermal conductivity	High (1)	Low (4)	Medium (2)	Low (3)
Optical reflectivity	High (1)	Low (3, 4)	Medium (2)	Low (3, 4)
Chemical reactivity	High (1)	Low (4)	Low (2, 3)	Low (3, 4)

[a] 1 = highest, 4 = lowest.

atoms, which integrate and smooth the complex behavior of subatomic particles, interact according to mathematically simple physical laws. A guiding operative principle in either case is that stable bonding configurations are the outcome of minimizing the energy that resides in electrons and atoms. In both cases compromises must be struck between attractive and repulsive forces, the resultant of which yields very precise solid-state atomic positions and well-defined bond energies.

Knowledge of atomic electron energy levels and transitions between them has been capitalized upon in assorted electron (AES) and X-ray spectrometers to identify atoms. These instruments have been indispensable tools in characterizing surface layers and interfaces in all classes of materials, particularly in microelectronics technology. Together with high-resolution electron microscopy (discussed in the next chapter), these methods have provided much of our information on the structural and bonding character of atoms within solids.

Generally speaking, hard, high-strength materials are the legacy of strong primary bonds. These are found particularly in ionic and covalent solids as well as in many metals. As we shall see later in the book, high electrical and thermal conduction and optical reflectivity go hand in hand with the availability of large numbers of mobile electrons, and metals have these in the greatest profusion.

There are, of course, materials that defy classification or exhibit anomalous behavior in particular composition and temperature ranges. Excellent examples are the high-temperature ceramic superconductors (e.g., $YBa_2Cu_3O_7$), which are outwardly ionic but behave as metals. They have a schizophrenic nature as do semiconductors that exhibit attributes of both ionic and metallic character. In design work, engineers must anticipate typical or expected property trends within material categories. But one must also be particularly alert to deviations from expected behavior; for example, diamond, a covalent solid, conducts heat better than any metal. This is what imparts excitement to the subject of materials.

Additional Reading

R. J. Borg and G. J. Dienes, *The Physical Chemistry of Solids,* Academic Press, Boston (1992).
L. Smart and E. Moore, *Solid State Chemistry: An Introduction,* Chapman & Hall, London (1992).
J. Waser, K. N. Trueblood, and C. M. Knobler, *Chem One,* 2nd ed., McGraw–Hill, New York (1980).

QUESTIONS AND PROBLEMS

2-1. a. How many atoms are there in a pure silicon wafer that is 15 cm in diameter and 0.5 mm thick?

b. If the wafer is alloyed with 10^{16} phosphorus atoms per cubic centimeter what is the atomic fraction of phosphorus in silicon?

2-2. Aluminum has an atomic density of 6.02×10^{22} atoms/cm^3. What is the mass density?

2-3. How many grams of Ni and Al are required to make 1 kg of the compound Ni_3Al?

2-4. What energy is associated with the absorption of the smallest quantum of vibrational energy in a typical solid?

2-5. A helium–neon laser beam is rated at 0.005 W and emits light with a wavelength of 632.8 nm. How many photons are emitted per second?

2.6. a. What is the de Broglie wavelength associated with an electron traveling at 1×10^6 m/s?

b. What is the de Broglie wavelength associated with a 3000-lb car traveling at 55 mph?

2-7. a. The wavelength of a copper X ray is 0.154 nm. What is the momentum associated with it?

b. When an X ray is emitted, the Cu atom recoils much like a rifle discharging a bullet. What momentum would be imparted to a free Cu atom upon X-ray emission?

c. What is the recoil velocity of the atom?

2-8. a. What is the expected ionization energy of the $3s$ electron in Na?

b. The actual ionization energy of Na is 5.2 eV. How do you account for the difference between the two values?

2-9. Calculate the energy and wavelength of the photon emitted when an electron in a titanium atom falls from the $n = 3$ to the $n = 1$ state.

The next several problems are based on the electron energy levels (in eV) for the elements listed in the following table and identified in Fig. 2-7.

Element	K	L_1	L_2	L_3	M_1	$M_{2,3}$	$M_{4,5}$
Cr	5,989	695	584	575	74.1	42.5	2.3
Cu	8,979	1,096	951	931	120	73.6	1.6
Mo	20,000	2,865	2,625	2,520	505	400	229
W	69,525	12,100	11,544	10,207	2820		

2-10. a. Create an energy level diagram for Cu.
   b. What is the energy of the photon emitted in the $M_3 \to K$ electron transition in Cu?
   c. In what range of the electromagnetic spectrum (visible, infrared, X-ray, etc.) does this photon lie?
   d. Repeat parts (b) and (c) for the $M_{4,5} \to M_{2,3}$ transition.

2-11. a. Will electrons that travel at velocities of 4.7×10^7 m/s have enough kinetic energy to eject the K electron from Cr? Can they eject Mo K electrons?
   b. Does a photon with a wavelength of 0.161 nm have enough energy to eject a K electron from Cu? Can it eject the Cr K electron?

2-12. The metals listed in the previous problem are all used commercially as targets in X-ray generating tubes (see Section 3.4.1.).
   a. What is the photon wavelength corresponding to the $L_3 \to K$ electron transition in each metal? This transition gives rise to the so-called K_{α_1} X ray.
   b. X-ray tubes with Cu, Cr, Mo, and W targets were mixed up in a laboratory. To identify them they were operated sequentially and the K_{α_1} wavelengths were measured with an EDX system. The first tube tested yielded a wavelength of 0.0709 nm. What is the target metal?

2-13. Moseley's law of atomic physics suggests that the energy of K_{α_1} X rays varies as $(Z - 1)^2$, where Z is the atomic number of the element. Plot the K_{α_1} X-ray energies for Ti, Cr, Cu, Mo, and W versus $(Z - 1)^2$ so that a straight line results. Based on your plot what are the energy and wavelength of K_{α_1} X rays in Sn?

2-14. Fluorescent X-ray analysis from an automobile fender revealed a spectrum with lines at 5.41 keV (intense) and 5.95 keV (less intense) and a weaker line at 8.05 keV. Interpret these findings.

2-15. During the Renaisance the white pigment used in oil paints was lead oxide. In the 19th century zinc oxide was used as well. In more recent times titanium oxide has been the preferred choice. A painting suspected of being a forgery is examined by EDX methods and yielded 75-keV $K\alpha_1$ X rays from a region painted white. Pending further investigation what, if anything, can you infer about the painting's age? (*Hint*: See Problem 2-13.)

2-16. The chemical composition of a series of binary Cu–Ti alloys is calibrated by measuring the relative intensity of fluorescent X rays emitted. For pure Cu the rate of X-ray emission is measured to be 1562 (photon) counts per second (cps), whereas 2534 cps was detected from pure Ti. The alloy yields a rate of 656 cps for Cu plus 1470 cps for Ti X rays. What is the overall alloy composition? (In all cases the same sample and measurement geometry were employed and a linear composition calibration is assumed.) Is the composition measured in weight or atomic percent?

2-17. Elemental analysis of the heavy metals by EDX methods is virtually independent of what phase (solid, liquid, gas) or state of chemical bonding (metallic, ionic, covalent) is involved. Why?

2-18. A common form of the potential energy of interaction between atoms is given by $U = -A/r^6 + B/r^{12}$, where A and B are constants.

a. Derive an expression for the equilibrium distance of separation in terms of A and B.

b. If $r_0 = 0.25$ nm, what is the ratio of B to A?

c. Derive an expression for the energy at the equilibrium separation distance in terms of A.

2-19. Consider a one-dimensional linear material consisting of ions of alternating charge separated by distance a as shown.

$$- \underline{\quad a \quad} + \underline{\quad a \quad} - \underline{\quad a \quad} + \underline{\quad a \quad} - \underline{\quad a \quad} +$$

Starting with any ion derive an expression for the electrostatic attractive energy between it and its two nearest-neighbor ions. Then add the repulsive energy between it and the two next nearest-neighbor ions. Continue in this manner and show that the sum of resulting terms is given by $U = (-2 \ln 2 \, q^2)/(4\pi\varepsilon_0 a)$.

2-20. In cesium chloride the distance between Cs and Cl ions is 0.356 nm and $n = 10.5$. What is the molar energy of a solid composed of Avogadro's number of CsCl molecules?

2-21. For the Li–F molecule what is the magnitude of the force of attraction between the two ions if the equilibrium separation distance is 0.201 nm? What is the magnitude of the repulsive force between these ions?

2-22. Ionic solids LiF and NaBr have the same structure as NaCl. For LiF, $n = 5.9$ and $r_0 = 0.201$ nm, whereas for NaBr, $n = 9.5$ and $r_0 = 0.298$ nm.

a. Which of these materials is expected to have a higher modulus of elasticity?

b. Calculate the molar energy of ionic interactions for both materials.

2-23. A cation of Mg^{2+} and an anion of O^{2-} are brought together.

a. Write an expression for the Coulomb energy of attraction.

b. If these two ions come to rest at the equilibrium distance of 0.201 nm, what is the value of the Coulomb energy?

c. If MgO has a molar energy of 603 kJ calculate the values of both n and B.

2-24. a. Opposite sides of a rocksalt crystal are pulled to extend the distance between neighboring ions from $r_0 = 0.2820$ nm to $r = 0.2821$ nm. Similarly, during compression, ions are squeezed to within a distance of 0.2819 nm. Compare the value of the tensile (or extension) force developed with that of the compressive force reached. Are they the same?

b. Repeat the calculation if the final distances are 0.2920 nm for extension and 0.2720 nm in compression.

2-25. Atoms on the surface of a solid make fewer bonds with surrounding atoms than do interior atoms. Sketch, in schematic fashion, the interatomic potential of surface and interior atoms as in Fig. 2-12.

2-26. Provide a reasonable physical argument for each of the following statements.

    a. The higher the melting temperature of the solid, the greater the depth of the potential energy well.

    b. Materials with deep energy wells are likely to have a more symmetrical potential energy curve.

    c. Materials with high melting points tend to have low coefficients of thermal expansion.

    d. Materials with high melting points tend to have large moduli of elasticity.

2-27. a. In a 1-cm^3 cube of Au metal what is the electron momentum in the x direction for the state $n_x = 3.83 \times 10^7$, $n_y = 1$, $n_z = 0$?

    b. What is the electron momentum in the y direction?

    c. How does the answer to (a) change for a 1000-cm^3 cube of Au?

2-28. Rationalize the decrease in magnitude of the energy band gap with increasing atomic number for elements in the fourth column of the Periodic Table.

2-29. In what ways are electrons in an isolated copper atom different from electrons in a copper penny?

2-30. Quantum well structures consisting of layers of very thin semiconductor films have been synthesized and used in advanced electronic devices. They have the property of trapping electrons in a well much like a "particle in a box." In a one-dimensional, infinitely high, 5-nm-wide well what is the electron energy for the $n = 1$ state?

2-31. How narrow would the quantum well in Problem 2-30 have to be before the electron energy for the $n = 1$ state were equal to that for the $n = 1$ ground state for the electron in the hydrogen atom? (Note that the electron energy is increased by squeezing its domain.)

2-32. What is so special about the electronic structure of carbon that enables it to form more than a million organic compounds with hydrogen, oxygen, nitrogen, and sulfur?

2-33. State whether ionic, covalent, metallic, or van der Waals bonding is evident in the following solids. (Where applicable distinguish between intramolecular and intermolecular bonding.)

a. Mercury   b. KNO_3 c. Solder   d. Solid nitrogen   e. SiC
f. Solid CH_4 g. Aspirin   h. Rubber   i. Na_3AlF_6 j. PbTe   k. Snow

2-34. Is it likely that a stable solid will have an interatomic potential energy of interaction where the attractive part of the potential is steeper than the repulsive part? Why?

3

STRUCTURE OF SOLIDS

3.1 INTRODUCTION TO CRYSTAL STRUCTURE

Although the previous chapter developed models of solids based on the **electronic structure** of atoms, it was largely done without assumptions as to how atoms were physically arranged or positioned relative to each other. In this chapter an almost opposite viewpoint is adopted: the **physical structure** is our primary concern and the electronic structure is of secondary interest. Of course, at a deeper level the two types of structures are intimately coupled. The richness in the diversity of materials properties is attributable to the countless combinations of different atomic species, their electronic constitution, and the way atoms are geometrically arranged within solids.

It is convenient to subdivide the structure of solids into two broad categories: **crystalline** and noncrystalline or **amorphous.** Many solids have a crystalline structure that is described by an ordered geometric array of atoms that stretches endlessly in all directions in a repetitive fashion. We do not think of materials like structural steel and solder as being crystalline because they are not clear, transparent, faceted, sparkling, angular, and so on. Despite external appearances to the contrary, materials like metals are, in fact, crystalline based on the scientific criterion of an orderly *internal* arrangement of atoms. On the other hand, atoms are not positioned in predictably ordered arrays in amorphous solids like silica glass and most hydrocarbon polymers.

In this chapter we are concerned with a description of the structure of crystalline solids. Although crystals have been known, and some even fondled for thousands of years, it is only in the last century that they have been classified. To create actual crystal structures one must first imagine a three-dimensional array of points in space distributed such that each has identical neighbors. There are only 14 ways to arrange points in space having this geometrical property. These special point arrays are known as **Bravais point lattices.** It must be appreciated that the space of any one of these point lattices can be delineated into cells in a number of arbitrary ways. The lines drawn in Fig. 3-1 have been found to be the most convenient way to outline **unit cells** of points that can be used to distinguish each of the Bravais lattices. Table 3-1 summarizes the lengths and interaxial angular relationships among the 14 unit cells that can be further classified as belonging to one of seven crystal systems. The unit cell dimensions, a, b, and c are known as **lattice parameters** or constants, and these can be measured with high precision by the X-ray diffraction techniques that are described later in the chapter.

TABLE 3-1   **SPACE LATTICES AND CRYSTAL GEOMETRIES**

Crystal system (Bravais lattice)	Axial lengths and interaxial angles	Examples
Cubic (simple cubic, body-centered cubic, face-centered cubic)	Three equal axes, three right angles $a = b = c$, $\alpha = \beta = \gamma = 90°$	Au, Cu, NaCl, Si, GaAs
Orthorhombic (simple orthorhombic, base-centered ortho-rhombic, body-centered orthorhombic, face-centered orthorhombic)	Three unequal axes, three right angles $a \neq b \neq c$, $\alpha = \beta = \gamma = 90°$	Ga, Fe$_3$C
Tetragonal (simple tetragonal, body-centered tetragonal)	Two of the three axes equal, three right angles $a = b \neq c$, $\alpha = \beta = \gamma = 90°$	In, TiO$_2$
Hexagonal (simple hexagonal)	Two equal axes at 120°, third axis at right angles $a = b \neq c$, $\alpha = \beta = 90°$, $\gamma = 120$	Zn, Mg
Rhombohedral (simple rhombohedral)	Three equal axes equally inclined, three equal angles $\neq 90°$ $a = b = c$, $\alpha = \beta = \gamma$	Hg, Bi
Monoclinic (simple monoclinic, base-centered mono-clinic)	Three unequal axes, one pair of axes not at 90° $a \neq b \neq c$, $\alpha = \gamma = 90°$, $\beta \neq 90°$	KClO$_3$
Triclinic	Three unequal axes, three unequal angles $a \neq b \neq c$, $\alpha \neq \beta \neq \gamma$	Al$_2$SiO$_5$

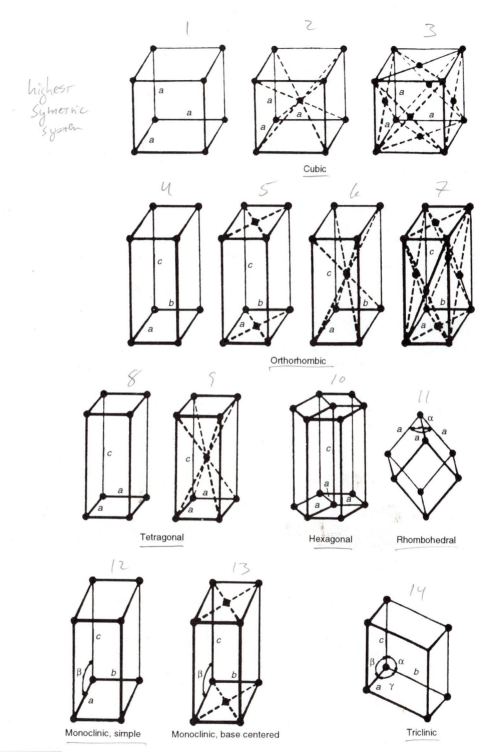

highest
symetric
system

1   2   3

Cubic

4   5   6   7

Orthorhombic

8   9   10   11

Tetragonal   Hexagonal   Rhombohedral

12   13   14

Monoclinic, simple   Monoclinic, base centered   Triclinic

FIGURE 3-1 The 14 Bravais point lattices.

When a **single** atom or **group** of two or more atoms is assigned to each lattice point a physically real crystalline solid emerges. The atom groupings must retain the same orientation at each lattice point to preserve the crystal geometry and symmetry. All of the structural information pertaining to the perfect placement of about 10^{22} atoms per cubic centimeter can now be conveniently condensed into what can be revealed by a single equivalent unit cell. A good portion of the chapter is devoted to the placement of atoms within unit cells of Bravais lattices and the geometric implications of doing so.

The focus is primarily on elemental solids but occasional reference is made to the crystal structure of simple compounds. A discussion of the structure of noncrystalline hydrocarbon polymers and inorganic silica glasses, as well as the more complex crystal structures of ceramic materials, is deferred to Chapter 4.

3.2. COMMON CRYSTAL STRUCTURES

3.2.1. Cubic Structures: Metals

Cubic crystal systems are the only ones we consider at any length in this book because they are the simplest to deal with. A more compelling reason for considering cubic structures, however, is that a majority of the important engineering materials crystallize in this geometry. For example, approximately 40 or so elemental metals have cubic crystal structures. Another 25 crystallize in hexagonal structures and they are treated in Section 3.2.4. The simple cubic (SC) structure is virtually never seen, although polonium metal at ~10°C is an exception. Instead, the body-centered cubic (BCC) and face-centered cubic (FCC) structures of Fig. 3-1 are the only ones commonly observed. In each structure the cube edge a, or lattice parameter, serves to differentiate one metal from another.

To visualize better the process of forming crystalline substances let us first consider a unit cell in the BCC Bravais point lattice. If individual iron atoms are now exactly placed at each lattice point such that $a = 0.2867$ nm, the material known as metallic iron would be produced. Iron atoms are represented as spheres in Fig. 3-2A, but just as there are no connecting lines in actual crystals there are no spheres. Each sphere represents an Fe nucleus surrounded by a complement of 24 core electrons (i.e., $1s^2, 2s^2, 2p^6, 3s^2, 3p^6, 3d^6$), plus two additional $4s$ valence electrons. Other BCC metals include Na, K, Cr, Ta, V, and W. We speak of there being only two atoms per unit cell, on the average, in BCC metals. One atom in the center is wholly contained within the cell. Each of the eight corner atoms is shared by eight other cells so that in total, only one atom ($8 \times \frac{1}{8} = 1$), belongs to the unit cell. Thus the center atom plus eight corner atom octants contribute a total of two atoms.

Similarly, gold metal is the result of placing Au atoms at each site of an FCC Bravais point lattice having a lattice parameter of 0.4079 nm (Fig. 3-2B).

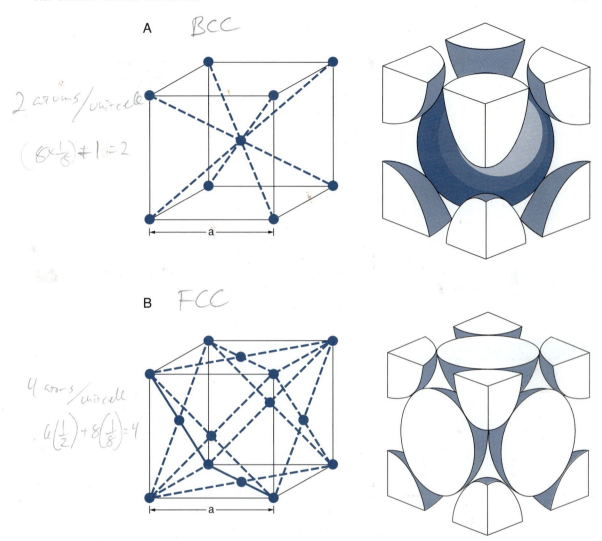

A BCC

2 atoms/unitcell

$(8 \times \frac{1}{8}) + 1 = 2$

B FCC

4 atoms/unitcell

$6(\frac{1}{2}) + 8(\frac{1}{8}) = 4$

FIGURE 3-2 Illustration of the placement of (A) iron atoms on a body-centered cubic point lattice and (B) gold atoms on a face-centered cubic point lattice.

There are four atoms per unit cell in FCC metals; three come about because atoms centered within the six square faces are each shared by two neighboring cells ($6 \times \frac{1}{2} = 3$) and, as before, one is contributed by eight corner atoms. Metals that exhibit FCC structures include Al, Cu, Ag, Ni, Pt, and Fe (from 910° to 1400°C). Interestingly, the inert gases Ar, Ne, Kr, and Xe also crystallize in FCC structures at low temperatures.

An abbreviated list of the crystal structures of metals and their lattice parameters is found in Table 3-2. More complete details of the structure of the elements can be found in Appendix A. These limited crystallographic data enable calculation of a considerable amount of information about the geometry of atomic packing provided atoms are assumed to be hard spheres.

1. **Atomic radii.** In BCC structures, atoms of radius r touch along the cube body diagonal. Thus, as shown in Fig. 3-3, $r + 2r + r$ or $4r = a\sqrt{3}$, and $r = a\sqrt{3}/4$. As $a = 0.2867$ nm, the calculated Fe atom radius is 0.1241 nm. This value is considerably larger than the ionic radii of Fe^{2+} and Fe^{3+}, which are 0.087 and 0.067 nm, respectively. In the case of FCC lattices, atoms touch along the cube face diagonal so that $r + 2r + r$ or $4r = a\sqrt{2}$, and $r = a\sqrt{2}/4$. For Au where $a = 0.4079$ nm, $r = 0.1442$ nm, which again is larger than the ionic radius of Au^{1+} (0.137 nm).

2. **Atomic packing factor.** The atomic packing factor (APF) is defined as the ratio of the volume of atoms, assumed to be spheres, to the volume of the unit cell (see Fig. 3-3). For BCC structures there are two atoms per cell. Therefore, APF $= 2 \times \frac{4}{3}\pi r^3/a^3$. But $r = a\sqrt{3}/4$ and after substitution APF $= \sqrt{3}\pi/8$ or 0.680, a number that is independent of atomic sphere size.

In the case of FCC structures, APF $= 4 \times \frac{4}{3}\pi r^3/a^3$. As $r = a\sqrt{2}/4$, APF $= 0.740$. This demonstrates that of the two structures, FCC is more densely packed.

TABLE 3-2   **LATTICE CONSTANTS OF SELECTED METALS AT ROOM TEMPERATURE**

Body-centered cubic		Face-centered cubic	
Metal	a (nm)	Metal	a (nm)
Chromium	0.2885	Aluminum	0.4050
Iron	0.2867	Copper	0.3615
Molybdenum	0.3147	Gold	0.4079
Potassium	0.5247	Lead	0.4950
Sodium	0.4291	Nickel	0.3524
Tantalum	0.3298	Platinum	0.3924
Tungsten	0.3165	Silver	0.4086
Vanadium	0.3023	Palladium	0.3891

Hexagonal close-packed			
Metal	a (nm)	c (nm)	c/a ratio
Cadmium	0.2979	0.5617	1.890
Zinc	0.2665	0.4947	1.856
Ideal HCP			1.633
Beryllium	0.2286	0.3584	1.568
Cobalt	0.2507	0.4069	1.623
Magnesium	0.3210	0.5211	1.623
Titanium	0.2951	0.4685	1.587
Zirconium	0.3231	0.5148	1.593

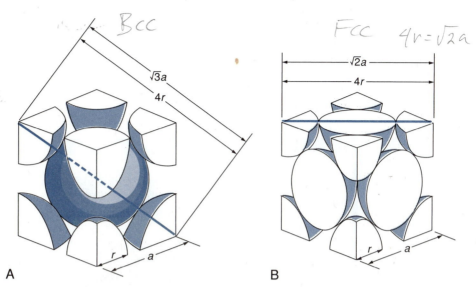

(A) BCC unit cell illustrating the geometry of contact between hard sphere atoms. Contact of atoms is made along the cube diagonal. Atomic packing factor = 0.680. (B) FCC unit cell illustrating the geometry of contact between hard sphere atoms. Contact of atoms is made along the cube face diagonal. Atomic packing factor = 0.740.

3. **Atomic planar density.** By analogy to APF we may define the atomic planar density (APD) as the ratio of the projected area of atoms in a plane to the total area of the plane (Fig. 3-4). To obtain APD only atoms whose centers lie in the plane are counted. The cube face in the BCC lattice contains the equivalent of a single atom because each of the four corner atoms is shared by four similar squares (i.e., $4 \times \frac{1}{4} = 1$). Therefore, $APD_{BCC} = 1 \times \pi r^2/a^2$, or 0.589, when accounting for the connection between r and a.

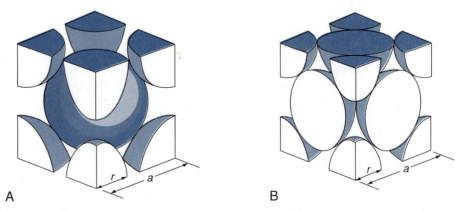

(A) Atomic planar density in the BCC lattice. On indicated plane APD = 0.589. (B) Atomic planar density in the FCC lattice. On indicated plane APD = 0.785.

In the case of the FCC cube face, there are the equivalent of two atoms. Hence $APD_{FCC} = 2 \times \pi r^2/a^2$, or 0.785. Of the two structures, this plane is more densely packed with atoms in FCC. If the FCC plane is rotated by 45° the atomic arrangements on the two planes appear to be similar. But, whereas atoms touch in FCC, they do not in BCC, and this accounts for the higher APF in the former.

EXAMPLE 3-1

a. What is the fractional volume change in iron as it transforms from BCC to FCC at 910°C? Assume that $a(BCC) = 0.2910$ nm and $a(FCC) = 0.3647$ nm.

b. What is the volume change if Fe atoms of fixed radius pack as hard spheres?

ANSWER   a. The volume (V) per atom is equal to volume per unit cell/atoms per unit cell. For BCC Fe, V (BCC) per atom $= a^3(BCC)/2$. Similarly for FCC Fe, $V(FCC)$ per atom $= a^3(FCC)/4$. After substitution, $V(BCC)/atom = (0.2910)^3/2 = 0.01232$ nm^3; $V(FCC)/atom = (0.3647)^3/4 = 0.01213$ nm^3. Therefore, $\Delta V/V = (0.01213 - 0.01232)/0.01232 = -0.0154$ or -1.54%, a result consistent with the abrupt contraction observed when BCC Fe transforms to FCC Fe.

b. If atoms pack as hard spheres, $a(BCC) = 4r/\sqrt{3}$ and $a(FCC) = 4r/\sqrt{2}$. Therefore,

$$\Delta V/V = \{[4r/\sqrt{2}]^3/4 - [4r/\sqrt{3}]^3/2\}/[4r/\sqrt{3}]^3/2$$

$$= (5.657 - 6.158)/6.158 = -0.0810, \text{ or } -8.10\%.$$

Note that the two values for $\Delta V/V$ differ considerably. The results for part b depend solely on geometric packing. They are independent of the lattice parameter and, therefore, of the particular metal. Obviously, atoms in real metals and iron, in particular, are not hard spheres of fixed radius. Other metals like manganese and plutonium also exist in both BCC and FCC crystal forms of different lattice parameters.

4. Linear atomic density.   Lines drawn through the centers of neighboring atoms will also pierce the centers of other atoms periodically spaced along the line. The linear atomic density (LAD) is defined as the ratio of the number of atomic diameters included along a line or direction to the length of the line (Fig. 3-5). Quite simply, LAD is the inverse of the distance between neighboring atomic centers and has units of atoms/length. LAD values will differ along the same direction in different crystals, as well as along different directions within the same crystal. For example, along the cube edge in BCC crystals, LAD $= 1/a_{BCC}$, but along the cube face diagonal, LAD $= 1/\sqrt{2}a_{BCC}$. In FCC crystals the respective LAD values are $1/a_{FCC}$ and $2/\sqrt{2}a_{FCC}$. As seen in Chapter 7, material deformation under stress usually occurs along directions of highest linear atomic density.

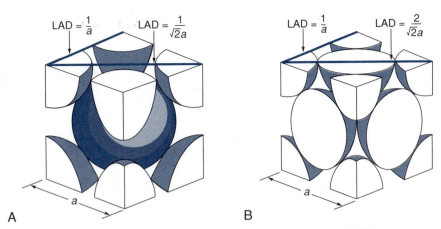

A

B

FIGURE 3-5 (A) Linear atomic density in the BCC lattice. (B) Linear atomic density in the FCC lattice.

3.2.2. More Complex Cubic Structures

3.2.2.1. Atoms of the Same Type

Earlier it was mentioned that in some crystal structures two or more atoms substitute for each Bravais lattice point. This is the case for solid-state silicon which is derived by placing two identically oriented Si atoms at each and every FCC lattice point as shown in Fig. 3-6A. One of the Si atoms sits precisely at the lattice point, which may be taken as the origin of the repeating two-atom motif. The other is displaced from it by $a/4$ (a is the lattice parameter) in each of the rectangular coordinate directions. Each Si atom in this crystal structure now has four nearest neighbors in tetrahedral coordination. In Section 2.4.4 it was noted that the four covalent bonds radiating out from carbon in diamond were configured in the same manner; hence the name *diamond cubic* for this

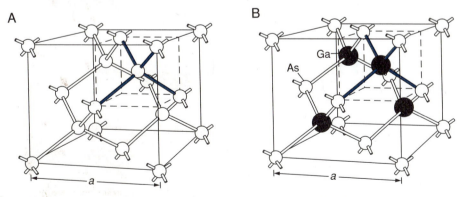

A

B

FIGURE 3-6 (A) Model of the crystal structure of the silicon diamond cubic lattice. The diamond cubic lattice may be viewed as two interpenetrating, but displaced FCC lattices. (B) Model of the crystal structure of GaAs. Both gallium and arsenic atoms populate separate FCC lattices which interpenetrate one another.

type of structure. Germanium and gray tin are the other two important members of the group IV elements with diamond cubic structures. The covalent nature of the bond weakens sufficiently by the time the metal lead is reached in the Periodic Table. This change in electronic character is also reflected by a different crystal structure, for example, FCC.

3.2.2.2. Dissimilar Atoms

If we consider the same FCC Bravais lattice but now substitute a *two-atom molecule* motif for each point, many very important materials can be generated. One can think of this structure as consisting of two interpenetrating FCC lattices, one for each atom. For example, when a GaAs molecule is located at each point, the As atom at the cell origin and the Ga atom at a distance $a/4$ away in each of the coordinate directions, the gallium arsenide crystal structure is produced (Fig. 3-6B). It is also known as the zinc blende structure, and can obviously also be generated by reversing the roles of Ga and As. Other important materials that crystallize with this same structure include the semiconductors InP, GaP, InSb, CdTe, ZnS, and InAs. The carbides TiC, ZrC, TaC, and VC; the nitrides TiN, ZrN, and VN; and the hydrides TiH and ZrH also exhibit this structure.

EXAMPLE 3-2

Calculate the atomic packing factor for Si and GaAs assuming the atoms (or ions) are hard spheres.

ANSWER   For Si there are eight atoms per unit cell: four associated with the original FCC point lattice and four additional atoms contained within each cell. Therefore, $\text{APF(Si)} = 8 \times \frac{4}{3}\pi r^3 / a^3$. If it is assumed that the Si atoms touch along the tetrahedral covalent bonds, $a\sqrt{3}/2 = 4r$ and $r = a\sqrt{3}/8$. After substitution, $\text{APF(Si)} = \pi\sqrt{3}/16 = 0.340$.

In the case of GaAs there are four Ga and four As atoms per unit cell and they possess atomic radii r_{Ga} and r_{As}, respectively. The total volume of the unit cell occupied by atoms is $4 \times \frac{4}{3}\pi(r_{Ga}^3 + r_{As}^3)$. If Ga and As atoms touch along the covalent bond then $2(r_{Ga} + r_{As}) = a\sqrt{3}/2$, or $a = (4/\sqrt{3})(r_{Ga} + r_{As})$. Therefore,

$$\text{APF(GaAs)} = \frac{\frac{16}{3}\pi(r_{Ga}^3 + r_{As}^3)}{\left[\frac{4}{\sqrt{3}}(r_{Ga} + r_{As})\right]^3}$$

$$= \frac{\pi\sqrt{3}}{4}\frac{(r_{Ga}^3 + r_{As}^3)}{(r_{Ga} + r_{As})^3}.$$

The atomic packing factor for GaAs depends on the radii for both Ga and As, that is, $r_{Ga} = 0.135$ nm and $r_{As} = 0.125$ nm. After substitution,

$$\text{APF(GaAs)} = (\pi/4)\sqrt{3}(0.135^3 + 0.125^3)/[0.135 + 0.125]^3 = 0.342.$$

These calculations suggest that both Si and GaAs are loosely packed compared with the FCC metals. This partially accounts for the low density of the former. Further, the values of APF for all monoatomic diamond cubic structures are identical irrespective of atomic size; if two different atoms (ions) are involved, APF values differ depending on atomic radius.

3.2.2.3. NaCl and CsCl

Let us again consider the FCC Bravais lattice. If we now faithfully replace each point by a NaCl molecule motif, we can generate the common rocksalt structure depicted in Fig. 3-7. In a slight structural twist, the Cl atom (ion) is positioned at the cell origin while the Na atom (ion) is located at the center of an original FCC cube edge (or vice versa). Each atom has six nearest neighbors of the other kind. Many oxides (e.g., MgO, BeO), fluorides (e.g., LiF, NaF), and chlorides (e.g., KCl) crystallize with rocksalt structures.

The least complex cubic structure containing two atoms is based on the simple cubic Bravais lattice. Although there are virtually no monoatomic simple cubic materials, there are a number of examples where a two-atom motif decorates a simple cubic Bravais lattice. The CsCl structure shown in Fig. 3-8 repeats a pattern of Cl (or Cs) at the cube origin and Cs (or Cl) at the cube

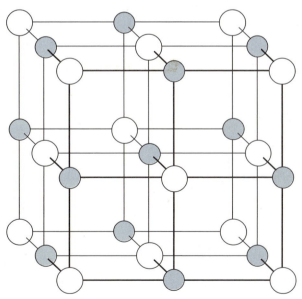

FIGURE 3-7 Rocksalt crystal structure. In NaCl the sodium and chlorine atoms populate separate FCC lattices which interpenetrate one another. Unlike GaAs, however, the orientation of the NaCl motif differs.

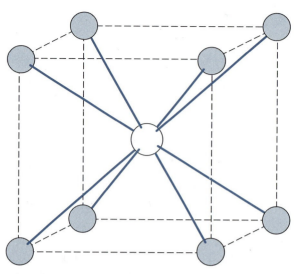

FIGURE 3-8 Crystal structure of cesium chloride. Two interpenetrating simple cubic lattices, one for Cs^+ and the other for Cl^-, generate this structure.

center. Each atom is coordinated with eight atoms of the other kind in what is known as cubic coordination. Of course CsCl is an ionic solid and if ions touch along the cube diagonal then $a\sqrt{3} = 2r_{Cs+} + 2r_{Cl-}$. Other CsCl-type structures are exhibited by halide salts (e.g., TlI, RbCl) and certain ordered binary metal alloys (e.g., CuZn or β brass, AlNi, BeCu).

3.2.3. Coordination of Ions

Is there any rhyme or reason to the way ions are coordinated in ionic solids? Why were eight Cl^- ions coordinated to each Cs^+ ion in CsCl, but only six Cl^- ions to each Na^+ ion in NaCl? Similar differences are observed in the coordination of metal and oxygen ions in ceramic (oxide) materials. Clearly, the model of pairwise interaction between ions in an ionic "molecule," so successfully used to calculate binding energies in Chapter 2 (Example 2-5), is of limited use in predicting crystal structure.

Linus Pauling, twice a Nobel laureate, has formulated useful rules to predict structural coordination of ions in ionic materials in terms of ionic size. Specifically, we wish to understand how anions are coordinated about a single cation in such structures assuming that they are larger than cations. With no loss in generality, what follows also holds if cations are larger than anions. It will be shown that a coordination number N_c, defined as the number of anions surrounding the cation, is dependent on the ratio of cation to anion radii (i.e., r_c/r_a). Assuming a hard sphere ionic model the central cation cannot remain

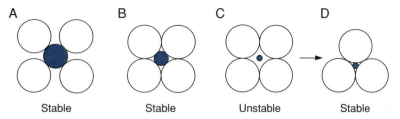

A   B   C   D

Stable   Stable   Unstable   Stable

FIGURE 3-9 Stable and unstable coordination between ions of different size. As the anion size remains constant, the cation successively shrinks in size (A)–(C), until an unstable configuration (C) is reached. This rearranges to a stable configuration (D).

in contact with all the surrounding anions if the ion size disparity is too great. Under these conditions the structure would tend to fly apart because of electrostatic repulsion between anions; a smaller N_c would be favored. This is apparent in Fig. 3-9 where stable and unstable ionic coordination configurations are indicated. On the other hand, if the cation grows too large, electrostatic energy will be reduced if greater numbers of anions participate in bonding, that is, if N_c rises. Between these extremes a given N_c will be stable.

The stability criterion assumes r_c/r_a must be greater than some critical value determined by the condition that anions touch each other and the cation simultaneously. Structures that are stable as a function of r_c/r_a are shown in Fig. 3-10. Up to an r_c/r_a ratio of 0.155 a linear structure is predicted, whereas a triangular **planar** ($N_c = 3$) array is stable at $0.225 > r_c/r_a > 0.155$. Beyond these ratio values three-dimensional **tetrahedrons** ($N_c = 4$), **octahedrons** ($N_c = 6$), **cubes** ($N_c = 8$), and **close-packed face-centered cubes** or **cuboctahedrons** ($N_c = 12$) become progressively more stable. Of these, tetrahedrons and octahedrons often appear in ceramic crystals; these polyhedra typically consist of a small cation contained within a cage of either four or six larger oxygen anions.

EXAMPLE 3-3

Calculate the critical values of r_c/r_a for octahedral and cubic coordination.

ANSWER   In octahedral coordination the ions assume the planar geometry shown in Fig. 3-9B, at the octahedron midsection. Therefore, $2r_a\sqrt{2} = 2r_a + 2r_c$. Solving, $r_c/r_a = \sqrt{2} - 1 = 0.414$. For r_c/r_a values greater than this critical value, octahedral coordination is stable.

In the case of cubic coordination, ions touch along the cube diagonal. Therefore, $2r_a\sqrt{3} = 2r_a + 2r_c$. Solving, $r_c/r_a = \sqrt{3} - 1 = 0.732$. For r_c/r_a values greater than this critical value cubic coordination is stable.

Linear
$N_c = 2$

Triangular
$N_c = 3$

Tetrahedron
$N_c = 4$

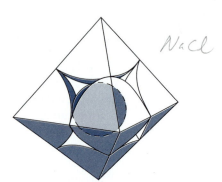

NaCl

Octahedron
$N_c = 6$

$r_{Cs} > r_{NA}$

CsCl

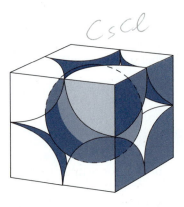

Cube
$N_c = 8$

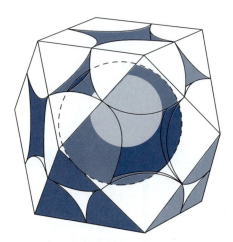

Cuboctahedron
$N_c = 12$

FIGURE 3-10 Coordination of dissimilar ions according to the Pauling scheme. Stability ranges for particular ionic coordination are linear, $r_c/r_a > 0$; triangular, $r_c/r_a > 0.155$; tetrahedral, $r_c/r_a > 0.225$; octahedral, $r_c/r_a > 0.414$; cubic, $r_c/r_a > 0.732$; and cuboctahedral, $r_c/r_a > 1$.

EXAMPLE 3-4

Predict values of N_c for NaCl and CsCl.

ANSWER Values for r in nm are r (Na$^+$) = 0.098, r (Cs$^+$) = 0.165, and r (Cl$^-$) = 0.181.

For NaCl, r (Na$^+$)/r (Cl$^-$) = 0.098/0.181 = 0.544. Therefore, N_c = 6. For CsCl, r (Cs$^+$)/r (Cl$^-$) = 0.165/0.181 = 0.912. Therefore, N_c = 8. These values are entirely consistent with the observed structures.

3.2.4. Hexagonal Structures

That some 25 elements crystallize with a **hexagonal** structure has already been noted. But most of them do not exhibit the simple hexagonal lattice suggested by the Bravais classification. Rather, metals like Zn, Mg, Be, Ti, and Zr crystallize in a **hexagonal close-packed** (HCP) structure, in which a two-atom motif is associated with each Bravais lattice point. These HCP materials are every bit as closely packed as FCC metals and bear a close relationship to them. To see how let us consider the hard sphere model of these structures and the close-packed planes represented in Fig. 3-11. On this plane atoms touch each other the way billiard balls do when racked on a pool table. For convenience, this plane is designated as A. Next, the structure is built up by depositing a second layer (B) of atoms which nest in the interstices of the first plane. But there are two interpenetrating sets of interstices, and once one is selected and totally occupied, the second set must be completely unoccupied.

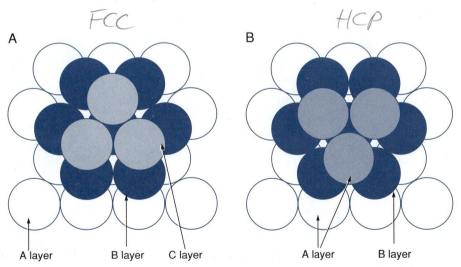

A layer   B layer   C layer   A layer   B layer

FIGURE 3-11 (A) Plan view of the hard sphere ABCABC... stacking characteristic of the close-packed plane in the FCC lattice. (B) Plan view of the ABABAB... stacking of close-packed planes leading to the HCP structure.

What happens on the third plane of atoms is quite critical, and nature allows a choice of two distinct possibilities.

1. In FCC materials the third plane of atoms sits directly above the set of nests or hollows *not* chosen in atomic layer B. Therefore, this top row is displaced horizontally from both planes A and B and is designated as the C layer. The fourth row, however, is identical to the first or A layer and the fifth to the second or B layer, and so on. This ABCABC... stacking (Fig. 3-11A) then continues in this sequence endlessly, and in the process a perfect FCC lattice is generated (Fig. 3-12A).

2. In HCP materials the third plane is identical to layer A and so the pattern of repetition is ABABABA... (Fig. 3-11B). Since atoms are as efficiently packed in HCP as in FCC lattices, the value of APF is 0.740. The structure has two lattice parameters: the dimension in the close-packed plane is a, and the distance between A–A planes is c (Fig. 3-12B). If atoms were hard spheres, a bit of geometry would show that $c/a = 1.633$.

Though the difference between FCC and HCP structures appears to be little more than an accident in stacking, property differences can be quite profound. As we shall see in Chapter 7, mechanical deformation of metals depends not only on the number of distinct close-packed planes, but on the number of close-packed directions in them. In HCP metals all close-packed planes are parallel so only one need be considered. In it lie three close-packed directions

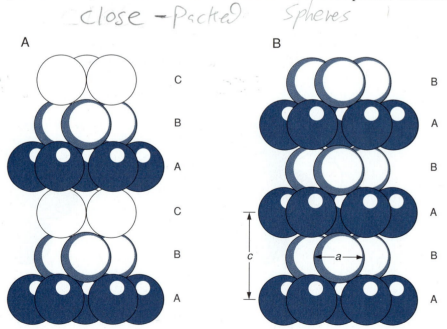

A

C
B
A
C
B
A

B

B
A
B
A
B
A

FIGURE 3-12 (A) Side view of a hard sphere model of the FCC structure. (B) Side view of a hard sphere model of the HCP structure.

(high LAD) set at 120° to each other. FCC lattices, however, can be cut in four distinct (nonparallel) ways, each yielding a set of close-packed planes. The same three close-packed directions exist and so FCC metals have four times the number of *plane–direction* combinations (12) as HCP metals (3). This is why aluminum is more ductile or deformable than zinc. It also accounts for differences in electronic structure.

Nonmetals also crystallize in hexagonal structures, for example, graphite (see Fig. 2-17B), tellurium, and selenium (see Fig. 4-17).

3.3. ATOM POSITIONS, DIRECTIONS, AND PLANES IN CRYSTAL STRUCTURES

3.3.1. Atomic Positions

Rather than continue to refer to "origin of unit cell," "cube diagonal," or "cube face" when speaking about atom positions, directions, and planes in a crystal, we have developed a shorthand notation to identify these geometric features. The notation is simply based on coordinate geometry and can be understood in most cases without resorting to analytic geometry or vector analysis. As the need arises, however, we will borrow some simple results from these latter mathematical methods. In what follows we strip away the atoms that were so carefully placed on Bravais lattice points in the previous section and just consider the Bravais points themselves.

Let us consider the rectangular coordinate axes shown in Fig. 3-13 and populate the space with SC, BCC, and FCC point lattices. Each point is given three indices to denote its x, y, and z coordinates in the space. Cube corners have a triad of integers, for example $(0, 0, 0)$, the origin, and $(1, 0, 0)$, $(0, 1, 0)$, $(0, 0, 1)$, $(1, 1, 0)$, and so on. Cube center indices are denoted by $(\frac{1}{2}, \frac{1}{2}, \frac{1}{2})$ whereas cube face-centered atoms are described by $(0, \frac{1}{2}, \frac{1}{2})$, $(\frac{1}{2}, 0, \frac{1}{2})$, $(\frac{1}{2}, \frac{1}{2}, 0)$, $(1, \frac{1}{2}, \frac{1}{2})$, and so on. From the standpoint of crystallography, SC is characterized by $(0, 0, 0)$, BCC by $(0, 0, 0)$ and $(\frac{1}{2}, \frac{1}{2}, \frac{1}{2})$, and FCC by $(0, 0, 0)$, $(0, \frac{1}{2}, \frac{1}{2})$, $(\frac{1}{2}, 0, \frac{1}{2})$, and $(\frac{1}{2}, \frac{1}{2}, 0)$. These numbers are consistent with the respective one, two, and four atoms per unit cell noted earlier. Translation of either the single-, two-, or four-point pattern throughout space will yield the indicated point lattices. In Fig. 3-13 only the unit cube or cell is considered and all coordinates are positive. For neighboring cells negative coordinates may arise. For example, BCC lattice cube centers in three adjacent cells might have coordinates of $(\frac{1}{2}, -\frac{1}{2}, \frac{1}{2})$, $(\frac{1}{2}, \frac{1}{2}, \frac{1}{2})$, and $(\frac{1}{2}, \frac{3}{2}, \frac{1}{2})$.

3.3.2. Directions

Directions are vectors that connect any two lattice points in a prescribed sense. The indices of a direction are given by the differences in x, y, z coordinates of the two points in question. For example, the direction connecting the outer

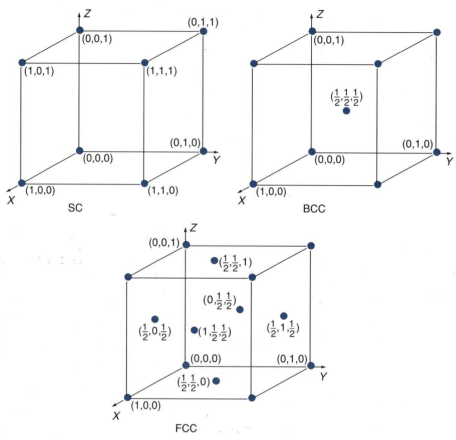

FIGURE 3-13 Coordinate positions of points within SC, BCC, and FCC unit cells relative to the indicated origin of the x, y, and z axes.

cube centers noted at the end of the previous paragraph is $(\frac{1}{2}, \frac{3}{2}, \frac{1}{2})$ minus $(\frac{1}{2}, -\frac{1}{2}, \frac{1}{2})$ or $(0, 2, 0)$. By convention, the resultant direction components are reduced to smallest whole numbers and placed in brackets without commas; the so-called *Miller indices* for the direction are [010]. For the antiparallel direction pointed toward the negative y axis the indices are [0$\bar{1}$0]. The minus sign makes crystalline directions different mathematically, but not physically.

A number of directions in cubic lattices are displayed in Fig. 3-14. Notice that the cube diagonal has [111] indices. But [$\bar{1}$11], [1$\bar{1}$1], and [11$\bar{1}$] are equivalent directions as are the oppositely oriented [$\bar{1}\,\bar{1}$1], [11$\bar{1}$], [$\bar{1}$1$\bar{1}$], and [$\bar{1}\,\bar{1}$1] directions. This collection of physically equivalent directions constitutes a family denoted by ⟨111⟩. That negative indices arise is the consequence of arbitrarily positioning the center of the coordinate axes. Note that any lattice point can serve as the center of coordinates.

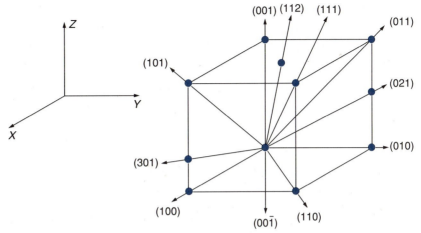

FIGURE 3-14   Miller indices of directions in cubic lattices. After C. R. Barrett, W. D. Nix and A. S. Tetelman, *The Principles of Engineering Materials,* Prentice–Hall, Englewood Cliffs, NJ (1973).

Many phenomena in materials occur preferentially along specific crystallographic directions. For example, there are stronger magnetic effects along [100] than along the [111] direction in iron. Similarly, when a crystal of MgO is stressed, deformation effects are observed along the [110] direction. Transport of electrical charge and heat as well as diffusion of atoms frequently occurs more readily in certain crystal directions. In such cases we speak of **anisotropic** behavior. This is to be contrasted with **isotropic** behavior where properties are independent of crystallographic direction. Although numerous lattice directions are possible, a surprisingly small number ever play a significant role in governing material properties. Those that do have low-number indices which correspond to directions densely populated by atoms (i.e., high LAD).

3.3.3. Planes

Like points and directions, crystallographic planes are identified by three Miller indices or numbers. Any three points define a plane; any two directions will also define a plane in which both lie. But, a single direction can lie in many different planes. The following simple recipe can be used to identify uniquely a given plane in cubic (and noncubic) crystals:

1. Express the intercepts of the plane on the three coordinate axes in number of unit cell dimensions.
2. Take reciprocals of these numbers.
3. Reduce the reciprocals to smallest integers by clearing fractions.

The resulting triad of numbers placed in parentheses without commas, (hkl), is known as the Miller indices of the plane in question. A number of common as well as uncommon planes in cubic lattices are shown in Fig. 3-15. Any time a plane passes through the origin, the above recipe will not work. In such a case it must be remembered that the origin of the coordinate axes can be arbitrarily shifted to any other lattice point. Another alternative is to translate the plane parallel to itself until intercepts are available. By either of these means it is always possible to have the involved plane slice through the unit cell. In the case of cube faces one intercept (e.g., x) is 1 and the other two intercepts extend to infinity. Therefore, $1/1 = 1$, $1/\infty = 0$, $1/\infty = 0$, and the planar indices are (100). Other (100)-type planes that are physically equivalent have (010), (001), ($\bar{1}$00), (0$\bar{1}$0), and (00$\bar{1}$) indices. These six planes constitute a family that is identified as {100}, and similarly for other planes.

Many phenomena in materials are associated with or occur preferentially on specific crystallographic planes. For example, brittle cleavage cracks readily nucleate and grow on (100) planes in MgO crystals. Silicon oxidizes more

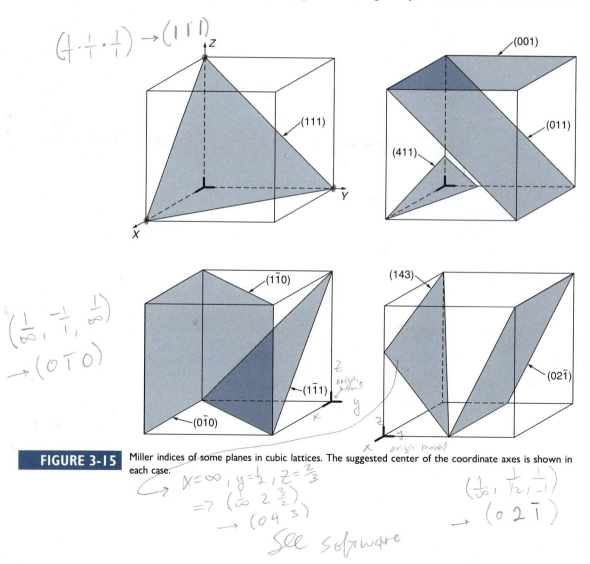

FIGURE 3-15 Miller indices of some planes in cubic lattices. The suggested center of the coordinate axes is shown in each case.

rapidly on the (111) plane than on the (100) plane. Crystal growth is frequently favored on one plane relative to others. Microelectronic (e.g., transistor) and optoelectronic (e.g., laser) device fabrication can be reliably accomplished only on specific semiconductor planes. Regardless of application, the important planes are usually the most atomically dense ones (i.e., high APD). Such planes are also ones of low index.

3.3.4. Theorems from Analytic Geometry and Vector Algebra

Several simple theorems that will prove useful when dealing with directions and planes in cubic systems are derived in this section. We start by considering the arbitrary (hkl) plane in the cubic system drawn relative to the cell cube depicted in Fig. 3-16. The intercepts are seen to be a/h, a/k, and a/l. From analytic geometry the equation of the plane can be expressed algebraically by $x/(a/h) + y/(a/k) + z/(a/l) = 1$ in an x, y, z coordinate system. A line drawn from the origin normal to the plane is, in fact, the interplanar spacing d_{hkl}. This plane normal has direction cosines α_1, α_2, and α_3 given by $\cos \alpha_1 = d_{hkl}h/a$, $\cos \alpha_2 = d_{hkl}k/a$, and $\cos \alpha_3 = d_{hkl}l/a$. Analytic geometry states that $\cos^2\alpha_1 + \cos^2\alpha_2 + \cos^2\alpha_3 = 1$. Substituting, $\cos^2\alpha_1 + \cos^2\alpha_2 + \cos^2\alpha_3 =$

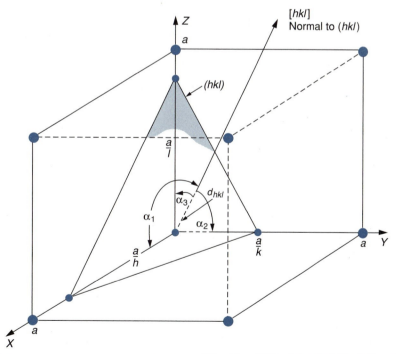

FIGURE 3-16 Geometry used to demonstrate that $d_{hkl} = a/(h^2 + k^2 + l^2)^{1/2}$ and that the Miller indices of the normal to a plane are identical to those of the plane itself.

$d^2_{hkl}(h^2 + k^2 + l^2)/a^2 = 1$, or

$$d_{hkl} = a/(h^2 + k^2 + l^2)^{1/2}. \tag{3-1}$$

This handy relationship, which is true only for cubic systems, will be used subsequently.

It is a simple matter now to determine the indices of the normal to the (hkl) plane. The coordinates of the point lying in the plane that is intercepted by the normal are $(d_{hkl} \cos \alpha_1/a, d_{hkl} \cos \alpha_2/a, d_{hkl} \cos \alpha_3/a)$. Therefore, the indices of the direction from the origin to the point in question are $[d_{hkl} \cos \alpha_1/a - 0, d_{hkl} \cos \alpha_2/a - 0, d_{hkl} \cos \alpha_3/a - 0]$ or $[d_{hkl} \cos \alpha_1/a, d_{hkl} \cos \alpha_2/a, d_{hkl} \cos \alpha_3/a]$. From the definition of direction cosines and Eq. 3-1 it is easy to show that $\cos \alpha_1 = h/(h^2 + k^2 + l^2)^{1/2}$, $\cos \alpha_2 = k/(h^2 + k^2 + l^2)^{1/2}$, and $\cos \alpha_3 = l/(h^2 + k^2 + l^2)^{1/2}$. Substitution for d_{hkl} (Eq. 3-1) and for the direction cosines yields the direction $[h/(h^2 + k^2 + l^2), k/(h^2 + k^2 + l^2), l/(h^2 + k^2 + l^2)]$. Finally by multiplying through by $(h^2 + k^2 + l^2)$, the direction $[hkl]$ emerges.

We have just proved the theorem that in cubic crystals the indices of the direction normal to plane (hkl) are $[hkl]$; both direction and plane have identical indices!

Next let us consider the angle ϕ between two arbitrary directions $[h_1 k_1 l_1]$ and $[h_2 k_2 l_2]$. The components of unit vectors representing these two directions are simply the direction cosines or $[h_1/(h_1^2 + k_1^2 + l_1^2)^{1/2}, k_1/(h_1^2 + k_1^2 + l_1^2)^{1/2}, l_1/(h_1^2 + k_1^2 + l_1^2)^{1/2}]$ and $[h_2/(h_2^2 + k_2^2 + l_2^2)^{1/2}, k_2/(h_2^2 + k_2^2 + l_2^2)^{1/2}, l_2/(h_2^2 + k_2^2 + l_2^2)^{1/2}]$. From the simple properties of vector dot products,

$$\cos \phi = [h_1 k_1 l_1] \cdot [h_2 k_2 l_2] \tag{3-2}$$

$$= \frac{h_1 h_2 + k_1 k_2 + l_1 l_2}{\sqrt{h_1^2 + k_1^2 + l_1^2} \sqrt{h_2^2 + k_2^2 + l_2^2}}.$$

3.3.5. Indices in Hexagonal Crystals

Although the three Miller indices are sufficient to identify planes and directions in any Bravais lattice, it is customary to consider four axes and corresponding indices in hexagonal crystals. We note in Fig. 3-1 that close-packed (111) or **basal** planes (also A–A layers in Fig. 3-11B) bound the top and bottom of the hexagonal prism. In the bottom basal plane, three [110]-type directions, set 120° from one another, radiate out from the center point of the coordinate system. Each of these three axes (a_1, a_2, a_3) intercepts a couple of prism side planes at distance a. The fourth axis lies in the vertical direction and intercepts the top basal plane at distance c.

As examples, the intercepts of a prism side plane (**prism** plane) are $a_1 = 1$, $a_2 = \infty$, $a_3 = -1$, $c = \infty$, and after reciprocals are taken the indices are $(10\bar{1}0)$; directions that outline the hexagonal perimeter of the basal plane, and which also lie in the prism plane, have $[11\bar{2}0]$ indices. This plane and direction are shown in Table 7-3. Note that the indices are always such that the sum of the first two is equal to the negative of the third index.

EXAMPLES 3-5

a. What are the d_{hkl} spacings of the four most widely separated planes in FCC Pt which has a lattice parameter of 0.3924 nm?

b. Indicate indices of at least four directions that lie in the (111) plane.

c. What is the angle between any two neighboring tetrahedral bonds in the diamond cubic structure?

ANSWERS   a. Reproducing Eq. 3-1 we have $d_{hkl} = a/(h^2 + k^2 + l^2)^{1/2}$. The largest d_{hkl} spacings occur for the smallest values of $(h^2 + k^2 + l^2)^{1/2}$. The (100), (110), (111), and (200) planes yield the four smallest respective denominators of 1, $\sqrt{2}$, $\sqrt{3}$, and 2. Therefore, the required d_{hkl} spacings are 0.3924 nm, 0.3924/$\sqrt{2}$ = 0.2775 nm, 0.3924/$\sqrt{3}$ = 0.2266 nm, and 0.3924/2 = 0.1962 nm.

b. Directions that lie in the (111) plane must be perpendicular to the normal to this plane or [111]. We seek directions [hkl] such that the dot product with [111] vanishes (cos 90° = 0). Therefore, $h + k + l = 0$, and examples of combinations that satisfy this requirement are [$\bar{1}$10], [01$\bar{1}$], [$\bar{2}$11], [5$\bar{3}$2], and so on. Note that one or more of the indices must be negative. But, if the (11$\bar{1}$) plane had been selected instead, the indices could be all positive, e.g., [101].

c. Tetrahedral bonds lie along [111]-type directions, which we will assume to be [111] and [$\bar{1}$$\bar{1}$1]. Therefore, by Eq. 3-2

$$\cos \phi = \frac{\{(1)(-1) + (1)(-1) + (1)(1)\}}{\sqrt{(1^2 + 1^2 + 1^2)}\sqrt{(-1^2 + (-1)^2 + 1^2)}} = -\frac{1}{3}$$

The angle whose cosine is $-\frac{1}{3}$ is $\phi = 109.5°$.

3.4. EXPERIMENTAL EVIDENCE FOR CRYSTAL STRUCTURE

This chapter has provided very elaborate geometric models of atomic positions within all classes of solids. Not only are the structures of materials specified with absolute certainty, but stated interatomic spacings are supposedly precise to small fractions of atomic dimensions. How do we know it is all correct? Virtually all of the evidence has come from X-ray diffraction methods and this is the first topic treated in this section. However, a number of other techniques of more recent origin have also yielded important structural information at the atomic level. Transmission electron microscopy, field ion microscopy, and scanning tunneling microscopy have provided exciting new ways to examine atomic details in thin-film materials and on the surfaces of bulk materials. These microscopy methods are indispensable in modern scientific work in materials and, for this reason, are discussed later in this section.

3.4.1. X-Ray Diffraction

Assume we did not know better and were asked to estimate very roughly the distance between atoms in a crystalline solid of atomic weight M and density ρ. A reasonable place to start is to assume that individual atoms are contained within a unit cell or cube of side a. In a volume of M/ρ there are Avogadro's number (N_A) of atoms. Therefore, the volume per atom is $M/\rho N_A$, and dimension a is equal to the cube root of this or $a = (M/\rho N_A)^{1/3}$. Substituting typical values—$M = 50$ g/mol and $\rho = 5$ g/cm^3—yields $a = [50/(5 \times 6.02 \times 10^{23})]^{1/3} = 2.55 \times 10^{-8}$ cm, or 0.255 nm. In the elementary theory of optics it was shown that physical **diffraction** effects arose from a periodic grating if photons had a comparable, but somewhat smaller, wavelength than the grating spacing. But crystals are essentially an atomic grating, and diffraction effects can therefore be expected at a wavelength of say ~0.1 to 0.2 nm. X rays have wavelengths in this range of the electromagnetic spectrum, hence their widespread use in diffraction experimentation on crystals.

In Fig. 3-17 we see two parallel X rays of wavelength λ, from a much wider beam, impinging on a crystal surface at angle θ. Parallel to the surface is a row of crystal planes, separated by distance d_{hkl}, that extends into the crystal. We assume that what happens at the top layers also occurs at the deeper planes reached by other, more penetrating X rays. The condition for constructive interference of the diffracted beam is that the two exiting rays be in phase, or that the path difference between ray 2 and ray 1, or $AB + BC$, be an integral number (n) of wavelengths. But simple geometry reveals that $AB = BC = d_{hkl}\sin\theta$. Therefore, the condition known as **Bragg's law** emerges as

$$n\lambda = 2d_{hkl}\sin\theta. \tag{3-3}$$

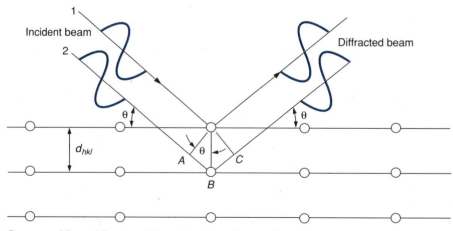

| **FIGURE 3-17** | Geometry of Bragg diffraction of X rays from atoms lying in the reflecting plane.

Satisfaction of Bragg's law results in a diffracted beam of high intensity that can easily be detected with a radiation counter or with photographic film. If the conditions of Eq. 3-3 are not rigorously met then the diffracted intensity is zero. Nothing in this formula restricts its use to cubic crystal structures although this will be our only application in the book.

Bragg's law is truly one of the most important relationships in materials science, and it is capitalized upon in many experimental arrangements (not only X-ray diffraction) to yield crystallographic information about solids. Basically two fundamental types of crystallographic information are sought. The first has to do with the size and shape of unit cells and the geometric orientation of these cells relative to the external surfaces. How the two or more atoms in the motif are distributed at each lattice point, and within a unit cell, is the second concern. The latter often poses a formidable experimental challenge and requires complex analysis of the intensities of the diffracted beams. We consider only the first and simpler of these applications.

Selection of λ and θ determines the basis of two broadly different experimental X-ray techniques based on Bragg's law:

1. λ *has a fixed wavelength (monochromatic), and θ is variable.* In one configuration a thin layer of crystalline powder of the material in question is spread on a nondiffracting planar substrate and exposed to the X-ray beam. Because there will always be some crystallites favorably oriented with respect to the beam each of the planar spacings, that is, $d_1, d_2, d_3, \ldots$, will diffract at the different angles $\theta_1, \theta_2, \theta_3, \ldots$. Therefore, provision must be made to experimentally detect diffraction at as many angles as possible. By this technique d spacings are directly measured, and from them lattice parameters can be extracted.

2. λ *is variable (white radiation), and θ is fixed.* Orientations of large single crystals can be determined this way. Surfaces of such crystals may bear little angular relation to the internal crystal structure but they can be oriented relative to it by this so-called Laue method.

The experimental geometry and equipment employed in the powder diffraction method is illustrated in Fig. 3-18. X rays are generated by directing an electron beam of high voltage and current (typically 35 kV, 20 mA) at a metal (e.g., Cu, Mo) target anode situated inside an evacuated X-ray tube. We have already studied how X rays can be generated from atoms in Section 2.3.1. In this case the incident electron beam first creates holes in the K shell of the target atoms. These are filled by electrons descending from the L and M shells to create K_α and K_β X rays, respectively. The more intense K_α X rays are used, and after collimation and filtering, they impinge on the specimen. A typical diffractometer trace is shown in Fig. 3-19 and reveals a number of peaks with varied intensities as a function of angle 2θ.

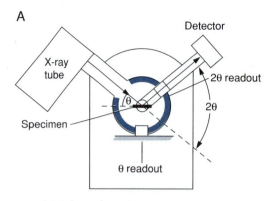

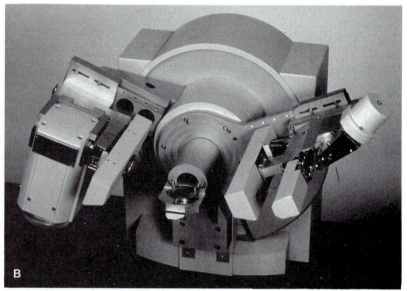

FIGURE 3-18 (A) Schematic of the diffraction geometry for the powder method including positions of X-ray tube, specimen, and detector. (B) Photograph of a modern X-ray diffractometer. Courtesy of North American Philips Corporation.

Before analyzing this diffraction pattern we assume for simplicity that the discussion is limited solely to cubic materials. Combination of Eqs. 3-1 and 3-3 yields, for $n = 1$,

$$\sin^2\theta = \frac{\lambda^2(h^2 + k^2 + l^2)}{4a^2}. \tag{3-4}$$

For two different planes $(h_1 k_1 l_1)$ and $(h_2 k_2 l_2)$, diffracting at angles θ_1 and θ_2, a proportionality can be drawn given by

$$\sin^2\theta_1/\sin^2\theta_2 = (h_1^2 + k_1^2 + l_1^2)/(h_2^2 + k_2^2 + l_2^2). \tag{3-5}$$

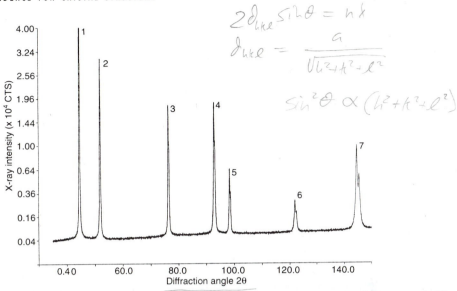

$$2d_{hkl}\sin\theta = n\lambda$$

$$d_{hkl} = \frac{a}{\sqrt{h^2 + k^2 + \ell^2}}$$

$$\sin^2\theta \propto (h^2 + k^2 + \ell^2)$$

FIGURE 3-19 Diffractometer trace of a cubic metal plotted as the intensity of the diffracted signal versus angle 2θ. Peak numbers refer to Example 3-6, where the identity of this metal is revealed. Courtesy of B. Greenberg, North American Philips Corporation.

It is easy enough to read the angles from the diffractometer trace and obtain d_{hkl}; but which planes do they correspond to? If the actual values for h, k, and l were known the lattice parameter could be determined from Eq. 3-1. But, in general we do not know the diffracting planes and additional information is required.

It turns out that not every plane diffracts, an unexpected result that can be capitalized upon during analysis of diffraction patterns. For example, let us consider diffraction from the (100) plane of the BCC structure shown in Fig. 3-20. The rays from neighboring (100) planes (i.e., 1 and 3), a distance a apart, are in phase but are 180° out of phase with the ray scattered from the center atom of plane 2. As there are equal numbers of type 1 and 2 planes, the net diffracted intensity vanishes for this plane in BCC materials; however, the (200) planes with interplanar spacing $a/2$ do diffract because the intensity cancellation will not occur for the second-order reflection ($n = 2$). Thus, complex wave phase relations between diffracted X rays within unit cells will cause there to be permitted as well as missing reflections. Table 3-3 lists the planes that do and do not diffract in SC, BCC, and FCC lattices. Note that in simple cubic lattices there are no missing reflections and all planes diffract. In BCC, $h^2 + k^2 + l^2$ must equal 2, 4, 6, 8, 10, and so on, whereas in FCC, $h^2 + k^2 + l^2$ has values of 3, 4, 8, 11, 12, and so on. This corresponds to h, k, l values that are all even or odd in both cases.

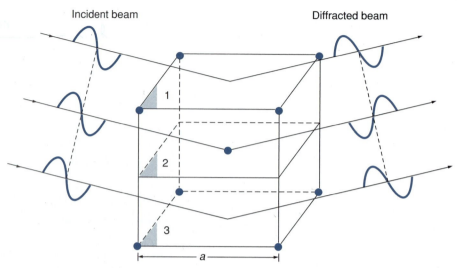

Incident beam   Diffracted beam

FIGURE 3-20 Reflected X rays from (100) planes of the BCC unit cell. As a result of adding the amplitudes from neighboring diffracted rays, the (100) reflection vanishes.

TABLE 3-3 **DIFFRACTING PLANES IN CUBE CRYSTALS**

1. Rules governing the presence of diffraction peaks

Lattice	Reflection present	Reflection absent
Simple cubic	Every (hkl) plane	None
BCC	$(h + k + l)$ = even	$(h + k + l)$ = odd
FCC	(h, k, l) all odd or all even	h, k, l mixed even and odd

2. Miller indices of diffracting planes in BCC and FCC

Cubic planes {hkl}	Sum $h^2 + k^2 + l^2$	Diffracting planes	
		BCC	FCC
{100}	1		
{110}	2	110	
{111}	3	—	111
{200}	4	200	200
{210}	5	—	—
{211}	6	211	—
{220}	8	220	220
{221}	9	—	—
{310}	10	310	—
{311}	11	—	311
{222}	12	222	222

EXAMPLE 3-6

Identify the cubic metal element that gave rise to the diffractometer trace shown in Fig. 3-19. Copper K_α radiation was used with $\lambda = 0.15405$ nm.

ANSWER To identify the metal, values for 2θ must be read from the trace and halved so that both $\sin \theta$ and $\sin^2\theta$ can be evaluated. Then from Eqs. 3-3 and 3-4, d_{hkl} can be calculated and $h^2 + k^2 + l^2$ ratios obtained, respectively. Finally, a is extracted from Eq. 3-4. The results are best tabulated as follows:

Line	2θ	θ	$\sin \theta$	d_{hkl} (nm)	$\sin^2\theta$	Ratio $(h^2 + k^2 + l^2)$	(hkl)	a (nm) (Eq. 3-4)
1	44.52	22.26	0.3788	0.2033	0.1435	3	(111)	0.3522
2	51.93	25.97	0.4379	0.1760	0.1918	4	(200)	0.3518
3	76.37	38.19	0.6182	0.1246	0.3822	8	(220)	0.3524
4	93.24	46.62	0.7268	0.1060	0.5283	11	(311)	0.3515
5	98.43	49.22	0.7572	0.1018	0.5734	12	(222)	0.3524
6	121.9	60.96	0.8742	0.0881	0.7644	16	(400)	0.3524
7	144.6	72.28	0.9526	0.0809	0.9074	19	(331)	0.3525

The key to the analysis is identifying the crystal structure. Because the ratio of $\sin^2\theta$ for the first two diffraction lines is $0.1435/0.1918 = 0.7482$ or ~ 0.750 ($= \frac{3}{4}$), the structure is FCC. Other lines are consistent with diffraction in FCC materials. The average value of a is determined to be 0.3522 nm and corresponds well to the accepted lattice parameter of nickel, $a = 0.3524$ nm.

3.4.2. Seeing Atoms

Seeing is believing, and the scientific as well as nonscientific skeptics among us will not be entirely happy until they actually see atoms. Inferring their positions indirectly to a fraction of an atomic dimension with the use of a ruler to measure diffraction patterns may still leave lingering doubts. Imaging atoms directly, however, offers the fascinating experience of visualizing phenomena in a unique way. In the past few decades remarkable progress has been made in exposing atoms to view. Two of the techniques developed for this purpose are **field ion microscopy** and **scanning tunneling microscopy**.

3.4.2.1. Field Ion Microscopy

Invented in the 1950s by E. Muller, field ion microscopy (FIM) employs a metal-tip specimen that is etched to a very fine point having a radius of curvature (r) of 50–100 nm. Viewing the tip head on, one sees that atoms in the form of tiny bumps are geometrically arranged over the crystalline terrain of a roughly hemispherical surface. The specimen is inserted into a surprisingly simple microscope, shown schematically in Fig. 3-21A, whose chamber is pumped to a very high vacuum ($\sim 10^{-11}$ atm). Other than the metal-tip specimen, there is a television-like fluorescent screen for viewing the image and a means of introducing a pure imaging gas, e.g., helium.

During operation, a high positive voltage ($\sim$20 kV) is applied to the tip after vacuum has been attained. Helium is then introduced and becomes ionized in the vicinity of the tip. An extraordinarily high electric field of several hundred million volts per centimeter exists there. Under these conditions an electron is stripped from the gas and enters one of the protruding atoms where the electric field is concentrated. The resulting He^{+} ion is strongly repelled from the positive tip and flies off in a line to impact the screen and light it up. Ionization occurs

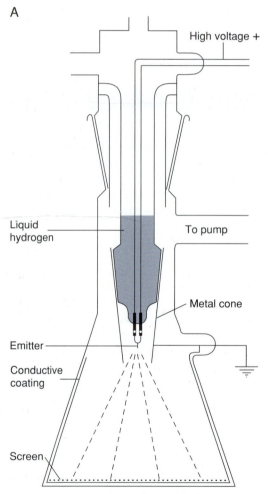

A

High voltage +

Liquid hydrogen

To pump

Metal cone

Emitter

Conductive coating

Screen

FIGURE 3-21 (A) Schematic of a field ion microscope employing helium ions. (B) Field ion microscope image of a tungsten tip. From E. W. Muller, *Advances in Electronics and Electron Physics*, Vol. 13, Academic Press, New York (1960).

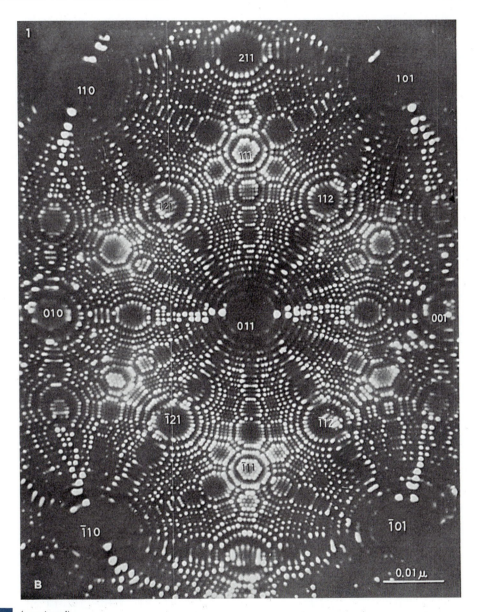

FIGURE 3-21 (*continued*).

all over the specimen surface which is projected, via He ions, across the flat screen to produce the image. The surface of a tungsten tip imaged in this manner is reproduced in Fig. 3-21B. Note that atoms are arranged around circular stepped terraces that outline specific crystal planes. Geometric magnifications of 2.4 million times (the tip–screen distance divided by the tip radius) are attained in FIM. Disadvantages of FIM include its limited applicability to certain metals, the necessity to cool specimens to cryogenic temperatures (i.e., ~20 K), the high vacuum requirement, and the possibility of an electric field-modified surface. Each of these disadvantages is dispelled in scanning tunneling microscopy.

3.4.2.2. Scanning Tunneling Microscopy

Like FIM, scanning tunneling microscopy (STM) has the capability of forming images of atoms at high magnifications. In common, it also uses a metal needle or probe, whose tip is etched to almost an atomic radius. But this tip serves as a conductive electrode and is not the specimen being observed. In operation the tip is brought to within 1 nm of the specimen surface and rastered across it, in atomic scale increments, along both x and y directions (Fig. 3-22). An electron conduction current (known as a **tunneling** current) that is exponen-

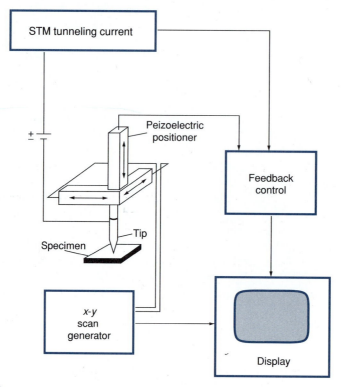

FIGURE 3-22   Schematic of a scanning tunneling microscope.

tially dependent on the narrow vacuum gap distance between tip and specimen flows when a voltage is applied between them. In one mode of operation a feedback loop maintains a constant tunneling current by adjusting the tip height. The feedback signal necessary to do this is monitored as the tip scans across the atomic bumps, and is converted into an image such as the one shown for a (111) Si surface in Fig. 3-23 (see color plate). Interestingly, the exposed **surface** atoms do not assume the crystallographic positions associated with the **bulk** crystal lattice (111) plane. At the free surface covalent bonds are necessarily cut. Rather than dangle vertically upward they bend over and link with cut bonds from neighboring atoms. In this way the pair of involved electrons form a low-energy surface covalent bond. In the process, surface atoms move into new positions on the **reconstructed** Si surface.

An allied technique known as **atomic force microscopy** (AFM) also senses the specimen surface topography with a fine probe tip. In this case a flexible cantilever beam supports the tip, which is either attracted to or repelled from the surface atoms. The forces responsible are attractive and van der Waals in nature for tip–specimen distances greater than 1 nm. For narrower gaps, electron cloud overlap in both tip and specimen results in repulsive forces. Sensitive force measurements during surface scanning are then converted into topographic images, and atomic resolution can be achieved in some cases. The ability to image insulators is a distinct advantage of AFM relative to STM methods.

3.5. DEFECTS IN CRYSTALLINE SOLIDS

Perfect crystal structures with all atoms situated in predictable lattice sites are rarely realized. Only in specially grown, oriented semiconductor single crystals is this ideal closely approximated in an engineering material over appreciable lattice dimensions. Polycrystalline solids consisting of numerous small single crystals or grains of random orientation invariably form during production and subsequent processing of materials. In them, errors abound in the perfect placement of atoms and the stacking of planes. Such effects tend to be magnified at the interfaces where grains fit together. At the outset a distinction should be made between such crystallographic defects and gross manufacturing defects and flaws like cracks and porosity. The latter are dealt with in Chapters 8, 10, and 15. Here our concern is with lattice defects having atomic size dimensions. What are the implications of such defects? How harmful are they? What properties are influenced by them? These questions are addressed in various ways in subsequent chapters with the benefit of additional concepts and knowledge of material properties. But first it is necessary to introduce crystallographic defects from a structural point of view, and that is what is done next.

3.5.1. Point Defects

A good way to appreciate *point* defects is through Fig. 3-24. All of the defects shown roughly extend over a single lattice point. They can be divided into two categories: those intrinsic to the pure matrix, and those generated by either the intentional or accidental introduction of foreign atoms.

3.5.1.1. Vacancies

These much studied intrinsic defects exist in all classes of crystalline materials and are represented by a missing atom from a lattice site (Fig. 24A). As they are granted a probability of existence from an energy standpoint, vacancies cannot be eliminated from the lattice. It will be shown in Section 5-8 that an

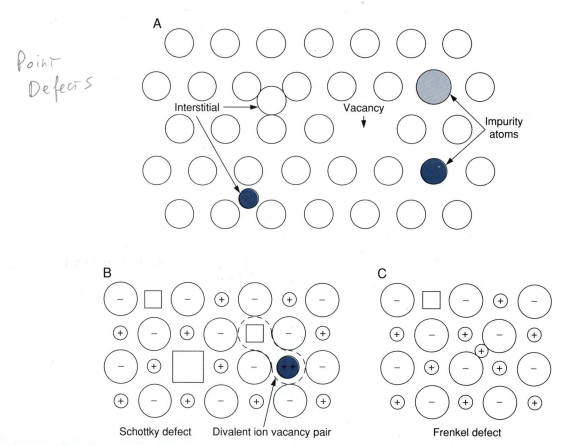

Point
Defects

Interstitial   Vacancy

Impurity
atoms

Schottky defect   Divalent ion vacancy pair   Frenkel defect

FIGURE 3-24   (A) Point defects in an elemental matrix. Shown are vacancies, interstitials, and impurity atoms. (B) Schottky defect consisting of a cation–anion vacancy pair. Also shown is a divalent ion–cation vacancy association. (C) Frenkel defect consisting of a cation interstitial–cation vacancy pair. Vacancies are shown as squares in B and C.

equilibrium vacancy concentration (C_V) varying exponentially with temperature is predicted to exist. This dependence is given by

$$C_V = \exp(-E_V/kT), \qquad (3\text{-}6)$$

where C_V is the fractional concentration relative to all lattice sites, E_V is the energy required to form a vacancy in units of eV/atom or J/atom, k is the Boltzmann constant ($k = 8.62 \times 10^{-5}$ eV/K or 1.38×10^{-23} J/K), and T is the absolute temperature. Alternately, on a per mole basis the gas constant R is equivalently used, where $R = N_A k$. Common values for R are 1.99 cal/mol-K = 8.31 J/mol-K. As an example, taking a typical value of E_V equal to 1.0 eV (1 eV/atom = 96.53 kJ/mol), at a temperature of 1273 K (1000°C), $C_V = \exp[-1.0/(8.62 \times 10^{-5} \times 1273)] = 1.1 \times 10^{-4}$. This means that about one lattice site in 10,000 will be vacant; however, at room temperature $C_V = 1.2 \times 10^{-17}$, a negligibly small concentration.

The presence of a vacancy can cause a local relaxation or redistribution of atoms and electrons surrounding it. This accounts for their importance in phenomena involving atomic motion or diffusion in solids. Lattice atoms or impurity atoms that are completely surrounded by nearest-neighbor lattice atoms are not mobile. But, if there is an adjacent vacancy, then the two can exchange places and atomic motion is possible. This subject is addressed again in Section 6.3 when mass transport in solids is discussed.

3.5.1.2. Interstitials

Atoms that take up positions between regular lattice sites are known as *interstitials* (Fig. 3-24A). The name is derived from sites that are interstices between the atoms. In this defect, atoms leave lattice sites and squeeze into relatively open positions between them. In FCC structures, for example, there are octahedral as well as tetrahedral interstitial sites where cages of either six or four large lattice atoms, respectively, surround the smaller interstitial atom. The center of the FCC unit cell ($\frac{1}{2}, \frac{1}{2}, \frac{1}{2}$) is an example of the former, whereas the ($\frac{1}{4}, \frac{1}{4}, \frac{1}{4}$) site is an example of the latter.

3.5.1.3. Point Defects in Ionic Compounds

In alkali halides and metal oxides where positive and negative ions populate lattice sites, point defect structures are more complex. Maintenance of charge neutrality is the reason. To visualize the issues involved let us consider Fig. 3-24B which depicts an electrically neutral lattice composed of monovalent (positive) cations and (negative) anions. Creation of a cation vacancy means the absence of a single positive charge or an effective negative lattice charge. This defect cannot exist by itself because the lattice is no longer electrically neutral. Therefore, a negative ion vacancy is created at the same time and the associated pair (positive ion vacancy + negative ion vacancy) is known as a **Schottky defect.** The fraction of Schottky defects relative to all molecules in the material is given by a formula similar to that for vacancies (Eq. 3-6);

however, the energy is halved because a pair of vacancies is involved. Another defect, shown in Fig. 3-24C, consists of a cation that hops into an interstitial site. The resulting cation vacancy–interstitial atom is known as a **Frenkel defect.** This defect occurs in AgCl.

Alloying alkali halides can complicate the defect picture. For example, if $CaCl_2$ is added to NaCl the two Cl^- ions occupy two normal anion sites, but the single Ca^{2+} ion can occupy only one cation site (Fig. 3-24B). To compensate for its extra charge an additional cation vacancy must be created. In such a case we must distinguish between two types of cation vacancies: those produced **intrinsically** simply by heating and described by Eq. 3-6, and those introduced **extrinsically** by alloying. Either way these defects play an important role in both material and charge transport processes in all types of ionic solids. Their influence is further discussed in Chapters 5 and 11. And as we shall later see, semiconductor doping has features in common with alloying ionic compounds.

3.5.2. Dislocations

Dislocations are defects that extend along a *line* of atoms in a crystalline matrix. They exist in all classes of solids but are most easily visualized in cubic lattices. There are two fundamental types of dislocations: the edge and screw. An **edge dislocation** can be imagined to arise by cutting halfway into a perfect crystal lattice, spreading the cut apart, and then inserting an extra half-plane of atoms. The resulting edge dislocation defect, denoted by a perpendicular symbol $\perp$, is the line of atoms at the bottom of the inserted plane. It is shown in Fig. 3-25A together with one of its chief attributes, the **Burgers vector.** If a closed loop clockwise traverse is made about a perfect lattice, then the endpoint (f) coincides with the starting point (i). A similar traverse around a region containing the core of an edge dislocation will not close, and the direction

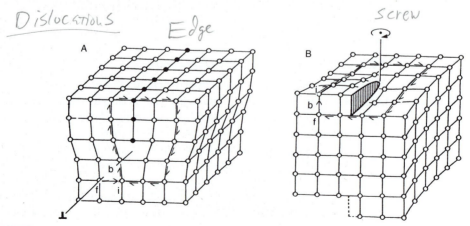

| FIGURE 3-25 | Atomic positions surrounding edge (A) and screw (B) dislocations in a crystal lattice. Burgers vector, **b**, lies perpendicular to the edge dislocation line and parallel to the screw dislocation line. |

connecting the endpoint to the initial point is known as the Burgers vector **b**. The vector sense is arbitrarily assigned by convention as indicated, and its magnitude is one lattice spacing. Furthermore, the Burgers vector is *perpendicular* to the dislocation line. A single dislocation in germanium, imaged by high-resolution transmission electron microscopy methods, is shown in Fig. 3-26 together with the Burgers circuit that defines it.

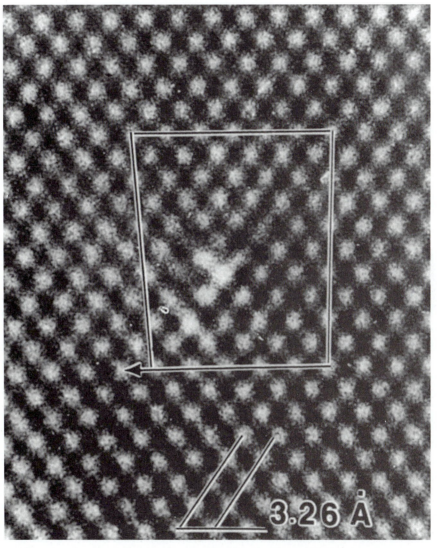

FIGURE 3-26   Electron microscope lattice image of an edge dislocation in germanium. The image is of the (111) plane. From A. Bourret and J. Desseaux, *Journale de Physique C* **6**, 7 (1979).

As before, the second type of dislocation can be imagined to arise by first making a cut halfway into the lattice. Then one-half is sheared up, the other down until a total relative displacement of one atomic spacing occurs. The resulting **screw dislocation** is shown in Fig. 3-25B together with the Burgers vector that defines it. Making a clockwise circuit about the axis of the dislocation is like going down a spiral staircase. The closure error or Burgers vector is parallel to the screw axis. In this case **b** is *parallel* to the dislocation line. Screw dislocations are often seen in the structure of crystals deposited from the vapor phase onto suitable substrates that serve as templates for growth. Substrate defects, especially screw dislocations, are perpetuated into the growing film by atomic extensions of the spiral ramp.

It frequently happens that a single continuous dislocation line can acquire mixed edge and screw character merely by turning a 90° angle corner in the crystal. A view of such a mixed dislocation structure is depicted in Fig. 3-27A. The emergence of edge and screw dislocations on crystal surfaces suggests a way to decorate them and perhaps distinguish individual dislocation types. Chemical etching is expected to be greater where dislocations emerge because atoms are more reactive there relative to atoms on defect-free surfaces. In addition, the resulting attack reflects the crystal orientation. Starting at the dislocation core the etching action laterally widens the base of the pyramidal pit as it deepens the apex. Eventually etch pits become large enough to be seen optically as in Fig. 3-27B. Destructive corrosive pitting of metals occurs by a similar mechanism.

Dislocation defects play a very vital role in influencing the mechanical behavior of crystalline solids. Section 7.4 is devoted to (1) a comprehensive examination of properties of individual as well as large numbers of interacting dislocations, (2) how dislocations are generated, and (3) how they respond to applied stresses. Much of the current interest in dislocations focuses on semiconductor materials, which are required to be defect free. Strategically located dislocations in crystalline regions that are only atomic dimensions in size can seriously impair electrical functions and render microelectronic devices useless.

3.5.3. Grain Boundaries

Virtually all of the materials considered until now have been **single crystals.** In them the ordered atomic stacking extends over macroscopic dimensions. All sorts of precious and semiprecious gems found in nature are essentially single crystals. Research in the solid state is often conducted on single crystals, as free of defects as possible, to assess intrinsic material properties. This usually necessitates growth by special techniques described in Sections 6.4.3 and 12.5.2. Single crystals are not laboratory curiosities, however, and they are grown for commercial electronic (silicon, quartz), magnetic (garnets), optical (ruby), and even mechanically functional (monocrystal metal turbine blade) applications. They can weigh 50 kg and be 30 cm in diameter. The reason for using single

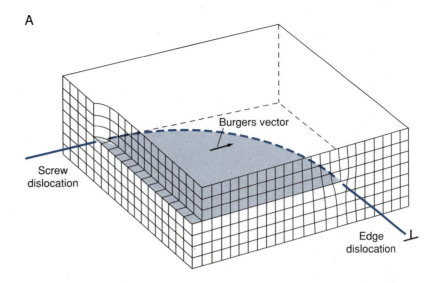

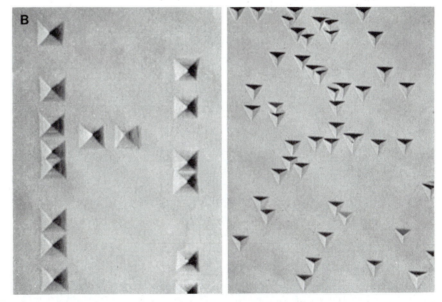

FIGURE 3-27 (A) Curved dislocation line with mixed edge and screw components that emerge on mutually perpendicular surfaces. (B) **Left:** Pyramidal dislocation etch pits on the cleaved (100) surface of LiF; **right:** triangular etch pits on the cleaved (111) surface of CaF_2. From W. G. Johnston, *Progress in Ceramic Science*, Vol. 2, Pergamon Press, New York (1962).

crystals is to eliminate the adverse influence of **grain boundaries.** These *surface or area* defects are the interfaces that separate individual single-crystal grains from one another.

In the hierarchy of defects, grain boundaries are the most structurally complex because they embody aspects of dislocations, point defects, and displaced atoms in a complex admixture. The simplest model of a grain boundary involves tilting two adjacent single-crystal grains relative to each other by a small angle ϕ as illustrated in Fig. 3-28A. When the crystals are welded together at the interface (Fig. 3-28B), a column of isolated edge dislocations is stacked vertically a distance L apart, where L is essentially b/ϕ. We speak of **subgrain** boundaries when very small angles are involved. For larger and arbitrary misorientations between grains, the geometric structural patterns of the resulting high-angle boundaries can be difficult to discern. Careful crystallographic analysis of grain boundaries in solids has revealed the common appearance of a high-angle-tilt

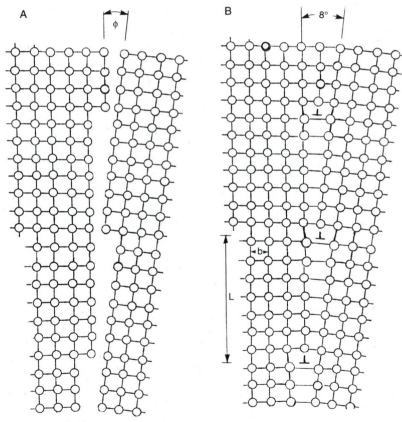

FIGURE 3-28 Dislocation structure of a small-angle-tilt grain boundary. (A) Separate crystals. (B) Crystals joined to form a tilt boundary. At a tilt angle of $\phi = \sim 8°$, $L = 7b$.

angle boundary with $\phi = 36.9°$. The reason for the stability of this boundary is that every fifth atom in it is precisely coincident with the lattice geometry of both adjoining crystals. Known as a **coincidence lattice site boundary** (Fig. 3-29A), a high-resolution image of it is shown in niobium (Fig. 3-29B). There are also twist grain boundaries composed of screw dislocations but they are more complex structurally.

As grain boundaries are relatively open structurally, atoms attached to the boundary tend to be more energetic than those within the bulk. This is apparent in the atomic energy displacement curve (e.g., Fig. 2-12) where larger average atomic spacings mean higher energies. Therefore, grain boundaries are the preferred location for chemical reactions (e.g., etching, corrosion) as well as the solid-state mass transport effects discussed in Chapter 6 (e.g., diffusion, atomic segregation, phase transformation, precipitation). Furthermore, grain boundaries are often stronger than the bulk at low temperatures and help resist

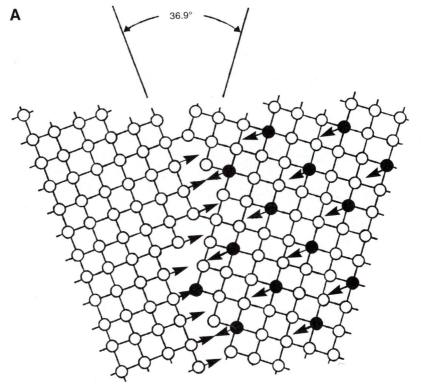

FIGURE 3-29 (A) Model of a coincidence lattice–site boundary. Darkened atoms in grain on right lie in a geometric pattern that coincides with lattice points of left grain. (B) High-resolution electron microscope image of a 36.9° coincidence lattice–site grain boundary in niobium. Notice that Nb atoms are dark. Courtesy of G. H. Campbell, Lawrence Livermore National Laboratory.

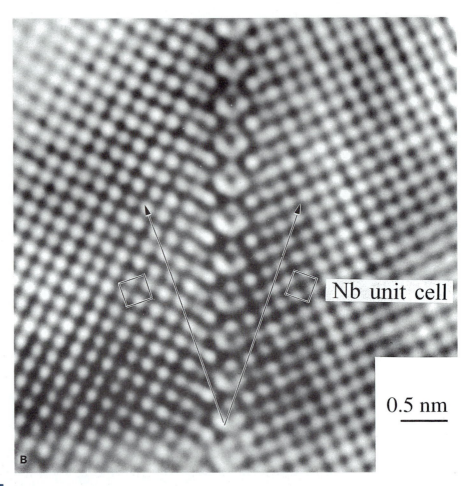

FIGURE 3-29 (*continued*).

the processes of deformation under stress. But at elevated temperatures, they are weaker than the bulk and when the material breaks, the fracture often propagates along a grain boundary path. The dividing line is approximately half the melting temperature (in degrees Kelvin), above which atomic motion and reactions are favored at grain boundaries, weakening their integrity. A more detailed discussion of these issues is presented in Chapters 7, 9, and 10, dealing with mechanical properties.

Great efforts are made to eliminate any hint of grain boundaries in semiconductor materials. An exception is Si solar cells, where cheaper-to-produce polycrystalline cells are often suitable for the application. In addition, other surface defects known as **stacking faults** are occasionally observed in single-crystal Si. Since they occur in other solids as well they are worth brief comment. As already noted, in perfect FCC structures atomic layers are ordered in sequential ABCABCABC... fashion. But, if the planar ordering is disrupted slightly and the sequence is ABCBABCABC... (by adding an extra B row) or is ABCACABC... (by withdrawing a B row), then stacking faults are created. Frequently nucleated at impurities such as oxides, stacking faults in Si fan out and, when they intercept the surface, create defects that are visible in the shape of triangles or squares. The geometric boundaries separate stacking faults from perfect material.

3.6. STRUCTURAL MORPHOLOGIES AND HOW THEY ARE REVEALED

3.6.1. Introduction

The common method for producing large volumes of engineering materials necessarily involves the simultaneous operation of numerous growth centers. In casting, for example, molten metals transform to solid crystallites all over the mold walls as well in the melt interior. Details of solidification are considered later in Sections 5.6.4 and 8.2. Typically, solid treelike projections known as **dendrites** first form at the mold wall. They jut into the melt, growing preferentially along a prominent crystallographic direction, for example, [100]. Secondary arms then sprout and grow in mutually perpendicular directions as dendrites thicken. With further melt consumption, dendrites impinge against one another, and a polycrystalline mass is produced. All the while heat transfer, impurity, and fluid flow effects play important roles in the evolution of resulting grain sizes, shapes, and crystallographic orientations. Dendrites and the development of the cast grain structure in rectangular molds are displayed in Fig. 3-30.

Collectively, the geometric features displayed constitute the structural **morphology,** also commonly known as the **grain structure.** Each grain has a particular crystallographic orientation that extends to the limits of the enveloping grain boundary. Materials in bulk or thin-film form, prepared by various methods (casting, consolidation of powders, mechanical forming, deposition from the vapor or liquid phase, etc.) and treated subsequently by thermal or mechanical means, display grain structures that reflect the prior processing and varied geometric constraints. And this is true of all classes of materials. *It is fair to say that creation and control of optimal grain structures is one of the primary concerns of materials scientists and engineers.* Chapters 5 and 6 are largely dedicated to the underlying issues involved in accomplishing this. But first it is helpful to appreciate nature's morphological designs and the ways in which they are revealed.

A

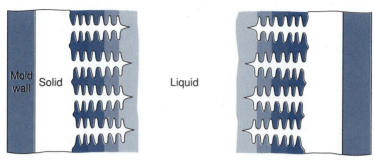

B

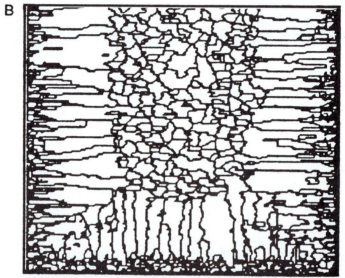

C

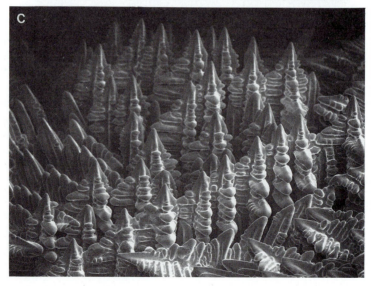

3.6.2. Grain Structure and Topology

Much can be learned about grain structure through optical examination. Sometimes grains are large enough to be seen with the unaided eye; zinc grains on galvanized steel sheet are an example. But usually grain sizes are very small, necessitating viewing with the aid of a microscope. Preparing specimens for optical metallographic observation first requires preparation of a flat, mirror-like surface through a series of grinding and polishing steps, with care taken to eliminate all scratches. Bulk metals, ceramics, semiconductors, and polymers are opaque and therefore cannot be viewed in transmitted light the way thin biological specimens can. Rather, they must be illuminated from above in a metallurgical microscope as shown in Fig. 3-31. Optical magnifications of ~1500× are possible. Revealing the grain structure, however, requires that the grain boundaries be delineated by etching with a suitable chemical solution that preferentially attacks them; light impinging at grain boundaries is then scattered, making them appear dark. Beautiful collections of microstructures have been published and reproducing a few of them cannot pretend to be representative. Nevertheless, a number of material structures obtained by optical microscopy are reproduced in Fig. 3-32, immediately revealing their artistic attributes.

Grain structures are generally tip sections of underlying icebergs. When individual grains of metal are separated from one another by treatment with liquid mercury or gallium, polyhedral shapes like those depicted in Fig. 3-33A have been observed. The resemblance to individual soap bubbles in a froth (Fig. 3-33B) is striking. Even in human fat tissue and vegetable cells there is a grain topology that parallels that of the inorganic materials dealt with here. Topological features of metal and soap grain polyhedra have been analyzed with the following interesting conclusions:

1. The average grain has about 13 faces (F).
2. The average polygon face has very nearly 5 sides or edges.
3. Each grain has about 23 corners (C).

According to topological considerations for stacked polyhedral grains sharing faces and corners,

$$F/2 - C/4 = 1. \qquad (3\text{-}7)$$

Measured values of $F = 14.50$ and $C = 24.85$ for brass and $F = 12.48$ and $C = 20.88$ for an Al–Sn alloy are in excellent agreement with Eq. 3-7 as simple substitution will show; soap bubbles parameters are in perfect agreement.

FIGURE 3-30   (A) Depiction of dendrites growing into a melt. After D. Apelian. (B) Computer simulation of grain structure in a rectangular ingot. Actual ingot cross sections look quite similar. Reprinted from *Acta Metallurgica*, **40**, F. Zhu and W. Smith, 683 and 3369, Copyright (1992), with permission from Elsevier Science Ltd., Pergamon Imprint, The Boulevard, Langford Lane, Kidlington OX51GB, UK. (C) Scanning electron microscope image of a cast dendritic structure in a nickel base superalloy (150×). Courtesy of G. F. Vander Voort, Carpenter Technology Corporation.

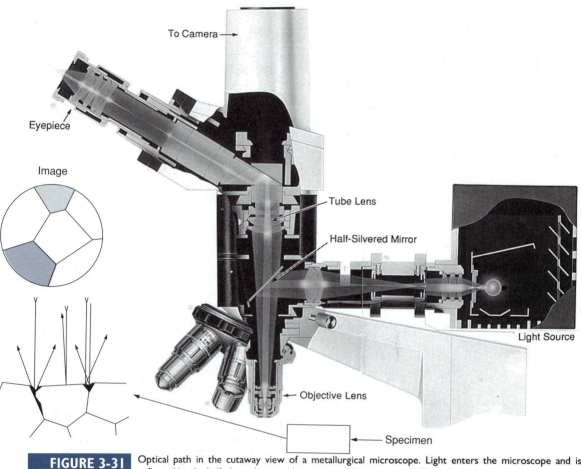

FIGURE 3-31   Optical path in the cutaway view of a metallurgical microscope. Light enters the microscope and is reflected by the half-silvered mirror down toward the specimen. Reflected light from the specimen passes through both the objective and eyepiece lenses. The image magnification is a product of the magnifying power of these two lens systems. Contrast within grains and at grain boundaries is illustrated in the inset. Courtesy of Olympus Optical Company, Ltd.

The topological analysis just presented is too complex for routine characterization of grain structures. Two- rather than three-dimensional information is normally available from microscopes. Therefore, a simpler measure that characterizes grain structure is **grain size.** The American Society for Testing and Materials (ASTM) has defined the grain size (n) as a direct function of the number of grains (N) observed per square inch at $100\times$ magnification. Explicitly,

$$N = 2^{n-1}. \tag{3-8}$$

Application of this formula is made in Example 3-7. Grain sizes and statistical parameters defining structural features and geometries are now commonly determined by computerized image analysis methods.

3.6.3. Scanning Electron Microscopy

Scientific disciplines are identified by the tools they use. In materials science and engineering the **scanning electron microscope** (SEM) is perhaps the most widely used tool. It enables structural characterization at magnifications ranging from a few times to $\sim 150,000\times$. Unlike optical microscopy, sample preparation is not normally required. By adding the ability to perform rapid elemental X-ray microanalysis (see Section 2.3.1), excellent structural and chemical analysis capabilities have been integrated into a single instrument. But this does not even begin to indicate the importance of the SEM in science and technology. More than any other analytical instrument our substantial industrial investment in the microscopic world of microelectronics, optoelectronics, and microbiology has been possible because of the SEM. Without it, advances in these fields would be impossible.

FIGURE 3-32 Optical micrographs of materials. (A) α brass alloy containing 70 wt% Cu–30 wt% Zn (100×). (For parts B and C, see color plates.) (D) Snowflake. Note the hexagonal symmetry (see Fig. 2-20).

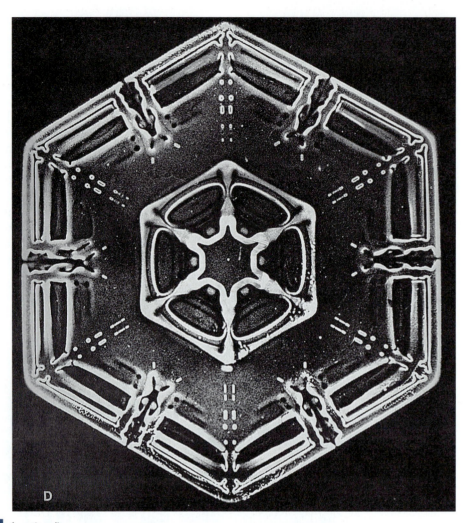

FIGURE 3-32 *(continued)*.

A photograph of a modern scanning electron microscope appears in Fig. 2-5; a schematic of its components and operation is shown in Fig. 3-34. Very briefly, electrons are emitted from a heated (cathode) filament and drawn to the anode by virtue of the ~30 kV potential difference applied between them. Electrons travel down the evacuated microscope column where successive condenser and objective lenses reduce the beam spot size to several nanometers in diameter. Then it impinges on the specimen, converting its energy into electronic excitation of the surface layers. Depending on the magnification required, scanning coils cause the finely focused electron beam to raster across

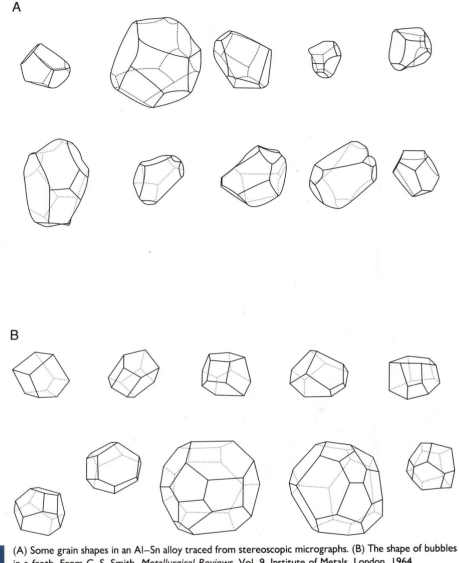

FIGURE 3-33 (A) Some grain shapes in an Al–Sn alloy traced from stereoscopic micrographs. (B) The shape of bubbles in a froth. From C. S. Smith, *Metallurgical Reviews*, Vol. 9, Institute of Metals, London, 1964.

a preselected rectangular area A_S of the specimen. In concert, low-energy (secondary) electrons are emitted from the specimen surface and collected by a detector. External to the SEM there is a cathode ray or TV tube whose electron beam also rasters, but across a screen of fixed area, A_{CRT}. The two raster scans are coupled and the SEM signal modulates the intensity of the TV image. A greater secondary electron signal is emitted from sharp edges and corners of

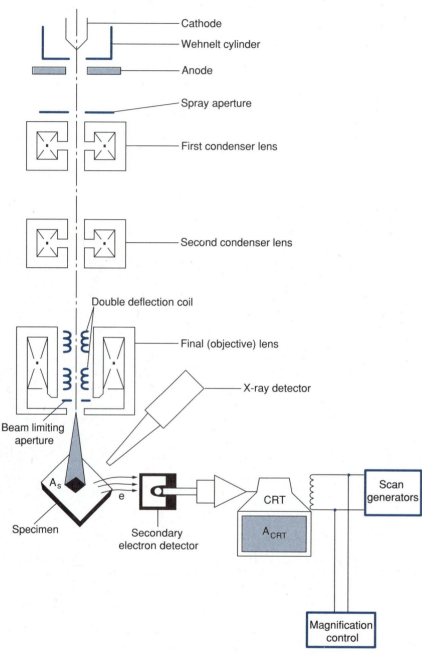

FIGURE 3-34 Schematic of the scanning electron microscope. From J. I. Goldstein, D. E. Newbury, P. Echlin, D. C. Joy, C. Fiore, and E. Lifshin, *Scanning Electron Microscopy and X-Ray Microanalysis,* Plenum Press, New York, 1981.

the topography than from flat surfaces. This source of contrast, together with the great depth of focus, is responsible for the dramatic images SEMs are capable of yielding. The image magnification M is simply given by

$$M = A_{CRT}/A_S. \tag{3-9}$$

Observation of grain morphology, fracture surfaces of materials, polymer composites, integrated circuit topography, sintered powder products, and all aspects of failure analysis are some of the materials, devices, and applications suited to SEM investigation. A potpourri of SEM images in Fig. 3-35 illustrates some of these capabilities.

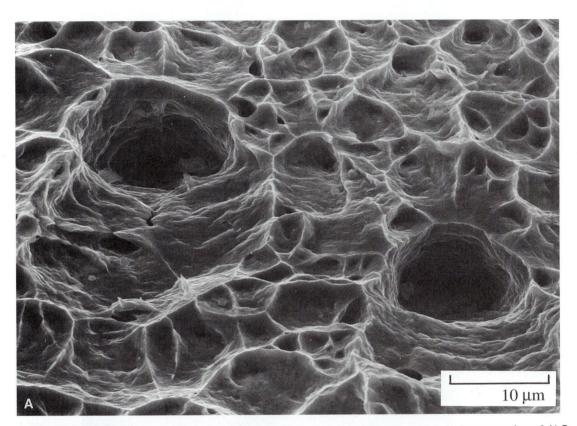

FIGURE 3-35 Assorted SEM images. (A) Ductile fracture surface of stainless steel. (B) Fracture surface of Al_2O_3. Courtesy of R. Anderhalt, Philips Electronic Instruments Company, a Division of Philips Electronics North America Corporation. (C) Polymer composite of carbon fibers in an epoxy matrix. From L. C. Sawyer and D. T. Grubb, *Polymer Microscopy*, Chapman & Hall, London (1987). (D) "SEM-Ant-ICs." An ant with an IC chip in its mandibles. Courtesy of Philips Electronic Instruments Company, a Division of Philips Electronics North America Corporation.

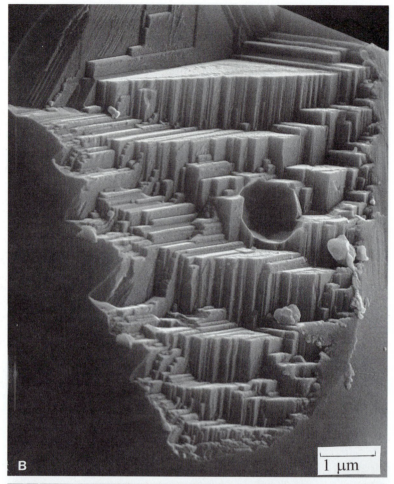

B

1 µm

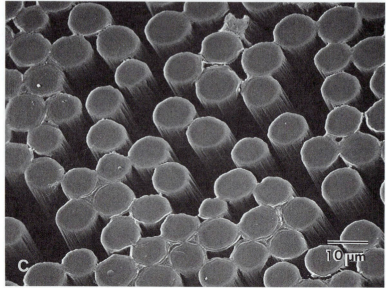

C

10 µm

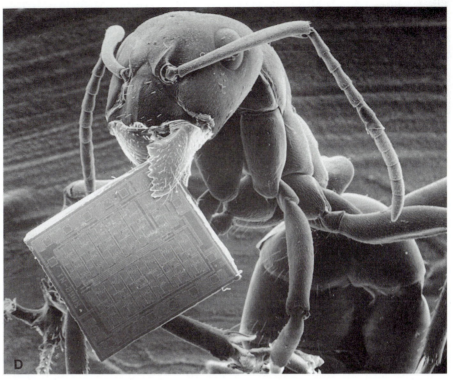

FIGURE 3-35 (*continued*).

3.6.4. Transmission Electron Microscope

The **transmission electron microscope** more closely resembles an optical microscope than an SEM because the latter builds images bit by bit through sequential scanning, whereas the TEM receives the image all at once. But, as shown schematically in Fig. 3-36, it is more instructive to compare TEM optics with those of the common slide projector. The lamp is like the electron source, and the slide is the thin specimen whose image is magnified by the projector (projection) lens. Transmission **optical** microscopes used to view thin biological specimens have a similar arrangement of lenses. Unlike the SEM which images only the material surface, the entire thickness of the TEM specimen is sampled by the impinging electron beam. But because electrons can typically penetrate only through ~100 nm to ~1 μm of a solid with the range of operating voltages available (100 kV normally, up to 1 MV in special TEMs), materials must be thinned to these thicknesses prior to observation. This requirement makes the TEM a less routine analytical tool than, say, the SEM. Rather, the TEM is primarily a research tool; one such high-performance instrument is

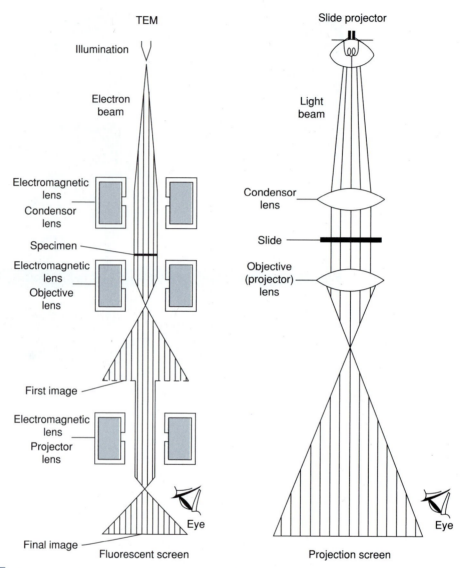

FIGURE 3-36   Comparison of the electron and optical imaging systems in the transmission electron microscope (*left*) and slide projector (*right*). All components in the TEM are enclosed in an evacuated column.

shown in Fig. 3-37. And a marvelous instrument it is because it not only provides high-resolution images of structures, but yields crystallographic information as well.

Electrons in the TEM possessing kinetic energies (KE) of 100 keV (1.6×10^{-14} J) travel down the column axis with a velocity of $v = (2KE/m_e)^{1/2}$ or

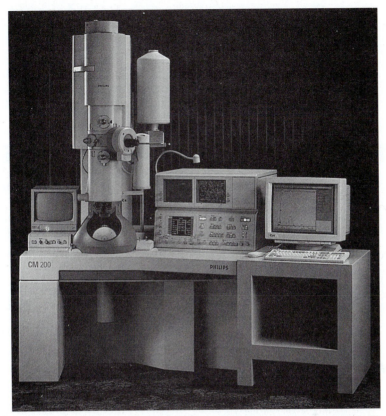

FIGURE 3-37   Modern transmission electron microscope. In addition to excellent imaging and diffraction capabilities this instrument can perform elemental analysis over nanometer-size areas. The energy-dispersive (EDX) fluorescent X-ray spectrum of the imaged region is shown on the right-hand monitor. Courtesy of Philips Electronic Instruments Company, a Division of Philips Electronics North America Corporation.

$[2(10^5 \text{ eV} \times 1.60 \times 10^{-19} \text{ J/eV})/9.11 \times 10^{-31} \text{ kg}]^{1/2} = 1.87 \times 10^8$ m/s. Through the de Broglie relationship (Eq. 2-1) this translates into an electron wavelength of $\lambda = h/mv$, or $(6.62 \times 10^{-34} \text{ J-s})/(9.11 \times 10^{-31} \text{ kg})(1.87 \times 10^8$ m/s) $= 3.89 \times 10^{-12}$ m, or 0.00389 nm. Therefore, diffraction effects similar to those obtained with X rays are possible at small diffraction angles. Image contrast occurs when the beam illuminates and scatters from lattice irregularities, for example, dislocations, grain boundaries, interfaces, precipitates, and distortion due to stress. The short wavelength confers the high resolution of the instrument, enabling features smaller than 0.2 nm, such as crystal planes, to be resolved. This compares with more than 2 nm in a good SEM. By employment of high-resolution **lattice imaging** techniques, atom positions can be inferred, and this is how the stunning electron micrographs of Figs. 3-26 and 3-29B were obtained.

One of the important advances in TEM techniques enables *cross sections* to be imaged. This method is indispensable for viewing through the side dimension of coatings and films that are already thin in plan view. The tedious thinning required is rewarded by the high-resolution cross-sectional views of interfaces and layers in all kinds of materials and microelectronic devices. A number of TEM images illustrating the techniques described above are shown in Fig. 3-38.

EXAMPLE 3-7

Determine the ASTM grain size number for the structure in Fig. 3-38A.

ANSWER In Fig. 3-38A there are approximately 28 "half-grains" cut by the photograph edges contributing $\frac{1}{2} \times 28 = 14$ grains. In addition there are ~30 interior grains for a total of 44 grains. These are contained in an area of 3 in. $\times$ 4 in. = 12 in.^{2}. This is comparable to an area of $12(100/72,500)^2 = 2.28 \times 10^{-5}$ in.^{2} at 100$\times$. Therefore, by Eq. 3-8,

$$n = 1 + \ln N/\ln 2 = 1 + \ln(44/2.28 \times 10^{-5})/\ln 2 = 21.9.$$

Note the extremely small grain dimensions in thin polycrystalline films.

3.6.5. The AFM Revisited

This section closes with the atomic force microscope image, shown in Fig. 3-39, of a deposited array of polycrystalline Si pillars each measuring 1.5 $\times$ 1.5 $\times$ 0.5 μm. For comparison the human hair is about 75 μm in diameter. Typical bacteria and red blood cells are several times larger than the size of the pillar. Imaging of structures always reveals surprises; one never knows what will turn up when the microscopic world is probed.

3.7. PERSPECTIVE AND CONCLUSION

The structure of matter telescopes a hierarchical order extending from subatomic particles (<0.01 nm in size), to atoms (~0.1 nm in diameter) in solids that are positioned in orderly lattice arrays (with features of roughly the same size), to tiny crystallites 10–100 nm, to grains of engineering metals more than

---

FIGURE 3-38 Images of materials obtained in the transmission electron microscope. (A) Plan view of a thin aluminum film (72,500$\times$). Courtesy of D. A. Smith. (B) TEM cross-section of an indium phosphide semiconductor diode laser structure. Courtesy of R. Hull, AT&T Bell Laboratories. (C) Electron diffraction pattern of polycrystalline iron film. Courtesy of S. Nakahara, AT&T Bell Laboratories. (D) High-resolution lattice image of ceramic superconductor (see Fig. 4-24). Courtesy of T. M. Shaw, IBM Corporation.

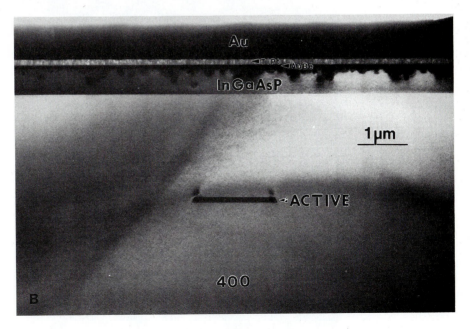

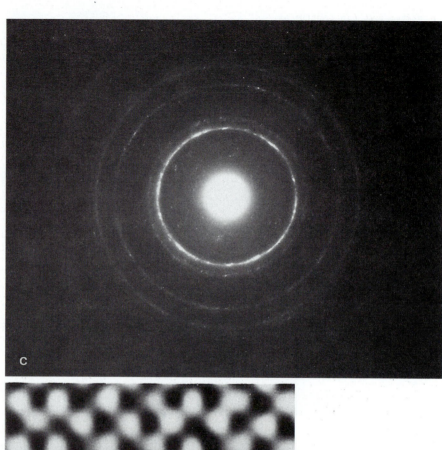

C

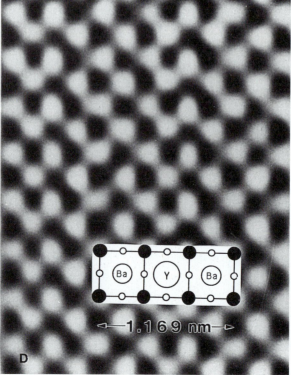

←— 1.169 nm —→

D

FIGURE 3-39   Atomic force microscope image of etched silicon pillars. Courtesy of J. E. Griffith, AT&T Bell Laboratories.

10,000 nm, to aggregates of grains in engineering structures yet thousands of times larger. Increasingly the microscopic world has become the arena of attention where technological progress has enabled materials processing, fabrication, and characterization at unprecedented dimensional levels.

It is found that the diversity of crystalline solids can be divided into only 14 categories or Bravais lattices. The four major classes of solids can each enumerate representatives with different Bravais lattices among their number, and conversely, there are examples of different types of materials having the same crystal structure. Fortunately, many important engineering materials crystallize as cubic solids. At the level of atomic lattices, coordinate geometry suffices to fix imporant crystallographic planes and directions. Periodicity or endless repetition of atomic stacking makes it possible to employ both X-ray and electron diffraction methods to reveal the details of lattice geometry. Atoms are never directly seen by employing X-ray techniques. Rather, we totally rely on Bragg's law, which in turn necessitates measurement of angles where diffraction occurs. By using a ruler and straight edge, diffraction data can unerringly yield values of interatomic spacings with great accuracy. It is paradoxical that when atoms are actually imaged by field ion or scanning tunneling microscopy, the resultant atomic dimensions cannot be determined to within much better than a percent or so. In contrast, X-ray diffraction yields values that are precise to better than 0.01%. The disadvantage of X-ray methods is their limited ability to probe and provide structural information over microscopic dimensions. Transmission electron microscopy overcomes this shortcoming, enabling very high spatial resolution to be achieved. Because diffraction angles cannot be measured as precisely with the TEM as with X rays, techniques based on the latter remain the standard source of lattice parameter information.

Direct observation of structures has revealed the ubiquitous presence of crystallographic defects. They are classified into point (vacancy), line (dislocation), and surface (grain boundary) categories. All of them play lesser or greater roles, acting either singly or collectively, in influencing virtually all material properties. In particular, dislocations and grain boundaries are effective in disrupting perfect lattices and often have an important beneficial influence on the development of certain mechanical properties. On the other hand, defects of any kind cannot be tolerated in semiconductor devices. Easiest to observe are grain boundaries; most difficult are vacancies, but even these elusive entities have been imaged by both FIM and STM.

Although materials *science* perhaps has a bias toward crystallographic descriptions of structure, materials *engineering* stresses structural morphology. The optical and scanning electron microscopes have thus become important engineering tools. They do not sample structural information at the atomic level and, therefore, frequently see "forests rather than trees" of structural information. All too often little of the internal atomic geometry is recognizable in the external structural morphology; but sometimes there is a strong resemblance. The most complete structural description of solids requires a creative synthesis of the two approaches and knowledge of the limitations of each.

Additional Reading

C. Barrett and T. B. Massalski, *Structure of Metals,* 3rd ed., McGraw–Hill, New York (1966).
B. D. Cullity, *Elements of X-Ray Diffraction,* 2nd ed., Addison–Wesley, Reading, MA (1978).
G. Thomas and M. J. Goringe, *Transmission Electron Microscopy of Materials,* Wiley, New York (1979).
G. F. Vander Voort, *Metallography, Principles and Practice,* McGraw–Hill, New York (1984).
D. B. Williams, A. R. Pelton, and R. Gronsky, *Images of Materials,* Oxford University Press, New York (1991).

QUESTIONS AND PROBLEMS

3-1. a. Cesium metal has a BCC structure with a lattice parameter of 0.6080 nm. What is the atomic radius?

b. Thorium metal has an FCC structure with a lattice parameter of 0.5085 nm. What is the atomic radius?

3-2. a. Rhodium has a lattice parameter of 0.3805 nm and the atomic radius is 0.134 nm. Does this metal have a BCC or FCC structure?

b. Niobium has a lattice parameter of 0.3307 nm and the atomic radius is 0.147 nm. Does this metal have a BCC or FCC structure?

3-3. Calculate the atomic packing factor for GaAs assuming the structure is composed of Ga and As ions. Compare your answer with that for Example 3-2.

3-4. What is the atomic planar density (APD) on the (100) and (111) planes of Si?

3-5. Calculate APD for the (111) plane of copper. What is the linear atomic density along the [$\bar{1}$10] direction in this plane?

3-6. Can you suggest a reason why silicon (111) planes oxidize more rapidly than (100) planes?

3-7. Titanium undergoes an allotropic phase change from HCP to BCC upon heating above 882°C. Assume that in HCP Ti, $a = 0.295$ nm and $c = 0.468$ nm, while in BCC Ti, $a = 0.332$ nm. What is the fractional volume change when HCP Ti transforms to BCC Ti?

3-8. Based on atomic weights and structural information show that gold and tungsten essentially have the same density. Calculate the density of each.

3-9. Show that the ideal c/a ratio in the HCP structure is 1.633.

3-10. Demonstrate that the densities of FCC and ideal HCP structures are identical if sites are populated by atoms of the same size and weight.

3-11. Calculate the theoretical density of beryllium from the known structure.

3-12. Calculate the percentage volume change when FCC γ iron ($a = 0.365$ nm) transforms to BCC δ iron ($a = 0.293$ nm) at 1394°C.

3-13. Cobalt exists in an FCC form with $a = 0.3544$ nm. What is the theoretical density of the FCC form of Co? What is the theoretical density of HCP Co?

3-14. CaO has a rocksalt structure with a lattice parameter of 0.480 nm. Determine the theoretical density of CaO and the number of atoms per unit cell.

3-15. Calculate the density of GaAs if the lattice constant is 0.5654 nm.

3-16. What are the Miller indices of planes a, b, and c in Fig. 3-40?

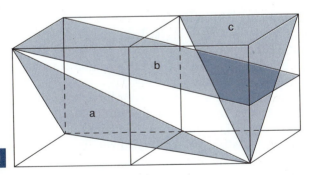

FIGURE 3-40

3-17. What are the Miller indices of directions 1, 2, and 3 in Fig. 3-41?

3-18. Atoms of a body-centered tetragonal metal are arranged in a square array on the (001) plane with a lattice constant of 0.460 nm. On the (100) and (010) planes, atoms are arranged in a rectangular array with lattice constants of 0.460 and 0.495 nm. Sketch the atomic positions on the (110) plane and indicate the dimensions.

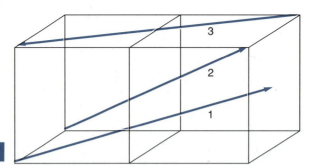

FIGURE 3-41

3-19. In a two-dimensional flatland there are *five* distinct surface point lattices where each point has the same surroundings. Can you draw and characterize them?

3-20. Explain why we can determine the lattice contants of materials with extraordinary precision by X-ray diffraction, but with considerably less precision and accuracy using electron diffraction methods.

3-21. Select all of the directions that lie in the (111) plane of a cubic crystal:
a. [111]   b. [$\bar{1}$11]   c. [100]
d. [110]   e. [$\bar{1}$12]   f. [$\bar{1}$01]
g. [3$\bar{2}$1]   h. [$\bar{2}$11]   i. [5$\bar{2}\,\bar{3}$]
j. [102]

3-22. For a cubic system select all the planes in which direction [011] can lie:
a. (101)   b. (100)   c. (3$\bar{1}$1)
d. (111)   e. (1$\bar{1}$1)   f. (201)
g. (11$\bar{2}$)   h. (200)   i. ($\bar{1}$10)
j. (011)

3-23. a. Determine the Miller indices of the plane that passes through the three coordinate points $(0, 0, 1)$, $(\frac{1}{2}, 1, \frac{1}{2})$, and $(1, \frac{1}{2}, \frac{1}{2})$ within a cubic lattice.
b. What are the coordinates of the intercepts on the x, y, and z axes?
c. What are the Miller indices of the direction connecting the last two points of part a?

3-24. A plane intercepts the x, y, and z coordinate axes of a cubic lattice at points whose coordinates are (100), (020), and (003).
a. What are the Miller indices of the plane?
b. Consider a parallel plane contained wholly within the unit cube. What are its intercepts on the x, y, and z axes?
c. If the lattice parameter is a, what is the distance between neighboring planes having these indices?

3-25. Titanium K_α X rays can be generated in a SEM (described in Section 2.3) or by using an X-ray tube (described in Section 3.4.1). What are the similarities and differences between these two kinds of X-ray sources?

3-26. Distinguish among the following three major applications of X rays: (a) lattice parameter determination, (b) identification of elements in a material, (c) medical imaging of bones and teeth. In each case indicate the physical phenomenon responsible for the X-ray signal, spectrum, or pattern.

3-27. Extend Table 3-3 to include the next three diffracting planes for both BCC and FCC.

3-28. It is desired to determine the unknown energy of an intense beam of X radiation. The beam is directed at the (110) plane of a copper crystal and a strong diffracted beam is detected at an angle of 23.55° with respect to the crystal surface. What are the wavelength and energy of the unknown radiation?

3-29. Powder diffraction from a pure FCC metal yields the following d_{hkl} spacings in nm: 0.2088, 0.1808, 0.1278, 0.1090, 0.1044, 0.09038, 0.08293, and 0.08083. What is the value of the lattice parameter? Identify the metal.

3-30. Diffraction from a pure cubic metal powder using CuK_α radiation yielded seven peaks with the following $\sin^2\theta$ values: 0.1118, 0.1487, 0.294, 0.403, 0.439, 0.583, and 0.691. Index the lines and determine the lattice parameter. What is the metal?

3-31. Calculate the first four 2θ diffractometer angles where diffraction peaks would be expected if the specimen were polycrystalline chromium and CuK_α radiation were employed.

3-32. A thin film of AlAs ($a = 0.56611$ nm) is deposited on a thick GaAs ($a = 0.56537$ nm) substrate. X rays whose wavelength is 0.15405 nm impinge on the layered structure, yielding diffraction peaks from both substances. What is the angular separation of the two (111) peaks? (Special diffractometers are required to resolve the peak separation.)

3-33. Does the topological equation $F/2 - C/4 = 1$ (Eq. 3-7) hold for a grain structure composed of cubes? What about tetrahedra?

3-34. Suggest possible reasons why the external geometric forms of some solids reflect the internal crystalline structure, while there is no such correlation in other solids.

3-35. A specimen 1 cm³ in volume contains grains that can be imagined to be spherical in shape with an average diameter of 20 μm. The crystal structure is cubic with $a = 0.2$ nm.
   a. How many grains are there in the specimen?
   b. What is the ratio of the number of atoms on grain surfaces to those in the grain interior?
   c. Approximately how many atoms will lie on a planar surface 1 cm² in area?
   d. Repeat part b for a nanocrystalline sample of the same material and size having a grain size of 20 nm.

3-36. Under a metallurgical microscope the 5-cm-diameter circular field of view reveals that there are 24 whole grains plus another 20 grains cut by the circumference.
   a. If the magnification were 100×, what would the ASTM grain size be?
   b. What would the ASTM grain size be if the magnification were 500×?

3-37. A plane crystal surface contains emergent screw dislocations. Atoms from the vapor phase condense sequentially on this surface and circular growth spirals are observed. Why?

3-38. A thick (100) silicon substrate with a lattice constant of 0.5431 nm is dislocation free. Then a layer of germanium ($a = 0.5657$ nm) is deposited on top with the same orientation. Careful examination of the planar interface between these semiconductors reveals the presence of dislocations. Why?

4

POLYMERS, GLASSES, CERAMICS, AND NONMETALLIC MIXTURES

4.1. INTRODUCTION

This chapter introduces two of the four important classes of engineering materials: polymers and ceramics. These include some of the oldest as well as some of the newest materials used by humans. Fired clays, glasses, and ceramic materials that contain mixtures of glassy and crystalline constituents were produced and used even before historical records were kept; synthetic polymers, on the other hand, date only to the last century. The former materials are *inorganic* in nature and generally produced at high temperatures; in contrast, polymers have *organic* compositions and are formed at low temperatures. Nevertheless, both display some common structural features. For example, ceramics are usually crystalline but can be disordered or amorphous; polymers are usually amorphous but can exhibit crystallinity. The closest parallels exist between amorphous inorganic glasses and polymers. In addition, both classes of materials are poor conductors of electricity and best suited to play important roles as insulators and dielectrics. Their mechanical properties range from hard and brittle ceramics to very soft and extendable rubbery polymer materials. These electrical and mechanical properties, as well as the methods used to process polymers and ceramics into useful shapes, are dealt with in subsequent chapters. Although ordered crystalline states of matter have been the focus of the book until now, this chapter deals with noncrystalline materials and mixtures of both for the first time.

Polymeric materials have made and continue to make great inroads into applications traditionally met by metal usage. As Fig. 4-1A indicates, polymer production and usage have grown at a far more rapid rate than production and use of metals in the last quarter of a century. Due largely to their ease of processing, high strength-to-weight ratios, low density (typically ~1.5 Mg/m³ compared with ~7.5 Mg/m³ for metals), and competitive costs, polymers and polymer-based composite materials continue to find expanded usage, particularly in automobiles and aircraft. Polymers have also replaced glass and ceramic materials in food containers, dinnerware, and assorted kitchen and toilet accessories. Chapters 7 and 9 address the mechanical properties of polymers and composites and reasons for their increasing selection in engineering design.

Ceramic materials have a number of attractive advantages relative to other materials. These include high melting points, great hardness, low densities, and chemical and environmental stability. However, ceramics are severely handicapped by a lack of toughness. Their intrinsic inability to absorb shock loading causes them to fracture in a brittle manner rather than deform in a ductile fashion the way metals do. Although they do not normally compete with metals, ceramics stand poised to potentially replace metals in selected applications, for example, high-temperature gas turbine engines and cutting tools. Ceramic materials still retain their traditional usage in pottery, bricks,

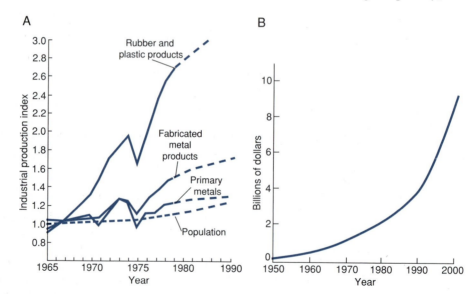

FIGURE 4-1   (A) Industrial production indices for metals and polymeric materials compared with growth in population. Data are normalized to 1.0 in 1967. *Source*: Federal Reserve System. Adapted from W. D. Compton, *Materials Technology*, Spring 1981. (B) Approximate annual value of advanced ceramics in billions of dollars. In the mid-1990s, the total market for glass plus ceramic products roughly estimated to be 30 billion dollars. *Source*: F. H. Norton, *Elements of Ceramics*, 1st and 2nd eds., Addison–Wesley, Cambridge (1952, 1974), and American Ceramic Society.

tiles, whiteware, and kitchen as well as toilet fixtures. But, in addition, there are exciting new high-tech uses for so-called new or advanced ceramics in assorted mechanical, electronic, magnetic, and optical applications. As Fig. 4-1B shows, markets for these high added value materials (discussed in Section 4.6.2) are growing rapidly; in contrast, traditional ceramics industries are expanding more slowly. Although a $10 billion advanced ceramics industry in a decade is impressive, it is still only a very small percentage of the total metal and polymer markets.

As structural considerations are prerequisite to understanding properties, the structure half of "structure–property" relationships is emphasized in this chapter.

Both ceramics and polymers are clearly distinct from metals and from one another. Treating these materials together in this as well as several subsequent chapters sharply emphasizes their different as well shared properties. Nevertheless, organization into separate sections also ensures that the individual attributes of each class of material are preserved and emerge undiminished.

4.2. INTRODUCTION TO POLYMERS

The origin of synthetic polymers can be traced to the American Wild West, of all places. In the mid-19th century, towns in Colorado attracted a motley crowd of adventurers, gamblers, saloon ladies, and other characters intent on acquiring wealth easily and quickly. In an effort to provide saloon patrons with enough billiard balls, John Hyatt sought a substitute for the scarce and expensive ivory spheres that had been used until that time. His early efforts resulted in balls consisting of a core of ivory dust bonded with shellac and an exterior coating of the somewhat unstable collodion. The latter frequently exploded when the balls collided, prompting every gunman in the saloon to instinctively draw his six-shooter. In 1868 Hyatt successfully adapted a material developed some 14 years earlier by Alexander Parkes in England. It consisted of a mixture of cellulose nitrate and camphor. This first synthetic plastic known as celluloid was widely used until recently in making toys, particularly dolls. Its dangerous flammability spawned substitutes, and when Baekeland patented the process for making Bakelite in 1907 the Age of Plastics formally arrived.* It only required an additional 80 years or so until the total volume of synthetic polymer products manufactured in the world exceeded that of metals produced.

A **polymer** is a very large molecule containing hundreds to many thousands of small molecular units or **mers** linked together into either chainlike or network structures. Although our concern here is with *synthetic* organic polymers, the definition is broad enough to include inorganic polymers based on silicon. The word *synthetic* is important because polymers are contained within natural

---

* R. A. Higgins, *The Properties of Engineering Materials*, R. E. Krieger, Huntington, NY (1977).

animal (wool, leather), insect (silk), and vegetable (wood, cotton) products. Cellulose and lignin are natural polymers found in vast quantities in wood. Some natural polymers are so well designed by nature that they have not yet been replaced by the synthetic variety. Polymer macromolecules are different from the other molecules studied in chemistry in at least two ways:

1. The mer building blocks each contain several atoms so that very large molecular weights (e.g., thousands to millions of grams per mole) are involved.
2. Individual polymer molecules do not all have a well-defined molecular weight. For example, in chainlike molecules there may be a factor of 10 or more difference in length and molecular weight between the shortest and longest macromolecules.

Of the polymer materials used in engineering practice **plastics** are, by far, produced in greatest amount and account for some 90% of U.S. tonnage. It is common to subdivide plastics into two main categories: **thermoplastics** and **thermosets.** Thermoplastics melt on heating and are processed in this state by a variety of extrusion and molding processes. Polyethylene, polyvinyl chloride, polypropylene, and polystyrene, in that order, are the four most widely produced thermoplastics. Thermosets (or resins) include phenolic resins, epoxies, unsaturated polyesters, and polyurethanes—substances that cannot be melted and remelted like thermoplastics, but irreversibly set instead. Thermoplastic production is approximately six times that of thermosets in tonnage and value.

Rubbers or **elastomers** form another group of polymeric materials. The natural rubber industry based on latex predated the introduction of plastics and originated before it was recognized that rubbers were polymeric materials. In addition to natural rubber, important synthetic elastomers include polybutadiene, styrene–butadiene, polyisoprene, and silicone rubber.

Appreciable added value is associated with polymers as they are derived primarily from petroleum. A ton of polyethylene currently costs about 1200 dollars, some ten times that of crude oil. Energy costs in polymer production are more than for steel, less than for aluminum, and comparable to those for many commodities. In use for only a century polymers will be with us for a very long time, especially because many of their raw materials can be synthetically derived from renewable resources.

Although the properties of specific classes of polymers differ widely, they are all composed of long molecules that have a spine or backbone of covalently bonded carbon atoms. Hydrogen atoms and hydrocarbon groups are typically distributed along and around the spine. Individual molecular strands are then held together laterally by weak van der Waals secondary bonds, as in thermoplastics, or by stronger covalent bonds that crosslink them together, as in thermosets. Weak polymer bonds decompose at a point not much above room temperature. That is why thermoplastics can be readily melted and formed at low temperatures. It is also the reason why these materials tend to be mechanically unstable and deform or stretch with time at relatively low temperatures. In these respects polymers broadly differ from metals and ceramics.

4.3. POLYMER CHEMISTRY AND STRUCTURE

4.3.1. Thermoplastics

4.3.1.1. Addition and Condensation Polymerization Reactions

Discussion of some aspects of the organic chemistry of hydrocarbons is necessary to understand both the reactions to form polymers and the resulting molecular structures. Consider the monomer molecule of ethylene C_2H_4 shown in Fig. 4-2A. A gaseous product of petroleum cracking, ethylene contains strong covalent **intra**molecular bonds between the C=C and C—H atom pairs. The **unsaturated** double bond between carbon atoms (that contains a total of four shared electrons) is particularly important in polymer formation. Ethylene can be activated by an oxidizing initiator (e.g., OH-), derived from hydrogen peroxide (H_2O_2), which attaches to one end of the ethylene molecule, altering its electronic structure by leaving a single inner **saturated** bond (two electrons) between C atoms. At the other end an energetic half-bond (one electron) dangles outward poised to capture another electron. Individual

$$\text{HO—}\underset{\underset{\text{H}}{|}}{\overset{\overset{\text{H}}{|}}{\text{C}}}\text{—}\underset{\underset{\text{H}}{|}}{\overset{\overset{\text{H}}{|}}{\text{C}}}\text{-}$$

entities are unstable and quickly attach to another ethylene monomer, once again shifting the unpaired electron to the end of the now larger molecule. With suitable catalysts and appropriate reactor temperatures and pressures, these active units continue to link to one another, propagating a chain of many prior monomers—reduced to mer units—within a gigantic linear polymer molecule. Chain reaction polymerization is ended when another -OH radical terminates the chain. The overall reaction, exclusive of the chain ends, can be written as

$$n \;\; \underset{\underset{\text{H}}{|}}{\overset{\overset{\text{H}}{|}}{\text{C}}}{=}\underset{\underset{\text{H}}{|}}{\overset{\overset{\text{H}}{|}}{\text{C}}} \longrightarrow \{{-}\underset{\underset{\text{H}}{|}}{\overset{\overset{\text{H}}{|}}{\text{C}}}{-}\underset{\underset{\text{H}}{|}}{\overset{\overset{\text{H}}{|}}{\text{C}}}{-}\}_n, \tag{4-1}$$

where the quantity in braces is the mer unit. A dense, intertwined collection of long molecular chains constitutes the bulk polymer.

We have just synthesized polyethylene (PE), the most widely produced thermoplastic, by an **addition or chain reaction polymerization** mechanism. A low-

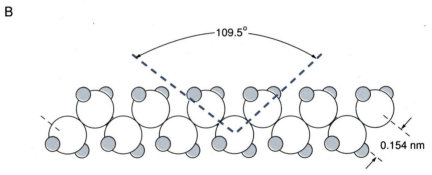

A

Activated ethylene + Ethylene monomer ⟶ Polymerized ethylene

$$-\overset{\displaystyle H}{\underset{\displaystyle H}{C}}-\overset{\displaystyle H}{\underset{\displaystyle H}{C}}-\overset{\displaystyle H}{\underset{\displaystyle H}{C}}-\overset{\displaystyle H}{\underset{\displaystyle H}{C}}- \; + \; n\left[\begin{matrix} H & H \\ | & | \\ C & = C \\ | & | \\ H & H \end{matrix}\right] \longrightarrow \left[\begin{matrix} H & H \\ | & | \\ C & - C \\ | & | \\ H & H \end{matrix}\right]_{n+2}$$

Polyethylene

B

—109.5°—

0.154 nm

FIGURE 4-2   (A) Steps in the polymerization of polyethylene. (B) A portion of the long-chain polyethylene molecule showing the angular geometry.

density form is used to make trash can liners, whereas containers are fabricated from a high-density form. The reason the reaction proceeds is due to a reduction in overall energy. For example, Table 2-1 reveals that C=C and C—C bonds have respective dissociation energies of 620 kJ/mol (6.42 eV/bond) and 340 kJ/mol (3.52 eV/bond). For every involved molecule one of the former bonds is replaced by two of the latter bonds. The overall dissociation energy difference for the chemical reaction per molecule is 2 (340) − 620 = 60 kJ (0.62 eV). As it takes more energy to dissociate the two C—C bonds than the one C=C bond, the products have lower energy than the reactants. Energy is reduced during polymerization and the difference is released as heat.

In forming the polyethylene macromolecule the number of mers that join together depends on the **degree of polymerization** (DP). If the average molecular weight M_{PE} of polyethylene were known, DP could be readily calculated. For example, the C_2H_4 mer has a mass of 28.04 g/mol, so that if $M_{PE} = 100,000$ g/mol, then DP $= 100,000/28.04 = 3566$. Commercial polymers have DP values of $\sim 10^3$ to 10^5.

The chemistry of polymerization of ethylene can be repeated in other molecules having the generic formula C_2H_3R, where R is an atomic (e.g., H, Cl) or molecular (e.g., CH_3, C_6H_5) species that readily forms a single covalent bond with carbon. When polymerized the very important class of thermoplastic polymers known as **polyvinyls** is formed. Structures of the major polyvinyl and other thermoplastic polymers are listed in Table 4-1 together with some properties and uses.

Condensation or step polymerization is another important type of reaction that can lead to formation of thermoplastics. The important polymer nylon is produced this way via the following chemical reaction.

$$H_2N-(CH_2)_6-N-\underset{H}{\overset{}{|}}\overbrace{H + HO}-\underset{O}{\overset{\|}{C}}-(CH_2)_4-\underset{O}{\overset{\|}{C}}-OH \longrightarrow$$

$$H_2N-(CH_2)_6-\underset{H}{\overset{}{\underset{|}{N}}}-\underset{O}{\overset{\|}{C}}-(CH_2)_4-\underset{O}{\overset{\|}{C}}-OH + H_2O \qquad (4\text{-}2)$$

hexamethylenediamine + adipic acid $\longrightarrow$
nylon (monoamide) + water

Here two different organic molecules combine to form a molecule of interest through elimination of water (H + HO). Once formed, each nylon molecule still has reactive groups at each end like those of the original precursors, and they can undergo repeated reactions to extend the polymer chain. Note that no initiators are required as in addition polymerization; rather, condensation monomers are intrinsically reactive and each molecule has an equal probability of reacting.

Similarly, when organic acids with terminal

$$-\underset{OH}{\overset{}{\underset{|}{C}}}=O$$

groups react with alcohols containing $-OH$ groups, the organic ester

$$-\underset{O-}{\overset{}{\underset{|}{C}}}=O$$

linkage binds the original molecule remnants, and water is rejected. (Note the analogy to the reaction between inorganic acids and bases.) In this way commercially important thermoplastic polyesters like polyethylene terephthalate (PET), polybutylene terephthalate (PBT), Dacron, and Mylar are produced.

TABLE 4-1   COMPOSITION AND USES OF THERMOPLASTICS

Thermoplastic	Composition of repeating unit	T_G (K)	Uses
Polyethylene (PE) (partly crystalline)	H | —C— | H	270	Tubing, film, sheet, bottles, packaging, cups, electrical insulation
Polyvinyl chloride (PVC) (amorphous)	H H | | —C—C— | | H Cl	350	Window frames, plumbing piping, phonograph records, flooring, fabrics, hoses
Polypropylene (PP) (partly crystalline)	H H | | —C—C— | | H CH_3	253	Same uses as PE, but lighter, stiffer, more resistant to sunlight
Polystyrene (PS) (amorphous)	H H | | —C—C— | | H (benzene ring)	370	Inexpensive molded objects, foamed with CO_2 to make insulating containers, toughened with butadiene to make high-impact polystyrene (HIPS), packaging
Polytetrafluoroethylene Teflon (PTFE) (amorphous)	F | —C— | F		High-temperature polymer with very low friction and adhesion characteristics, nonstick cookware, bearings, seals
Polymethylmethacrylate Lucite (PMMA) (amorphous)	H CH_3 | | —C—C— | | H $COOCH_3$	378	Transparent sheet, aircraft windows, windscreens
Thermoplastic polyesters Examples: polyethylene terephthalate (PET), Mylar, Dacron	—O—C—C—O—C—(benzene ring)—C— with H,H on the C—C and =O on the carbonyl carbons		Fiber, films
Nylon (partly crystalline when drawn)	See Eq. 4-2	340	Textiles, rope, gears, machine parts

Note: (benzene ring symbol) = benzene ring

4.3.1.2. Molecular Weights

Previously it was mentioned that molecular weights in polymers are not fixed; rather measurement reveals that few polymer chains are either very short or very long; rather molecular sizes are statistically distributed, as we shall see, depending on the degree of polymerization. Two popular ways of defining the average molecular weight have evolved. The first is a **number average relative molecular weight** M_n defined as

$$M_n = \Sigma f_i M_i, \qquad \Sigma f_i = 1, \qquad (4\text{-}3)$$

where f_i represents the fraction of all molecular chains that have a molecular weight M_i. Thus, fraction f_1 has molecular weight M_1, fraction f_2, molecular weight M_2, and so on. The second is the **weight average relative molecular weight** M_w defined as

$$M_w = \Sigma w_i M_i, \qquad \Sigma w_i = 1, \qquad (4\text{-}4)$$

where w_i represents the fractional weight of polymer chains. Thus, a fraction of the total polymer weight, w_1 has molecular weight M_1, weight fraction w_2 has molecular weight M_2, and so on. The ratio M_w/M_n is referred to as the heterogeneity or polydispersity index (PDI) and it always has a value greater than one. Polymer compositions with PDI values considerably larger than unity have large numbers of small molecules that adversely affect thermal stability.

EXAMPLE 4-1

The number and weight fractions of molecular weights in the indicated range for a batch of polyethylene are tabulated below and plotted in Figs. 4-3A and B. Calculate: (a) the number average molecular weight, (b) the weight average molecular weight, and (c) the degree of polymerization.

ANSWER The indicated products and sums defined in Eqs. 4-3 and 4-4 are carried out and entered in the table.

M Range	M_i	f_i	w_i	$f_i M_i$	$w_i M_i$
0–5,000	2,500	0.01	0.00	25	0
5,000–10,000	7,500	0.02	0.01	150	75
10,000–15,000	12,500	0.06	0.03	750	375
15,000–20,000	17,500	0.13	0.09	2,275	1,575
20,000–25,000	22,500	0.21	0.18	4,725	4,050
25,000–30,000	27,500	0.27	0.28	7,425	7,700
30,000–35,000	32,500	0.23	0.26	7,475	8,450
35,000–40,000	37,500	0.07	0.15	2,625	5,625
		$\Sigma = 1.00$	$\Sigma = 1.00$	$\Sigma = 25,450$	$\Sigma = 27,850$

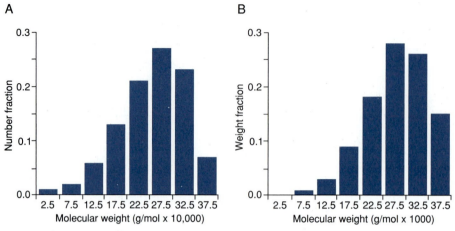

A   B

FIGURE 4-3 Distribution of polymer molecular weights: (A) Number fraction. (B) Weight fraction.

> a. M_n = 25,850 g/mol.
> b. M_w = 27,850 g/mol.
> c. The degree of polymerization DP = $M_n/28.04$ or 25,450/28.04 = 908 mers/mol.

Average polymer molecular weights are very influential in affecting many polymer properties (Fig. 4-4) including softening temperature and strength. Low-molecular-weight chains form relatively few intermolecular bonds and slide apart too easily. Therefore, they soften at low temperatures and have low cohesiveness or strength. Higher values of molecular weight impart more desirable properties but molding the polymers becomes more difficult.

4.3.1.3. Linear Chain Geometry and Motion

The chain length, an important characteristic of linear polymers, is, like molecular weights, also statistically distributed. A polyethylene molecule with 1000 carbon atoms has about the same length-to-thickness ratio as a piece of string that is a couple of meters long. The C—C spacing has been accurately determined to be 0.154 nm but the chain is not linear. Instead it zigzags as shown in Fig. 4-2B with a 109.5° angle between neighboring C atoms. Therefore, in the above example the actual length of a 907-mer chain of polyethylene is (907)(0.154) sin(109.5/2) = 114 nm, or 0.114 μm.

Polypropylene and other vinyl polymers exhibit other structural features that profoundly influence properties. In Fig. 4-5A all of the CH_3 groups are attached on the same side of the chain, yielding a so-called **isotactic** structure. The CH_3 groups alternate regularly on either side of the chain in the **syndiotactic** structure of Fig. 4-5B. In contrast, the **atactic** structure shown in Fig. 4-5C

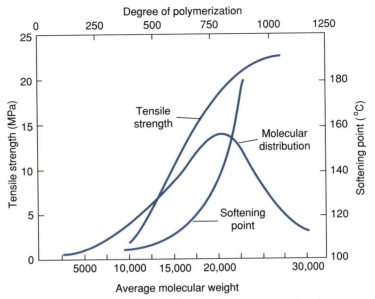

FIGURE 4-4 Effect of the degree of polymerization or molecular weight on the softening temperature and tensile strength of polyethylene. A typical weight fraction distribution of molecular weights is shown. (The tensile strength is the maximum pulling force per unit area a sample of polyethylene can sustain without breaking.)

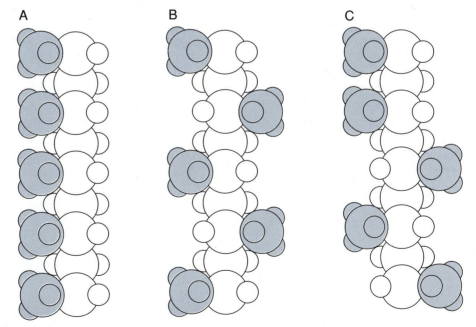

FIGURE 4-5 Models of polypropylene polymer chain molecules showing placement of side groups. Large spheres are carbon atoms; small spheres are hydrogen atoms. (A) Isotactic. (B) Syndiotactic. (C) Atactic.

has a random arrangement of CH_3 pendant groups. Tacticity or **stereoregularity** defines distinct molecular configurations that cannot be changed by simple molecular rotation. The iso- and syndiotactic forms of polypropylene can be packed more densely and regularly, giving the bulk polymer a partially crystalline character. Because of the better molecular fit these chains are stiffer and more heat resistant than those in amorphous atactic polypropylene.

Linear chains have considerable motional freedom despite the restriction that the C—C spacing and bond angle are fixed. In the liquid phase, a physical state important in processing, absorbed thermal energy is reflected in molecular **translation** as a whole, **rotation** about C—C chain bond links, and **vibration** of individual bonds. Due to continued molecular buffeting, great chain flexibility, and weak intermolecular bonding forces, the analogy to a "stirred pot of spaghetti" or "can of worms" is apt. The situation in the solid state is not as clear. In some polymers the molecules align in a reasonably ordered crystalline-like fashion in the sense meant in Chapter 3; other polymer structures are amorphous with no long-range crystalline order. In either case vibrational motion persists to low temperatures, whereas rotational modes of motion are restricted, and large molecular translations are impeded due to now stronger intermolecular bonding forces. Lowering the temperature limits the motional freedom still further.

4.3.1.4. Viscosity

The totality of internal molecular motions is manifested macroscopically by a material property known as **viscosity** (η). It is measured in units of Poise (P), where 1 P = 0.1 Pa-s. Materials with low viscosity (e.g., in water $\eta \approx 10^{-4}$ Pa-s) offer little resistance to flow or deformation when acted on by external forces. More viscous materials (e.g., molasses) flow slowly and when the viscosity reaches high enough levels ($\eta \approx 10^{12}$ Pa-s) the former liquid becomes a rigid solid (glass) that is effectively immobile. In Chapter 7 viscous deformation of polymers is treated quantitatively, and in Chapter 8 the role of viscosity in processing polymers is noted. Importantly, the coefficient of viscosity depends strongly on temperature in a complex way. A widely used approximate expression for η, the Williams–Landel–Ferry equation, is given by

$$\eta = \eta_0 \exp - \frac{17.4(T - T_G)}{[51.6 + (T - T_G)]}, \qquad (4-5)$$

where $\eta_0 = 10^{12}$ Pa-s, and T and T_G are the polymer and glass transition temperatures (in K), respectively. The glass transition temperature, a material property that depends on the polymer in question, is defined in Section 4.4.2. Clearly, when $T = T_G$, $\eta = 10^{12}$ Pa-s.

Large increases in viscosity can also occur with time during the synthesis of vinyl polymers at *constant* temperature as schematically indicated in Fig. 4-6. When polymerization nears completion the magnitude of the van der Waals forces between adjacent molecules rises greatly. This elevates the viscosity to

Nature of molecule   Change in viscosity

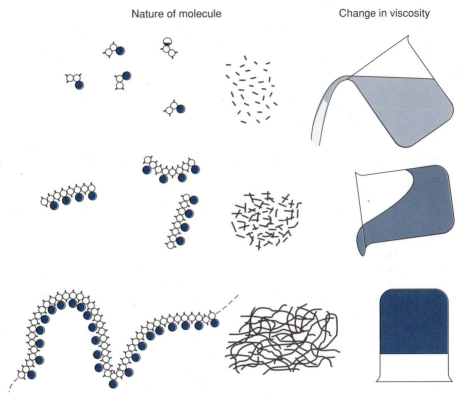

FIGURE 4-6 Stages in the polymerization of a vinyl polymer and the relative changes in viscosity they produce. From R. A. Higgins, *The Properties of Engineering Materials*, R. E. Krieger, Huntington, NY (1977).

the point where the prior liquid cannot be poured; instead, it effectively behaves as a solid.

4.3.1.5. Copolymers

When two or more kinds of mers are mixed within a single polymer chain then copolymers are produced. Synthesis of copolymers has an obvious analogy to alloying in metal systems and similarly yields materials that potentially combine the beneficial properties of each component. Copolymer structures are schematically depicted in Fig. 4-7 and at least three types—**alternating**, **random**, and **block**—can be distinguished. In addition, linear chains can interact by **grafting** to one another (Fig. 4-7D).

Several important copolymers incorporate polystyrene as one of the ingredients. Polystyrene by itself is rather inflexible and brittle at room temperature, a condition caused largely by the steric hindrance of the benzene rings. But when copolymerized with some 3–10% of the stretchy rubber polybutadiene,

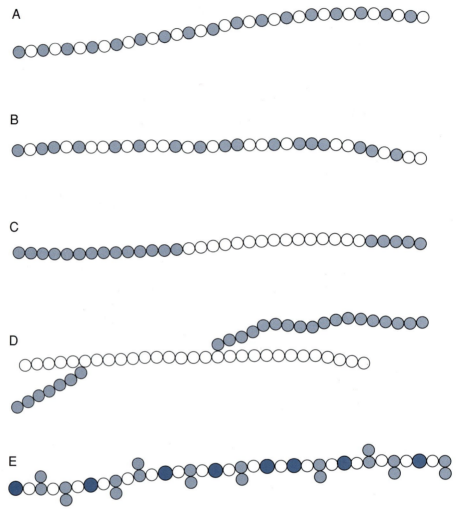

FIGURE 4-7   Types of copolymers: (A) Alternating. (B) Random. (C) Block. (D) Graft. (E) Terpolymers.

a high-impact-strength polystyrene (HIPS) results. If butadiene additions can improve styrene, what about improving HIPS by further ternary alloying? This is what is done in the ABS family of thermoplastics. **A**crylonitrile monomer is copolymerized with **B**utadiene and **S**tyrene to yield **ABS** polymers with improved properties. Drain and vent piping in buildings and assorted auto and appliance parts capitalize on the individual attributes of the components in this **terpolymer** (Fig. 4-7E). Acrylonitrile confers heat and chemical resistance, butadiene imparts the ability to withstand impact loading, and styrene provides part rigidity and processing or shaping ease.

The branching in graft copolymers can also occur in homopolymers. Interestingly, polymerization processes can be controlled to produce only linear chains [e.g., high-density (HD) polyethylene] or branched chains [e.g., low-density (LD) polyethylene]. Branching, like atacticity, strongly hinders alignment of neighboring chains, inhibiting crystalline ordering. Thus, LDPE is typically 50% crystalline, whereas HDPE is more crystalline ($\sim$80%). As we shall subsequently see, the extent of crystallinity influences the polymer softening temperature, a property that in turn affects mechanical properties and processing behavior.

4.3.2. Thermosets and Elastomers

Unlike linear thermoplastic chains there are materials, not considered to be polymers, that contain extended two- or even three-dimensional networks of chemical bonds. For example, graphite and boron nitride are composed of **regular** or ordered, two-dimensional crystalline networks, whereas some diamond-like materials possess a three-dimensional noncrystalline structure. In thermosetting polymers and elastomers, however, the entire network is interconnected through primary bonds in a **nonregular** (noncrystalline) fashion. Distinct molecules do not exist. Two principal types of networks have been identified: (1) those in which small molecules link linear chains and (2) those in which small molecules (including short polymer chains) react directly to produce chain branching.

Elastomers belong to the first category. Polyisoprene, the hydrocarbon that constitutes raw natural rubber, is an example. It contains unsaturated C=C bonds, and when vulcanizing rubber, sulfur is added to promote crosslinks. Two S atoms are required to fully saturate a pair of —C=C— bonds and link a pair of adjacent molecules (mers) as indicated in the reaction

$$
\begin{array}{ccc}
\text{H} \ \text{CH}_3 \ \text{H} \ \ \text{H} & & \text{H} \ \text{CH}_3 \ \text{H} \ \ \text{H} \\
\ \ | \ \ \ \ | \ \ \ | \ \ \ \ | & & \ \ | \ \ \ \ | \ \ \ | \ \ \ \ | \\
-\text{C}-\text{C}=\text{C}-\text{C}- & & -\text{C}-\text{C}-\text{C}-\text{C}- \\
\ \ | \ \ \ \ \ \ \ \ \ \ | & & \ \ | \ \ \ \ | \ \ \ | \ \ \ \ | \\
\text{H} \ \ \ \ \ \ \ \ \ \text{H} & & \text{H} \ \ \ \ \text{S} \ \ \text{S} \ \ \text{H} \\
& +\,2\text{S} \longrightarrow & \\
\text{H} \ \ \ \ \ \ \ \ \ \text{H} & & \text{H} \ \ \ \ | \ \ \ | \ \ \ \ \text{H} \\
\ \ | \ \ \ \ \ \ \ \ \ \ | & & \ \ | \ \ \ \ | \ \ \ | \ \ \ \ | \\
-\text{C}-\text{C}=\text{C}-\text{C}- & & -\text{C}-\text{C}-\text{C}-\text{C}- \\
\ \ | \ \ \ | \ \ \ | \ \ \ | & & \ \ | \ \ \ \ | \ \ \ | \ \ \ \ | \\
\text{H} \ \text{CH}_3 \ \text{H} \ \ \text{H} & & \text{H} \ \text{CH}_3 \ \text{H} \ \ \text{H}
\end{array}
\qquad (4\text{-}6)
$$

Without vulcanization, rubber is soft and sticky and flows viscously even at room temperature. By crosslinking about 10% of the sites, the rubber attains mechanical stability while preserving its flexibility. Hard rubber materials contain even greater sulfur additions.

Exposure to oxygen also promotes a similar crosslinking of polymer molecules containing double bonds to create

$$-\overset{|}{\underset{\underset{|}{O}}{C}}-\overset{|}{\underset{\underset{|}{O}}{C}}-$$

structures. In contrast to beneficial vulcanization with sulfur, oxygen "vulcanization" of rubber molecules raises a reliability issue. Oxygen crosslinks embrittle elastomers, causing them to lose some of their elasticity.

Examples in the second category include the network polymers listed in Table 4-2 which are commonly known as thermosets. They are generally formed by mixing and reacting two precursors together, for example, a resin and a hardener in the case of epoxies. Like certain thermoplastics, a number of important thermosets (e.g., phenol–urea and melamine–formaldehyde) undergo a **condensation reaction** where H_2O is rejected. For example, when phenol (C_6H_5OH) and formaldehyde (CH_2O) react, polymerization occurs through loss of H from the former and O from the latter as indicated by the reaction

2 phenol + 1 formaldehyde = 1 phenol–formaldehyde + H_2O.   (4-7)

The reaction occurs at random H sites (*), leaving a CH_2 bridging group (from the formaldehyde) to link two phenol molecules together at the severed bonds. There are typically three or so such bridges per phenol group that extend randomly in all directions. The result is an irregular or amorphous three-dimensional network of phenol groups welded to one another through flexible CH_2 bridges (Fig. 4-8). Because there are two or more functional groups per precursor, branched rather than linear structures result. This product, commercially known as Bakelite, is still widely used. Similar amorphous networks arise in other thermosets which are used as adhesives, coatings, and

| TABLE 4-2 | COMPOSITION AND USES OF THERMOSETTING POLYMERS |

Thermoset	Composition of repeating unit	Uses
Phenolics ⠀Phenol–formaldehyde ⠀Bakelite (amorphous) ⠀See Eq. 4-7, Fig. 4-8	$\begin{array}{c} OH \\ | \\ -C_6H_2-CH_2- \\ | \\ -CH_2 \end{array}$	Electrical insulation
Epoxy (amorphous)	$-O-\!\!\bigcirc\!\!\overset{\overset{CH_3}{|}}{\underset{\underset{CH_3}{|}}{C}}\!\!\bigcirc\!\!-O-CH_2-\overset{\overset{OH}{|}}{CH}-CH_2-$	Fiberglass matrix, adhesives
Polyester (amorphous)	$-\overset{\overset{O}{\|}}{C}-(CH_2)_m-\overset{\overset{O}{\|}}{C}-O-\overset{\overset{CH_2OH}{|}}{\underset{\underset{CH_2OH}{|}}{C}}-$	Fiberglass composites; cheaper than epoxy
Melanine–formaldehyde	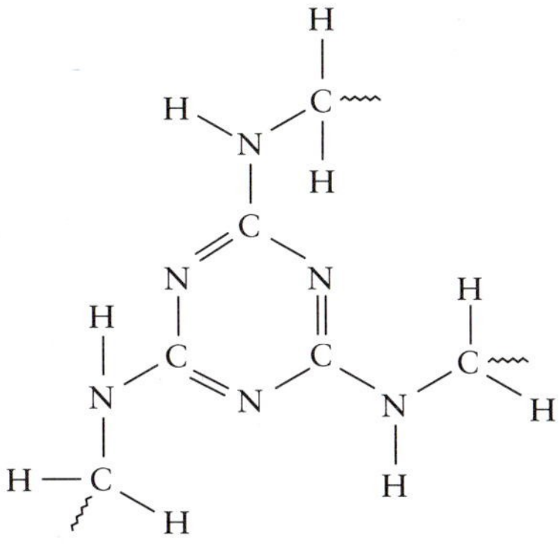	Molded dinnerware, adhesives and bonding resins for wood, flooring, and furniture (usually cellulose-filled)

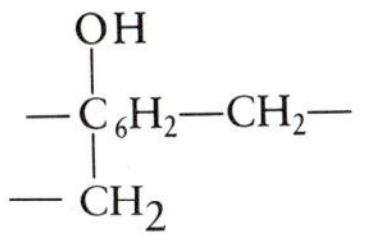

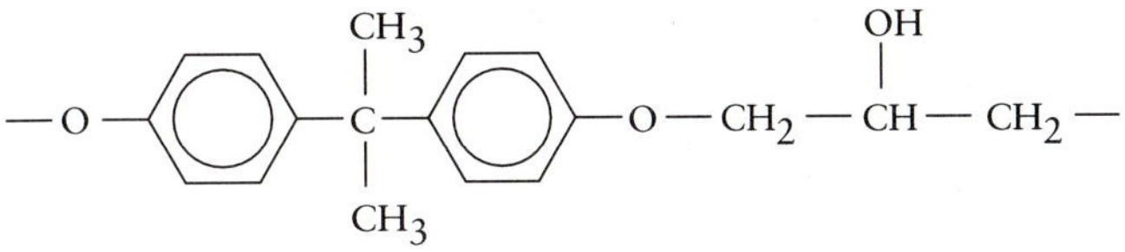

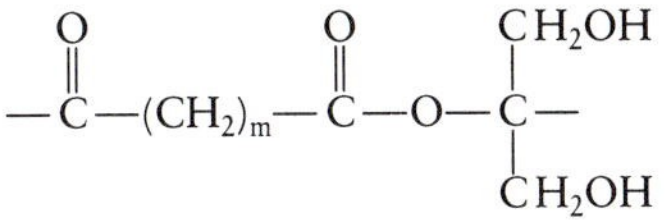

and similar units
randomly connected
by a variety of links

matrices for composites, for example, epoxies. On heating, secondary bonds in the structure may break or melt, but the overwhelming number of primary bonds are resistant to heat. Therefore, thermosets do not melt or flow viscously, but rather decompose or char if the temperature is high enough.

⠀⠀⠀Elastomers or rubbers also have amorphous structures. But unlike thermosets that form amorphous three-dimensional networks, elastomers are based on irregular linear chains that are occasionally crosslinked. When elastomers

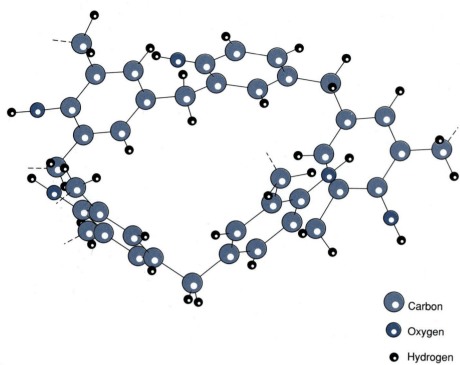

○ Carbon

○ Oxygen

○ Hydrogen

FIGURE 4-8   Molecular model of the network structure of phenol–formaldehyde or Bakelite.

are extended the chains tend to straighten and some crystallinity is induced. The common rubbers listed in Table 4-3 are based on the structure

$$\{-\underset{\underset{\text{H}}{|}}{\overset{\overset{\text{H}}{|}}{\text{C}}}-\underset{\underset{\text{R}}{|}}{\text{C}}=\underset{\underset{\text{H}}{|}}{\text{C}}-\underset{\underset{\text{H}}{|}}{\overset{\overset{\text{H}}{|}}{\text{C}}}-\}_n$$

where R is an atomic or molecular unit, for example, H, Cl, CH$_3$.

EXAMPLE 4-2

A rubber contains 55% butadiene, 35% isoprene, 6% sulfur, and 4% carbon. What fraction of possible crosslinks is involved in vulcanization? Assume one sulfur atom is required to crosslink each mer.

ANSWER   For butadiene (C$_4$H$_6$),

4 C atoms × 12 g/mol + 6 H atoms × 1 g/mol = 54 g/mer.

TABLE 4-3	COMPOSITION AND USES OF ELASTOMERS (RUBBERS)

Elastomer	Composition of repeating unit	Uses
Polybutadiene	$-\overset{\overset{\displaystyle H}{\mid}}{\underset{\underset{\displaystyle H}{\mid}}{C}}-\overset{\overset{\displaystyle H}{}}{\underset{\underset{\displaystyle H}{}}{C}}=\overset{}{\underset{\underset{\displaystyle H}{}}{C}}-\overset{\overset{\displaystyle H}{\mid}}{\underset{\underset{\displaystyle H}{\mid}}{C}}-$	Tires, moldings; amorphous except when stretched
Polyisoprene (natural rubber)	$-\overset{\overset{\displaystyle H}{\mid}}{\underset{\underset{\displaystyle H}{\mid}}{C}}-\overset{}{\underset{\underset{\displaystyle H}{}}{C}}=\overset{}{\underset{\underset{\displaystyle CH_3}{}}{C}}-\overset{\overset{\displaystyle H}{\mid}}{\underset{\underset{\displaystyle H}{\mid}}{C}}-$	Tires, gaskets; amorphous except when stretched
Neoprene	$-\overset{\overset{\displaystyle H}{\mid}}{\underset{\underset{\displaystyle H}{\mid}}{C}}-\overset{}{\underset{\underset{\displaystyle H}{}}{C}}=\overset{}{\underset{\underset{\displaystyle Cl}{}}{C}}-\overset{\overset{\displaystyle H}{\mid}}{\underset{\underset{\displaystyle H}{\mid}}{C}}-$	Oil-resistant rubber used for seals
Silicone rubber	$-O-\overset{\overset{\displaystyle CH_3}{\mid}}{\underset{\underset{\displaystyle CH_3}{\mid}}{Si}}-O-\overset{\overset{\displaystyle CH_3}{\mid}}{\underset{\underset{\displaystyle CH_3}{\mid}}{Si}}-O-$	Thermal and electrical insulation components and coatings, foam rubber

For isoprene (C_5H_8),

$$5 \text{ C atoms} \times 12 \text{ g/mol} + 8 \text{ H atoms} \times 1 \text{ g/mol} = 68 \text{ g/mer}.$$

Atomic weight of sulfur = 32. Therefore, the fraction of crosslinks = $(6/32)/(55/54 + 35/68) = 0.122$.

4.4. POLYMER MORPHOLOGY

4.4.1. Crystalline Growth

We now consider the chain architecture in bulk polymers at several levels of microstructural organization beyond individual molecules. The forms and inner structures observed in optical and electron microscopes characterize polymer morphology. It is the subtle interplay at all of these structural levels that determines polymer properties. Polymers conceal their internal morphology very effectively but much can be learned by melting a small thermoplastic bead between glass slides. Optical observation of the solidified material between crossed polarizers reveals the presence of spherulites (Fig. 4-9A). High in the structural hierarchy of polymers, spherulites are also observed in crystalline minerals, and provide the first clue that some bulk polymers are crystalline. The individual small platelike lamellar single crystals of polyethylene in Figs. 4-9B and C exhibit more convincing evidence of crystallinity.

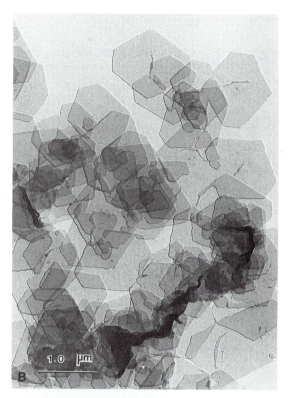

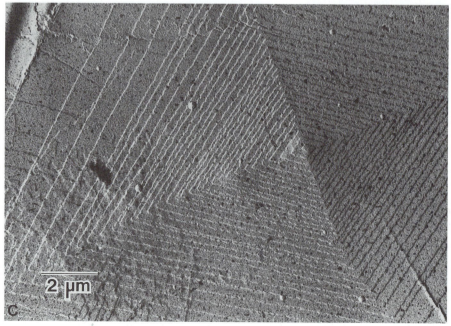

D

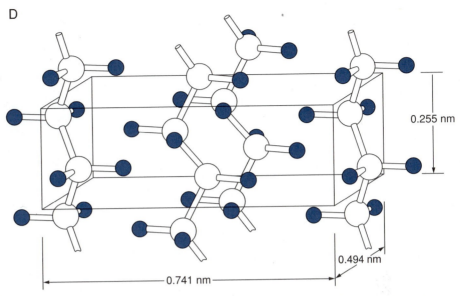

0.255 nm

0.494 nm

0.741 nm

FIGURE 4-9 Structure of polyethylene from the macroscopic to the microscopic level. (A) Spherulites of polyethylene. (B) Electron microscope image of small single-crystal plates of polyethylene grown from xylene solution. From L. C. Sawyer and D. T. Grubb, *Polymer Microscopy*, Chapman & Hall, London (1987). (C) Electron microscope detail of large single-crystal polyethylene plates. Courtesy of A. J. Lovinger, AT&T Bell Laboratories. (D) Unit cell of polyethylene. (Not to scale.)

X-ray diffraction methods have demonstrated that many polymers are crystalline. It is, however, clear that they possess none of the classic structural perfection of metal crystals whose atoms and unit cells are predictably and beautifully ordered over very long distances. Instead, polymer crystals start with good geometric intentions, and ordered unit cells (Fig. 4-9D) can be identified within groups of chain molecules arranged in parallel bundles. Long strands of molecules then fold like computer paper whose stack width is much less than the overall bundle length. With repeated folding and spreading something like a shaggy rug composed of polymer is woven over two dimensions as depicted in Fig. 4-10A. Through the thickness the ∼100-carbon-long crystalline segments are well ordered, but where they repeatedly loop over in up and down U bends, bundles occasionally tear. This frays the otherwise geometric (quasicrystalline) rug pattern, leaving an amorphous tangle of loose molecular ends that interfaces with the surrounding melt or solution.

The shaggy crystalline platelets often organize themselves in spherulites by first attaching at some central crystalline nucleus (Fig. 4-10B). These plates then extend radially outward, executing complex growth mechanisms and patterns that resemble a spherical spark shower at a fireworks display on July 4th. Initial ordered stacking of platelets is short-lived, however, because amorphous polymer melt gets trapped in between. As the spherulite continues

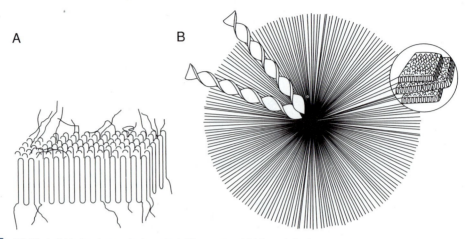

FIGURE 4-10   (A) Chain-folded polymer single crystal. The continual folding of the linear molecule onto itself creates thin carpetlike plates that extend over micrometer-sized domains. The frayed edges are an amorphous component of the crystalline structure. (B) Model of spherulite structure. Both planar and twisted crystalline plates grow radially outward from the central nucleus.

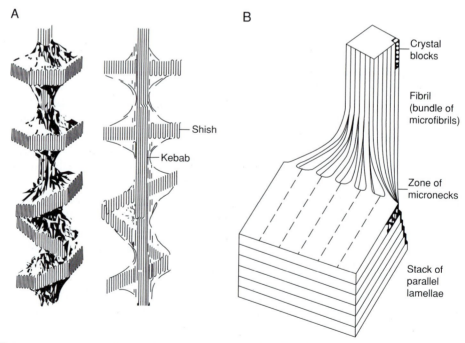

FIGURE 4-11   Polymer structures produced under flow conditions: (A) Shish kebab. (B) Fibrillar.

to grow the plates splay or fan apart until eventually they butt against other spherulites. The resulting grainlike struture is exhibited by a number of thermoplastics (e.g., polyethylene, nylon, polystyrene). Interestingly, spherulitic growth in polymers parallels that of graphite flakes in iron castings (see Fig. 3-32C).

Polymers usually crystallize under melt flow conditions rather than under the quiescent circumstances that lead to spherulitic growth. In such cases elongated crystals align along the flow direction and a fibrous microstructure results. The "shish kebab" structure shown in Fig. 4-11A is typical, where the shish is the elongated crystals in the rows, and the kebab, the overgrown crystalline plates along the vertical spine. A more extreme alignment of crystals occurs if a polymer is pulled or drawn cold (below the melt temperature). The stacked lamellae of the spherulites are then forcibly aligned as shown in Fig. 4-11B into an elongated fibrillar structure. X-ray diffraction patterns then typically reveal varying degrees of crystallinity ranging, perhaps, to a maximum of 95%. Unavoidable molecular entanglements prevent highly crystallizable linear polymers from achieving higher levels of perfection.

Orienting and extending chain crystals are very desirable from an engineering standpoint because both the modulus of elasticity and the strength of polymers are enhanced this way. Fiber-reinforced composites, tires, and industrial belts are some of the applications for high-modulus fibers.

4.4.2. Amorphous Growth

We have already alluded to the fact that polymers assume amorphous forms at the molecular level. To place amorphous polymers within the broad context that includes the melts from which they as well as other amorphous solids (e.g., silica glass, metal glasses) form, it is useful to refer to Fig. 4-12. This important representation displays the volume (V) of the polymer in question as a function of temperature. The liquid is structurally disordered but as the melt is cooled below the melting point (T_M) it does not crystallize the way a metal does. Metals usually undergo an abrupt volume decrease at T_M, and unless cooled in an extraordinarily rapid fashion they solidify as crystalline solids (**1**). (Some metals also expand on solidification; see Section 5.3.3.) Crystalline polymers emulate metals to some extent, but because there is a spread in molecular chain lengths the melting point sharpness is blurred (**2**).

By increasingly interfering with the stereoregularity of the molecules through atacticity, crosslinking, and addition of random copolymer branches, any tendency to crystallize is totally suppressed. The melting point disappears and instead the polymer melt solidifies as an amorphous glass over a range of temperatures (**3,4**). Volume contraction occurs because atomic rearrangements allow molecules to pack more efficiently. Higher polymer density (less volume) is the result of lower cooling rates because molecules have more time to achieve tighter packing. A more or less linear decrease in volume occurs until the

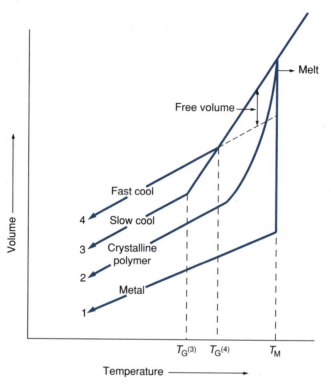

FIGURE 4-12 Volume change as a function of temperature in (1) metals, (2) partially crystalline polymers, (3) amorphous polymers cooled at low rates, and, (4) amorphous polymers cooled at high rates.

glass transition temperature (T_G) is reached. Typically, $T_G = 0.5$ to $0.75\ T_M$. Properties often undergo large property changes in the vicinity of $T = T_G$. Above T_G the amorphous polymer is rubbery and extends readily; below it brittle, glasslike behavior is observed.

During cooling of the melt the molecules of the polymer are free to translate, rotate, extend, bend, twist, and so on, with a freedom proportional to the difference in temperature between T and T_G. At elevated melt temperatures there is a maximum amount of so-called "free-volume" available for molecules to execute their motions in. The free volume at any temperature is the difference between the volume present for unconstrained molecular motion and that occupied by tightly packed motionless molecules. When T_G is reached, this free volume and the last vestiges of the melt disappear. The molecules, now effectively immobilized, solidify as a rigid, solid glass. Below T_G there are no further atomic rearrangements, and the volume change with temperature reflects the reduced amplitude of thermal vibrations. Molecular chains no longer uncoil when loaded; this and the less deformable carbon spine conspire to embrittle the glassy polymer. A similar transition between readily stretchable

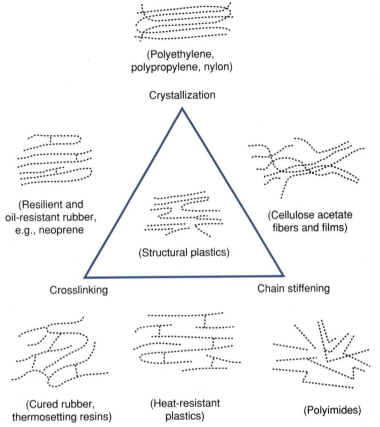

(Polyethylene,
polypropylene, nylon)

Crystallization

(Resilient and
oil-resistant rubber,
e.g., neoprene

(Structural plastics)

(Cellulose acetate
fibers and films)

Crosslinking

Chain stiffening

(Cured rubber,
thermosetting resins)

(Heat-resistant
plastics)

(Polyimides)

FIGURE 4-13 Strengthening polymers by crystallization, crosslinking, and chain stiffening. After H. F. Mark, *Scientific American* 217(3), 149 (1967).

(ductile) and brittle (nonductile) behavior occurs in some engineering metals (including steel, see Section 7.7.2) which, however, are crystalline in both states. For this reason, T_G is to polymers what the ductile–brittle transition temperature is to metals.

4.4.3. Designing Polymer Structures with Improved Properties

A fitting summary of this chapter's introduction to polymer structure is found in the triangle of Fig. 4-13. Three basic ways to impart mechanical strength and thermal resistance to polymers are represented at the corners: (1) **crystallization**, (2) **crosslinking**, and (3) **chain stiffening**. Combinations of two or, better yet, three of these strengthening mechanisms have proven effective in achieving useful polymer properties. For flexible chain molecules the only options for strengthening are crystallization and crosslinking (e.g., rubber).

Strengthening by chain stiffening is accomplished by hanging bulky side groups that prevent the chain from bending. As an example, the large benzene rings attached to the carbon spine serve to stiffen polystyrene, a polymer that is neither crystalline nor crosslinked. Lastly, in polyethylene terephthalate (PET) the combination of moderately stiff chains and crystallization raises the melting point and strengthens the polymer.

Later in Chapter 9 another very important approach to polymer strengthening—creating composites—is discussed. Polymer and polymer composite processing and mechanical properties are subjects dealt with in Chapters 7, 8, and 9. Additional references to polymeric materials are scattered throughout the remainder of the book.

4.5. INORGANIC GLASSES

Glass is one of our most versatile materials and also one of the oldest. Obsidian, a common volcanic glass, was widely used for arrowheads, spearheads, and knives in the Stone Age. The earliest example of an object completely crafted of glass dates to ~7000 BC. Today 700 different glass compositions are in commercial use. These are fabricated into tens of thousands of products having combinations of properties for countless uses.

4.5.1. Structure and Composition

4.5.1.1. Silica Glass

If vitreous (glassy) silica (SiO_2) were examined at the atomic level it would appear to be a random three-dimensional network of SiO_4 groups as envisioned in Fig. 4-14A. Each group consists of an ionic tetrahedral structure consisting of four O^{2-} ions surrounding a smaller Si^{4+} ion. These tetrahedra interlock in the sense that each O belongs to two adjacent groups and are thus linked to form one giant (polymerized) molecule. In contrast, the regularity of crystalline SiO_2 is schematically depicted in Fig. 4-14B. Corresponding X-ray diffraction patterns of *glassy* silica and cristobalite, a crystalline form of SiO_2, are compared in Fig. 4-15. Both patterns peak at a common diffraction angle, suggesting approximately the same atomic surroundings, coordination numbers, and interatomic distances. Actually, the average Si–O distance in the glass is 0.162 nm compared with 0.160 nm in crystalline silicates. Other models of glass structure assume it to be little different from that of the melt from which it formed.

Extending what had been learned about SiO_2, Zachariasen in 1932 postulated a number of conditions that must be fulfilled before an oxide can be a glass former:

1. Each oxygen atom must be linked to no more than two cations.
2. The number of oxygen atoms around each cation must be small, that is, three or four.

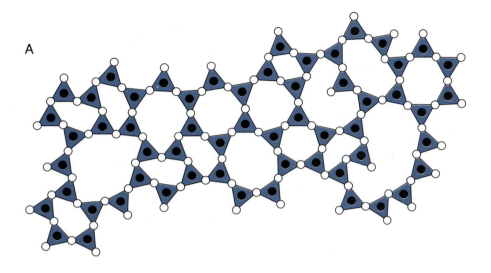

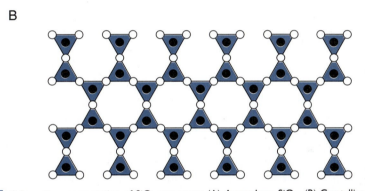

FIGURE 4-14   Schematic representation of SiO_2 structures: (A) Amorphous SiO_2. (B) Crystalline SiO_2.

3. The oxygen polyhedra must share corners, not faces or edges, to form a three-dimensional network.
4. At least three corners must be shared.

Silica is the classic exemplar of these rules. Glasses based on SiO_2 are nearly universal solvents and virtually all elemental oxides dissolve in them to a considerable degree. Broadly speaking, the various metal oxides indicated in Table 4-4 are classified into one of three categories: **glass formers, intermediates, and modifiers**. Interestingly, all of the good glass formers have high bond strengths. They also possess bonds that are estimated to have 50% or more covalent character (e.g., Si—O, Ge—O, B—O, etc). Strong bonds mean that deformation and flow will be difficult, and that melts of pure glass formers are likely to be viscous. Modifier cations open the network by bonding ionically to the anions. Adding a molecule of Na_2O releases two Na^+ ions, which attach

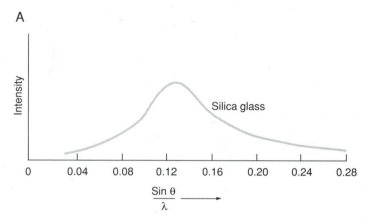

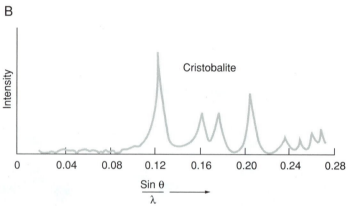

FIGURE 4-15   X-ray diffraction patterns of SiO_2: (A) Amorphous SiO_2. The single broad diffraction peak is indicative of the spread in atomic spacings and the breakdown in long-range crystalline ordering. (B) Cristobalite. Sharp diffraction peaks indicate the presence of crystalline SiO_2.

to the oxygens at the tetrahedron corners making them nonbridging. These low-energy bonds cause the structure to open, thereby lowering the viscosity and facilitating melt flow or fluidity characteristics. This means the glass can be worked at lower temperatures. Electrical conductivity and thermal expansion are both increased by modifier additions. Intermediate oxides straddle both worlds. Though not capable of forming a glass they assume both network and modifier sites in the glass. A schematic representation of a glass with modifiers and intermediates is shown in Fig. 4-16.

Another classification of oxides emphasizes their functional nature. In this scheme the common oxides are divided into glass formers (as before), **fluxes,** and **stabilizers.** Fluxes are added to lower the melting and working temperatures by decreasing the viscosity. Common fluxes are Na_2O, K_2O, and B_2O_3. Stabiliz-

TABLE 4-4	ROLE, COORDINATION, AND BOND STRENGTHS OF METAL IONS IN OXIDES

M^{x+} in MO (valence)	Coordination number	Single-bond strength (kcal/mol)	M^{x+} in MO (valence)	Coordination number	Single-bond strength (kcal/mol)
Glass former			Modifiers		
B (3)	3	119	Y (3)	8	50
Si (4)	4	106	Sn (4)	6	46
Ge (4)	4	108	In (3)	6	43
Al (3)	4	101–79	Mg (2)	6	37
P (5)	4	111–88	Li (1)	4	36
V (5)	4	112–90	Pb (2)	4	36
As (5)	4	87–70	Zn (2)	4	36
Sb (5)	4	85–68	Ba (2)	8	33
Zr (4)	6	81	Ca (2)	8	32
			Cd (2)	4	30
Intermediates			Na (1)	6	20
Ti (4)	6	73	K (1)		13
Zn (2)	2	72			
Cd (2)	2	60			
Al (3)	6	67–53			
Zr (4)	8	61			

From J. R. Hutchins and R. V. Harrington, *Glass: Encyclopedia of Chemical Technology*, 2nd ed., Wiley, New York (1966).

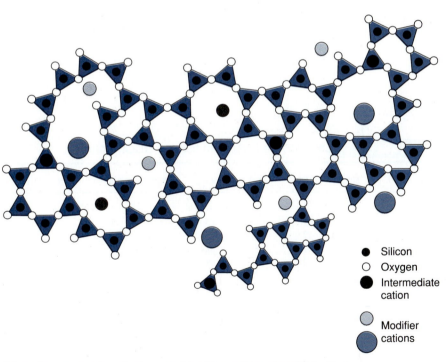

● Silicon
○ Oxygen
● Intermediate cation
◐ Modifier cations

FIGURE 4-16	Schematic representation of a glass containing intermediate and modifier cations.

ers (e.g., CaO, MgO, and Al_2O_3) are added to prevent crystallization and improve chemical durability.

A number of the important silicate glass compositions are listed in Table 4-5. Soda-lime glasses, typically containing (by weight) 72% SiO_2, 15% $Na_2O + K_2O$, 10% $CaO + MgO$, 2% Al_2O_3, plus 1% miscellaneous oxides, account for 90% of the tonnage of all glass produced. With minor composition variations, products include plate glass, containers, and light bulb and TV tube envelopes. Borosilicate glasses such as Pyrex exhibit low thermal expansion; containing ~13% B_2O_3, they are widely used in laboratory glassware. In addition, there are aluminosilicate and lead glasses. The former are chemically durable and resistant to crystallization and elevated temperature degradation. High-lead silica glasses are used for radiation shielding, optical, and decorative art (lead crystal) applications.

4.5.1.2. Nonsilicate Glasses

There are a great many kinds of nonsilicate glasses. Polymers, the most important of these, have already been discussed at length. Later (Chapter 5) we shall see that even some metal alloys can be made glassy through extremely rapid cooling of melts. Amorphous silicon and germanium thin films are readily deposited, even at relatively elevated temperatures. Glass-forming oxides rather than SiO_2 are the basis of several commercial compositions. They include an 83% B_2O_3 glass for high X-ray transmission applications and an 80% GeO_2 optical glass.

Thus far our discussion of glasses has focused only on oxygen anions. Are other nonmetals also the basis for glass-forming systems? Interestingly, sulfur,

TABLE 4-5 **COMPOSITION (wt%) AND USES OF SILICATE GLASSES**

Glass[a]	SiO₂	Al₂O₃	B₂O₃	Na₂O	K₂O	CaO	PbO	MgO	Applications
0010	63	1		7	7		22		Lamp tubing
0080	73	1		17		5		4	Lamp bulbs
1720	62	17	5	1		8		7	Ignition tubes
7070	71	1	26	1	1				Low-loss electrical
7740	81	2	13	4					General
7900	96	0.3	3						High temperature
8363	5	3	10				82		Radiation shielding
Window	72	1		14		10		3	
Fiberglass	54.5	14.5	8.5	0.5		22			Composites
Borosilicate	81	2.5	12	4.5					Low expansion, chemical industry

[a] The four-digit codes refer to Corning glasses. Viscosities of these glasses are given in Fig. 8-33.
From J. R. Hutchins and R. V. Harrington, *Glass: Encyclopedia of Chemical Technology*, 2nd ed., Wiley, New York (1966).

selenium, and tellurium, also group VI elements from the Periodic Table, all form glasses. As these elements are a couple electrons shy of completing their outer shell complement they share electrons and form covalent bonds. Crystalline Se and Te solids are hexagonal and consist of continuous spiral chains aligned along one axis of the crystal as illustrated in Fig. 4-17A. Other crystalline modifications show "puckered" or chainlike rings, usually consisting of eight atoms (Fig. 4-17B). A fraction of these open in the melt and, with moderately rapid cooling, can be supercooled to form elemental glasses. These group VI elements are the base of the **chalcogenide** glasses which also contain assorted metals. Some of them behave like semiconductors but commercial applications have not yet been found.

4.5.2. Solidification of Glassy Melts

Definitions of glass stress that they are products of melts cooled to a rigid condition without crystallizing or that they have attained such a high degree of viscosity as to be rigid for all practical purposes. But glass is not simply a

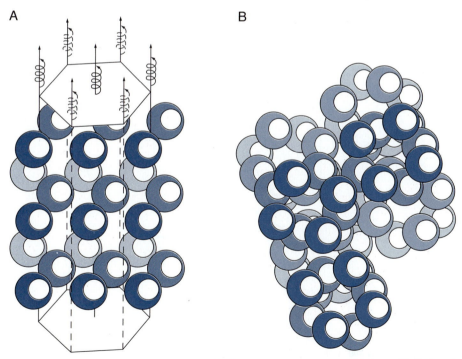

A   B

FIGURE 4-17 Crystalline forms of selenium and tellurium: (A) Hexagonal structure with spiral chains. (B) Puckered ring structure. From R. J. Charles, *The Nature of Glass*, GE Report No. 67-C-317 (1967).

supercooled liquid, as can be illustrated in the already familiar (e.g., Fig. 4-12) volume (V)–temperature (T) plot of Fig. 4-18. A supercooled liquid is the linear extension of the liquid thermal expansion line extrapolated to low temperature. Somewhere in the vicinity of T_G, the glass transformation temperature, there is a bend in the curve and the viscosity reaches values of 10^{11} to 10^{12} Pa-s. The slope, dV/dT, no longer follows the (supercooled) liquid line but rather parallels the thermal expansion behavior for the glassy solid. It is generally thought that the slope change is due to kinetic factors or limited by sluggish atomic motion in the melt. The viscosity becomes so large that structural change cannot be detected in the time scale of measurement. Melts cooled at a lower rate have more time to relax and the supercooled liquid persists to lower temperature, resulting in a denser glass. More rapid cooling raises T_G, and an approximate glass transformation range between these two T_G values can be associated with a particular glass.

The dominant role viscosity plays in processing and treatment of glass is addressed in Chapter 8.

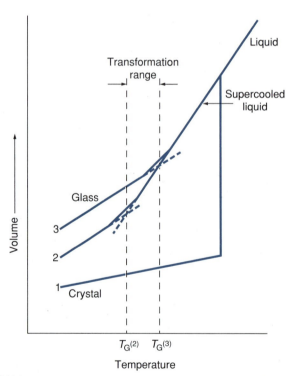

FIGURE 4-18 Volume–temperature relationships in (1) a crystal, (2) a slowly cooled glass, and (3) a rapidly cooled glass. Other properties of glass like refractive index and heat content (see Section 5.1) show similar changes as a function of temperature.

4.6. CERAMICS: AN INTRODUCTION

4.6.1. Traditional Ceramics

With such a long history of use, ceramics have been defined in terms that stress both their artistic and scientific attributes. One definition of ceramics is "the art that deals with the design and fabrication of objects made from fired clay." The earliest known clay figures have been dated at 22,000 BC and may be considered the root of an artistic tree that has borne fruit up to the present day. Scientifically, ceramics are most broadly defined as inorganic nonmetallic solids. This definition considerably extends the meaning of the original Greek word κεραμοσ (*keramos*), referring to pottery or earthenware. In common, ceramics include various hard, brittle, corrosion-resistant materials made by firing clay and other minerals consisting of one or more metals and one or more nonmetals; oxygen is invariably present in the form of oxides. Because many inorganic materials are broadly considered to be ceramics it is instructive to classify them according to the traditional roles they have played and continue to play.

4.6.1.1. Stone

Stone and rock are naturally occurring ceramics that have always been and continue to be important building materials. The popular ones are granite [a mixture of feldspar (aluminosilicate containing sodium, potassium, or calcium), quartz, and mica (aluminosilicate)], limestone, and marble. Structurally, they are usually agglomerates of crystallites held together by an intergranular glassy network. The latter two stones can support compressive loads of 700 kg/cm^2, or 10,000 lb on every square inch of area, and granite twice as much. (Mechanical properties will be put in a more understandable context in Chapter 7.) Their brittle nature facilitates quarrying and polishing, while their low coefficients of thermal expansion ($\sim 8 \times 10^{-6}$ C^{-1}), relatively low density (~ 2.5–2.7 g/cm^3), and low tendency to absorb water are beneficial attributes in structural applications.

4.6.1.2. Clay Products

Ceramics have traditionally meant fired **whiteware** and **structural clay** products. The ingredient common to all of these is clay, which varies widely in chemical, mineralogical, and physical characteristics. Clay basically consists of electrically neutral aluminosilicate layers in the form of tiny crystalline platelets that readily slide over each other. In combination with water they form plastic masses that are easily shaped. Whiteware includes porcelain, sanitary ware, electrical insulators, and dishes. They are made from mixtures of clay, feldspar, and flint. Porcelain, an important member of this group, can be ideally thought of as a complex mixture of K_2O, SiO_2, and Al_2O_3 that is fired at very high temperature ($\sim 1400°C$), yielding a translucent product. Fired porcelain structures typically contain a glassy alkali matrix in which tiny crystals of quartz and mullite ($3Al_2O_3 \cdot 2SiO_2$) are embedded.

Structural clay products include the bricks, tiles, and sewer pipes commonly employed in the construction industry. Other examples of similar compositions include stoneware and earthenware such as terra-cotta. Clays of different composition are used in these products and firing temperatures are lower than for whiteware. For both decorative purposes and to make fired clay impervious to water penetration, surface glazes (glasses) are generally applied.

4.6.1.3. Glasses

The previous section of this chapter may have given the impression that glasses and ceramics are distinct materials. Silica glasses are, however, generally classified as a subgroup of ceramics. And like glass, ceramics can even sometimes be made optically transparent. One interesting category of glass has unambiguous attributes of ceramics; these are the **glass–ceramics**, and Corning cookware is perhaps the most visible example. They are formed as a glass but, in the ensuing processing treatments, are converted to a fine-grained crystalline ceramic that is dense, is heat resistant, and has a high thermal shock resistance. More will be said about this glass-to-ceramic transformation in Section 6.5.3.

4.6.1.4. Refractories

These ceramics largely provide the thermal insulation, crucibles, and hardware in all kinds of casting operations, as well as elevated temperature processing and heat treatment furnaces. Whether they line the immense blast furnaces used in steel making or line molds for casting turbine blades, they must withstand the ravages of direct contact with molten metals and glasses (e.g., slags) without decomposing or cracking. For these applications, refractories are commonly sold in brick form but they can also be purchased in rolls, powder, and loose form. Refractory oxides used for these general purposes most commonly include fireclays containing alumina–silicates with varying Al_2O_3–SiO_2 proportions, magnesium oxide-rich compositions, relatively pure silica, and zirconia (ZrO_2). A problem of concern in refractories is their porosity, which may exceed 20% of the volume. High-density ceramics are stronger and are not as susceptible to chemical attack as the more porous varieties. Specialized refractories processed from pure Al_2O_3, ZrO_2, BeO, and other compounds can be made relatively pore free for special applications.

4.6.1.5. Abrasives

Abrasive ceramic materials are critical in the grinding, lapping, and polishing operations that enable the production of parts requiring high-dimensional tolerances. The sharp abrasive grains must necessarily be harder and tougher than the material being cut so that they do not round off or fracture during contact. Materials used as abrasives include silicon carbide, tungsten carbide, alumina, silica, and industrial diamond dust. They are most commonly bonded to grinding wheels or to paper and cloth (e.g., sandpaper), but are also used in loose form as well as embedded in pastes and waxes.

4.6.2. The New Ceramics

In the past several decades there has been a rapidly growing interest in what are called **new**, **fine**, **advanced**, or **engineering** ceramic materials. The functions, properties, and applications of these ceramics are conveniently displayed in a condensed format in Fig. 4-19. A glance at the applications immediately reveals the enormous economic potential of these high-tech materials. Some of these engineering ceramics are not new. But, many others are, as are the applications

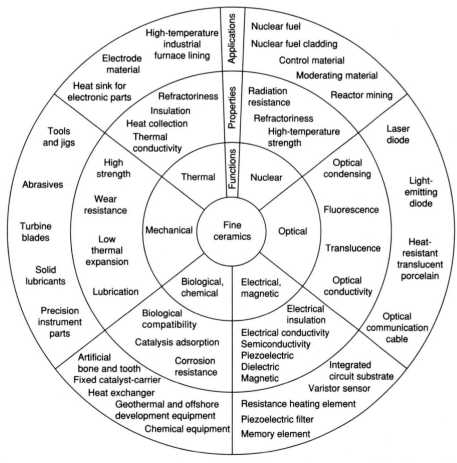

· FIGURE 4-19 Functions, properties, and applications of the new ceramics. From the Fine Ceramics Office, Ministry of International Trade and Industry, Tokyo.

and the methods used to process and characterize them. It is instructive to compare the conventional and new ceramics with the aid of Fig. 4-20 which draws distinctions in cartoon form. The conventional and new ceramic materials do not generally compete with one another in the marketplace; rather, they coexist in largely independent applications. Although some of the details are reserved for later discussion, the following differences are readily apparent.

4.6.2.1. Materials

Conventional ceramics are very heterogeneous physical and chemical mixtures of oxide compounds and glasses, containing relatively large impurity

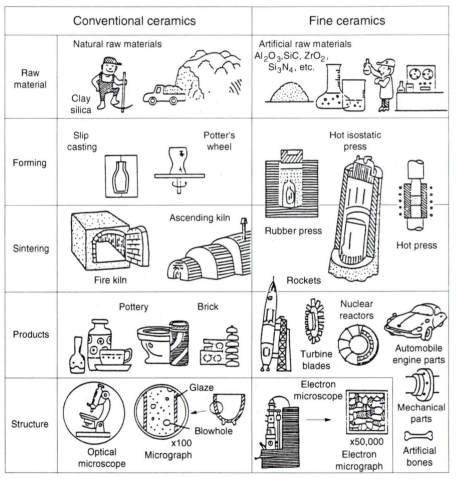

	Conventional ceramics	Fine ceramics
Raw material	Natural raw materials — Clay, silica	Artificial raw materials Al_2O_3, SiC, ZrO_2, Si_3N_4, etc.
Forming	Slip casting — Potter's wheel	Hot isostatic press — Rubber press — Hot press
Sintering	Ascending kiln — Fire kiln	Rockets
Products	Pottery — Brick	Nuclear reactors — Turbine blades — Automobile engine parts
Structure	Optical microscope — Micrograph ×100 — Glaze — Blowhole	Electron microscope — Electron micrograph ×50,000 — Mechanical parts — Artificial bones

FIGURE 4-20   Contrast between the new and conventional ceramics. From the Fine Ceramics Office, Ministry of International Trade and Industry, Tokyo.

concentrations. The new ceramics are frequently composed of single compounds such as carbides (e.g., SiC, TiC) and nitrides (e.g., Si_3N_4, TiN, BN) whose stoichiometry, purity, and grain size must be well controlled. Therefore, conventional ceramics rely on the time-tested mines, sources of clays, and quarries. Fine ceramics, on the other hand, are often synthesized in laboratory-like environments from pure or well-controlled precursor raw materials.

4.6.2.2. Forming and Processing

Traditional methods of producing ceramicware like casting, pressing, and throwing clay on a potter's wheel are still widely practiced. And firing in large, poorly temperature-controlled kilns is also common. At the other extreme many of the new ceramics must be in the form of single crystals which are carefully grown from the melt. Hot-pressing powders in well-controlled furnaces to achieve high-density consolidation is another of the techniques used to fabricate the new ceramic materials. Not only are bulk ceramics of interest, but there is increasing demand for thin films to coat assorted substrates. Methods like chemical vapor deposition (see Section 5.2.2) that are spinoffs from microelectronic processing have been adapted to deposit ceramic films and coatings from suitable vapors. A more detailed discussion of how green (unfired) ceramic bodies are formed and then densified is deferred to Section 8.7.

4.6.2.3. Applications

It is in the applications that remarkable differences between conventional and new ceramics are evident. A sampling of these together with the materials involved are listed in Table 4-6. Space constraints do not allow a treatment of all these applications, but many are discussed further throughout the book. It should be noted that not all of the applications have been commercially realized. The ceramic engine, whose components are shown in Fig. 1-11, is still undergoing development. Nevertheless, the potential gain in energy efficiency and fuel savings as a result of higher engine temperatures and lighter weight is a worthy quest. In this case, the hope is that ceramics will replace metals.

The Space Shuttle orbiter shown in Fig. 4-21 also relies on the new ceramics for reusable thermal insulation tiles composed largely of SiO_2 and Al_2O_3 fibers. These tiles are often coated with high-thermal-emissivity borosilicate glass layers. Some 30,000 tiles with their low-heat-transfer characteristics protect the underlying aluminum frame from heating appreciably. The orbiter nose must withstand reentry temperatures as high as 1450°C and is therefore made of a very special ceramic, a carbon/carbon composite (see Section 9.8.3). In these and other applications (e.g., capacitors, sensors, optical materials, high-temperature superconductors, insulating magnets) there are no viable substitutes for ceramics.

TABLE 4-6 **APPLICATIONS FOR THE NEW CERAMIC MATERIALS**

Application	Material
1. Machinery	
Heat-resistant structural materials	Si_3N_4, SiC, cordierite $(2MgO \cdot 2Al_2O_3 \cdot 5SiO_2)$, ZrO_2 (transformation toughened), mullite $(3Al_2O_3 - 2SiO_2)$
Heat engine components	Si_3N_4, SiC
Diesel engine	
Turbine engine	
High-temperature bearings	
High-temperature gas and molten metal transport pipes	
Stirling combustion engine components	
Ceramic heat exchanger	Cordierite
Wear resistant materials	SiC, Al_2O_3, ZrO_2, TiC
2. Cutting tools	WC, Al_2O_3, TiC, TiN, cubic BN, diamond
3. Nuclear	
Fuel	UO_2
Reactor components	SiC, B_4C, Al_2O_3
Dosimeters	CaF_2, K_2SO_4, LiF
4. Electronics	
Integrated circuit substrates	Al_2O_3, AlN, BeO, SiC
Capacitors	$SrTiO_3$, $BaTiO_3$
Thermistor (temperature sensor)	$(NiMn)_3O_4$, $KTaNbO_3$
Piezoelectrics	PZT (lead zirconate titanate), $LiNbO_3$
Ferroelectrics	$LiTaO_3$, $Pb(TiZr)O_3$
Magnetics	Ferrites $(NiZn)Fe_2O_4$, garnets
5. Sensors	
Oxygen	Y-doped ZrO_2
Humidity	Ti-doped $MgCr_2O_4$
Hydrocarbon gas	Doped SnO_2
6. Optical	
Lasers	Ruby (Cr-doped Al_2O_3), Nd-doped yttrium iron garnet
Transparent windows	Al_2O_3, MgF_2, ZnSe
Electrooptical materials	$LiNbO_3$, KH_2PO_4
7. Medical and bioengineering	
Bone, artificial joints, tooth replacements	Apatite hydroxide (for bone meal), Al_2O_3 for implantation

Reprinted from S. Musikant, *What Every Engineer Should Know about Ceramics*, p. 36, courtesy of Marcel Dekker, Inc.

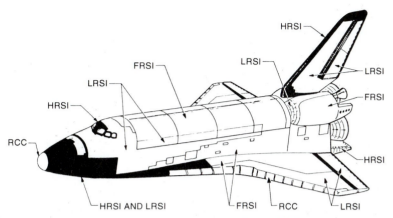

FIGURE 4-21 Thermal protection system for the Columbia Space Shuttle. HRSI, high-temperature reusable surface insulation (1260°C); LRSI, low-temperature reusable surface insulation (540°C); FRSI, felt reusable surface insulation (430°C); RCC, reinforced carbon/carbon composite (1450°C). Temperatures that must be withstood are indicated. Reprinted from S. Musikant, *What Every Engineer Should Know about Ceramics*, p. 149, courtesy of Marcel Dekker, Inc.

4.6.2.4. Structure

Conventional ceramics are inhomogeneous even at levels that can sometimes be seen with the naked eye. But the new ceramics require electron microscopes and other modern analytical tools to assess structural and chemical variations.

Depending on the method of preparation, ceramics can be either crystalline or glassy. This is especially true of thin films and coatings. But, for the most part, the new ceramic materials are crystalline, and in some applications (e.g., lasers, optics) single crystals are essential. For this reason the next section primarily, though not exclusively, emphasizes crystalline ceramics. Consideration of ionic groupings is a good place to start.

4.7. STRUCTURE OF CERAMICS

4.7.1. Crystalline Ceramics

The varied and complex structures of ceramics derive from the fact that two or more atoms (ions) of different size are usually involved and the condition that charge neutrality must be maintained. An extensive treatment of these structures is not possible and only a few will actually be presented in figures. Instead an extensive compilation of ceramic oxide structures is provided in Table 4-7. When properly understood this table contains useful information for visualizing some of the important representatives that were already mentioned

TABLE 4-7	**SIMPLE IONIC STRUCTURES GROUPED ACCORDING TO ANION PACKING**			
Anion packing	**Coordination number of M and O**	**Cation sites**	**Structure name**	**Examples**
Cubic close-packed	$6:6\ MO$	All octahedral	Rocksalt	NaCl, MgO, FeO, NiO, CaO, SrO, MnO, BaO
Cubic close-packed	$4:4\ MO$	Half tetrahedral	Zinc blende	ZnS, BeO, SiC
Cubic close-packed	$4:8\ M_2O$	All tetrahedral	Antifluorite	Li_2O, Na_2O, K_2O, Rb_2O
Cubic close-packed	$12:6:6\ ABO_3$	One-fourth octahedral (B)	Perovskite	$CoTiO_3$, $SrTiO_3$, $SrSnO_3$, $SrZrO_3$
Simple cubic	$8:8\ MO$	All cubic	CsCl	CsCl, CsBr, CsI
	$8:4\ MO_2$	Half cubic	Fluorite	ThO_2, CeO_2, PrO_2, UO_2, ZrO_2, HfO_2
Hexagonal close-packed	$4:4\ MO$	Half tetrahedral	Wurtzite	ZnS, ZnO, SiC
Distorted cubic	$6:3\ MO_2$	Half octahedral	Rutile	TiO_2, GeO_2, SnO_2, PbO_2, VO_2, NbO_2
Hexagonal close-packed	$6:4\ M_2O_3$	Two-thirds octahedral	Corundum	Al_2O_3, Fe_2O_3, Cr_2O_3, Ti_2O_3

From W. D. Kingery, H. K. Bowen, and D. R. Uhlmann, *Introduction to Ceramics*, 2nd ed., Wiley, New York (1976).

previously or will be discussed in this and subsequent chapters. Both a review of Section 3.2.3 and practice will help to convert word descriptions into pictures. As examples, the simpler structures are selected.

4.7.1.1. Rocksalt

The position of cations and anions is familiar and shown in Fig. 3-7. In the case of MgO, for example, O ions are considered to populate FCC Bravais lattice points. The (interstitial) Mg ions are octahedrally coordinated so that six anions surround each cation and vice versa.

4.7.1.2. Zinc Blende

This structure, exhibited by GaAs, was also reproduced earlier in Fig. 3-6. In the case of BeO, oxygen atoms populate all the FCC Bravais lattice points. Be, however occupies only half of the tetrahedral sites. Four O ions surround each Be ion, and vice versa. Other materials exhibiting this structure include ZnS and SiC.

Charge neutrality is easily attained in these examples because ions with the same valence magnitude are present in equal numbers. This restriction is removed in the next few examples.

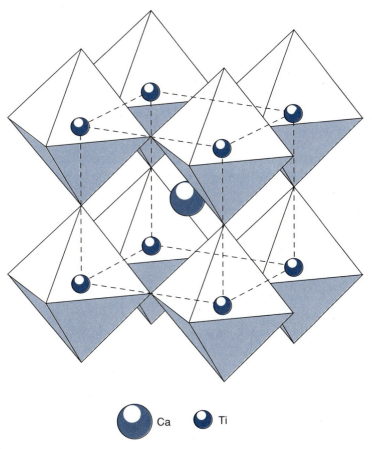

Ca   Ti

FIGURE 4-22   Perovskite structure of CaTiO$_3$.

4.7.1.3. Perovskite

Three ions are involved in these structures, whose idealized cubic unit cell is shown in Fig. 4-22. In CaTiO$_3$ the central ion is Ca^{2+} (the larger of the two cations), and it lies in 12-fold coordination with O^{2-} ions and in 8-fold coordination with Ti^{4+}. Meanwhile, each Ti^{4+} ion is octahedrally coordinated to 6 O^{2-} ions. It is instructive to account for both the stoichiometry and the charge balance in this unit cell. The one Ca^{2+} ion is wholly contained within the cell. Each of the corner Ti^{4+} ions is shared by 8 identical cells so there is effectively only $8 \times \frac{1}{8} = 1$ Ti atom. Lastly, each of the 12 O^{2-} ions is shared by 4 cells, and therefore contribute a total of 3 O atoms. The sum of the positive charge $2+ + 4+ = 6+$ balances the negative charge ($3 \times 2- = 6-$). Furthermore, the centers of the positive and negative charges coincide, unlike the situation in BaTiO$_3$, a slightly distorted perovskite. As we shall see in

Chapter 11, this separation of charge creates an electric dipole and a structure with interesting electrical properties.

EXAMPLE 4-3

a. For UO_2 and B_2O_3 predict the value of the coordination number, N_c.

b. Suggest a crystal structure for UO_2 and show that the charge balances in it.

ANSWER   a. Ionic radii are $r(U^{4+}) = 0.105$ nm, $r(B^{3+}) = 0.023$ nm, and $r(O^{2-}) = 0.132$ nm.

For UO_2, $r(U^{4+})/r(O^{2-}) = 0.105/0.132 = 0.795$. Therefore, from Fig. 3-10, $N_c = 8$.

For B_2O_3, $r(B^{3+}/r(O^{2-}) = 0.023/0.132 = 0.174$. Therefore, $N_c = 2$. For B_2O_3, N_c is experimentally found to be 3. The discrepancy stems from uncertainty in ionic radii and contributions from covalent bonding.

b. For UO_2, a CsCl-like cubic structure is suggested with U^{4+} located in the cube center surrounded by 8 O^{2-} ions at the cube corners. In such a cubic cell the U^{4+} ion contributes an effective charge of $+\frac{1}{2}e$ to each bond with O^{2-}; for 8 bonds the total charge is $+4e$. Each O^{2-} ion has a charge of $-2e$ and is shared by 8 neighboring cubes. Therefore, the total negative charge within the cell is $8 \times \frac{1}{8} \times (-2e) = -2e$. The charge imbalance means that the CsCl structure is not correct.

If the U^{4+} occupied *every other* cube center position the charge would balance. This is what occurs in the fluorite structure and ceramic materials exhibiting this structure include CaF_2, ThO_2, ZrO_2, and UO_2. The fluorite structure is shown in Fig. 4-23 with one complete CsCl-like cell indicated at the extreme right.

EXAMPLE 4-4

Shown in Fig. 4-24 is the crystal structure of the YBaCuO superconductor, which may be thought of as three stacked $CaTiO_3$ cells. Neglecting the vacant sites what is the chemical formula for this material?

ANSWER   There are obviously 1 Y atom and 2 Ba atoms per cell. Each of the 8 Cu atoms at the corner of the cell is shared by 8 other cells. Therefore, $8 \times \frac{1}{8} = 1$ Cu atom. Each of the other 8 Cu atoms on the cell edges is shared by 4 other cells and they contribute $8 \times \frac{1}{4} = 2$ Cu atoms. Altogether there are 3 Cu atoms.

Each of the 8 O atoms on the large rectangular faces is shared by 2 cells, each yielding $8 \times \frac{1}{2} = 4$ atoms. The O atom at each pyramid apex is shared

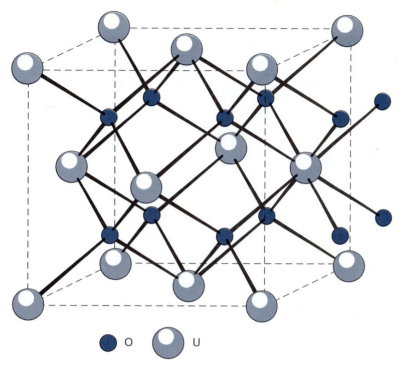

O U

FIGURE 4-23 Fluorite structure of UO_2.

by 4 cells and there are 8 such atoms. Thus, the number of O atoms from this source is $8 \times \frac{1}{4} = 2$. Finally, there are an additional 4 O atoms located on the top and bottom planes which are each shared by 4 cells, so that $4 \times \frac{1}{4} = 1$ O atom. Altogether there are 7 O atoms.

Therefore, the chemical formula is $YBa_2Cu_3O_7$.

4.7.2. Silicates

After carbon, silicon forms more compounds than any other element. The earth's crust is largely composed of silicon–oxygen compounds, or silicates, and many of them find engineering uses in glass, cement, brick, and electrical insulators. They all have structures that are based on the fundamental tetrahedral SiO_4^{4-} building block, the orthosilicate anion. By linking to one another across the bridging oxygen atoms one-, two-, and three-dimensional subunits can be constructed. These connect to one another through intervening cations such as Al^{3+}, Mg^{2+}, Ca^{2+} and Fe^{2+} to create a diverse array of rather complex crystalline structures.

The simplest silicate is the linear chain of SiO_4 units shown in Fig. 4-25A. Because the bridging oxygen is shared by two silicon atoms the formula changes

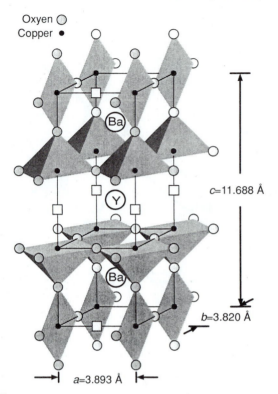

Oxen ◯
Copper ●

c=11.688 Å

b=3.820 Å

a=3.893 Å

FIGURE 4-24   Crystal structure of the $YBa_2Cu_3O_7$ high-temperature superconductor (see Fig. 3-38D).

from SiO_4 to SiO_3; the mineral enstatite, $MgSiO_3$, is an example. Two chains can then link together to form the double-chain inosilicate structure (Fig. 4-25B), an example of which is anthophyllite, $Mg_7Si_3O_{22}(OH)_2$. With the lateral addition of more chains, sheet silicates form where the triangular faces of the SiO_4 tetrahedra all lie in a common plane. Such sheet silicates (Fig. 4-25C) are the basis of the phyllosilicate minerals, which are important constituents of igneous rocks and schists. Members include the mica group (e.g., muscovite) and the clay mineral group (e.g., kaolinite). A much simplified depiction of the bonding of silica sheets in these minerals by metal ion bridges is shown in Fig. 4-26. Asbestos essentially consists of such sheets bridged by magnesium ions. The more detailed structure of the layered clay mineral pyrophyllite is displayed in Fig. 4-27, with tetrahedral and octahedral sites noted. Kaolinite crystals are photographed in Fig. 4-28, revealing the actual platelet morphology. The slippery and soapy feeling of such clay minerals is a consequence of the weak van der Waals forces between layers. Mica sheets, on the other hand, are bonded ionically via K^+ ions, but the ionic bonds are not sufficiently strong to prevent cleavage separation of sheets.

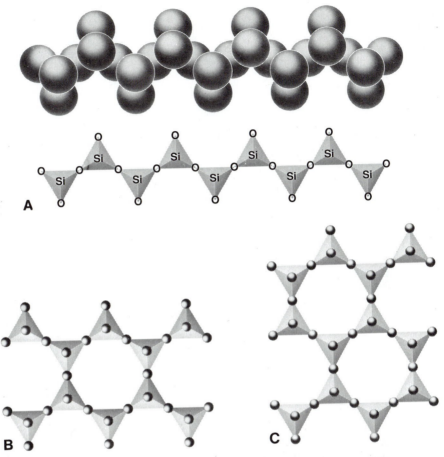

FIGURE 4-25 (A) A single chain of SiO_4^{-4} units. Each singly bonded oxygen, one in the plane of the page and the other behind it, has a single charge. Every silicon is joined by two bridging oxygens. (B) Double-chain silicate structure (inosilicate). (C) Sheet silicate structure (phyllosilicate). After R. J. Borg and G. J. Dienes, *The Physical Chemistry of Solids*, Academic Press, San Diego, CA (1992).

4.8. CEMENT AND CONCRETE

4.8.1. Introduction

Concrete and reinforced concrete are two of the most important civil engineering materials. Their tonnage consumed far exceeds that of steel, wood, and polymers combined. They are the essential ingredient in some of the largest structures built in this century such as high-rise buildings and airport runways. Many large concrete structures have to restrain water from rivers as in dams, whereas others must withstand ocean forces and marine environments. As an

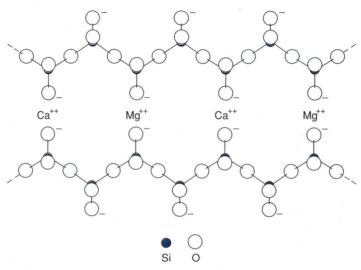

FIGURE 4-26   Schematic representation of the bonding of silicate chains by bridging metal ions. After R. J. Borg and G. J. Dienes, *The Physical Chemistry of Solids,* Academic Press, San Diego, CA (1992).

important example of the latter consider their use in ocean platforms for oil drilling. Up until 1972 the petroleum industry used steel for such purposes. Then a bold decision was made to use concrete as the principal structural material in the Ekofisk oilfield in the North Sea. From then until 1989 some 20 concrete platforms, containing about 2×10^6 m^3 ($\sim 5 \times 10^9$ kg) of high-

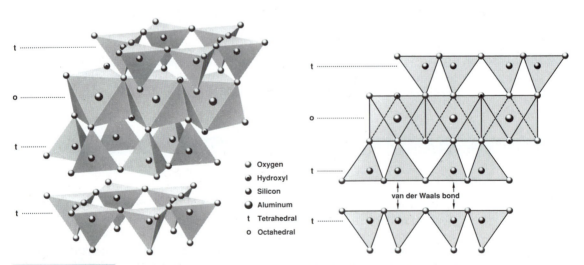

FIGURE 4-27   Structure of the clay mineral pyrophyllite. After R. J. Borg and G. J. Dienes, *The Physical Chemistry of Solids,* Academic Press, San Diego, CA (1992).

FIGURE 4-28 SEM image of kaolinite platelets. Courtesy of R. Anderhalt, Philips Electronic Instruments Company, a Division of Philips Electronics North America Corporation.

quality concrete, were built in the North Sea in water depths ranging from 70 m to more than 200 m. The world's largest platform is shown in Fig. 4-29. Evidently, our long and varied engineering experience with cement and concrete has engendered sufficient confidence in the strength and environmental degradation resistance of these materials in such challenging applications.

Adhering to the pattern set earlier in this chapter, we are concerned chiefly with the structure and composition of concrete materials.

4.8.2. CEMENT

Concrete is composed of cement, aggregates (sand, gravel, crushed rock), and water. The water is important because it makes the concrete moldable and is needed for the hydration reactions that harden the cement. As the cement hardens it binds the aggregate and gives concrete the stonelike quality of bearing high compressive loads. Cement, a generic term for concrete binder, is the key ingredient. The most important binder for concrete is Portland cement and it will be discussed exclusively. Portland cement is produced from an initial mixture of 75% limestone ($CaCO_3$), 25% clay, assorted aluminosilicates, and iron and alkali oxides. This mixture is ground and fed into a rotary kiln (a

FIGURE 4-29 Gullfaks C, the world's largest offshore concrete oil drilling platform. Water depth = 216 m, concrete volume = 240,000 m^3, production capacity = 245,000 bbls/day. Courtesy of J. Moskenes, Norwegian Contractors.

large cylindrical rotating furnace) together with powdered coal. Progressive reactions at temperatures extending to 1800 K break down clays, decompose the limestone to yield quicklime or CaO, and then fuse them to produce clinker (pellets) 5–10 mm in size. After cooling, the latter is mixed with 3–5 wt% gypsum (CaSO$_4$) and ground to the powder that is Portland cement.

The final powder size influences the setting rate; finer sizes mean a higher specific surface area and faster hydration and setting rates. Portland cement is composed of the following identifiable compounds:

Calcium oxide (lime)	$CaO = C$
Silicon oxide (silica)	$SiO_2 = S$
Aluminum oxide (alumina)	$Al_2O_3 = A$
Water	$H_2O = H$

Some three-quarters by weight is composed of tricalcium silicate (C_3S) and dicalcium silicate (C_2S). Different proportions of these and other ingredients are blended to produce special cements that either set slowly or more rapidly, liberate less heat of hydration, or are intended to resist degradation by water containing sulfates and other compounds. Wet cement is workable only until it sets, and this period establishes the pot life of the mix. The typical time dependence of heat evolution and hardening of cement is shown in Fig. 4-30. After a large initial burst of heat there are much smaller secondary peaks that coincide with the start of solidification. During the setting period cement has no strength, but with solidification, strength rises.

Chemical reactions that are responsible for heat evolution and hardening are complex and not entirely understood. Cement does not harden by a drying reaction, but, conversely, requires water; surprisingly, the reactions continue for years. Several important hydration reactions occur at different rates:

$$C_3A + 6H \longrightarrow C_3AH_6 + \text{heat} \qquad \text{(hours)}$$
$$2C_3S + 6H \longrightarrow C_3S_2H_3 + 3CH + \text{heat} \qquad \text{(days)}$$
$$2C_2S + 4H \longrightarrow C_3S_2H_3 + CH + \text{heat} \qquad \text{(months)}$$

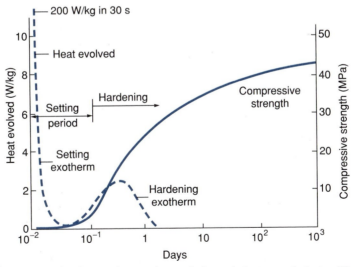

FIGURE 4-30 Time-dependent heat evolution and strength change during cement hydration. (The compressive strength is the maximum load per unit area a sample of cement can sustain without fracturing.) After G. Weidmann, P. Lewis, and N. Reid. *Materials in Action Series: Structural Materials*, Butterworths, London (1990).

The first reaction causes the cement to set; the second and third reactions cause hardening. Some 70% of the structure is taken up by $(CaO)_3(SiO_2)_2(H_2O)_3$ or $C_3S_2H_3$ which is hydrated $(CaO)_3(SiO_2)_2$. Despite the long reaction times, the hydrated tricalcium sulfate product is in the form of very small particles (less than 1 μm in size) and is known as **tobomorite gel**. Gels are swollen polymer-like networks of high viscosity, and they coat the surfaces of the unreacted cement grains, slowing hydration reactions of the latter. Hardening and strengthening occur with the growth and intertwining of these spinelike gels. The increasing silicate content with time suggests the corner-to-corner joining of $(SiO_4)^{4-}$ tetrahedra:

$$2(SiO_4)^{4-} \longrightarrow Si_2O_7^{6-} + O^{2-}$$

These combinations can give rise to higher-order silicate groupings, essentially producing a polymer. Hardening is associated with the production of these higher-order silicates.

4.8.3. Concrete

Concrete is the first of a number of synthetic composites or engineered materials mixtures discussed in the book. In addition to cement, concrete contains aggregate consisting of sand (less than 2 mm in diameter) and larger particles of gravel and crushed rock. A high proportion of fine aggregate is required for smooth finishes. The typical ratio of aggregate to cement is 5 : 1, whereas that of sand to gravel is about 3 : 2. Aggregates contain particles with a distribution of sizes that enable smaller ones to fit neatly into the spaces between larger ones. Cement then flows into the remaining spaces.

For many applications steel bars are embedded in the concrete to reinforce it. Concrete, like other ceramics, is weak when pulled in tension but strong when compressed. Steel, which is strong in tension, is positioned in those parts of the concrete likely to be so stressed. In that way steel supports tensile loads and concrete bears the compressive loads.

4.9. PERSPECTIVE AND CONCLUSION

This chapter has been one of startling material contrasts. First, there is age. Ceramics and silica glasses are among the oldest materials used by humans, whereas polymers are among the newest. But, surprisingly, even among ceramics the new or fine varieties are truly modern, state-of-the-art materials. This represents a vivid example of art being supplanted by science; nevertheless, there is still much room for art. Second, one group of materials is organic in nature, the other inorganic. Raw materials and sources for them thus differ.

Third, there are extremes in atomic bonding. Organic polymers are low-temperature materials reflecting the dominant role of van der Waals bonding. Ceramics are composed of strong ionic and covalent bonds that decompose or melt only at extremely elevated temperatures. Lastly, there is the huge gulf in what is meant by crystallinity. Ceramic crystals, even those found in nature, are paradigms of beautifully ordered structures; in fact the science of crystallography was originally devoted to unfolding their internal atomic arrangements. For many of these ceramic crystals, oxgyen atoms (or anions) fill the close-packed structures that were introduced in Chapter 3. And in between the oxygen atoms are the cations located in tetrahedral or octahedral sites. Crystalline polymers, on the other hand, exhibit a very imperfect crystallinity and structures are disordered or even amorphous at the edges of crystals.

Despite these distinctions there are important similarities between glasses and polymers. Both types of materials exist in amorphous forms. But even among amorphous structures there are differences. In polymers, individual molecules can be flexible linear chains or they can be interconnected in three-dimensional networks. This latter configuration is the one that silicate glasses exhibit. But, whereas tetrahedral units are linked via bridging oxygen atoms in silicates, there are no simple rules that structurally or chemically characterize network polymer units.

Both glasses and polymers display similar volume–temperature crystallization behavior. A given amount of amorphous material can have different volumes or, equivalently, differing densities, depending on the cooling rate of the melt. Low cooling rates allow a more complete structural relaxation of liquid to solid, resulting in a denser solid. Both materials are characterized by a similar temperature-dependent viscosity; this important property reflects the ease of molecular motions and macroscopic deformation under applied loads. In the liquid state the viscosity (η) is low, facilitating forming operations, whereas in the solid state η rises to 10^{12} Pa-s. Removed from either of these states is one in between, corresponding to the glass transition temperature (T_G). Above T_G polymers tend to be rubbery and extend readily; below it they are brittle and crack easily. Mechanical properties will thus depend on the relative position of T_G with respect to the use temperature. Materials other than polymers and glass can also be "amorphized," for example, certain metal alloys and semiconductors. The trick is to cool or condense them rapidly enough from the liquid or vapor state, respectively, so that something resembling a supercooled liquid can be retained. So far they can be produced only in thin sections (films, coatings, foils, and powder) because heat can be more easily extracted from such geometries.

Our formal introduction to electronic and material structure has ended. Concepts presented in the last three chapters will, to a greater or lesser extent, permeate the rest of the book. Interestingly, the next chapter, dealing with thermodynamics, makes virtually no reference to either atomic electrons or specific crystal structures.

Additional Reading

C. Hall, *Polymer Materials*, Macmillan Press, London (1981).

W. D. Kingery, H. K. Bowen, and D. R. Uhlmann, *Introduction to Ceramics*, 2nd ed., Wiley, New York (1976).

S. Musikant, *What Every Engineer Should Know about Ceramics*, Dekker, New York (1991).

D. W. Richerson, *Modern Ceramic Engineering*, Dekker, New York (1992).

F. Rodriguez, *Principles of Polymer Systems*, 2nd ed., McGraw–Hill, New York (1982).

QUESTIONS AND PROBLEMS

4-1. Hydrogenation of acetylene yields ethylene according to the reaction

$$C_2H_2 + H_2 \longrightarrow C_2H_4$$

Calculate the energy change in the reaction per mole, and indicate whether energy is released or absorbed.

4-2. The end-to-end distance of a stretched long-chain polymer molecule is much larger than the end-to-end distance of the unrestrained molecule. Why?

4-3. Write the formula for the condensation reaction between urea and formaldehyde to produce thermosetting urea–formaldehyde if urea has the structure

$$H-N-C-N-H$$
$$\ \ \ \ |\ \ \ \ \|\ \ \ \ |$$
$$\ \ \ \ H\ \ \ \ O\ \ \ \ H$$

4-4. What is the degree of polymerization of a polystyrene sample that has a molecular weight of 129,000?

4-5. A certain rubber is composed of equal weights of isoprene ($C_4H_5CH_3$) and butadiene (C_4H_6).

a. What is the mole fraction of each of the rubber components?

b. How many grams of sulfur must be added to 2 kg of this rubber to crosslink 1% of all the mers? *Note:* 1 S atom crosslinks 2 mers.

4-6. If the degree of polymerization of polyvinyl chloride is 729, what is the molecular weight of the polymer?

4-7. An electrical terminal block composed of thermosetting phenol–formaldehyde (Bakelite) weighing 0.1 kg is compression molded. If this part can be thought of as a single large crosslinked molecule, what is its molecular weight? *Note:* A mole of polymer contains 6.02×10^{23} mers.

4-8. A polypropylene polymer has equal numbers of macromolecules containing 450, 500, 550, 600, 650, and 700 mers. What is the mass average molecular weight in amu?

4-9. A copolymer of polyvinyl chloride and vinyl acetate contains a ratio of 19 parts of the former to 1 part of the latter. If the molecular weight is 21,000 g/mol what is the degree of polymerization? *Note:* The vinyl acetate mer contains 4 C, 6 H, and 2 O atoms.

4-10. Plasticizers in polymers behave like modifiers in silica glasses. Speculate on the behavior of plasticizers.

4-11. In crystalline form a polymer melts at 240°C but in amorphous form it has a glass transition temperature of 90°C. The coefficient of thermal expansion of the liquid polymer is three times that of the solid polymer. If the polymer volume expands linearly with temperature in both states, what is the ratio of the free volumes at 100°C and 200°C?

4-12. a. What is the approximate glass transition temperature of a polymer that has a viscosity of 10^9 Pa-s at 75°C?

   b. At what temperature will the viscosity of this polymer be 10^7 Pa-s?

4-13. A kilogram of vinyl chloride polymerizes to polyvinyl chloride.

   a. What bonds are broken and what bonds are formed?

   b. How much energy is released in the process?

4-14. For atactic polypropylene the temperature (T in °C) dependence of the specific volume V_s (in units of mL/g) is given by

$$V_s \text{ (L)} = 1.137 + 1.4 \times 10^{-4} \, T \qquad \text{(for amorphous solid at low temperature)}$$

$$V_s \text{ (H)} = 1.145 + 8.0 \times 10^{-4} \, T \qquad \text{(at high temperature)}$$

   a. What is the glass transition temperature of this polymer?

   b. What is the polymer density at 25°C?

   c. Suppose $V_s \text{ (L)} = 1.15 + 1.4 \times 10^{-4} \, T$. Would this behavior signify a faster or slower cooling rate from the melt?

4-15. Plastic foams have many uses including padding, flotation devices, and insulation. They consist of large volumes of entrapped gas that can reside either in interconnected open cells or in isolated closed cells.

   a. Are open or closed cells more desirable from the standpoint of a water flotation device?

   b. A polymer has a specific gravity of 1.11 and is foamed to a density of 0.07 g/cm^3. What is the percentage expansion during foaming?

4-16. Which rubber, polyisoprene or polybutadiene, is more likely to be susceptible to atmospheric oxygen degradation?

4-17. The earth's crust is almost 50% oxygen by weight and some 96% oxygen by volume. Geologists say that the earth's crust is solid oxygen containing a few percent impurities. Assess the truth of this statement by roughly comparing the ionic sizes and weights of the most abundant elements on earth.

4-18. Distinguish between the following pairs of terms:

   a. Traditional and new ceramics

   b. Glasses and ceramics

   c. Refractories and abrasives

   d. Glass formers and modifiers

   e. Silicate and chalcogenide glass

4-19. A 100-kg charge of glass contains 75.3 kg of SiO_2, 13.0 kg of Na_2O, and 11.7 kg of CaO.

   a. What is the molar percentage of each oxide?

   b. What is the empirical formula of the glass?

4-20. It is desired to make 100 g of the $YBa_2Cu_3O_x$ high-temperature superconductor by compounding Y_2O_3, Cu_2O, and BaO.
   a. How many grams of each ingredient must be added?
   b. What is the value of x?
   Note: Additional oxidation is carried out to raise the oxygen level.

4-21. What compositional and structural differences distinguish the new and structural ceramic materials?

4-22. The glass–ceramic composition β-spodumene or $Li_2O \cdot Al_2O_3 \cdot 4SiO_2$ is compounded and 10 mol% TiO_2 is added to it.
   a. Calculate the weight percentage of lithia, alumina, and silica in β-spodumene.
   b. What is the weight percentage of titania in this material?

4-23. a. Is $BaTiO_3$, which has a structure like $CaTiO_3$, simple cubic, BCC, or FCC? Why?
   b. What is the lattice parameter of $BaTiO_3$ if its density is 6.02 Mg/m^3?
   c. Which ions appear to contact each other in $BaTiO_3$?

4-24. The lattice constant of CeO_2, which has a fluorite structure, is 0.54 nm. What is the density of this oxide?

4-25. A soda-lime glass composition specifies 70 wt% SiO_2, 15 wt% Na_2O, and 15 wt% CaO. If the raw materials for melting are SiO_2, Na_2O, and $CaCO_3$, what weight of each ingredient is required to produce 100 kg of glass? (At an elevated temperature the carbonate decomposes, releasing CO_2 gas bubbles that promote melt stirring.)

4-26. Calculate the predicted cation coordination numbers for the following oxides:
   a. BeO   b. TiO_2 c. SiO_2 d. Al_2O_3.

4-27. Calculate the predicted cation coordination numbers for the following compounds: a. MgF_2 b. FeS   c. AlF_3 d. NaF.

4-28. a. When a glass modifier like Na_2O is added to SiO_2 does the oxygen-to-silicon (O : Si) ratio increase or decrease?
   b. If the O : Si ratio reaches 3, glasses tend to crystallize. Why?

4-29. Clay minerals like pyrophyllite cleave readily. Suggest a reason why.

4-30. Calculate the approximate density of $YBa_2Cu_3O_7$, the high-temperature superconductor.

4-31. Distinguish between the following terms:
   a. Cement and concrete
   b. Gypsum and lime
   c. Setting and hardening of cement
   d. Concrete and reinforced concrete

4-32. To make concrete pavements, a mix of 1 : 2 : 3 by volume of cement : sand : aggregate is required. In addition suppose the water–cement ratio (by weight) necessary for this cement is 0.5. If the average specific gravities of the cement, sand, and aggregate are 3, 2.6, and 2.5 Mg/m^3, respectively, what quantities (by weight) of the four ingredients are needed to mix 1 m^3 of concrete? Assume all of the solids are dry.

5

THERMODYNAMICS OF SOLIDS

The concept of equilibrium is an important one in science. Statics, a common course of study, deals with the balance of forces that prevails in mechanical structures and systems when they are in equilibrium. When such a state exists it means the structure or system has no tendency to move or change and, in effect, has minimized its mechanical or potential energy. We may view the materials we are concerned with in this book as chemical systems that are occasionally relatively pure, but more frequently contain two or more different elements. Several questions then arise with respect to chemical compared with mechanical systems: What do we mean by chemical equilibrium? Are there comparable forces that equilibrate in chemical equilibrium? Is there a corresponding energy that is minimized at equilibrium? Fully developed answers to these questions are given in chemistry and thermodynamics courses, but here we can only stress some highlights in a qualitative way.

A good place to start is with two famous quotations by the German physicist Clausius: "Die Energie der Welt ist konstant." "Die Entropie der Welt strebt einem maximum zu." (The energy of the world is constant. The entropy of the world strives toward a maximum.) Although Clausius had global notions in mind, what if the world we are interested in is some solid we deal with in this book? Energy in this sense deals broadly with the interactions between atoms and specifically with the bond energies discussed in Chapter 2. There we noted that a balance of attractive and repulsive forces led to a minimum energy in a single bond. The concept of entropy is, however, more subtle. It is common to view entropy as a measure of the state of disorder or randomness in the system. A quantitative statistical measure of entropy is related to the number of possible configurations atoms can assume; the larger this number the greater the randomness and entropy. The creative synthesis of these opposing natural tendencies—of lowering internal energy by bringing atoms together, while simultaneously increasing entropy by randomizing their location—is a main cornerstone of thermodynamics. Here, we state without proof that a single function known as the free energy (G) combines both internal energy (E) and entropy (S) attributes within systems as

$$G = E + PV - TS, \qquad (5\text{-}1)$$

where T is the temperature in degrees Kelvin. The additional term, a product of the pressure (P) and volume (V), is small in solids. G is temperature dependent, reflecting the strong influence of T in Eq. 5-1 and the generally slight temperature dependence of both E and S.

In this book we are always concerned with **changes** in energy between initial (i) and final (f) states that are at **constant** temperature and pressure. The latter conditions imply that enthalpy (H), defined as $H = E + PV$, rather than internal energy E, is strictly more appropriate; because the difference between E and H is small, we use both energy terms interchangeably. With these facts taken into consideration the equation of interest is

$$\Delta G = \Delta H - T\Delta S, \qquad (5\text{-}2)$$

where ΔH and ΔS are the respective changes in enthalpy and entropy (usually on a molar basis).

A very important consequence of the laws of thermodynamics is that spontaneous reactions occur at constant T and P when $\Delta G = G_f - G_i$ is minimized, or when ΔG is negative. Note that neither the sign of ΔH nor the sign of ΔS taken individually determines reaction direction; rather it is the sign of the *combined* function ΔG that is crucial. For example, during the condensation of a vapor to form a solid there is a reduction in entropy ($\Delta S < 0$). This is due to the many atomic configurations in the vapor and the markedly reduced number in the ordered solid. But the enthalpy (energy) decrease on forming the solid more than offsets the entropy decrease, and the net change in ΔG is negative. Systems thus naturally proceed to reduce G_i successively to a still lower, more negative value G_f, until it is no longer possible to reduce G any further. When this happens, $\Delta G = 0$ and the system is at rest or in equilibrium; there is no longer a driving force for change.

Thermodynamic driving forces include the gradients or spatial derivatives of temperature, pressure, and chemical potential. When the temperature in all parts of the system is the same there are no temperature gradients or driving forces for heat flow. This constitutes a state of **thermal equilibrium.** Similarly, mechanical equilibrium implies a balance of pressures. If pressure is not equili-brated there will be bulk flows and deformations of matter in parts of the system. **Chemical equilibrium** implies no tendency for chemical change. The chemical potential, a type of thermodynamic concentration, is the most difficult of the variables to understand. It is defined as the free energy per unit mass of a given atomic species. If there is a gradient in chemical potential (~ effective atomic concentration), then a driving force is established that causes individual atoms to diffuse. (The subject of diffusion is addressed in more detail in Chapter 6.) In summary, systems are only in true **thermodynamic equilibrium** when all gradients, and the driving forces they create, vanish.

The remainder of this chapter is devoted to a number of applications in which thermodynamic issues play a major role. Of these topics the greatest emphasis is placed on binary phase diagrams. To preview this important subject within the context of the above concepts it is instructive to consider Fig. 5-1. A binary system is schematically depicted in which there are a number of possible configurations of the same number of A and B atoms (components). What are the thermodynamic implications, if any, of these different physical distributions of atoms? In Fig. 5-1A initial blocks of pure A and B atoms are brought together. Next these atoms are allowed to mix together freely as they might do if they were melted together (Fig. 5-1B). The entropy of this randomly mixed system (or **liquid solution**) is larger relative to case A. Upon solidification of the melt one possible atomic configuration resembles that of the prior liquid solution. In this so-called **random solid solution** there does not appear to be any great tendency for A atoms to bond to A neighbors or for B atoms to couple to B neighbors (Fig. 5-1C). Rather, both A and B atoms show little preference in selecting their neighbors, but the energy of the solid is lower than that of the liquid. This state is completely different from the one shown in Fig. 5-1D. Here, A atoms clearly prefer to bond to A atoms, and likewise for B atoms. Relative to case B the entropy has fallen, but relative to case A it has risen. This segregation of A and B atoms into pure clusters is apparently due to A–A and B–B bonds having lower energies than A–B bonds. Rather than a solution, atoms have separated into a so-called **two-phase mixture**. Surprisingly, although concentration variations and gradients are evident, the system is actually in equilibrium. This is a case that shows the absence of any tendency toward homogenization by atomic diffusion. Although there are concentration gradients, there are, in fact, no chemical potential gradient driving forces. State D also has a somewhat higher surface energy than state A. This is due to the larger total perimeter length and interfacial surface area associated with clusters relative to the linear boundary separating A and B atoms in case A. Finally, in marked contrast to case D there is the solid-state configuration shown in Fig. 5-1E. It resembles a **compound**, but in comparison to case C we might also describe it as an **ordered solid solution**. In this configuration the energy reduction due to favored A–B bonds offsets the lowered entropy due to or-dering.

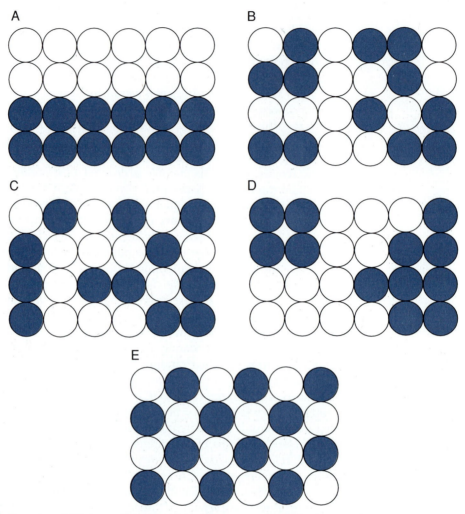

Illustration of differing configurations of A and B atoms within binary alloys. (A) Separate blocks of A and B atoms. (B) A melt of A and B randomly distributed. (C) Random solid solution of A and B. (D) Separation of A- and B-rich regions or phases. (E) Ordered compound of A and B.

It should be appreciated that interest in thermodynamics is wonderfully interdisciplinary in nature. There are many approaches to the subject and each can be understood on different levels. At the most fundamental level all of these approaches actually converge. Mechanical engineering stresses the **macroscopic** nature of matter and the thermodynamic approach taken makes no assumption about the atomic constitution of the working medium (e.g., steam) of the system. Electrical engineering frequently adopts this point of view in energy conversion applications. Materials science, physics, and chemistry, however, usually stress the **microscopic** statistical nature of atomic phenomena, illustrated, for example, in Fig. 5-1. When binary atomic distributions exist in stable thermodynamic equilibrium, it is certain that the system has achieved minimum free energy. Chemical thermodynamics is also very much concerned with free energy as we shall see in the next section.

5.2. CHEMICAL REACTIONS

5.2.1. Oxide Formation Reactions

There are many instances where chemical reactions play an important role in materials science and engineering. The oxidation of metals, high-temperature reactions in polymer mixtures, corrosion, and the deposition of hard ceramic coatings from gases are but a small sample of the examples that can be profitably described by the thermodynamics of chemical reactions. According to the rules of thermochemistry that apply to chemical reactions at equilibrium, we recall that the energies (enthalpies, free energies) of reactants are additive when account is taken of the number of moles involved, and similarly for the products. The **net** energy difference (ΔH, ΔG) is what remains when the energies of the products are subtracted from those of the reactants. To see how this works, consider the oxidation of metal M by the reaction

$$M + O_2 = MO_2, \qquad \Delta G^\circ_{MO_2}. \tag{5-3}$$

If MO_2 is produced under standard conditions (e.g., pure M, 1 atm O_2), then the change in free energy of oxide formation per mole, ΔG°, is given by $\Delta G^\circ = G_{MO_2} - G_M - G_{O_2}$. In elementary chemistry it is shown that the equilibrium constant of a reaction (K_{eq}) is related to ΔG° through the important equation

$$K_{eq} = \frac{[MO_2]}{[M]P_{O_2}} = \exp(-\Delta G^\circ / RT). \tag{5-4}$$

The terms in brackets are the effective concentrations, P_{O_2} is the oxygen pressure, T is the absolute temperature, and R is the gas constant. ($R = 8.314$ J/mol-K or 1.987 cal/mol-K.)

Extensive tables of ΔG° values exist not only for oxides, but for other compounds formed from elemental constituents. Data for oxides have been conveniently graphed in Fig. 5-2, providing a handy collection of thermodynamic information which is used in the following illustrative problems.

EXAMPLES 5-1

a. Thin-film Al conducting stripes are brought into intimate contact with insulating layers of SiO_2 in integrated circuits. Is there a thermodynamic tendency for them to react at 400°C?

b. What is the maximum O_2 pressure to which a Ni melt at 1460°C can be exposed before it tends to oxidize?

c. What is the reason for the negative slope of ΔG^0 versus T for oxidation of carbon whereas the slope for the other metals is positive?

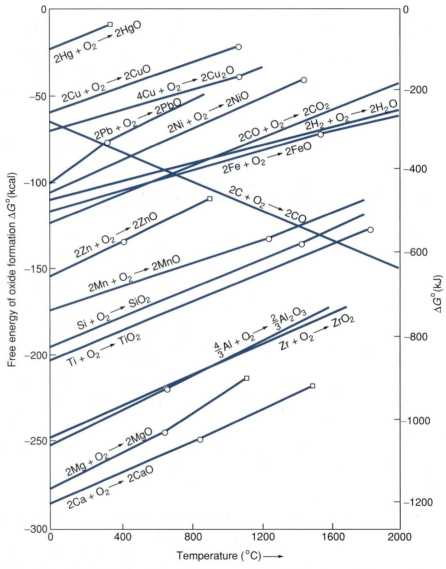

FIGURE 5-2 Free energy of metal oxide formation versus temperature. ○, Melting point; □ boiling point (1 atm). From A. G. Guy, *Introduction to Materials Science,* McGraw–Hill, New York (1972).

ANSWERS   a. We must consider the reactions (from Fig. 5-2)

$$\tfrac{4}{3}Al + O_2 = \tfrac{2}{3}Al_2O_3 \qquad \Delta G^\circ_{Al_2O_3} = -235 \text{ kcal/mol at } 400°C$$
$$Si + O_2 = SiO_2 \qquad \Delta G^\circ_{SiO_2} = -180 \text{ kcal/mol at } 400°C$$

Eliminating O_2 through subtraction of these two equations yields

$$\tfrac{4}{3}Al + SiO_2 = \tfrac{2}{3}Al_2O_3 + Si$$
$$\Delta G^\circ = \Delta G^\circ_{Al_2O_3} - \Delta G^\circ_{SiO_2}$$
$$= -235 - (-180) = -55 \text{ kcal/mol (or } -230 \text{ kJ/mol).}$$

Because ΔG° is negative the reaction is thermodynamically favorable. Whether reaction will, in fact, occur is dependent on kinetic factors. In this case reaction rates are too sluggish to produce troublesome effects.

b. Assume that molten nickel oxidizes according to the reaction

$$2Ni + O_2 = 2NiO \qquad \text{and} \qquad \Delta G^\circ_{NiO} = -40 \text{ kcal/mol at } 1460°C \text{ (1733 K).}$$

In Eq. 5-4 the effective standard concentrations of both metal and oxide are usually taken to be unity. If the same can also be assumed for Ni and NiO, then $[NiO] = 1$ and $[Ni] = 1$. Thus,

$$\ln P_{O_2} = \Delta G^\circ/RT = (-40{,}000 \text{ cal/mol})/(1.99 \text{ cal/mol K})(1733 \text{ K}) = -11.6.$$

Therefore, $P_{O_2} = \exp -11.6 = 9.17 \times 10^{-6}$ atm.

c. If Eq. 5-2 is plotted as ΔG versus T, a straight line results provided ΔH and ΔS are not temperature dependent but constant. This generally appears to be the case in Fig. 5-2. The slope of ΔG versus T is $-\Delta S$. In the case of carbon two molecules of CO form for each O_2 consumed. This corresponds to a gas volume expansion and an entropy *increase* ($\Delta S > 0$). During oxidation of all the metals there is a *reduction* in gas volume as O_2 reacts and is consumed. Therefore, $\Delta S < 0$, and the slopes are positive.

Because its free energy will fall below that for any metal, provided the temperature is high enough, carbon is a universal reducing agent. In the form of charcoal it has enabled metals to be extracted from oxide ores since antiquity.

As a generalization, the metal of an oxide that has a more negative ΔG° than a second oxide will reduce the latter, and be oxidized in the process.

5.2.2. Chemical Vapor Deposition Reactions

There are a number of important applications in which thin films or coatings are synthesized from high-temperature gas mixtures. The objective of these chemical vapor deposition (CVD) processes is to coat selected substrates with the required materials. As an example consider the TiN, TiC, and Al_2O_3 compound coatings deposited on tungsten carbide metal machining and cutting tools. (The gold color commonly seen on cutting tools and on tungsten carbide tool inserts is due to a TiN coating.) Hundreds of millions of tool bits worldwide have been coated with either single or sequentially deposited multiple layers of these materials. Each coating has a hardness double that of hardened tool steel, a hard material in its own right. Only 5- to 10-μm (5000- to 10,000-nm)-thick layers are deposited. This coating thickness is enough to bestow considerable wear resistance to the underlying tool so that it can last 5 to 10 times longer than uncoated tools before requiring sharpening. A reactor used to deposit all three coatings is shown schematically in Fig. 5-3. Input gases react at the substrate surface forming the compound deposit of interest while gaseous products are swept out of the reactor. The chemical reactions relied upon are

$$2TiCl_4(g) + N_2(g) + 4H_2(g) = 2TiN(s) + 8HCl(g) \qquad 1000°C \qquad (5\text{-}5a)$$

$$TiCl_4(g) + CH_4(g) = TiC(s) + 4HCl(g) \qquad 1000°C \qquad (5\text{-}5b)$$

$$2AlCl_3(g) + 3CO_2(g) + 3H_2(g) = Al_2O_3(s) + 3CO(g) + 6HCl(g) \qquad 1000°C$$
$$(5\text{-}5c)$$

Each reaction takes place at roughly 1000°C under approximate equilibrium conditions. Also, all of the reactants and products, with the exception of the coating denoted by (s), are gaseous; this is a general feature of CVD reactions. Several coated tools are reproduced in Figs. 5-4A and B together with the structure of a modern multilayer coating on a WC tool substrate (Fig. 5-4C).

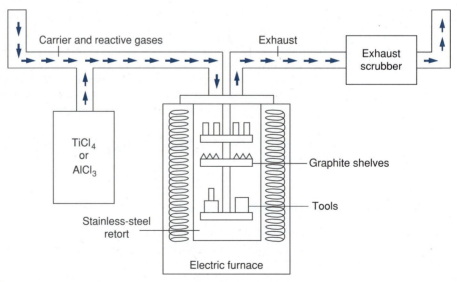

FIGURE 5-3   Schematic of chemical vapor deposition reactor used to deposit hard coatings of TiN, TiC, and Al$_2$O$_3$.

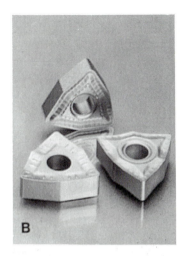

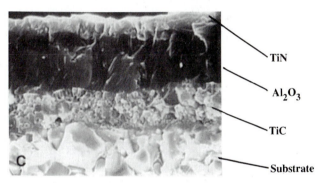

TiN

Al$_2$O$_3$

TiC

Substrate

FIGURE 5-4 (A) Assorted cutting tools coated with TiN. Courtesy of Multi-Arc Scientific Coatings. (B) Multilayer coated lathe cutting tool inserts. (C) Scanning electron microscope image of a trilayer coating, deposited by CVD methods, on a WC cutting tool substrate (3500×). The TiC is hard and bonds well to the WC. Protection of the tool against excessive heating is provided by the hard but insulating Al$_2$O$_3$ layer. Finally, the hard, wear-resistant upper TiN layer is chemically stable and offers low friction in contact with the workpiece being machined. Courtesy of S. Wertheimer, ISCAR.

EXAMPLE 5-2

An important CVD reaction employed in the fabrication of Si microelectronic devices is

$$SiCl_4 + 2H_2 = Si(s) + 4HCl \qquad 1500°K \qquad (5\text{-}6)$$

Published thermodynamic data on the free energy of formation of SiCl$_4$ and HCl from the elements at 1500°K reveal

$$Si + 2Cl_2 = SiCl_4 \qquad \Delta G° = -441 \text{ kJ/mol} \qquad (5\text{-}7a)$$
$$\tfrac{1}{2}H_2 + \tfrac{1}{2}Cl_2 = HCl \qquad \Delta G° = -105 \text{ kJ/mol}. \qquad (5\text{-}7b)$$

What are $\Delta G°$ and the equilibrium constant for the reaction of Eq. 5-6 at 1500°K?

ANSWER The additive nature of energy and the simple rules of thermochemistry yield

$$\Delta G° \text{ (Eq. 5-6)} = 4\Delta G° \text{ (Eq. 5-7b)} - \Delta G° \text{ (Eq. 5-7a)}$$
$$= 4(-105) - (-441) = +21 \text{ kJ}$$

Therefore, the equilibrium constant is $K_{eq} = \exp - (21,000/8.32 \times 1500) = 0.19$. Crystal growth from the vapor phase is normally carried out as close as possible to $\Delta G° = 0$ or $K = 1$.

The deposition of the Si film takes place on a prior single-crystal wafer of Si. Unlike the polycrystalline structure of the hard coatings, the Si deposit is a high-quality single-crystal film. It is known as an **epitaxial film** (from the Greek *epi* and *taxis*, which means "arranged upon") because the deposit uses the substrate as a template on which to extend crystal growth. Epitaxial film growth is addressed again in Section 12.5.2.

5.3. SINGLE-COMPONENT SYSTEMS

5.3.1. Variable Pressure and Temperature (Phase Diagrams)

By **single-component systems** we mean materials possessing one chemical species that is retained in all of its phases over the broad range of temperatures (T) and pressures (P) of interest. For example, a single type of either atom (carbon) or molecule (H_2O) can constitute a single component. Thus, carbon can exist as graphite, diamond, liquid carbon, or carbon vapor, and H_2O as water, ice, or steam. Each of these recognizable homogeneous forms (solid, liquid, gas) constitutes a **phase**. A convenient way of visualizing regions of phase stability in single-component systems as a function of T and P is through phase diagrams. The widely reproduced phase diagram for H_2O is shown in Fig. 5-5. We know from experience that water, ice, and steam are individually stable over broad temperature and pressure ranges. The phase diagram, determined from careful experiment, broadly maps these limits of single-phase stability. In addition, the lines depict the special set of conditions under which two phases coexist in equilibrium. Thus, the boundary between liquid and vapor defines a set of pressure-versus-boiling point data, whereas that between the vapor and solid defines corresponding vapor pressure-versus-sublimation temperature data. Under a very restricted set of conditions, namely, 0.0075°C and 0.006 atm, all three phases of H_2O coexist in equilibrium at the triple point.

The vapor pressure–temperature data just referred to are extremely important in thin-film and coating deposition technologies. A handy representation

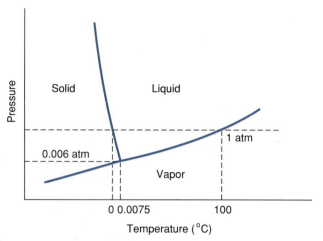

FIGURE 5-5 Pressure–temperature phase diagram for H_2O. From C. R. Barrett, W. D. Nix, and A. S. Tetelman, *The Principles of Engineering Materials*, Prentice–Hall, Englewood Cliffs, NJ (1973).

of such thermodynamic information is provided in Fig. 5-6 for a number of elements. Note that the vapor pressure for each element rises exponentially with temperature. This is not an entirely unexpected dependence because evaporation may be viewed as a reaction where the vapor (V) and condensed liquid (L) or solid (S) phases are in equilibrium. Therefore, the vaporization reaction

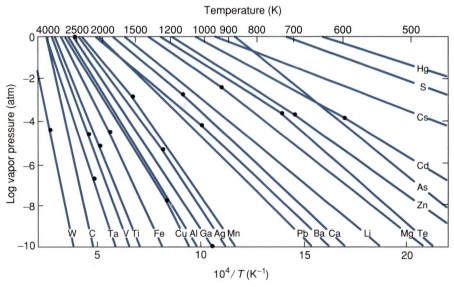

FIGURE 5-6 Vapor pressure–temperature data for selected elements. From C. H. P. Lupis, *Chemical Thermodynamics of Materials*, North-Holland, Amsterdam (1983).

for component A (from the liquid) and the corresponding equilibrium relationship are given by

$$A_L = A_V, \qquad P_A/[A_L] = \exp(-\Delta G_{vap}/RT). \qquad (5\text{-}8)$$

Substitution of $\Delta G_{vap} = \Delta H_{vap} - T\Delta S_{vap}$ (Eq. 5-2) finally yields

$$P_A = [A_L] \exp(\Delta S_{vap}/R) \exp(-\Delta H_{vap}/RT) = P_0 \exp(-\Delta H_{vap}/RT). \qquad (5\text{-}9)$$

In this expression ΔH_{vap} may be associated with the latent heat of vaporization
and P_0, a constant, is equal to $[A_L] \exp(\Delta S_{vap}/R)$.

The phase diagram for carbon (Fig. 5-7) exhibits two triple points and
contains two different solid phases, diamond and graphite. One of the most
exciting quests in the history of materials science has been the synthesis of
diamond. This diagram suggests that graphite might possibly be converted to
diamond provided high enough pressures could be maintained at elevated
temperatures. Long experimentation was rewarded with success in 1954, when
the General Electric Corporation announced a process to produce diamond
crystals. Typical conditions required pressures of 54,000 bars (53,500 atm),
temperatures of 1430°C, and, importantly, a liquid Ni solvent from which
dissolved carbon could separate out as diamond. Thermodynamic equilibrium
is essentially maintained in this method for making diamond.

A no less exciting achievement of the 1980s has been the synthesis of thin
films of diamond from the vapor phase under much less forbidding processing
conditions. Readily accessible pressures of less than 1 atm and temperatures
of 1000°C or less are required. One way to prepare such films is to reduce

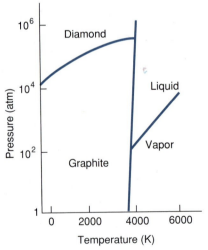

FIGURE 5-7   Pressure–temperature phase diagram of carbon. From A. L. Ruoff, *Materials Science*, Prentice–Hall,
Englewood Cliffs, NJ (1973).

methane with atomic hydrogen generated in an energetic gaseous discharge or plasma. Similar discharges occur in neon signs. In this way the formation of tetrahedral sp^3 bonded carbon (see Section 2.4.4) is promoted. Examples of the resulting faceted diamond jewels are shown in Fig. 5-8. This surprisingly simple production method means that the equilibrium conditions of the phase diagram have been bypassed in favor of **metastable** synthesis. Metastability denotes a nonequilibrium state of existence that in some cases can be virtually permanent, for example, glass. If heated, glass may crystallize and more closely approach thermodynamic equilibrium. In the present case metastable diamond is retained at room temperature.

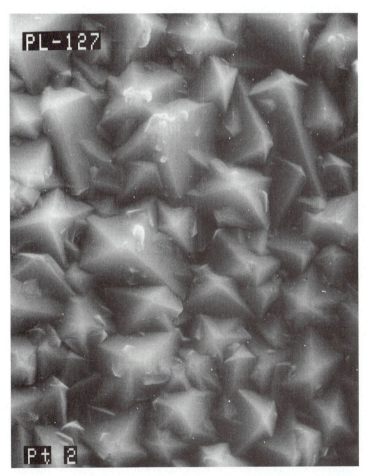

FIGURE 5-8 Diamond crystals grown by low-pressure CVD methods. Courtesy of T. R. Anthony, GE Corporate Research and Development.

5.3.2. The Gibbs Phase Rule

The celebrated *phase rule* derived by Josiah Willard Gibbs provides an interesting way to analyze not only one-component three-phase systems, but systems containing an arbitrary number of components and phases. This rule surprisingly connects the number of degrees of freedom (F) in a system to the number of components (C) and phases (ϕ) it contains, as

$$F = C + 2 - \phi. \tag{5-10}$$

Though deceptively simple in appearance, the Gibbs phase rule is arguably the most important linear algebraic equation in all of science (whole books have been written on it). By degrees of freedom or variance, we mean the number of variables (here T and P) that can be **independently** varied without altering the phase state of the system. Thus, water ($C = 1$) will remain a liquid ($\phi = 1$) even though both temperature and pressure ($F = 2$) can be independently raised or lowered over a reasonable range of values.

When water (liquid) and steam (vapor) coexist at 100°C and 1 atm, $\phi = 2$ and $F = 1$. Now assume that P is increased independently, as in a pressure cooker. The temperature must then rise in a specific, **dependent** fashion along the vaporization line to continue to preserve the water–steam equilibrium, albeit at different values of P and T. Finally, at the critical triple point, $\phi = 3$ and $F = 0$. Any change in either temperature or pressure will destroy the very precarious three-phase equilibrium and push the system into one of the single-phase fields.

5.3.3. Constant Pressure

Most frequently we are interested in systems at atmospheric pressure. Under such conditions one variable is fixed by the equation $P = 1$ atm. The variance is thus reduced by one so that the Gibbs phase rule now reads

$$F = C + 1 - \phi. \tag{5-11}$$

Thus, for one component ($C = 1$) at the melting point, both liquid and solid are present ($\phi = 2$), so that $F = 0$. This is why melting occurs at a fixed temperature.

Interestingly, some materials experience a volume expansion whereas others undergo a volume contraction upon melting. Typically the volume change is a few percent for both types of behavior. Most metals belong to the former category. Their crystal structures are tightly packed. In the case of FCC metals each atom has 12 nearest neighbors. On melting, crystalline packing deteriorates, fewer neighbors surround each atom, and separations between them extend beyond the prior solid-state lattice spacings. The resulting volume expansion is turned to a contraction upon solidification. Such shrinkage is a particularly troublesome issue during casting where it can lead to defects if not compensated for (see Section 8.2.5). In contrast, those materials that have relatively open network structures containing stereospecific bonds frequently undergo a volume *contraction* upon heating and a corresponding *expansion* upon solidification. The excited atoms in the liquid state can now reduce their energy by forging new lower-energy bonds of shorter length. Examples of such materials include diamond cubic silicon, ice, rhombohedral bismuth, and orthorhombic gallium. In each case the solid structure is relatively loosely packed.

Another important phenomenon exhibited by certain single-component materials in the solid state is **polymorphism** or **allotropy**. The involved materials assume different crystallographic structures as the temperature changes. Iron is the outstanding engineering example exhibiting allotropy. Because it undergoes polymorphic changes we are able to harden steel by heat treatment. From below room temperature up to 910°C, α-Fe, a body-centered cubic phase ($a = 0.2866$ nm), is stable. At 910°C, α-Fe transforms to face-centered cubic γ-Fe ($a = 0.3647$ nm), and it, in turn, is stable until 1390°C. From this temperature up to its melting point (1534°C), iron exists as δ-Fe, another body-centered cubic form ($a = 0.2932$ nm). All of these phase transformations, like melting, are reversible. By Eq. 5-1, G falls with increasing T and stable phases are those that minimize the free energy as illustrated graphically in Fig. 5-9. This accounts for the existence of α-Fe below 910°C where γ-Fe and liquid Fe are metastable.

Another important example of polymorphism occurs in SiO_2 or silica. Crystalline packing of silica SiO_4^{4-} tetrahedra occurs in quartz, tridymite, and cristobalite phases, each of which exists in several modifications. Quartz trans-

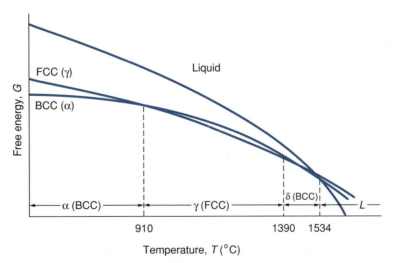

FIGURE 5-9 Free energy of phases in iron versus temperature. Phases with the lowest free energy are thermodynamically stable. After C. R. Barrett, W. D. Nix, and A. S. Tetelman, *The Principles of Engineering Materials*, Prentice–Hall, Englewood Cliffs, NJ (1973).

forms slowly into stable tridymite at 867°C. Tridymite remains stable until 1470°C when it transforms to cristobalite. These crystal forms have densities in the neighborhood of 2.65 g/cm³, while that for amorphous fused quartz is only 2.2 g/cm³.

5.3.4. Thermal Energy Absorption in Solids

5.3.4.1. Heat Capacity

One of the important attributes of a solid is its energy. When heat flows into a solid its temperature rises as the absorbed thermal energy (enthalpy at fixed pressure) increases. As H is a function of temperature (T) it is then possible to define the derivative of H with respect to T as the heat capacity (c_p) of the material:

$$c_p = dH/dT. \tag{5-12}$$

Because H is not necessarily linear with respect to T, c_p is generally a function of temperature. Liquids and gases also have heat capacities. Therefore, if the heat capacity of materials is measured experimentally as a function of temperature, H can be extracted through mathematical integration. By making use of other thermodynamic equations and identities, values for entropies and free energies can also be determined. This building block approach enables a consistent set of thermodynamic data for materials to be assembled for use in assorted thermochemical calculations.

Returning to our one-component solid we may well ask how thermal energy is absorbed. The core electrons are too energetic to be affected, and the valence electrons absorb relatively little energy. This leaves the atomic ion cores as the major absorber of energy. What is the probability, $P(E_i)$, that atoms can absorb a given energy E_i at a temperature T? Statistical thermodynamic considerations provide the answer to this important question. The result, given without proof, is

$$P(E_i) = \exp(-E_i/kT), \tag{5-13}$$

where k is the Boltzmann constant ($k = R/N_A$). Of the form given earlier in Eq. 5-4, the exponential is known as the Maxwell–Boltzmann factor. This equation is easily one of the most important in the book and certainly one worth remembering. Since temperature constitutes such an important variable in materials science it is significant that Eq. 5-13 accurately predicts the exponential increase in rate with temperature of a host of phenomena governed by the thermally induced motion of atoms. A partial list of examples includes rates of chemical reactions, diffusive motion of atoms, viscous flow and deformation of polymers, solid-state phase transformations, sintering of ceramic and metal powders, elevated temperature degradation, and ionic conduction in ceramics. The value of E_i differs but, remarkably, ranges from perhaps a fraction of an electron-volt per atom (1 eV/atom = 96,500 kJ/mol) to only a few electron-volts per atom in each of these solid-state processes.

This equation can also be profitably used to describe thermal energy absorption by atoms in a solid. The latter may be viewed as being held together by a three-dimensional collection of coupled springs much like the inner spring of a mattress or trampoline. A difference, however, is that the springs are in perpetual motion. Atoms sit at the spring nodes and when thermal energy is absorbed they vibrate at higher frequencies and with greater amplitude. A spring's potential energy is half the product of the spring constant and the square of its displacement (i.e., $\frac{1}{2}k_S(r - r_0)^2$ by Section 2.5.1); its kinetic energy is the same in magnitude on average. Their sum is basically the E_i of Eq. 5-13, which is the link between the spring's mechanical and thermal energy (i.e., kT).

This classical picture is a bit simplistic. We have already seen that electrons absorb energy in discreet quanta and so do atoms. But rather than photon waves for electrons, quantized vibrational waves (phonons) are associated with lattice atoms; their energies, $E = nh\nu$, were already noted in Eq. 2-3, where ν is the vibrational frequency. What average energy can then be expected in such a complex situation? No less a person than Albert Einstein solved this problem, and for a mole of atoms, he showed that the energy $E (\sim H)$ is

$$E = 3N_A h\nu/2 + 3N_A h\nu/\{\exp(h\nu/kT) - 1\}. \tag{5-14}$$

To obtain the specific heat of the solid, this expression needs to be differentiated with respect to T. At a high temperature, where $h\nu/kT$ is small, the problem is simplified because $\exp(h\nu/kT) - 1$ can be expanded as a series and approximated by $1 + h\nu/kT + \cdots + -1 = h\nu/kT$. Therefore, $E = 3N_A h\nu/2 + 3N_A kT$. Finally, by virtue of Eq. 5-12, differentiation yields $c_p = 3N_A k$ or $3R$. Two features of this quantum calculation (note the presence of Planck's constant h) are worth noting:

1. Irrespective of material the specific heat of a solid approaches $3R$ or very nearly 25 J/mol (6 cal/mol) at elevated temperatures. This feature is illustrated in Fig. 5-10 for a number of different materials and is known as the law of Dulong and Petit.

2. Heat capacity differences between materials are reflected primarily at lower temperatures where ν does not cancel from the expressions for E and c_p. Each material has a different characteristic value of ν. In the very hard and stiff diamond lattice, atoms vibrate at a high frequency of $\sim 2.75 \times 10^{13}$ Hz. For soft lead, however, the atomic vibrational frequency is low and approximately equal to 1.83×10^{12} Hz.

5.3.4.2. Blackbody Radiation

Another phenomenon based on absorption of thermal energy is blackbody radiation. If enough heat is absorbed by a solid and it gets sufficiently hot, it begins to emit electromagnetic energy from the surface, usually in the infrared

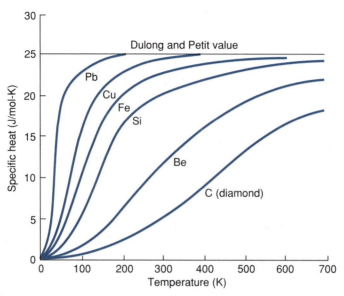

FIGURE 5-10 Heat capacity versus temperature for various solids. The upper limiting value of c_p is that predicted by Dulong and Petit. From C. Newey and G. Weaver, *Materials in Action Series: Materials Principles and Practice*, Butterworths, London (1990).

and visible regions of the spectrum. According to the formula given by Planck, the power density (P) radiated in a given wavelength (λ) range varies as

$$P(\lambda) = C_1\lambda^{-5}/\{\exp(C_2/\lambda T) - 1\} \quad \text{W/m}^2, \tag{5-15}$$

where C_1 and C_2 are constants. The mathematical similarity between Eqs. 5-14 and 5-15 is a reason for introducing this phenomenon here. As the temperature is increased the maximum value of P shifts to lower wavelengths. This accounts for the fact that starting at 500°C, a heated body begins to assume a dull red coloration. As the temperature rises it becomes progressively red, orange, yellow, and white. The total amount of heat power emitted from a surface, integrated over all directions and wavelengths, depends on temperature as T^4. This dependence is known as the Stefan–Boltzmann law.

5.4. INTRODUCTION TO BINARY PHASE DIAGRAMS

5.4.1. Phase Diagrams and How They Are Obtained

When two or more components (elements) react or are alloyed the number of possible product phases at different temperatures and pressures is limitless. Although one-element phase diagrams can comfortably fit in a book of about 100 pages, binary or two-element systems would necessitate a shelf of books to completely account for their reactions. A library would probably be required for all of the possible ternary and higher-order systems. With the exception of several ternary alloy systems considered in Chapter 13 in connection with optical properties, these complex materials are not treated in this book.

In binary systems we are simply interested in knowing what to expect when two elements, of initially known proportions, are mixed and brought to temperatures of interest where they come to thermodynamic equilibrium. To simplify matters, only fixed pressure ($P = 1$ atm) is considered so that the Gibbs phase rule is $F = 3 - \phi$ when $C = 2$. Because it is impossible for F to be negative, three phases, at most, can coexist in equilibrium at fixed pressure. Whatever phases are present, however, can have differing melting temperatures, chemical compositions, and solubilities. Therefore, a graphic representation, that is, a binary phase diagram, that reflects this information would be very handy. In what follows the steps required to experimentally generate such a binary phase diagram by thermal analysis are described.

Consider the copper–nickel system. A series of alloys composed of different proportions by weight of Cu and Ni are prepared (e.g., 25Cu–75Ni, 50Cu–50Ni, 75Cu–25Ni), placed in separate crucibles, and then melted. A thermocouple (temperature-sensing device) is immersed in each melt and the temperature is carefully recorded as a function of time as the alloys cool from high temperatures. An extremely important assumption is that the crucible contents are in **thermodynamic equilibrium** at all times. This condition frequently means long exposure to elevated temperature coupled with low cooling rates. The individual cooling curves obtained are displayed in Fig. 5-11 together with those for pure Cu and pure Ni. In each case heat is lost to the surroundings by a combination of mechanisms that depend largely on the thermal properties (e.g., thermal conductivity, heat capacity, heat transfer coefficients) of the phases contained within the melt. At critical temperatures where new phases appear and others disappear, the cooling kinetics are strongly altered. Evolution of the latent heat of solidification and changes in the amounts and compositions of the phases present serve to pinpoint these singular temperatures; above and below them different cooling rates are frequently manifested.

Differences in cooling behavior are most pronounced in the pure metals. Both above and below the melting temperature (T_M), liquid and solid metal, respectively, cool in similar ways with time. But at T_M there is a thermal arrest that marks the transformation of liquid to solid. As there are one component and two phases, liquid (L) and solid (S), the phase rule predicts zero degrees of freedom, a condition satisfied by an **invariant** or fixed temperature.

Things are a bit more complicated in the alloys for there are no thermal arrests during solidification; instead, three regions can be distinguished. The upper and lower behaviors apparently describe cooling of either a liquid or a solid, respectively, as before. In these cases there is one phase but two components, so that $F = 2$. One of the degrees of freedom is taken up by a temperature drop as observed. In between, some combination of both liquid and solid exists

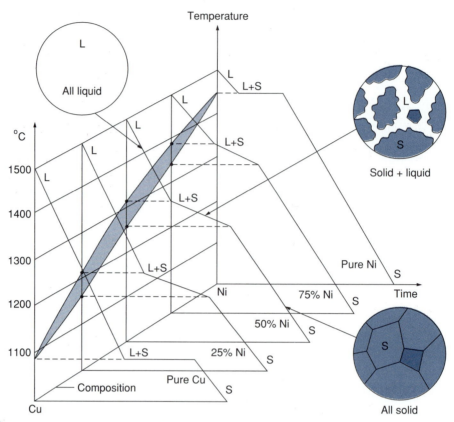

Temperature

FIGURE 5-11   Temperature–time cooling curves for a series of Cu–Ni alloys.

in the crucible and the cooling behavior indicates at least one degree of freedom. If the crucible contents were stirred in this region the presence of solid pieces would be felt, but continued stirring would not dissolve them.

When corresponding sets of critical temperatures so determined are connected together for all of the metals and alloys, the resulting **equilibrium phase diagram**, depicted in Fig. 5-12, emerges for the Cu–Ni system. In the experimental technique just described, thermal property response was the vehicle used to locate critical temperatures. Although thermal analysis is the most widely employed method for this purpose, other properties sensitive to phase transformations (e.g., electrical, optical, magnetic, lattice parameter) have been used. The resulting experimentally determined phase diagrams have been compiled in books, that serve as the first references consulted when synthesizing or heat-treating materials.

The microstructure of an **equilibrated** binary Cu–Ni alloy, shown in Fig. 5-13, is not very different from that of the pure metals it comprises. Other

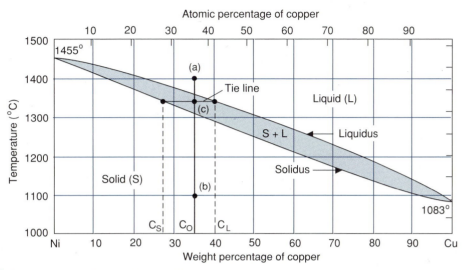

FIGURE 5-12 Equilibrium phase diagram of the Cu–Ni system.

binary combinations exhibiting similar solid solution solidification behavior include Ag–Au, Ag–Pd, Al$_2$O$_3$–Cr$_2$O$_3$, and Ge–Si, a system that is discussed further in Section 5.4.3.

5.4.2. Analysis of Binary Phase Diagrams

Phase diagrams compress a great deal of chemical and physical information about the involved element pairs into a convenient form. On the horizontal axis composition is plotted, with pure components represented at the extreme ends and alloys in between. Temperature is plotted on the vertical axis so that the intersection of composition and temperature, a point, represents a single state of the system, for example, a crucible containing a specific composition raised to the indicated temperature. In Cu–Ni alloys the possible states of the system contain either one or two phases. The former are labeled either L (liquid) or S (solid); the latter are confined to the lens-shaped region labeled L + S.

5.4.2.1. Single Phase

If a state exists in a single-phase field (in this or any other phase diagram) the **chemical** and **physical** analyses of what is present in thermodynamic equilibrium are very simple. Chemical analysis means just that—the percentage by weight or by atoms of each constituent component. Let us consider a 65 wt% Ni and 35 wt% Cu alloy (close in composition to Monel metal, which is used in petroleum refineries) and heat it to 1400°C (state a, Fig. 5-12). The alloy would melt and if we could make a chemical analysis at this temperature it would, not surprisingly, contain 65 wt% Ni and 35 wt% Cu. Physically, we

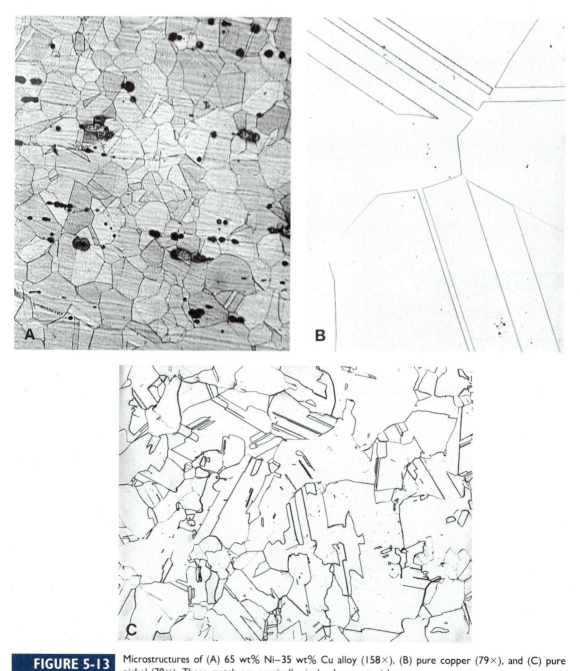

FIGURE 5-13 Microstructures of (A) 65 wt% Ni–35 wt% Cu alloy (158×), (B) pure copper (79×), and (C) pure nickel (79×). These metals are nominally single phase materials.

know that 100% of what would be present would be liquid. Similarly, if the original alloy were cooled to 1100°C (state b) we know the chemical analysis would be identical (i.e., 65 wt% Ni and 35 wt% Cu), and that physically, 100% of what would be present would be solid.

In any single-phase field of any binary equilibrium phase diagram the rules are:

1. The chemical composition is the same as that of the original alloy.
2. Physically, the alloy is either 100% homogeneous liquid or 100% homogeneous solid.
3. Single phase-alloy liquids and solids are generally stable over some range of composition and temperature.

5.4.2.2. Two-Phase Mixture: Tie Line Construction and Chemical Composition

If the system state point falls in a two-phase (L + S) field, a two-phase **physical mixture** exists. The chemical and physical analyses are now more complicated and follow rules different from those of the single-phase field. Importantly, both analyses center around the construction of a **tie line.** This line is drawn through the state point (c) and extends to the left and right until it ends at the single-phase boundaries on either side. Such a tie line is shown at 1340°C for the initial 65 wt% Ni–35 wt% Cu alloy. Tie lines are very significant because they connect two phases in **thermodynamic equilibrium.**

Mechanical equilibrium prevails because these phases coexist at the same constant pressure (1 atm).

Thermal equilibrium is assured because the tie line is drawn horizontally and not at an angle; the connected phases are thus at the same temperature.

Chemical equilibrium is also maintained but it is not as easy to visualize this. Basically the sloped lines on phase diagrams represent solubility limits. We all know, for example, that water at a given temperature dissolves as much sugar as the solubility limit allows. Below this limit, sugar dissolves in the one-phase liquid solution; above it, sugar is rejected (precipitates) out of solution to create a two-phase solid plus liquid mixture. The same considerations hold in the hot liquid and solid phases of the Cu–Ni system. The boundary line between the S and L + S fields is known as the **solidus,** and that between the L + S and L fields, the **liquidus.** Within the L + S field, chemical equilibrium requires that the composition of the solid at the solubility limit is given by the intersection point of the tie line with the solidus curve (read off on the horizontal axis). Similarly, the corresponding composition of the liquid is given by the intersection of the tie line with the liquidus curve. For example, in the 65 wt% Ni–35 wt% Cu alloy at 1340°C the solid has a composition of 27 wt% Cu–73 wt% Ni, whereas the liquid has a composition of 40 wt% Cu–60 wt% Ni. At these chemical compositions the free energy is minimized; slight increases or decreases beyond these solubility limits raise the overall free energy of this alloy relative to its minimum value.

5.4.2.3. Two-Phase Mixture: Lever Rule and Relative Phase Amounts

We are only halfway through our analysis of the two-phase mixture. Now that we know the **chemical analysis** of each phase let us perform a **physical analysis** to determine the relative proportions or weights of each phase present. The analysis is not unlike algebra word problems dealing with mixtures. Consider a portion of the phase diagram containing an alloy of A and B atoms with an initial composition C_0 expressed in terms of wt% B. Assume that two phases, α and β, coexist in equilibrium and the **chemical** composition of phase α is C_α and the composition of phase β is C_β. (It is common to identify solid phases by Greek letters.) Thus, phase α contains C_α wt% of component B and $(100 - C_\alpha)$ wt% of component A. The problem is to determine the fraction of the mixture by weight that is phase α (i.e., f_α) and the fraction that is phase β (i.e., $f_\beta = 1 - f_\alpha$). On the basis of W grams of total alloy, conservation of the mass of B atoms requires

$$WC_0 = WC_\alpha f_\alpha + WC_\beta (1 - f_\alpha). \tag{5-16}$$

The term on the left represents the total weight of B atoms in the alloy, and the terms on the right the weights of B atoms partitioned, respectively, to phases α and β. Solving for f_α and f_β yields the lever rule, which is expressed by

$$f_\alpha = (C_\beta - C_0)/(C_\beta - C_\alpha) \qquad \text{and} \qquad f_\beta = (C_0 - C_\alpha)/(C_\beta - C_\alpha). \tag{5-17}$$

The lever rule is so named because the fraction of a phase (f_α) multiplied by its "lever arm" ($C_0 - C_\alpha$) is equal to the product of f_β and lever arm ($C_\beta - C_0$). Thus, whichever phase (boundary) the initial composition is closest to is present in the greatest amount. When applied to our 65 wt% Ni–35 wt% Cu alloy at 1340°C, the fraction of solid is $f_S = (C_L - C_0)/(C_L - C_S) = 40\text{-}35/40\text{-}27 = 0.385$. For the liquid $f_L = (C_0 - C_S)/(C_L - C_S) = (35\text{-}27)/(40\text{-}27) = 0.615$.

1. Draw a horizontal tie line. (Tie lines are drawn only in two-phase fields. They make no sense in a one-phase field.)
2. The chemical composition of the two phases is given by the ends of the tie line, extended vertically down to and read off the horizontal axis.
3. The physical composition or weight fraction of each phase within the two-phase mixture is based on the tie line and given by the lever rule (Eq. 5-17).
4. Quantitative phase analyses cannot be made if the system state lies exactly on a line or boundary that separates phases. One must be either slightly above or slightly below the line to analyze the system by the above rules.

These are the rules. The rest is commentary and detail.

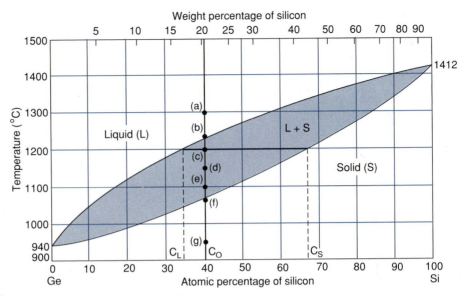

FIGURE 5-14 Equilibrium phase diagram of the Ge–Si system.

5.4.3. Analysis of an Equilibrium Cooled Alloy

We now have all the tools required to chemically and physically analyze a binary alloy system. Rather than the Cu–Ni system, which has a restricted L + S field (narrow solidification range) that makes tie line constructions difficult to read, let us consider the Ge–Si system of Fig. 5-14, which has a similar but wider two-phase field. Both phase diagrams exhibit **solid solution solidification**. The single-phase binary alloy liquid gives way to a **solid solution** upon cooling. Binary alloy liquids have many features in common with those which we imagine aqueous solutions possess. Atoms are in an agitated state of motion, sliding past each other in an amorphous medium with a density slightly different from that of the solid. These liquids are chemically homogeneous throughout and are single-phase materials because the individual randomly mixed atoms cannot be distinguished or separated. Solid solutions are similar, except that the **solute** atoms are randomly mixed in a crystalline matrix of **solvent** atoms.

EXAMPLE 5-3

An alloy containing 6×10^{22} atoms of Ge and 4×10^{22} atoms of Si is heated to 1300°C and cooled to 950°C.

a. What is the alloy composition in atomic and weight percentages?

b. At temperatures of 1300, 1230, 1200, 1150, 1100, 1070, and 950°C, what phases are present? What are their chemical compositions? How much of each phase is physically present? Assume equilibrium is attained at each temperature.

ANSWER   a. The atomic percentage of Ge is

$$[6 \times 10^{22}/(6 \times 10^{22} + 4 \times 10^{22})] \times 100 = 60 \text{ at.\%}$$

Therefore, 40 at.% Si is present. The weight of the Ge atoms is

$$6 \times 10^{22} \, M_{Ge}/N_A = 6 \times 10^{22} \times 72.6/6.02 \times 10^{23} = 7.24 \text{ g}$$

The weight of the Si atoms is

$$4 \times 10^{22} \, M_{Si}/N_A = 4 \times 10^{22} \times 28.1/6.02 \times 10^{23} = 1.87 \text{ g.}$$

The wt% Ge = $[7.24/(7.24 + 1.87)] \times 100 = 79.5$. Therefore, there is 20.5 wt% Si as noted on the upper weight percentage silicon scale. Both weight and atomic percentage notations are widely used as measures of composition.

b. The included table is self-explanatory and has the required answers.

Temperature (°C)	Phases	Chemical composition (at.% Si)	Phase amounts (atomic fraction)	Comment
a. 1300	L	40	1.0	All liquid
b. 1230	L	~40	~1.0	Solidification
	S	72	~0	just beginning
c. 1200	L	34	(67-40)/(67-34) = 0.82	Use of Eq. 5-17
	S	67	0.18	Use of Eq. 5-17
d. 1150	L	25	(56-40)/(56-25) = 0.52	Use of Eq. 5-17
	S	56	0.48	Use of Eq. 5-17
e. 1100	L	16	(47.5-40)/(47.5-16) = 0.24	Use of Eq. 5-17
	S	47.5	0.76	Use of Eq. 5-17
f. 1070	L	12	~0	Solidification
	S	~40	~1	almost complete
g. 950	S	40	1.0	All solid

An idealized picture of the solidification events during slow cooling of this alloy would first start with an amorphous melt. At 1230°C tiny Si-rich dendrites (see Section 3.6.1) would begin to form at the mold wall and perhaps in the bulk of the melt as well. With a drop in temperature the dendrites continue to grow by thickening and lengthening. Surrounding the dendrites is the remaining Ge-rich liquid which continually shrinks in amount until it effectively disappears at 1070°C, leaving grain boundaries behind.

5.5. ADDITIONAL PHASE DIAGRAMS

5.5.1. The Binary Eutectic Phase Diagram

Many pairs of elements (e.g., Bi–Cd, Sn–Zn, Ag–Cu, Al–Si) solidify in a manner similar to that of Pb–Sn, whose **eutectic phase diagram** is shown in Fig. 5-15. The eutectic is so named for the critical constant temperature or **isotherm** that dominates the phase diagram. Below this temperature (183°C in Sn–Pb) for any composition lying within the span of the isotherm (19.2–97.5 wt% Sn), liquid is no longer stable. Above the eutectic temperature, however, it is as though portions of two independent solid solution systems exist—one on the Pb side yielding Pb-rich solid solutions, the other on the Sn side giving rise to Sn-rich solid solutions. Their liquidus lines converge at a critical point, the eutectic composition (61.9 wt% Sn). This means that the liquid phase of every alloy containing between 19.2 and 97.5 wt% Sn will eventually reach the eutectic composition (61.9 wt% Sn) upon cooling from an elevated temperature to slightly above 183°C. Liquid at this very special eutectic composition then decomposes at 183°C into a mixture of two solid phases: α, a Pb-rich

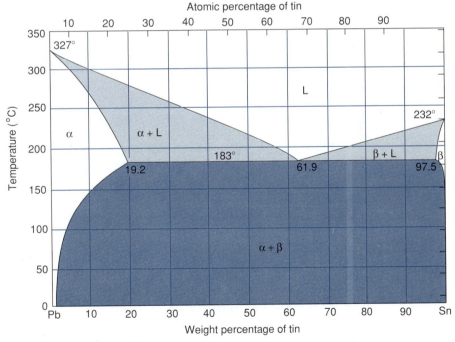

FIGURE 5-15   Equilibrium phase diagram of the Pb–Sn system.

solid solution, and β, a Sn-rich solid solution. The equation describing the decomposition of a single-phase liquid into a two-phase solid mixture at the eutectic temperature

$$L \rightarrow \alpha + \beta \qquad (5\text{-}18)$$

may be taken as a definition of the generic eutectic solidification reaction. According to the phase rule, $C = 2$, $\phi = 3$ (L, α, and β); therefore, $F = 2 - 3 + 1 = 0$. Zero degrees of freedom means that the reaction is invariant; both the temperature and phase compositions remain constant during the transformation of eutectic liquid to $\alpha + \beta$. Consider now the cooling behavior of an alloy of exactly eutectic composition (61.9 wt% Sn–38.1 wt% Pb). Until the melt reaches 183°C there is the usual temperature drop. But then there is a thermal arrest at the eutectic temperature until the eutectic reaction is complete. This is followed by cooling to room temperature by the two-phase solid. If one did not know better the solidification cooling behavior of a pure metal with a melting point of 183°C might be imagined. The two-phase $\alpha + \beta$ eutectic microstructure that develops is generally characteristic of the alloy system. It often appears as a lamellar array of alternating plates of the two phases, but sometimes as a dispersion of one phase in a matrix of the other.

The rules of phase analysis in single- and two-phase fields were given previously and they will be applied once again in the example below. First, however, a few additional details of eutectic solidification should be noted. When compositions on the Pb-rich side begin to solidify, the solid α phase (in equilibrium with L) that increases in amount as the temperature drops is known as the **proeutectic** or **primary** α phase. Similarly, the solid proeutectic or primary β phase separates from the melt in Sn-rich alloys. After the eutectic reaction is completed, the eutectic mixture of $\alpha + \beta$ adds its presence to the already solidified larger crystals of proeutectic α (or β). Further cooling is accompanied by rejection of solute from proeutectic phases to satisfy solubility requirements. But solid-state reactions are often sluggish, and structural and chemical change practically ceases once the melt has totally solidified.

EXAMPLE 5-4

a. For the alloys 5 wt% Sn–95 wt% Pb, 40 wt% Sn–60 wt% Pb, and 70 wt% Sn–30 wt% Pb, identify the phases present and determine their compositions and relative amounts at 330, 184, and 25°C.

b. For a 1-kg specimen of the 70 wt% Sn–30 wt% Pb alloy, what weight of Pb is contained within the eutectic microstructure at 182°C?

ANSWER   The answers are tabulated below using the tie line and lever rule in all two phase regions.

Temperature (°C)	Phases	Chemical composition (wt% Sn)	Phase amounts (weight fraction)	Comment
		5 wt% Sn–95 wt% Pb		
330	L	5	1.0	All liquid
184	α	5	1.0	All solid
25	α	~2	(100-5)/(100-2) = 0.97	(Eq. 5-17)
	β	~100	(5-2)/(100-2) = 0.03	No eutectic micro-structure
		40 wt% Sn–60 wt% Pb		
330	L	40	1.0	All liquid
184	α	19.2	(61.9-40)/(61.9-19.2) = 0.51	Proeutectic α
	L	61.9	(40-19.2)/(61.9-19.2) = 0.49	Eutectic liquid
25	α	~2	(100-40)/(100-2) = 0.61	(Eq. 5-17)
	β	~100	(40-2)/(100-2) = 0.39	

Or alternatively,

182	α	19.2	(61.9-40)/(61.9-19.2) = 0.51	Proeutectic α
	$\alpha + \beta$	61.9	(40-19.2)/(61.9-19.2) = 0.49	Eutectic mixture (Eq. 5-17)
		70 wt% Sn–30 wt% Pb		
330	L	70	1.0	All liquid
184	β	97.5	(70-61.9)/(97.5-61.9) = 0.23	Proeutectoid β
	L	61.9	(97.5-70)/(97.5-61.9) = 0.77	Eutectic liquid
25	α	~2	(100-70)/(100-2) = 0.31	(Eq. 5-17)
	β	~100	(70-2)/(100-2) = 0.69	

Or alternatively,

182	β	~97.5	(70-61.9)/(97.5-61.9) = 0.23	Proeutectic β
	$\alpha + \beta$	61.9	(97.5-70)/(97.5-61.9) = 0.77	Eutectic mixture (Eq. 5-17)

b. In 1 kg of the 70 wt% Sn–30 wt% Pb alloy the eutectic mixture weighs 0.77 kg. Of this, 61.9% is Sn and 38.1% is Pb. Therefore, the amount of Pb = (0.77)(.381) = 0.29 kg.

Schematic microstructures of these alloys are sketched in Fig. 5-16 for the indicated alloys. They should be compared with actual room temperature microstructures of several alloys in the Pb–Sn system (Fig. 5-17). The one containing 63 wt% Sn–37 wt% Pb is common electrical solder; the 50 wt% Sn–50 wt% Pb alloy was used in plumbing. (Recently, the use of Pb in plumbing has been outlawed.)

5.5.2. Side-by-Side Eutectics

More complex phase diagrams can arise through the addition of solid solution, eutectic, and yet other solidification features in horizontal or vertical

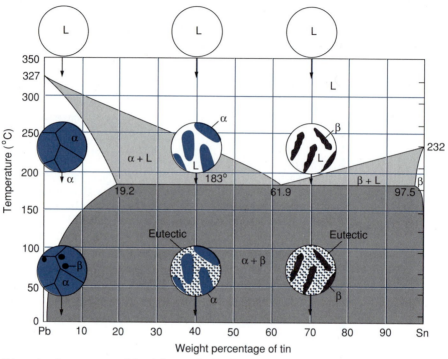

FIGURE 5-16 Schematic microstructures of three alloys in the Pb–Sn system at temperatures of 330, 184, and 25°C. The eutectic microstructure is a mechanical mixture of $\alpha + \beta$.

configurations. Phase analysis in such systems follows rules laid down earlier. Tie lines never extend beyond two-phase fields so that no more than one eutectic at a time is ever considered. Phase diagrams are like road maps; on a map of the greater New York metropolitan area one is not concerned about the neighboring states of New Jersey and Connecticut when motoring through Manhattan. In this spirit consider the Ga–As phase diagram depicted in Fig. 5-18. This system is a very important one in semiconductor technology and contains two eutectic systems stacked side by side. The eutectic compositions are so close to the respective terminal axes that they cannot be resolved on the scale of the phase diagram. A dominant feature in the phase diagram is the central compound GaAs, containing equiatomic amounts of Ga and As. It is regarded as a terminal component in the Ga–GaAs eutectic phase diagram on the left half or Ga-rich side. GaAs is also shared in common with As to

FIGURE 5-17 Optical microstructures of Pb–Sn alloys (96×). (A) 30 wt% Sn–70 wt% Pb. Dark etching, proeutectic Pb-rich α phase surrounded by light etching eutectic. (B) 50 wt% Sn–50 wt% Pb. Dark etching dendritic α phase within the light etching eutectic matrix. (C) 63 wt% Sn–37 wt% Pb. Two-phase eutectic microstructure. (D) 70 wt% Sn–30 wt% Pb. Light etching, dendritic, proeutectic Sn-rich β phase surrounded by eutectic.

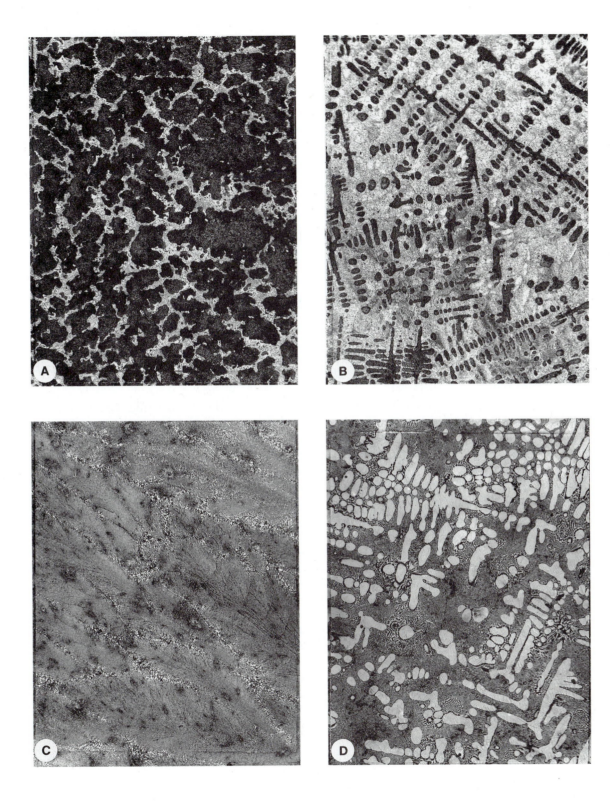

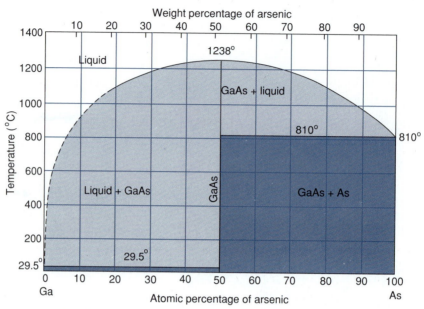

FIGURE 5-18   Equilibrium phase diagram of the Ga–As system.

generate the As–GaAs binary eutectic on the As-rich side. Compounds on phase diagrams often have well-defined stoichiometries, in this case 50 at.% Ga–50 at.% As, and are therefore denoted by narrow vertical lines.

One of the important single-crystal growth techniques employed in the fabrication of compound semiconductor lasers and diodes is known as **liquid phase epitaxy**. The technique can be simply understood in terms of this phase diagram. Consider a Ga-rich melt containing 10 at.% As that is cooled slowly just below 930°C so that it enters the two-phase L + GaAs field. A small amount of solid GaAs separates from the melt. If a single-crystal GaAs wafer substrate is placed in the melt, it will provide an accommodating surface template for the rejected GaAs to deposit on. In the process thin single-crystal epitaxial layers are grown. Readers will recognize that the crystals they grew as children from supersaturated aqueous solutions were essentially formed by this mechanism.

The combination of the ceramic materials silica (SiO_2) and corundum (Al_2O_3) yields another binary system of interest in which eutectics are joined horizontally. As shown in Fig. 5-19 one eutectic forms between SiO_2 and γ (mullite) and the other between γ and Al_2O_3. Mullite is a compound with nominal composition $3Al_2O_3 \cdot 2SiO_2$, but it apparently exists over a broader stoichiometric range than does GaAs. Alloy mixtures of these oxides are important because they are used to make refractory bricks that line high-temperature industrial furnaces.

It is interesting to note the passage from one-phase to two-phase to one-

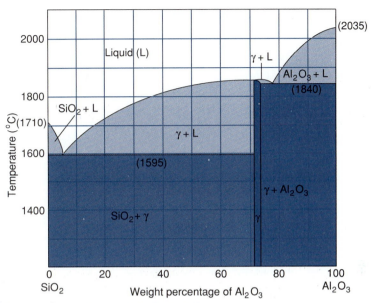

FIGURE 5-19 Equilibrium phase diagram of the Al_2O_3–SiO_2 system.

phase to two-phase to one-phase regions, and so on, across any constant temperature on the phase diagram. This 1–2–1–2–1–...–1 rule, rooted in thermodynamics, is readily illustrated by all phase diagrams.

EXAMPLE 5-5

The phase diagrams depicted in Fig. 5-20 are all thermodynamically impossible as drawn. Provide good physical arguments that specifically indicate what is wrong.

ANSWER   a. The hump in the lens-shaped L + S region is incorrect. Below T_1 tie lines correctly intersect L and S fields. But above T_1 it is possible to draw tie lines that intersect only liquid fields on either side and not L + S as labeled.

b. A slanted eutectic isotherm is a contradiction in terms. It means the system is not in thermal equilibrium. Furthermore, horizontal tie lines to the slanted eutectic line connect phases that are inconsistent with the given labeling of the phase fields.

c. On the eutectic isotherm four phases—α, β, γ, and L—appear to coexist in equilibrium. But this is contrary to the phase rule (by Eq. 5-11, $F = 2 + 1 - 4 = -1$) because the maximum number of phases in a two-component system is 3.

d. The gap in composition along the eutectic isotherm is incorrect. It implies a variable composition of the eutectic liquid and negative degrees of freedom.

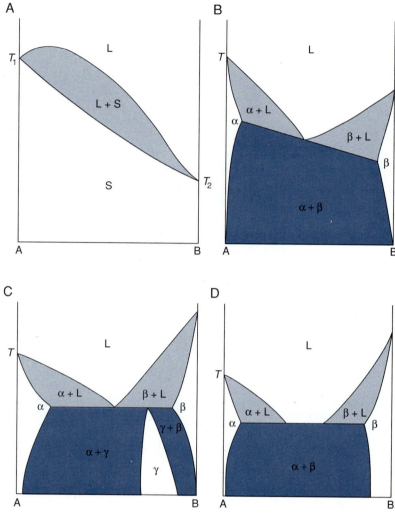

FIGURE 5-20 Four thermodynamically impossible phase diagrams.

5.5.3. Peritectic Solidification

5.5.3.1. Pt–Re

Not uncommon in the catalog of solidification sequences is the invariant **peritectic** reaction, which can be written in the generic form

$$L + \beta \rightarrow \alpha. \tag{5-19}$$

Peritectic reactions often occur when there is a large difference in melting point between the two components. On the phase diagram the peritectic reaction is characterized by the prominent isotherm (zero degrees of freedom), and appears as an upside-down eutectic. There are few examples of alloy systems that exhibit only peritectic behavior over the complete temperature range. One is the platinum–rhenium system, whose phase diagram is shown in Fig. 5-21. Above the peritectic temperature of 2450°C, solid solution solidification prevails in Re-rich alloys. Below this temperature down to the melting point of Pt, liquid exists for Pt-rich compositions which also solidify as solid solutions. This is in contrast to the eutectic isotherm below which no liquid exists.

Only alloys with compositions between ~43 wt% Re and ~54 wt% Re participate in the peritectic reaction. For any alloy in this range a liquid containing ~43 wt% Re and a solid β containing ~54 wt% Re coexist just above 2450°C. The proportions depend on the initial composition and can be calculated from the lever rule (Eq. 5-17). If, for example, we consider the alloy with the critical peritectic composition, that is, 46 wt% Re, then just above 2450°C, $f_\beta = (46\text{-}43)/(54\text{-}43) \times 100 = 27\%$. Thus, 27% of the alloy is solid β, and

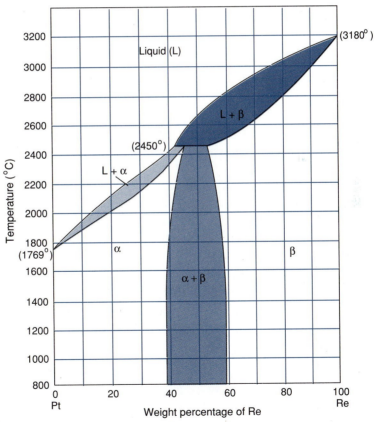

FIGURE 5-21 Equilibrium phase diagram of the Pt–Re system.

the rest is liquid. Given sufficient time, α phase will consume the prior liquid and β, as Eq. 5-19 suggests. But, as the temperature drops, β will be rejected from α and both phases will persist down to lower temperatures. Unlike eutectics, peritectics have no particularly distinguishing microstructural features.

The issue of equilibrium is an important one in peritectic solidification. As peritectic reactions are so sluggish they do not readily proceed to completion unless maintained at high temperature for long times. With practical cooling rates the peritectic reaction may be bypassed, yielding nonequilibrium phases that may not appear on the phase diagram.

5.5.3.2. Cu–Zn

One glance at Fig. 5-22 reveals a binary system with complex solidification behaviors, containing a number of phase features not yet discussed. The cop-

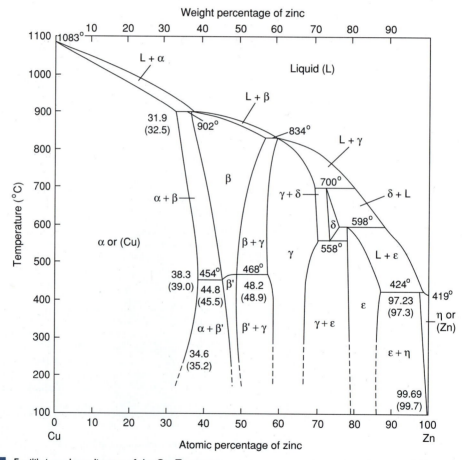

FIGURE 5-22 Equilibrium phase diagram of the Cu–Zn system.

per–zinc system includes several commercially important coppers and brasses among its compositions, for example, gilding metal (95Cu–5Zn), cartridge brass (70Cu–30Zn), and Muntz metal or β brass (60Cu–40Zn). A cascade of peritectic solidification behaviors dominates the high-temperature portion of the phase diagram for all compositions; actually there are five peritectic isotherms as well as a 558°C **eutectoid**-type reaction, whose features are discussed in the next section.

5.5.4. Fe–Fe₃C Phase Diagram

The best known and arguably most important phase diagram in metallurgy is the one shown in Fig. 5-23 between Fe and the compound Fe₃C containing 6.67 wt% C. From it much can be learned about plain carbon steels and cast irons, two of the most widely used classes of structural metals. Figure 5-23 is not a true *equilibrium* diagram, however, because Fe₃C can decompose into graphite (C) and Fe at elevated temperatures. Nevertheless, carbon steel struc-

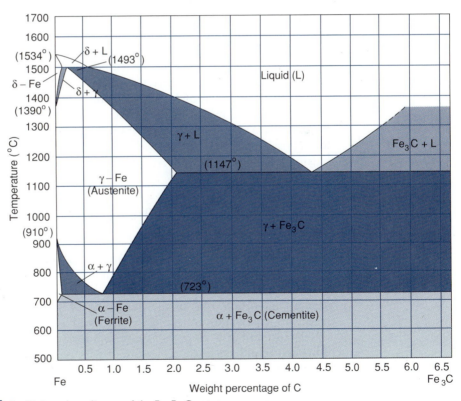

FIGURE 5-23   Equilibrium phase diagram of the Fe–Fe₃C system.

tures are not about to collapse because Fe_3C is, for all practical purposes, stable indefinitely at ambient temperatures.

It is instructive to proceed down the weight percentage C axis and indicate some of the ferrous materials commercially employed. Containing less than 0.1% carbon are the wrought irons used by railroads and in shipbuilding and decorative architectural applications (iron gratings, fences, etc.). Wrought irons also contain several percent slag (mixtures of oxides) strung out in the matrix as a result of mechanical processing. The transition to materials known as steels occurs beyond roughly 0.1 wt% C. Plain carbon steels range from ~0.1 to about 1.5 wt% C and are identified by the designation *SAE** or *AISI†* *10XY*. The last two digits refer to carbon content in hundredths of a percent.

Structural steel (1015 or 1020 containing 0.15 or 0.20 wt% C) is produced in the greatest tonnage. From roughly 0.20 to 0.60 wt% C (1020–1060) there are the medium-carbon steels, and above this, the high-carbon steels. Machine parts and tools are made from steels within these compositions. Steel containing 0.8 wt% C is at the critical eutectoid composition and is known as eutectoid steel. **Hypoeutectoid** steels contain less than 0.8 wt% C, and **hypereutectoid** steels have more than this amount of carbon. For carbon levels between ~2 and 4 wt% C we speak of cast irons. The Fe–Fe_3C phase diagram exhibits all of the solidification and transformation behaviors already discussed and one we did not yet introduce. **Solid solution** freezing is evident between the γ, $\gamma + L$, and L phase fields at elevated temperatures, **eutectic** solidification occurs at 1147°C, and there is a **peritectic** reaction at 1493°C.

The new transformation feature, the **eutectoid**, occurs at the critical temperature of 723°C and resembles a eutectic. Instead of a liquid transforming as in the eutectic, a wholly solid-state transformation occurs in which one *solid* decomposes into two different *solid* phases. The generic, invariant eutectoid reaction is

$$\gamma \text{ (solid)} \rightarrow \alpha + \beta, \tag{5-20}$$

which, in steel, is explicitly exemplified by

$$\gamma\text{-Fe (austenite)} \xrightarrow{723°C} \alpha\text{-Fe (ferrite)} + Fe_3C \text{ (cementite)}. \tag{5-21}$$

Austenite (FCC γ-Fe) dissolves as much as 2 wt% C at 1147°C, whereas **ferrite** (BCC α-Fe) can hold only a maximum of a hundredth as much or less. Partitioning of carbon between these phases at the transformation temperature plays an important role in the hardening of steel by nonequilibrium heat treatment. This important subject is discussed again in Chapter 9.

Reference to Fig. 5-24 reveals that room temperature equilibrium structures of hypoeutectoid steels are a mixture of α-Fe and Fe_3C phases. Upon cooling γ-Fe into the two-phase α-Fe $+$ γ-Fe field, **proeutectoid** α-Fe nucleates at the

---

* Society of Automotive Engineers.
† American Iron and Steel Institute.

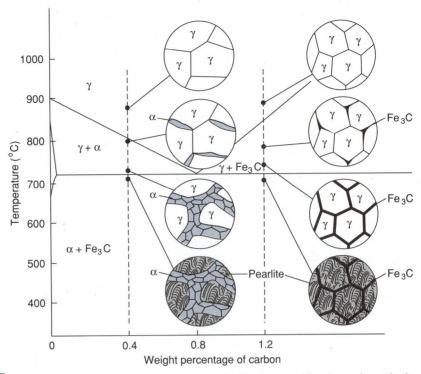

FIGURE 5-24 Equilibrium phase transformations in hypoeutectoid and hypereutectoid steels together with schematic microstructures.

prior γ grain boundaries. The situation is not unlike rejection of a proeutectic phase from a prior liquid in a eutectic system. Just above the eutectoid isotherm, the lever rule (Eq. 5-17) indicates that the amount of γ-Fe present is $(C_0 - 0.02)/(0.8 - 0.02)$, where C_0 is the initial hypoeutectoid steel composition. Below 723°C all of this γ-Fe transforms to a two-phase mixture of α-Fe + Fe$_3$C known as **pearlite.** Just as eutectic two-phase mixtures have a characteristic appearance, so does pearlite, whose microstructure is shown in Fig. 5-25. It consists of plates of Fe$_3$C (also known as **cementite**) embedded in an α-Fe matrix, so that overall, the eutectoid appears to have a fingerprint-like lamellar structure. Slowly cooled hypoeutectoid steels then contain varying proportions of ferrite and pearlite depending on steel composition.

The situation in hypereutectoid steels is complementary to that in hypoeutectoid steels. Now the proeutectoid phase is cementite, and phase fields to the right of the eutectoid composition, extending to 6.67 wt% C, are relevant. Just above 723°C there is a two-phase mixture of γ-Fe + Fe$_3$C, and again austenite transforms to pearlite when the steel cools below this temperature. The microstructure now consists of pearlite colonies, surrounded by a network of Fe$_3$C that delineates the prior austenite grain boundaries where cementite

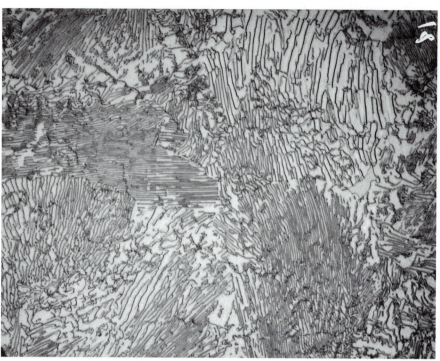

FIGURE 5-25 Microstructure of pearlite in eutectoid steel (970×). Courtesy of G. F. Vander Voort, Carpenter Technology Corporation.

nucleates. Because cementite is a hard, brittle phase it outlines potential paths for crack propagation. A coarse Fe_3C network is, therefore, viewed with concern. Heating high-carbon steels to just below the eutectoid temperature for long periods causes Fe_3C, in both pearlite and proeutectoid phases, to ball up as in Fig. 5-26. These cementite spheres are now surrounded by a softer, but more malleable, ferrite grain matrix and resemble fat globules floating in soup. Importantly, hard-to-machine tool steels are softened by this spheroidizing heat treatment. It will be clear from Section 5.7 that reduction of the Fe_3C plate surface area (or energy) is the driving force behind the spheroidization process.

5.6. STRUCTURE AND COMPOSITION OF PHASES

5.6.1. Equilibrium Phases: Solid Solutions

5.6.1.1. Substitutional Solid Solution

Binary phase diagrams usually contain an assortment of single-phase materials known as **solid solutions** and these have already been introduced in Sections 5.1 and 5.4.3. In the substitutional solid solution alloy the involved solute and

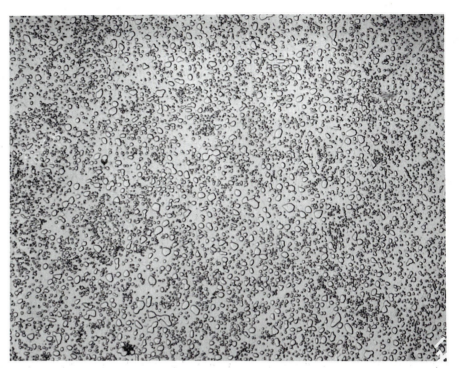

FIGURE 5-26 Microstructure of spheroidite in high-carbon steel (970×). Courtesy of G. F. Vander Voort, Carpenter Technology Corporation.

solvent atoms are randomly mixed on lattice sites. The single-phase solid alloys that extend across the entire phase diagram in the Cu–Ni and Ge–Si systems are good examples of random substitutional solid solutions. Both atoms are randomly distributed on FCC lattice sites in Cu–Ni and on diamond cubic sites in Ge–Si. In either case it makes no difference which component one starts with; the atoms mix in all proportions. A *necessary* condition for single-phase solid solution formation across the entire phase diagram is that both components have the same crystal structure. It is not a *sufficient* condition, however, because there are combinations like Ag–Cu (both FCC) and Fe–Mo (both BCC) that do not form an extensive range of solid solutions. Instead **terminal solid solutions,** so named because they appear at the ends of the phase diagram, form. The terminal α and β phases in the Pb–Sn diagram (Fig. 5-16) are examples. They are both substitutional solid solutions and display the limited solubility often exhibited by such phases. Introduction of foreign atoms into the lattice, whether by design (as dopants or solutes) or accident (as impurities), will always create dilute solid solutions which are often substitutional in nature. In fact, all "pure" materials are in effect dilute terminal solid solutions because it is thermodynamically impossible to remove all impurities.

The lattice parameter of substitutional solid solutions is usually an average of the interatomic spacings in the pure components weighted according to the atomic fractions present. This observation is known as Vegard's law. It predicts, for example, that the lattice parameter of a 25 at.% Cu–75 at.% Ni alloy will be approximately 0.25 × 0.3615 nm + 0.75 × 0.3524 nm = 0.355 nm. (See Table 3-2.)

5.6.1.2. Interstitial Solid Solution

When undersized alloying elements dissolve in the lattice they sometimes form interstitial solid solutions. Important examples include carbon and nitrogen in BCC iron, the former being the α phase on the Fe–Fe$_3$C phase diagram (Fig. 5-23). Rather than sit on substitutional sites, these atoms occupy the interstice in the center of the α-Fe cube faces, e.g. $(\frac{1}{2}, 0, \frac{1}{2})$. In FCC γ-Fe, carbon atoms assume the large interstitial site in the center of the unit cell. Based on a hard sphere model it is not too difficult to show that considerably less room is available for carbon to squeeze into in the α-Fe interstitial site relative to the γ-Fe site. More severe lattice bond straining in the former means that less carbon will dissolve interstitially in α-Fe compared with γ-Fe, in accord with solubility limits defined by the Fe–Fe$_3$C phase diagram.

5.6.1.3. Intermediate Solid Solution

Unlike terminal solid solutions that extend inward from the outer pure components, intermediate phase fields are found *within* the phase diagram. Two phase regions border either side of an intermediate phase. In the case of Cu–Zn (Fig. 5-22), α and η are terminal phases, and β, γ, δ, and ε are intermediate phases. These single phases are generally stable over a relatively wide composition range.

5.6.1.4. Ordered Solid Solution

Atoms within certain solid solutions can, surprisingly, order and give every appearance of being a compound. The resulting structure is schematically depicted in Fig. 5-1E and actually realized in 50 at.% Cu–50 at.% Zn β'-brass. In this alloy we can imagine a CsCl-like structure (Fig. 3-8) populated by atoms of Cu and Zn. Above 460°C the ordering is destroyed and a random solid solution forms. The order–disorder transformation, interestingly, bears a close resemblance to the loss of magnetism exhibited by magnets that are heated above the Curie temperature (see Chapter 14).

5.6.2. Compounds

Another very common type of solid phase that appears in phase diagrams is the compound. Elementary inorganic chemistry provides a basis for understanding compounds in many binary systems. For example, we know that salt compounds form when elements, drawn from opposite ends of the Periodic

Table, react. Further, such compounds possess well-defined stoichiometries and melting points. Cementite (Fe_3C) in Fig. 5-23 is such a compound. The oxides of iron, magnetite (Fe_3O_4), and wustite (FeO), shown in Fig. 10-19, are also compounds, but they exist over a range of oxygen content and correspondingly have varying melting points. Among the most technically important compounds are those formed between column 3A and 5A elements, for example, GaAs (Fig. 5-18). These compound semiconductors play an important role in optoelectronics and are discussed further in Chapter 13. A closer look at GaAs reveals that it can be either slightly Ga-rich or slightly As-rich, although this is not visible on the scale of the diagram. In general, most of the compounds we are concerned with have a slightly variable stoichiometry.

Although the above compounds are neither unusual nor unexpected from the standpoint of elementary chemistry, the same is not true of **intermetallic compounds.** These are composed of two metals that can, for example, stem from the same column of the Periodic Table (e.g., Na_2K), represent two different columns (e.g., $BaAl_2$), or be derived from the collection of transition metals (e.g., $MoNi_4$), The stoichiometry of these as well as the Ni_3Al intermetallic, a compound that plays an important role in commercial Ni base alloys, is not readily predictable from normal chemical valences. In this sense more complex metallic bonding issues govern the phase stability of these materials. For our purposes the following attributes of intermetallic compounds are noteworthy:

1. They often have a reasonably well defined stoichiometry (e.g., $NiAl_3$; see Fig. 5-27).
2. Just as frequently some are stable over a range of composition (e.g., Ni_3Al, NiAl).
3. Some compounds (e.g., GaAs, NiAl) melt **congruently** or maintain their composition right up to the melting point.
4. Many other compounds melt **incongruently** and undergo phase decompositions and chemical change. Examples are Ni_2Al_3 and $NiAl_3$, which display peritectic reactions at the appropriate isotherms (Fig. 5-27).

5.6.3. Some Phase Trends

Careful examination of many binary phase diagrams has led to guidelines that enable a qualitative prediction of the extent to which one metal dissolves in another when alloyed. The general relevant factors and trends they produce were summarized by Hume-Rothery, a noted British metallurgist, as follows:

1. *Atomic size.* Typically, if the two atoms differ by less than ~15% in radius, extensive solubility can be expected. For example, in Cu–Ni, a system that displays unlimited solubility, $(r_{Cu} - r_{Ni})/r_{Cu} \times 100 = 2.3\%$. Large solute atoms extend the bonds of smaller surrounding host atoms; alternately smaller

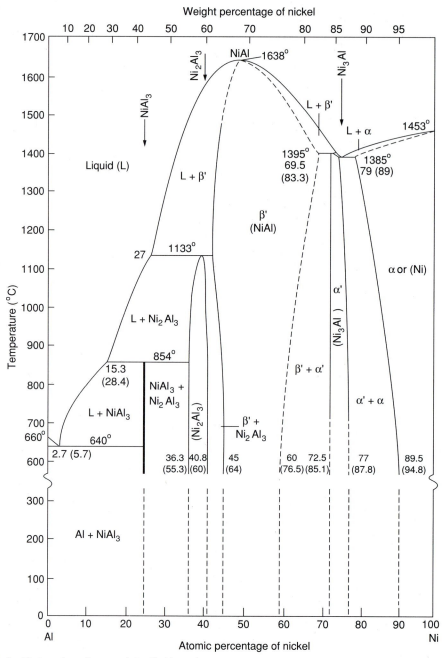

FIGURE 5-27   Equilibrium phase diagram of the Al–Ni system.

solute atoms cause a corresponding relaxation of bond lengths. In general, large differences in atomic size lead to limited solubility because strained bonds cause an unfavorable rise in their energies.

2. *Valence.* The tendency toward compound formation is greater the more electropositive one atom is and the more electronegative the other is. Under these conditions very limited solubility can also be expected. Greater solubility occurs when there is little valence difference between components.

3. *Relative valence.* Other things being equal (which they never are), a metal of low valence dissolves more readily in one of higher valence than vice versa.

5.6.4. Nonequilibrium Phases and Structures

Equilibrium is an ideal that is not often achieved in practice. Even though phase diagrams provide information on phase composition and relative amounts, they provide few clues as to their microstructure and morphology or appearance. This is all the more true of nonequilibrium phases. As properties of materials frequently hinge very strongly on phase size, shape, and distribution, as well as composition, it is important to appreciate the connections between processing and structure. Prolonged heating and slow cooling lead to equilibrium microstructures. But, relatively simple thermal treatments, such as rapid cooling from elevated temperatures, can often yield nonequilibrium microstructures and phases.

5.6.4.1. Metal Glasses

Nonequilibrium cooling of polymer and glass melts that yield amorphous structures has already been discussed in Chapter 4. An extreme example of **nonequilibrium** solidification in alloys occurs in the commercial process for the production of **metal glasses** (**Metglas**). The nonequilibrium or metastable structures in these materials evolve under conditions in which thermodynamic equilibrium is suppressed. The intent is to be as far removed from equilibrium as possible, a state achieved by quenching liquid metal alloys on cryogenically cooled surfaces. When this happens, metastable and even amorphous (noncrystalline) alloys are produced. Cooling rates of a million degrees centigrade per second are required to retain the character of the prior amorphous liquid; such rates are actually attained in shapes of small dimension (e.g., powder, thin foils less than 0.05 mm thick) because heat can be readily extracted from them. The reason that such high cooling rates are required to amorphize metals is the low melt viscosity and resulting high atomic mobility. Metal atoms do not interfere with one another when seeking to enlarge solid nuclei the way polymer and glass molecules do in viscous melts.

Continuous production of thin metal sheet, shown in Fig. 5-28, involves pouring the molten alloy on a chilled roller and removing the solidified product. A conceptually similar process is used during the very widely employed continu-

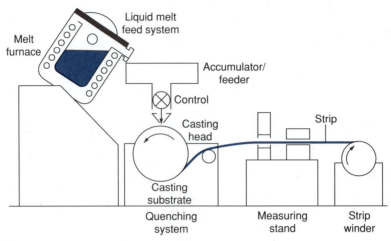

FIGURE 5-28   Diagram of the melt quenching apparatus used to produce metal glasses. Courtesy of Allied Metglas Products.

ous casting of metals. Only certain alloy compositions will yield amorphous phases. Important binary examples include combinations of transition metal and metalloid atoms (e.g., Fe–Si, Pd–Si, Fe–B). The alloys are generally strong and brittle, and some compositions display useful soft magnetic properties (e.g., Fe–B).

5.6.4.2. Cored Dendritic, Cast Structures

An important example of nonequilibrium solidification occurs during casting of metals where the luxury of very slow cooling in molds is neither feasible nor desirable. Rather, alloys are cooled relatively quickly in the foundry (but not quenched!) so that they can be processed further. What happens under these conditions can be understood with reference to Fig. 5-29. During cooling of the A–B solid solution alloy, the first solid nucleus that forms is richer in A than the overall alloy. As the temperature drops, more and more A and B atoms condense as the solid dendrite now reaches macroscopic dimensions. At each stage chemical equilibrium requires that the alloy solid have a uniform composition throughout, meaning thorough intermixing of atoms. This happens easily enough in the liquid phase but not in the solid state. Atomic diffusion, a sluggish process requiring the hopping and squeezing of atoms in between and past one another through lattice sites, does not occur readily in solids. And as the temperature continues dropping, equilibrium becomes harder and harder to achieve. Atoms are frozen in place as diffusion is further limited at low temperatures. Besides, there is physically more solid to homogenize. As a result, successive layers of increasingly richer B content surround the initial dendrite. The overall composition of the solid follows the nonequilibrium

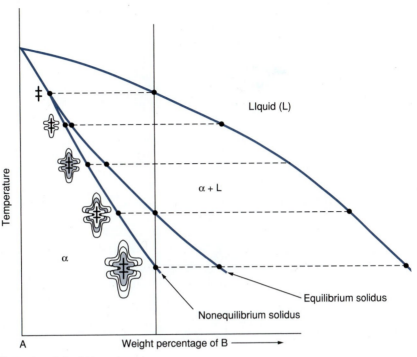

FIGURE 5-29 Illustration of dendritic coring.

solidus line rather than the equilibrium solidus line. The microstructure is said to exhibit **coring.**

Upon solidification of Cu–Ni alloys such inhomogeneous cored, dendritic microstructures develop as shown in Fig. 5-30. Undesirable properties (e.g., corrosion, lack of ductility) stem from such microstructures and they are, therefore, modified through subsequent thermomechanical processing such as hot rolling and recrystallization, which simultaneously homogenize and bring the alloys closer to equilibrium. Structural evolution considerations are generally distinct from thermodynamic ones, however, so that a better understanding of these will have to wait until some of the involved issues are addressed in Chapter 6 (also see Fig. 8-11).

5.6.4.3. Thin-Film Superlattice

As another example of nonequilibrium microstructures, this time in the Ge–Si system, consider Fig. 5-31. Shown with atomic resolution is a synthesized structure known as a **superlattice,** consisting of alternating thin-film layers of Si and a solid solution alloy of 40 at.% Ge–60 at.% Si. It was produced through careful sequential vapor deposition of the involved components an atom layer at a time. There are no visible dislocation or grain boundary defects

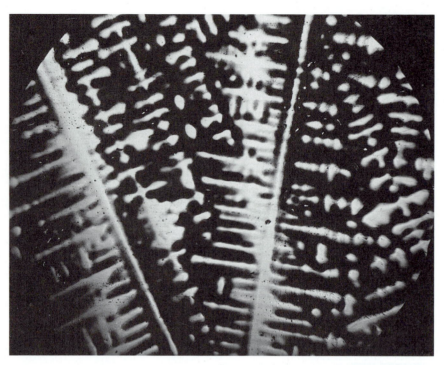

FIGURE 5-30   Dendritic cored microstructure of Monel metal (63 wt% Ni–32 wt% Cu + Fe, Si) at 50×. The light etching center of the dendrite is Ni rich; the darker surroundings are Cu rich.

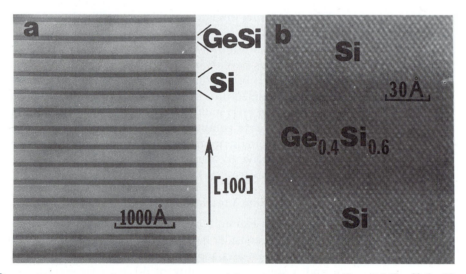

FIGURE 5-31   (A) TEM cross section of a superlattice consisting of alternating layers of pure silicon and 40 at.% Ge–60 at.% Si. (B) High-resolution lattice image of superlattice crystal. This composite structure is an uncommon two-phase, perfect single crystal. Courtesy of J. C. Bean and R. Hull, AT&T Bell Laboratories.

so the superlattice is a single crystal even though it contains atomically sharp interfaces between layered phases. The structure is in metastable equilibrium but can persist indefinitely at room temperature. Increasingly, new advanced solid-state electronic devices employ similar artificially tailored structures to achieve desired functions. But if heated to elevated temperatures, atoms will intermix and a single uniform random solid solution would replace the super-lattice.

5.7. THERMODYNAMICS OF SURFACES AND INTERFACES

5.7.1. Atoms on Surfaces

Every condensed phase has a surface that is exposed not only to gases but also to other condensed phases in contact with it. All sorts of processes involving chemical reactions and atom movements occur at surfaces and interfaces (e.g., oxidation, corrosion, soldering, sintering or bonding of particles). A large fraction of atoms reside at surfaces in materials like films, coatings, and fine powders. Atoms on surfaces are more energetic than atoms in the interior of a solid. Quite simply there are fewer atoms to bond to and restrain them. Therefore, they have a greater tendency to "escape" the solid than subsurface atoms, which are effectively surrounded on all sides. The bonding curve (see Fig. 2-12) for a pair of surface atoms would lie at higher energy and be displaced to a slightly larger equilibrium spacing relative to the response for a comparable pair of interior atoms.

Interestingly, the atomic "escaping tendency" also depends on the actual *geometry* of the surface. In order of increasing reactivity are atoms on concave, planar, and convex surfaces. These respective cases are shown in Figs. 5-32A–C, and reflected by a decreasing number of interatomic bonds between the atom and the surface. It is not difficult to quantify these tendencies. First, a surface (free) energy γ has to be introduced. A sufficient definition for our purposes is the difference in energy per unit area between atoms in a surface layer relative to the same number of atoms in a layer embedded well below the surface. For practical purposes γ is known as the surface tension and values of it have been measured for both liquids and solids of many different materials. They typically range from 0.1 to 1 J/m^2. The reason that liquid mercury balls up is that surface tension effects convert some energetic surface atoms to less energetic bulk atoms during the accompanying reduction in surface area.

Let us consider a material having a spherical surface of radius r and total surface energy E_s equal to $4\pi r^2\gamma$. If the volume of an atom is Ω, then the total number of atoms (n) in the sphere of volume $4\pi^3/3$ is $4\pi r^3/3\Omega$. A convenient measure of the atomic escaping tendency or reactivity is μ, also known as the **chemical potential.** It can be defined as the rate at which surface energy changes

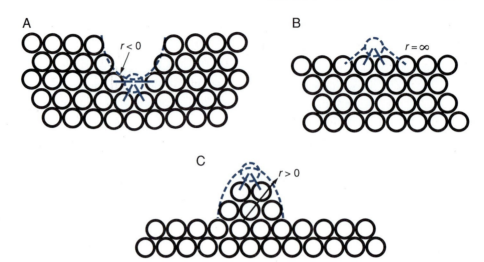

FIGURE 5-32 Atoms on (A) concave surface, (B) planar surface, and (C) convex surface. On the convex surface atoms are most reactive; on the concave surface they are least reactive.

with the addition or withdrawal of a number of atoms. Mathematically,

$$\mu = dE_s/dn = d(4\pi r^2\gamma)/d(4\pi r^3/3\Omega) = 8\pi r\gamma dr/(4\pi r^2/\Omega)\ dr$$

or

$$\mu = 2\Omega\gamma/r. \qquad (5\text{-}22)$$

Small particles have large values of μ and are thus reactive. Furthermore, for a single surface, μ is numerically largest for atoms on convex surfaces ($r > 0$) and smallest for atoms on concave surfaces ($r < 0$). In between is the flat surface where $r = \infty$ and $\mu = 0$. If particles of different size were in contact and could freely exchange atoms, then the larger particle would grow at the expense of the smaller one. Free energy is lowered in the process as equilibrium is approached.

EXAMPLE 5-6

a. What is the energy reduction that occurs when spherical particles of radii r_1 and r_2 coalesce into a single particle?

b. What fraction of the total initial surface energy is reduced when two spherical powders of equal size coalesce?

ANSWER   a. The respective particle volumes are $4\pi r_1^3/3$ and $4\pi r_2^3/3$, whereas the volume of the coalesced sphere of radius r_c is $4\pi r_c^3/3$. Because mass (volume) is conserved, $4\pi r_c^3/3 = 4\pi r_1^3/3 + 4\pi r_2^3/3$, or $r_c = (r_1^3 + r_2^3)^{1/3}$. The surface energy reduction is therefore $(4\pi r_1^2\gamma + 4\pi r_2^2\gamma) - 4\pi r_c^2\gamma$, with r_c given above.

b. The fractional energy reduction $(\Delta E/E)$ is given by $(4\pi r_1^2\gamma + 4\pi r_2^2\gamma - 4\pi r_c^2\gamma)/(4\pi r_1^2\gamma + 4\pi r_2^2\gamma)$. As $r_1 = r_2 = r$, $\Delta E/E = \{1 - (2r^3)^{2/3}/2r^2\} = 1 - (2^{2/3}/2) = 0.206$ or 20.6%. Note that $\Delta E/E$ is independent of r.

Later, in Chapter 6, the *rate* at which such particles sinter together is addressed.

5.7.2. Surface Tension Effects

Why does solder sometimes ball up but flow nicely when properly fluxed? A clue to the answer can be found by considering Fig. 5-33, where a liquid (L), for example, a droplet of solder, or more generally any liquid or solid material, is in contact with a planar substrate. The surface tension is just that: a *tension,* or pull, that acts along a thin *surface* layer that is the interface between two phases. There are three such surface tensions in the system. The first is the tension at the liquid–vapor interface γ_{L-V}, and the second is that between the liquid and substrate γ_{L-S}; lastly, there is the tension between the substrate and vapor, γ_{S-V}. If the droplet is in thermodynamic equilibrium, the surface tension forces that act on it must balance. The arrows drawn in Fig. 5-33 are the directions in which these forces pull to reduce the areas of the indicated interfaces. Simple static equilibrium of forces in the horizontal direction demands

$$\gamma_{S-V} = \gamma_{L-S} + \gamma_{L-V}\cos\theta, \qquad (5\text{-}23)$$

where θ is known as the wetting angle.

When θ is 0, the liquid is stretched out into a film that effectively wets or covers the surface, and $\gamma_{S-V} = \gamma_{L-S} + \gamma_{L-V}$. On the other hand, when the liquid dewets, it agglomerates into a sphere and $\theta = 180°$. Reducing γ_{L-S} is the key to enhancing wetting and is accomplished through alloying, fluxing, or the use of wetting agents.

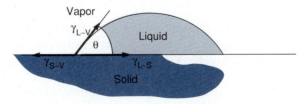

FIGURE 5-33   A liquid drop on a solid surface pulled by the various interfacial tensions.

5.8. THERMODYNAMICS OF POINT DEFECTS

What does thermodynamics say, if anything, about point defects in solids? To the extent that they behave like chemical entities they can be described in terms of the established rules of equilibrium thermochemistry. Vacancies are the defects most amenable to analysis and are the only ones discussed here. Let us start with a perfect lattice of N atoms which is taken to be the initial state. A number of vacancies (n_v) is created by extracting atoms from the lattice interior and placing them on the surface. Through the use of Eq. 5-2 we can calculate the free energy change relative to the initial state. First, energy E_v is expended in creating each vacancy and thus the total value of ΔE is $n_v E_v$.

The ΔS term is more complicated because entropy is associated not only with the internal nature of atoms and vacancies (due to lattice vibrations), but also with their locations. It makes a difference if all the vacancies are concentrated in the upper quarter, the upper half, or the lower tenth of the solid, or are distributed randomly, or in any one of an almost infinite number of other ways. Let us first start with a vacancy-free lattice; because there is only one geometric distribution, the entropy is minimum and may be taken to be zero. Then N atoms and n_v vacancies are randomly distributed on $N + n_v$ sites and the entropy can be obtained using the following statistical recipe. Consider an empty lattice of $N + n_v$ sites. The first atom can be introduced in any one of $N + n_v$ ways. If two atoms are sequentially introduced they can be accommodated in $(N + n_v)(N + n_v - 1)$ *indistinguishable* ways. Exchange of the two atoms halves this; so the number of *distinguishable* ways or combinations of placement is $(N + n_v)(N + n_v - 1)/2$. For three atoms, $(N + n_v)(N + n_v - 1)$ $(N + n_v - 2)/2 \times 3$, and so on. For N atoms the total number of *distinguishable* distributions is $(N + n_v)!/N! \, n_v!$. The symbol $N!$, or N factorial, means $N(N - 1)(N - 2)(N - 3) \dots 1$. A cornerstone of the science of statistical mechanics is that the statistical or configurational entropy is given by

$$S = k \ln\{(N + n_v)!/N! \, n_v!\}, \qquad (5\text{-}24)$$

where k is the Boltzmann constant. The total free energy change on creating n_v vacancies, neglecting PV (Eq. 5-1), is therefore

$$\Delta G = n_v E_v - Tk \ln\{(N + n_v)!/N! \, n_v!\}. \qquad (5\text{-}25)$$

Thermodynamic equilibrium implies minimization of G or $d\Delta G/dn_v = 0$. Rather than evaluate this mathematically, it is instructive to illustrate it graphically. This is done in Fig. 5-34, where the quantities ΔE and $T\Delta S$ and the sum ΔG are plotted versus n_v at two different temperatures, T_1 and T_2 ($T_2 > T_1$). At the higher temperature it is clear that the number of vacancies in equilibrium is higher because the minimum is pushed to the right. There is no reason why the vacancy concentration should not be governed by the laws of chemical equilibrium for the reaction $M_a(N) \rightarrow M_a(N - n_v) + n_v$, where M_a refers to

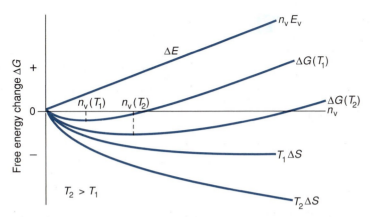

FIGURE 5-34 Plots of internal energy, entropy, and free energy as a function of the vacancy concentration. The minimum of the free energy curve graphically yields the equilibrium vacancy concentration. After C. R. Barrett, W. D. Nix, and A. S. Tetelman, *The Principles of Engineering Materials*, Prentice–Hall, Englewood Cliffs, NJ (1973).

the matrix atoms. The thermally activated concentration of vacancies given by Eq. 3-6, $C_v = \exp(-E_v/kT)$, is then not unexpected.

Values for E_v have been measured in metals by heating them to different elevated temperatures and measuring their concentrations through thermal expansion and X-ray techniques. Alternatively, they have been retained at low temperature through very rapid cooling; their numbers have then been estimated by noting the amount by which they have raised the electrical resistance of metals. Either way, values of E_v have been obtained for many metals. As already noted in Section 3.5.1, values of 1 eV per vacancy (96,500 J/mol) are typical. This is certainly true of the noble metals where $E_v = 0.98$, 1.1, and 1.0 eV for gold, silver, and copper, respectively.

Point defects in alkali halide and metal oxide compounds are also amenable to similar thermodynamic analysis. Let us consider the formation of a Schottky defect (see Section 3.5.1), a metal cation (M)–oxygen anion (O) vacancy combination, by a reaction that removes the pair of ions from the lattice and places them on the surface:

$$(M^{2+}-O^{2-})_{\text{lattice}} \rightarrow M_v + O_v + (M^{2+}-O^{2-})_{\text{surface}}. \tag{5-26}$$

The equilibrium constant is then given by

$$(M_v)(O_v) = \exp(-E_{sv}/kT), \tag{5-27}$$

where E_{sv} is the energy required to form an **intrinsic** Schottky vacancy (v) pair. Because of charge neutrality considerations, $(M_v) = (O_v)$, and either concentration is thus equal to $\exp(-E_{sv}/2kT)$ by taking the square root. Schottky disorder in oxides is characterized by E_{sv} values of several electron-volts. For example, in BeO, MgO, and CaO values of ~6 eV have been measured. With such a high energy, the intrinsic point defect concentration will be negligible in ceramics except at extremely elevated temperatures. Later, in Section 11.6.1, we shall see that the presence of impurities can generate large numbers of **extrinsic** point defects, even at low temperatures.

It is interesting to note that, unlike vacancies, dislocations are not thermodynamic defects. Since many atoms are distributed along the line the internal energy is large; simultaneously the entropy is low because dislocation lines are constrained to assume few crystallographic orientations or configurations. As a result ΔG is large. Therefore, although dislocations can be eliminated from crystals, resulting in a free energy reduction, vacancies cannot be so removed.

5.9. PERSPECTIVE AND CONCLUSION

Thermodynamics is very definite about denying the impossible and defining limits for behavior that is possible. But, for a possible event, thermodynamics is noncommittal as to the extent to which it will occur, or even whether it will occur at all. For example, amorphous SiO_2 should crystallize and Fe_3C should thermodynamically decompose into iron and graphite at 300 K, but both can be retained indefinitely in metastable equilibrium. This is true of many other solid-state materials and reactions whether they are chemical in nature or involve physical transformations of state. Only reactions that lower the free energy of the system can proceed; and if the free energy of the system reaches a minimum and cannot be lowered further, true thermodynamic equilibrium is attained. Minimization of free energy is a cardinal principle used over and over in this book to establish the configuration of material states in equilibrium. In this chapter this one principle was expressed in different but equivalent forms and applied in the following ways:

1. $\Delta G°$ *must be negative.*   For chemical equilibria (oxidation and CVD processes) a negative free energy change is required for the reaction to proceed as written. Shape changes that minimize surface energy also necessarily reduce free energy. The derivation of many important thermodynamic laws, for example, the Gibbs phase rule, stem from free energy minimization.

2. $K_{eq} = \exp(-\Delta G°/kT)$.   The equilibrium constant defines magnitudes of concentrations or pressures for both chemical reactions and physical transformations. Examples include the vapor pressure above condensed elemental phases and the equilibrium concentration of vacancies.

3. $d\Delta G/dn_v = 0$.   When the free energy can be expressed as a function of the composition of a chemical species, for example, point defects or vacancies, this condition leads to the equilibrium concentration. This formalism also holds for the determination of the chemical potential of atoms on curved surfaces.

4. *Phase diagrams.*   Phase diagrams are simply graphic displays of equilibrium solubilities in single- or multicomponent systems. The equilibrium phase

compositions and amounts for any state are those that minimize the free energy of the system. In particular, the tie line construction is rooted in thermodynamic equilibrium.

Questions of reaction rates and conditions under which equilibrium can be expected are not addressed by thermodynamics. The next chapter, on kinetics, does focus on such issues and it is to them that we now turn.

Additional Reading

ASM Handbook. Vol. 9: *Metallography and Microstructures*, 8th ed., ASM International, Materials Park, OH (1985).
A. H. Cottrell, *An Introduction to Metallurgy*, Edward Arnold, London (1967).
M. Hansen and K. Anderko, *Constitution of Binary Alloys*, 2nd ed., McGraw–Hill, New York (1958).
C. H. P. Lupis, *Chemical Thermodynamics of Materials*, North-Holland, New York (1983).
R. A. Swalin, *Thermodynamics of Solids*, 2nd ed., Wiley, New York (1972).

QUESTIONS AND PROBLEMS

5-1. An initially solid material becomes a liquid as the temperature is raised above the melting point. Explain the probable signs of the enthalpy, entropy, and free energy changes. What are the signs of these quantities as the liquid transforms back to the solid?

5-2. In Thermit welding processes aluminum reacts with iron oxide, releasing molten iron as the weld metal. If the reaction can be written as

$$2Al + 3FeO = Al_2O_3 + 3Fe,$$

what is the free energy change per mole of Al during reaction at 1600°C?

5-3. What is the minimum temperature at which it would be thermodynamically possible for carbon to reduce silica according to the reaction

$$SiO_2 + 2C = Si + 2CO?$$

Assume the effective concentrations are unity and the pressure is 1 atm.

5-4. The free energies of oxides are usually lower than those of the corresponding metals. Nevertheless, metals do not rapidly oxidize and we can comfortably rely on their long-term stability. Why?

5-5. Hydrogen is often used to reduce metal oxides (MO) at elevated temperatures. If $2M + O_2 = 2MO$ (ΔG°_{MO}) and $2H_2 + O_2 = 2H_2O$ ($\Delta G^\circ_{H_2O}$), write an expression for the equilibrium constant of the reaction $H_2 + MO = M + H_2O$ in terms of ΔG_{MO}, ΔG_{H_2O}, and the pressures of H_2 and H_2O.

5-6. Aluminum is melted at 1000°C. What is the maximum oxygen partial pressure that can be tolerated to eliminate the possibility of any Al_2O_3 formation? Practically, much higher oxygen pressures can be present without appreciable oxidation. Why?

5-7. From the data given in Fig. 5-5 determine the heat of vaporization of H_2O.

5-8. Determine the heat of vaporization for aluminum using Eq. 5-9.

5-9. On the basis of Fig. 5-5 show that ice floats on water. (The reason that aquatic life in lakes can survive winters is that water at 4°C is denser than ice and sinks to the bottom.)

5-10. A mole of lead and a mole of diamond are heated from ~4 to 700 K at atmospheric pressure. Based on Fig. 5-10 what is the approximate ratio of the heat absorbed by C (diamond) relative to that by Pb?

5-11. When the temperature of a two-phase mixture of any Cu–Ni alloy is raised, the Ni content of both the liquid and solid phases increases. Does this violate the conservation of mass?

5-12. A solder manufacturer wishes to make a batch of solder having the eutectic composition. On hand is a supply of 250 kg of electrical solder scrap containing 60 wt% Sn–40 wt% Pb and 1250 kg of plumbing solder scrap containing 60 wt% Pb–40 wt% Sn. If all of the scrap is to be melted, how much pure Sn must be added to achieve the desired 61.9 wt% Sn composition?

5-13. Gallium arsenide crystals are grown from Ga-rich melts. Can you mention one reason why As-rich melts are not used for this purpose?

5-14. A 90 wt% Sn–10 wt% Pb alloy is cooled from 300 to 0°C.
a. What is the composition of the alloy in terms of atomic or molar percentage?
b. Draw a cooling curve for this alloy.
c. Upon cooling of this alloy, the degrees of freedom change from 2 to 1 to 0 to 1. List possible temperatures that correspond to these four conditions.
d. In what ways does the equilibrium microstructure of this alloy differ from that of the 90 wt% Pb–10 wt% Sn alloy?

5-15. Consider a binary system composed of components A and B with the following features (all compositions in at.%):
1. Melting point of A = 1500°C and melting point of B = 750°C.
2. Melting point of intermediate phase A_2B = 1250°C.
3. Zero degrees of freedom at a composition of 20B at 500°C.
4. Zero degrees of freedom at a composition of 50B at 1000°C.
5. A single solid phase of composition 50B at 900°C.
6. Isotherms at 500 and 1000°C.
Sketch a possible phase diagram based on this information.

5-16. The Al_2O_3–Cr_2O_3 phase diagram resembles that for Cu–Ni. The melting point of Al_2O_3 is 2040°C, whereas that of Cr_2O_3 is 2275°C.
a. Sketch the phase diagram.
b. How many components are there in this system?
c. How many degrees of freedom are there in the liquid, liquid + solid, and solid phase regions?
d. It is desired to grow a ruby single crystal with the composition 22 wt% Cr_2O_3. On your phase diagram indicate the melt composition and temperature for growth.

5-17. Suppose you have a large quantity of gallium amalgam (Ga–Hg alloy) scrap. The alloy is liquid at room temperatures and has much less value than the isolated pure metals. Suggest a physical method to separate these metals if they are initially present in equal amounts by weight.

5-18. a. In the Ga–As binary system perform total chemical and physical composition analyses for an 80 at.% Ga–20 at.% As alloy at 1200, 1000, 200, and 29°C.

b. What is different about the two two-phase fields labeled liquid + GaAs?

5-19. It is desired to pull an alloy single crystal of composition 78 at.% Ge–22 at.% Si from a binary Ge–Si melt.

a. At what temperature should the crystal be pulled?

b. What melt composition would you recommend?

c. As the crystal is pulled what must be done to ensure that its stoichiometry is kept constant?

5-20. Consider the 50 wt% Pt–50 wt% Re alloy.

a. This alloy is heated to 2800°C and cooled to 800°C. Sketch an equilibrium cooling curve for this alloy.

b. Perform complete equilibrium chemical and physical phase analyses at 2800, 2452, 2448, and 1000°C.

5-21. Make enlarged sketches of the phase regions that surround the two highest isotherms in the Cu–Zn phase diagram (see Fig. 5-22).

a. What is the name of the solidification behavior displayed in both cases?

b. Which lines on this phase diagram represent the solubility limit of Zn in the α phase?

c. Which lines on this phase diagram represent the solubility limit of Zn in the β phase?

d. Which lines on this phase diagram represent the solubility limit of Cu in the β phase?

e. Is it possible for the solubility of a component to drop as the temperature is raised?

Give an example in this phase diagram.

5-22. Perform quantitative physical and chemical phase analyses at temperatures corresponding to the four states indicated for the 0.4% C, and 1.2% C steels in the Fe–Fe$_3$C system of Fig. 5-24. Do your analyses correspond to the sketched microstructures?

5-23. Up until the Middle Ages iron was made by reducing iron ore (e.g., Fe$_2$O$_3$) in a bed of charcoal (carbon). Steel objects were shaped by hammering the resulting solid sponge iron that was reheated to elevated temperatures using the same fuel. As the demand for iron and steel increased, furnaces were made taller and larger. They contained more iron ore and charcoal that were heated for longer times. An interesting and extremely important thing occurred then. *Liquid* iron was produced! Why? (Molten iron issuing from the bottom of the furnace could be readily cast, a fact that helped to usher in the Machine Age.)

5-24. Give a good physical reason for the following statements:
   a. Elements that are soluble in all proportions in the solid state have the same crystal structure.
   b. Unlike substitutional solid solutions, interstitial solid solutions do not range over all compositions.

5-25. Distinguish between intermediate phases and compounds. Contrast metals and nonmetallic inorganic materials with respect to their tendency to form intermediate phases and compounds in binary phase diagrams.

5-26. a. A 40 wt% Al_2O_3–60 wt% SiO_2 melt is slowly cooled to 1700°C. What is the proeutectic phase and how much of it is there?
   b. Furnace bricks are made from this composition. Is the melting point of these bricks higher or lower than that of pure silica bricks?
   c. What is the highest melting temperature attainable for bricks containing both SiO_2 and Al_2O_3?

5-27. Gold and silver dissolve in all proportions in both the liquid and solid states. The same is true for gold and copper. Does this mean that silver and copper will form solid solutions in the solid state? Check your answer by consulting a book of phase diagrams, for example, M. Hansen, *Constitution of Binary Alloys*, McGraw–Hill, New York (1958).

5-28. A 10-g ball bearing sphere is made of a steel containing 1.1 wt% C. Suppose the ball is austenitized and slowly cooled to room temperature.
   a. What is the weight of pearlite present?
   b. What is the total weight of ferrite present?
   c. What is the total weight of cementite present?
   d. What is the weight of cementite present in the pearlite?
   e. What is the weight of cementite present as the proeutectoid phase?

5-29. Cast iron weighing 1 kg and containing 3.5 wt% C (remainder Fe) is melted and slowly cooled from 1300°C. Perform a complete phase analysis (a) just below 1147°, (b) at 900°C, and (c) just below 723°C. (d) Sketch the expected microstructure at 500°C. (e) In true thermodynamic equilibrium cementite decomposes to graphite and iron. What weight of graphite would be present if all of the cementite decomposed?

5-30. The copper–oxygen system is a simple eutectic on the copper-rich side. The eutectic temperature is 1065°C and the eutectic composition contains 0.39 wt% O. Tough pitch copper, a product produced on a large scale for electrical applications, contains a few hundredths of a percent of oxygen. Sketch the microstructure of such a copper containing 0.04 wt% O if the solubility of O in Cu is nil, and (a) the eutectic mixture consists of Cu_2O particles dispersed in a nearly pure Cu matrix, or (b) the eutectic forms near the grain boundaries of primary Cu.

5-31. The fire assaying of precious metal (Au, Ag) content in alloys containing Cu and other base metals dates to antiquity, but is still the preferred method for such analysis by the U.S. Mint. In this technique known as *cupellation*, the alloy is dissolved in molten lead, a solvent for these metals. The lead and other base metals are then oxidized at high temperature and effectively removed as their oxides are absorbed into the bone ash crucible (cupel). After complete oxidation of the lead a ball of pure unoxidized Au, Ag, or Ag–Au remains (shades of alchemy)! The solubility of Pb in the precious metals is nil. Based on this information roughly sketch features of the high-temperature Pb–precious metal phase diagram applicable in the cupellation process.

5-32. What vacancy concentration is present in gold (a) at 25°C? (b) At 1060°C? (c) Suppose Au is heated to 1062°C and then quenched so rapidly that all of the vacancies are retained at 25°C. By what fraction is the density of Au reduced?

5-33. Suppose instead of three phases, that is, solid, liquid, and vapor, equilibrated with respect to surface tensions (Fig. 5-33), we consider three grains of the same phase at a triple-joint junction. Along grain boundaries between neighboring grains there is an interfacial tension γ or force, and in equilibrium these forces balance.

a. If γ is identical between grains, sketch the force balance at the triple point. What angle exists between adjacent tensions?

b. When the three individual tensions are different, a general result of static equilibrium is $\gamma_1/\sin \theta_1 = \gamma_2/\sin \theta_2 = \gamma_3/\sin \theta_3$, where the angle θ_1 lies between the directions γ_2 and γ_3, and so on. If $\gamma_1 = 1.0$ J/m^2, $\gamma_2 = 1.0$ J/m^2, and $\gamma_3 = 1.2$ J/m^2, solve for θ_1, θ_2, and θ_3.

5-34. Consider a sphere of copper that weighs 10 g.

a. If the surface energy is 1.4 J/m^2, what is the energy associated with the surface atoms?

b. The original sphere is broken down into fine spherical grains of powders each measuring 100 nm in diameter. What is the total surface energy of all of the powder grains?

5-35. Zone refining is a process that was used to purify early semiconductor materials. Consider a long rod of solid polycrystalline silicon, containing uniformly distributed small amounts of impurities in solid solution. A narrow zone of the rod is melted at one end and the molten zone is slowly translated to the other end of the rod. At the advancing zone edge new impure solid is melted, and at the trailing edge liquid solidifies.

a. By considering that impurities typically lower the melting point of Si, what is true of the purities of solid and liquid phases in equilibrium?

b. Why is the resulting rod purer at the initially melted end and less pure at the far end?

6

KINETICS OF MASS TRANSPORT AND PHASE TRANSFORMATIONS

We noted in Chapter 5 that when materials are not thermodynamically equilibrated, forces naturally arise to drive them toward equilibrium. This takes **time,** which now becomes a central variable just as it is in the subject of kinetics and dynamics of mechanical systems. In our context, kinetics deals with questions of how long it will take compositional and structural change to occur in materials systems. We first consider systems in which chemical concentrations are not uniform throughout the volume. When this happens the resulting concentration gradients are the driving forces that cause time-dependent mass transport or diffusion of atoms. Quite frequently, concentration gradients are intentionally imposed during processing or manufacturing to controllably alter the subsurface composition of materials. Examples that will be considered later include diffusion of (1) carbon or nitrogen into mild steels to harden their surfaces, (2) phosphorus or boron into silicon to produce semiconductor devices, and (3) oxygen into SiO_2 thin films during semiconductor device processing. Another diffusion-controlled process treated is powder metallurgy. Here mass transport effects are relied upon to cause metal or ceramic powders to bond or sinter to one another. Surprisingly, atoms diffuse even though powders are chemically homogeneous.

It is interesting to observe changes in these and other processes on a **microscopic** level. Unlike our notion of a geometrically ordered array of immobile

249

atoms, solids can, depending on temperature, become very animated places. At first glance individual atoms appear to hop around randomly, oblivious to the presence of other migrating atoms. One wonders how all this apparently uncorrelated atomic motion can push the system toward equilibrium, but this is precisely what happens. It is the sheer numbers of atoms involved and the statistical weight they carry that cause **macroscopic** change. Manifestations of change include reduction of gross concentration gradients in the matrix, local accumulations of atoms that evolve into new phases or precipitates, and smoothing of previously rough surfaces.

Although diffusional effects are often capitalized upon in processing they are quite troublesome when they cause degradation of materials in service. Exposure to elevated temperatures can accelerate otherwise dormant mass transport processes. What is important in materials degradation is the distance over which change occurs and the time it takes to occur. Turbine blades at elevated temperatures might be able to tolerate diffusion-induced oxide layers that are a few thousandths of an inch thick. (*Note:* 0.001 in. = 25.4 μm.) On the other hand, integrated circuit (IC) interconnections and semiconductor contacts have features that are at most only a few micrometers in size. Such dimensions are uncomfortably small compared with the extent of diffusional reaction needed to render circuits inoperable; a turbine blade may be operable at 1000°C whereas failure may readily occur at 35°C in an IC. Therefore, the nature of the involved materials, their dimensions, the temperature of reaction, and the time of exposure must all be considered when assessing degradation effects caused by mass transport. Material damage and failure caused by diffusion are treated more fully in Chapters 10 and 15.

Finally, mass transport effects are also of interest because they play a central role in phase transformations of all kinds, high-temperature annealing of metals, electrical transport in ceramics, mechanical behavior at elevated temperatures, and assorted heat treatments to strengthen or toughen materials. Considerable attention is given in this chapter to phase transformations. These occur when phase boundaries in single- or multicomponent system phase diagrams are crossed. As we now suspect, the newly stable phases do not appear instantly, but rather need time for mass transport to adjust concentrations to new solubility demands. Morphologies or structural shapes of the new phases reflect the atomic motion during transformation.

Kinetics describes in a word the totality of time-dependent mass transport and phase transformation phenomena in solids. Common to all of the subjects discussed in this chapter is diffusion, the first subject treated.

6.2. MACROSCOPIC DIFFUSION PHENOMENA

6.2.1. Mathematics of Diffusion

Diffusion may be defined as the observed migration of atoms (or molecules) in a matrix under the influence of a concentration gradient driving force. In most cases the diffusing and matrix atoms differ. But they can be the same, for example, radioactive tracer atoms in a matrix of the same but nonradioactive atoms, causing a phenomenon known as **self-diffusion.** Either way, the fundamental equation that describes diffusion is known as Fick's law:

$$J = -D \; dC/dx. \tag{6-1}$$

Decreasing atomic concentration (C) in the positive x direction signifies a *negative* concentration gradient (dC/dx is negative in magnitude). But as indicated in Fig. 6-1, a *negative* gradient at some point x induces a positive flux of atoms, J, in the $+x$ direction. Hence the negative sign in Eq. 6-1. Similarly, if the concentration increased in the positive x direction, mass would flow in the opposite direction. The units of flux mirror those of concentration. For C in atoms/m^3, atoms/cm^3, g/m^3, mol/m^3, the corresponding units of J are atoms/m^2-s, atoms/cm^2-s, g/m^2-s, and mol/m^2-s. Irrespective of concentration units, values of D, the diffusion coefficient or diffusivity, are always expressed in units of (distance)^2/time, that is, m^2/s or cm^2/s. More will be said about D later in the chapter.

Consider what happens when atoms of one kind (e.g., A) originating from the surface ($x = 0$) diffuse into a matrix of B atoms (Fig. 6-1). At any given time more atoms will diffuse into a subsurface volume element through plane x than will leave through plane $x + dx$. Therefore, in the next time instant the difference between these atom fluxes must accumulate in the element,

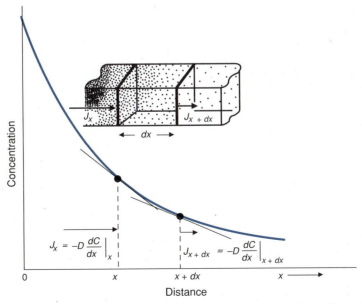

FIGURE 6-1 Illustration of the laws of diffusion. The direction of the concentration gradient and the diffusion flux it produces are indicated. When the mass fluxes into and out of a volume differ, then dC/dt is not zero.

raising the concentration of A there. Similar effects occur at each value of x throughout the volume, and when proper account is taken of instantaneous mass conservation, the governing equation of diffusion in one dimension is

$$\partial C(x, t)/\partial t = D \, \partial^2 C(x, t)/\partial x^2. \tag{6-2}$$

This partial differential equation describes **non-steady-state diffusion** under conditions where D is assumed to be constant. The objective is to find $C(x, t)$ and the consequences that flow from it. In **steady-state diffusion,** the same constant mass flux passes through every plane in the system and therefore a linear concentration profile exists. A more concise requirement for steady-state diffusion is $\partial C(x, t)/\partial t = 0$.

Let us apply Eq. 6-2 to diffusion into a semi-infinite matrix (one that extends from $x = 0$ to $x = \infty$). If atoms are supplied at the $x = 0$ boundary from an inexhaustible or **continuous** surface source of concentration C_s, and it is assumed that the initial concentration in the matrix is C_0, the solution to Eq. 6-2 is

$$[C(x, t) - C_0]/[C_s - C_0] = \text{Erfc}(x/\sqrt{4Dt}). \tag{6-3}$$

Known as the complementary error function, Erfc is mathematically defined as

$$\text{Erfc}(x/\sqrt{4Dt}) = 1 - \frac{2}{\sqrt{\pi}} \int_0^{x/2\sqrt{Dt}} \exp(-z^2) \, dz \tag{6-4}$$

Like trigonometric functions, values for the complementary error function are tabulated as a function of the argument $x/\sqrt{4Dt}$. For small numerical values of the argument, $\text{Erfc}[x/\sqrt{4Dt}] \sim 1 - x/\sqrt{4Dt}$. In Fig. 6-2, a wide range of values for Erfc and the related gaussian function (discussed below) are displayed graphically. A couple of examples illustrate the use of this figure.

EXAMPLE 6-1

Steel gears containing 0.20 wt% C are exposed to a carbon-containing gas atmosphere that maintains a surface concentration of carbon equal to 1.0 wt%. How long will it take the carbon concentration to reach a level of 0.5 wt% at a distance 0.1 cm below the surface? Assume $D = 5.1 \times 10^{-6}$ cm^2/s.

ANSWER Substitution in Eq. 6-3 yields $[0.5 - 0.2]/[1.0 - 0.2] = 0.375$ or $\text{Erfc}(x/\sqrt{4Dt}) = 0.375$. From Fig. 6-2 the argument, whose complementary error function is 0.375, is estimated to be 0.63. Therefore, $x/\sqrt{4Dt} = 0.63$. Substituting for x and D,

$$t = \frac{0.1 \text{ cm}^2}{(0.63)^2 \times 4 \times 5.1 \times 10^{-6} \text{ cm}^2/\text{s}} = 1235 \text{ s or } 20.6 \text{ min.}$$

Because the diffusion zone is small compared with typical gear dimensions, the assumption of an "infinite" matrix is reasonable, validating the use of Eq. 6-3.

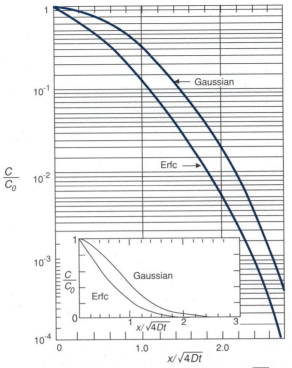

FIGURE 6-2 Normalized gaussian and Erfc curves of C/C_0 versus $x/\sqrt{4Dt}$. Both logarithmic and linear (inset) scales are shown. From W. E. Beadle, J. C. C. Tsai, and R. D. Plummer, *Quick Reference Manual for Silicon Integrated Circuit Technology*, Wiley (1985). Copyright © 1985 by AT&T; reprinted by permission.

EXAMPLE 6-2

Silicon solar cells are made by diffusing phosphorus into the surface of a silicon wafer doped with an initial uniform concentration of boron C_B. The purpose of this treatment is to create a junction at a distance below the surface where the concentration of phosphorus C_P reaches the boron concentration, that is, $C_P = C_B$. What is the junction depth if $C_B = 10^{16}/cm^3$ and the wafer surface concentration of P is maintained at 10^{20} atoms/cm³? Diffusion of P is carried out for 15 min at 1000°C where $D = 1.2 \times 10^{-14}$ cm²/s. A schematic of the relevant concentration profiles involved is provided in Fig. 6-3.

ANSWER We assume diffusion of phosphorus is independent of the presence of boron and is described by Eq. 6-3. At the junction depth x_j, $C_P(x_j, t) = C_B$. Because there is no phosphorus in the wafer initially, $C_0 = 0$. After substitution, $10^{16}/10^{20}$ or $10^{-4} = \text{Erfc}(x_j/\sqrt{4Dt})$. Reference to Fig. 6-2 shows that when $x_j/\sqrt{4Dt} = 2.73$, Erfc 2.73 $= 10^{-4}$. Therefore, $x_j = 2.73 \times [4 \times 1.2 \times 10^{-14}$

cm^2/s × 900 s]$^{1/2}$ = 1.79 × 10^{-5} cm, or 179 nm. Junction depths must be shallow for light to penetrate Si and illuminate the junction. (See Section 13.4.3 for discussion of solar cells.)

These two illustrative problems represent extremes in values of the complementary error function. High concentration levels mean small arguments and values of Erfc close to 1. Large values of the argument render the magnitude of Erfc and the relative concentrations very small. The results obtained for time and distance in these problems are quite sensitive to the values of D, and these were selected from data given in Sections 6.2.2. and 6.3.2.

The simplest mathematical solution to Eq. 6-2 is the gaussian

$$C(x, t) = \frac{S}{\sqrt{\pi D t}} \exp\left(\frac{-x^2}{4Dt}\right), \tag{6-5}$$

which describes situations where a **finite,** rather than continuous, surface concentration S (in units of atoms/cm^2) is present. Such finite sources are used in diffusional doping of semiconductors. As atoms enter a Si wafer the amount left on the surface decreases with time. For example, at $x = 0$, $C(0, t) = S/\sqrt{\pi D t}$.

A comparison between the time evolutions of both types of diffusional profiles is shown in Fig. 6-4. Irrespective of diffusion geometry all concentration profiles, whether from finite or continuous sources, are mathematical superpositions of gaussian-like behavior. The same bell-shaped curve is employed in

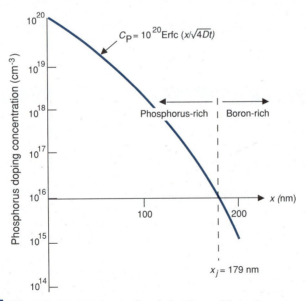

FIGURE 6-3   Schematic concentration profile of phosphorus diffusing from a continuous source into silicon doped with boron.

probability theory and the statistics of random events (see Chapter 15). Why atomic diffusion is random is developed later in Section 6-3. The range over which atoms diffuse from a point source is estimated to be the spread of the gaussian profile. On this basis, when $C(x, t)/C (x = 0, t) = 1/e$, the diffusional spread is given by

$$x = 2\sqrt{Dt}. \tag{6-6}$$

This very handy rule of thumb connects the expected diffusional distance to the pertinent values of D and t irrespective of whether Erfc or gaussian diffusion applies. Later the strong exponential dependence of D on temperature T [i.e., $D = D_0 \exp(-\Delta E_D/RT)$, with D_0 and ΔE_D constant] will be developed. Thus, **position** x, **temperature** T, and **time** t, three key variables in this chapter, are intimately packaged in this one useful equation.

6.2.2. Diffusional Treatments

Carburization of steel is easily the most well known and widely used diffusional surface treatment. The intent is to boost the C concentration at the surface sufficiently so that it can be made harder by a subsequent quenching and tempering heat treatment (described in Section 9-2). The result is a case hardened layer at the surface that can be made visible metallographically. In the carburization process, carbon-rich gases, such as methane and CH_4–CO–H_2 mixtures, are made to flow over low- or medium-carbon steels (0.1–0.4 wt% C) maintained at temperatures of ~900°C. Pyrolysis at the metal surface releases elemental carbon which diffuses into austenite or γ-Fe, a phase that can dissolve

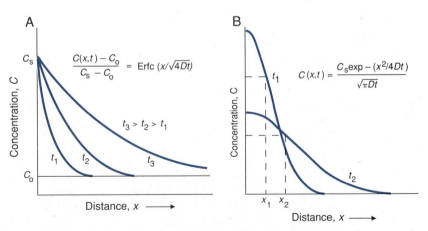

FIGURE 6-4 (A) Sequence of concentration profiles generated by diffusion from a constant surface source. A complementary error function solution governs the mass transport. (B) Gaussian diffusion profiles from a finite surface source. Average diffusant concentrations and distances are dotted in. A quadrupling of the diffusion time ($t_2 = 4t_1$) results in a doubling of the diffusional penetration ($x_2 = 2x_1$).

about 1.25 wt% C at this temperature. A surface layer or **case** typically 1 mm thick is enriched to about 1 wt% C. After heat treatment, a hardened, wear-resistant case consisting of a hard, metastable phase of steel known as **martensite** will then surround the softer but tougher interior. Many automotive parts, machine components, and tools such as gears, cam-shafts, and chisels require this combination of properties and therefore are carburized. Some carburized parts are schematically reproduced in Fig. 6-5. In designing carburization treatments the value for the diffusion coefficient of carbon in α-iron may be taken as $D_{\alpha}(C) = 0.004 \exp(-19.2 \text{ kcal/mol}/RT)$ cm^2/s. Similarly, in γ-iron the diffusion coefficient of carbon is approximately $D_{\gamma}(C) = 0.12 \exp(-32.0 \text{ kcal/mol}/RT)$ cm^2/s. Note that these D values depend on steel composition.

Even harder steel surfaces can be produced by nitriding or carbonitriding. In nitriding, ammonia pyrolysis at 525°C provides the nitrogen that penetrates the steel. After 2 days, case layers extending 300 μm deep can be expected. Typical components that are nitrided include extrusion, deep-drawing, and metal-forming tools and dies. Through low-temperature diffusional processing such machined parts avoid thermal distortion. The resulting surfaces retain high hardness and are resistant to wear as well as corrosion without requiring subsequent heat treatment.

Other commercial diffusional processes involve the introduction of the elements aluminum, boron, silicon, and chromium into the surface layers of metals being treated. The corresponding aluminizing, boronizing, siliciding, and chromizing treatments yield surfaces that are considerably harder or more resistant to environmental attack than the original base metal. For example, coatings based on Al have been used for decades to enhance the resistance of materials to high-temperature oxidation, hot corrosion, particle erosion, and

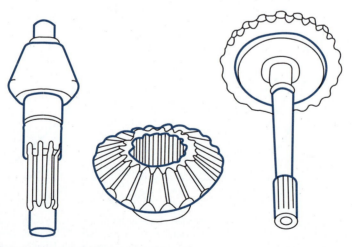

FIGURE 6-5 Collection of carburized steel components.

wear. Aluminized components find use in diverse applications: nuclear reactors, aircraft engines, and chemical processing and coal gasification equipment. Metals undergoing aluminizing treatments include nickel and iron base super-alloys, heat-resistant alloys, and a variety of stainless steels.

6.2.3. Oxidation of Silicon

Oxidation is a process in which oxygen chemically reacts with a metal surface to produce a metal oxide. Thickening of the oxide is usually accomplished by the diffusion of either oxygen or metal, or both, through the existing oxide. Oxidation can lead to the deterioration of metal surfaces during elevated temperature service. This aspect of oxidation related to materials reliability and failure is deferred to Chapter 10, where its connection to corrosion is developed. But not all oxidation is harmful, however. In fact, one of the key steps in the processing of integrated circuits involves the controlled oxidation of silicon to form a high-quality SiO_2 thin film by the reaction $Si + O_2 \rightarrow SiO_2$. From a mass transport standpoint, the reaction between oxygen and Si is quite instructive and worth discussing because of its general applicability to other oxidation (and sulfidation, nitriding, etc.) processes.

To extend the thickness of an already present oxide layer it is assumed that at least three sequential steps must occur:

1. Oxygen has to be transported from the bulk of the flowing gas, through a stagnant gas boundary layer, to arrive at the outer surface of the SiO_2. (The wafer surfaces are assumed to lie parallel to the gas flow.)
2. Then oxygen, probably in atomic form, must diffuse through the entire SiO_2 film to arrive at the Si–SiO_2 interface.
3. Chemical reaction is required to convert the Si substrate to SiO_2.

Considering only transport and reaction in the solid, steps 2 and 3 can be modeled by

$$J_2 = D(C_0 - C_i)/x_0 \tag{6-7a}$$

and

$$J_3 = K_s C_i. \tag{6-7b}$$

Mass flux J_2 is based on Eq. 6-1, where C_0 and C_i are the respective oxygen concentrations at the gas and SiO_2 interfaces, respectively, D is the diffusivity of oxygen in SiO_2, and x_0 is the SiO_2 thickness at any time. The rate of chemical reaction at the interface is proportional to both the oxygen concentration C_i and the rate constant K_s, so that the amount of oxide formed per unit time and area is flux J_3. During steady-state oxide growth $J_2 = J_3$, and, therefore, a bit of simple algebra gives $C_i = C_0(1 + K_s x_0/D)^{-1}$. But J_3 is also proportional to the rate at which oxide grows, that is, $J_3 = N_0 \, dx_0/dt$. The oxide growth rate is dx_0/dt, and N_0 is the number of oxidant molecules incorporated into a unit volume of oxide film ($N_0 \sim 10^{22}/cm^3$). Therefore,

$$\frac{dx_0}{dt} = \frac{K_s C_i}{N_0} = \frac{K_s C_0}{N_0(1 + K_s x_0/D)} \tag{6-8}$$

Two important limiting cases can be recognized in this equation. In the first case oxygen diffusion through the oxide is rapid (D large), but reaction with Si is slow (K_s small). Because oxidation is hindered or limited by the speed of the interfacial chemical reaction we speak of **reaction rate-limited oxide growth.** In this regime $1 > K_s x_0/D$, and direct integration of what remains of Eq. 6-8 yields

$$x_0 = (C_0 K_s/N_0)t \qquad \text{(reaction rate limited)}. \tag{6-9}$$

At the other extreme, diffusion of oxygen through the SiO_2 is sluggish (D small), but when it reaches the interface, reaction with Si is rapid (K_s large). In this case we speak of *diffusion-limited growth.* Now $K_s x_0/D > 1$, and the

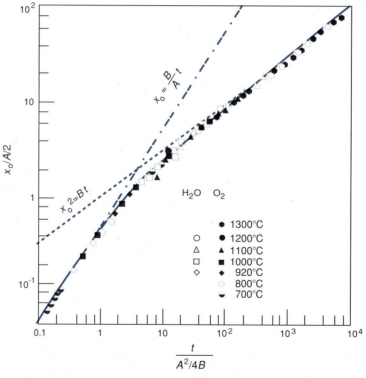

FIGURE 6-6   Experimental determinations of silicon oxidation kinetics in O_2 and H_2O ambients, plotted in dimensionless coordinates. Limiting cases of reaction rate and diffusion-controlled oxidation are indicated, where $A = 2D/K_s$ and $B = 2C_0D/N_0$. From A. S. Grove, *Physics and Technology of Semiconductor Devices,* Wiley, New York (1967).

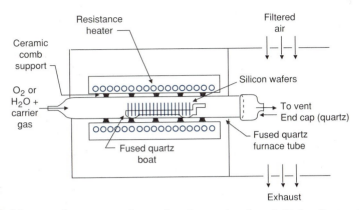

differential equation that has to be solved is $dx_0/dt = C_0 D/N_0 x_0$. After variables are separated direct integration yields

$$x_0^2 = 2(C_0 D/N_0)t \qquad \text{(diffusion limited)}. \qquad (6\text{-}10)$$

Oxidation kinetics data have been obtained in this very extensively studied system using both dry oxygen and steam ambients. Some of these results are plotted in dimensionless coordinates in Fig. 6-6 so that all of the experimental determinations fall on one line. At short times (x_0 small) oxide growth exhibits a linear time dependence characteristic of a reaction-limited process. For longer times, parabolic growth is followed in accord with diffusion-limited kinetics. These results are in excellent agreement with the simple model presented here.

Oxidation of Si is generally carried out at temperatures of 1100 to 1200°C in tubular reactors (Fig. 6-7) where oxygen flows axially. **Diffusional doping** of silicon is carried out in virtually identical furnaces at similar temperatures. Gases such as B_2H_6, PH_3, and AsH_3 are used as dopant sources and are fed into the reactor. Together with oxygen corresponding B-, P-, or As-rich silica glass layers deposit in a CVD-like process (see Section 5.2.2) on the Si wafer surface. From there the required dopant (either B, P, or As) diffuses into the underlying semiconductor. Additional aspects of semiconductor processing are addressed in detail in Chapter 12.

6.3. ATOM MOVEMENTS AND DIFFUSION

In this section the focus is on diffusion phenomena from a **microscopic** rather than **macroscopic** point of view. In particular, we wish to trace atom movements and determine how they are influenced by temperature.

6.3.1. The Diffusion Coefficient

Figure 6-8 shows a model for atomic diffusion between neighboring planes within a crystalline matrix. For atoms to jump from one lattice site (Fig. 6-8A) to another, two conditions must be fulfilled:

1. A site (e.g., a vacancy) must first be available for atoms to move into.
2. The atom must possess sufficient energy to push aside and squeeze past surrounding lattice atoms to reach an activated state (Fig. 6-8B). Then it readily falls into the neighboring site (Fig. 6-8C) while the vacancy moves to the left.

The language of diffusion is similar to that used in describing chemical reactions. We speak of surmounting an *energy barrier* along the *reaction path* coordinates. Both examples hinge on the thermodynamic probability P that atomic events succeed in producing reaction. They are based on similar, but slightly modified, Boltzmann-type formulas that were introduced previously in Eq. 5-13:

$$P = \exp(-\Delta G^*/RT). \tag{6-11}$$

The free energy difference, ΔG^* (per mol), between normal and activated states is known as the **activation free energy.** Small ΔG^* values and high temperatures increase P exponentially and thus strongly enhance the prospects for atomic motion.

It should come as no surprise that the diffusion coefficient introduced earlier has a form that is proportional to P, namely,

$$D = D_0 \exp(-\Delta E_D/RT) \quad \text{cm}^2/\text{s}. \tag{6-12}$$

To derive this formula in an approximate way let us consider the rate at which atoms jump into the vacancy, a distance a away. We have already noted that atoms vibrate in place with frequency ν (typically $\sim 10^{13}$ s^{-1}). Every so often a vibrating atom will acquire enough thermal energy to ascend to the activated state. The rate (r) at which this happens per second is given by the product of ν and P or

$$r = \nu \exp(-\Delta G^*/RT) \quad \text{s}^{-1}. \tag{6-13}$$

This fundamental step, a precursor to atomic diffusion, must ultimately be consistent with Eq. 6-6. Therefore, for a nearest-neighbor diffusive jump we must associate x with a, and the jump time (t) with r^{-1}. Substitution yields $D = \frac{1}{4}a^2 r$. In three dimensions, D is equal to $\frac{1}{12}a^2 r$ or $\frac{1}{12}a^2\nu \exp(-\Delta G^*/RT)$. Making use of Eq. 5-2, we obtain $D = \frac{1}{12}a^2\nu \exp(\Delta S^*/R) \exp(-\Delta E^*/RT)$, where ΔS^* is the **activation entropy.** Comparison with Eq. 6-12 implies that $D_0 = \frac{1}{12}a^2\nu \exp(\Delta S^*/R)$, and that ΔE_D, the **activation energy** for diffusion, is equal to ΔE^*. For typical values of ν, $a = 2.5 \times 10^{-8}$ cm, and $\exp(\Delta S^*/R) =$

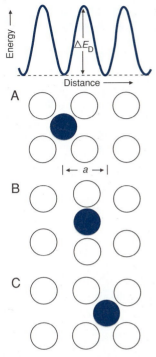

FIGURE 6-8 Model of atomic diffusion in a lattice. In exchanging with a vacancy, the diffusing atom must surmount an energy barrier as shown.

10, the preexponential factor D_0 is equal to $\sim$0.005 cm^2–s. Both D_0 and ΔE_D are constant. Each is dependent on the nature of the diffusant and matrix atoms, and the specific diffusion path.

Over the years values of D have been experimentally determined in a large number of metallic, semiconductor, and ceramic systems in both liquid and solid states. In most cases D is extracted after fitting measured concentration profiles to mathematical solutions (e.g., Eqs. 6-3 and 6-5) for the experimental geometry in question. One popular experimental geometry involves the so-called "diffusion couple" consisting of two different materials joined at a planar interface. This configuration often arises in practice, for example, clad metals and layered semiconductors, and simulates interdiffusion reactions in these structures. Most of the available data are obtained from measurements in the bulk lattice where diffusion occurs on substitutional sites (Fig. 6-9). Occasionally, as for carbon and nitrogen in α-Fe, transport of these small atoms occurs via the network of interstitial sites.

Diffusivity values are invariably reported as a function of temperature as indicated in the next section.

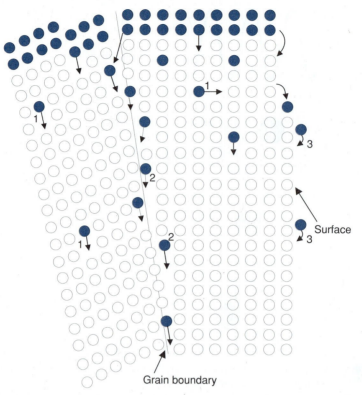

FIGURE 6-9   Atomic transport mechanisms in crystalline solids. Illustrated are (1) bulk diffusion due to vacancy–atom exchange, (2) grain boundary diffusion, and (3) surface diffusion.

6.3.2. Effect of Temperature: Arrhenius Behavior

That temperature plays an extremely important role in kinetic processes in solids should be evident by the ubiquitous presence of the Boltzmann factor (e.g., Eqs. 6-12 and 6-13). A universal way of graphically representing the effect of temperature is through the **Arrhenius plot.** Experimental values of the temperature-affected variable, for example, diffusivity and rate of reaction, are plotted logarithmically versus $1/T$ (K). In this way the highly nonlinear, thermally activated response is linearized and made amenable to simple analysis. An example is the diffusion coefficient of (interstitial) nitrogen in α-iron. Measured experimental values span some 16 orders of magnitude over a temperature range of about 1000°C (Fig. 6-10) and yet an excellent linear fit to the data is obtained.

Arrhenius plots of diffusivity data have been accumulated for metal, semiconductor, and ceramic systems as shown in Figs. 6-11A, B, and C, respectively. In the case of metals, typical substitutional diffusivity values in BCC or FCC

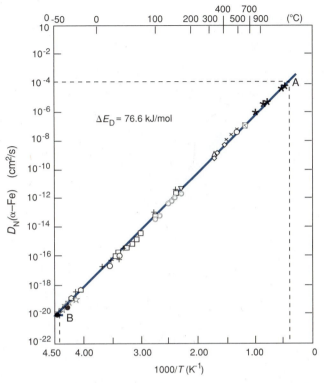

FIGURE 6-10 Arrhenius plot of the diffusion coefficient of nitrogen in iron. Note the reversed sense of the abscissa. After D. N. Beshers, in *Diffusion*, American Society of Metals, Metals Park, OH (1973).

alloy matrices of melting point T_M are plotted. This body of information can be briefly summarized as follows:

1. In solids, values for ΔE_D range from ~0.2 to several electron-volts per atom (1 eV/atom = 23.1 kcal/mol = 96.6 kJ/mol).

2. In many materials ΔE_D values scale directly with the melting point (T_M) of the matrix. For example, in FCC metals the bulk self-diffusion diffusion coefficient can, to a good approximation, be written as $D = 0.5 \exp(-17.0 T_M / T)$ cm^2/s, so that $\Delta E_D = 17.0 R T_M$. To achieve equivalent diffusional penetrations in ceramics relative to metals, higher temperatures would generally be required.

3. Although not indicated in Fig. 6-10, the diffusion coefficients of atoms in liquids of virtually all kinds (metals, oxide slags, polymer melts, aqueous solutions, etc.) only span the narrow range between 10^{-4} and 10^{-6} cm^2–s, irrespective of temperature.

4. Much less frequently are measurements reported for atomic diffusion through grain boundaries, dislocation cores, or surface paths (Fig. 6-9). The

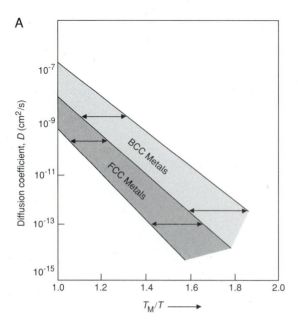

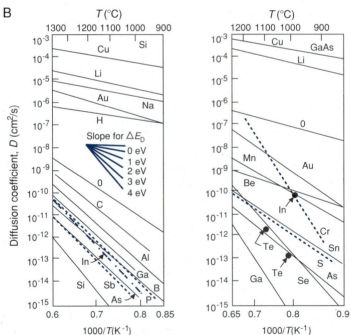

FIGURE 6-11 (A) Approximate range of diffusivity versus temperature for substitutional diffusion in BCC and FCC alloys. T_M is the melting point of the alloy. Adapted from O. D. Sherby and M. T. Simnad, *Transactions of the American Society for Metals* **54,** 227 (1961). (B) Diffusion of dopants in Si and GaAs. From S. M. Sze, *Semiconductor Devices: Physics and Technology,* Wiley (1985). Copyright © 1985 by AT&T; reprinted by permission. (C) Diffusion in oxide materials. *Note:* CSZ refers to stabilized ZrO_2. From W. Rhodes and R. E. Carter, *Journal of the American Ceramic Society* **49,** 244 (1966). Reprinted by permission of the American Ceramic Society.

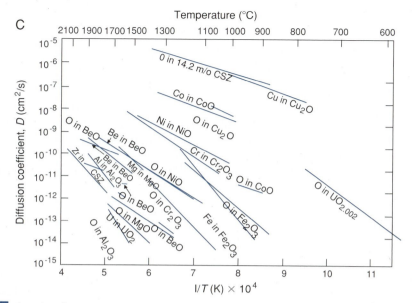

FIGURE 6-11 (*continued*).

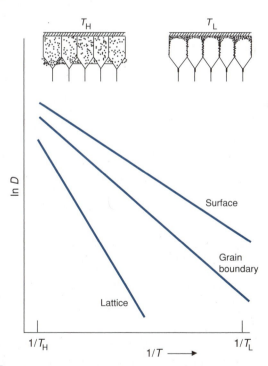

FIGURE 6-12 Schematic dependence of lattice, grain boundary, and surface diffusion on temperature. For operative lattice and grain boundary diffusion mechanisms, diffusional penetration is primarily through the bulk of the grains at high temperature (T_H). At low temperature (T_L), diffusional penetration of grain boundaries is dominant as shown in the above structural model.

looser atomic packing facilitates atomic motion along these short-circuit paths. As a result, measured activation energies for such transport are roughly half those required for diffusion through the bulk lattice.

With respect to item 4 it should be noted that although short-circuit diffusion is rapid, the path cross-sectional areas are small. Therefore, little net transport occurs and it is swamped by the amount of diffusion through the broad avenues of the bulk lattice at elevated temperatures. However, the reverse is true at low temperatures, as shown in Fig. 6-12. Lattice diffusion is virtually completely halted, but the mass transport that does occur is concentrated in grain boundaries. Though diminished, the amount of diffusion may still be sufficient to cause damage. This is why short-circuit mass transport at low temperatures is viewed as a reliability issue in microelectronic devices (see Chapter 15).

EXAMPLE 6-3

Calculate a value for the activation energy of nitrogen diffusion in α-iron.

ANSWER   From Eq. 6-12,

$$\ln D = \ln D_0 - \Delta E_D/RT \qquad (6\text{-}14a)$$

or

$$\log D = \log D_0 - \Delta E_D/2.303RT. \qquad (6\text{-}14b)$$

The latter equation is of the form $y = mx + b$, the algebraic representation of a line, where y is associated with $\log D$, x with $1/T$, and b with $\log D_0$. The slope m, defined as $\triangle(\log D)/\triangle(1/T)$, is equal to $-\Delta E_D/2.303R$ and, therefore, $\Delta E_D = -2.303mR$. From Fig. 6-11, selecting points A and B, $m = [-4 - (-20)]/(0.0004 - 0.0044) = -4000$. Since $R = 8.314$ J/mol-K, $\Delta E_D = -2.303 \times 8.314 \times (-4000) = 76.6$ kJ/mol or 18.3 kcal/mol.

EXAMPLE 6-4

Is Cu predicted to diffuse more rapidly in Ni at 1000°C than Ni in Cu at the same temperature? Assume dilute alloys in each case.

ANSWER   Both Ni and Cu matrices have FCC structures with respective melting points of 1726 K and 1356 K. At 1000 + 273 = 1273 K, $T_M/T = 1726$ K/1273 K = 1.36 in the case of Cu in Ni. For Ni in Cu, $T_M/T = 1356$ K/1273 K = 1.07. Reference to Fig. 6-11A suggests a midrange diffusivity value of $\sim 7 \times 10^{-12}$ cm²/s for diffusion of Cu in Ni and a value of $D = \sim 8 \times 10^{-10}$ cm²/s for the more rapid diffusion of Ni in Cu.

Measured values are 2×10^{-11} cm^2/s and 6×10^{-10} cm^2/s, respectively, indicating the approximate nature of the calculation.

6-4. NUCLEATION

Irrespective of whether single- or multicomponent systems are involved, phase transformations occur in a material system when boundary lines on phase diagrams are crossed. Out of the now unstable prior phases, new materials form and grow until they entirely consume the former. This takes time. There is usually a sequence of steps that occur, first involving germination or **nucleation** of stable clusters of the new phase. Nucleation is then followed by a period of phase **growth.** The dividing line between these two processes is not sharply defined, however, and often it is not clear when nucleation ends and growth begins. Furthermore, both processes frequently occur simultaneously in different parts of the same system.

It is important to understand nucleation and growth effects because they conspire to establish the times required for phase compositions and structural morphologies to evolve during transformations. Structure–property linkages make it imperative to control both nucleation and growth if materials are to be creatively synthesized or modified.

6.4.1. Homogeneous Nucleation

6.4.1.1. Thermodynamics of Nucleation

Let us start by considering the early stages of solidification of molten metal and determine the energy changes that characterize the liquid–solid (L–S) phase transformation. At the moment we deal with only one of many identical spherical solid nuclei of radius r that form in the interior of the parent liquid phase. **Homogeneous nucleation** of solid is said to occur because nuclei form without benefit of interaction with the mold walls. The driving force for nucleation is derived from the (negative) bulk chemical free energy difference (ΔG_V) for the reaction L $\rightarrow$ S, where $\Delta G_V = G_S - G_L$. But, as the nucleus forms, an interface develops between the solid and liquid. New solid–liquid bonds mean an increase in system **surface energy** that must, in essence, be provided by the **chemical energy** of transformation. The net free energy change (ΔG_N), plotted in Fig. 6-13, is given by the sum of these two contributions:

$$\Delta G_N = \tfrac{4}{3}\pi r^3 \, \Delta G_V + 4\pi r^2 \gamma. \qquad (6\text{-}15)$$

In this equation, γ, the surface free energy or surface tension, is *positive* and has units of energy per unit area. On the other hand, ΔG_V represents ΔG on a per unit volume basis and has a *negative* magnitude because $G_L > G_S$. Clearly if liquid is stable so that $G_L < G_S$, there can be no nucleation of solid.

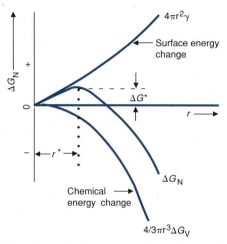

FIGURE 6-13 Contributions to the free energy change associated with the homogeneous nucleation of a solid spherical nucleus in a liquid.

The opposing influences of these two terms on ΔG_N are depicted in Fig. 6-13. For small values of r (i.e., $r < 1$), $r^2 > r^3$, whereas for large values of r, the inequality reverses. Thus, ΔG_N initially rises with r, goes through a maximum, and then falls. Solving for r after setting $d\,\Delta G_N/dr = 0$ yields the critical radius of the stable nucleus (r^*) as

$$r^* = -2\gamma/\Delta G_V. \qquad (6\text{-}16)$$

Substituting r^* in Eq. 6-15 enables the corresponding critical value of the net free energy—the barrier to nucleation—to be evaluated as

$$\Delta G_N^* = \frac{16\pi\gamma^3}{3(\Delta G_V)^2}. \qquad (6\text{-}17)$$

Our nucleation model suggests that spherical solid clusters form under initially favorable conditions, but if they do not grow larger than r^*, they redissolve in the melt. Like certain fish species that lay many eggs only to see a few embryos survive, it is those clusters favored by chance that attain a size larger than r^*; they then achieve thermodynamic stability and become nuclei. Their growth is favored because ΔG_N is reduced as they grow larger.

The energies involved in nucleation deserve a closer look. Through the use of Eq. 5-2 we can write

$$\Delta G_V = \Delta H_V - T\Delta S_V. \qquad (6\text{-}18)$$

In the case of solidification it is reasonable to associate ΔH_V with the latent heat of fusion ΔH_F. At the melting point, T_M, it is further assumed that

$\Delta S_V = \Delta H_F/T_M$. Substitution of these terms yields

$$\Delta G_V = \Delta H_F(T_M - T)/T_M = \Delta H_F \,\Delta T/T_M. \qquad (6\text{-}19)$$

As a result, ΔG_V depends linearly on the extent of melt supercooling (ΔT), which is defined as the difference between the melting point and the actual melt temperature. Likewise, the energy barrier for nucleation (ΔG_N^*) also depends on the amount of supercooling. The dependence is large and the implications important as we shall soon see.

6.4.1.2. Nucleation Rate

Until now only a single nucleus has been considered, but it is well known that many usually form. Therefore, we must address two questions: How many nuclei form in a given volume per unit time? What is the temperature dependence of this nucleation rate? The population of nuclei is critically dependent on the probability of acquiring sufficient energy (i.e., ΔG_N^*) to create nuclei. Following our prior experience with the thermodynamics of vacancies (see Section 5-8 and Eq. 3-6) we might expect that a Boltzmann factor would govern the equilibrium concentration (C_N) of solid nuclei generated within the melt interior:

$$C_N = C_0 \exp(-\Delta G_N^*/RT) \qquad (6\text{-}20)$$
$$= C_0 \exp - \frac{16\pi\gamma^3}{3RT(\Delta H_F \Delta T/T_M)^2} \quad \text{nuclei/cm}^3.$$

This expression combines Eqs. 6-11, 6-17, and 6-19. The only term not previously defined is C_0, the total concentration of nuclei that could potentially form. Unlike vacancies, however, the closer one gets to the melting point, the fewer the nuclei there are.

Lastly, we are interested in the nucleation rate $\dot{N}$ (nuclei/cm³-s), which is essentially proportional to the product of thermodynamic and kinetic factors. The first is C_N and the second is the rate at which atoms attach to nuclei by diffusion. A very approximate formula that accounts for these two contributions is

$$\dot{N} = A\{C_0 \exp(-\Delta G_N^*/RT)\}\{\nu \exp(-\Delta E_D/RT)\} \quad \text{nuclei/cm}^3\text{-s}, \quad (6\text{-}21)$$

where A is a constant. The results are graphically depicted in Fig. 6-14 and surprisingly reveal that the maximum nucleation rate occurs well below the transformation temperature. This effect also occurs in solid-state transformations and reflects the temperature dependencies of the two terms in braces. Nucleation is sluggish near T_M and enhanced at low temperatures, whereas diffusion exhibits a contrary behavior, being larger the higher the temperature. It is therefore a case of multiplying terms that are alternately large and small, yielding a small net value of $\dot{N}$ at both the highest and lowest temperatures; the product is largest at temperatures in between.

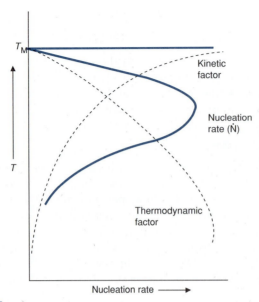

FIGURE 6-14   Schematic variation of the nucleation rate of a transformation as a function of temperature. Just below the transformation temperature, $\dot{N}$ is small because the nucleation energy barrier is very high (thermodynamic factor is small). Similarly, $\dot{N}$ is small at low temperature because diffusion is limited (kinetic factor is small).

EXAMPLE 6-5

a. Suppose a nickel melt can be supercooled 300°C below its melting point of 1452°C. If the liquid–solid surface energy is 2.55×10^{-5} J/cm^2, and the latent heat of fusion is 301 J/g, what is the size of the nucleus of critical size? The density of Ni is 8.9 g/cm^3.

b. What is the ratio of the nucleation rate expected with 250°C supercooling relative to 300°C supercooling? Assume the contribution due to diffusion in the liquid can be neglected.

ANSWER   a. From Eq. 6-19,

$$\Delta G_V = \Delta H_F \, \Delta T / T_M = (301 \text{ J/g})(8.9 \text{ g/cm}^3)(300 \text{ K})/(1452°C + 273 \text{ K})$$
$$= 466 \text{ J/cm}^3.$$

Substitution in Eq. 6-16 (recalling that ΔG_V is physically negative in sign) yields $r^* = 2(2.55 \times 10^{-5})/466 = 1.09 \times 10^{-7}$ cm or 1.09 nm. A spherical nucleus of this size contains approximately 660 Ni atoms.

b. For 250°C supercooling, where $T = 1475$ K,

$$\Delta G_V = (301)(8.9)(250)/(1725) = 388 \text{ J/cm}^3.$$

Therefore by Eq. 6-17,

$$\Delta G_N^* = 16\pi (2.55 \times 10^{-5})^3/3(388)^2 = 1.85 \times 10^{-18}\,\text{J}.$$

Similarly for 300°C supercooling, where $T = 1425$ K,

$$\Delta G_N^* = 16\pi (2.55 \times 10^{-5})^3/3(466)^2 = 1.28 \times 10^{-18}\,\text{J}.$$

From Eq. 6-21, and employing k rather than R,

$$\frac{\dot{N}\,(250°C)}{\dot{N}\,(300°C)} = \frac{\exp(-1.85 \times 10^{-18})/[(1.38 \times 10^{-23})(1475)]}{\exp(-1.28 \times 10^{-18})/[(1.38 \times 10^{-23})(1425)]} = 10^{-25}.$$

The ratio is vanishingly small. This problem illustrates the fact that although nucleation theory is very far from being quantitatively correct, it qualitatively indicates the profound effect temperature can have on nucleation.

6.4.2. Heterogeneous Nucleation

Homogeneous nucleation occurs only rarely. It is virtually always the case that nucleation processes are catalyzed by a heterogeneity such as an accommodating substrate surface. The liquid cap wetting a flat substrate in Fig. 5-33 may be viewed as an example of **heterogeneous nucleation.** Because of the bonding across the substrate interface, less energy is expended in creating the cap heterogeneously than in creating a spherical nucleus homogeneously. Nuclei that solidify from a melt onto a mold wall are similar. Because of the effective reduction in surface energy, the critical nucleus is smaller and the rate of formation is greater relative to that obtained by homogeneous nucleation. Heterogeneous nucleation is difficult to analyze not only because of the geometry, but also because of the often unknown interactions between the nucleus and substrate.

Many phase transformations take place heterogeneously in the solid state. Grain boundaries are favored sites for such reactions because excess interfacial energy is available there for nucleation of new phases. This is why proeutectoid phases like α-Fe and Fe_3C, as well as two-phase mixtures of pearlite, nucleate at prior austenite grain boundaries. Precipitates and inclusions in the matrix also serve as heterogeneous nucleation sites.

6.4.3. Implications of Nucleation

6.4.3.1. Making Ice

The formulas that have been developed can help us qualitatively understand a number of important commonly observed phase transformation phenomena. We all know, for example, that when water is slightly below 0°C ice does not readily form. Few crystals form, but those that do nucleate tend to grow large. Because supercooling (ΔT) is very small very little chemical driving force exists

(ΔG_V small), and the large barrier to nucleation keeps $\dot{N}$ small. Next consider the problem of water nucleation during a drought. There is often enough moisture in the atmosphere, and the temperature is low enough, but it still does not rain. To induce nucleation of ice, it is common practice to seed clouds with AgI. The reason this promotes rain production is that both ice and AgI have hexagonal crystal structures with very similar lattice parameters. In AgI, ice has a low-energy accommodating surface template for heterogeneous nucleation. Ice that forms then melts to rain as it warms in falling to earth.

6.4.3.2. Growing Single Crystals

The preceding suggests that a good way to grow large single crystals of materials is to cool melts just below T_M. If one prior nucleus, a seed crystal, is carefully placed in the melt and we are lucky, other nuclei will not form. Rather, solid deposits on the seed instead, causing it to grow while retaining its original crystallographic orientation. A crystal puller, schematically shown in Fig. 6-15, capitalizes on the effect, lifting and rotating the oriented solid as it forms. In the process a crystal–melt interface is continually left behind for further ordered atomic attachment. All of the bulk single crystals employed in

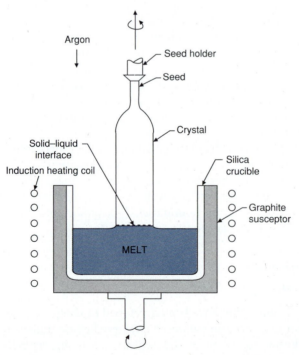

FIGURE 6-15   Apparatus to grow single crystals of Si from the melt by the Czochralski method (see Section 12.5.2). From S. M. Sze, *Semiconductor Devices: Physics and Technology*, Wiley (1985). Copyright © 1985 by AT&T; reprinted by permission.

electronics technology are basically grown from the melt in this or similar ways, including silicon and gallium arsenide for semiconductor and optoelectronic devices (see Chapters 12 and 13) and ruby and garnet crystals for lasers. Creative variants of the single-crystal growth process are tailored to the specific material, its shape, the temperature involved, and the crystal perfection required. Examples include growth of monocrystalline superalloy turbine blades for jet engines (see colorplate Fig. 8-2) and crystals grown in the Space Shuttle to eliminate defects caused by convective flow of the liquid. In common, every single-crystal growth method suppresses all nuclei save one, and promotes as high a growth rate of it as possible.

6.4.3.3. Grain Refinement

At the other extreme, fine-grained materials are often desired in castings. When melts solidify at large levels of supercooling, nucleation theory predicts high nucleation rates and tiny crystal formation. Practically, however, dendrites often grow readily even when ΔT is large. Dendritic cast structures have inferior mechanical properties because the grains tend to be long and narrow, chemically inhomogeneous, and not readily deformable. A more desirable equiaxed, finer-grain structure can be produced with the use of small additions of **inoculants** and **grain refiners** to melts. Commercially, titanium and boron are added to aluminum and zirconium to magnesium. In the latter case, grain refinement occurs by a peritectic reaction in which Mg nucleates around the Zr, resulting in spherical grains rather than dendrites. The close lattice match between Mg and Zr reduces ΔG_N^* (as in ice on AgI), even when the amount of supercooling is low.

6.4.3.4. Nucleation of Thin Films from the Vapor Phase

Liquid–solid and solid–solid transformations are not the only ones of interest. Vapor–liquid and vapor–solid transformations are responsible for the condensation of moisture and frost, respectively. The deposition of thin films, materials that serve critical functions in many technologies, often proceeds from the vapor phase. For example, films for mirrors are prepared by thermally evaporating metals like Al and Ag from a molten source, and collecting the condensed vapor on glass substrates placed within the same vacuum chamber. What drives the vapor–solid transformation in such cases is the **supersaturation** or excess partial pressure of metal atoms above the equilibrium vapor pressure value. Atom-by-atom deposition of thin films results in **heterogeneous** nucleation of films. The picture of small nuclei with spherical caps on a flat substrate models the earliest stage of thin-film formation. With further deposition the islandlike nuclei grow and coalesce, resulting in a connected network with channels in between. Finally, the channels fill in and the film is continuous. The sequence of events leading to the nucleation and growth of a thin silver film deposited on a NaCl substrate is shown in Fig. 6-16.

The resulting film morphology depends on the magnitude of nucleation and growth effects. If many nuclei form under high evaporation rates, they will

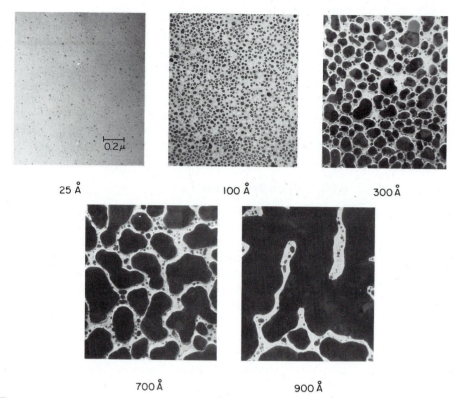

25 Å   100 Å   300 Å

700 Å   900 Å

FIGURE 6-16   Sequence illustrating the nucleation and growth of Ag films on a NaCl crystal substrate. Film thicknesses are indicated. From R. W. Vook, *Int. Metals Rev.* **27**, 209 (1982).

have little opportunity to grow when the substrate temperature is low. A fine-grained film results. But if the substrate temperature is unusually low, atoms are frozen in position where they land. As surface diffusion to equilibrium crystal lattice sites is suppressed, a disordered or even amorphous film can be produced. On the other hand, if there are few nuclei but they are given the opportunity to grow at high temperature, a coarse-grained deposit results. These same simple trends summarize, in a nutshell, the extremes of structural evolution in all sorts of phase transformations, including those treated in the next section.

6.5. KINETICS OF PHASE TRANSFORMATIONS

Phase transformations from unstable to stable phases are complicated because the entire solid does not transform all at once in unison. Rather, one region starts ahead of others due to some favorable local composition, temperature, defect, variation, or bias in the system. Then nucleation occurs elsewhere, and later somewhere else, while growth continues independently in regions of prior nucleation. The situation is very much like rain droplets falling on a pond. Where rain impinges on the pond, surface wave nuclei are created. The circular ripples grow outward and begin to impinge on one another. In solids, the transformation would be complete after all of the prior untransformed matrix disappeared due to growth and impingement processes. Not only is the sequential time dependence of events complex, but so too are the geometric shape and spatial distribution of new phases.

An important equation derived by M. Avrami over half a century ago applies to such transformations. In it the fractional amount of transformation (f) varies with time (t) as

$$f = 1 - \exp(-Kt^n). \tag{6-22}$$

The constant K, which accounts for a complex combination of nucleation and growth rate factors, is thermally activated, that is, is proportional to the Boltzmann factor, $\exp(-E/RT)$. (Here, as elsewhere in the book, E replaces the more correct ΔE.) In addition, n is a number related to the geometric shape of the growing transformed phase. In the case of spheres $n = 4$. For short times ($t \to 0$), $f \to 0$. At longer times the reaction nears completion and f approaches unity as the exponential term vanishes. It is important to know the quantities n and E (the activation energy) that govern the time dependence of solid-state phase transformations if they are to be carried out efficiently. Several nucleation and growth transformations of scientific and technical interest are treated next to illustrate the use of the Avrami equation.

6.5.1. Amorphous-to-Crystalline Transformation

Amorphous materials tend to transform to their crystalline counterparts of lower free energy provided the necessary kinetic processes occur. In this regard, the amorphous-to-crystalline transformation of cobalt disilicide ($CoSi_2$), a metal alloy used in integrated circuits, has recently been extensively studied and serves as a model example. Thin amorphous films were first deposited and subsequently heated to elevated temperatures to crystallize them. An electron microscope image of such a film transformed at 150°C is reproduced in Fig. 6-17, revealing spherical crystals, typically 1 μm in size, immersed in the amorphous matrix. A value of f can be ascertained by determining the fraction of the total volume occupied by crystals. If the transformation is followed as a function of time and f is continually measured, curves of the type shown in Fig. 6-18 are obtained at different temperatures. The curves are excellent fits to the Avrami equation.

The S-shaped curve is common to many diverse phenomena.* In our example

---

* The technology "learning curve" exhibits a similar behavior. Learning is difficult initially and little progress is made. With greater experience, a period ensues when bugs are eliminated and barriers bypassed, enabling rapid advances. In time there is little left to learn and further progress is difficult to achieve.

FIGURE 6-17   TEM image of a partially transformed amorphous $CoSi_2$ film. The varied dark and light geometric patterns indicate that the circular grains are crystalline. From K. N. Tu, J. W. Mayer, and L. C. Feldman, *Electronic Thin Film Science,* Macmillan, New York (1992). Courtesy of K. N. Tu, IBM Corp.

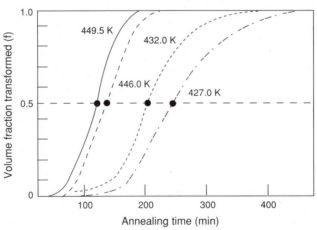

FIGURE 6-18   Kinetics of the amorphous to crystalline transformation in $CoSi_2$ at several temperatures. The indicated points represent the $t_{1/2}$ values used in Example 6-6. From K. N. Tu, J. W. Mayer, and L. C. Feldman, *Electronic Thin Film Science,* Macmillan, New York (1992).

there is an incubation period before any appreciable amount of transformation occurs. Then nucleation is rapid and simultaneous growth results in a quickening of the reaction. Finally, as the transformed crystals butt up against each other, f saturates and rises very slowly at the end of the transformation.

EXAMPLE 6-6

Assuming that Eq. 6-22 describes the data of Fig. 6-18 and that $n = 4$, calculate the value of E for the transformation.

ANSWER In Eq. 6-22, let $K = K_0 \exp(-E/RT)$. Therefore, $\ln(1 - f) = -[K_0 \exp(-E/RT)]t^4$. In Fig. 6-18 data points of $t_{1/2}$, corresponding to $f = \frac{1}{2}$, or 50% transformation at different values of T, are now selected. Taking natural logarithms of both sides, $\ln(-\ln \frac{1}{2}) = \ln K_0 - E/RT + 4 \ln t_{1/2}$. This can be rewritten as $\ln t_{1/2} = E/4RT + \frac{1}{4}C$, where $C = \ln(-\ln \frac{1}{2}) - \ln K_0$. The last equation is in the Arrhenius form (Eq. 6-14) except for the sign of the slope. A plot of $\log t_{1/2}$ versus $1/T$ (K), shown in Fig. 6-19, yields a straight line with a slope of 3870. Equating this to $E/2.303 \times 4R$, $E = 2.303 \times 4 \times 8.31 \times 3870 = 296{,}000$ J/mol. Therefore, $E = 296$ kJ/mol or 3.07 eV/atom.

This energy reflects contributions from nucleation and three-dimensional growth of crystalline $CoSi_2$.

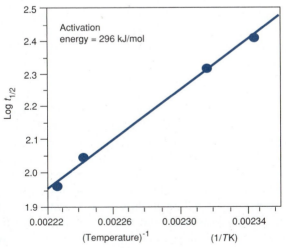

FIGURE 6-19 Arrhenius plot of log $t_{1/2}$ versus $1/T$ (K) for the transformation of amorphous to crystalline $CoSi_2$.

6.5.2. Nonequilibrium Transformation of Steel

In Section 5.5.4 the **equilibrium** phase transformation of austenite to ferrite and cementite was discussed. As with all phase transformations carried out under equilibrium conditions, rates of cooling are very low and time is never a variable of concern. But when austenite is *rapidly* cooled its transformation can be suppressed; it can be retained in metastable equilibrium just like super-cooled glass, quenched metal glasses, or amorphous $CoSi_2$ discussed above. Quenching austenite from elevated temperature is the first step of a very import-ant heat treatment that steel undergoes to harden it. Details are given later in Section 9.2. Here we are concerned with the controlled phase transformation of austenite by the method to be described.

Let us consider plain carbon eutectoid steel (AISI 1080 containing 0.8 wt% C), a model alloy because there are no proeutectoid phases (α-Fe or Fe_3C) to contend with. Austenite transforms only to pearlite. A large number of thin specimens, each the size of a dime, are first austenitized at a temperature of say 885°C. Sufficient time is allowed to transform the mixture of ferrite and cementite to austenite. Then one by one a number of these are quickly brought to a transformation temperature T_t below the eutectoid temperature of 723°C. According to Fig. 5-23, γ-Fe is no longer stable below 723°C and must trans-form to α-Fe and Fe_3C. Suppose T_t is 400°C. The first specimen is kept at 400°C for 1 second and then rapidly quenched in cold water. For the second specimen the time at 400°C might be 4 seconds prior to quenching. This procedure is repeated for the remaining specimens, with times suitably chosen to track the transformation of γ-Fe to α-Fe + Fe_3C until it is certain that austenite is no longer present. Now another T_t is chosen, for example, 700°C, and a new set of austenitized specimens is sequentially transformed **isothermally** and quenched. This procedure is repeated until it is deemed that a sufficient matrix of temperature–time treatments has been carried out. All of the speci-mens are then metallographically prepared for optical microscopy to estimate the relative amounts of phases present. In the days prior to computerized image analysis one can imagine how tedious this chore was.

A typical sequence of microstructures that tracks the transformation of eutectoid steel at 400°C is schematically reproduced in Fig. 6-20. Each structure contains varying proportions of the dark etching lamellar pearlite (two-phase mixture of ferrite + carbide, or F + C) and a light etching acicular or needlelike **martensite** phase (shown as white). Martensite does not appear on the equilib-rium Fe–C phase diagram, indicating that this phase is thermodynamically metastable. It is the product that austenite transforms to when very rapidly quenched. More will be said about martensite in Chapter 9, but here it serves the useful function of telling us how much austenite did *not* transform at temperature T_t. By now it should be clear that the Fe–Fe_3C phase diagram is no longer useful in describing the response of steel to rapid quenching heat treatments.

When the percentage of pearlite formed is plotted versus time, the transformation kinetics shown in Fig. 6-20 is typically obtained at any T_t. Comparison with Fig. 6-18 reveals that the percent increase in pearlite with time appears to reflect the nucleation–growth processes mathematically described by the Avrami equation. It is now a simple matter to determine the times required to initiate and to complete the austenite–pearlite transformation. The former time is arbitrarily associated with the appearance of 1% pearlite and the latter time with 99% pearlite. After plotting of the pair of times for each transformation temperature, the so-called **time–temperature transformation** or **TTT** curve displayed in Fig. 6-21 emerges.

Why it takes a long time to form pearlite at both high and low temperatures, but a short time in between, is no longer a mystery. The **C** shape of the TTT curve again underscores the effect of coupled nucleation and growth effects that control phase transformations. Just below 723°C *nucleation* of pearlite is very sluggish because there is little thermodynamic driving force for its decomposition. But, at the same time atomic *diffusion* and *phase growth* rates are large at high temperatures. Because the nucleation rate ($\dot{N}$) is small and the growth rate ($\dot{G}$) is large, the $\dot{N}\dot{G}$ product is small, meaning that it takes a long time to transform austenite. Conversely, at temperatures below the nose of the curve, the driving force for austenite decomposition is large while atomic

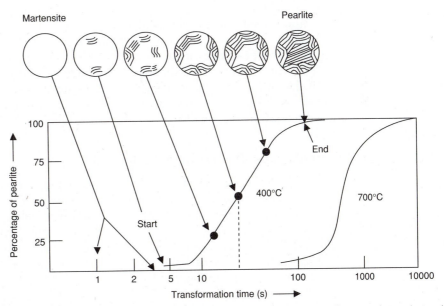

FIGURE 6-20 Time sequence of pearlite plus martensite microstructures that evolve as a function of time during the transformation of 1080 steel at 400 and 700°C. The percentage pearlite in the microstructure tracks the kinetics of transformation.

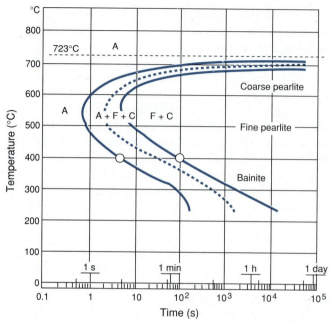

FIGURE 6-21   A TTT curve for 1080 (eutectoid) steel. Times for start and end of transformation at 400°C are noted.

diffusion rates are small. The product of a large $\dot{N}$ and small $\dot{G}$ is again small, accounting for the long time required for austenite to transform. At intermediate temperatures pearlite forms with greatest ease (see Fig. 6-14).

TTT curves graphically summarize complete sets of $f–T–t$ data and have been determined for other unstable systems undergoing nucleation–growth phase transformations. The ones for steel are the most important of these and are discussed in Section 9.2.

6.5.3. Catalyzed Crystallization of Glass to Glass–Ceramic

One of the important advances in ceramic engineering has been the development of **glass–ceramic** materials. Corning (Pyroceram) cookware is perhaps the best known glass–ceramic product. Guided missile radomes that act as lenses for high-frequency radio waves, machinable ceramics, and photosensitive glasses (Fotoceram) capable of storing three-dimensional images are other examples of these remarkable materials. Some commercial glass–ceramics products are shown in Fig. 6-22.

FIGURE 6-22   Glass–ceramics products. (A) Machinable glass components. (B) Transparent β-quartz solid solution Visions cookware. (C) Assorted black and white Corningware. Courtesy of Corning Incorporated.

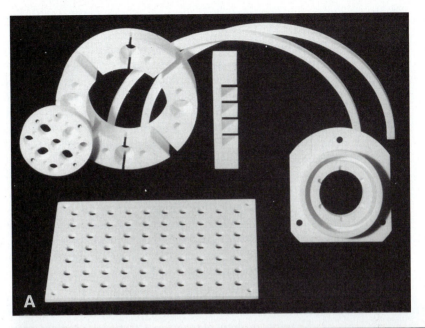

In the manufacture of transparent glass great care is taken to avoid crystallization, which can be triggered by the presence of precipitates. But small concentrations of second phases in the form of crystals, gas bubbles, and droplets of immiscible glass have long been intentionally introduced to make opaque or translucent glasses. In addition, colloidal particles (of the order of ~1 nm) have been accidentally and intentionally used to color glass since antiquity. The rare, beautiful gold ruby, copper ruby, and silver yellow glasses of the Middle Ages, made using closely held secret recipes, are some unusual examples that involved the introduction of small quantities of the indicated metals into the glass melt. Interestingly, some of the glass–ceramic compositions listed in Table 6-1 also contain metals that act as catalysts or nucleating agents for crystalline phases. What distinguishes these modern materials from the prior glasses is composition and the unique processing they are given. The process for manufacturing glass–ceramic articles can be understood with reference to the thermal history displayed in Fig. 6-23, and involves the following steps:

1. Conventional melting at temperature T_M, making sure the catalyst is completely dissolved.

2. Shaping molten *glass* into a transparent glass article by any forming method, for example, blowing, pressing, drawing, or casting.

3. Cooling glass to the temperature of optimum nucleation, T_N. Here tiny crystallites, approximately a few tens of nanometers in size, nucleate with the assistance of the catalyst. Alternately, the glass can be cooled still further to

TABLE 6-1	GLASS–CERAMIC COMPOSITIONS AND APPLICATIONS

Glass composition	Crystal phases	Catalysts	Properties and applications
1. Pyroceram $MgO–2Al_2O_3–SiO_2$	$2MgO \cdot 2Al_2O_3 \cdot 5SiO_2$ (cordierite)	TiO_2	Transparent to radar Radomes for missiles
2. Pyroceram $Li_2O–2Al_2O_3–SiO_2$	$Li_2O \cdot 2Al_2O_3 \cdot SiO_2$ (β-eucryptite) $Li_2O \cdot 2Al_2O_3 \cdot 4SiO_2$ (β-spodumene)	TiO_2 TiO_2	Low expansion Corning cookware
3. Fotoceram $Li_2O–2Al_2O_3–SiO_2$	$Li_2O–SiO_2$, SiO_2 (lithium metasilicate)	Au, Ag, Cu, Pt, Ce	Photosensitive, photoetchable ceramics
4. Macor (Fluormica)	$KMg_3AlSi_3O_{10}F_2$	Internally nucleated fluormica crystals	Machinable glass

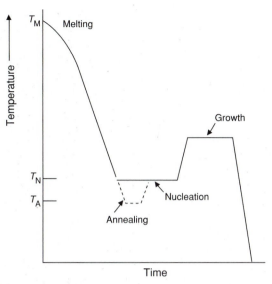

FIGURE 6-23 Sequence of thermal treatments required for the controlled nucleation and growth of crystalline phases in glass–ceramics.

the annealing temperature, T_A (to remove stress), or even to room temperature prior to treatment at T_N. The article is typically held at T_N (T_N is about 50°C higher than T_A) for an hour or until a maximum number of *crystalline ceramic* nuclei form.

4. Raising the temperature at a rate of a few degrees a minute to T_G, the crystal growth temperature. The article is maintained there until growth of the ceramic crystalline phase(s) is complete.

5. Cooling to room temperature completes the process.

The microstructure in Fig. 6-24 demonstrates the extent of the glass-to-ceramic transformation. Unlike the other nucleation–growth phase transformations considered above, the nucleation and growth temperatures here are distinct, effectively decoupling these processes.

6.6. GENERALIZED SOLID-STATE KINETICS

In this section we consider the transport of atoms in systems where *generalized* driving forces exist. These cause the chemical reactions, diffusion in concentration gradients, solidification, and solid-state phase transformations we have discussed in multicomponent systems. In addition, other processes in single-component (pure material) systems occur under different driving forces. Included are grain boundary motion, the sintering of particles, and diffusion in

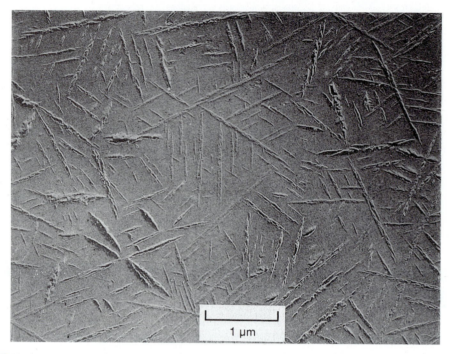

FIGURE 6-24 Microstructure of Fotoform, as revealed by electron microscopy. Exposure to light during processing allows development of a continuous dendritic pattern of lithium metasilicate, a structure that is etchable. Courtesy of G. H. Beall, Corning Incorporated.

an electric field (electromigration). In every one of these examples atomic motion is involved. Do Newton's laws apply to such motion, or are other factors at play?

When driving forces exist over atomic dimensions of a crystal, the periodic array of lattice energy barriers assumes a tilt or bias as shown in Fig. 6-25. In fact, the slope of G with distance ($\Delta G/a$), or the free energy gradient, is caused by and is physically equal to the applied **driving force** (F). Values of $F = \Delta G/a$ are in force per mole or force per atom units. This tilting is unlike the situation in Fig. 6-8, where the energy hills and valleys were level, and lattice sites were energetically equivalent. Now atoms in site 2 have a lower energy than those in site 1 by an amount ΔG. Therefore, the energy barrier to atomic motion from 1 to 2 is lower than that from 2 to 1. In the former case the rate at which atoms move to the right is given by Eq. 6-13:

$$r_{12} = \nu \exp(-\Delta G^*/RT) \quad \text{s}^{-1}. \tag{6-23}$$

Similarly

$$r_{21} = \nu \exp[-(\Delta G^* + \Delta G)/RT] \quad \text{s}^{-1}. \tag{6-24}$$

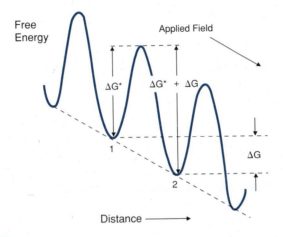

FIGURE 6-25 Potential energy of atoms in a periodic lattice: (A) In the absence of a force. (B) In the presence of a force.

Extensions of these two formulas can be applied to all of the above generalized reactions provided proper account is taken of the origin and magnitude of ΔG.

6.6.1. ΔG Large (Chemical Reactions)

In the case of chemical reactions where large energy changes accompany breaking of former bonds and creation of new ones, ΔG is large. Typically, ΔG values of 1 to 10 eV/atom or $\sim 10^5$ to 10^6 J/mol can be expected. For the chemical reaction A $\rightarrow$ B, the forward reaction rate is proportional to $C_A \cdot r_{AB}$, and the reverse reaction rate to $C_B \cdot r_{BA}$, where C's are the concentrations and r's, the reaction rates. The net reaction rate, r_{net}, is given by the difference of these two quantities. With the help of Eqs. 6-23 and 6-24,

$$r_{net} = C_A \cdot r_{AB} - C_B \cdot r_{BA}$$
$$= C_A \nu \exp(-\Delta G^*/RT) - C_B \nu \exp[-(\Delta G^* + \Delta G)/RT]. \quad (6\text{-}25)$$

When the forward and reverse reaction rates proceed equally, $r_{net} = 0$. Therefore,

$$C_B/C_A = \exp(\Delta G/RT) \qquad [\text{or } K_{eq} = C_B/C_A = \exp(-\Delta G^0/RT)], \qquad (6\text{-}26)$$

where $\Delta G = G_A - G_B$. But in thermochemistry we have defined free energy changes as the difference between product and reactant energies, that is, $\Delta G^0 = G_B - G_A = -\Delta G$. Substitution yields the familiar equation of chemical equilibrium shown in brackets in Eq. 6-26, and noted in Eq. 5-4.

6.6.2. ΔG Small

Returning to Eqs. 6-23 and 6-24 we note that for a general reaction between states 1 and 2

$$r_{net} = r_{12} - r_{21} = \nu \exp(-\Delta G^*/RT)[1 - \exp(-\Delta G/RT)]. \qquad (6\text{-}27)$$

If it can be assumed that ΔG is small compared with RT, a series expansion of the term in brackets yields $1 - \exp(-\Delta G/RT) \approx 1 - (1 - \Delta G/RT + ...)$, or $\Delta G/RT$. But, noting that $\Delta G = Fa$,

$$r_{net} = \nu\, [\exp(-\Delta G^*/RT)]\, \Delta G/RT = \nu[\exp(-\Delta G^*/RT)]aF/RT. \qquad (6\text{-}28)$$

To see whether $\Delta G/RT < 1$ is physically justified we note the following:

1. For diffusion in a chemical concentration gradient, $\Delta G \sim 10^3$ J/mol.
2. For solidification (with a small degree of supercooling), $\Delta G \sim 10\text{--}100$ J/mol.
3. For crystallographic phase changes, $\Delta G \sim 1\text{--}10$ J/mol.
4. For grain and precipitate growth and sintering of powders, $\Delta G = 0.01\text{--}1$ J/mol.

On the other hand, the value of RT at 1000 K is about 8300 J/mol and thus, in general, $\Delta G/RT < 1$. This means that Eq. 6-28 describes the processes noted reasonably well. Upon multiplication of both sides by a, the left-hand side is converted to the atomic velocity V, as V (m/s) $= a$ (m) $\cdot r_{net}$ (s^{-1}).

Finally, after collecting all terms we have basically derived the **Nernst–Einstein equation** as

$$V = ar_{net} = [a^2\nu \exp(-\Delta G^*/RT)]F/RT \qquad \text{or} \qquad V = DF/RT. \qquad (6\text{-}29)$$

(In Section 6.3.1 it was shown that the term in brackets is essentially the diffusivity D.) Larger values of D and F cause larger atomic velocities and displacements. The physical processes introduced above can be modeled by Eq. 6-29. In all cases small energy changes, corresponding to bond shifting, or breaking and re-forming in a single material, are involved. This contrasts with chemical reactions where high-energy primary bonds break down in reactants and re-form in products.

The Nernst–Einstein equation is important because it quantitatively connects the energy gradient or force in the system to the atomic motion it induces. Unlike Newton's laws of motion in bulk mechanical systems, atoms do not accelerate under a constant force, but migrate at constant velocity. Also, motion is very much dependent on temperature now and, to a first approximation, is independent of atomic mass. Because a slightly biased atomic hopping or diffusion between lattice sites occurs, the presence of D in the formula is not unexpected.

6.6.3. Application to Sintering and Grain Growth

6.6.3.1. Sintering

This chapter concludes with applications of the above analyses to two important processes used to fabricate materials and control grain size. The two are usually independent but in ceramics they are often related. Sintering, an important part of the powder metallurgy process to produce shaped parts (discussed in Chapter 8), involves heating a compact of pressed powders. The purpose is to densify it through elimination of porosity at particle contacts as illustrated in Fig. 6-26A. Fine powders of all classes of materials are amenable to sintering processes. What is significant is that consolidation and densification occur without resorting to melting. Rather, temperatures of only ~0.5 to $0.8T_M$ (K) are required. In this process necks form where particles meet, as seen in Fig. 6-26B. Let us therefore consider identical particles of a pure metal in contact and heat them to temperatures where mass transport readily occurs. It was demonstrated in Section 5.7.1 that atoms on convex material surfaces have a higher chemical potential or *effective concentration* than atoms on concave surfaces. Therefore, atoms have a tendency to be transported from the (convex) spherical surface to the narrow crevice at the (concave) interface between particles. As the circular neck volume grows in size, the total surface area (or surface energy) of both particles is reduced in accordance with thermodynamic dictates (see Section 5.7.1).

Although the macroscopic energetics of sintering has been accounted for, the exact mass transport mechanism remains uncertain. Have atoms diffused through the bulk lattice, along the surface, or through grain boundary paths to fill in the neck region? Alternatively, have the particles physically flattened by viscous flow the way glass does at elevated temperatures? Or have atoms simply thermally evaporated from the particle surface only to deposit at the neck? And what difference does it make? Actually, knowledge of the detailed atomic transport mechanisms and process variables is of considerable practical importance because it enables more efficient control of sintering and greater product yield.

The early time kinetics of neck radius (r_N) growth for the different transport mechanisms during sintering depends on the powder particle radius (r_p), temperature, and time (t). By equating the volume increase in the neck to the involved mass transport, theoretical considerations have suggested the relationship

$$(r_N)^n = K r_p^m t. \tag{6-30}$$

K is a temperature-dependent constant and values of n and m for the different transport mechanisms are entered in Table 6-2. Different transport mechanisms may dominate sintering at specific times and temperatures. For example, surface

A

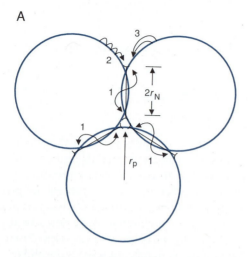

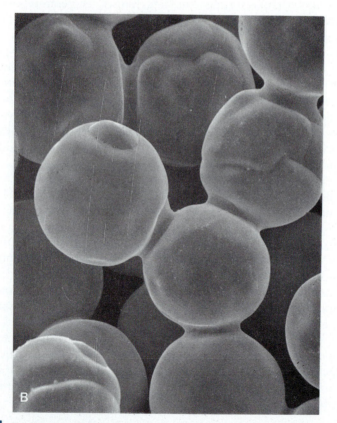

(A) Sintering of spherical particles illustrating mass transport filling of pores by (1) bulk diffusion, (2) surface diffusion, and (3) evaporation–condensation mechanisms. Bulk diffusion of atoms out of necks into pores reduces the interparticle distance and results in shrinkage. (B) SEM image of 33-μm-diameter Ni spheres sintered at 1030°C for 6 hours. Courtesy of Professor R. M. German, Pennsylvania State University.

TABLE 6-2	SINTERING MECHANISMS		
Transport mechanism		**n**	**m**
Bulk diffusion		5	2
Grain boundary diffusion		6	3
Surface diffusion		7	3
Viscous flow		2	1
Evaporation–condensation		3	1

diffusion may occur initially only to be supplanted by bulk diffusion later in the process. To determine values for n and m, the sintering geometry of individual spherical particles to flat plates has been measured. It is much easier, however, to measure the shrinkage or volume reduction in a standard rectangular bar. In such a case shrinkage is directly proportional to the ratio of the sintered neck area to particle cross-sectional area, $(r_N/r_p)^2$. With a bit of algebra, Eq. 6-30 is transformed to

$$\text{shrinkage} \sim (r_N/r_p)^2 = K^{2/n}r_p^{2(m-n)/n}t^{2/n}. \tag{6-31}$$

Experimental evidence for the sintering of Al_2O_3 powders is shown in Fig. 6-27, where the slope of the displayed lines is equal to $2/n$.

6.6.3.2. Grain Growth

Grains in solids tend to grow larger when heated to high temperatures for long times. If an energy is ascribed to grain boundaries, then the likely driving

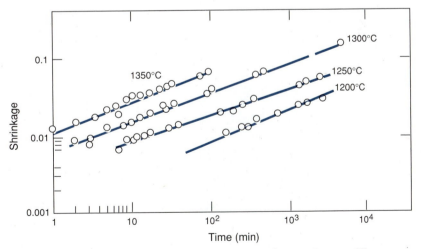

| FIGURE 6-27 | Fractional shrinkage in length of Al_2O_3 powder compacts as a function of time at different temperatures. From R. L. Coble and J. E. Burke, *Progress in Ceramic Science*, Vol. 3, Pergamon Press, Oxford (1963). |

force for this phenomenon is the reduction in interfacial area between grains. It is possible to estimate the kinetics of grain migration and growth through a simple application of formulas already introduced. Consider the curved grain boundary between neighboring grains in Fig. 6-28. Atoms on the boundary of convex spherical grain 1 have a higher chemical potential than otherwise identical atoms situated in concave grain 2. As a result the boundary moves toward the center of curvature. The driving force per mole is G/d, or μ/d per atom, where d can be taken as the grain boundary width (i.e., 0.5 nm). As the velocity of grain motion is $V = dr/dt$, by combining the Nernst–Einstein relationship (Eq. 6-29) and the definition of μ (Eq. 5-22),

$$\frac{dr}{dt} = \frac{2\gamma\Omega D}{dRTr}.$$

(6-32)

Simple integration after separating variables yields

$$r_\text{f}^2 - r_\text{i}^2 = \frac{4\gamma\Omega Dt}{dRT},$$

(6-33)

where r_f and r_i are the final and initial grain radii. Parabolic or $t^{1/2}$ kinetics of grain size growth are predicted in this case. Experimental grain growth data for brass (Cu–Zn), sintered Y_2O_3 doped with ThO_2, and sintered Al_2O_3 compacts are displayed in Fig. 6-29. It is interesting to note that irrespective of material or grain size, the growth laws are similar.

The phenomena of sintering and grain growth were treated separately, but both occur simultaneously to different extents. In ceramic powder compacts rapid densification through sintering occurs at low temperatures, but there is little grain growth. Paradoxically, at high temperatures grain growth is rapid but there is little densification.

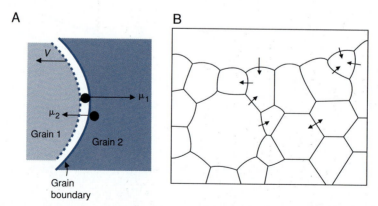

FIGURE 6-28   (A) Mechanism of grain boundary migration. When atoms on surfaces of convex grains jump into the neighboring concave grains, the grain boundary effectively moves to the left. (B) During grain growth, boundary segments move toward their center of curvature.

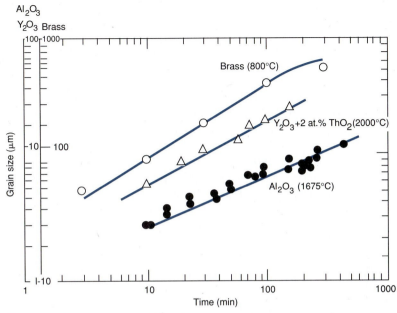

FIGURE 6-29 Grain growth kinetics in brass, thoria-doped yttria, and alumina. After J. E. Burke, R. L. Coble, P.J. Jorgensen, and R. C. Anderson.

6.7. PERSPECTIVE AND CONCLUSION

Atom movements and reactions occur in solids to drive them toward thermo-dynamic equilibrium. The resulting time-dependent mass transport processes produce changes in *chemical composition,* in *physical amounts,* and in the *geometry and size* of phases present.

Diffusion

Kinetic change in solids invariably occurs by diffusional mass transport. Atomic diffusion can occur in response to externally imposed concentration gradients or to internal composition variations due to inhomogeneous process-ing. The former is readily modeled by the diffusion equation together with appropriate initial and boundary conditions. Simple solutions in terms of com-plementary error and gaussian functions enable the concentration of the diffus-ing species to be determined as a function of position and time. There are two things to remember about diffusion:

1. The distance over which appreciable diffusion can be expected is $\sim\sqrt{4Dt}$.
2. The diffusion coefficient varies with temperature as $\sim\exp(-E_D/RT)$.

Phase Transformations

Phase transformations may be divided into several categories. The least complex involves a change in physical state, for example, solidification, with no alteration of phase chemical composition. Decomposition from a single phase to a mixture of two phases (e.g., austenite to α-Fe + Fe_3C), each with different composition and structural morphology, is at the other extreme of complexity. Concurrent heterogeneous nucleation and diffusional growth processes make phase transformations complex events. In the conversion of glass to glass–ceramics these effects are decoupled, enabling finer tuning of the transformation. The generic similarity of nucleation–growth phenomena in the crystallization of amorphous films and the transformation of steel should not go unnoticed. Similar transformations occur in the precipitation hardening treatment for nonferrous metals and in recrystallization effects that occur in previously deformed metals, two phenomena that are treated in Chapter 9. The Avrami equation provides a fruitful way to analyze the kinetics for *all* such nucleation–growth transformations.

Surface Energy Effects

The desire to minimize surface energy is the driving force to reduce surface and interfacial area and to alter surface geometry. But the mass transport driving force that induces geometric change is extremely small in practice, for example, typically a millionth of that which drives chemical reactions. We must be patient with solids. With long, high-temperature heat treatments sufficient mass transport occurs to cause fine particles to sinter or grains to grow in size.

There are a number of ways in which reactions in solids generally differ from those in liquids and gases. Atoms interfere much more effectively with each other's movements in solids than in liquids or gases. Therefore, diffusion coefficients are lowest in solids, highest in gases. As a result change will occur much more slowly in solids, an hour or more being a characteristic time. In the chemistry laboratory reactions commonly occur within seconds to minutes in liquids, but with explosive speed in gases. The time factor is of course dependent on temperature. Furthermore, the characteristic dimensions of space affected by reaction are generally much larger in gases and liquids than in solids. Convection phenomena in response to unavoidable density and temperature variations in fluids cause matter to be transported in swirls that spread atoms over large volumes. Such convective transport cannot occur in solids. Other differences stem from the heterogeneous nature of solid materials. The latter may contain several phases, precipitates, grain boundaries, and so on, each with a different degree of reactivity. Localized reactions may then be expected in solids and this is how they tend to degrade (e.g., corrode) in service.

Additional Reading

R. J. Borg and G. J. Dienes, *Solid State Diffusion*, Academic Press, Boston (1988).

D. A. Porter and K. E. Easterling, *Phase Transformations in Metals and Alloys*, 2nd ed., Van Nostrand Reinhold (UK), Berkshire (1992).

P. G. Shewmon, *Diffusion in Solids*, McGraw–Hill, New York (1963).

J. D. Verhoeven, *Fundamentals of Physical Metallurgy*, Wiley, New York (1975).

QUESTIONS AND PROBLEMS

6-1. Show that the gaussian diffusion profile (Eq. 6-5) is a solution to the non-steady-state diffusion equation (Eq. 6-2) by direct substitution and differentiation.

6-2. Phosphorus diffuses into the surface of a 1-μm-thick layer of silicon under *steady-state* conditions. At the source surface the concentration is 10^{20} P atoms/cm^3, whereas at the other surface the concentration is maintained to be zero. At 1060°C what is the atomic flux at the source surface? What is the atomic flux at the other surface?

6-3. Suppose steady-state diffusion of carbon occurred through a 0.1-mm-thick sheet of α-Fe at 800°C. If the carbon concentrations maintained at the surfaces of the sheet were 8 at.% and 2 at.%, what is the flux of carbon in units of C atoms/cm^2-s? *Hint:* First calculate the number of Fe atoms per cubic centimeter.

6-4. a. Write expressions for values of $D_\alpha(C)$ and $D_\gamma(C)$ in units of m^2/s with ΔE_D in units of kJ/mol.

b. At what temperature are both diffusion coefficients equal?

c. At the α-Fe to γ-Fe transformation temperature what is the ratio of $D_\gamma(C)/D_\alpha(C)$?

6-5. During the carburization of a gear, the complementary error function is used to model the diffusion. Do steady-state diffusion conditions prevail? Explain. Do steady-state diffusion conditions normally prevail during diffusion of dopants into a silicon wafer 0.05 cm thick? Explain.

6-6. Gases permeate solid materials of thickness δ with a steady-state flux J [cm^3 gas (STP)/cm^2-s] given by $J = -P_g \{\exp(-E_g/RT)\}(p^{1/2} - p_0^{1/2})/\delta$. Here p and p_0 are the pressures (atm) on either side of the solid, and P_g and E_g are constants. Consider a 20-cm-diameter, 140-cm-long steel cylinder with a 0.15-cm wall thickness containing hydrogen gas under a pressure of 150 atm. The tank is maintained at a temperature of 0°C.

a. What is the flux of hydrogen that flows out to the ambient through each square centimeter?

b. After 1000 hours what will the H$_2$ pressure in the tank be?

c. How will the answers to parts and and b change if the tank is maintained at 40°C? *Note:* $P_g = 2.9 \times 10^{-3}$ cm^3 (STP)/cm-s-atm$^{1/2}$ and $E_g = 8400$ cal/mol.

6-7. A continuous carbonitriding diffusion source introduces both carbon and nitrogen into steel surfaces simultaneously. Show that the ratio of the penetration depth of carbon (x_C) to that of nitrogen (x_N) is given by $x_C/x_N = [D(C)/D(N)]^{1/2}$. Evaluate the magnitude of the ratio at 500°C and at 900°C.

6-8. A reaction proceeds 10,000 times more rapidly when the temperature is raised from 800 to 1300°C.

   a. What is the reaction rate activation energy?

   b. If the temperature is decreased to 600°C, by what factor is the reaction rate decreased relative to that at 800°C?

6-9. Using the data plotted in Fig. 6-10B determine both D_0 (in units of cm²/s and m²/s) and ΔE_D (in units of kJ/mol, kcal/mol, and eV) for the diffusion coefficient of Be in GaAs.

6-10. It is desired to create a p–n junction 500 nm beneath the silicon surface by diffusing boron into a wafer initially doped with 10^{15} atoms of phosphorus/cm³. If the diffusivity of B in Si is 1×10^{-13} cm²/s, what continuous surface source composition is required for a half-hour diffusion treatment?

6-11. A 1010 steel is carburized by a source that maintains a surface carbon content of 2.0 wt% C. It is desired to produce a 0.80 wt% C concentration 0.05 cm below the steel surface after a 4-hour treatment. At what temperature should carburization be carried out?

6-12. Write an expression that approximates the diffusivity of Mo in a dilute BCC Fe alloy at any temperature.

6-13. A 50 : 50 copper–nickel alloy casting has an inhomogeneous structure in which the interdendritic spacing between Cu-rich and Ni-rich regions is 0.0015 cm. It is desired to homogenize the structure in an hour-long diffusional anneal.

   a. What effective atomic diffusivity is required?

   b. Estimate the annealing temperature needed.

6-14. a. During a 2-hour-long gas carburization of steel at 950°C the furnace temperature varied by ±3% of the set point. By what percent will the diffusion depth x vary if it is assumed that $x^2 = 4Dt$?

   b. If there is a ±3% variation in diffusion time, by what percent will the diffusion depth vary?

6-15. Whereas *carburization* leads to a desirable hardening of the steel surface, *decarburization* or loss of carbon softens the steel and is undesirable.

   a. Schematically sketch carbon concentration profiles during decarburization of a steel that initially contains a uniform 1% C composition. (For comparison, see Fig. 6-4A.)

   b. Suggest an equation involving error functions that quantitatively describes decarburization of this steel.

   c. Mention one possible way a steel can lose its surface carbon.

6-16. In face-centered cubic metals with melting point T_M (K), the following expressions describe the diffusivity of atoms along paths indicated by the following transport mechanisms:

Lattice or bulk diffusion	$D_L = 0.5 \exp(-17.0 T_M/T)$ cm²/s
Grain boundary	$D_b = 0.3 \exp(-8.9 T_M/T)$ cm²/s
Surface	$D_s = 0.014 \exp(-6.5 T_M/T)$ cm²/s.

a. For gold what is the activation energy for lattice diffusion?

b. What is the activation energy for grain boundary diffusion?

c. For which mechanism is the diffusivity highest at 1000°C?

d. For which mechanism is the diffusivity highest at 30°C?

6-17. At elevated temperatures copper atoms continually exchange places with vacancies in the copper matrix, undergoing self-diffusion.

a. Using data in Problem 6-16 calculate approximately how far Cu atoms diffuse in a second at 1000°C.

b. How far would Cu atoms diffuse along grain boundaries in the same time?

6-18. Suppose a finite surface source containing both phosphorus and lithium atoms is applied to the surface of a silicon wafer. Ten times as much P is present as Li and a diffusional anneal is carried out at 1200°C for 5 hours. After codiffusion, at what distance beneath the wafer surface will the concentration of P equal that of Li?

6-19. Calcium ions diffuse into silica with an activation energy of 53.7 kcal/mol, whereas for sodium ions the activation energy is 22.8 kcal/mol. If a constant source containing both ions is maintained at the surface of an integrated circuit encapsulated with silica, what is the ratio of penetration of Na^+ to Ca^{2+} ions after a 400°C treatment. Assume D_0 is the same for both ions.

6-20. During oxidation what thickness of silicon is consumed for every 100 nm of SiO_2 formed?

6-21. An activation energy of 330 kJ/mol was measured for the thermal oxidation of a chromium surface to Cr_2O_3. Does this activation energy correspond to the diffusional migration of Cr in Cr_2O_3 or O in Cr_2O_3?

6-22. Blocks of pure tin and lead are butted against one another to create a planar interface. This couple is heated to 175°C and held there for a sufficient time to promote solid-state diffusion.

a. Sketch the resulting diffusion profile across this couple.

b. Suppose the diffusion temperature were 200°C, which is below the melting points of both components. Sketch the diffusion profile now.

Hint: Consult the Pb–Sn phase diagram.

6-23. A specific formula describing reaction rate-limited growth of SiO_2 is $x_0 = 950[\exp(-2.0 \text{ eV}/RT)]t$ μm. Similarly, a formula describing diffusion-limited growth of SiO_2 is $x_0^2 = 5.9 \times 10^6 [\exp(-1.24 \text{ eV}/RT)]t$ μm^2. (*Note:* t = time in hours.)

a. Derive expressions for the *velocity* of oxide growth for both transport mechanisms.

b. For which growth mechanism is the velocity initially greater?

c. What is the ratio of diffusion to reaction rate growth velocities after a 10-hour oxidation at 1200°C?

6-24. Chemists use the rough rule of thumb that the rate of reaction doubles for every 10°C rise in temperature above 25°C. For this to be true what activation energy is involved?

6-25. a. During diffusion-limited SiO_2 growth the diffusivity is raised by a factor of 4. By what factor is the oxide thickness increased for the same oxidation time?

       b. By what factor must the oxidation time be increased at constant temperature to increase the oxide thickness by a factor of 3?

       c. Repeat parts a and b if reaction rate-limited growth prevails.

6-26. What is the ratio of the slopes of the two limiting lines characterizing oxidation of silicon (Fig. 6-6). Is this ratio consistent with linear and parabolic oxide growth rates?

6-27. A solid cube-shaped nucleus forms homogeneously within a liquid where the chemical free energy difference between the phases is ΔG_V and the surface energy is γ. Derive expressions for the critical nucleus size and energy barrier for nucleation.

6-28. A spherical nucleus forms during a solid-state transformation and it is determined that $\gamma = 0.2$ J/m^2 and $\Delta G_V = -6.5 \times 10^6$ J/m^3. Assume homogeneous nucleation.

       a. What is the radius of the nucleus of critical size?

       b. What is the magnitude of the nucleation energy barrier?

6-29. Why do new phases usually nucleate in the grain boundaries of the prior unstable grains during solid-state transformations?

6-30. The nucleation rate (nuclei/m^3-s) on gold particle catalysts within a certain glass–ceramic matrix at 800 K is observed to be $\dot{N} = 5 \times 10^{11}$ nuclei-m^{-3}. For this material, $\Delta G_V = -1.74 \times 10^{10}$ J/m^3, $\Delta E_D = 6.6 \times 10^5$ J/mol, $\gamma = 1.2$ J/m^2, and $\nu = 10^{13}$ s^{-1}. If ΔG_V is temperature independent over a small range, calculate $\dot{N}$ at 900 K. (*Hint:* Although this is a clear case of heterogeneous nucleation use the homogeneous nucleation formulas.)

6-31. TTT curves are not limited to steels. Sketch and label a schematic TTT curve for the transformation of amorphous to crystalline CoSi$_2$. Approximately where on this curve would the data shown in Fig. 6-18 lie?

6-32. Fused quartz tubing is often the enclosure for high-temperature systems used to process, anneal, or heat-treat materials. With use it is observed that the initially transparent tubing becomes opaque. Suggest a reason for this effect.

6-33. Catalysts effectively lower the activation energy barrier for gas-phase chemical reactions whose rates are speeded up as a result. In the process they are not used up.

       a. Represent the free energy variation as a function of reaction coordinate for a chemical reaction without a catalyst.

       b. Schematically indicate how the behavior of part a changes when a catalyst is used.

       c. In time the catalyst may become poisoned and lose some of its effectiveness. Sketch the behavior of a poisoned catalyst.

6-34. During solidification a free energy difference of $\Delta H_F \, \Delta T / T_M$ (Eq. 6-19) exists across an interfacial distance $\delta = 0.5$ nm between the solid and supercooled liquid. Silicon has a heat of fusion of 50 kJ/mol, melting point of 1420°C, and estimated atomic diffusivity of 10^{-11} m^2/s. At what approximate velocity can single crystals be pulled from the melt under a 1°C supercooling driving force? *Note:* In practice silicon crystals are grown at rates of several millimeters per minute.

6-35. Explain how forces that make macroscopic bodies move are different from forces that make the atoms within them move.

6-36. Data for the shrinkage of Al_2O_3 compacts during sintering are given in Fig. 6-27. What is the likely mass transport mechanism for sintering?

6-37. How closely do the grain growth data for brass (Cu–Zn), sintered Y_2O_3 doped with ThO_2, and sintered Al_2O_3 in Fig. 6-29 fit parabolic kinetics?

6-38. With reference to Fig. 6-27 calculate the activation energy for sintering of alumina by comparing shrinkages at different temperatures but at a fixed time (e.g., 100 min).

6-39. Suppose the diffusion coefficient that controls grain growth in brass has a value of 10^{-14} m^2/s at 800°C. What driving force, in units of J/mol-m, must be operative?

7 MECHANICAL BEHAVIOR OF SOLIDS

7.1. INTRODUCTION

Whether it was the need to support lofty cathedrals, make swords that did not break, or meet current performance standards in aircraft engines, the history of engineering is a continuous saga of grappling with the strength and mechanical properties of materials. No branch of engineering is immune from such concerns. In electrical engineering, for example, the packaging of microelectronic chips has raised a host of issues related to mechanical reliability and failure of substrates, metal solder joints, polymer boards, contacts, and connectors. Although some components may be minuscule in size compared with structures dealt with by civil or mechanical engineers, they are subject to the same limitations imposed on elastic and plastic phenomena.

This chapter is concerned with the many issues and facets related to the mechanical behavior of solids. At one extreme the subject matter is concerned with **elasticity** or deformation that is recoverable. In this regime materials deform (e.g., elongate or shorten) *linearly* with applied load. Elastic behavior forms the basis for virtually all structural and machine design. The reason we can be so confident of the mechanical integrity of our engineering structures and components is that elastic deformation phenomena are predictable to a high degree of accuracy. Design loads are such as to keep dimensional changes small, and upon unloading, the material springs back and regains its original shape without any apparent damage.

At the other extreme are **plasticity** effects induced at levels of loading above the limit of elastic response. Permanent deformation occurs during plastic loading and the material does not recover its shape upon unloading. Slight extension, large-scale stretching, and, finally, breakage into two or more pieces are stages in the way materials behave in the plastic regime. Plastic deformation effects are, unfortunately, *nonlinear*, depend on prior processing, and are not always predictable. Virtually all materials exhibit manifestations of plastic deformation under suitable conditions. Metals and polymers deform plastically to the greatest extent, and glasses and ceramics, to the least. Depending on treatment or operating conditions, the same material might stretch plastically, or it might fracture without warning.

The plastic regime is important to engineers for two main reasons. It enables an appreciation of (1) how components are manufactured and (2) how some of them will fail in use. Plasticity is capitalized upon in all sorts of materials forming or processing operations (e.g., rolling, extrusion, forging). Even structures designed to operate in the elastic range may stretch beyond original specifications and slowly change shape over extended periods of use, or may even fail altogether through fracture. Treatment of these two engineering aspects of plasticity is reserved for Chapters 8 and 10, respectively. The important issues of strengthening and toughening materials are the subject of Chapter 9. But first, this chapter introduces the mechanical response of all classes of engineering materials to various kinds of loading. How the deformation is manifested and evaluated on both macroscopic and microscopic levels is discussed.

7.2. ELASTIC BEHAVIOR

7.2.1. Elements of Elasticity

An appropriate way to start a discussion of elastic behavior is to define stress. When forces are applied to the surface of a body they act directly on the surface atoms. The forces are also transmitted to the internal atoms via the network of bonds that permeates the solid. To quantify the idea of stress consider the rectangular bar of initial length l_0 with a square cross section of area A_0 shown in Fig. 7-1A. If forces (F) are applied normal to the end surfaces, the bar stretches in tension by an amount Δl, which is known as the **displacement**. Simultaneously, the bar contracts in each of the two short dimensions. Referring to the indicated coordinate axes, the applied normal or **tensile stress** is defined by

$$\sigma_z = F/A_0 \quad \text{(in units of N/m}^2\text{, Pa, lbs}_f/\text{in}^2 \text{ or psi).} \qquad (7\text{-}1)$$

The corresponding normal or **tensile strain** is defined by

$$e_z = \Delta l/l_0 \quad \text{(in dimensionless units of m/m, cm/cm, in./in., etc).} \qquad (7\text{-}2)$$

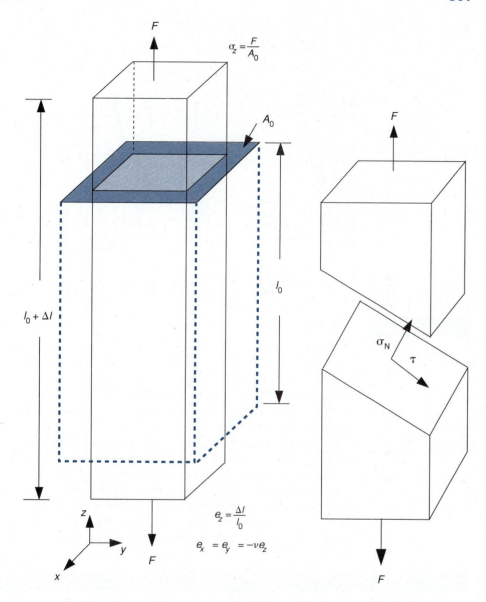

$$\sigma_z = \frac{F}{A_0}$$

$$e_z = \frac{\Delta l}{l_0}$$

$$e_x = e_y = -\nu e_z$$

A

B

FIGURE 7-1 (A) Tensile force applied to a bar (dashed) and resultant elastic deformation. (B) Arbitrary section revealing stress distribution through bar. Both tensile and shear stresses exist on exposed surface of free body. (C) Solid cube acted on by shear stresses and the resultant shear distortion.

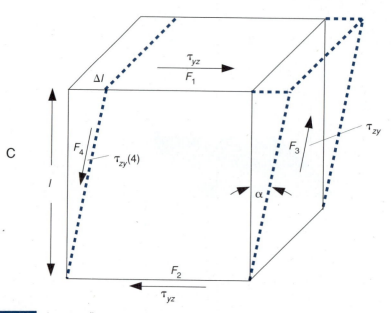

FIGURE 7-1   (*continued*).

Normal stresses σ_x and σ_y and corresponding strains e_x and e_y in the remaining two coordinate directions can be similarly defined. During elastic deformation the strain must be small; displacements (Δl), however, may be large depending on the size (l_0) of the loaded structure.

As the bar is in static equilibrium it can be arbitrarily cut to expose internal surfaces as indicated in Fig. 7-1B. The resultant force that must be applied to prevent the free body from moving can be resolved into forces that lie both normal and parallel to the cut surface. When divided by the area over which each force acts, a tensile stress (σ_N) arises from the former, whereas **shear stress** (τ) is generated by the latter. Thus, external forces acting on a body in equilibrium establish an internal state of stress throughout its volume.

EXAMPLE 7-1

If the tensile stress on the bar of Fig. 7-1 is $\sigma_z = 150$ MPa (21,800 psi), what is the magnitude of the shear stress on planes oriented 45° to the axis of the bar?

ANSWER   The applied force resolved in the plane is $F \cos 45°$ and the area over which the force acts is $A_0/\cos 45°$. The shear stress, defined as the ratio of the resolved force to the resolved area, is therefore given by $\tau = F/A_0$

$\times \cos^2 45° = F/2A_0 = \sigma_z/2$. After substitution, $\tau = 150/2 = 75$ MPa or 75 MN/m^2 (= 10,900 psi).

For this geometry the magnitude of the shear stress is half that of the tensile stress, a result that will be used again in Section 8.3.1.

Like tensile stresses, compressive stresses are the result of normal forces. But because they are directed *into* rather than *away* from the surface, convention assigns a negative sign to compressive stresses. For linear elastic deformation there is a relationship connecting the tensile or axial strains of the above bar to the lateral or transverse contractions (strains), namely, $e_x = -ve_z$ and $e_y = -ve_z$. The quantity v is a material constant known as **Poisson's ratio**, and for many substances it has a value of approximately 0.3. Note that even though no forces are applied in the x or y directions, there are, nevertheless, components of compressive strain in these directions.

Elastic shear stresses on the solid cube of Fig. 7-1C distort it into the shape shown. Note that a single shear force (F_1) on the upper surface directed in the positive y direction disrupts the mechanical equilibrium. An oppositely directed *equal* shear force (F_2) on the lower surface is required to restore the force balance in this direction. But now there is a tendency for the body to tip or rotate in the clockwise direction. Therefore, an oppositely directed couple in the form of antiparallel forces (F_3, F_4) must be applied to the remaining set of surfaces to achieve a balance of moments. Two subscripts are required to specify shear stresses. The first defines the direction of the force; the second indicates the direction normal to the surface of the plane on which the forces are applied. In Fig. 7-1C the shear stresses shown are τ_{yz} and τ_{zy}, and it is left as an exercise to show that when there is an equilibrium of moments, $\tau_{yz} = \tau_{zy}$. In more complicated three-dimensional states of stress τ_{xz} ($= \tau_{zx}$) and τ_{xy} ($= \tau_{yx}$) may also have to be considered. Shear stresses induce shear strains identified as γ_{yz} ($= \gamma_{zy}$) defined by $\gamma_{yz} = \Delta l/l = \tan \alpha$, where α is the distortion angle. As elastic strains are necessarily small, $\tan \alpha \approx \alpha$, and γ_{yz} can be effectively replaced by α.

7.2.2. Visualizing Stress States

It is useful to be able to recognize the types and orientations of stresses operative in structures or components subjected to the common loading conditions of Fig. 7-2. In each case a volume element is outlined and stresses are indicated. Examples of **tensile** and **compressive** loading in hooks and columns are shown in Figs. 7-2A and B. The metal casing of the gas cylinder shown in Fig. 7-2C is subjected to **biaxial** tensile loading under the influence of the pressurized gas. Hoop (σ_h) or circumferential stresses tend to stretch the diameter, whereas the longitudinal stresses (σ_l) act to extend the cylinder. The membrane of a blown-up spherical balloon is stretched in *balanced* biaxial tension (Fig. 7-2D) unlike the unbalanced biaxial tension of the previous gas tank.

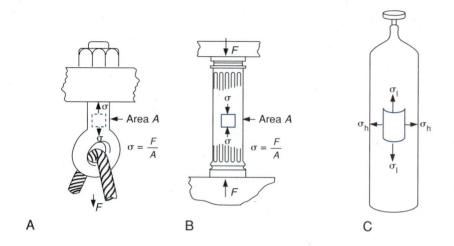

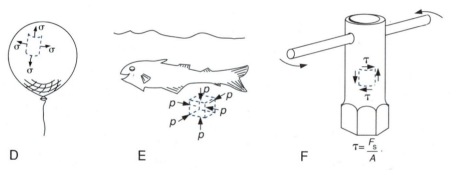

FIGURE 7-2   Visualizing common states of stress. (A) Simple tension. (B) Simple compression. (C) Unbalanced biaxial tension. (D) Balanced biaxial tension. (E) Hydrostatic pressure. (F) Shear stress. From M. F. Ashby and D. R. H. Jones, *Engineering Materials 1: An Introduction to Their Properties and Applications*, Pergamon, Oxford (1980).

Objects submerged in water are acted on by **hydrostatic** forces (Fig. 7-2E). Uniform pressure or compressive stresses act over the entire surface of the body, tending to compress its volume. The idea of uniformly densifying an object by shrinking its volume is an appealing way to improve material properties. Through hydrostatic extrusion and compaction in pressurized fluids and gases, the void volume in castings and powder compacts has been controllably reduced. The result is stronger materials that better resist cracking. Lastly, **shear** stresses are shown arising from applied torques or twisting forces (Fig. 7-2F).

7.2.3. Hooke's Law

The most important equation defining elastic phenomena is known as Hooke's law. It simply draws a linear proportionality between the strains produced in a body to the stresses that cause them, or between the stresses and the strains that produce them. In isotropic materials, Hooke's law is independent of direction and for normal components can be expressed by

$$\sigma_j = E e_j, \qquad (7\text{-}3)$$

where $j = x, y, z$. The quantity E is the important material constant known as **Young's modulus of elasticity.** (In Chapter 2 it was given the symbol E_y, but here Young's modulus is defined as E.) For the same strain, materials with larger values of E develop higher stress levels than materials having lower values of E. Alternatively, for the same applied stress the former materials deform less than the latter materials. In the case of combined stresses, a situation common in practice, strains in a given direction are additive. Therefore, Hooke's law can be expanded to the form

$$e_x = (1/E) \{\sigma_x - \nu (\sigma_y + \sigma_z)\}, \qquad (7\text{-}4)$$

and similarly for e_y and e_z.

For shear stresses, Hooke's law is appropriately written in the form

$$\tau_{xy} = G \gamma_{xy} \qquad \text{(similarly for other components)}, \qquad (7\text{-}5)$$

where G is known as the **shear modulus.** Of the three elastic constants that have been introduced, ν, E, and G, only two are independent, as they are related by elasticity theory: $E = 2(1 + \nu) G$. Therefore, only two material constants are necessary to specify elastic stresses and strains in materials whose properties are isotropic or the same in all directions.

7.2.4. Elastic Moduli

It was previously noted in Section 2.5 that the elastic modulus at the atomic level was proportional to the curvature of the interatomic potential-versus-distance curve or to the stiffness of the bonds. Since stiffer materials are associated with higher melting points (T_M), a correlation between E and T_M may be expected. On this basis, in order of increasing modulus, we have the polymers, metals, ceramics, and covalent solids. Values of Young's moduli and Poisson ratios for a number of materials are entered in Table 7-1, and for the most part materials follow these property trends. It must, however, be noted that there are exceptions to this ordering. Thus, E for tungsten, a BCC metal with a significant amount of covalent bonding, is greater than E for most ceramics and covalent materials; however, the latter two groups of materials are generally stiffer than iron, copper, aluminum, and their alloys. Towering

TABLE 7-1   ELASTIC PROPERTIES FOR SELECTED ENGINEERING MATERIALS AT ROOM TEMPERATURE[a]

| Material | Elastic modulus, E | | Shear modulus, G | | Poisson's ratio, ν | Density, ρ (Mg/m^3) | E/ρ (GN-m/Mg) |
	10^6 psi	GPa	10^6 psi	GPa			
Metals							
Aluminum alloys	10.5	72.4	4.0	27.6	0.31	2.7	26.8
Copper alloys	17	117	6.4	44	0.33	8.9	13.1
Nickel	30	207	11.3	77.7	0.30	8.9	23.2
Steels (low alloy)	30.0	207	11.3	77.7	0.33	7.8	26.5
Stainless steel (8–18)	28.0	193	9.5	65.6	0.28	7.9	24.4
Titanium	16.0	110	6.5	44.8	0.31	4.5	24.4
Tungsten	56.0	386	22.8	157	0.27	19.3	20
Ceramics							
Alumina (Al$_2$O$_3$)	53	390	—	—	0.26	3.9	100
Zirconia (ZrO$_2$)	29	200	—	—	—	5.8	34.5
Quartz (SiO$_2$)	13.6	94	4.5	—	0.25	2.6	36.2
Pyrex glass	10	69.0	—	—	—	—	—
Fireclay brick	14	96.6	—	—	—	—	—
Covalent							
Diamond	145	1000	—	—	—	3.51	285
Silicon carbide	65	450	—	—	—	2.9	155
Titanium carbide	55	379	—	—	—	7.2	52.6
Tungsten carbide	80	550	31.8	—	0.22	15.5	35.5
Plastics							
Polyethylene	0.058–0.19	0.4–1.3	—	—	0.4	0.91–0.97	~0.96
PMMA	0.35–0.49	2.4–3.4	—	—	—	1.2	~2.4
Polystyrene	0.39–0.61	2.7–4.2	—	—	0.4	1.1	~3.2
Nylon	0.17	1.2	—	—	0.4	1.2	~1
Other materials							
Concrete–cement	6.9	45–50	—	—	—	2.5	~19
Common bricks	1.5–2.5	10.4–17.2	—	—	—	—	—
Rubbers		0.01–0.1	—	—	0.49	—	—
Common wood (∥ grain)	1.3–2.3	9–16	—	—	—	0.4–0.8	~20
Common wood (⊥ grain)	~0.1	0.6–1.0	—	—	—	0.4–0.8	~1.3

[a] Property values for ceramics and polymers vary widely depending on structure and processing. Ranges and average values are given. Data were taken from many sources.

over all E values is the one for diamond. As diamond is also the hardest material known, one suspects a connection between these two properties. This subject is discussed further in Section 7.7.1.

The modulus values are basically structure *insensitive*, meaning that they are affected only slightly by small alloy additions, grain boundaries, dislocations, or vacancies. For larger alloying, phase and particle or fiber additions (as in composites) the modulus value is generally given by a complex weighted average of the moduli for individual constituents. The magnitude of the modulus is largest at the lowest temperatures. As the temperature increases, E falls slightly in a roughly linear fashion up to approximately $0.5 T_M$, and then more rapidly as T_M is approached. Just prior to melting, E is typically somewhat less than half its value at absolute zero. A decreasing modulus with increasing temperature was predicted in Section 2.5.

In isotropic polycrystalline materials the elastic moduli, weighted over the different directions, reduce to only two in number, E and G. But many materials are not isotropic. For example, single crystals with anisotropic atomic bonding have more than the two elastic moduli. The number of moduli rises proportionately as the unit cell increasingly deviates from cubic symmetry. Thus, polycrystalline iron has two, single crystal iron has three (i.e., E_{100}, E_{110}, and E_{111}), and crystalline quartz has six distinct elastic constants. With so many elastic moduli to contend with, the question of how to measure them arises. Young's modulus is most precisely measured by acoustic methods that rely on the fact that the speed of sound in a solid (v_s) is related to E and density (ρ) as $v_s \sim (E/\rho)^{1/2}$. In the popular pulse-echo technique an oscillator crystal is bonded to one surface of an oriented single crystal of the material in question. In radarlike fashion the oscillator both launches and detects a sound pulse after it reflects from the opposite crystal surface. An oscilloscope measures the time required to traverse the crystal length twice, enabling v_s to be determined.

7.3. PLASTIC DEFORMATION OF METALS

7.3.1. The Tensile Test

The tensile strength is the most widely recognized and quoted measure of a material's ability to support load. To obtain information on the tensile behavior of metals, bar or plate specimens are prepared in accordance with American Society for Testing and Materials (ASTM) standards as to shape and dimensions. The specimen is more massive at the ends and has a reduced section in between known as the *gauge length*. It is then gripped in a tensile testing machine (Fig. 7-3) and loaded in uniaxial tension until it fails. What is recorded is the instantaneous specimen load (F) and extension (Δl) as the crosshead moves down at a constant rate of straining, that is, $d \Delta l/dt = $ const.

7.3.1.1. Elastic Stresses and Strains

Since the centrally located gauge length (l_0, typically 5.08 cm) has the smallest specimen cross-sectional area, the stress is highest there. This defines the region where tensile properties are measured. The so-called **engineering stress** (σ_e) at any time during the test is related to the quotient of the instantaneous force and the original cross-sectional area of the specimen (A_0): $\sigma_e = F/A_0$. Correspondingly, the **engineering strain** (e) is defined, as previously, by $e = \Delta l/l_0$.

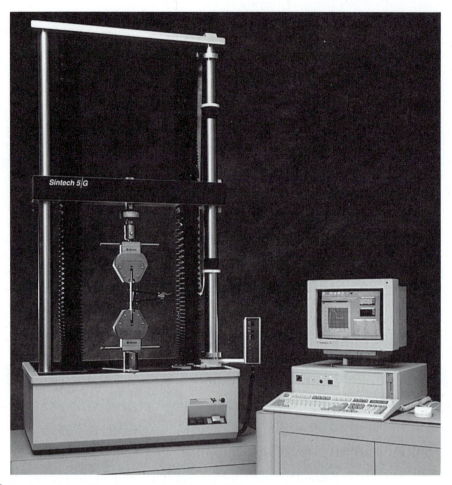

FIGURE 7-3 Tensile testing machine. Courtesy of P. S. Han, MTS Systems Corporation.

The *engineering* stress–strain tensile behavior of mild steel is shown in Fig. 7-4. At low applied load levels stress varies linearly with strain, and the proportionality or slope of the line (i.e., $\Delta\sigma/\Delta e$) in the elastic regime yields **Young's modulus**, E. The response is reversible; upon unloading, the original elastic curve is retraced. With further loading, however, the response is no longer linear, thus defining the onset of plastic deformation. Now there is much more strain per unit stress increment than before. Upon unloading now the load drop is parallel to the original elastic line and a permanent plastic strain offset occurs. The plastic regime has been entered.

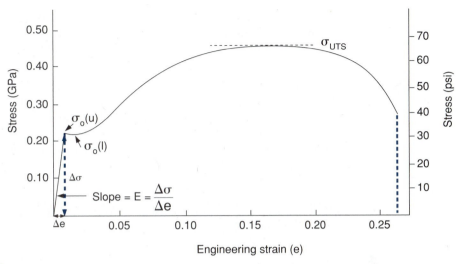

Engineering tensile stress–strain curve for mild steel.

7.3.1.2. Plastic Stresses and Strains

Just where elastic stresses end and plastic ones begin is taken as the **yield stress**, σ_0. In some materials the exact elastic–plastic transition is difficult to pinpoint. The stress level required to achieve 0.2% *plastic* strain (or $e = 0.002$) is then defined as the so-called **offset yield stress**. It can be obtained by drawing a line parallel to the elastic line at 0.2% strain and noting the stress level intersected on the stress–strain curve. Peculiar to carbon steel, however, is a marked upper yield point, $\sigma_0(u)$, followed by plastic flow at a constant lower yield stress level, $\sigma_0(l)$. These effects have been traced to the interaction of carbon and dislocations as described in Section 7.4.3.3. Following this, a region of nonlinear plastic behavior ensues, where the steel actually gets stronger with continued plastic strain. The specimen usually undergoes reasonably uniform plastic stretching in this range as the stress continues to rise.

The next notable happening is that the load reaches a maximum value F_{max} resulting in the **ultimate tensile stress or strength** ($\sigma_{UTS} = F_{max}/A_0$). Continued loading reveals a surprising drop in load, and shortly thereafter the specimen breaks, terminating the test. The fact that the specimen breaks at a less than *maximum load* does not mean that it has not broken at the *maximum stress* level. Associated with the drop in engineering stress after σ_{UTS} has been reached, it is observed that the tensile specimen has undergone localized **necking** or thinning. When the fractured halves are examined it is clear that the final cross-sectional area (A_f) is less than A_0. If the load at fracture (F_{frac}) were divided by A_f, the **true** ultimate tensile stress would exceed σ_{UTS} (i.e., $F_{frac}/A_f > F_{max}/A_0$). Necking or thinning is an undesirable feature displayed by metals during plastic forming operations (e.g., deep drawing) and is discussed again in Section 7.3.3.

7.3.1.3. Ductility

This important property of interest is also provided by the tensile test. Measures of ductility include percentage elongation and percentage reduction in cross-sectional area ($\% RA$). These are respectively defined by

$$\% \text{ elongation} = \frac{l_f - l_0}{l_0} \times 100; \qquad \% RA = \frac{A_0 - A_f}{A_0} \times 100, \qquad (7\text{-}6)$$

where the subscript f refers to final dimensions. Although l_f can usually be determined by fitting together the broken halves of a tensile bar, A_f is not measured as easily.

7.3.1.4. Resilience and Toughness

A couple of additional mechanical properties that are defined with reference to the area under a stress–strain curve are resilience and toughness. Both properties have the same units of stress times strain or (force/area) × (distance/distance) or (energy/volume), that is, *energy density*. Resilience is a measure of the ability of materials to absorb and release *elastic* strain energy. The modulus of resilience, E_R, a material property, is defined as the area under the *elastic* portion of the stress–strain curve. This triangular area has a value of $\frac{1}{2}\sigma_0 e_0$, where e_0 is the elastic strain at the yield stress. Alternatively, making use of Eq. 7-3,

$$E_R = \int_0^{\sigma_0} \sigma \, de = \frac{1}{E} \int_0^{\sigma_0} \sigma \, d\sigma = \sigma_0^2/2E \qquad \text{or} \qquad \tfrac{1}{2} E e_0^2. \qquad (7\text{-}7)$$

Toughness is defined as the area under the *entire* stress–strain curve (elastic + plastic) up to fracture and is, therefore, also a measure of ductility. A material's toughness reflects its ability to absorb energy and depends on test conditions (e.g., strain rate) and defects within the specimen (e.g., notches). Ceramic materials usually have high resilience (because of large σ_0) but low toughness because the plastic region is severely restricted. Metals are tough because of the generally large plastic region, but may have low resilience. As it is such an important goal to toughen engineering materials, much of Chapter 9 is devoted to this subject.

7.3.1.5. Fracture Surfaces

Lastly, there is the appearance of the tensile bar fracture surfaces. A typical cup–cone fracture for ductile metals is shown in Fig. 7-5A, where the 45°

FIGURE 7-5 (A) Cup-cone fracture in mild steel. (B) Brittle fracture of steel. Courtesy of G. F. Vander Voort, Carpenter Technology Corporation. (For part C, see color plate.)

angular lip at the specimen circumference indicates that shearing stresses play a role here. On the other hand, the fracture surfaces of brittle materials (Figs. 7-5B and C) display no evidence of shear. To confound matters, the same steel can exhibit either ductile or brittle fracture depending on the test temperature and treatment history. Though not easily quantified like the other tensile properties noted above, the morphology or appearance of the fracture surface is nevertheless extremely important. It often provides clues as to the origin and type of failure suffered by components and structures; this important subject is addressed again in Section 10.5.

7.3.2. True Stress–True Strain Behavior

For reasons that will become apparent, it is more desirable to specify tensile stress and strain values that are tied to instantaneous area (A_i) and length (l_i) dimensions rather than to initial dimensions that change during the test. Therefore, rather than the *engineering stress* (σ_e)–*engineering strain* (e) behavior of the previous section, we now recast tensile behavior in terms of *true stress* (σ_t)–*true strain* (ε) variables. These are respectively defined by

$$\sigma_t = F/A_i \tag{7-8}$$

$$\varepsilon = \int_{l_0}^{l_i} dl/l = \ln(l_i/l_0). \tag{7-9}$$

Since $l_i = l_0 + \Delta l$, $\varepsilon = \ln(1 + \Delta l/l_0)$ or $\varepsilon = \ln(1 + e)$. Thus the true strain can be calculated once the engineering strain is known. Furthermore, it is possible to express true stress in terms of initial dimensions plus measured engineering quantities. To do this we first note that the volume is preserved during deformation. Thus, equating the initial and instantaneous volumes, that is, $A_0 l_0 = A_i l_i$, and combining the previous relationships, $A_i = (A_0 l_0)/l_i = (A_0 l_0)/(l_0 + \Delta l)$. Finally, $\sigma_t = F/A_i = (F/A_0)(1 + e)$ or $\sigma_t = \sigma_e(1 + e)$, a result that is strictly valid during uniform deformation (before necking).

True stress–true strain quantities enjoy a physical significance that the corresponding engineering stress–strain quantities lack. This is illustrated in the following examples.

EXAMPLE 7-2

a. A specimen of initial length l_0 is first plastically extended to $2l_0$. Then it is further stretched until the final length is $3l_0$. Calculate both the engineering and true strains after each stage of deformation and cumulatively.

b. The specimen is compressed from l_0 to $\frac{1}{2}l_0$. Again, compare both strains.

ANSWER   a. After the first deformation the engineering strain is $e_1 = (2l_0 - l_0)/l_0 = 1$, and after the second deformation it is $e_2 = (3l_0 - 2l_0)/2l_0 = \frac{1}{2}$. The total engineering strain is $e = (3l_0 - l_0)/l_0 = 2$, but the sum of the strains is $e_1 + e_2 = 1 + \frac{1}{2} = 1.5$.

After the first deformation the true strain is $\varepsilon_1 = \ln\frac{2}{1} = \ln 2$, and after the second deformation $\varepsilon_2 = \ln\frac{3}{2} = \ln 3 - \ln 2$. The total true strain is $\varepsilon = \ln\frac{3}{1} = \ln 3$. Note that $\varepsilon_1 + \varepsilon_2 = \ln 2 + \ln 3 - \ln 2 = \ln 3$. *Therefore, true strains are additive, while engineering strains are not.*

b. The engineering compressive strain is $e = (\frac{1}{2}l_0 - l_0)/l_0 = -\frac{1}{2}$, whereas the true compressive strain is $\varepsilon = (\ln\frac{1}{2})/1 = -\ln 2$. Thus, compressing a specimen to half its length is equivalent to stretching it to double its length, insofar as true strain is concerned. But this is not the case for engineering strains.

The following example will additionally illuminate the quantitative distinctions between true and engineering values of stress and strain in the tensile test.

EXAMPLE 7-3

The following load elongation data were recorded during the tensile testing of a copper bar measuring 0.505 in. (0.0128 m) in diameter, with a gauge length of 2 in. (0.0508 m).
a. What is the value of Young's modulus?
b. What is the yield stress?
c. What is the ultimate tensile stress?
d. What is the percentage elongation at fracture?
e. Calculate true stress–true strain values up to onset of necking.

Tensile test data				Derived stress–strain values					
Load		Elongation		Engineering stress (σ)–strain (e)			True stress (σ_t)–strain (ε)		
lb	N	in.	10^{-5} m	psi	MPa	in./in. or m/m	psi	MPa	in./in. or m/m
0	0	0	0	0	0	0	0	0	0
1700	7,550	0.001	2.54	8,500	58.6	0.0005	8,500	58.6	0
3340	14,800	0.002	5.08	16,700	115	0.001	16,700	115	0.001
3500	15,500	0.004	10.2	17,500	121	0.002	17,500	121	0.002
4200	18,600	0.10	254	21,000	145	0.05	22,000	152	0.049
5000	22,200	0.2	508	25,000	172	0.1	27,500	189	0.095
6200	27,500	0.4	1020	31,000	214	0.2	37,200	256	0.182
7200	32,000	0.6	1520	36,000	248	0.3	46,800	322	0.262
7800	34,600	0.8	2030	39,000	269	0.4	54,600	376	0.336
7500	33,300	1.00	2540	37,500	258	0.5	↓ Necking to failure		
6000	26,600	1.05	2670	30,000	207	0.525			

ANSWER a. The first three data points correspond to elastic loading. As $E = \Delta\sigma/\Delta\varepsilon$, $E = (16,700 - 0)/(0.001 - 0) = 16.7 \times 10^6$ psi or 115 GPa.
b. The offset yield stress occurs (approximately) at a strain of 0.002. Therefore, the yield stress is 17,500 psi or 121 MPa.
c. The ultimate tensile stress is the stress level at the maximum load or 39,000 psi (269 MPa).
d. The percentage elongation to failure by Eq. 7-8 is $[(l_f - l_0)/l_0] \times 100 = [(2 + 1.05 - 2)/2] \times 100 = 52.5\%$
e. Calculated true stress–strain values are entered in the table above. As a sample calculation we consider the load = 6200 lb, elongation = 0.4 in. data point, for which $e = \Delta l/l_0 = 0.4/2 = 0.2$ in./in. or 0.2 m/m. As $\varepsilon = \ln(1 + e)$, $\varepsilon = \ln(1 + 0.2) = 0.182$. To calculate the true stress, the formula $\sigma_t = (F/A_0)(1 + e)$ given above can be used. Substitution yields $\sigma_t = (6200/0.2)(1 + 0.2) = 37,200$ psi or 256 MPa. (Note that $A_0 = \pi(0.505)^2/4 = 0.2$ in.^{2}

or 1.29×10^{-4} m^2.) Beyond necking σ_t and ε cannot be calculated because local test bar dimensions are unknown. Both engineering and true stress–strain curves based on these results are plotted in Fig. 7-6A.

7.3.3. Necking

The true stress–true strain curve exhibits a plastic loading response that rises continuously, reflecting strengthening or **work hardening** as the material strains. This behavior has a form that can often be approximated by a relatively simple mathematical equation. A power law expression for the stress–strain relationship,

$$\sigma_t = K\varepsilon^n, \qquad (7\text{-}10)$$

is often used, where K is a constant with units of stress, and n, a dimensionless constant, is known as the **strain hardening coefficient.** By plotting tensile test data in the form $\ln \sigma_t$ versus $\ln \varepsilon$, the value of n can be extracted. This is done in Fig. 7-6B for the data of Example 7-3 and the slope yields a value of $n = 0.5$. Values for K and n are given in Table 7-2 for a number of other metals, whose true stress–true strain behaviors are plotted in Fig. 7-7.

Let us consider a simple consequence of Eq. 7-10 by calculating the true strain at necking. Prior to necking the strain is uniform. In this regime as the load F rises, $dF > 0$, and the test bar strengthens through strain hardening. But then a local constriction occurs and the cross-sectional area decreases at a more rapid rate than the rate of hardening. When this happens the load that can be supported decreases ($dF < 0$). Thus, at the necking instability the tensile load goes through a maximum, an effect that can easily be described mathematically. Instantaneously, $F = \sigma_t A_i$, and at necking, $dF = 0$. Differentiation yields

$$A_i d\sigma_t + \sigma_t dA_i = 0 \qquad \text{or} \qquad d\sigma_t/\sigma_t = -dA_i/A_i. \qquad (7\text{-}11)$$

Physically, the rate of strain hardening precariously balances the rate of area contraction. By constancy of volume, $V = A_i l_i = $ constant. Therefore, $dV = 0 = A_i\, dl_i + l_i\, dA_i$, or $dA_i/A_i = -dl_i/l_i$. But, $dl_i/l_i = d\varepsilon$, and by Eq. 7-11, $d\sigma_t/d\varepsilon = \sigma_t$. This last equation holds only at the necking strain ε_n. Through simple substitution of Eq. 7-10, which holds for all values of strain including ε_n, we obtain $nK\varepsilon_n^{n-1} = K\varepsilon_n^n$. Therefore, the condition for necking is

$$\varepsilon_n = n, \qquad (7\text{-}12)$$

indicating that necking will initiate at a critical value of strain equal to n.

Thus, the stages of tensile deformation for ductile materials include the following:

1. Elastic strain, for ε ranging from 0 to $\ln(1 + \sigma_t/E)$.
2. Uniform plastic strain, for ε ranging from $\ln(1 + \sigma_t/E)$ to n (Here, $dF > 0$, and by differentiating as above, $d\sigma_t/\sigma_t > -dA_i/A_i$.)

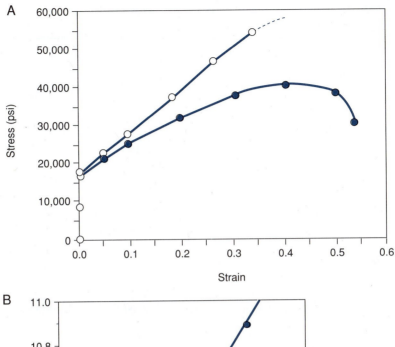

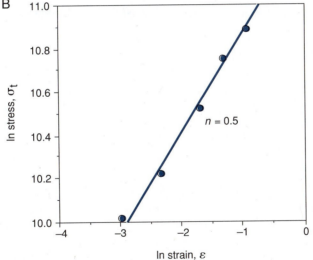

FIGURE 7-6 (A) Engineering and true stress–true strain behavior for annealed copper, ○, true; ●, engineering. Note true stress continues to rise until fracture. (B) Plot of ln σ_t versus ln ε. The slope of the plot yields the strain hardening coefficient n.

3. Nonuniform necking and thinning, for ε extending from n to the strain at failure. (After necking, $dF < 0$ and $-dA_i/A_i > d\sigma_t/\sigma_t$.)

An important characteristic of metals is that they become harder as they plastically deform in stages 2 and 3. For each increment of plastic strain the stress must be progressively increased. We say that metals work or **strain**

| TABLE 7-2 | TYPICAL VALUES OF K AND n (EQ. 7-10) FOR METALS AT ROOM TEMPERATURE |

Materials[a]	K (MPa)	n
Aluminum alloys		
1100, annealed	180	0.20
2024, hardened	690	0.16
6061, annealed	205	0.20
6061, hardened	410	0.05
7075, annealed	400	0.17
Brass (Cu–Zn)		
70–30, annealed	895	0.49
85–15, cold-rolled	580	0.34
Bronze, annealed	720	0.46
Cobalt–base alloy	2070	0.50
Copper, annealed	315	0.54
Molybdenum, annealed	725	0.13
Steel		
low-carbon, annealed	530	0.26
1045, hot-rolled	965	0.14
4135, annealed	1015	0.17
4135, cold-rolled	1100	0.14
4340, annealed	640	0.15
52100, annealed	1450	0.07
302 stainless, annealed	1300	0.30
304 stainless, annealed	1275	0.45
410 stainless, annealed	960	0.10

[a] Compositions of metal alloys are given in Tables 9–1 to 9–3.
From S. Kalpakjian, *Manufacturing Processes for Engineering Materials*, 2nd ed., Addison–Wesley, Reading, MA (1991).

harden. This property is evident because metals are straightened only with difficulty after they have been bent with ease. Far from being a disadvantage, work hardening is relied upon in all kinds of metal-forming operations like extrusion, drawing, and stretch forming. In these processes we want to prolong the region of uniform plastic strain and avoid, at all costs, the onset of necking that would lead to localized thinning and failure. Thus, a large n value is desirable, or, equivalently, a strongly work hardening material is required.

Tensile behavior of metals is very much a function of temperature as indicated in Fig. 7-8. Yield (as well as ultimate tensile) stresses are reduced at elevated temperatures because, as we shall see later, deformation involves thermally activated atomic flow; ductility is, however, enhanced at elevated temperatures. Both properties are capitalized upon in many mechanical forming operations, the former by reducing the loads required and the latter by enabling metals to stretch more without breaking.

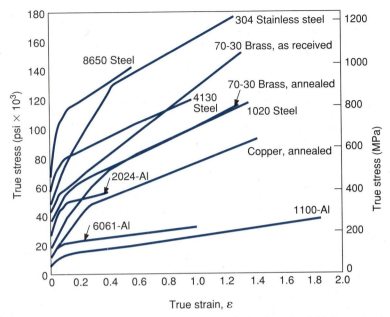

FIGURE 7-7 True stress–true strain curves for a number of metals. On the scale of this figure the elastic regions are not shown. Intercepts on the stress axis are the approximate yield stresses. Alloy compositions are given in Tables 9-1 and 9-2. After S. Kalpakjian, *Manufacturing Processes for Engineering Materials*, 2nd ed., Addison-Wesley, Reading, MA (1991).

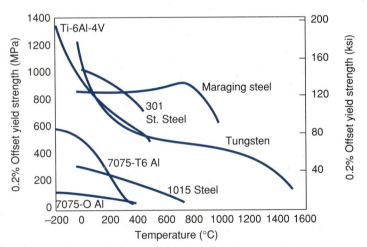

FIGURE 7-8 Temperature dependence of the yield stress for various metals. Note the following compositions in wt%: Maraging steel—18 Ni, 8 Co, 5 Mo, rem Fe; 301 stainless—17 Cr, 7 Ni, rem Fe; 1015 steel—0.15 C, 0.5 Mn, rem Fe; 7075—5.6 Zn, 2.5 Mg, 1.6 Mg, rem Al. O refers to softened, and T6 to heat-treated conditions. From *Metals Handbook,* Vol. 8, 9th ed., American Society of Metals, Metals Park, OH (1985).

7.3.4. Deformation of Single Crystals

The discussion to this point has referred to polycrystalline materials. No reference was made to the microscopic nature (e.g., atomic structure, grain structure, size and crystallographic orientation, dislocation and grain boundary distribution) of the involved material. To isolate the intrinsic lattice deformation characteristics from the complicating influence of grain boundaries, deformation studies on large single crystals have been performed over the years. The first step of course is to acquire or grow a single crystal, a process described in Section 6.4.3.

When such crystals (Fig. 7-9A) are pulled in tension, deformation occurs with the formation of numerous visible lines or surface markings known as **slip bands**. A high-magnification look at the surface topography reveals that the bands are, in turn, composed of microstep displacements or slip lines that form only on specific crystallographic planes known as **slip planes**. These

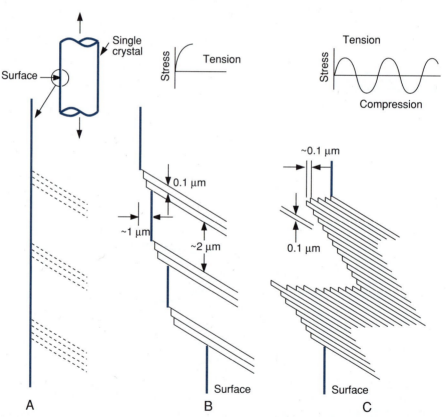

FIGURE 7-9 (A) Incipient slip lines on single-crystal surface prior to plastic deformation. (B) Slip bands on the surface of a crystalline solid plastically deformed in tension. (C) Slip band structure produced by alternating tensile and compressive stresses. Unlike case B, microcrack-like surface intrusions form.

closely packed planes are stepped perhaps a hundred nanometers apart, and displaced from one another some ten times as much as shown in Fig. 7-9B. With greater levels of plastic deformation, the density of both steps and bands rises. It is as though the single crystal behaved like a deck of cards that sheared under load to create a stepped terrain on the surface.

When the applied loading is cyclic or alternately tensile and compressive as indicated in Fig. 7-9C, slip bands can extend outward as well as inward, leaving incipient surface cracks behind. These processes are addressed again in Section 10.7 in connection with fatigue effects.

What is unique about the microscopic details of plastic deformation phenomena in crystalline solids is the following:

1. Slip occurs through the action of *shear stresses* on specific crystallographic planes. The planes that glide under stress are usually ones of densest atomic packing (high APD) and largest interplanar spacing.
2. Slip planes are displaced along the specified crystallographic directions that lie in them. These slip directions are invariably the ones of highest linear atomic density (LAD). The slip plane and slip direction combination constitutes the **slip system**, and predominant slip systems are listed for various metals and nonmetals in Table 7-3.
3. Plastic deformation initiates on the slip plane when the resolved shear stress reaches a critical value along the most favorably oriented slip direction.
4. These same properties hold for plastic deformation by compressive loading.

What is meant by the resolved shear stress can be understood from Fig. 7-10. A cylindrical single crystal of cross-sectional area A is loaded uniaxially

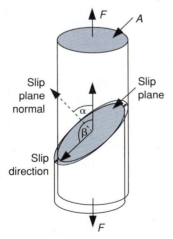

FIGURE 7-10   Activated slip plane and slip direction in a single crystal loaded uniaxially in tension.

TABLE 7-3   SLIP SYSTEMS IN CRYSTALS

Structure and material	Slip plane	Slip direction	Number of slip systems	τ_{CRSS} (MPa)	Slip geometry
FCC	$\{111\}$	$\langle 1\bar{1}0 \rangle$	$\{4\} \times \langle 3 \rangle = 12$		
Ag (99.99%)				0.58	
Cu (99.999%)				0.65	
Ni (99.8%)				5.7	
Diamond cubic	$\{111\}$	$\langle 1\bar{1}0 \rangle$	$\{4\} \times \langle 3 \rangle = 12$		
Si, Ge					
BCC					
Fe (99.96%)	$\{110\}$	$\langle 1\bar{1}1 \rangle$	$\{6\} \times \langle 2 \rangle = 12$	27.5	
Mo	$\{110\}$	$\langle 1\bar{1}1 \rangle$		49.0	
HCP					
Zn (99.999%)	(0001)	$[11\bar{2}0]$	$\{1\} \times \langle 3 \rangle = 3$	0.18	
Cd (99.996%)	(0001)	$[11\bar{2}0]$		0.58	
Mg (99.996%)	(0001)	$[11\bar{2}0]$		0.77	
Al_2O_3, BeO	(0001)	$[11\bar{2}0]$			
Ti (99.99%)	$(10\bar{1}0)$	$[11\bar{2}0]$	$\{1\} \times \langle 3 \rangle = 3$	13.7	
Rocksalt	(110)	$[1\bar{1}0]$	$\{6\} \times \langle 1 \rangle = 6$		
LiF, MgO					

After H. W. Hayden, W. G. Moffatt, and J. Wulff, *The Structure and Properties of Materials,* Vol. III, Wiley, New York (1965).

in tension by force *F*. By Example 7-1 we know that tensile forces induce shear stresses. With respect to the loading axis the normal to the slip plane makes an angle α, whereas the slip direction makes an angle β. Thus, the component of the force *resolved* in the slip plane along the slip direction is shearing force $F \cos \beta$, and the area over which it acts is $A/\cos \alpha$. The resolved shear stress τ_{RSS} is, by definition, the ratio of the two quantities, or

$$\tau_{RSS} = (F \cos \beta)/(A/\cos \alpha) = (F/A) \cos \alpha \cos \beta, \qquad (7\text{-}13)$$

where F/A is the applied tensile stress. When τ_{RSS} just initiates plastic flow, it assumes a value known as the **critical resolved shear stress,** τ_{CRSS}. Values for it, entered in Table 7-3, are unique to particular materials for given slip systems.

EXAMPLE 7-4

A nickel single crystal grown along the [210] axis has an initial diameter of 1 cm. What force is required to induce the onset of plastic deformation in this crystal if it is pulled along the growth axis?

ANSWER   Table 7-3 indicates that slip occurs on (111) planes, in $[1\bar{1}0]$ directions, and that $\tau_{\text{CRSS}} = 5.7$ MPa. Making use of Eq. 3-2 and noting that the indices of the direction normal to the (111) plane are [111], we obtain

$$\cos \alpha = \frac{[(1)(2) + (1)(1) + (1)(0)]}{\sqrt{1^2 + 1^2 + 1^2} \cdot \sqrt{2^2 + 1^2 + 0^2}} = \frac{3}{\sqrt{15}}$$

$$\cos \beta = \frac{[(1)(2) + (-1)(1) + (0)(0)]}{\sqrt{1^2 + (-1)^2 + 0^2} \sqrt{2^2 + 1^2 + 0^2}} = \frac{1}{\sqrt{10}}.$$

Therefore, from Eq. 7-13,

$$F = \tau_{\text{CRSS}} A / \cos \alpha \cos \beta = \frac{5.7\pi (0.01)^2}{4 (3/\sqrt{15}) \cdot (1/\sqrt{10})}$$

$$= 1.83 \times 10^{-3} \text{ MN}.$$

7.3.5. Twinning

In addition to slip, twinning is another mechanism that helps to account for plastic deformation in metals, intermetallic compounds, crystalline polymers, and ceramic materials. During twinning, atoms undergo a coordinated shear displacement in the **twin direction** such that the deformed lattice is brought into a mirror-image orientation relative to the undeformed lattice. Each atom is displaced slightly relative to its neighbors, but the overall macrostrain may be considerable. Twinning enables orientations favorable for slip to be exposed, enabling further, larger amounts of deformation by the latter mechanism. In twinning, two mirror or **twin planes** bound the deformed matrix in between as shown in Fig. 7-11. Thus, although it is possible to polish slip steps away removing traces of plastic flow, twin boundary lines (T_1 and T_2) on the surface cannot be so eliminated because they reflect changes in the bulk crystallography. Twin planes and directions in BCC metals are (112) and [111], respectively; the reverse, namely, (111) and [112], is the case in FCC metals.

Whether metals deform by slip or twinning depends on the stress level required to activate each mechanism. This, in turn, is dependent on a number of intrinsic (e.g., crystal structure, number of independent slip systems) and extrinsic (e.g., impurities, temperature) factors. Deformation twinning is not normally observed in FCC metals because the stress required for slip is so much lower. In BCC metals slip is generally operative at elevated temperatures

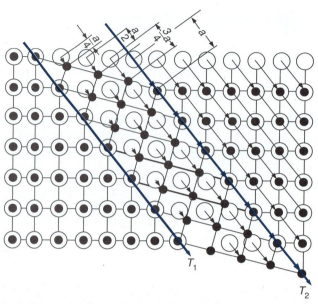

Schematic view of atom displacements during twinning. Open circles represent the perfect atomic lattice and smaller black circles assume the atom positions of the twinned lattice. T_1 and T_2 represent both the twinning planes and the direction of twinning.

but twinning may be favored at reduced temperatures. On the other hand, HCP metals exhibit extensive twinning behavior because so few slip systems are available.

Twins possessing the same crystallography described above are also commonly observed in metals like copper (see Fig. 5-13) and brass (see Fig. 3-32A). They originate not from deformation, but from heating deformed metals to elevated temperatures, and are known as **annealing twins**.

7.4. ROLE OF DISLOCATIONS

7.4.1. Theoretical Stress for Deformation of Crystals

Dislocation or line defects, introduced in Chapter 3, play a central role in our understanding of the microscopic nature of plastic deformation in crystalline solids. Paradoxically, dislocations explain not only why materials are strong, but also why they are weak. To see why, consider a portion of a perfect lattice containing a slip plane as shown in Fig. 7-12. How much shear stress is required to initiate the minimum amount of plastic deformation, that is, a one-atom slip step on an initially flat crystal surface? A rigid displacement of the top half of the crystal relative to the bottom half that severs all of the bonds in the plane simultaneously is one way to achieve such slip. In such a

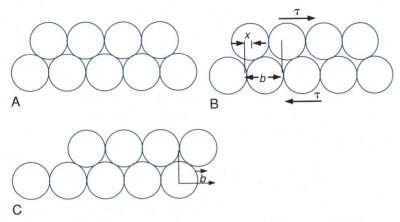

FIGURE 7-12 Model of rigid slip of crystal planes in a perfect lattice. (A) Atoms at rest. (B) Application of shear stress and displacement of atoms. (C) Atomic positions after a unit of slip.

mechanism, atoms on top collectively glide or conformally roll over those on the bottom until they nest in the next set of equilibrium sites. It takes force initially to stretch bonds elastically, break them plastically, and then re-form them between neighboring atoms. Because the process has the lattice periodicity, it makes sense to describe slip mathematically by the periodic function

$$\tau = \tau_0 \sin(2\pi x/b). \qquad (7\text{-}14)$$

Here τ and x are the shear stress and displacement, respectively, τ_0 is the maximum shear stress value attained, and b is the atom size (also Burgers vector). The quantity x/b is essentially the shear strain, and for small strain values, $\sin(2\pi x/b) \approx 2\pi x/b$. But, Hooke's law also holds for small strains so that

$$\tau = Gx/b = \tau_0\, 2\pi x/b. \qquad (7\text{-}15)$$

Through elimination of x/b the maximum shear stress is simply estimated to be $\tau_0 = G/2\pi$. If typical values for the shear modulus are substituted, the magnitude of τ_0 would be a factor of $\sim$100 to $\sim$1000 or more times larger than experimentally measured values of τ_{CRSS}. The conclusion that can be drawn is that plastic slip deformation does not occur by the rigid displacement of atomic planes in unison; such a mechanism requires exertion of too much force. Conversely, materials like thin whisker filaments contain no dislocations and exhibit the theoretical maximum strength because they can deform only by rigid displacement of neighboring planes.

7.4.2. Dislocation Model of Plastic Deformation

Dislocation defects provide a convenient model for slip deformation and explain why it can occur at relatively low stress levels. This is evident in Fig.

7-13, where an edge dislocation in an otherwise perfect crystal is acted on by shear stresses (τ) acting across opposing crystal faces. In the absence of load, the dislocation core atom forms stable bonds equidistant between the two neighboring atoms (**1** and **2**) below the slip plane; however, the applied forces upset this balanced state and the core atom moves slightly to shorten the forward bond (to **2**) and extend the trailing bond (to **1**) lengths. The latter bond breaks as the dislocation moves one atomic distance to the right, where it is again poised (between **2** and **3**) to repeat this stress-biased motion. Continued loading causes the dislocation to undulate in a wavelike manner, alternately breaking and re-forming bonds. The analogy to caterpillar motion is a vivid, self-explanatory image of dislocation motion. Finally, the dislocation reaches the crystal surface where it produces a step whose height is equal to Burgers vector, **b**. This is evidence that slip occurs, and once the dislocation leaves the crystal, a perfect lattice is left behind. Similarly, shear stresses applied to a crystal containing a screw dislocation cause it to glide on the slip plane. Much like a zipper opening, a slip step eventually appears on the surface parallel to **b**.

An appreciation of the diminished force required for dislocation motion can be gained by considering the problem of moving a large rug across a floor. If one edge is lifted and pulled, the remaining rug hugging the floor reluctantly moves, restrained by the weight-induced, rug–floor friction. The situation is like that of slip in the absence of dislocations. Raising a bump or high wrinkle near one edge is akin to introducing a dislocation in the rug. If the bump is pushed in while the edge is held fixed, the wrinkle moves to the adjacent region. With successive holding and pushing maneuvers the bump migrates to the opposite edge causing a net rug displacement. Successive incremental motion is a slow but steady way to translate the rug; the one-time pull is quick but requires lots of muscle.

Forces required for dislocation motion vary widely in different materials. In metals, dislocations navigate through a very mobile electron sea, nudging atoms aside slightly during passage. Any disruption of the electron–ion bonding charge distribution is instantly smoothed and the lattice is barely disturbed. That is why the stress required for dislocation motion in metals is only $\sim 10^{-4}$ to 10^{-3} E. In covalently bonded materials, however, dislocations have to break very strong bonds. Rather than a sea, dislocations can be imagined to move through a thick forest where passage requires that treelike bonds be uprooted. The stress for plastic flow is now $\sim 10^{-3}$–10^{-2} E. Ionic solids fall somewhere in between; during dislocation motion, ions of like charge periodically closely approach one another. Electrostatic repulsion then forces them apart, serving to discourage dislocation motion and plastic deformation. The combination of low dislocation densities and few slip systems in nonmetals makes them inherently less ductile.

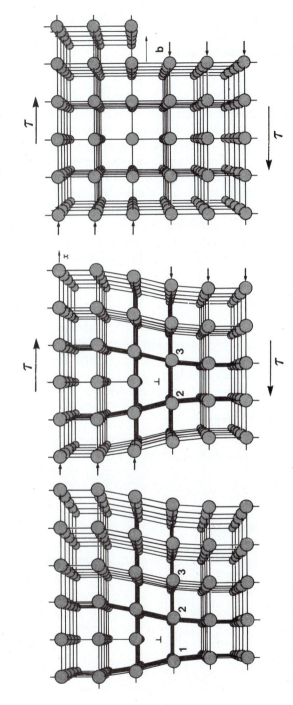

FIGURE 7-13 Atom movements associated with edge dislocation motion across a stressed crystal. When the dislocation emerges at the surface a step is produced. After A. G. Guy, *Essentials of Materials Science*, McGraw–Hill, New York (1976).

7.4.3. Properties of Individual Dislocations

7.4.3.1. Stress Field

Dislocation behavior and its influence on the properties of materials stem from an admixture of complex geometric and physical attributes. Although they are massless entities, dislocations surprisingly exhibit a host of dynamical effects. They move in response to applied forces and they interact with one another by attracting or repelling one another. One key to their mechanical behavior is the elastic stress field that surrounds them. The elastic stress field around edge dislocations has a complex spatial distribution. Above the slip plane the lattice is compressed along the axis of Burgers vector; below the slip plane the lattice is expanded.

In screw dislocations the deformation is pure shear and relatively easy to analyze because the shear strain (γ) and stress (τ) are simply given by

$$\gamma = b/2\pi r \tag{7-16a}$$

$$\tau = Gb/2\pi r, \tag{7-16b}$$

in cylindrical coordinates (Fig. 7-14). To derive these formulas we recall that a circular traverse of radius r around the screw axis results in a Burgers vector displacement of magnitude b (see Fig. 3-25B). The strain is therefore the elastic displacement (b) per circular length ($2\pi r$) over which the displacement occurs; the stress is G times as large as the strain.

7.4.3.2. Strain Energy

Following Eq. 7-7 we would expect the elastic **shear strain energy per unit volume** of a screw dislocation to be $\frac{1}{2}G\gamma^2$ or $\frac{1}{2}G(b/2\pi r)^2$. To roughly obtain

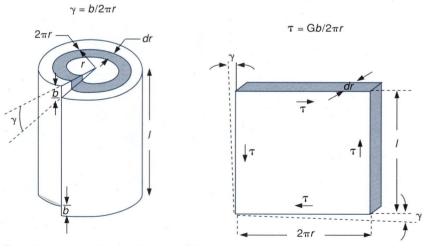

FIGURE 7-14   Elastic distortion associated with a screw dislocation.

the strain energy *per unit length* (E_d) of the dislocation line, we must multiply by the area (πr^2) surrounding the dislocation core or axis, yielding $\sim Gb^2/8\pi$. It is common to refer to E_d as the dislocation line energy and approximate its magnitude by $\sim Gb^2$. Importantly, E_d is proportional to the square of the Burgers vector.

Energy always seeks a minimum value, and in the case of dislocations this can be achieved by essentially reducing the magnitude of the Burgers vector, as well as the length of the dislocation line. Decomposition into two or more partial dislocations is a common mechanism for reducing strain energy. To see why let us consider slip for FCC metals on the (111) plane in the [110] direction as shown in Fig. 7-15. The shortest slip distance is half the cube face diagonal, a length of $a/\sqrt{2}$, that by definition is equivalent to the Burgers vector. Therefore, $b = (a/2)$ [110]. But physically, it appears that if slip occurred via a two-step zigzag process within atomic valleys, first along [$21\bar{1}$] and then along [121] directions, less energy would be expended than for [110] dislocation motion.

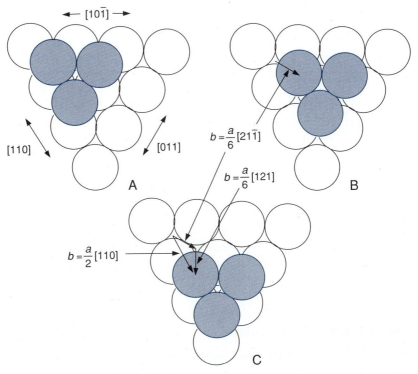

FIGURE 7-15 (A) Atomic view of slip along the [110] direction in the (111) plane of an FCC crystal. Activation of successive (B) a/6 [$21\bar{1}$] and (C) a/6 [121] partial dislocations facilitates plastic deformation.

EXAMPLE 7-5

Show that the dislocation defined by $b = (a/2)$ [110] is unstable relative to two partial dislocations of the type $(a/6)$ [211].

ANSWER   The dislocation reaction that must be considered is

$$(a/2)[110] \rightarrow (a/6)[21\bar{1}] + (a/6)[121]. \tag{7-17}$$

The energy of each dislocation is $\sim Gb^2$. Therefore, the energy of the $(a/2)$ [110] dislocation is $Ga^2 [1^2 + 1^2 + 0^2]/2^2$ or $Ga^2/2$. Similarly, the energy for each of the partial dislocations is $Ga^2/6$. The reaction of Eq. 7-17 is energetically favorable as written because $Ga^2/2 > 2Ga^2/6$.

7.4.3.3. Interaction with Impurity Atoms

Individual dislocations interact with impurity atoms in the lattice in an interesting way. Undersized atoms find it comforting to seek out and segregate to the relatively open space of the dislocation core where matrix atoms do not squeeze them. Elastic strain energy is reduced in this way. Carbon segregation to dislocations in iron is an example of this effect which has been identified to be the cause of the yield point phenomenon in mild steels. This feature is shown in the stress–strain curve of Fig. 7-4. If, soon after yielding, the steel is retested in tension, there is no upper yield point and no subsequent load drop. If, however, the specimen is aged for a long time at room temperature (or heated to a hundred degrees or so for a short time) and then retested, the yield point and load drop reappear!

The reason for this odd behavior has to do with carbon-dislocation interactions. Initially dislocations are decorated by carbon atoms, forming what are known as **Cottrell atmospheres** (so named for A. H. Cottrell, a prominent British metallurgist). Loading tears dislocations away from the restraining influence of the carbon atoms, causing a load drop. Now dislocation propagation occurs at the decreased stress levels of the lower yield point. Immediate retesting does not allow sufficient time for carbon to respond to the pull of the dislocations and diffuse interstitially to reestablish the atmospheres. But given sufficient temperature or time, carbon resegregates and the yield point reappears.

7.4.4. Behavior of Many Dislocations

7.4.4.1. Dislocation Multiplication

Until now we have considered the behavior of only isolated dislocations. There are examples where single or very few dislocations can have significant engineering implications. A dislocation, critically positioned at a transistor in an integrated circuit, can adversely affect device reliability by altering charge transport; however, typical mechanical behavior and phemomena require the collective response of immense numbers of dislocations that are distributed throughout the crystalline matrix. Dislocations generally enter crystalline solids during growth and subsequent processing, but their densities can be considerably enlarged through further plastic deformation. Mechanisms for generating dislocations have been proposed and experimentally observed. A popular one known as the Frank–Read source is depicted schematically in Fig. 7-16. An initial edge dislocation line pinned at its extremities is required. Impurity atoms, precipitates, grain boundaries, and intersecting dislocations can serve to hold the ends fixed. An applied shear stress in the slip plane then forces the dislocation line to bow into an arc. With continued stress, the loop opens, closes in on itself, and then pinches. Finally, a complete loop of new dislocation is generated, leaving a pinned length of edge dislocation behind to start the dislocation breeding process over again.

7.4.4.2. Dislocation Density

The dislocation density, ρ_d, a type of concentration, is measured by counting the number of dislocation lines that thread a unit area of surface (i.e., $\#/m^2$);

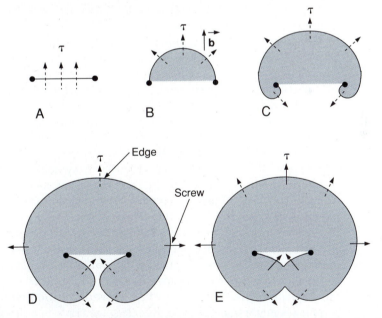

FIGURE 7-16 Sequence depicting dislocation multiplication by the Frank–Read mechanism. The pinned edge dislocation (A) acted on by the applied shear stress first bows into a semicircular loop (B). Portions of the loop parallel to Burgers vector have a screw dislocation character. Unlike edge components which travel *parallel* to **b**, screw dislocations glide *perpendicular* to **b**, and serve to expand the loop (C). Eventually the loop is pinched as screw dislocations of opposite sense (D) glide toward each other and annihilate one another (E).

ρ_d is also defined in terms of the total dislocation length per unit volume (i.e., m/m^3). Dislocations can be imaged in the electron microscope (Fig. 7-17) and made visible optically on crystal surfaces where they emerge (see Fig. 3-27). As a result, dislocation densities generated during growth totaling roughly $10^3/cm^2$ (or 10^3 cm/cm^3) are observed in undeformed ionic and covalent materials. In metals, however, densities of $\sim 10^8/cm^2$ are found. But the number increases greatly with plastic strain, and densities as high as $10^{12}/cm^2$ are not uncommon in heavily deformed metals. Dislocations on the surface of such a deformed metal are close enough for their individual stress fields to overlap.

7.4.4.3. Dislocation Interactions

With so many dislocations around they are bound to get in each other's way. This generally makes it more difficult for a particular dislocation to move. Restricted dislocation motion means that a higher stress level is required for further plastic straining. The material work hardens; it becomes harder and stronger but, simultaneously, less ductile. This phenomenon is observed whenever metals are plastically deformed by any suitable means, for example, by rolling, hammering, tensile stretching beyond the elastic regime. Such cold deformation or working of metals is an important way to shape as well as strengthen them.

Much can be learned about such strengthening by understanding the nature of interactions between a pair of dislocations. In Fig. 7-18A two dislocations of the *same sign* lie on a slip plane. Additive superposition of overlapping stress fields further compress the lattice above the slip plane. This raises the strain energy above that which exists around individual dislocations. The system reacts to reduce the strain energy by forcing the dislocations apart. A cause of work hardening in metals is this repulsion between dislocations. To see why, consider a group of dislocations of the same sign gliding on a common slip plane until the leading one hits an obstacle and is immobilized. Grain boundaries are such obstacles because slip planes are not crystallographically continuous from grain to grain. Unable to switch slip planes, the trailing dislocations pile up and locally increase the strain energy. An effective barrier or **back stress** is established, repelling other dislocations from approaching the pileup as shown in Fig. 7-18C (and Fig. 7-17A). A higher load is now required to force dislocations to move and produce more strain; the material work hardens.

Continued loading may simply force many dislocations of the same sign to impinge on one another in pileups. By summing Burgers vectors on a single slip plane a kind of superdislocation is produced that may be thought of as an incipient microcrack (Fig. 7-19A). How dislocation motion on intersecting slip planes might nucleate a similar cleavage crack in a brittle solid is modeled in Fig. 7-19B.

When two dislocations of *opposite sign* share the same slip plane then strain energy is reduced through attraction (Fig. 7-18B). In this case they move toward one another until dislocation annihilation ultimately leaves the lattice defect

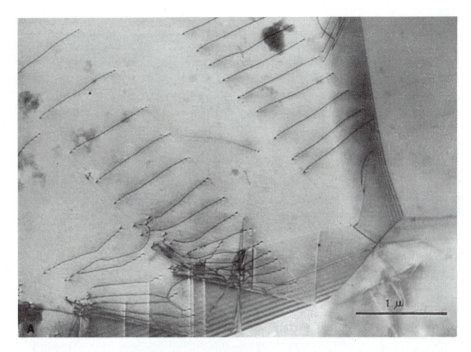

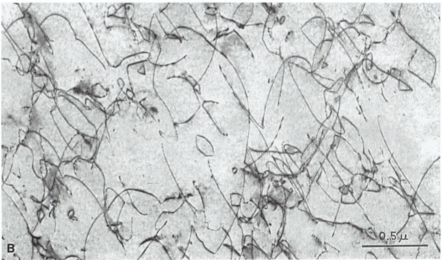

FIGURE 7-17 (A) Dislocation pileups in stainless steel imaged with a TEM operating at 1000 kV. Picture taken by G. Dupouy. (B) Dislocations in a 1.5-μm-thick foil of an Fe–Mn–N alloy imaged with a TEM operating at 800 kV. Picture taken by K. F. Hale and M. H. Brown.

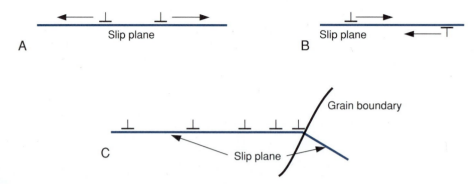

FIGURE 7-18   Interaction between edge dislocations lying in a slip plane. (A) Dislocations with parallel Burgers vectors repel one another. (B) Dislocations with antiparallel Burgers vectors attract and annihilate each other. (C) Pileup of dislocations at a grain boundary.

free. Rather than a horizontal array on a single slip plane, consider a number of dislocations distributed on a series of parallel slip planes. For dislocations of the *same sign* minimum strain energy calls for them to be vertically stacked. This is the configuration noted earlier (see Fig. 3-28) in grain boundaries of low crystallographic misorientation. One way dislocations can rise vertically is by the mechanism of **climb**. If atoms constituting the core diffuse away, then the dislocation effectively climbs one lattice spacing above the slip plane. Such a mechanism, further treated in Section 10.6, helps to explain the role dislocations play during elevated temperature creep of materials.

7.4.4.4. Dislocation Dynamics

Is there a way to deal with the seemingly intractable **microscopic** behavior of such a large number of dislocations? And can dislocation behavior be correlated to or even predict **macroscopic** plastic response and mechanical behavior? To address these knotty questions a dislocation dynamics approach has been adopted with some success. The spirit of the approach is analogous to the well-established science of statistical mechanics where the known properties of individual atoms are statistically summed according to the laws of physics; the objective is to model bulk material behavior.

A few of the key equations used in dislocation dynamics are relatively simple:

$$\dot{\varepsilon} = \tfrac{1}{2}\,\rho_d\,bv \tag{7-18}$$

$$\rho_d = \rho_0 + A\varepsilon \qquad (\rho_0,\, A \text{ are constants}) \tag{7-19}$$

$$v = v_0\tau^m \qquad (v_0,\, m \text{ are constants}) \tag{7-20}$$

The first equation states that the strain rate $\dot{\varepsilon}$ (strain/time) is proportional to the product of the *mobile* dislocation density and velocity; only when dislocations move does the material strain. This equation is physically analogous to

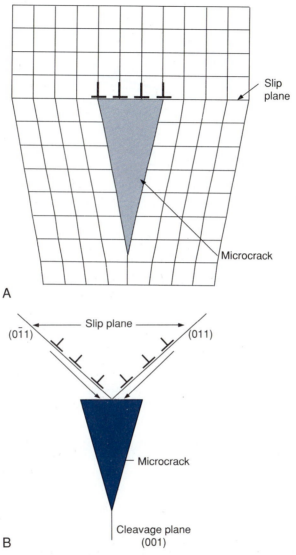

Slip plane

Microcrack

A

Slip plane

$(0\bar{1}1)$ (011)

Microcrack

B

Cleavage plane
(001)

FIGURE 7-19 Dislocation models for microcrack formation. (A) A "super" dislocation composed of closely squeezed individual edge dislocations. (B) Microcrack formed by gliding dislocations accumulating at intersecting (011) slip planes.

Eq. 11-1, connecting current (charge/time) density to charge concentration and velocity. Experimental observations are the basis of the remaining two equations. Dislocation densities have been measured to multiply linearly with strain, and dislocations glide with velocities (v) that depend strongly on some power of the applied shear stress τ. In these equations, the variables stress,

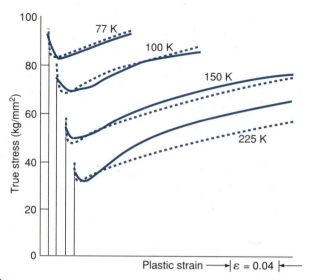

FIGURE 7-20 Calculated tensile stress–strain curves (dashed lines) for steel derived using dislocation dynamics concepts. Experimental curves (solid lines) are shown for comparison. After H. Conrad.

strain, and time are linked; they are also related in a tensile test. Therefore, by integrating the elastic and plastic responses in tension for both the testing machine and specimen, load–time (or stress–strain) behavior has been predicted for different materials. For example, a comparsion between calculated and measured stress–strain curves in mild steel is shown in Fig. 7-20.

7.5. MECHANICAL BEHAVIOR OF POLYMERS

7.5.1. Young's Modulus

Understanding Young's modulus (E) is a key to appreciating the mechanical properties of polymers. Unlike the case for metals and ceramics, E for polymers is often time dependent so that it is, in effect, a **viscoelastic** modulus. After load is applied, polymers not only instantaneously stretch, but also viscously elongate further at a slower rate as polymer molecules find new avenues for relaxing or uncoiling. A more or less constant stress coupled with a time-dependent, but recoverable strain means a time-dependent modulus, $E(t) = \sigma/\varepsilon(t)$. More will be said about this subject in Section 7.5.3.

In addition, we are reasonably confident that E for metals and ceramics will vary only by a very small and predictable amount over a several hundred degree range about room temperature. The same is not true of polymers. This is immediately apparent in Fig. 7-21, where E for a typical linear polymer is plotted against temperature; as much as a huge 10^6-fold variation in E may

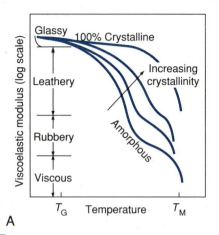
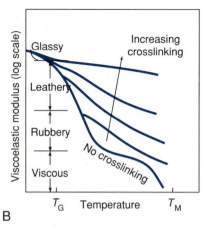

FIGURE 7-21   Typical temperature-dependent changes in the viscoelastic modulus of long-chain polymers. (A) Effect of degree of crystallinity. (B) Effect of extent of crosslinking. From S. Kalpakjian, *Manufacturing Processes for Engineering Materials*, 2nd ed., Addison–Wesley, Reading, MA (1991).

occur over a relatively small temperature range (e.g., ~200°C). As before, T_G and T_M are the glass transition and melting temperatures. Together with Fig. 4-12, this is one of the most important representations of polymer behavior. And just like it, glassy and viscous regimes of polymer mechanical behavior exist at the temperature extremes, with leathery and rubbery behavior in between. Note that very large changes in modulus can also occur by increasing the extent of either crystallinity or crosslinking.

7.5.1.1. Glassy Regime

For values of $T/T_G < 1$ the polymer behaves like a relatively brittle glass. The free volume in the structure has just about been eliminated and the polymer molecules are either densely packed amorphous tangles or collections of misoriented crystals permeated by an amorphous matrix. Applied stresses (σ) stretch the strong covalent bonds of the chain spine, as well as the weaker secondary bonds between chains. The resultant strain, ε, reflects the differential stiffness of both kinds of bonds weighted by their respective concentrations in the polymer. Adding strain contributions from both yields

$$\varepsilon = \sigma[f/E_p + (1 - f)/E_s], \qquad (7\text{-}21)$$

where E_p and E_s are the moduli associated with primary chain and secondary bonds, respectively, and f is the overall fraction of the former in the polymer. By definition, the modulus of the polymer is $E = \sigma/\varepsilon$ and, therefore,

$$E = [f/E_p + (1 - f)/E_s]^{-1}. \qquad (7\text{-}22)$$

As Example 7-6 suggests, a noncrosslinked polymer has a typical E value of 2–3 GPa. If the fraction of strong chain bonds increases, so does E, and values of up to 10 GPa occur. A dramatic increase in E is possible if the polymer fibers are drawn. Then the fraction of covalent bonds oriented in the loading direction increases so that f is effectively >0.95. In **orientation strengthened** nylon, values of E can then approach 100 GPa, which is close to that for copper.

7.5.1.2. Rubbery Plateau

The glassy region gives way to the rubbery plateau above T_G. A thousandfold drop in E occurs, a consequence of the partial breaking of secondary bonds. Molecular chains do not yet slither in an unrestricted way as they do in the melt. Rather they are now loosely entangled with one another; under loading, chains extend but then wriggle back when the load is removed. If there is a dilute concentration of covalent crosslinks that do not break, the polymer becomes a true elastomer. It can now elastically extend a few hundred percent and snap back when the load is withdrawn. But, if there is a high concentration of crosslinks, the covalent bonds associated with them will create a three-dimensional network of relatively short linear molecular segments. No longer are there long linear molecules to stretch large amounts in a unidirectional way.

Surprisingly, in elastomers, unlike other materials, E increases slightly as the temperature rises. This is caused by the increased random thermal motion of polymer segments, which makes it more difficult to orient coiled chains. The higher uncoiling load required means a higher elastic modulus.

7.5.1.3. Viscous Regime

Above about $1.5\,T_G$ all of the secondary bonds break and the modulus suffers another large decline. Linear polymers become viscous liquids and undergo so-called Newtonian flow. Applying a shear stress (τ) to such fluids causes them to shear or flow at a strain rate of $\dot{\gamma}$ (or $d\gamma/dt$) that is determined by the coefficient of viscosity, η, such that

$$\tau = \eta\dot{\gamma}. \tag{7-23}$$

Viscosity has units of poise P ($1\ \mathrm{P} = 10^{-1}$ Pa-s) and *drops* with temperature in a thermally activated manner (see Eq. 4-5); its role in processing polymers is treated in Chapter 8.

EXAMPLE 7-6

Young's modulus for diamond, which is composed totally of primary chain bonds, is 10^3 GPa. For a soft wax, composed entirely of secondary bonds, E is 1 GPa. Estimate E for a linear polymer composed of a 50–50 mixture of primary and secondary bonds.

ANSWER In diamond ($f = 1$), $E_p = 10^3$ GPa, whereas for wax ($f = 0$), $E_s = 1$ GPa. Therefore, $E = [f/10^3 + (1 - f)/1]^{-1}$. For the linear polymer, $f = 0.5$, and after substitution, $E = [0.5/10^3 + 0.5/1]^{-1} = 2$ GPa.

7.5.2. Strength and Toughness

As with metals, much can be learned about the strength and toughness of polymers and elastomers from tensile tests. Below about $0.75\,T_G$ they become quite brittle and lose most or all of their plastic extension. [Tennis balls cooled to liquid nitrogen temperatures (~77 K) shatter like glass when bounced.] The tensile behavior shows a linear elastic range followed by a tiny region of plastic deformation prior to fracture. Small surface cracks or surface scratches due to machining, abrasion, atmospheric attack, and other processes become dangerous flaws because they concentrate stress and lead to brittle fracture. Although they are brittle, glassy polymers display heterogeneous plasticity which makes them considerably tougher than silica glass. Brittle fracture will be discussed again in Chapter 10 within the context of failure of materials.

Above T_G typical stress–strain curves for linear polymers display the behavior shown in Fig. 7-22. After an elastic region, polymers often briefly undergo yielding at the upper yield point, followed by very lengthy uniform extension at the lower yield stress level. The behavior is similar to that of mild steel pulled in tension, and like it, the polymer also necks. Thinning initiates at one point and the resulting neck propagates along and may extend the bar to a

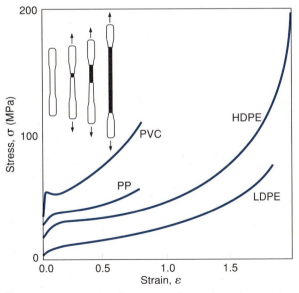

FIGURE 7-22 True stress–strain curves for several polymers. Inset shows uniform necking extension.

strain of a few hundred percent. During this period chains straighten and align in a process called **drawing**. Proof that significant strengthening has occurred in the drawing direction is the fact that fracture does not occur in the much reduced cross section. Interestingly, the test bar also whitens and becomes more opaque in the drawn region. This phenomenon has been attributed to the formation of microvoids and is related to **crazing**, a phenomenon discussed further in Section 10.5.5. Polymers that do not draw well in tension sometimes develop crazes within the deformed matrix. The drawn matrix fibers surround and interconnect successive crazes, which link up to become nuclei for fracture cracks. Crazes appear white and can often be seen when polymers are bent to and fro.

Individual polymers display the entire spectrum of tensile phenomena discussed above by tuning to the appropriate test temperature as indicated in Fig. 7-23. The transition from the strong, brittle version of PMMA at 40°C to the tougher, but weaker, form at elevated temperatures is evident.

Mechanical properties of a number of thermoplastics and thermosets are entered in Table 7-4.

7.5.3. Springs and Dashpots

Between the extremes of elastic behavior and fracture, the mechanical behavior of thermoplastic polymers is often modeled by a mechanical analog involv-

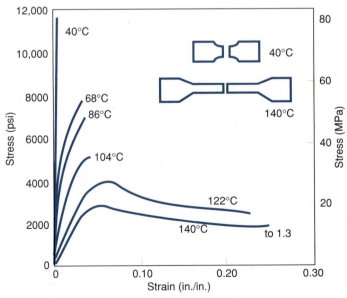

FIGURE 7-23   Stress–strain curves for polymethyl methacrylate at different test temperatures. After T. Alfrey, *Mechanical Behavior of Polymers*, Wiley–Interscience, New York (1967).

ing a combination of springs and dashpots. Springs are perfectly elastic and respond instantaneously to load application or removal. The force (F)–extension (x) characteristics of a spring are linear, that is, $F = k_s x$, where k_s is the spring constant. Dashpots are devices that cushion the uncomfortable bounciness of springs by slowing down their speed of response. A flat metal plate constrained to move through oil is such a dashpot. Because viscous flow is involved, the *rate* of plate displacement $\dot{x}$ is proportional to F. Therefore, $F = \eta\dot{x} = \eta\, dx/dt$, where again, η is the coefficient of viscosity.

All materials that deform plastically, not only polymers, respond to applied forces as though they were a complex admixture of springs and dashpots. They are said to exhibit **viscoelastic** behavior. A simple example modeled by a spring and dashpot in *series* is the Maxwell model, shown in Fig. 7-24. The instant a constant load F is hung on the combination, the spring elongates elastically. Slower to respond is the dashpot, which extends linearly with time to yield

| **TABLE 7-4** | **MECHANICAL PROPERTIES OF POLYMERS** |

Polymer	E (GN/m^2)	ρ (Mg/m^3)	$\sigma_0{}^a$ (MN/m^2)	T_G (°C)
Thermoplastics				
Polyethylene (PE)				
High-density	0.56	0.96	30	
Low-density	0.18	0.91	11	−20
Polyvinyl chloride (PVC)	2.0	1.2	25	80
Polypropylene (PP)	1.3	0.9	35	0
Styrene				
ABS^b	2.5	1.2	50	80
Polystyrene (PS)	2.7	1.1	50	100
Polycarbonate (PC)	2.5	1.2–1.3	68	150
Polyethylene terephthalate (PET)	8	0.94	135	67
Polymethyl methacrylate (PMMA)	2.8	1.2	70	100
Polyesters	1.3–4.5	1.1–1.4	65	67
Polyamide	2.8	1.1–1.2	70	60
Polytetrafluoroethylene (PTFE)	0.4	2.3	25	120
Thermosets				
Epoxies	2.1–5.5	1.2–1.4	60	(107–200)
Phenolics	18	1.5	80	(200–300)
Polyesters (glass-filled)	14	1.1–1.5	70	200
Polyimides (glass-filled)	21	1.3	190	350
Silicones (glass-filled)	8	1.25	40	300
Ureas	7	1.3	60	80
Urethanes	7	1.2–1.4	70	100

^{a} σ_0 = yield strength, T_G = glass transition temperature (maximum value).
^{b} ABS, acrylonitrile–butadiene–styrene.
After B. Derby, D. A. Hills, and C. Ruiz, *Materials for Engineering*, Longman Scientific and Technical, London (1992).

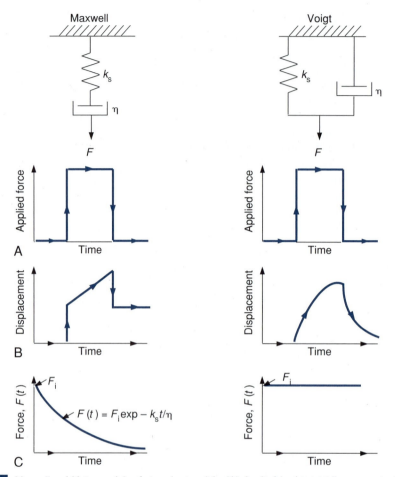

Maxwell and Voigt models of viscoelastic solids. (A) Applied load is initially zero, raised to a constant level for a fixed time, and then unloaded. (B) Resultant time-dependent displacement. (C) Time-dependent force on viscoelastic components if the load (F_i) is initially applied and the strain in each solid is maintained constant thereafter.

the indicated displacement or *strain*. When the load is removed the spring snaps back but not the dashpot. A permanent deformation or set results.

The Voigt model for a viscoelastic solid, shown in Fig. 7-24, involves a *parallel* combination of a spring and dashpot. This time when a constant load is applied, the spring is restrained from immediately extending by the dashpot. As a result, the material responds by stretching in a nonlinear way. Provided the temperature and loads are appropriately high, all materials display similar **creep**like behavior.

Now, suppose a load (F_i) is initially applied to the Maxwell spring–dashpot combination, which is constrained thereafter to have a fixed total strain or displacement. With time, the load relaxes as shown by the following analysis. If x_s and x_d are the spring and dashpot displacements, the defining equations are

$$x_{total} = x_s + x_d = \text{constant}; \quad \text{by differentiating,} \quad 0 = \dot{x}_s + \dot{x}_d$$

$$F_s = k_s x_s; \quad \text{by differentiating,} \quad \dot{x}_s = \dot{F}_s/k_s$$

$$F_d = \eta \dot{x}_d; \quad \text{or} \quad \dot{x}_d = F_d/\eta.$$

Furthermore, $F_s = F_d = F$, because both elements are in series. Therefore,

$$\dot{x}_s + \dot{x}_d = 0 = \dot{F}/k_s + F/\eta \quad \text{or} \quad dF/dt + k_s F/\eta = 0.$$

Upon integrating the last equation and noting that F_i is the initial force, we obtain

$$F = F_i \exp(-k_s t/\eta). \tag{7-24a}$$

With little loss in generality, F can be replaced by σ, the stress, and k_s by E, Young's modulus, so that

$$\sigma(t) = \sigma_i \exp(-Et/\eta). \tag{7-24b}$$

This problem physically corresponds to the relaxation in strength of a polymer bolt tightened to a fixed strain and initial stress level σ_i. Note that the effective viscoelastic modulus defined by $\sigma(t)/\varepsilon$ is time dependent. An exponential drop in stress occurs with time. In the case of a polymer where $E = 3$ GPa and $\eta = 10^{13}$ Pa-s, the fractional stress change is $\sigma(t)/\sigma_i = \exp(-0.0003 t)$. Half the initial stress will be reached in only 2307 seconds. Stress relaxation can be accelerated with temperature because of its effect in lowering viscosity. Torqued steel bolts also exhibit stress relaxation. But it takes much longer for them to lose their strength or loosen because the ratio E/η is much smaller in steel than in polymers due to an immensely larger effective viscosity.

7.6. MECHANICAL BEHAVIOR OF CERAMICS AND GLASSES

7.6.1. Introduction

At first glance ceramics and glasses have a number of outstanding mechanical property attributes. In particular, ceramics have high moduli of elasticity, are hard, and resist softening. Silica-based glasses are also hard and strong in compression. And because these materials are composed of light atoms their density (ρ) is low. Thus, E/ρ is large (Table 7-1), an attractive feature in lightweight mechanical design applications. Nevertheless, the use of these materials is severely restricted in structural applications because of their fatal tendency to fracture in a brittle manner under tension.

There are *intrinsic* as well as *extrinsic* reasons for the brittle behavior of these materials. It has already been mentioned that passage of dislocations through crystalline ceramics means rupturing very strong interatomic bonds. Therefore, dislocations are not prevalent and their absence generally limits plastic deformation. Factors that adversely affect their mechanical behavior stem from the surface and interior cracks and notches that are unavoidable in processing or that develop in use. Because they cannot be easily melted, ceramics are formed into shape by pressing and sintering powders. In consolidating powders, sharp crevices at internal pores can then behave as incipient cracks.

As the mechanical behavior of these materials is severely limited by macroscopic cracks and pores, much of the subsequent discussion is devoted to the implications of stressing matrices containing such defects.

7.6.2. Cracks and Stress Concentration

Surface and interior cracks in objects under load can greatly alter local stress distributions relative to the case of flaw-free material. This intuitive result is illustrated in Fig. 7-25, where a body (plate) containing elliptical-shaped surface and interior holes or cracks is pulled by uniaxial forces, F. The undisturbed force lines in the defect-free portion of the plate mean that the stress distribution is uniform there at a level σ_a. But, the force lines are concentrated in the vicinity of cracks; stess is magnified there because there are effectively more force lines per unit area. A **stress concentration factor**, k_σ, can be defined as the ratio of amplified local to background stress, that is, $k_\sigma = \sigma/\sigma_a$. For flat elliptical surface cracks the maximum value of k_σ is at the crack tip and given by

$$k_\sigma = 2(c/\rho)^{1/2}. \qquad (7\text{-}25)$$

Here c is half the length of the major axis and ρ is the radius of curvature at the crack tip. It is apparent that longer and sharper cracks raise k_σ. Although cracks can exist within the bulk, they are more commonly found at the surface. Furthermore, they pose a much greater hazard in brittle ceramics and glasses than they do in ductile metals. In metals, the elevated stress ahead of the crack causes local plastic deformation, which strengthens the matrix. It is this effect that blunts the advance of the crack. But no such mechanism to arrest crack growth exists in brittle materials. The simultaneous magnification of stress at a crack tip coupled with a lack of dislocations and plastic deformation to absorb or dissipate the strain energy buildup is a disastrous combination. Quite simply put, metals are tough and ceramics and glasses are not: the former plastically yield before fracture; the reverse is true for the latter.

On this basis we can understand why both ceramics and glasses respond differently in tension and compression. Young's modulus measured either way is virtually identical, but the fracture strength in compression is typically 5 to 10 or more times that in tension. Flat cracks are closed shut in compression but tend to be pulled open in tension. Glasses can be melted and made dense, but atmospheric attack can enlarge natural surface flaws into cracks in the presence of tensile stress. Water can penetrate these surface cracks and weaken

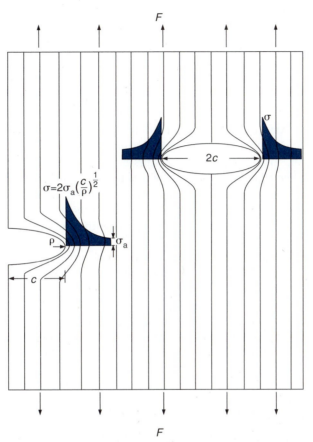

$$\sigma = 2\sigma_a \left(\frac{c}{\rho}\right)^{\frac{1}{2}}$$

FIGURE 7-25   Stress concentration around internal and surface cracks in a uniformly stretched solid. Parallel lines denote the presence of uniform stress.

the glass by converting Si—O bonds to Si—OH bonds. This is the mechanism that appears to account for **static fatigue** in glass, a subject that is discussed again in Section 10.5.1.

In view of their brittle nature the question of mechanical testing of ceramics and glasses arises. Unless they are properly shaped they cannot be easily gripped and pulled in tension because failure will occur preferentially in the grips. Therefore, these materials are often fashioned into rectangular or circular bars and loaded in three-point bending. The bars rest on two widely spaced pins while a central pin is driven into the surface of the bar. **Transverse rupture strength** (or **modulus of rupture**), the maximum tensile stress at the beam surface prior to fracture, is thus frequently quoted for these materials. As brittle materials usually undergo little deformation beyond the elastic regime, interest in their mechanical properties centers on fracture. For this reason an introduction to the theory of fracture mechanics follows.

A more extensive treatment of fracture phenomena including actual failure mechanisms and analyses of case histories is reserved for Chapter 10. There, fracture of glass will be a particular focus of attention.

7.6.3. Brittle Fracture

In 1920 A. A. Griffith formulated an elegantly simple model to account broadly for the fracture of brittle materials and glass filaments in particular. Preexistent flaws or cracks were assumed to play a leading role in influencing strength. The approach and results have survived as the basis of the actively researched branch of study known as **fracture mechanics.** With suitable modification, fracture mechanics is applicable to all classes of materials and it is therefore worth presenting aspects of this important subject. It is assumed that a flat elliptical crack of length $2c$ exists within a large plate oriented as shown in Fig. 7-26A, where an applied tensile stress σ is directed perpendicular to its axis. The process of crack formation *releases* strain energy in an assumed circular disk volume of $2\pi c^2$, for a plate of unit thickness. Earlier it was shown that the elastic strain energy per unit volume is $\sigma^2/2E$ (Eq. 7-7). Therefore, the reduction of strain energy is equal to $-\pi c^2 \sigma^2/E$. Simultaneously, crack surface or cohesive energy associated with the breaking of bonds between atoms and creation of fresh crack surface area must be considered. It is this term that reflects the toughness of the material. Because the total crack area is $2c + 2c$, or $4c$, and it possesses an effective surface energy per unit area γ, the energy contribution from this source is $4c\gamma$. Therefore, total energy, E_T is

$$E_T = -\pi c^2 \sigma^2/E + 4c\gamma. \qquad (7\text{-}26)$$

Development of this formula is not unlike that for homogeneous nucleation (see Eq. 6-15). The criterion for rupture assumes passage from unbroken to broken states via a continuous decrease in system energy. When the crack is

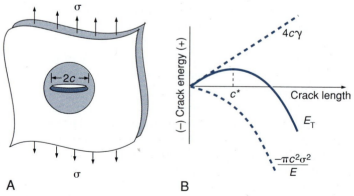

A

B

FIGURE 7-26 (A) Griffith crack model geometry. (B) Crack energy as a function of crack length. The critical crack length is reached when the rate of strain energy release equals the rate at which surface energy is created as the crack opens.

on the verge of growth the rate of decrease of elastic strain energy precariously balances the rate at which surface energy increases. The recipe for determining critical values of σ and c is to set $dE_T/dc = 0$. Thus, $2\pi c\sigma^2/E = 4\gamma$, and the critical stress ($\sigma = \sigma^*$) that just causes crack instability is given by

$$\sigma^* = (2E\gamma/\pi c)^{1/2}, \quad \text{or equivalently,} \quad \sigma^*(\pi c)^{1/2} = K_{1C}. \quad (7\text{-}27)$$

Alternatively, a given stress level σ will cause instability in cracks that are larger than a critical size (c^*), given by $c^* = (1/\pi)(K_{1C}/\sigma)^2$. A graphical model demonstrating the energy trade-offs is displayed in Fig. 7-26B. The constant K_{1C} is equal to $(2E\gamma)^{1/2}$ and has the peculiar units of psi-in.$^{1/2}$ or MPa-m$^{1/2}$. K_{1C}, pronounced "kay one see," is known as the **critical stress intensity factor** or simply the **fracture toughness**. At any noncritical stress level σ, the **stress intensity** is defined as

$$K = \sigma(\pi c)^{1/2}. \quad (7\text{-}28)$$

Therefore, when $K < K_{1C}$, the crack is stable. But, when the reverse is true, the crack is unstable and will open rapidly. Equation 7-27 is modified slightly when surface cracks or cracks with different geometries are involved.

Values of K_{1C} for a number of materials are entered in Table 7-5. They were experimentally determined by methods discussed in Section 7.7.2. The high values of K_{1C} for metals is due to the plastic zone at the crack tip. Among the important facts gleaned from this table are:

1. Glasses and metal oxides have the lowest values of K_{1C}. Metals are much tougher because K_{1C} are typically 10–100 times larger than for these nonmetals.
2. K_{1C} for some of the toughened ceramics are much higher than for the usual ceramics and glasses but are still well below the level for metals.

These results can be understood by noting that the total area under the tensile stress–strain curve (toughness) is a measure of K_{1C}. For both polymers and ceramics this area is restricted; in polymers high ductility is offset by low strength, while in ceramics high strength is offset by low ductility. Metals, however, possess high strength that persists to large strains; this is a recipe for high K_{1C}.

EXAMPLE 7-7

What is the ratio of the critical crack sizes in aluminum relative to aluminum oxide when stressed to their maximum respective elastic stresses?

ANSWER   At the maximum elastic stress, $\sigma = \sigma_0$ (the yield stress), and $c = (1/\pi)(K_{1C}/\sigma_0)^2$. The required ratio is

$$\frac{c(\text{Al})}{c(\text{Al}_2\text{O}_3)} = \frac{\{(1/\pi)(K_{1C}/\sigma_0)^2\}_{\text{Al}}}{\{(1/\pi)(K_{1C}/\sigma_0)^2\}_{\text{Al}_2\text{O}_3}}.$$

From Table 7-5, for Al, K_{1C} = 45 MPa-m$^{1/2}$ and σ_0 = 260 MPa, and for Al$_2$O$_3$, K_{1C} = 3.7 MPa-m$^{1/2}$ and σ_0 = 270 MPa. Upon substitution,

$$\frac{c(\text{Al})}{c(\text{Al}_2\text{O}_3)} = 160.$$

The different response to flaw size of metals relative to ceramics can be qualitatively understood from this example. High values of K_{1C} in the former relative to the latter, coupled with comparable σ_0 values, make critical crack dimensions a hundred times larger in metals than in ceramics. Large cracks in the more defect-tolerant metals can often be detected during inspection and possibly repaired by sealing them shut. But, the much smaller flaws of critical size in ceramics are more effectively hidden and easier to overlook, a combination that makes ceramics fracture prone in service.

The important subject of brittle fracture continues to bedevil engineers and requires a multifaceted approach in order to design against it. A frequent mistake is to think that high-yield-strength materials will provide effective protection against brittle fracture. Quite the opposite is often true as shown in Fig. 7-27. The counter trends that strength and fracture toughness display are not only seen among different classes of materials; they also hold in a single material, for example, an alloy heat-treated to different strength levels. An important lesson that must be learned is that *the price paid for high strength (yield stress) is often low fracture toughness, and vice versa.*

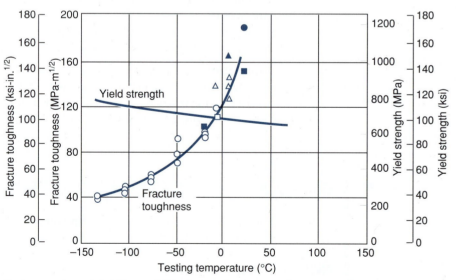

FIGURE 7-27   Fracture toughness and yield strength trends in stainless steel as a function of testing temperature. Note that high yield strengths imply low fracture toughness values, and vice versa. (See also Fig. 9-13.) From *Metals Handbook*, Vol. 8, 9th ed., American Society of Metals, Metals Park, OH (1985).

| TABLE 7-5 | TOUGHNESS OF MATERIALS (FRACTURE TOUGHNESS AND IMPACT ENERGY) |

Material	K_{IC}		Yield stress		Impact energy (Charpy)	
	MPa-m$^{1/2}$	ksi-in.$^{1/2}$	MPa	ksi	J	ft-lb
Metals						
Aluminum	45	41	260	38	27	~20
2024, hardened	26	24	455	66	14	~10
2024, annealed	44	40	345	50		
7075, hardened	26	22	495	72		
Ti-6A1-4V	55	50	1035	150	54	~40
Mild steel	140	127	220	32	203	~150
4340, hardened	60.4	55	1515	220	18	13
4340, annealed	99	90	860	125	106	78
Maraging steel	~133	~120	1700	247	54	~40
Pressure vessel steel	200	182	500	72		

			Modulus of rupture			
			MPa	ksi		
Ceramics						
Soda glass	0.7		70	10		
Al$_2$O$_3$	3.7		270	40		
SiC	3.1		170	25		
Si$_3$N$_4$	4					
ZrO$_2$, stabilized	12					
Concrete	0.2					

					Izod (ft-lb)[a]	
Polymers						
Polyethylene (high density)	~2				0.4–14	
PMMA	~1.3					
Polystyrene	~1				0.2–0.5	
Polystyrene (high impact)					0.5–4	
Polycarbonate	2				12–16	
PVC					0.4–20	
Epoxy	~0.5				0.2–3.0	
Epoxy (fiberglass reinforced)	~40				2–30	

[a] Impact energy values for polymers are for notched Izod tests and are in units of ft-lb/in. Izod and Charpy values are not easily compared.

Sources. R. W. Hertzberg, *Deformation and Fracture Mechanics of Engineering Materials*, 3d ed., Wiley, New York (1989); C. T. Lynch, *Practical Handbook of Materials Science*, CRC Press, Boca Raton, FL (1989); M. F. Ashby and D. R. H. Jones, *Engineering Materials 2: An Introduction to Microstructures Processing and Design*, Pergamon, Oxford (1986).

7.7. MECHANICAL TESTING OF MATERIALS

In this section a number of nondestructive as well as destructive mechanical testing methods are discussed together with an interpretation of the information they convey.

7.7.1. Hardness

By providing information on strength, ductility, and toughness, tensile testing yields the broadest profile of a material's response to loading. Nevertheless, hardness testing is the most widely performed of the mechanical tests. The reason is the ease and speed of the test and the ability to nondestructively probe the hardness of surfaces without the need for special specimen preparation. This makes it ideal for quality-control purposes during manufacturing. As suggested by Fig. 7-28, hardness tests are performed by pressing a hard steel ball or

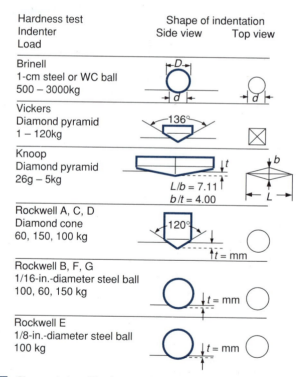

FIGURE 7-28   Characteristics of hardness testing methods for applied load F. Brinell hardness, $\text{BHN} = 2F/\pi D \, [D - (D^2 - d^2)^{1/2}]$. Vickers hardness, $H_V = 1.72 \, F/d^2$. Knoop hardness, $H_K = 14.2 \, F/L^2$. Rockwell hardness scales A, C, D, R_A, R_C, $R_D = 100 - 500t$. Rockwell hardness scales B, F, G, E, R_B, R_F, R_G, $R_E = 130 - 500t$. After H. W. Hayden, W. G. Moffatt, and J. Wulff, *The Structure and Properties of Materials*, Vol. 3, Wiley, New York (1965).

diamond (cone or pyramid) indenter into the surface of the material under a specified compressive load. The Brinell, Vickers, and Knoop hardness numbers that are recorded reflect the stress, or load per unit surface area of indentation. For the popular Rockwell tests, hardness values (HRA, ..., HRG) are related to the penetration depth, with smaller penetrations yielding higher hardness numbers.

Hardness is a complex property. It is related to the strength of interatomic bonding forces and apparently depends on more than one variable. This suggests that a combination of high cohesive energy and short bond length could correlate with high hardness. Measures of cohesive energy per unit volume are plotted against hardness in Fig. 7-29 and the correlation appears to be valid for nonmetallic solids. The wide range in hardness exhibited by different materials is noteworthy. Hardened steels with Vickers (H_V) hardnesses of perhaps 900 are softer than virtually all of the materials displayed in this figure. We

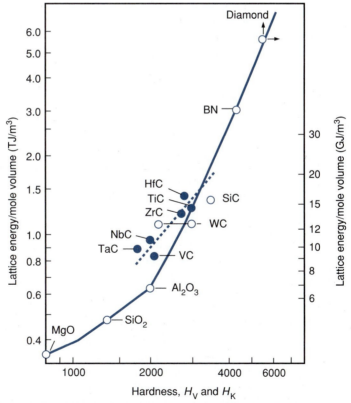

FIGURE 7-29 Relationship between hardness and cohesive energy of solids. The heat of sublimation per molar volume is taken as a measure of cohesive energy. From M. Ohring, *The Materials Science of Thin Films*, Academic Press, Boston (1992).

generally think of metals as being hard, but compared with covalent and ionically bound solids, they are not hard. Hardness values in metals correlate well with their tensile strengths. For example, an approximate equation connecting the two variables for steel is σ_{UTS} (psi) $= 500 H_B$, where H_B is the Brinell hardness. In addition, theory has shown that the hardness value (H) is related to the yield stress of metals through the formula $H = 3\sigma_0$ and this relationship is approximately observed in practice.

As noted in Section 5.2.2, a number of the very hardest materials (e.g., diamond, BN, TiC, TiN, Al_2O_3) are finding increasing commercial use in cutting tools, forming dies, and bearings, applications that require resistance to wear. For these purposes, thin coatings ($\sim 5-10$ μm thick) are deposited, usually by CVD methods. To assess process variables and coating properties it is often necessary to measure hardness without sampling the underlying substrate. For this and similar applications, **micro**hardness testing methods have evolved from the conventional **macro**hardness testing of bulk solids. With the use of much smaller indenters and loads, the hardness of coatings as well as microscopic phases (e.g., inclusions, precipitates) can be measured. The latest development involves **nano**indentation hardness testing which has been devised to test coatings and thin films typically less than 1 μm thick. Yet smaller loads cause indentations less than 100 nm deep.

7.7.2. Notch and Fracture Toughness

7.7.2.1. Notch Toughness

Many mechanically functional materials contain surface notches, pores, and cracks of varying size and geometry that stem from design (e.g., sharp corners), manufacturing (e.g., casting, welding), or service (e.g., atmospheric corrosion). As we have already seen in Section 7.6.2, cracks concentrate stress and influence toughness and fracture in varied ways for different materials. The widely used Charpy impact test is a standard way to assess toughness quantitatively in notched specimens. In this test a standard bar specimen, with a square cross section and a V-shaped notch cut into it, is strained very rapidly to fracture by means of a swinging, pendulum-type hammer (Fig. 7-30). The difference between the initial and final potential energies of the hammer (measured by the initial and final hammer heights) is the impact energy absorbed by the specimen. Brittle materials like glass, polymers, ceramics, and some metals cannot tolerate cracks of any size; they are **notch sensitive,** have low toughness, and absorb little energy. On impact, these materials frequently snap into two pieces that can usually be neatly fitted together.

Tough materials, on the other hand, absorb strain energy at the crack tip through plastic deformation of the surrounding matrix. In effect, the specimen traverses all stages of the stress–strain response, from yielding to work hardening to fracture, in a split second. The fractured halves cannot be fitted together,

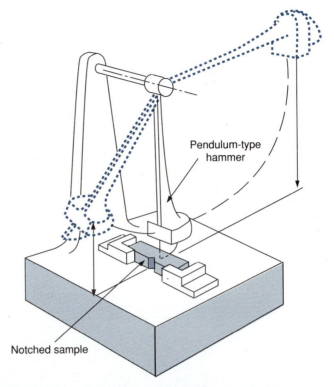

Pendulum-type
hammer

Notched sample

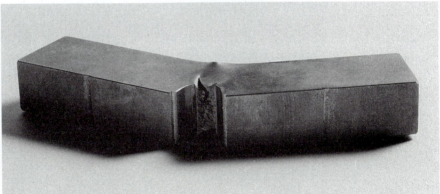

FIGURE 7-30 Schematic of the Charpy impact testing machine. From C. R. Barrett, W. D. Nix, and A. S. Tetelman, *The Principles of Engineering Materials,* Prentice–Hall, Englewood Cliffs, NJ (1973). Inset shows a ductile Charpy specimen after impact. Courtesy of P. S. Han, MTS Systems Corporation.

and for really tough materials the hammer is effectively stopped as the specimen severely distorts but does not break. Ductile metals are such materials and they are not notch sensitive. Impact energies for a number of engineering materials are entered in Table 7-5 where they may be compared with fracture toughness values. By and large, impact energies roughly scale directly with K_{1C} values.

7.7.2.2. Fracture Toughness

To determine K_{1C} values of materials, specimens in the form of plates are fashioned containing machined cracks of known length as shown in Fig. 7-31A. As tensile loads are applied the specimen halves open, and the resulting crack extension is continuously monitored (Fig. 7-31B). Noting the critical crack size and stress necessary to induce fracture, use of Eq. 7-27 yields the value for K_{1C}. Although it might not be suspected from Charpy impact testing, where a specimen of standard dimensions is used, the toughness of a material depends on its size. This is demonstrated schematically in Fig. 7-32, where the K_{1C} dependence on plate thickness is shown. Thin plates exhibit shear lip or

FIGURE 7-31   (A) Metal and ceramic fracture toughness specimens. (B) Experimental arrangement used to obtain K_{1C} values. Inset shows fractured specimen. Courtesy of P. S. Han, MTS Systems Corporation.

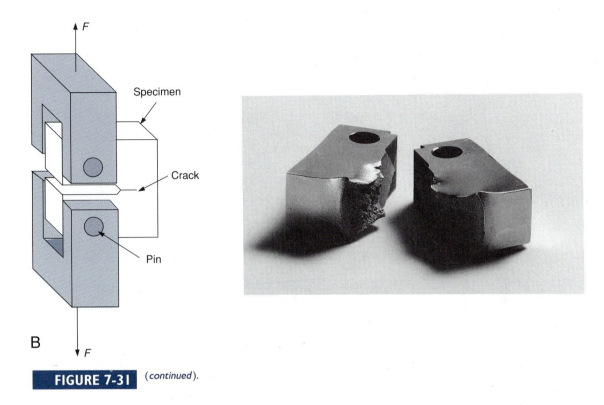

B

FIGURE 7-31 *(continued).*

slant fracture surfaces under so-called **plane stress** conditions. The shear stress-induced plastic deformation is associated with considerable toughening as reflected in the large K_{1C} values. Thick plates, on the other hand, exhibit a flat fracture surface. In this **plane strain** geometry, plastic deformation is constrained, and low, relatively constant K_{1C} values are obtained over a broad thickness range. In between, a mixed fracture mode is evident. More conservative engineering practice often mandates the selection of minimum K_{1C} values, and these are the ones usually reported.

7.7.2.3. Notch Toughness versus Fracture Toughness

Since notch or impact energy toughness is related to fracture toughness, it is worth comparing their similarities and differences. Both use notched or cracked specimens that are tested to fracture. For the purposes of engineering analysis and design, however, K_{1C} values generally have greater utility than impact energies. Whether a crack of given dimensions constitutes a hazard can be immediately assessed if both K_{1C} and the stress in its vicinity are known. A similar assessment is not possible based simply on knowledge of the impact energy. The potential for fracture is exaggerated in the standard Charpy test because of the notch and very rapid rate of deformation. On the other hand,

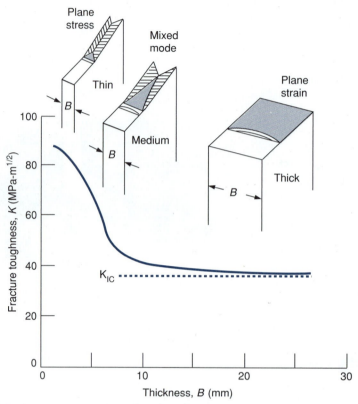

FIGURE 7-32   Fracture toughness as a function of plate specimen thickness. *Note:* Plane strain conditions generally prevail when plate thickness B exceeds 2.5 $(K_{IC}/\sigma_0)^2$. After H. O. Fuchs and R. I. Stephens, *Metal Fatigue in Engineering,* Wiley, New York (1980).

fracture toughness testing is often done at low strain rates, and K_{1C} values are independent of loading rate. Lastly, a variant of fracture toughness testing involving cyclic loading makes it possible to assess the potential for fatigue failure, a subject addressed at the end of the chapter as well as in Section 10.7.

7.7.2.4. Transition Temperature

The important issue of the temperature dependence of toughness remains to be addressed. Steels often display an alarming drop in notch toughness resistance at the so called **transition temperature**. This is shown in Fig. 7-33 for several types of ship steels used during World War II, when all too many ships suffered brittle fracture of varying severity (see Fig. 1-7). In them steels underwent a ductile-to-brittle transition at temperatures too close to room temperature for comfort. This is one reason among others that, even today, structures like storage tanks, oil rigs, and even ships fracture catastrophically more often in winter than in summer.

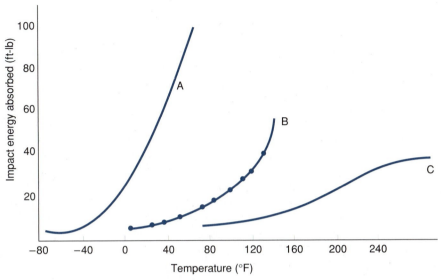

Charpy impact energies of various World War II ship steels as a function of test temperature. (A) Deck plates from S. S. Warrior. This steel is toughest, having high impact energies and the lowest transition temperature. (B) Impact energies on fractured deck plates. Each point is an average of 30 or more Charpy tests. (C) Deck plates at corner hatches of forward port on the S. S. Simon Willard. This steel is least tough, having low impact energies and the highest transition temperature. From Welding (1947).

Transition temperatures can be lowered through alloying. *Reducing* the carbon content of steel has a potent effect in this regard; it also raises the magnitude of the (absorbed) impact energy as well. Alloying with nickel is particularly beneficial in toughening steel and lowering the transition temperature. In general, alloying elements in steel affect toughness in complex ways depending on composition, matrix structure, and heat treatment. Ductile face-centered cubic metals are generally tough and do not exhibit a transition temperature.

7.7.3. Creep

At temperatures of about half the melting point ($0.5\ T_M$) and above, metals (and ceramics) undergo time-dependent plastic straining when loaded. This phenomenon is known as creep and it can occur at stress levels less than the yield strength! Extension of an involved component may eventually produce a troublesome loss of dimensional tolerance or even ultimately lead to catastrophic rupture. Turbine blade creep due to the inhospitable temperatures of a jet engine is an often-cited example. High-pressure boilers and steam lines, nuclear reactor fuel cladding, and ceramic refractory brick in furnaces are components and systems that are also susceptible to creep effects.

Creep tests are performed on round tensilelike specimens that are stressed by fixed suspended loads, while being heated by furnaces that coaxially surround them. The typical response obtained in a creep test is shown in Fig. 7-34, where the specimen elongation or strain is recorded as a function of time. Not unlike creep in the polymer (Fig. 7-24), there is an initial **elastic** extension or strain the instant load is applied. Then a viscouslike plastic straining ensues in which the creep strain rate ($\dot{\varepsilon} = d\varepsilon/dt$) decreases with time. This **primary creep** period then merges with the **secondary creep** stage where the strain rate is fairly constant. Alternately known as **steady-state creep**, this region of **minimum creep** normally occupies most of the test lifetime. The specimen undergoes a viscouslike extension in this range. Finally the strain rate increases rapidly in the **tertiary creep** stage, leading to rupture of the specimen. Creep may be thought of as a competition between the rate of work hardening, $\dot{H}$, and the rate of thermal softening or recovery, $\dot{S}$ (discussed in Section 9.4.2). The three successive creep stages can then be respectively

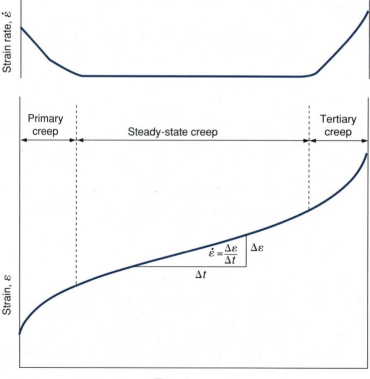

FIGURE 7-34   Stages of plastic strain as a function of time during creep testing at temperatures above $0.5\,T_M$. The corresponding creep strain rate is plotted above.

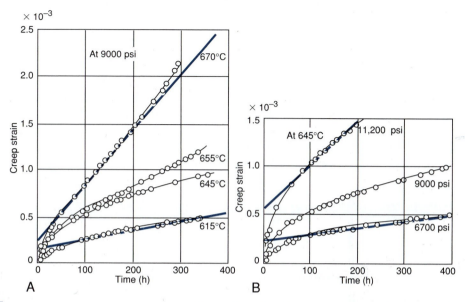

FIGURE 7-35 Creep strain versus time test results in a 0.5 wt% Mo, 0.23 wt% V steel. (A) Constant stress, variable temperature. (B) Constant temperature, variable stress. From A. J. Kennedy, *Processes of Creep and Fatigue in Metals*, Wiley, New York (1963).

characterized by $\dot{S} < \dot{H}$, $\dot{S} = \dot{H}$, and $\dot{S} > \dot{H}$, where tertiary creep additionally involves microfracture and stress instabilities due to necking. Thus, initially, hardening wins out over softening and creep straining is inhibited. At the end the reverse is true, and the creep strain rate accelerates largely because of the reduction in specimen area. In between, the two processes are roughly balanced.

To better understand the creep response of materials, engineers perform two types of tests. The first aims to determine the steady-state creep rate over a suitable matrix of stress and temperature. Thus tests are performed at the *same stress* but at *different temperatures*, as well as at the *same temperature* but at *different stresses*, as shown in Fig. 7-35. Specimens are usually not brought to failure in such tests; accurately predicting their extension is of interest here. The second, known as the creep rupture test, is conducted at higher stress and temperature levels to accelerate failure. Such test information can either be used to estimate short-term life (e.g., turbine blades in military aircraft) or be extrapolated to lower service temperatures and stresses (e.g., turbine blades in utility power plants) to predict long-term life.

7.7.3.1. Steady-State Creep

Much engineering design is conducted on the basis of steady-state creep data. As the creep strain rate is accelerated by both temperature and stress, a useful phenomenological equation that succinctly summarizes the data of Fig. 7-35 and reflects the role of these variables is

$$\dot{\varepsilon} = A\sigma^m \exp(-E_c/RT). \qquad (7\text{-}29)$$

A and m are constants and the exponential term is the ubiquitous Boltzmann factor that signifies the thermally activated nature of creep. Further, the creep activation energy is E_c, and m, a constant, typically ranges in value from 1 to 7 (see Section 10.6). In view of the fact that relatively elevated temperatures are involved, one may be certain that atomic diffusion is involved in some way. In fact, E_c is often close to the activation energy for bulk diffusion.

EXAMPLE 7-8

Determine E_c and m for the data shown in Fig. 7-35.

ANSWER   The steady-state strain rate is the slope of the strain–time response. Limiting the calculation to only the maximum and minimum slopes for the superimposed lines, the following results are tabulated:

Test	Temperature (°C)	Stress (psi)	Stain rate (h^{-1})
1.	670 (943 K)	9,000	$(2.5 - 0.28) \times 10^{-3}/(370 - 0) = 6 \times 10^{-6}$
2.	615 (888 K)	9,000	$(0.56 - 0.2) \times 10^{-3}/(400 - 0) = 9 \times 10^{-7}$
3.	645 (918 K)	11,200	$(1.5 - 0.6) \times 10^{-3}/(200 - 0) = 4.5 \times 10^{-6}$
4.	645 (918 K)	6,700	$(0.5 - 0.25) \times 10^{-3}/(400 - 0) = 6.25 \times 10^{-7}$

To determine E_c, data from tests 1 and 2 are used. From Eq. 7-29,

$$\dot{\varepsilon}(1)/\dot{\varepsilon}(2) = [\exp(-E_c/RT_1)]/[\exp(-E_c/RT_2)]$$

or

$$E_c = [RT_1T_2/(T_1 - T_2)] \times \ln [\dot{\varepsilon}(1)/\dot{\varepsilon}(2)].$$

After substitution,

$$E_c = [8.31 \text{ J/mol-K } (943 \times 888/55)] \times \ln[6 \times 10^{-6}/9 \times 10^{-7}] = 240 \text{ kJ/mol.}$$

Similarly, using test data 3 and 4, $\dot{\varepsilon}(3)/\dot{\varepsilon}(4) = (\sigma_3/\sigma_4)^m$. Taking logs $m = \ln[\dot{\varepsilon}(3)/\dot{\varepsilon}(4)]/\ln(\sigma_3/\sigma_4)$. After substitution,

$$m = \ln(4.5 \times 10^{-6}/6.25 \times 10^{-7})/\ln(11,200/6700) = 3.84.$$

Knowing E_c and m enables one to evaluate $\dot{\varepsilon}$ at any other temperature and stress level.

7.7.3.2. Creep Rupture

Creep or stress rupture test data are normally plotted as stress versus rupture time (t_R) on a log–log plot as shown in Fig. 7-36. Each data point represents

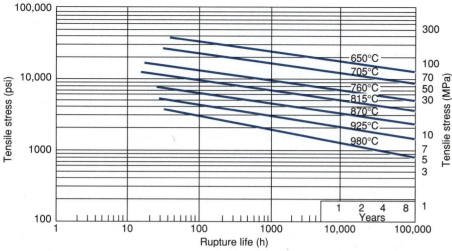

FIGURE 7-36 Creep-rupture properties of Incoloy alloy 800, an iron-base alloy containing 30 wt% Ni and 19 wt% Cr. *Source*: Huntington Alloys Inc.

one test at a specific temperature and stress level. Given sufficient testing at different temperatures, a complete profile of the material response is obtained. It is a good assumption to believe that creep rupture data will be thermally activated. Therefore, high-temperature stress-induced mass transport processes (bulk and grain boundary diffusion, grain growth, precipitation, etc.) will contribute to the evolution of damage. Specific creep damage mechanisms are further explored in Section 10.6.

7.7.4. Fatigue

It has been estimated that more than 80% of all service failures can be traced to fatigue. Included in this category are components and structures that are subjected to lengthy repeated or cyclic loading. By cyclic loading we mean almost any reasonably periodic stress–time variation, for example, axial tension–compression, reversed bending, and reversed torsion or twisting. To complicate matters, variable stress magnitude and frequency may also be involved. In an insidious manner, undetected flaws or incipient cracks grow to macroscopic dimensions through incremental propagation during each stress cycle. Unnoticed until the very end, when the greatly reduced section is unable to support the load, the component quickly undergoes fracture without warning; in essence a crack of critical size is reached. The failure mode just described is known as **fatigue**. This sequence of events has been repeated all too often in components of rotating equipment such as motor and helicopter shafts, train wheels and tracks, pump impellers, ship screws and propellers, and gas turbine disks and blades. Even surgical prostheses implanted into the human body are not immune and have suffered fatigue failure in service. What is difficult to design against is the fact that under cyclic loading, failure can occur significantly below the designed static stress levels.

Failure of materials undergoing repeated cyclic loading is measured employing a fatigue testing machine. In a traditional version (Fig. 7-37A), a bar specimen that narrows in diameter toward the middle is mounted horizontally and rotated at high speed with a motor. A hanging load tilts the grips so that reversed bending moments are transmitted to the specimen, stressing its surface alternately in tension and compression during rotation. Sinusoidal loading about a zero mean stress level is most often used in testing (Fig. 7-37B). For the case of arbitrary sinusoidal loading, maximum (σ_{max}), minimum (σ_{min}), mean (σ_m), and stress range (σ_r) values are defined in Fig. 7-37C. The number of rotations to failure is counted at a specific specimen stress amplitude. Then a new specimen is mounted, a different load is chosen yielding a new peak stress level, and the number of cycles to failure is determined once again. In this tedious manner sufficient data points are accumulated to generate an S–N (or stress–number of cycles) curve.

Amid a typically large scatter band, the S–N response for assorted materials is displayed in Fig. 7-38. Importantly, below a stress level known as the **endurance** or **fatigue limit** (the curve horizontal), most steels can rotate indefinitely

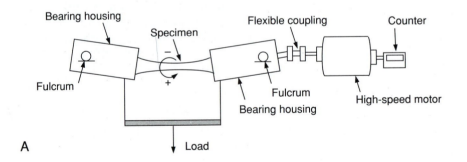

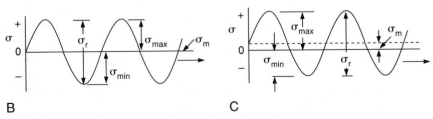

| **FIGURE 7-37** | (A) Moore fatigue testing machine. (B) Symmetrical sinusoidal loading about $\sigma_m = 0$. (C) Asymmetrical sinusoidal loading about $\sigma = \sigma_m$. *Note:* $\sigma_m = 1/2(\sigma_{max} + \sigma_{min})$ and $\sigma_r = \sigma_{max} - \sigma_{min}$. |

(i.e., $\geqslant 10^6$ cycles). This fact gives hope to motor manufacturers; they design for shaft stresses below the fatigue limit, which is typically one-third to one-half the ultimate tensile strength. Some high-strength steels and many nonferrous metals (e.g., Al and Cu), on the other hand, do not exhibit a well-defined endurance limit. Polymers display similar fatigue characteristics. To guard against fatigue, stress levels should not exceed $\frac{1}{3}\sigma_{UTS}$. Determination of reliable fatigue strength values necessitates the use of highly polished specimens. They must be free of scratches or notches which can serve as crack nuclei that effectively lower endurance limits.

When the applied stress causes unconstrained deformation, stress–life (S–N) tests and the information they convey are appropriate. But, in most practical applications, engineering components are constrained, particularly at stress

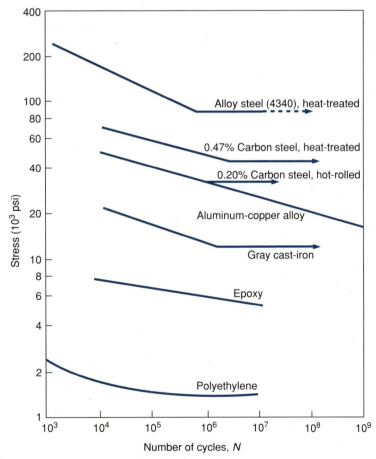

FIGURE 7-38 S–N data for various metals and polymers. Data were taken from several sources.

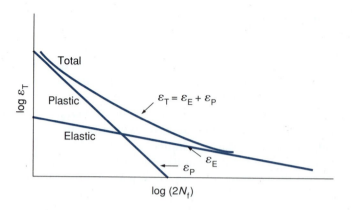

FIGURE 7-39 Typical variation of number of stress cycles to fatigue failure as a function of strain amplitude.

concentrations. For such cases it is often more useful to plot fatigue data in terms of *strain–life* variables. The ordinate is usually taken as *total* strain amplitude, ε_T, and the abscissa is the number of cycles to failure, N_f, or, more commonly, the number of strain reversals to failure $(2N_f)$. A schematic plot of $\log \varepsilon_T$ versus $\log 2N_f$ is shown in Fig. 7-39, and is often used to describe low cycle fatigue. The total strain can be broken down into the sum of elastic (ε_E) and plastic (ε_P) strain contributions as indicated. An equation commonly used to describe the overall behavior is $\varepsilon_T = \varepsilon_E + \varepsilon_P$, or

$$\varepsilon_T = (\sigma_f/E)(2N_f)^b + \varepsilon_f(2N_f)^c, \tag{7-30}$$

where σ_f and ε_f are defined as the fatigue strength and fatigue ductility coefficients, respectively. Constants b and c are the respective slopes of the ε_E and ε_P lines in Fig. 7-39. It is observed that at *short lives*, ε_P dominates and, therefore, *ductility or toughness controls performance.* At *longer lives*, ε_E dominates and *performance is controlled by strength.*

EXAMPLE 7-9

The low cycle fatigue properties of a certain steel are given by Eq. 7-30, where $\sigma_f/E = 0.005$, $\varepsilon_f = 0.07$, $b = -0.08$, and $c = -0.7$.
a. What is the value of the transition fatigue life (i.e., value of $2N_f$ when $\varepsilon_E = \varepsilon_P$)?
b. What is the total strain amplitude at the transition fatigue life?

ANSWER   a. Equating $\varepsilon_E = \varepsilon_P$, $0.005(2N_f)^{-0.08} = 0.07(2N_f)^{-0.7}$. Therefore, $(2N_f)^{0.62} = 0.07/0.005$, and solving, $2N_f = (14)^{1/0.62} = 70.6$.
b. Substituting $2N_f = 70.6$, $\varepsilon_T = 0.005(70.6)^{-0.08} + 0.07(70.6)^{-0.7} = 0.0071$.

Modern approaches to fatigue testing involve the introduction of prior cracks of known length and shape. Under a programmed cyclic loading regimen the progressive extension of the crack is monitored optically. The results of this experimental approach are incorporated into a fracture mechanics framework in Section 10.7.2. Viewing fatigue from the standpoint of fracture mechanics should be contrasted with the classical S–N approach. The former is more realistic because it accounts for unavoidable initial flaws. Furthermore, as we shall see later, the fracture mechanics approach has a predictive capability regarding the number of stress cycles to failure.

As with fracture and creep, the detailed microscopic mechanisms of fatigue are treated in Chapter 10 to round out our understanding of damage manifestations.

7.8. PERSPECTIVE AND CONCLUSION

This chapter is essentially the first to deal with engineering properties. The mechanical behavior of solids is broadly divided into the elastic and plastic regimes. Elasticity theory and engineering simplifications of it form the basis of the engineering design of structures and machine parts. The tensile, compressive, and shear strains are always small in the elastic regime. Elastic tensile stresses exert a pull on atomic bonds, and the resultant atomic separation is manifested macroscopically as strain; both stress and strain vanish when the load is removed. The reason we can be so confident about stress analysis in the elastic range is the validity of Hooke's law or the linear connection between stress and strain. (Ohm's law linearly connecting voltage and current plays a similar singular role in the design and analysis of electrical circuits.) Young's modulus and Poisson's ratio are the only material constants of concern in isotropic elasticity, but there are more moduli to contend with when single crystals are involved. With computers it is now possible to determine accurately the stress and strain distribution throughout arbitrarily shaped bodies loaded elastically in almost any given way.

It is, however, the more complex plastic regime that is important in this book. Plastic deformation is relied on in materials-forming operations described in the next chapter. But, it is also the forerunner to mechanical failure, a subject dealt with in this as well as Chapter 10. Crystalline solids deform differently than amorphous solids. Attributes of plastic deformation in crystalline solids worth noting include the following:

1. *Deformation is severely constrained by the lattice geometry.* Slip, the vehicle of plastic deformation, is peculiar to the material in question and occurs only on certain atomically dense and widely spaced planes. These slip planes move relative to one another in directions of close packing only when a critical level of shear stress is exceeded.

2. *Dislocations play a well-defined but complex role in all stages of the plastic deformation process.*   They paradoxically explain why metals are generally weak and also why they grow stronger with deformation. When dislocations move freely, plastic flow occurs without much increase in stress. If dislocation motion is blocked by impediments (e.g., grain boundaries, precipitates, other dislocations), higher stress levels are required for continued deformation; the material work hardens.

3. *Plastic deformation is always facilitated at elevated temperatures.*   Even normally brittle crystalline ceramics and semiconductors will plastically deform if the temperature is sufficiently high.

Deformation phenomena in amorphous polymers cannot be summarized so easily. Crystalline dislocation mechanisms are out of the question. Rather, the intertwined long-chain molecules stretch, twist, uncoil, and so on, in response to the applied load. The resultant mechanical response resembles that of a collection of springs and dashpots connected in both series and parallel combination. In this way stress relaxation and creep of polymers can be readily modeled.

The questions of how to view and deal with failure and fracture of materials have not been entirely resolved. Oblivious to the presence of manufacturing defects it has been traditional to simply specify the yield stress as the maximum stress level in design. But, when flaws exist, safe stress levels may very well be overestimated. Selecting a high-yield-stress material could actually be a disastrous choice. A more rational approach is to employ fracture mechanics, a subject that developed from simple Griffith crack theory. On this basis crack size and the K_{1C} value are of critical concern.

The fatal tendency of glasses and ceramics to fracture in a brittle manner because of processing flaws dominates the mechanical behavior of these materials. There are fewer such defects or preexistent cracks in metals; and happily, because metals are not notch sensitive, their presence is not as troublesome. Metals have significantly higher values of K_{1C} and impact energy than ceramics, glasses, and polymers. But, there is always room for improvement. Methods have been devised to strengthen and toughen all of these materials, and they are described in Chapter 9.

Mechanical testing is performed on materials for a number of reasons, including:

1. To obtain material properties (Young's modulus, yield and tensile strength, percentage elongation, etc.) that can be used in engineering design.
2. To facilitate materials selection choices.
3. To ensure that products meet manufacturing specifications.
4. To discover how materials will respond under load to extremes in environmental conditions.

5. To understand and cope better with the effects of impact, elevated temperature (creep), and cyclic (fatigue) loading.
6. To assess the susceptibility of materials to brittle fracture.

By combining the results of macroscopic testing with models of deformation at the atomic level, a rather complete understanding of the mechanical behavior of solids has emerged.

Additional Reading

M. F. Ashby and D. R. H. Jones, *Engineering Materials 1: An Introduction to Their Properties and Applications,* Pergamon Press, Oxford (1980); *Engineering Materials 2: An Introduction to Microstructures, Processing and Design,* Pergamon Press, Oxford (1986).

T. H. Courtney, *Mechanical Behavior of Materials,* McGraw–Hill, New York (1990).

G. E. Dieter, *Mechanical Metallurgy,* 3rd ed, McGraw–Hill, New York (1986).

N. E. Dowling, *Mechanical Behavior of Materials,* Prentice–Hall, Englewood Cliffs, NJ (1993).

R. W. Hertzberg, *Deformation and Fracture Mechanics of Engineering Materials,* 3rd ed., Wiley, New York (1989).

QUESTIONS AND PROBLEMS

7-1. a. What is the final length of a 2-m-long bar of copper, 0.01 m in diameter, stressed by a 500-kg force?

b. If a steel bar of the same diameter has the same force applied to it, how long must it be to extend the same amount as the copper bar in part a?

7-2. Wire used by orthodontists to straighten teeth should ideally have a low modulus of elasticity and a high yield stress. Why?

7-3. The engineering stress–strain curve for a given metal can be well described by three straight lines, with stress σ given in psi and strain e in units of in./in.

Elastic range: $\sigma = 20 \times 10^6 \, e$ for $0 < e < 0.0025$.

Plastic range: $\sigma = 47{,}400 + 1.026 \times 10^6 \, e$ for $0.0025 < e < 0.1$

and $\sigma = 250{,}000 - 1.0 \times 10^6 \, e$ for $0.1 < e < 0.15$.

a. What is the value of Young's modulus?

b. What is the value of the yield stress?

c. What is the value of the ultimate tensile stress?

d. Calculate a value for the modulus of resilience.

e. Calculate a value for the toughness.

f. What is the percentage elongation.

7-4. Sketch the engineering stress–engineering strain curve of 70–30 brass (as received) from the known true stress–true strain curve depicted in Fig. 7-7.

7-5. For a hardened alloy steel the true stress–true strain behavior is $\sigma = 190\varepsilon^{0.3}$ ksi.

a. What is the true strain at necking?

b. What is the engineering strain at necking?

    c. What are the true and engineering stress levels at necking?

    d. What is the value of the 0.2% offset yield strength?

7-6. The true stress–true strain relationship for a work hardening metal is $\sigma = K\varepsilon^n$ and the metal is compressed to a strain of ε_1.

    a. What is the true compressive stress level achieved?

    b. The average stress up to strain ε_1 is defined by

$$\sigma_a = \int_0^{\varepsilon_1} \sigma \, d\varepsilon / \varepsilon_1.$$

    Grapically indicate the physical meaning of σ_a on a stress–strain curve.

    c. Show that $\sigma_a = K\varepsilon_1^n/(n + 1)$ by evaluating the integral.

7-7. a. A bar is pulled in tension to a certain plastic stress level and the machine is halted at a fixed crosshead separation. What will happen to the recorded load as a function of time? Why?

    b. The elastic modulus determined from the slope of the stress–strain curve in the elastic region is more accurate the lower the testing temperature. Why?

7-8. During the growth of silicon single crystals a small seed measuring 0.003 m in diameter is immersed in the melt and solid Si deposits on it. Small seeds minimize the likelihood of dislocation defects propagating into the growing crystal. What is the length of a 0.15-m-diameter crystal that can be supported by such a seed if the yield stress of silicon is 210 MPa?

7-9. Steel pipe employed for oil drilling operations is made up of many segments joined sequentially and suspended into the well. Suppose the pipe employed has an outer diameter of 7.5 cm, an inner diameter of 5 cm, a density of 7.90 Mg/m^3, and a yield strength of 900 MPa. Assume a safety factor of 2.

    a. What is the maximum depth that can be safely reached during drilling?

    b. What is the maximum elastic strain in the pipe and at what depth does it occur?

    c. At what depth does the elastic strain have a minimum value?

7-10. a. Write an expression for the maximum elastic strain energy per unit weight of a material.

    b. How much elastic strain energy (in units of J/kg) can be stored in 15 kg of steel if the yield stress is 300,000 psi or $0.01\,E$?

    c. A 12-V automobile battery weighs 15 kg and can deliver 150 A-h of charge. How much energy (J/kg) is stored by the battery?

7-11. A 1500-kg load is hung from an aluminum bar of dimensions $1 \times 0.01 \times 0.01$ m and stretches it elastically along the long axis. What is the change in volume of the bar?

7-12. A tensile bar has an initial 2.000-in. gauge length and a circular cross-sectional area of 0.2 in.^{2}. The loads and corresponding gauge lengths recorded in a tensile test are:

Load (lb)	Gauge length (in.)
0	2.0000
6,000	2.0017
12,000	2.0034
15,000	2.0042
18,000	2.0088
21,000	2.043
24,000	2.30
25,000	2.55
22,000	3.00

After fracture the bar diameter was 0.370 in. Plot the data and determine the following:

a. Young's modulus
b. 0.2% Offset yield stress
c. Ultimate tensile stress
d. Percentage reduction of area
e. True fracture stress.

7-13. Identically shaped tensile specimens of copper and Al_2O_3 with a diameter of 1.5 cm contain a 0.02-cm-deep V-groove machined into the surface along the bar circumference (perpendicular to the applied load). Explain, using schematic stress–strain curves, the difference in behavior when these bars are pulled in tension compared with those that have no grooves.

7-14. A single crystal of Mo shaped as a cylindrical rod with a 1-cm^2 cross section is pulled in tension. If the crystal axis lies along the [123] direction, at what applied load will crystal slip occur on the (110) plane in the [$1\bar{1}1$] direction?

7-15. A single crystal of LiF is cleaved in the shape of a rectangular prism whose faces are (100)-type planes. Slip systems in LiF involve {110}-type planes and ⟨110⟩-type directions, for example (011) and [$0\bar{1}1$].

a. Show that a compressive load applied normal to any crystal face equally activates four slip systems.
b. Dislocation etch pits in LiF have a pyramidal shape as shown in Fig. 3-27A. Although all pits have a square base, some pyramids are skewed while others are symmetric. Explain this observation.

7-16. Explain how dislocations have paradoxically rationalized why metals can be both mechanically weak and strong.

7-17. Show by means of a sketch the result of screw dislocation motion under applied shear stresses.

7-18. How have dislocations helped to explain the following phenomena?

a. Yield stress of mild steel
b. Structure of small-angle grain boundaries
c. Work hardening of metals
d. Microcrack formation in ceramics
e. Slip offsets on deformed crystal surfaces

7-19. A cube of copper 1 cm^3 in volume containing a dislocation density of 10^6 cm/cm^3 is plastically compressed in an adiabatic fashion (with no loss of heat). Copper has a heat capacity of 0.38 J/g-°C and a Burgers vector of 0.25 nm.

a. If the yield stress of Cu is $0.001E$, what is the elastic strain energy and temperature rise after elastic deformation?

b. If the dislocation density rises to 10^{12} cm/cm^3, what is the plastic strain energy and temperature increase after plastic deformation?

7-20. Estimate the dislocation density in the electron micrograph of Fig. 7-17A if the specimen is 1 μm thick.

7-21. During deformation in a tensile testing machine, metal is strained at a typical rate of 10^{-3} s^{-1}. During impact loading a strain rate of 10^4 s^{-1} is typical.

a. For the same initial defect density, what is the ratio of dislocation velocities in these two deformation processes?

b. At what velocities will dislocations travel if their initial density is 10^6 cm/cm^3 and $b = 0.25$ nm?

7-22. Show that the resulting strain in the Voigt model of a polymer is given by $\varepsilon = \sigma/E[1 - \exp(-Et/\eta)]$. Describe a physical situation in which this formula applies.

7-23. A polymer bolt is stressed to a level of 4000 psi and its length is held constant. After 30 days the stress has fallen to 2500 psi. How long will it take for the stress to fall to 1000 psi?

7-24. Ten minutes after a stress of 1000 psi was applied to a polymer rod with an elastic modulus of 1×10^5 psi, the strain was measured to be 0.005.

a. What is the viscosity if this polymer behaves like a Voigt solid?

b. After 1000 minutes what strain can be expected?

See Problem 7-22.

7-25. A stress of 2 MPa applied to a polymer decayed to one-tenth this value in 1 hour at 50°C. If the temperature rises to 70°C, E drops by a factor of 3. Estimate how long it will take for a similar stress relaxation at 70°C if the glass transition temperature for this polymer is 10°C. *Hint*: See Chapter 4.

7-26. Which of the following polymeric materials is expected to have a higher tensile strength and why?

a. Low-density polyethylene or high-density polyethylene

b. Network phenol–formaldehyde or linear vinyl chloride

c. Isotactic polypropylene or atactic polypropylene

d. Rubber at room temperature or rubber at 77 K

e. Alternating acrylonitrile–butadiene rubber copolymer or graft acrylonitrile–butadiene rubber copolymer

7-27. A spring (k_1) is in series with a parallel combination of a spring (k_2) and dashpot (η). Sketch and verbally describe the displacement of strain-versus-time behavior after a load is hung on this system.

7-28. a. A steel with a yield stress of 150 ksi contains a circular surface crack. What stress can be safely applied without local deformation occurring near the crack?

b. In the same steel an elliptical surface crack has a radius of curvature equal to 0.5 mm. How deep a crack can be tolerated without plastic deformation if the applied stress is 35,000 psi?

7-29. Estimate the yield stress of the material tested in Fig. 7-32.

7-30. Fracture toughness (K_{1C} in ksi-in.$^{1/2}$) and yield stress (σ_0 in ksi) are inversely related in metals. Suppose that in a high-strength tool steel $K_{1C} = 160$ ksi-in.$^{1/2}$ when $\sigma_0 = 150$ ksi, and $K_{1C} = 20$ ksi-in.$^{1/2}$ when $\sigma_0 = 250$ ksi. Other data suggest the following relationship over the range of variables: $K_{1C} = a + b/\sigma_0$, where a and b are constants. What value of K_{1C} can be expected at a yield stress of 190 ksi?

7-31. Big Bird Helicopter Company requires a stainless steel that is tough enough for use in the Tropics (120°F) as well as the Arctic ($-60°$F). They peruse this book and find the data in Fig. 7-27. At the designed use stress, the component can tolerate a surface flaw of 0.080 in. at 120°F. Manufacturing inspection equipment can only detect flaws that are larger than 0.050 in. Will a fracture unsafe situation arise at $-60°$F for the same loading?

7-32. What is the difference between the following pairs?
a. Stress intensity and critical stress intensity
b. Strength and toughness
c. Cracks in metals and cracks in ceramics
d. Elastic strain energy and surface energy associated with cracks
e. Transverse rupture strength and tensile strength

7-33. Figure 7-31 suggests that ceramic specimens for fracture toughness testing can be considerably thinner than corresponding metallic specimens. Why?

7-34. A plate of steel has a yield stress of 1000 MPa. The plate fractured when the tensile stress reached 800 MPa and it was therefore hypothesized that a surface crack was present. If the fracture toughness for this steel is 60 MPa-m$^{1/2}$, approximately what crack size is suggested?

7-35. A steel cylinder containing CO_2 gas pressurized to 1500 psi is being transported by truck when someone fires a bullet that pierces it, making a 1-cm hole.
a. Explain the conditions that might cause the steel to fracture violently with fragments flying off at high velocity.
b. Similarly, explain the conditions that would cause the cylinder to release the gas relatively harmlessly.

7-36. The equation describing steady-state creep of a material is given by $\dot{\varepsilon} = A\sigma^m \exp(-E_c/kT)$, where $m = 6$ and $E_c = 100$ kJ. The creep rate can be increased by raising either the stress level of 10 MPa or the creep temperature of 1000 K.
a. For a 1% change in stress calculate the percentage change in $\dot{\varepsilon}$.
b. For a 1% change in temperature calculate the percentage change in $\dot{\varepsilon}$.
c. Is the 1% increase in stress or temperature more effective in raising the creep strain rate?

Hint: In parts a and b calculate $d\dot{\varepsilon}/\dot{\varepsilon}$ versus $\dot{\varepsilon}^{-1}d\dot{\varepsilon}/d\sigma$ and $\dot{\varepsilon}^{-1}d\dot{\varepsilon}/dT$.

7-37. Steady-state creep testing of electrical solder wire yielded the indicated strain rates under the test conditions given.

Test	Temperature (°C)	Stress (MPa)	Strain rate (s^{-1})
1	22.5	6.99	2.50×10^{-5}
2	22.5	9.07	6.92×10^{-5}
3	46	6.99	2.74×10^{-4}

From these limited data determine the creep activation energy, E_c, and stress exponent, m.

7-38. Incoloy 800 tubes are selected for use in a pressurized chemical reactor. If they are designed to withstand wall stresses of 2000 psi at 900°C, predict how long they will survive.

7-39. Suppose the fatigue behavior of a steel is characterized by a two-line response when plotted on an S–log N plot, namely, (1) S (MPa) = 1000 − 100 log N, between $N = 0$ and $N = 10^6$ stress cycles, and (2) S (MPa) = 400, for $N > 10^6$ stress cycles.

a. Sketch the S–log N plot.

b. What is the value of the endurance limit?

c. How many stress cycles will the steel probably sustain prior to failure at a stress of 460 MPa?

d. Elimination of surface scratches and grooves changed the endurance limit to 480 MPa. (Assume plot 1 is not altered.) How many stress cycles will the steel probably sustain prior to failure at a stress of 470 MPa? How many stress cycles will the steel probably sustain prior to failure at a stress slightly above 480 MPa?

8 MATERIALS PROCESSING AND FORMING OPERATIONS

Engineering materials are useful only if they can be processed in quantity and fabricated in varied shapes, while meeting composition and property specifications. **Primary processing** is concerned with large-scale production of basic materials like steel and polymer precursors. This output is then sent either to **secondary processors** for alloying or blending or directly to the manufacturing industries (e.g., auto and plastic packaging companies), which additionally shape and assemble them into final products. Many variations on this theme exist and some prime producers also manufacture intermediate finished products for further assembly, for example, I beams for construction and huge rotors for steam turbines. Some companies and industries derive relatively little benefit, but others involved with semiconductors and electronics reap huge added value profits from additional processing.

In this chapter we focus on the processing associated with shaping and forming components in traditional primary and secondary manufacturing industries. A comparative approach is adopted, stressing similarities in processing methods for different materials, for example, die casting of metals and injection molding of polymers, and powder pressing and sintering of metal and ceramic powders. At the same time, differing competitive routes to production of the same component, for example, by forging, casting, or powder metallurgy, are compared. Qualitative process descriptions, engineering modeling of some of their features, product design, and elimination of processing defects are some of the additional issues addressed.

Materials processing issues underscore a vitally important and broad subject that could easily fill a book by itself. The subject is interdisciplinary in nature. To confront it our knowledge of materials must be integrated with fundamental scientific and engineering concepts drawn largely from the fields of mechanical and chemical engineering. Solid mechanics, strength of materials, heat and mass transfer, and fluid mechanics—a substantial part of the educational core baggage borne by engineers—all play important roles in processing. For simplicity, only elementary descriptions and applications of these subjects are considered. Although processing has been benignly neglected in this book until now, it is hoped a new perspective and appreciation of materials properties will emerge from the implications of the present chapter.

Let us start with casting or, more appropriately, solidification processing.

8.2. SOLIDIFICATION PROCESSING OF METALS

Casting is a viable process because liquid metals flow readily, have a high density, and are good conductors of heat. **Fluidity,** or the ability to flow, is inversely proportional to viscosity. Liquid metals are unique in having very low viscosities (e.g., $\eta = 10^{-3}$ P or 10^{-4} Pa-s) and therefore flow readily. Coupled with high densities, liquid metals easily fill molds, whereas less dense nonmetallic fluxes, drosses, and oxide slags float to the surface where they can be skimmed off. High heat transfer rates make it possible to complete solidification rapidly, making casting production rates economical. A wonderfully diverse collection of casting processes has been developed over thousands of years. New processes as well as modifications of older ones continue to appear. The reason, as we shall see, is that solidification processing is a very attractive way to shape not only metals, but other materials as well.

8.2.1. Expendable Mold Casting

8.2.1.1. Lost Wax Method

Lost wax casting employs expendable molds that are used just once because they are destroyed in the process of making a single impression. Much can be learned about casting by considering how the bronze sculpture **Dream 1,** shown in Fig. 8-1A, was made. Except for new metal compositions and improvements in waxes and mold materials, tooling, furnaces, and other aspects, the basic process steps have not changed since antiquity. First the sculptor created an exact wax replica or pattern of what the final bronze was envisioned to look like. Then the wax was essentially removed without a trace and completely

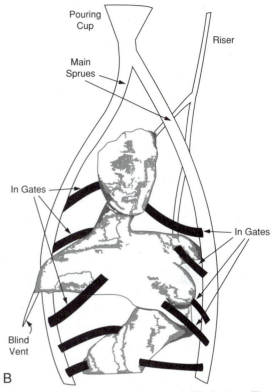

FIGURE 8-1	(A) *Dream I*, a bronze sculpture by Professor R. B. Marcus, New Jersey Institute of Technology. The sculpture is 24 in. high. (B) Schematic of the gating system used for this casting. The risers provide metal to feed shrinkage and enable gases to escape. Gates help reduce turbulent flow of the melt as it fills the mold. The blind vent ensures that voids do not form in corners.

replaced by bronze. This was not achieved by magical sleight of hand, but rather through the following time-honored series of steps:

1. Wax rods are initially attached to the wax pattern to create a type of plumbing system (Fig. 8-1B) which directs the flow of metal from the **pouring cup,** through a **sprue** and into **gates** that meter the molten metal flow into the casting. Provision must also be made to attach a wax **riser.** When filled, the riser provides a reservoir of molten metal that can flow into and fill casting cavities caused by metal shrinkage during the transformaton of liquid to solid bronze.

2. The wax and the gating system are either immersed in a plasterlike, water slurry of refractory oxide powder (e.g., SiO_2, CaO, ZrO_2, plus binders) or stuccoed over with this mold material until completely surrounded. To replicate wax copies or additional castings split or multipiece permanent molds with parting lines are used.

3. After drying, the entire mold assembly is inverted and placed in a furnace. The wax is melted, drained, and then burned out to remove the last vestiges. In the process the mold hardens and is capable of withstanding the shock of hot metal contact.

4. Molten metal is then poured into the empty cavity created by the lost wax. After solidification the mold is broken apart to free the casting. The solid gating and riser rods are sawed off and the metal surface undergoes a long finishing operation to enhance the beauty of the bronze and the artistic quality of the final sculpture.

Note that the original pattern and final sculpture are not solid but rather hollow shells with a wall typically about 0.5 to 0.75 cm thick. This not only reduces the sculpture weight but eliminates thermal and mechanical damage to the mold and lessens the probability of casting defects. **Lost wax** casting is of course not simply limited to creating bronze sculpture. All metals can be cast by this technique. In fact, it is the preferred method for producing some of the most critical, intricate, and costly shaped castings that are manufactured, for example, the prostheses in Fig. 1-10. The cobalt base alloys used to make these precise castings necessitate a high-temperature refractory ceramic (e.g., ZrO_2) mold material or investment. This is why lost wax casting is also known as **precision** or **investment casting.**

Among the most sophisticated of all castings are jet engine turbine components. Higher temperatures enable turbine blades to operate more efficiently but unfortunately accelerate thermal degradation processes due to creep. The first problem is ameliorated by incorporating fine internal passages that help cool them. To address the second problem we note (in Section 10.6) that grain boundaries are important sites for creep damage. Therefore, one strategy is to increase the casting grain size as shown in the turbine blades of Fig. 8-2 (see color plate). Ideally, single-crystal turbine blades with no grain boundaries are most immune to creep damage. This goal is achieved by first directionally solidifying the melt to induce columnar grain growth. One such grain is then isolated and constricted to form the seed onto which the single-crystal blade subsequently grows. Monocrystalline turbine blades have proven to be a significant technological advance. Other products commonly cast by lost wax techniques include tools, gold rings, and precious metal jewelry.

8.2.1.2. Sand Casting

Based on a weight criterion, more sand castings are produced than any other type. Cast iron as well as aluminum engine blocks, cast iron machine tool bases, steel machinery components of all types, stainless-steel pump housings, plumbing fixtures of all sizes, and huge bronze ship propellers are examples of some sand casting applications and the metals used. Wooden or plastic patterns of the desired shapes together with attached gates and risers are first embedded into a special casting sand contained within halves of a split mold.

This allows the pattern and its attached gates and risers to be easily removed to create the cavity, which is subsequently filled by molten metal, after the halves have been rejoined. A good idea of sand casting can be obtained from the exploded view of Fig. 8-3A.

In an important variant of sand casting known as **shell mold casting,** a thin shell (0.5–1 cm thick) of a sand–resin mixture is used as the mold. In this form split molds can be stored; when needed they are clamped together, backed by sand, and then filled with molten metal.

Another expendable mold process employs easily shaped and produced low-density polystyrene patterns of the objects to be cast. After they are embedded in sand, molten metal is poured in, rapidly vaporizing the polystyrene. Finally, metal completely displaces the original polystyrene pattern. Applications for such **full mold castings** have included aluminum cylinder heads, brake components, and manifolds for automobiles.

8.2.2. Permanent Mold Casting

8.2.2.1. Die Casting

Molds are used over and over again in permanent mold casting processes. The most widely practiced of these methods is known as die casting. Dies are extremely important manufacturing tools made of hard, tough alloy steels that enable components made of metals, polymers, and ceramics to be shaped and replicated. They not only are used as molds to contain molten metals and polymers, but find extensive applications in mechanical forming operations like extrusion, wire drawing, sheet metal forming, and the pressing of metal and ceramic powders. Because the process chamber and die steels cannot withstand lengthy elevated-temperature contact with molten metals, die casting is limited predominantly to the low-melting-point metals like zinc and aluminum alloys. In the widely used **hot chamber process,** shown in Fig. 8-3B, molten metal is first pressurized to anywhere from 14 to 35 MPa (2–5 ksi). It is then injected into the clamped, split die mold, via a gooseneck, and held under pressure until it solidifies. The die is then opened, the solidified part ejected and the cycle repeated. Production rates depend on the alloy used and shape and weight of the casting. Approximately 900 shots or injections of zinc per hour are possible.

A variant of die casting is the **cold chamber process.** Higher-melting-point metals like aluminum, magnesium, and some copper alloys can be cast this way by first ladling out metal into the shot chamber and then rapidly ramming it into the die cavity. Pressures employed range from 20 to 70 MPa (3–10 ksi) but can be even higher.

Parts made by die casting include lawnmower engine and motor housings, carburetors, hand tool housings and parts, appliance components, door handles, and zipper teeth. A collection of die castings is shown in Fig. 8-4. Most die castings weigh between 0.1 and 25 kg.

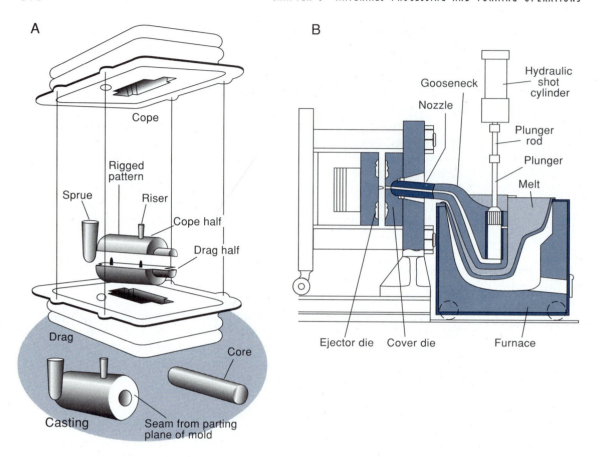

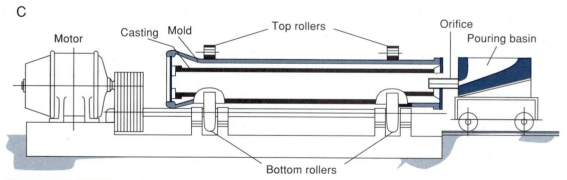

FIGURE 8-3　Casting methods. (A) Sand casting. From H. J. Heine, *Casting Design and Applications* (Fall, 1989). (B) Hot chamber die casting. (C) Centrifugal casting of pipes.

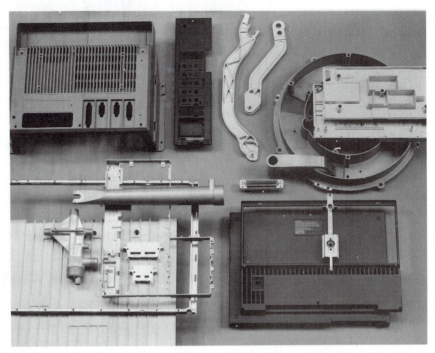

FIGURE 8-4 Assorted aluminum, magnesium, and zinc die castings. Courtesy of Chicago White Metal Casting, Inc.

8.2.2.2. Centrifugal Casting

As the name suggests, centrifugal casting (Fig. 8-3C) involves rotation to increase the inertial force that drives molten metal to fill mold cavities. By spinning permanent molds at ~1000 rpm, sounder pipes, cylinder liners, and hollow cylindrical parts of all metals can be cast while retaining good dimensional tolerances. Investment cast gold rings are often rotated about the vertical sprue axis during pouring to better fill intricately shaped molds.

Now that a number of expendable and permanent mold casting processes have been introduced, it is instructive to compare their relative merits. This is done in a self-explanatory way in Table 8-1.

8.2.3. Heat Transfer Considerations

It is important to know the temperature history at all locations within solidifying castings to identify hot spots that may become potential defect sites. There will be some discussion of how this is accomplished employing computer methods, but it is helpful first to have a physical feeling for the problem. Therefore, let us calculate the rate of solidification of the melt shown in Fig. 8-5. It is assumed that nucleation of solid from the slightly undercooled liquid

TABLE 8-1   **GENERAL CHARACTERISTICS OF CASTING PROCESSES**

Process	Metals cast	Weight (kg)		Dimensional accuracy[a]	Shape complexity[a]	Porosity[a]
		min	max			
Sand	All	0.05	No limit	3	1–2	4
Shell	All	0.05	100+	2	2–3	4
Investment	All (High melting point)	0.005	100+	1	1	3
Permanent mold	All	0.5	300	1	3–4	2–3
Die	Nonferrous (Al, Mg, Zn, Cu)	<0.05	50	1	3–4	1–2
Centrifugal	All	—	5000+	3	3–4	1–2

[a] Relative rating: 1 = best, 5 = worst. Variations can occur, depending on the evaluation methods used.

From S. Kalpakjian, *Manufacturing Processes for Engineering Materials,* 2nd ed. © 1991 by Addison–Wesley Publishing Company, Inc. Reprinted by permission of the publisher.

is not a problem and that the solid–liquid interface is linear. Interface motion depends on how fast heat is extracted from the melt. Initially, a relatively insulating mold contains liquid metal at its melting point T_M. As metal of density ρ_M freezes, its latent heat of fusion H (J/kg) is liberated at the interface between liquid (L) and solid (S). This heat quickly flows through the already

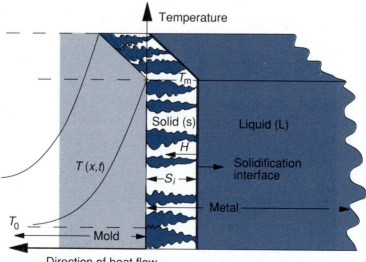

FIGURE 8-5   Model of solid metal solidification front advancing into a melt at T_M. The liberated latent heat flows through the solidified metal and into the mold.

solidified metal and enters the mold, raising its temperature. Concurrently, the amount of metal solidified increases with time. The position of the solidification front (S_i) as function of time (t) is given by

$$S_i = \frac{2}{\pi} \frac{(T_M - T_0)(\kappa c \rho_m)^{1/2} t^{1/2}}{\rho_m H}, \tag{8-1}$$

where κ, c, and ρ_m are the thermal conductivity (W/m-K), heat capacity (J/kg-K), and density (kg/m^3) of the mold material, respectively. (Derivation of Eq. 8-1 is left as an exercise in Problem 8-1.)

Conduction or diffusion of heat obeys the same physical laws as diffusion of atoms in solids. In fact, heat conduction in one dimension obeys equations similar to Eqs. 6-1 and 6-2 if C is replaced by T. Additionally, in Eq. 6-1 the mass flux J must be replaced by the heat flux Q (in units of J/m^2-s or W/m^2), and D by κ (in units of W/K-m-s). But in Eq. 6-2, the diffusion coefficient D is replaced by the thermal diffusivity α, which is equal to $\kappa/\rho_m c$; both D and α have identical units (m^2/s). Lastly, the ambient temperature is T_0. Note that S_i depends on the product of two factors, one containing metal dependent, and the other, mold dependent properties. The solidification front is seen to advance parabolically in time. This result should not be too surprising as the diffusional spreading of matter also follows a $t^{1/2}$ dependence.

Figure 8-5 reveals that S_i is equal to the volume of the solidified metal divided by the area of the planar solidification front. This result can be generalized to arbitrarily shaped solidified regions of volume (V) and (curved) surface area (A). Furthermore, if these quantities refer to the final casting volume and surface area, the time can be associated with the solidification time (t_s). Substitution of these terms in Eq. 8-1 yields Chvorinov's rule:

$$V/A = Ct_s^{1/2}; \qquad C = \text{a constant.} \tag{8-2}$$

This equation has been verified for both small and large castings in insulating molds.

8.2.4. Structural Implications of Solidification Rates

The heat flow analysis given above provides no clue as to the grain structure of the casting, an important consideration as already noted in Sections 3.6.1 and 8.2.1. Actually, dendritic growth and columnar grains are favored when the temperature gradient in the melt is small and the solid–melt interface velocity is large. When the opposite conditions prevail, as in crystal pulling (see Section 6.4.3), a planar, nondendritic solidification front develops.

In Section 9.4.2 we shall see that both higher strength and ductility of metals are associated with fine grain size. The same is true of dendritic castings, but instead of grain size the appropriate dimension of concern is the **secondary dendrite arm spacing**. Growing perpendicular to the main dendrite trunk are the secondary dendrite arms (see Fig. 5-30) and their interspacing can be reduced by cooling the melt more rapidly. In the process, undesirable composition variations due to coring are minimized. Unfortunately, heat transfer concerns limit the size of the casting whose properties can be enhanced this way. At extreme cooling rates of $\sim 10^6\,°C/s$ the arm spacing may be effectively reduced to atomic dimensions if amorphous structures result.

EXAMPLE 8-1

Suppose solid aluminum spheres 0.2 m in diameter solidify in 30 minutes. An engineer suggests casting solid hemispheres sequentially (followed by joining) as a time-saving measure. Is this course of action wise?

ANSWER   The constant C can be determined for this casting process by direct substitution of the quantities for the sphere. Therefore by Eq. 8-2,

$$C = t_s^{-1/2}(V/A) = (30^{-1/2})[(4\pi(0.1)^3/3)/(4\pi(0.1)^2)] = 6.09 \times 10^{-3}\ \text{m/min}^{1/2}.$$

Assuming C is the same for casting hemispheres, the time it will take each to solidify is

$$t_s = \frac{1}{C^2}\left(\frac{V}{A}\right)^2 = \frac{1}{(6.09 \times 10^{-3})^2}\frac{[2\pi(0.1)^3/3]^2}{[2\pi(0.1)^2 + \pi(0.1)^2]^2} = 13.3\ \text{minutes}.$$

The suggestion does not appear to be wise because it will only leave $30 - 2(13.3) = 3.4$ minutes, at most, for additional preparation of two molds, double pourings, and assembly. Besides, duplicate castings are wasteful of mold material.

8.2.5. Design of Castings

Chvorinov's rule suggests that sections of the casting having a large volume-to-surface area ratio will take a long time to solidify. Such hot spots should be avoided when designing castings because they are potential sites of porosity defects. When the melt within a hot spot solidifies it contracts, leaving a shrinkage cavity that must be refilled if the casting is to be sound. If, however, neighboring thin sections solidify around the hot spot, the flow of liquid metal may be choked off, depriving the cavity of metal. Such a problem might develop when casting a metal bowling pin upright, the orientation depicted in Fig. 8-6. After the mold is completely filled, grains bridge in the neck section, restricting metal flow to the bulbous base. A mass of liquid is trapped that cannot be fed, and when it solidifies a shrinkage cavity develops. The situation is remedied by casting the pin upside down. Sufficient liquid metal will be stored above in the riser to feed the shrinkage in the casting below. The desirable progressive **directional solidification** will thus result in a sound casting.

Although the preceding example demonstrated the need for orienting the casting properly, Fig. 8-7 illustrates a number of poorly designed casting features together with recommended alternate designs. A rough but useful way

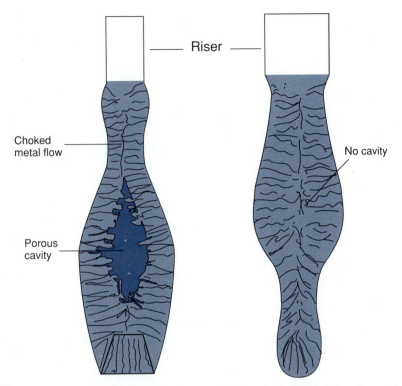

FIGURE 8-6 *Left:* Metal bowling pin cast upright. The melt flow from the riser is choked in the neck, leaving a cavity where feeding is prevented. *Right:* Bowling pin cast upside down. Directional solidification results in a sound casting.

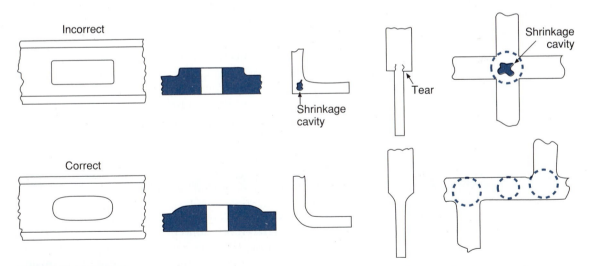

FIGURE 8-7 Incorrect and correct design geometries when casting the indicated sections. For the cross at the extreme right, spheres (circles) are inscribed to reveal casting hot spots. Shrinkage cavity formation and hot tearing effects can be minimized by modifying casting designs to eliminate hot spots, sharp corners, and nonuniform sections.

to locate hot spots at suspected sites is to find where the largest circle (sphere) can be inscribed. Large radii mean long solidification times because of the large volume-to-area ratio. On this basis, the design modifications shown reduce the probability of shrinkage defects. Sharp corners and abrupt changes in dimension should also be avoided to minimize stress concentrations. Differential thermal contraction and strength beween contiguous hotter and cooler regions of metal promote a tendency toward cracking and tearing.

8.2.6. Computers and Casting

We now consider the computer modeling of the temperature–time distribution throughout a complex shaped casting after pouring. The procedure is conceptually straightforward but rather involved in practice. First the casting, riser, and mold geometries are subdivided into a bricklike array of elements. Hundreds to hundreds of thousands of elements may be required depending on the complexity of the casting geometry. Changes in temperature of any given element caused by conduction, radiation, and convection heat transfer to it from neighboring elements are calculated by appropriate mathematical techniques. The temperature solutions are dovetailed together in three dimensions in a self-consistent manner, and the results are updated at desired time intervals. Temperature-dependent mold and metal thermal and heat transfer constants are folded into the analysis to make the simulation as accurate as possible. The power of the method is illustrated in Fig. 8-8A (see color plate), where the computer-calculated time sequence of the cooling behavior of the wheel reveals the presence of a hot spot. Note that on either side of the hot spot (void) the metal is cooler. The situation is improved by redesigning the rim to be lighter. As seen in Fig. 8-8B (see color plate), solidification now proceeds in the desired directional manner.

Computers have dramatically transformed the art of engineering design and manufacturing into a science. They enable the concurrent integration of computer-aided design (CAD) and manufacturing (CAM), involving stress (strength) and solidification analyses of the part to be cast, with the production of the required tooling (dies, molds). How this is all accomplished in the case of a lightweight automobile engine connecting rod is illustrated in Fig. 8-9.

8.3. MECHANICAL FORMING OPERATIONS

Unlike the casting of liquids that readily flow to fill molds by gravity, mechanical forming operations redistribute solid matter under the influence of applied forces. Examples of common bulk mechanical forming operations are schematically depicted in Fig. 8-10. Included are forging, rolling, extrusion, and drawing processes that one associates with the primary forming of metals. In contrast, sheet metal forming operations (treated in Section 8.3.6) are nor-

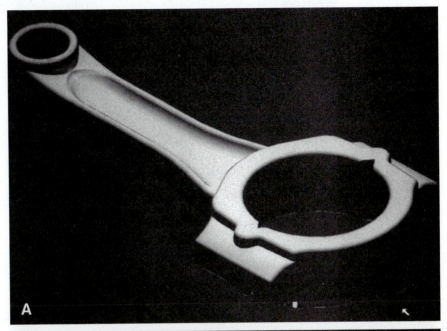

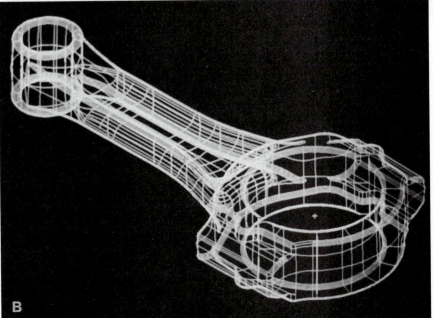

FIGURE 8-9 CAD/CAM methods used in producing a lightweight automobile engine connecting rod. (A) Shaded surface model. (B) Computerized "wireframe" model. (For parts C and D, see color plates.) (E) Computer-controlled cutter path for developing prototype tooling. Courtesy of G. Ruff, CMI International, Inc.

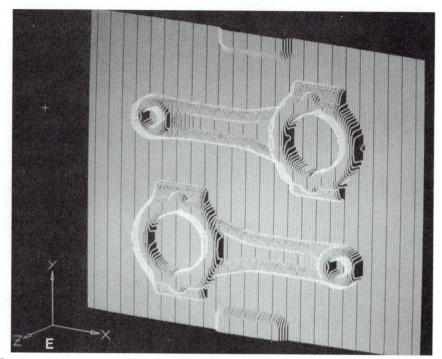

FIGURE 8-9 (continued)

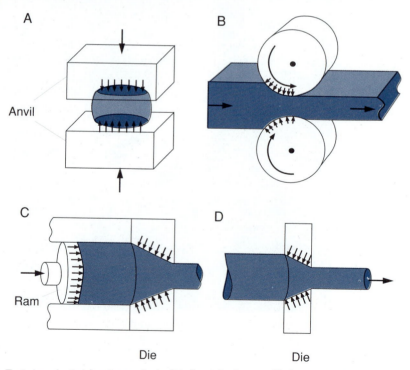

A

Anvil

B

C

Ram

Die

D

Die

FIGURE 8-10 Typical mechanical forming methods. (A) Open die forging. (B) Rolling. (C) Extrusion. (D) Drawing. Arrows show forces on workpiece.

mally carried out in secondary manufacturing. The dividing line is not so sharp, however, and there are examples of each forming process shown in Fig. 8-10 being practiced in secondary manufacturing environments. All bulk metal forming processes can and have been carried out at room temperature, but it is more common to practice them at elevated temperatures, especially when strong, large shaped objects are involved. In this way greater ductility is achieved and lower plastic flow stresses can be capitalized on to reduce the load-bearing requirements of the forming equipment.

Mechanical forming of metals often endows them with superior mechanical properties. The traditional route of metals processing typically includes ingot casting followed by either (hot) rolling, extrusion, or forging at elevated temperatures. This causes a cast structure that is dendritic, chemically inhomogeneous, and full of shrinkage voids or porosity to be broken up by the combined effects of stress and high-temperature diffusion. The prior casting of reasonable strength but low ductility is converted into a much tougher and homogeneous material. Then the metal is frequently shaped by low-temperature forming, for example, cold rolling, where the requisite strength, ductility, and uniform grain structure are conferred for final manufacturing operations, for example, deep forming of sheet metal. These processes are illustrated in Fig. 8-11 for Monel metal. Starting with the dendritic cored structure shown in Fig. 5-30, the sequence of thermomechanical treatments progressively alters the grain structure as shown.

What is true of each forming operation in Fig. 8-10 is that the required force is usually applied in one direction; however, it invariably happens that this unidirectional force actually stresses the workpiece not only in one, but in two approximately perpendicular directions simultaneously. For example, pushing the work through the die during extrusion creates a compressive reaction from the sidewalls. And the tension applied to a wire to reduce its cross-sectional area results in a similar compressive die wall reaction on the drawn metal. Even during forging where the work appears to be simply compressed, lateral stresses are frequently generated. Frictional contact at the anvil–workpiece interface establishes forces that restrain the work from expanding laterally, and causes bulging (see Fig. 8-10A).

8.3.1. Deformation due to Biaxial Stresses

Let us expand these concepts by considering the application of mutually perpendicular forces on a solid cube as shown in Fig. 8-12. For simplicity, only the two-dimensional state of stress in the projected square section is shown. In case (A) both stresses σ_x and σ_y are tensile and assumed equal (balanced biaxial tension). Under the action of σ_x alone the deformation displaces (very exaggeratedly) the two halves of the square such that overall it extends in the x direction and contracts in the y direction. If σ_y is now sequentially applied, the square will extend in the y direction and contract in the x

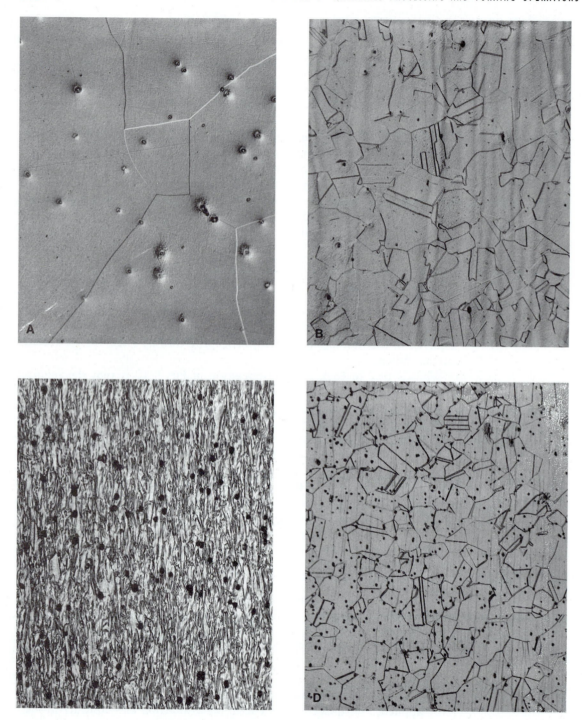

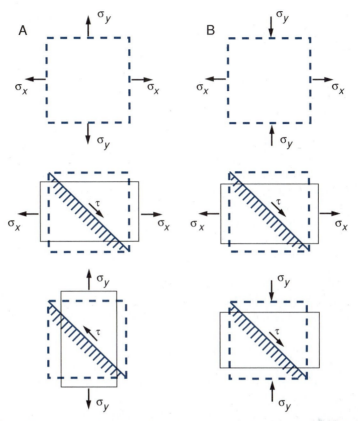

(A) Biaxial tension. In balanced biaxial tension or compression, superposition of the separate uniaxial stresses leads to zero net shear stress. (B) Body acted upon by mutually perpendicular tensile and compressive loads. Superposition of these uniaxial stresses leads to an enhancement of the shear stress components.

direction. A *superposition* of the two stresses will leave the square unchanged. In essence the shear stresses that act along the diagonals have vanished. If σ_x and σ_y become very large they will ideally tear the cube apart rather than plastically deform it.

Let us see what happens in case (B), where the sign of σ_y is reversed, so that it is compressive but with the same magnitude as σ_x. Clearly the deformation is doubled, as the amount produced by σ_x alone is now reinforced by a like

Microstructures resulting from the thermomechanical processing of Monel (nominally 63 wt% Ni, 32 wt% Cu, plus Fe and Si). The cast microstructure is shown in Fig. 5-30. Sequential treatments and microstructures: (A) Homogenized at 1120°C for 20 hours. (156×). (B) Hot forged between 650 and 1000°C (156×). (C) Cold rolled to 75% reduction in area (156×). (D) Annealed at 870°C for 1 hour (156×).

amount through the action of σ_y. Under uniaxial loading conditions it was shown in Example 7-1 that the *shear stress* on the diagonal plane has a magnitude equal to *half* that of the applied stress. Therefore, these physical results can be algebraically expressed through addition of the stresses by

$$\tfrac{1}{2}\sigma_x - \tfrac{1}{2}\sigma_y = \tau. \qquad (8\text{-}3)$$

Readers familiar with the elements of structural mechanics (or Mohr's circle construction) will recognize this result; in this case σ_x and σ_y are the so-called *principal stresses*.

For our needs we must realize that forming operations involve plastic and not elastic stresses, and that Eq. 8-3 is a formula that is derived from elasticity. Nevertheless, it also applies to plasticity. If a bar of metal is pulled in uniaxial tension we know that it will plastically deform when the yield or flow stress (σ_0) is reached; when this happens the plastic shear stress τ is equal to $\sigma_0/2$. Assuming τ, σ_x, and σ_y can be associated with the corresponding plastic stresses,

$$\sigma_x - \sigma_y = \sigma_0. \qquad (8.4)$$

This equation represents a criterion for the onset of plastic deformation and is known as the **Tresca condition.** Simply stated, plastic deformation can be expected when the difference between the principal (plastic) stresses is equal to the yield stress of the material. A practical implication of Eq. 8-4 is that the best way to promote plastic forming operations is to apply combined stresses that maximize the internal shear stresses. Mutually perpendicular applied tensile and compressive forces, produced by pulling and pushing the workpiece, will have the desirable effect of enhancing plastic deformation.

8.3.2. Forging

Consider the forging of a rectangular block of metal compressed between a pair of flat, parallel anvils as shown in Fig. 8-13. In practice, hammerlike impact loads are usually applied. If the yield stress is σ_0 we wish to know the compressive load required to deform the block. The presence of friction at the work–anvil interface means that tangential shear stresses exist such that $\tau/\sigma_y = f$, where f is the coefficient of friction. It also means that compressive stresses in the x direction, generated by friction forces, are distributed throughout the workpiece. The effect is cumulative. At the edges $x = \pm a$, the frictional forces are zero, but they build symmetrically to a maximum value at the center of the forging. A so-called **friction hill,** which becomes higher and steeper with increasing f, develops. The larger f is, the greater is the applied force necessary to plastically deform the block. Undesirable barrel-shaped forgings are a manifestation of work–die friction. Theory shows that the compressive stress in the y direction varies exponentially about $x = 0$ as

$$\sigma_y = -\sigma_0 \exp[2f(a - x)/h]. \qquad (8\text{-}5)$$

A

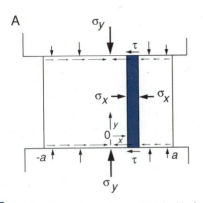

B

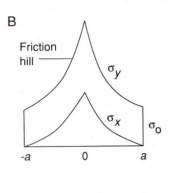

FIGURE 8-13 (A) Forging geometry in a rectangular block. Stresses acting on the interior shaded element are indicated. (B) Distribution of compressive stresses, σ_x and σ_y, versus distance. Note that vertical surfaces of the forging are stress free ($\sigma_x = 0$).

Similarly, σ_x, the compressive stress in the x direction, also reflects the symmetrical friction hill. The two stresses are related by Eq. 8-4 and their variation with x is indicated in Fig. 8-13B. Because the total area under the σ_y stress distribution is proportional to the total forging load, every effort is made to reduce f and σ_0. And since elevated temperatures reduce the yield stress, forging and other bulk metal working operations such as rolling and extrusion are carried out hot.

If f/h is sufficiently small, then the exponential term can be approximated by $1 + 2f(a - x)/h$. These are the first two terms of the series expansion for the exponential. In such a case the simpler linear formula

$$\sigma_y = -\sigma_0[1 + 2f(a - x)/h] \qquad (8\text{-}6)$$

can be used. To a good approximation, Eqs. 8-5 and 8-6 also hold for forging cylinders of radius r and height h, if r is substituted for a.

EXAMPLE 8-2

A 10,000-ton (8.88×10^7N) press is used to forge Cu alloy blocks that are 0.1 m high having a square cross section equal to 0.0225 m². An average flow stress of 600 MPa is maintained throughout the forging process under conditions where $f = 0.2$.

a. What load is required for forging?

b. If the coefficient of friction were reduced to 0.1, what load would be required?

c. What is the largest average flow stress this press can handle if the metal is in the form of a 0.15-m-high ingot with a square cross section measuring 0.10 m²? Assume $f = 0.1$.

ANSWER   a. As $a = 0.075$ m, $2fa/h = (2)(0.2)(0.075)/(0.10) = 0.3$. Because the exponent is less than 1, we can use Eq. 8-6. The load (P) required is the product of the average value of σ_y and base area. Calculation of the peak value of $\sigma_y = \sigma_0(1 + 2fa/h)$ yields $600(1 + 0.3) = 780$ MPa, whereas the minimum value of σ_y at $x = a$ is 600 MPa. The average value of σ_y is, therefore, $\frac{1}{2}(780 + 600) = 690$ MPa. Hence,

$$F = (690 \times 10^6 \text{ Pa})(0.0225 \text{ m}^2) = 1.55 \times 10^7 \text{ N}.$$

b. If $f = 0.1$, the average stress is half the sum of $600(1 + 0.15)$ and 600 or 645 MPa. The forging load is reduced to

$$F = (645 \times 10^6)(0.0225) = 1.45 \times 10^7 \text{ N}.$$

c. In this problem we must equate the press capacity (F_p) to the load sustained in the metal. Hence, $F_p = \sigma_0[(1 + 2fa/h + 1)/2](A)$, where $A = 0.1$ m^2, $a = 0.316$ m, $h = 0.15$, and $f = 0.1$. Solving, $8.88 \times 10^7 = \sigma_0[1 + (0.1)(0.316)/(0.15)](0.1)$ yields $\sigma_0 = 7.33 \times 10^8$ Pa, or 733 MPa.

8.3.3. Forging Practice and Products

At the outset a distinction between open and closed die forging should be noted. Open die forging is the type to which the previous section applies. Large shapes such as steam turbine rotors and military cannons are forged between open die anvils of simple shapes. A blacksmith's hammer and anvil would loosely approximate open dies; however, many additional finished products like automobile crankshafts, wrenches, hand tools, turbine disks, railroad wheels, and gears have, at one time or another, been forged within shaped dies that close around the metal within. In the case of the crankshaft shown in Fig. 8-14, the steel rod is first roughly forged to bulge the metal in the right places. Then it is progressively forged between intermediate closed blocker dies which shape it further until finishing dies can be used. After several strikes a faithful impression of both die faces is made. Directed metal flow in dies confers a type of graininess to the forging, making it stronger in directions parallel to the deformation. The crankshaft emerges but with a thin flare or **flash** of steel that spreads outward from the die parting line. After the flash is ground away the finished forged shaft emerges. Since the flash is thin (h small), Eq. 8-5 indicates that the forging load rapidly mounts. Simultaneously, the flash is sufficiently strengthened so that it prevents metal in the closed die from extruding out. Friction promotes metal filling of the cavity, and without it closed die forging would not be feasible.

Metals must be ductile to be forged, but for casting purposes there is no such restriction. In general, forgings exhibit greater toughness and ductility than equivalent castings. Nevertheless, cheaper castings compete with and have replaced forgings in a number of applications including some automotive connecting rods (see Fig. 8-9) and crankshafts. For function in a car a bent

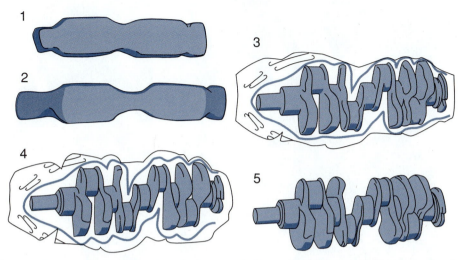

FIGURE 8-14   Stages in the closed die forging of a crankshaft. (1), (2) Preliminary roll forging. (3) Blocking in closed dies. (4) Finishing in closed dies. (5) Flash trimming. From *Metals Handbook*, 8th ed., Vol. 5, American Society of Metals, Metals Park, OH (1970).

forged crankshaft is as bad as a broken cast one. Therefore, the less tough, cast crankshafts sometimes serve just as well.

8.3.4. Rolling

Rolling of metals, a popular mechanical forming process, results in a reduction of the initial thickness as shown schematically in Fig. 8-10B. Work in the form of slabs or sheets is fed in at the left and the thinned material exits the rolls at higher speed. If the process is frozen in time, rolling is not unlike forging. A difference is that an asymmetrical stress distribution (friction hill) arises in the deformed metal because of the more complicated curved roll (anvil)–work geometry. The sheer volume of steel that passes between rolling mills dwarfs, by far, that of other metals. Irrespective of metal, however, the process is carried out at hot, warm, and cold temperatures. Hot refers to temperatures of $\geqslant 0.6 T_M$, where T_M is the melting point in degrees Kelvin. Recrystallization of the worked matrix can occur at such temperatures. (See Chapter 9.) Between $\sim 0.3\ T_M$ and $0.6 T_M$, the warm working regime is operative. Cold rolling is usually done at temperatures lower than $\sim 0.3 T_M$ and most frequently at room temperature. Normalizing to the melting point of the material means that temperatures for cold working of tungsten are comparable to those required for hot rolling steel. As the rolling temperature drops, the thickness of the stock that can be rolled must necessarily decrease. Also, thinner stock is rolled at higher roll velocities and reaches delivery speeds of over a mile a minute in the case of sheet steel or aluminum foil.

FIGURE 8-15 Microstructures of cold-rolled cartridge brass (70Cu–30Zn) starting with the undeformed metal of Fig. 3-32. (A) 15% reduction. (B) 30% reduction. (C) 50% reduction. (D) 75% reduction. 80×. Courtesy of G. F. Vander Voort, Carpenter Technologies Corporation.

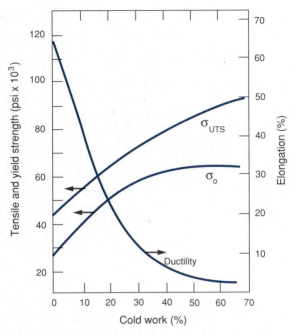

FIGURE 8-16 Effect of cold working on the yield stress, tensile stress, and ductility (percentage elongation) of cartridge brass (70Cu–30Zn).

If the initial stock has an equiaxed grain structure, cold rolling will elongate the grains and string out the grain boundaries in a distinctive manner. This is illustrated for cartridge brass in Fig. 8-15. The dislocation density grows enormously and the metal becomes harder. As shown in Fig. 8-16, strength properties increase significantly, but simultaneously, ductility measured by percentage reduction in area suffers with increasing amount of cold work. Surprisingly, these microstructural and mechanical property trends can be reversed by annealing; a metal approximating the original one prior to rolling can be regenerated through heating to elevated temperatures. Discussion of these important practical treatments is continued in the next chapter (see Section 9.4).

EXAMPLE 8-3

For a certain application, brass plate must have a tensile strength of 45,000 psi, a minimum ductility of 25% elongation, and a thickness of 0.5 in. The only plate stock available is 1.0 in. thick. Design a processing schedule that would enable the required specifications to be met.

ANSWER   The total percentage reduction required is $[(1.0$ in. $- 0.5$ in.$)/1.0$ in.$] \times 100 = 50\%$. If the plate were rolled 50% in one pass, the tensile strength would be close to 80,000 psi but the elongation would only be about 5%. To achieve a 25% elongation, the maximum amount of cold work that can be tolerated is only 20%. Fortunately, such a deformation will produce a more than sufficient strength of ~60,000 psi. Therefore, we must ensure that the final pass involves no more than a 20% reduction. Therefore, the thickness d must be $[(d - 0.5)/d] \times 100 = 20$. Solving, $d = 0.625$ in.

Thus, the required processing could involve a single reduction from 1 to 0.625 in. [i.e., $[(1.0 - 0.625)/1.0] \times 100 = 37.5\%$ reduction], followed by an annealing treatment to totally soften the brass to the original condition. This should be followed by the final 20% reduction. Alternately, several rolling passes could be initially employed to reduce the stock to 0.625 in., but the brass must then be fully annealed prior to the final 20% thickness reduction.

8.3.5. Extrusion and Drawing

Extrusion and drawing (Fig. 8-10) are normally treated together because they are analyzed in similar ways. In extrusion, materials are *pushed* through the die with a ram. Toothpaste flowing through the tube opening is an example; a shaped Playdoh extrusion is a better example. Like forging, extrusion is frequently employed to break up undesirable dendritic structures of large cast-ings. In drawing, material is reduced in diameter by *pulling* it through a die using grips. It is a simple matter to estimate the stress required for these processes if friction between the die and work can be neglected. Suppose a billet of length L_0 and area A_0 is extruded through a die by applying uniform pressure p. The external work, W, done by the ram is the product of the force, pA_0, and the length, L_0, over which it acts. Therefore, $W = pA_0L_0$. This work is equal to the plastic deformation energy expended in the billet,

$$A_0L_0 \int_{L_0}^{L} \sigma_0 \, d\varepsilon = A_0L_0 \int_{L_0}^{L} \sigma_0 \frac{dl}{l}.$$

Direct integration yields $\sigma_0 A_0 L_0 \ln(L/L_0)$, where the final length is L. Equating, $pA_0L_0 = \sigma_0 A_0 L_0 \ln(L/L_0)$, or

$$p = \sigma_0 \ln(L/L_0) = \sigma_0 \ln(A_0/A). \qquad (8\text{-}7)$$

This formula makes use of the constancy of volume $(A_0L_0 = AL)$, where A is the final billet area. In extrusion, compression is involved (p negative), whereas in drawing, tension is applied (p positive). The quantity A_0/A is known as the extrusion ratio R and has values typically ranging from 10 (for Ni superalloys) to more than 100 (for Al). Required deformation pressures are seen to increase with both the material yield strength and the extrusion ratio.

EXAMPLE 8.4

a. What is the percentage reduction of area (%RA) if $R = 10$? If $R = 100$?

b. What is the maximum percentage reduction attainable during extrusion for a non-work-hardening (perfectly plastic) material?

ANSWER   a. The %RA is defined by $100(A_0 - A)/A_0 = 100 (1 - 1/R)$. For $R = 10$, the %RA = 90%. If $R = 100$, the %RA = $100 (1 - 1/100)$ = 99%.

b. If the material does not work harden, the extrusion pressure cannot exceed σ_0. Therefore, $\sigma_0 = \sigma_0 \ln (A_0/A)$, or $\ln R = 1$. Thus, $R = e$ or 2.718, and the %RA is $100 (1 - 1/e)$ = 63.2%.

In actuality, friction is present and materials may harden during forming. Therefore, Eq. 8-7 is too simplistic. An approximate formula that is used instead is

$$p = K_e \ln R, \qquad (8\text{-}8)$$

where K_e is the extrusion constant, a quantity that incorporates many factors. Experimentally determined values of K_e for a number of metals, extruded at different temperatures, are given in Fig. 8-17. Virtually all materials, even normally brittle ceramics, can be extruded at sufficiently high temperatures because the compressive stresses of the process promote cohesion. In Fig. 8-18 the grain structure of extruded uranium dioxide fuel rods is depicted.

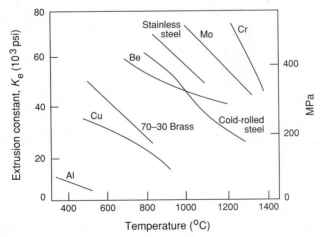

FIGURE 8-17   Extrusion constant as a function of temperature for assorted metals. After P. Lowenstein, ASTME Paper SP63-89.

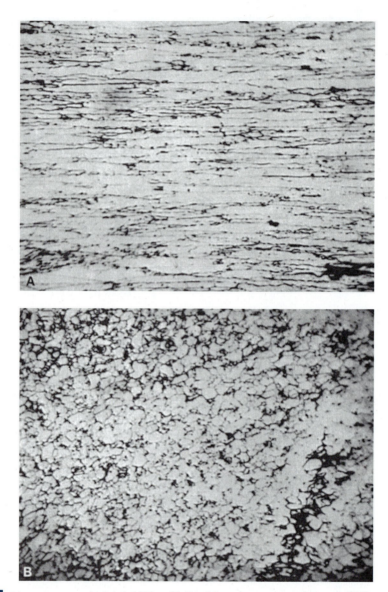

FIGURE 8-18 Microstructures of extruded UO_2: (A) Parallel to the extrusion direction. (B) Perpendicular to the extrusion direction. From Nuclear Metals, Inc.

Pressures of 1.03 to 1.38 GPa (150,000–200,000 psi) combined with temperatures above 1700°C are required before UO_2 flows plastically during extrusion.

A variant of extrusion, **coextrusion,** plays an important role in the technology of superconductors. Practical devices, for example, coils for superconducting magnets, require wire but unfortunately, virtually all of the commercially important superconductor materials are brittle and cannot be drawn without cracking. To overcome this problem, the hard, brittle materials are first incorporated into a softer, more ductile matrix. In the case of NbTi, holes are first drilled along the axis of an aluminum cylinder and cast rods of NbTi inserted. The composite is then extruded and the NbTi diameter shrinks (Fig. 8-19A). After the first extrusion the bundles are packed together in a billet for a second extrusion (Fig. 8-19B), and so on. The process is continued until a superconducting cable of required length and filament diameter is made. Similar methods have been adopted to fabricate fine-diameter, superalloy welding rods used in the repair of jet engine turbine blades. In this case, the matrix must be dissolved away to free the rods.

Among the defects encountered in extruded metals are surface cracking and internal cracking. The former are due to oxides and the latter to low-melting impurities (e.g., sulfur) which embrittle grain boundaries, a phenomenon known as **hot shortness**. A remedy for the latter in steel is to tie up S in the harmless form of MnS through small manganese additions. In addition, tailpipes (**extrusion defect**) invariably develop in the tail section or last portion to be extruded because the metal at the billet center is hotter and flows more rapidly than the metal at the outer edges. These pipes (tapered tubing) must be cut away and can account for as much as a 30% product loss. Reduction of friction, removal of surface scale, and elimination of temperature gradients help minimize the effects of this important defect.

Unlike extrusion, wire drawing tends to be a finishing operation that is generally, but not always, carried out at low temperatures. Wire and wire products are used in electrical applications, cables of all types, fencing, strings for muscial instruments, and other items. Because tensile rather than compressive stresses are involved in drawing, the metals must be intrinsically ductile and furthermore must work harden. If metals do not become stronger with deformation, they might fracture at the die exit because of their reduced cross section. Lead, for example, cannot be drawn into wire because it does not work harden at room temperature. Actually, Pb does harden, but immediately softens due to recrystallization processes (see Section 9.4), so there is no net strengthening.

8.3.6. Sheet Metal Forming

Sheet metal forming and working are among the most widely employed processes in manufacturing. Unlike the mechanical forming processes considered above where the ratio of the volume to surface area is large, the reverse is true in sheet forming. Automotive bodies, aircraft frames, major home appliance housings, and building fixtures are some of the arenas where large-area sheet products are used. But there are countless applications for smaller sheet metal parts, for example, washers, beer cans, kitchenware, and utensils.

Punches and **dies** are the basic shaping and cutting tools used in sheet forming. These tools are mounted in presses that provide the necessary forces

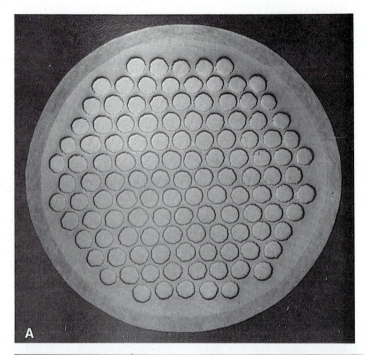

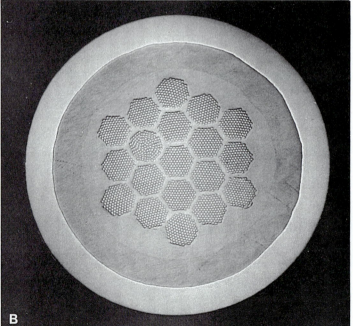

FIGURE 8-19 (A) Cross section of 121 NbTi rods in an extruded Al matrix. The initial diameter was 0.076 m and the extrusion ratio was 10 at a ram speed of 0.0085 m/s. (B) Cross section of a double-extruded superconductor structure containing 19 × 121 NbTi filaments. From AIRCO Superconductor.

to accommodate workpieces that must be sheared, simply bent, curved, deeply recessed, or impressed with a pattern in relief (coining). A couple of examples will illustrate sheet forming extremes. First consider the production of washers, perhaps the most widely produced sheet metal product. The tooling required, shown in Fig. 8-20, consists of successive (male) blanking and piercing punches and the mating (female) die. Provision must be made for feeding the sheet metal and clamping it. The basic forming operation involves shearing across a cylindrical surface. For example, in blanking a washer of thickness d with outer diameter r, a shearing force F_s (stress × shear surface area) between $\sigma_0 2\pi rd$ and $\sigma_{UTS} 2\pi rd$ is required. In practice it is typically found that

$$F_s = 0.7\sigma_{UTS}2\pi rd. \tag{8-9}$$

At the other extreme of sheet metal forming are deep drawing or cupping operations (Fig. 8-21). Here, a circular blank sheet is held down while a punch extends the metal into the die cavity. The relatively small clearance between punch and die guides the metal that forms the cup walls. Stretching (biaxial tension) and drawing (tension–compression) modes of deformation are operative in different portions of the sheet. Deep forming makes stringent demands on metal properties. Large values of the strain hardening coefficient n are necessary to forestall necking and tearing, as noted in Section 7.3.3. It is also

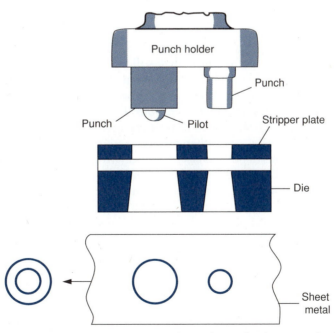

FIGURE 8-20 Schematic of the sheet metal process used to produce washers in a progressive die. From *Metals Handbook*, 8th ed., Vol. 4, American Society of Metals, Metals Park, OH (1969).

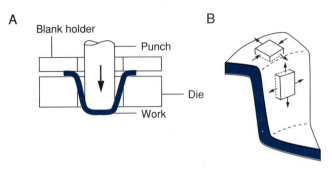

FIGURE 8-21   (A) Deep drawing of a cup. (B) Stress distribution in cup.

desirable that the metal be isotropic. Sheet metals, however, often exhibit a **preferred orientation** or **texture** that is a function of prior rolling and annealing processing. This means that a majority of indivdiual grains assume a more or less common crystallographic orientation relative to the sheet surface. Far from creating anything like a large, thin single crystal, preferred orientation produces a type of grain. The net effect is to superimpose the anisotropy of crystallographic slip on the deformation geometry. An undesirable waviness known as **earing,** which sometimes develops in the rim of formed cups, is a manifestation of this preferred texture.

8.3.7. Residual Stress

It appears paradoxical that solid bodies can be internally or residually stressed even when there are no externally applied forces. Yet this phenomenon of **residual stress** is a rather common occurrence in both solidification processing and mechanical forming operations. In fact, virtually every time materials are processed or heat treated at temperatures different from the use temperature, residual stresses are either produced or relieved. They occur in heat-treated glass, in films deposited electrolytically from solution or from the vapor phase, during welding, and in machining and assembly operations. In general, residual stresses arise from nonuniform changes in size or shape. Often it is not easy to visualize the source of the nonuniformity, much less understand or prevent its occurrence. Like a "jack in the box," residual stresses are sometimes released with disastrous consequences, as in the case of the Liberty Bell. Although they are generally harmful, sometimes residual stresses can be beneficial. By intentional introduction of residual compressive stresses through peening of the surface, materials are better able to withstand harmful applied tensile stresses that cause fracture.

To see how residual stresses are introduced into a casting consider the cooling of the initially stress-free ingot shown in Fig. 8-22A. As the outside of the ingot cools it contracts. The center is still hot, and being extended, it

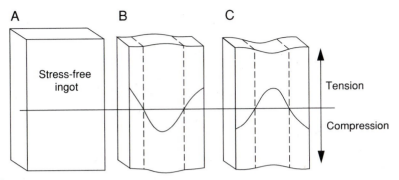

FIGURE 8-22 Illustration of how residual stresses develop in a casting.

opposes the surface contraction. Therefore the surface is stretched in tension while the core is compressed and plastically flows (Fig. 8-22B). With further cooling the core continues to contract thermally, but the surface now restrains it from shrinking. A reversal of stress ensues with the core stretched in tension while the surface winds up in compression (Fig. 8-22C). Such effects are accentuated at interfaces between solidifying sections of different sizes.

Residual stresses also occur in the absence of thermal processing. As an example consider the cold rolling of metals. During a light rolling pass of a plate, the surface fibers are stretched more than the inner metal, an effect shown in exaggerated fashion in Fig. 8-23A. Compatibility of the workpiece demands a straight cross section, however. Therefore, a compromise is struck

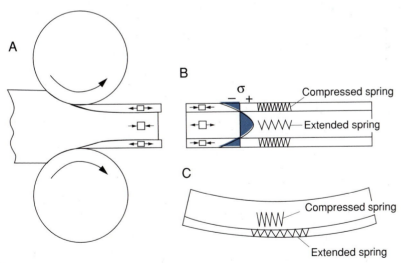

FIGURE 8-23 Development of residual compressive stresses in the surface of a lightly cold-rolled metal plate. If a thin layer is removed by machining, the plate will bow as shown.

and the surface fibers do not stretch as much as they would like to and wind up in compression. Meanwhile the interior is pulled in tension (Fig. 8-23B). In this as well as all cases of residual stress, the body is in mechanical equilibrium, and the tensile and compressive forces must balance. Now consider what happens when a strip of the rolled plate is clamped in a vice and a thin layer of metal is machined from one surface. After the plate is released, the machined surface bows distinctly concave upward (Fig. 8-23C), ruining any required tolerances. Bowing can be understood by imagining that both surfaces consist of compressed springs counterbalanced by a single stretched interior spring. Removal of one compressed spring upsets the equilibrium, causing the remaining outer spring to extend and the central spring to contract.

8.4. POWDER METALLURGY

8.4.1. Introduction

Some materials melt at too high a temperature to be easily cast, or are too hard and brittle to be plastically deformed into shape. In such cases salvation frequently comes from the powder processing route. This entails first producing powder of the material in question and then compacting it into the desired shape by pressing it in dies. Finally the compact is sintered at elevated temperatures to produce a densified, mechanically strong, finished part. Advantages of powder processing include the following:

1. Applicability to all classes of materials
2. Relatively low processing temperatures (significantly below T_M)
3. Attainment of highly homogeneous and uniform microstructures
4. Ability to create composites, mixed phases, and nonequilibrium metastable and microcrystalline structures
5. Production of precision parts with close tolerances
6. High-volume production rates, because the process is amenable to automaton
7. Absence of finishing operations like machining and grinding
8. Efficient use of material resources (no scrap)

With such an outstanding list of favorable attributes one suspects that the generally high costs of capital equipment (presses, furnaces), materials, and tooling will be disadvantages in those instances where competitive processing methods exist. Nevertheless, many metal parts are produced by the so-called **powder metallurgy (P/M)** route. Ceramic components are also fabricated this way but there does not seem to be an equivalent term for their processing.

Virtually every major industry and technology has applications requiring the use of metal and ceramic parts made by P/M. The collection of components displayed in Fig. 8-24 attests to the broad capabilities of this processing tech-

FIGURE 8-24   Assorted steel, stainless-steel, and brass powder metallurgy parts. Courtesy of Metal Powder Industries Federation.

nique. A very abbreviated list of applications includes automotive (gears, bushings, brake linings, spark plugs), aerospace (engine components), chemical (catalysts, filters), manufacturing (cutting tools, dies), and electrical (metal and ceramic magnets, contacts, relays, sputtering targets for thin film deposition). Additional technological applications use powders in the unsintered form, for example, abrasives, paint pigments, solder pastes, inks, and explosives. Making powder is a good place to start.

8.4.2. Powder Production and Properties

A number of processes including atomization, electrodeposition, and chemical reduction of oxides have been employed to make metal powders. Atomization is the most popular for metals. If molten gold is poured into water, the metal stream is broken up into shot ranging up to a few millimeters in size. This phenomenon is optimized in atomization processes (Fig. 8-25) to produce fine powder instead of shot. High-pressure jets of either gas (air, H_2, Ar) or

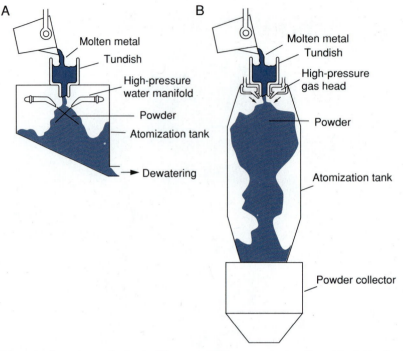

FIGURE 8-25   Schematic of powder production by (A) water and (B) gas atomization. From *Metals Handbook,* 9th ed., Vol. 7, American Society of Metals, Metals Park, OH (1984).

water are focused at molten metal streams issuing from a heated crucible. As a result the metal disintegrates in rapid order into liquid sheets, ligaments, and droplets which solidify into powder particles that are collected. The higher the kinetic energy of the H_2O or gas directed at the metal, the finer the resulting powder. Micrographs of the powder morphology of gas- and water-atomized stainless steel are reproduced in Figs. 8-26A and B.

Electrolytic powder is produced by electroplating metals onto cathodes they poorly adhere to at high current densities. Because the deposits do not stick they are easily brushed to the bottom of the plating tank and collected. In this way dendritic shaped copper, iron, titanium, and beryllium powders are produced. Crushed metal oxide particles, for example, WO_2 and MoO_2, are other sources of metal powder. When reduced by hydrogen at ~1000 to 1500°C, metal powder is liberated. In this way tungsten and molybdenum powders are produced at well over 1000°C below their melting points. Ceramic powder is usually obtained by pulverizing naturally occurring minerals of high purity. Such powder may not be suitable for some applications and may require further purification.

Powder size and shape are important variables because they influence packing, flow, and compressibility during compaction, as well as uniformity and

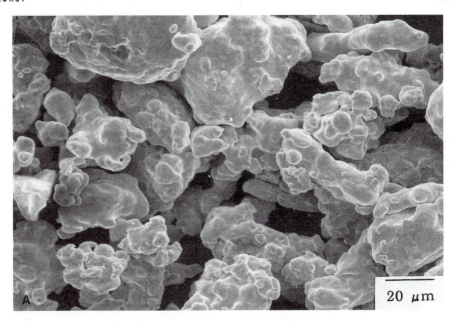

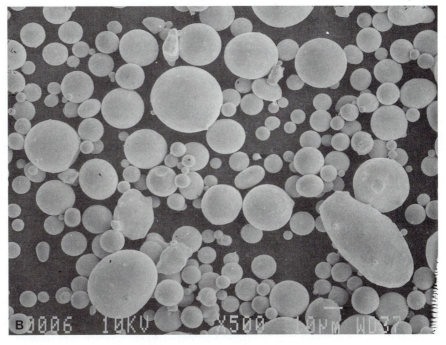

FIGURE 8-26 Morphology of stainless steel metal powder made by (A) water atomization and (B) gas atomization. Courtesy of Professor R. M. German, Pennsylvania State University.

homogeneity during sintering. Particle size is a key characteristic of powder. For a given process, particle sizes are usually statistically distributed and are measured by sieving through mesh screens with openings of different size; they typically range from ~30 to 200 μm in size. Ceramic powder can be comminuted or fractured to submicrometer (1 μm = 10^3 nm) dimensions by ball milling. In this process, steel balls continually tumble and wear the original particles by attrition upon compressive impact.

Powder shapes have descriptor names such as spherical, rounded, angular, spongy or porous, flake, polygonal, and dendritic. Both size and shape strongly influence powder friction. This property in turn determines powder flow characteristics during packaging, blending, and mixing, as well as during die filling. Not unexpectedly, smooth powder surfaces exhibit less friction. Uniform, fine, spherical powder, in particular, is in great demand for such applications as compounding solder (Sn–Pb) creams for microelectronic use and producing jet engine parts (superalloys, Ti alloys). Reasonably smooth and uniform spherical powder is produced by centrifugal atomization. In this process liquid metal streams impinge upon rapidly rotating disks, and droplets are spun off and collected as powder. The high cooling rates reduce the interdendritic arm spacing and overall compositional variations in the powder.

8.4.3. Sintering

Powders are pressed in shaped, multipiece die sets at high pressures, for example, 0.4 to 0.8 GPa (60–120 ksi), to compact them as much as possible. The resulting green compacts are like pharmaceutical pills that crumble under stress and mishandling. To eliminate porosity and strengthen them, the compacts are sintered at elevated temperatures in controlled atmospheres. The kinetics of sintering were considered at length in Section 6.6.3 and serve to design processing treatments. Sintering temperatures are obviously lower than melting temperatures, and may be hundreds to well over 1000°C lower.

It rarely happens that sintering is accomplished without enhancing its rate in some way. In one of the most common processes, **liquid phase sintering,** a liquid provides for enhanced material transport and more rapid sintering. Either low-melting powder blended into the compact or external infiltrants that penetrate the compact surface pores by capillary wicking action during sintering are the source of the liquid. For example, in the production of tungsten carbide, a very widely used cutting tool material for machining, liquid cobalt, produced at the sintering temperature, binds the WC particles together (Fig. 8-27). During sintering, the particles are first wet and may rearrange in the melt. Dissolution processes, followed by reprecipitation elsewhere, often result in better packing and densification. Similarly, infiltration of steel powder with copper base alloys is employed in the production of gears.

Beneficial changes occur in assorted properties as a function of sintering temperature as indicated schematically in Fig. 8-28. For densities ~99% of

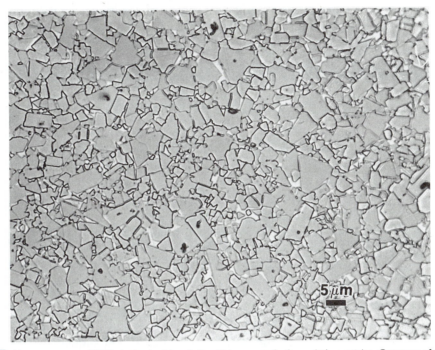

FIGURE 8-27   Microstructure of WC sintered with cobalt which surrounds the carbide particles. Courtesy of G. F. Vander Voort, Carpenter Technologies Corporation.

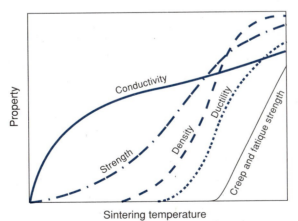

FIGURE 8-28   Schematic variation of P/M compact properties as a function of sintering temperature. After R. M. German, *Powder Metallurgy Science,* Metal Powder Industries Federation, Princeton, NJ (1984).

bulk values, mechanical properties resulting from P/M processing often exceed those in wrought (cast and mechanically worked) materials.

Polymer melts are molded or blown in the viscosity range 10^3 to 10^5 Pa-s (10^4–10^6 P), which compares with only 10^{-4} Pa-s for molten metals. Thus, pressure is generally required for polymers to fill molds, whereas metals need only be poured to be cast. A variety of extrusion and molding operations are used to process polymers. In particular, the following processing methods now enable thermoplastic products of complex shape to be produced on a very large scale: extrusion, injection molding, blow molding, compression molding, calendering, and thermoforming.

In addition there are processes employed to make reinforced polymers, and they are treated in the next chapter. Extrusion processing underlies both injection and blow molding of polymers and, therefore, is discussed first.

8.5.1. Extrusion

Extrusion of polymers resembles extrusion of metals. Lengths of tubular, sheet, and rod products possessing both simple and complex cross sections emerge from dies after pushing polymer feedstock through them. The details differ because metals do not melt during extrusion. Batch processing (one billet of metal at a time), unlike continuous extrusion of polymers, is another difference. In this sense polymer extrusion also resembles continuous casting of metals. (In continuous casting, metal is continually poured from a large, replenishable melt reservoir into an open, chilled mold, where it solidifies. Later, downstream, the solidified moving billet is cut into convenient lengths.)

Thermoplastic materials (resins) in pellet form are fed in at one end of the extruder as shown in Fig. 8-29A. Polymer feeds often contain additives, the most important of which are *lubricants,* which assist processing; *fillers and plasticizers* to modify mechanical properties; *flame retardants;* and *stabilizers* (antioxidants) to enhance degradation resistance. The polymer feed is conveyed down the extruder barrel by one (or more) long, rotating (and reciprocating) screw(s) where it is **compressed.** Through contact with the heated walls and the mechanical action of the screw, polymer **melting** occurs. In some systems polymer is pushed along by a piston. Next, melt **pumping** or **metering** under pressure through a tapered region and into a shaped die enables the product to exit. The extrudate must have a sufficiently high viscosity when it leaves the die to prevent it from deforming mechanically or even collapsing in an uncontrolled way. Therefore, water or air sprays are required to cool the product. The cooling rate must be controlled, however, because it can influence

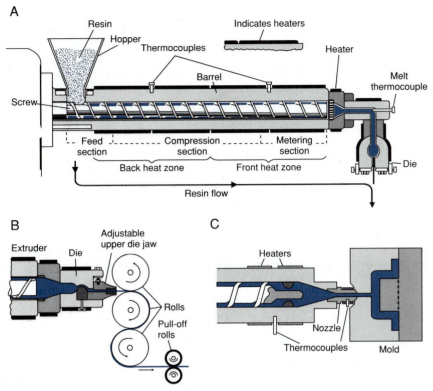

FIGURE 8-29 Polymer forming processes based on the use of a screw-type extruder: (A) Extrusion. (B) Extrusion of sheet. (C) Injection molding. Modified from R. J. Baird, *Industrial Plastics,* Goodheart–Willcox Co., South Holland, IL (1976).

the extent of crystallinity and affect mechanical properties. By use of an appropriate die and array of rollers, plastic sheet can be produced (Fig. 8-29B).

8.5.2. Injection Molding

To extend the parallels with metal processing, injection molding of shaped polymer parts closely resembles die casting. The process shown in Fig. 8-29C starts in the same way as an extrusion. Instead of being forced through a die into the unpressurized ambient, the melt is injected under pressure into a split die cavity. Injection molding is usually carried out at $\sim 1.5T_G$, where T_G is the glass transition temperature of the polymer. Pressures in the mold are typically maintained at somewhat higher levels than those employed in die casting, and range from 35 to 140 MPa (5–20 ksi). After the part cools below T_G (under pressure) the die opens and the part is ejected. The chamber is essentially a chemical reactor that allows the mixed and heated ingredients to blend, polymerize, and crosslink during simultaneous shaping. Fibers and particles can also be introduced, enabling the production of composites.

As the sizes of parts increase so do the magnitudes of the injection pressure, and the clamping force needed to keep the dies together. To address both problems *reaction injection molding* processes have been developed. They allow polymerization reactions to occur in the dies (mold), not the chamber. Injection viscosities are then reduced by more than two orders of magnitude with a corresponding drop in back pressure. Solid and foam urethanes are the chief reaction injection molded materials.

Injection molding is a high-production-rate process with typical cycle times ranging from seconds to several minutes depending on the type of polymer and the part size. Poor heat transfer properties of polymers basically limit the speed of the process. High injection pressures mean that good tolerances and surface finishes can be maintained. Among the parts produced by injection molding are containers, housings, plumbing fixtures, gears, telephone receivers, and toys. Parts comparable in size to those produced in die casting can be injection molded. That is why polymers have increasingly replaced metals in the above and other applications.

EXAMPLE 8-5

Suppose you wish to injection mold a set of plastic toy soldiers and must determine how many can be produced during one cycle. A 100-ton (8.9×10^5-N) press is available and the die pressure required to reproduce fine die features is 100 MPa ($\sim$15 ksi). Each soldier has an average projected area of 10 cm^2 (10^{-3} m^2) and is 0.75 cm thick. How many soldiers can be molded at one time?

ANSWER   If the parting plane divides the halves of the mold at the center of each soldier, the die clamping force required is the product of the pressure and die face area occupied by polymer. This force must be less than or equal to the press capacity. If n is the number of soldiers, then $(100 \times 10^6$ Pa$) \times n \times (10^{-3}$ m$^2) = 8.9 \times 10^5$ N. Solving, $n = 8.9$. For safety, only eight soldiers should be molded during one cycle. The thickness of the part is basically immaterial in this problem.

8.5.3. Blow Molding

The combination of extrusion and air pressure makes blow molding and the production of plastic bags and bottles possible. As depicted in Fig. 8-30A, a thin-walled tube of polymer is extruded vertically and expanded by blowing through the die until the desired film thickness is produced. Plastic wrap and

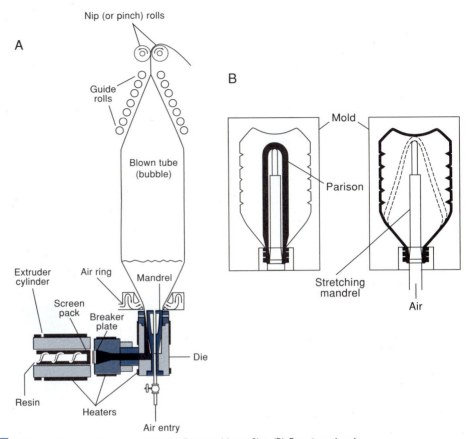

FIGURE 8-30 Polymer blow molding processes: (A) Forming blown film. (B) Forming a bottle.

bags are made continuously in variations of this process. In making beverage bottles a prior extruded tube, known as a *parison,* is pinched at one end and clamped (Fig. 8-30B) within a mold that is much larger than the tube diameter. It is then blown outward until the polymer extends to the mold wall, in a way that a balloon would if blown up within a bottle. A hot air blast at a pressure of 350 to 700 kPa (50–100 psi) is used for this purpose.

The plastic beverage bottle represents an interesting problem in materials design. For marketing purposes the bottle must be highly transparent. In addition it must be sufficiently strong (creep resistant) that it does not lose its shape on the shelf. Lastly, and importantly, it must be impermeable to CO_2 so that gas loss does not make the contents go flat. To meet these needs the polymer polyethylene terephthalate (PET) is selected and blow molded. But, if cooled rapidly, PET is amorphous. It is quite clear, but its relatively loose molecular structure makes it permeable to CO_2 and susceptible to creep. At the other

A

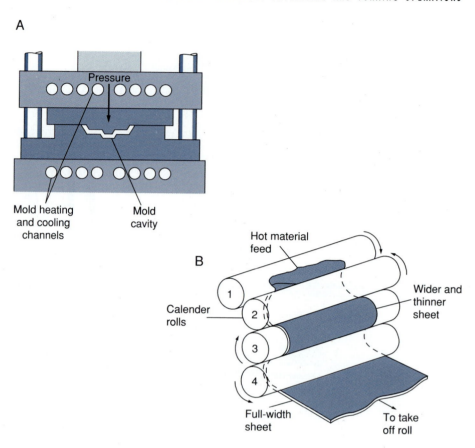

Pressure

Mold heating
and cooling
channels

Mold
cavity

B

Hot material
feed

Calender
rolls

Wider and
thinner
sheet

Full-width
sheet

To take
off roll

C

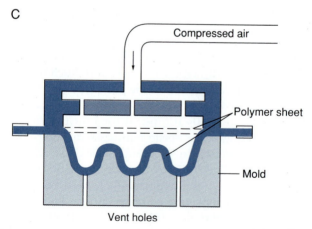

Compressed air

Polymer sheet

Mold

Vent holes

FIGURE 8-31 Additional polymer processing methods: (A) Compression molding. (B) Calendering. (C) Thermoforming. Modified from R. J. Baird, *Industrial Plastics*, Goodheart–Willcox Co., South Holland, IL (1976).

extreme of slow cooling, the polymer crystallizes. It is now strong and gastight, but opaque. The answer is first to cool the parison rapidly below T_G (340 K), where it is amorphous, and then to stretch it by blow molding at 400 K, a temperature high enough for plastic flow but too low for appreciable crystallization. Stretching a polymer induces additional crystallization and strength, and if controlled, no loss in clarity occurs.

8.5.4. Compression Molding

Where the expense of an extruder and injection molding dies is not warranted, compression molding of parts is practiced. In this process (Fig. 8-31A) premeasured volumes of polymer powder or viscous resin–filler mixtures are introduced into a heated multipiece die and then compressed with an upper plug. Thermosetting polymers and elastomers are shaped by compression molding and products include electrical terminal strips, dishes, and washing machine agitators.

8.5.5. Calendering

Polymer sheets are often made by calendering (Fig. 8-31B). In this process a warm plastic mass is fed through a series of heated rolls and stripped from them in the form of sheet. Because thermoplastics have a high strain hardening coefficient they undergo large uniform deformations, making sheet production possible.

8.5.6. Thermoforming

Thermoforming, shown in Fig. 8-31C, parallels methods used to form sheet metals. Sheet polymer (e.g., formed by calendering) is preheated and then laid over a mold having the desired shape. The applied air pressure is normally sufficient to make the sheet flow plastically and conformally cover the mold interior. Parts made this way include advertising signs, panels for shower stalls, appliance housings, and refrigerator linings.

8.6. FORMING GLASS

8.6.1. Scope

As with all materials that have been worked for a long time, many methods have evolved and been developed to form glass over the years. The important ones discussed below include pressing, blowing, casting, and rolling and float molding. These processing techniques are schematically depicted in Fig. 8-32.

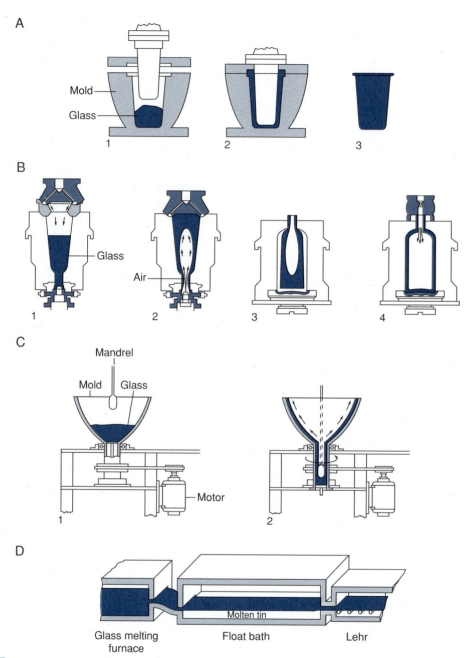

FIGURE 8-32 Processes to form and shape glass: (A) Pressing a glass container. (B) Blowing a bottle. (C) Centrifugal casting of a cathode-ray tube. (D) Plate glass manufacture by the Pilkington method. From J. R. Hutchins and R. V. Harrington, *Encyclopedia of Chemical Technology*, 2nd ed., Vol. 10, Wiley, New York (1966).

8.6.1.1. Pressing

An example of pressing involves the forming of glassware by compressing it in a mold with a plunger. The process (Fig. 8-32A) resembles closed die forging of metals and compression molding of polymers. Parts with diameters of 30 cm, depths of 10 cm, and weights of up to 15 kg are pressed using pressures of 0.69 MPa (100 psi).

8.6.1.2. Blowing

Hand blowing of glass is practiced today much as it was in antiquity. Glass is gathered on the end of an iron blowpipe and, with lung power or the use of compressed air, the glass blower shapes the "gathers." Artistic glassware is hand blown while the glass is continually rotated. The largest pieces reach weights of 15 kg, lengths of almost 2 m, and diameters of 1 m. Examples of machine-blown ware include glass containers and light bulb envelopes. Gobs of glass are delivered to the *blank* mold (1,2), where they are preformed by either blowing or pressing. After the blank is rotated, it is then blown in the *blow* mold (3,4) to finish the operation (Fig. 8-32B).

8.6.1.3. Casting

In casting, glass is poured into or on molds, tables, or rolls. One of the largest pieces ever cast is the Mount Palomar reflector, a cylinder measuring 5.08 m (200 in.) in diameter and 0.457 m thick. Other products routinely cast are radiation-shielding window blocks and television and cathode-ray tubes. The latter are centrifugally cast, and as the molten gobs are spun the glass flows up to create a uniform wall thickness (Fig. 8-32C). There is a 10^5-fold difference in viscosity between molten glasses and metals. The implication of this with respect to casting can be appreciated by comparing how honey and water pour.

8.6.1.4. Rolling and Float Molding

Both rolling and float molding are used to produce plate glass. In continuous processes, raw materials are fed in at one end of very large horizontal furnaces (with capacities of 1000 tons), where successive melting and refining of glass occur. At the other end the glass is fed into a pair of cooled rollers and the emerging ribbon of solid glass is then conveyed on rollers through an annealing lehr. Plate glass nearly 4 m wide and 1 cm thick can be rolled at rates greater than 6 m/min. The surfaces of rolled plate glass must be further ground and polished.

The revolutionary Pilkington float molding process (Fig. 8-32D) is now widely used to make plate glass. In this process soft glass is floated on the flat surface of a molten tin alloy that controls the lower glass surface temperature. The upper surface of the glass develops by gravity. As the temperature is reduced the glass stiffens and is conveyed to be annealed. This process produces a distortion-free plate glass comparable in flatness to that produced by rolling glass. A drawback of the process is incorporation of Sn into the glass.

All one has to do is look through plate glass windows of the last century to appreciate how far we have come in glass plate manufacture.

8.6.2. Viscosity

The single most important property of glass that influences these processes and subsequent heat treatments is viscosity (η). This thermomechanical property effectively combines a number of material attributes (flow resistance, fluidity, atomic diffusion) in a single constant. In inorganic glasses the temperature dependence of η is given by

$$\eta = \eta_0 \exp(E_\eta/RT), \tag{8-10}$$

where η_0 and E_η are characteristic constants that depend on composition, and RT has the usual meaning. Equation 4-5 expresses an analogous dependence applicable to polymers. The magnitude of the viscosity of glass is critical in all stages of its manufacturing and forming processes. Viscosity determines virtually all of the melting, forming, annealing, sealing, and high-temperature heat treatments of glass. Typical viscosity ranges for a number of operations are as follows:

Operation	Viscosity (Pa-s)
Glass melting and fining (bubble elimination)	5–50
Pressing	50–700
Gathering or gobbing for forming	50–1,300
Drawing	2,000–10,000
Blowing	1,000–3,000
Removal from molds	10^3–10^6
Annealing	10^{12}–10^{15}
Use	$10^{13.4}$–$10^{14.5}$

The temperatures at which these operations are carried out differ with glass composition and these are depicted in Fig. 8-33 for a number of commercial silica-based glasses. Each of the latter has a specific viscosity–temperature dependence and individual set of strain, annealing, softening, and working point temperatures. These are defined in the following ways:

Strain point. The temperature at which internal stresses are reduced significantly in a matter of hours. At this temperature the viscosity is taken as $10^{13.5}$ Pa-s.

Annealing point. The temperature at which the viscosity is about 10^{12} Pa-s and the internal stresses are reduced to acceptable commercial limits in a matter of minutes.

Softening point. At this temperature glass will rapidly deform under its own weight. For soda-lime glass this occurs at a viscosity of $10^{6.65}$ Pa-s.

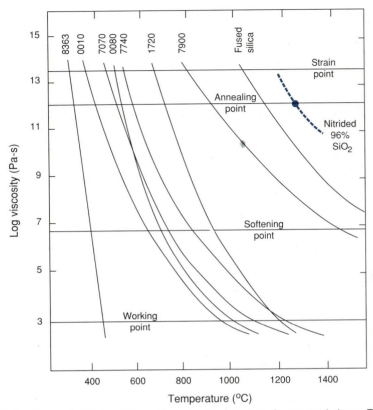

FIGURE 8-33 Coefficient of viscosity versus temperature for a number of commercial glasses. The four-digit codes refer to Corning glasses whose compositions are given in Table 4-5. From J. R. Hutchins and R. V. Harrington, *Encyclopedia of Chemical Technology*, 2nd ed., Vol. 10, Wiley, New York (1966).

Working point. At this temperature glass is soft enough for most of the common hot working processes. The working point corresponds to a viscosity of 10^3 Pa-s.

EXAMPLE 8-6

It is accurately determined that a glass has a strain point of 515°C and a softening point of 820°C. Estimate the lowest temperature at which this glass can be blown.

ANSWER In this problem we assume that Eq. 8-10 describes the viscosity of this glass as a function of temperature. Once η_0 and E_η are determined, the lowest glass blowing temperature, which corresponds to a viscosity of 3000

Pa-s, can be calculated. Using the two given pieces of information, $\eta = 10^{6.65}$ Pa-s at 820°C (1093 K) at the softening point and $\eta = 10^{13.5}$ Pa-s at 515°C (788 K) at the strain point, there are two simultaneous equations:

$$10^{6.65} = \eta_0 \exp[E_\eta/R(1093)]$$
$$10^{13.5} = \eta_0 \exp[E_\eta/R(788)].$$

Dividing the second equation by the first yields

$$10^{6.85} = \exp\frac{E_\eta}{8.314}\left(\frac{1}{788} - \frac{1}{1093}\right).$$

Solving, $E_\eta = 370$ kJ/mol. With knowledge of E_η, η_0 is calculated to be 9.27×10^{-12} Pa-s.

Finally, $3000 = 9.27 \times 10^{-12} \exp(370,000/8.314T)$. Solving for T, the minimum glass blowing temperature is 1332 K, or 1059°C.

8.7. PROCESSING OF CERAMICS

8.7.1. Forming the Green Body

With millennia of experience as background, many processes and variants of processes have been used to make **vitreous** (glass containing) ceramic products. They generally involve a number of common steps including preparing and mixing the raw ingredients (sands, clays) and binders (in a water base), forming or shaping the green body (e.g., on a potters wheel), drying it, firing the body to densify it, and finishing the product by grinding, cutting, polishing, and glazing. Shaping the green body is facilitated because the hydroplastic nature of the clay particles enables them to slide easily over one another. Upon firing, the glassy (vitreous) phase melts and coats the higher-melting clay particles, bonding them together. Most of the new ceramic (see Section 4.6.2) material processes are similar in spirit, but often there is no glassy phase. They start with powders that are generally purer. Additives to the composition may include binders (for green strength), lubricants (to reduce friction and release the body from molds), wetting and water retention agents, deflocculants (to control pH, electrostatic charge and particle dispersion), and sintering aids.

Green bodies of both traditional and new ceramics are formed by a number of methods that are enumerated below.

8.7.1.1. Slip Casting

First, a slip consisting of a suspension of clay or ceramic powder in water is prepared. This is cast into an absorbent mold as indicated in Fig. 8-34A. After sufficient water loss, the partially wet solid body is removed and dried further prior to firing. Vases as well as large and complex parts like Si_3N_4 turbine rotors have been reproduced this way.

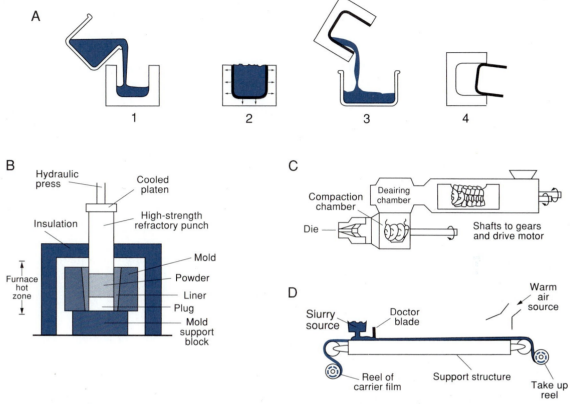

Several techniques used to process ceramics. (A) Slip or drain casting: (1) Pouring the clay/ceramic suspension. (2) Water absorption by plaster mold until a solid shell forms. (3) Pouring out the excess suspension. (4) Removal of body for firing. (B) Pressing. (C) Extrusion. (D) Tape forming. From S. Musikant, *What Every Engineer Should Know about Ceramics*, Marcel Dekker, New York (1991).

8.7.1.2. Pressing

The specially prepared dry powder is placed in a die and pressed under high uniaxial forces to make a green compact (Fig. 8-34B). The process is very much like pressing of metal powders in powder metallurgy dies (see Section 8-4). Only relatively small parts can be made in this way. An improvement over applying uniaxial compressive loads is to compact the powder **isostatically.** Here the part is enclosed in a flexible airtight rubber bag and immersed in a chamber filled with hydraulic fluid. Hydrostatic (uniform over all directions) pressure is applied during (cold) compaction, eliminating the complex pressure distribution throughout the die due to mold wall friction. Electrical insulators, bushings, magnetic components, and spark plug bodies are produced in this way for subsequent sintering.

Hot pressing combining compaction and sintering operations is also practiced to make a variety of highly dense ceramic components.

8.7.1.3. Extrusion

Extrusion of green bodies parallels the extrusion of polymers in that powder and screws (augers) are involved. Polymers, however, are extruded in the molten state whereas ceramic green body extrusions are created at room temperatures from pliable, powder–binder mixtures. This material is extruded through a die as shown schematically in Fig. 8-34C, and then sintered. Long tubes, rods, and honeycomb structures used to shield thermocouples or for heat exchanger applications are made in this way.

8.7.1.4. Tape Forming

Tape forming, shown in Fig. 8-34D, is used to make thin ceramic parts like alumina substrates for integrated circuit chips and special capacitors. A thin layer of slip is laid on a flat carrier which can be a paper sheet or polymer film. A doctor blade controls the slip thickness, resulting in a tape that can be stamped into small shapes. These can be stacked or coiled for subsequent sintering.

8.7.1.5. Injection Molding

In injection molding, a hybrid process, ceramic powder is mixed with thermosetting (organic) polymer binder and injected under pressure into a die or mold. The bodies are then ejected and sintered while new ones are replicated. Complex parts with excellent dimensional control can be made in this way, including many of those proposed for use in all ceramic engines.

8.7.2. Densification

8.7.2.1. Firing

As anyone who has made pottery knows, the dried green ware is next fired in a furnace to densify it into a hard body. Firing temperatures vary depending on the composition of the body, as the following brief listing indicates.

Ceramic	Approximate firing temperature (°C)
Porcelain enamels (on cast irons and steels)	650–950
Clay products (bricks, sewer pipes, earthenware, pottery)	1000–1300
Whitewares (porcelain, china, sanitary ware)	1000–1300
Refractories (alumina, silica and magnesia brick, silicon carbide)	1300–1700
Electronic and newer ceramics (aluminas, ferrites, titanates)	1300–1800

As a rough proposition highly fluxed bodies containing silica and alkali oxides are fired at lower temperatures than bodies richer in alumina and magnesia.

Note that the colored designs on the surface of ceramic ware often require application of an appropriately compounded glaze and a *second firing*. The glass that forms fills all the pores, making the body impervious to water penetration.

Reactions that occur during firing are generally complex. They have been studied in kaolin using microscopy, diffraction techniques, and calorimetry (to measure heat absorption). It is instructive to follow the sequence of events depicted in Fig. 8-35 as the temperature is raised. The kaolinite (clay) crystals remain intact until 450°C, when they decompose into a noncrystalline mass; simultaneously heat is absorbed and water is expelled. There is little shrinkage in the green body up to a temperature of ~600°C, although by this time most of the weight loss has occurred. Then at about 1000°C both alumina and mullite ($3Al_2O_3 \cdot 2SiO_2$) crystallize as heat is liberated; these reactions are accompanied by a large rate of shrinkage. At higher temperatures mullite crystals grow in size and are surrounded by the glassy (vitreous) phase that pulls the particles together by surface tension forces. At 1200°C cristobalite crystallizes from the surrounding silica-based glass, and again shrinkage and pore reduction occur at a high rate. The significant shrinkage that occurs must be accounted for in the original design of all ceramic products.

8.7.2.2. Sintering

Sintering has already been discussed on two prior occasions (see Sections 6.6.3 and 8.4.3). Only a few comments of relevance to ceramics are made here. Perhaps the chief problem associated with the sintering of ceramics is the retention of porosity. One consequence is a resultant product that is opaque because of efficient light scattering by pores. Sintering under pressure (hot

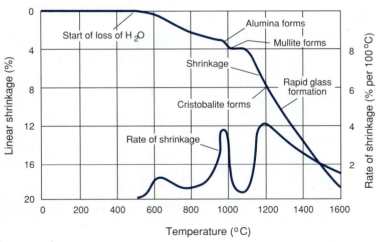

FIGURE 8-35   Processes that occur during the firing of kaolin as a function of temperature. From F. H. Norton, *Elements of Ceramics*, Addison–Wesley, Cambridge, MA (1952).

pressing) is often effective in collapsing pores. But there are systems in which pores are stubborn and cannot be easily removed because they exist in the interior of grains. If voids can be brought to grain boundaries, the resulting enhanced mass and vacancy transport there are effective in eliminating them. A strategy that will work is to limit exaggerated grain growth through suitable alloy additions. This prevents stable pores from being isolated too far from grain boundaries. In this way addition of MgO to Al_2O_3 eliminates porosity, yielding a translucent product that is used as an envelope for high-pressure sodium vapor lamps. In the same vein, 9 mol% La_2O_3 (lanthana)-strengthened Y_2O_3 (yttria) has been developed as an infrared window material (Fig. 8-36).

Together with the use of liquid-phase sintering aids to enhance pore closure, improved physical pressing and processing techniques have enabled production of denser ceramics. **Hot isostatic pressing** (HIP), schematically shown in Fig. 8-37, is one such method that goes well beyond simple isostatic pressing discussed earlier. The practice of HIPing has dramatically improved the quality not only of sintered ceramics and metals, but of cast metals as well. Parts to be HIPed are placed within a chamber that is first evacuated and then pressurized with inert gas to levels of ~3000 atm. Simultaneously, parts are heated to temperatures of up to ~2000°C. This combination of hot uniform squeezing from all directions shrinks pores and allows full density to be achieved in many cases; however, HIP equipment and operating costs are high.

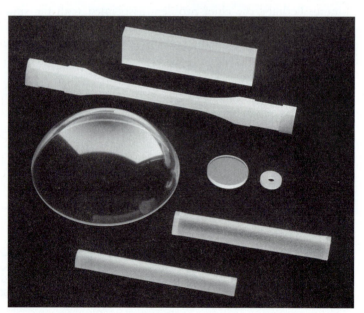

FIGURE 8-36   Pore-free sintered ceramic products. Tubes in the foreground are used in sodium vapor lamps and are made of translucent Al_2O_3. The clear dome, tensile bar, and other shapes are composed of Y_2O_3 strengthened with La_2O_3. Courtesy of W. H. Rhodes, Osram Sylvania.

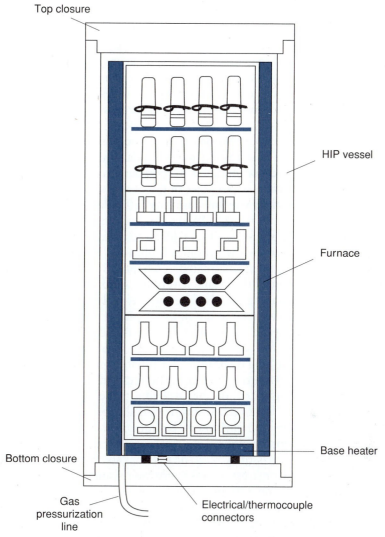

Top closure

HIP vessel

Furnace

Base heater

Bottom closure

Gas
pressurization
line

Electrical/thermocouple
connectors

FIGURE 8-37 Schematic of a HIP system. Courtesy of HIP Division, Howmet, Whitehall, MI.

8.8. PERSPECTIVE AND CONCLUSION

When viewed from the world of processing, materials are surprisingly similar. The *micro*scopic electronic and atomic structural distinctions between materials are not directly relevant and blur somewhat when it comes to issues of shaping or forming them. Rather, the *macro*scopically linked thermomechanical flow properties of materials are capitalized on. It is the characteristic temperatures (e.g., melting points, diffusion temperatures) and flow parameters (e.g., yield stress, viscosity) that are the important variables in shaping. If melting temperatures are easily attained, as they are in most metals, polymers, glasses, and semiconductors, then casting or solidification processing is practiced. Ceramic materials simply melt at inaccessible temperatures and, therefore, cannot be cast or otherwise be processed in the liquid state. This accounts for the extensive use of powder processing in ceramics and high-melting-point metals.

Because they are worked at the lowest as well as the highest viscosities, metals are singular with respect to materials processing. They are cast, mechanically formed, and consolidated by powder processing; no other class of material enjoys the same wide choice of possibilities. **Solidification processing** is possible because of their very low viscosity (10^{-3} Pa-s) compared with molten polymers and glasses ($\eta = \sim 10^3$ Pa-s). This combined with high density enables liquid metal to be poured easily and fill molds readily. Although heat transfer considerations determine the overall solidification rates, it is the combination of nucleation–growth effects treated in Chapter 6 that determines the size, shape, and composition of the grains within the solid. **Deformation processing** is another route to shaping metals. The most effective way to deform them is to ensure the presence of large amounts of plastic shear stresses induced by working above the yield stress. Both hot and cold plastic deformation are practiced, the former for large shapes and high-strength metals, the latter for thin stock and for finishing operations. Fortunately, metals are relatively plastic and flow sufficiently—during rolling to be reduced in thickness, during forging to fill dies, during extrusion and drawing to be reduced in area, and during sheet forming to be stretched into varied shapes.

Polymers and glasses can be liquefied but the viscosity is relatively high so they cannot be poured as easily as metals. Both are commonly shaped at viscosity levels of $\eta = \sim 10^3$ to 10^5 Pa-s, requiring simultaneous application of pressure or stress to make sure they assume the desired shape during molding or blowing operations. (Metals cannot normally be blown because viscosity levels never assume these intermediate levels.) The chief reason that polymers enjoy processing advantages relative to both metals and glasses is the very low temperatures required. For example, plastic bottles are typically blown at $\sim 100°C$ compared with $\sim 1000°C$ for glass bottles. In addition, polymer forming loads are often lower than those for metals. Having emphasized distinctions between polymers and metals, it is only fair that we point out similarities, for example, between injection molding and die casting, between compression molding and closed die forging, and extrusion in both materials.

Ceramics always require a two-step processing treatment: the first to form or mold them into a green product or compact, and the second to fire or sinter them. The difficulty of these steps depends on the ceramic in question. Traditional ceramics are generally easier to process than the new ceramics because the low-melting glassy phase helps to bind and consolidate particles on firing. For new ceramics, hot pressing and sintering routes must often be followed in the quest to achieve fully dense products. High melting points and compressive strengths, coupled with low ductility, limit conventional deformation processing of ceramics.

Computer-aided design of components and control of processing methods are increasingly the keys to achieving desirable properties. The aim is to control dimensional tolerances, chemical composition and uniformity, grain size and microscopic porosity, as well as eliminate macroscopic defects such as cracks. These are generic concerns common to very different materials that are often shaped in remarkably similar ways.

Additional Reading

L. Edwards and M. Endean, Eds., *Materials in Action Series: Manufacturing with Materials,* Butterworths, London (1990).

M. C. Flemings, *Solidification Processing,* McGraw–Hill, New York (1974).

R. M. German, *Powder Metallurgy Science,* 2nd ed., Metal Powder Industries Federation, Princeton, NJ (1994).

S. Kalpakjian, *Manufacturing Processes for Engineering Materials,* 2nd ed., Addison–Wesley, Reading, MA (1991).

J. A. Schey, *Introduction to Manufacturing Processes,* 2nd ed., McGraw–Hill, New York (1987).

QUESTIONS AND PROBLEMS

8-1. Consider metal maintained at T_M initially in contact with a semi-infinite long mold at temperature T_0. As solidification occurs, latent heat (H in J/g) is released and this heat flows into the mold. The problem is like that of diffusion into a semi-infinite matrix. Therefore, by Eq. 6-3, $[T(x, t) - T_0]/[T_M - T_0] = \mathrm{Erfc}[x/(4\alpha t)^{1/2}]$.

a. If the heat flux (Q) is defined as $-\kappa dT/dx$, consult a calculus book to determine Q at $x = 0$. (Answer: $Q = \kappa[T_M - T_0]/(\pi\alpha t)^{1/2}$.)

b. Show that the rate of heat production at the melt–solid interface is given by $Q_{L-S} = \rho H\, dS_i/dt$.

c. By equating the two heat fluxes and integrating, obtain Eq. 8-1.

8-2. Metal castings of a sphere and a cube of the same metal have the same surface area. What is the ratio of the solidification times for the sphere and the cube?

8-3. Castings of a sphere and a cube of the same metal have the same volume. What is the ratio of the solidification times for the sphere and the cube?

8-4. Copper is cast into two different insulating molds of the same rectangular geometry, under identical conditions. Mold 1 is sand with the following properties: specific heat = 1150 J/kg-K, thermal conductivity = 0.6 W/m-K, density 1400 kg/m³. Mold 2 is mullite investment with the following properties: specific heat = 750 J/kg-K, thermal conductivity = 0.5 W/m-K, density 1600 kg/m³. What is the ratio of the times required for these two castings to solidify?

8-5. Magnesium and aluminum are poured into identical sand molds at 25°C from their respective melting temperatures. For Mg, heat capacity = 1300 J/kg-K and latent heat of fusion = 360 kJ/kg. For Al, heat capacity = 1100 J/kg-K and latent heat of fusion = 390 kJ/kg. If it takes the Al casting 10 minutes to solidify, how long will it take the Mg casting to solidify?

8-6. Design the dimensions of a cylindrical riser with height twice the diameter that would be suitable to cast the bowling pin of Fig. 8-6. Assume the riser solidification time is 25% longer than that of the casting. The pin has a volume of 1000 cm^3 and a surface area of 700 cm^2.

8-7. What casting process would you recommend to make the following items?
   a. A large bronze ship propeller   b. Steel railroad wheels
   c. Aluminum sole plates for electric steam   d. Iron base for a lathe
      irons
   e. Aluminum engine block for a lawn   f. Steel pressure vessel tubes
      mower
   g. Nickel-base alloy turbine blades   h. Large brass plumbing
                                      valves

8-8. South American natives have made beautiful jewelry containing platinum–gold alloys for centuries. How do you account for this if these alloys, bits of which are found in nature, melt at much higher temperatures than could be reached in their furnaces?

8-9. Under uniaxial tension, plastic deformation initiates in a tensile bar at a stress level of 400 MPa. It is desired to deform a plate of this material plastically by applying biaxial stresses. If one of the stresses is compressive of magnitude 220 MPa, what is the minimum stress (magnitude and sign) that must be applied in the transverse direction?

8-10. A work hardening cylinder of annealed brass 5 cm in diameter and 5 cm in height is compressed. If the true stress–true strain relationship for this metal is $\sigma = 895\varepsilon^{0.49}$ MPa, calculate the load required to compress the cylinder to half its height if there is no friction between the metal and anvils.

8-11. Derive a formula for the total forging load L on a rectangular solid of cross-sectional area $2a \times w$ by directly integrating $\int \sigma_y \, dx$ using Eq. 8-5 for σ_y. Verify the answer to illustrative Example 8-2 using this formula.

8-12. A rectangular solid block of unit depth, $a = 0.2$ m, and $h = 0.15$ m is forged with $f = 0.3$.
   a. What is the fractional change in the maximum forging stress or pressure if the coefficient of friction changes by 10%?
   b. What is the fractional change in the maximum forging stress if the yield stress changes by 10%?
   (*Hint:* Determine $d\sigma_y/\sigma_y$ in terms of df and $d\sigma_0$.)

8-13. Show that if a steel plate enters the rolls at an initial velocity v_0 with thickness h_0 and width w, and leaves with velocity v_f, thickness h_f, and the same width, then $v_f > v_0$ (i.e., steel speeds up on the exit side).

8-14. If the diameter d of a wire is reduced 50% in length during drawing, what is the percentage reduction in area?

8-15. a. A material is reduced in area by 50% by extrusion. What is the extrusion ratio?

b. What percentage reduction in area does an extrusion ratio of 95 correspond to?

8-16. Circular disks of sheet steel 0.01 m thick and 0.15 m in diameter are punched out of a plate. If the tensile stress of the steel is 80,000 psi, what shearing force is required?

8-17. What problem do you foresee during deep drawing of metal cups:

a. If the punch used has too small a radius of curvature at the edge?

b. If there is too much clearance between punch and die?

c. If the sheet is held too loosely due to insufficient hold-down forces?

d. If the necking strain is exceeded?

e. If the sheet metal is strongly anisotropic?

8-18. Suggest possible ways of producing different powder sizes during atomization processes in metals.

8-19. Enumerate the similarities and contrast the differences in:

a. Pressing and sintering metal and ceramic powders.

b. Components made by casting and by powder metallurgy.

c. Shaping parts by forging and by powder metallurgy.

d. Extruding metals and polymers.

8-20. Why are polymers not normally shaped into parts by the powder metallurgy processes employed for metals and ceramics?

8-21. What pressure is required to extrude a bar of stainless steel at 1000°C from an initial diameter of 10 cm to 2.5 cm?

8-22. It is desired to make 4000 polymer gears per hour, each measuring 3 cm in diameter and 0.5 cm thick. What capacity press is required to injection mold them if a 120-MPa die pressure must be applied and the cycle time is 15 seconds?

8-23. Are thermosetting polymers suitable for injection molding processes? Explain.

8-24. A considerable amount of heat is generated by the interaction between the rotating auger and the churning polymer feedstock during extrusion. Suppose a 4-kW motor is used and 70% of the power goes into heating polyethylene fed at a rate of 100 kg/h. If the heat capacity of polyethylene is 1.4 J/g-°C, what will the temperature rise be in a half-hour? Assume no heat loss.

8-25. The following values were obtained for 7740 glass:

	Temperature		Viscosity
	°C	K	(Pa-s)
Annealing point	~580	853	10^{12}
Softening point	~830	1103	$10^{6.65}$
Strain point	~530	803	$10^{13.5}$
Working point	~1230	1503	10^3

a. From annealing and softening point data calculate E_η. Compare this value with that obtained from strain and working point data for this glass. (This shows that it is not wise to obtain E_η from only two data points.)

b. Determine a value for E_η by making an Arrhenius plot using all four sets of data.

8-26. Why is fused silica so much harder to shape than soda-lime glass?

8-27. Molten glass is compounded to flow easily during shaping or mold filling operations. Glazes and enamels are also glasses, but they must adhere to the fired ceramic or metal surface. Would you expect the compositions to differ in glasses as opposed to glazes and enamels? Mention one required property difference.

8-28. Crucibles used for melting 25 charges of silicon for single-crystal growth are made of the purest grade of fused silica and have the shape of a rounded cup with a flat bottom. The wall thickness is ~6 mm and the maximum crucible diameter is ~25 cm. Suggest a way to make these crucibles.

8-29. Tempered glass is strengthened by rapid cooling of the surface with an air blast after the glass is processed near the softening point. After cooling, the surface layers are under compression while the bulk of the interior is stressed in tension. Explain how such a state of residual stress develops.

8-30. a. A soda-lime glass has a viscosity of $10^{13.5}$ Pa-s at 570°C. If the activation energy for viscous flow is 400 kJ/mol, what is the expected viscosity at 670°C?

b. At what temperature will the viscosity be 10^9 Pa-s?

8-31. One way to strengthen surface layers of glass is to chemically diffuse in oversized alkali ions (e.g., K^+) that replace the original Na^+ ions. What state of stress develops at the surface and interior of a piece of soda-lime glass so treated?

8-32. Contrast structural and chemical differences in Al_2O_3 used for (1) substrates for microelectronic applications, (2) transparent tubes for sodium vapor lamps, and (3) furnace brick. Note any special processing requirements for these applications.

8-33. Why do ceramic bodies undergo much larger shrinkages than powder metal parts during high-temperature processing?

8-34. Porosity in ceramics is usually undesirable but there are applications where they are desired. Contrast these applications.

8-35. Why is thick ceramic ware more prone to cracking than thin ware during its processing?

8-36. A formula that describes the time (t in minutes)-dependent fractional shrinkage in length ($\Delta L/L$) of a sintered ceramic plate product is $\Delta L/L = A[\exp(-E/RT)]t^{2/5}$, where A is a constant, E is the activation energy for sintering, and RT has the usual meaning. At 1300°C and $t = 100$ minutes, $\Delta L/L = 0.05$; at 1200°C and $t = 100$ minutes, $\Delta L/L = 0.005$. Calculate A and E.

8-37. Assume that the formula $\Delta L/L = A[\exp(-E/RT)]t^{2/5}$ accurately describes the sintering of WC drawing dies. For this application the fractional shrinkage $\Delta L/L$ must be controlled to be 0.05 ± 0.003. If the sintering process is carried out at 1400°C for 60 minutes and $E = 420$ kJ/mol, what maximum variation in furnace temperature can be tolerated?
[*Hint:* Evaluate $(\Delta L/L)^{-1} d(\Delta L/L)/dT$.]

8-38. What is the probable method for fabricating the following ceramic objects?
a. A rare Ming dynasty vase
b. Bathroom wall tiles
c. Cups for a tea set
d. Spark plug insulation
e. A rotor for an all-ceramic engine

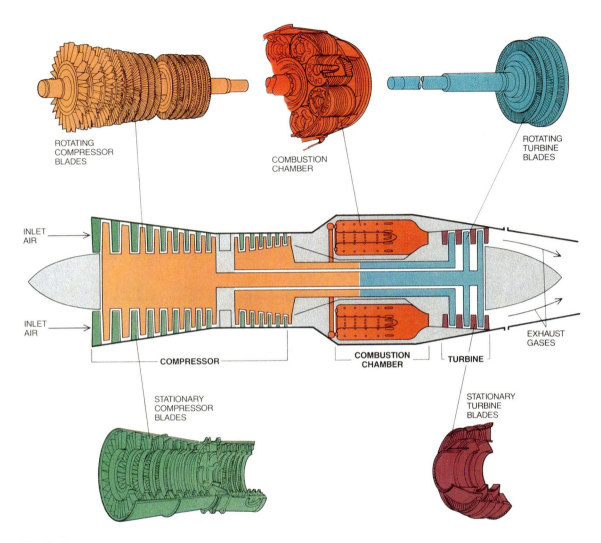

Fig. 1-9. Turbojet engine components. Air is pulled into the **compressor**, composed of a series of fan blades attached to rotating disks; surrounding these are stationary compressor blades that redirect the airflow between rotors. Titanium alloys are used in the compressor. In the **combustion chamber** the compressed air is mixed with fuel and ignited. The expanding, hot exhaust gases move with great velocity through the **turbine**, making its blades whirl at high speed to propel the engine forward. Nickel base and cobalt base superalloys are employed in these latter engine sections. Adapted from "Advanced Metals," by B. H. Kear. Copyright © 1986 by Scientific American, Inc. All rights reserved.

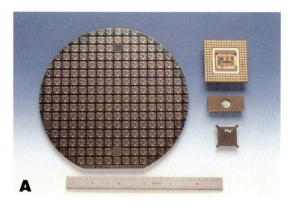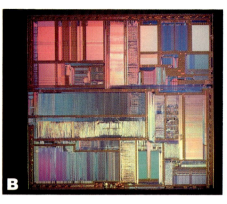

Fig. 1-14. (A) A 6-in. silicon wafer containing numerous 80386 microprocessor dies (chips). Shown are various packaging configurations for the chip. (B) The Pentium microprocessor chip. Measuring 1.8 × 1.7 cm, this chip contains 3.1 million transistors. Courtesy of INTEL Corporation. (For part C, see text.)

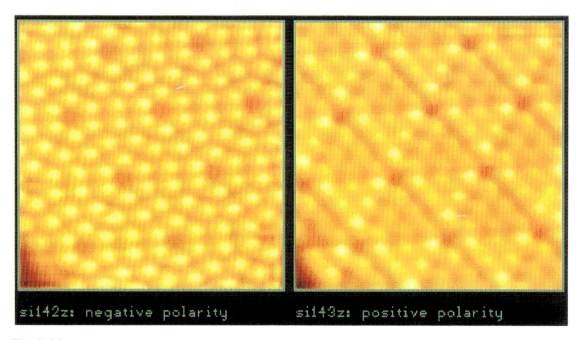

Fig. 3-23. Image of a (111) silicon surface obtained with a STM. **Left:** With negative polarity between tip and specimen. **Right:** With reversed polarity. Courtesy of Burleigh Instruments, Inc.

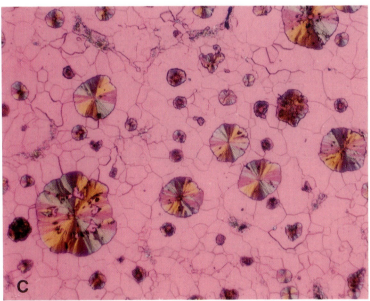

Fig. 3-32. Optical micrographs of materials. (For parts A and D, see text.) (B) Polycrystalline grain structure of a YBaCuO oxide superconductor viewed in polarized light. (C) Nodular graphite in etched ductile cast iron viewed in polarized light (164×). Micrographs are courtesy of G. F. Vander Voort, Carpenter Technology Corporation.

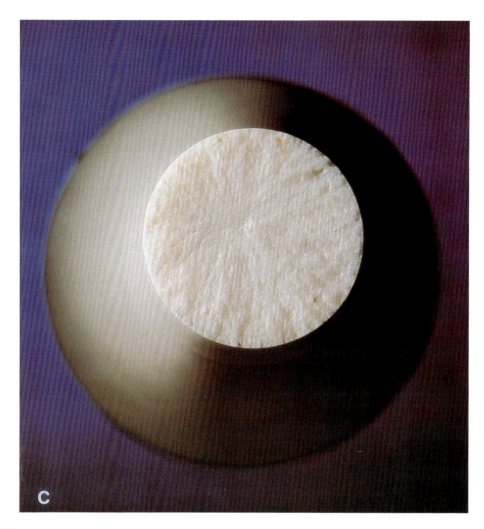

Fig. 7-5. (For parts A and B, see text.) (C) Brittle tensile fracture surface of Mg-PSZ, a toughened MgO–ZrO$_2$ ceramic. Courtesy of P. S. Han, MTS Systems Corporation.

Fig. 8-2. Investment cast turbine blades. From left to right are the equiaxed grain refined, directionally solidified polycrystalline, and single-crystal turbine blades, in the high-pressure (front row) and low-pressure (back row) jet engine sections. Courtesy of Howmet Corporation.

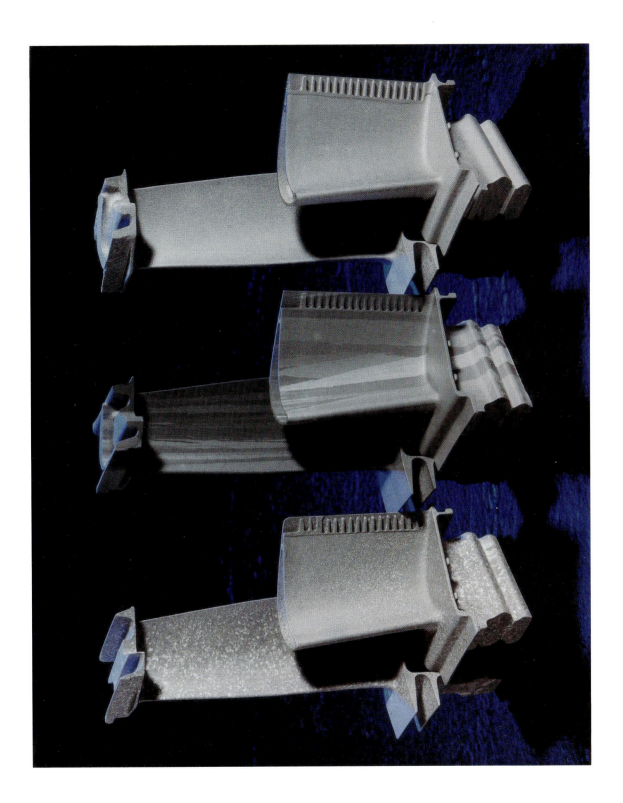

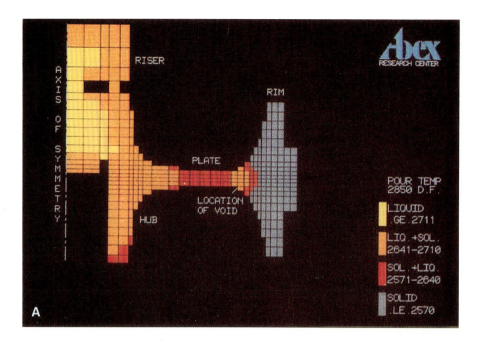

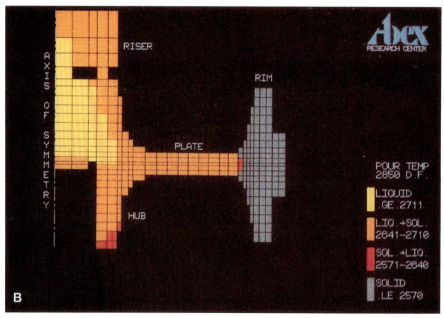

Fig. 8-8. Computer-modeled temperature distribution in wheel castings. (A) In the heavy rim section a liquid pocket is trapped between cooler metal on either side. When it solidifies a cavity may develop because this hot spot cannot be fed with metal. (B) The hot spot is eliminated in the case of a lighter rim. From M. K. Walther, *Modern Casting*, March 1984.

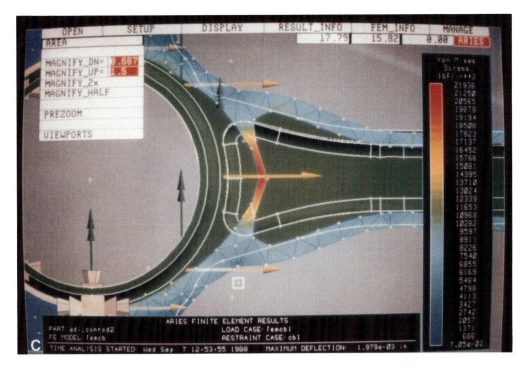

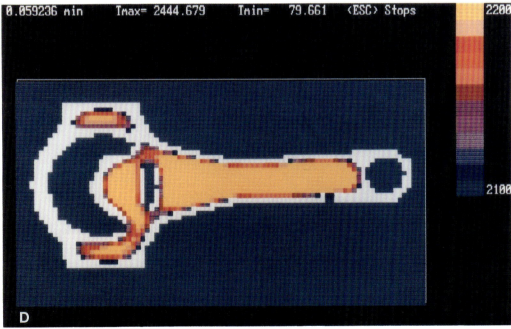

Fig. 8-9. CAD/CAM methods used in producing a lightweight automobile engine connecting rod. (For parts A, B, and E, see text.) (C) Finite element analysis (FEA) of stress distribution. (D) Two-dimensional solidification analysis.

Fig. 10-11. Case histories of corrosion and degradation. (For parts A–D, and F, see text.) (E) Ammonia corrosion of a pressure gauge valve. From E. D. D. During, *Corrosion Atlas: A Colllection of Illustrated Case Histories*, Elsevier, Amsterdam (1988).

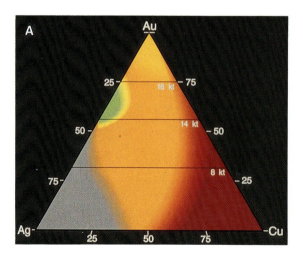

Fig. 13-6. (A) Color of ternary gold–silver–copper alloys. From *Gold Bulletin* **25,** 1 (1992). (For part B, see text.)

9

HOW ENGINEERING MATERIALS ARE STRENGTHENED AND TOUGHENED

9.1. INTRODUCTION

In this chapter we consider materials that are employed in mechanically functional engineering applications. Those materials that achieve commercial status have necessarily passed a tough development ordeal whose hurdles include (1) fulfillment of stringent mechanical property specifications (e.g., tensile strength, elongation, toughness); (2) development of reproducible processing and fabrication methods (e.g., casting, mechanical forming, joining); (3) extensive testing to ensure reliable operation in components and structures; and (4) competitive pricing. The surviving materials are typically but a fraction of potential candidates that for one reason or another were discarded and did not see commercialization.

Historically, metals have survived longest and been the materials of choice because they have represented the best compromise in terms of a basket of engineering properties that include stiffness (Young's modulus), yield strength, toughness, melting temperature, and cost. This can be seen in the bar graphs of Fig. 9-1, where the properties of metals, ceramics, and polymers are compared. If materials are scored 3 (best), 2, and 1 (worst) based on the maximum property value attained by any representative material, then the cumulative grades for the four properties considered are, roughly, 9 for metals, 10 for ceramics, and 5 for polymers. Surprisingly, ceramics score highest on this limited, and perhaps biased, test. They are the stiffest, strongest, and most heat

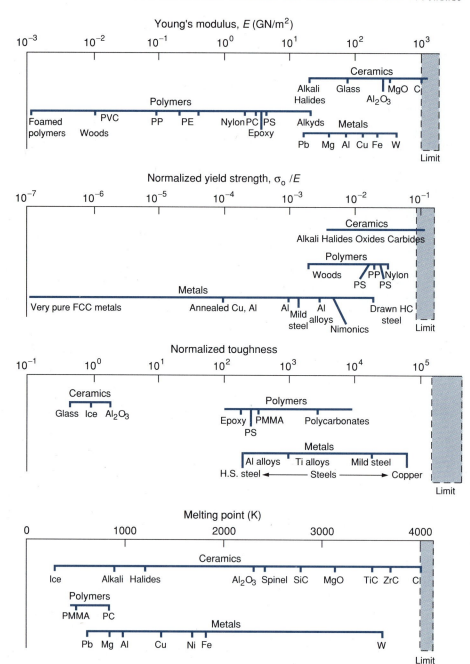

FIGURE 9-1 Property value ranges for metals, ceramics, and polymers. From M. F. Ashby and D. R. H. Jones, *Engineering Materials 1: An Introduction to their Properties and Applications*, Pergamon Press, Oxford (1980).

resistant materials. Their Achilles heel is toughness, which always raises the frightening specter of brittle fracture. Although metals occasionally suffer brittle fracture their outstanding attribute is toughness. If weight considerations are critical, as in aerospace applications, polymers rise to the top when strength-to-density comparisons are made. The latter is an example of the factors that must be taken into account in the increasingly complex arena of materials selection and utilization. Concerns of limited availability of strategic materials, the thrust toward energy independence, and an increasingly competitive world have exerted a strong impetus to tighten engineering design, improve performance, and lower costs. Many of these demands cannot be met by traditional materials. New approaches involving novel heat treatments, composites, and utilization of coatings are now important and will be discussed.

Metal properties and treatments have traditionally been the base against which new materials are compared, and for this reason the chapter initially emphasizes them. The remainder of the chapter is devoted primarily to polymers, composites, and ceramics. In each material category the more important representatives are introduced, followed by a description of the methods and strategies employed to develop optimal strength and toughness properties.

The closest thing to a universal recipe for strengthening materials is: *disrupt the perfection of the lattice or matrix and reduce the size of constituent phases.* Introduction of foreign atoms, dislocations, grain boundaries, and second-phase distributions will work, but unfortunately may impact toughness adversely. Practical methods for implementing this strengthening recipe include composition control, phase transformations, heat treatments, and thermomechanical processing techniques that combine thermally activated solid-state reactions (e.g., phase transformations, diffusion) with some sort of mechanical forming. A number of these treatments are generic in nature and, although initially developed in metals, are applicable to and practiced in different classes of materials.

9.2. HEAT TREATMENT OF STEEL

9.2.1. Background

Steels are our most important engineering materials because they enable the creation of the structures, tools, machines, and power plants that make all of our technological endeavors possible. Up until the beginning of the 20th century the only steels available were of the plain carbon variety. The practice of heating iron to elevated temperatures in a bed of glowing charcoal, holding it there for a while to absorb sufficient carbon to create a plain carbon steel, forging it into the required shape, and slowly cooling it to room temperature dates to antiquity. Carbon's importance is disproportionate to its small presence not only in plain carbon steel, but other grades of steel as well. A good basis for understanding plain carbon steels and some of the microstructures arising from simple thermal treatments is provided by the Fe–C (Fe–Fe$_3$C) phase diagram (Fig. 5-23). Reviewing the equilibrium properties of the Fe–C system at this point is, therefore, worthwhile. It should, however, be remembered that only equilibrium transformations, brought about by *slow cooling* of steel from elevated temperatures, were considered previously.

A dramatic improvement in the hardness and strength of the steel results from a simple heat treatment. Instead of being slowly cooled after it is forged, the piece must be again heated to an elevated temperature, *rapidly quenched* (in water or oil), and then tempered or reheated to a rather low temperature prior to use. By these means hardnesses and tensile strengths can often be more than doubled, a huge increase when one speaks about mechanical properties. And, very importantly, parts can be hardened after they have been formed to the final shape without appreciably distorting them during heat treatment.

Humankind owes a great debt to the dirty, soot-laden, ancient blacksmith(s) who, millennia ago, invented and passed on the secret of the magical process for hardening iron. Without means of controlling composition and temperature it was little wonder that the product of such a complex heat treatment frequently exhibited variable behavior in service as an agricultural tool or sword. Occasionally, however, blacksmiths refined the process, and then were able to reproducibly fashion legendary swords. The early treating of steel was an art and largely remained so until the individual steps were studied scientifically in the 1920s and 1930s. A milestone in our understanding of steel heat treatment occurred with the generation of experimental **time, temperature, transformation** (TTT) curves introduced in Section 6.5.2. It is a good idea to review this subject because the important implications of TTT curves with respect to strengthening steel by heat treatment will now be dealt with.

9.2.2. Isothermal Transformation of Austenite

9.2.2.1. Eutectoid Steel

Let us reconsider the TTT curve for eutectoid steel (Fig. 6-21) that is reproduced again in Fig. 9-2 for convenience. Recall that the **C** shape of the TTT curve underscores combined nucleation and growth effects that characterize the isothermal austenite–pearlite transformation. Suppose stable austenite (A), above the eutectoid critical temperature of 723°C, is rapidly cooled below it to temperature $T_1 = 475$°C and left to decompose. With time, the now unstable austenite transforms isothermally (at constant temperature) via solid-state reaction to ferrite (F) and iron carbide (C) along the horizontal path as shown. As long as the A + F + C field has not been entered, only A will be present. After about 1 second the boundary between A and A + F + C is crossed, causing pearlite (F + C) to nucleate and grow with time. For approximately the next 18 seconds the transformation of A to F + C proceeds slowly at first and then

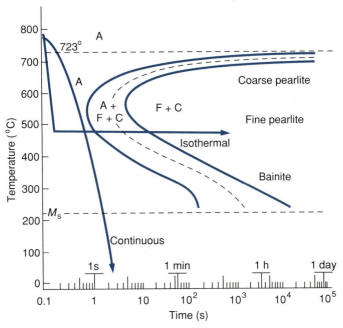

FIGURE 9-2 TTT curve for eutectoid steel with superimposed paths of isothermal and continuous cooling transformation of austenite.

more rapidly; finally, it tails off toward completion in the sigmoidal fashion described previously in Chapter 6. Once the transformation enters the F + C field there is no more austenite left and the steel will not appreciably change if taken to any lower temperature.

This same sort of isothermal transformation will occur at any temperature between ~220 and 723°C, but with some significant differences. Importantly, the specific isothermal transformation temperature has a significant bearing on both the microstructure and hardness (strength) of the pearlite that forms. At high temperatures the few F + C patches that nucleate grow rapidly, leading to a coarse pearlitic structure. The large interlammelar spacing between the α-Fe and Fe_3C phases of pearlite results in a relatively soft, low-strength steel. Dislocations encounter relatively few barriers to motion in such coarse structures. Later in the chapter it will be evident that mechanical strength is enhanced if phases in a two-phase structure are small in size, for then they more effectively restrict dislocation motion. Considerably harder and stronger steels are produced if the transformation occurs at successively lower temperatures. The many nuclei that cannot readily grow cause evolution of finer pearlitic microstructures possessing higher hardness and strength.

If the transformation temperature is reduced well below the nose of the curve (Fig. 9-2), austenite isothermally decomposes into **bainite**. This structure is named after E. C. Bain, an early investigator of this type of transformation. Even today the nature of the bainite transformation is not entirely understood. Like pearlite, bainite consists of ferrite and Fe_3C, but unlike pearlite the micro-structure is nonlamellar (Fig. 9-3A). In some ways bainite displays characteristics of martensite, which are discussed in the next section. Distinctions in bainitic structures occur depending on the temperature of formation. As might be expected, the finer lower bainites produced below ~350°C are harder than the upper bainites formed between ~350 and 450°C. Lower bainites contain fine precipitates of carbides distributed throughout the ferrite plates, yielding a hard and tough structure.

9.2.2.2. Noneutectoid Steel

Initially austenitized noneutectoid carbon steels convert to either ferrite or Fe_3C when isothermally transformed above 723°C, depending upon whether the involved steel is hypoeutectoid or hypereutectoid in composition, respectively. This accounts for the respective A + F and A + C fields in the TTT diagrams for these steels as shown in Fig. 9-4. Below the critical eutectoid temperature the transformation of A to A + F or A + C proceeds first, followed by a subsequent A to F + C transformation, as before, for any austenite that remains.

9.2.3. Transformation of Austenite to Martensite

Life is too short to wait for steel to transform isothermally. Suppose that while eutectoid steel was isothermally transforming the process were suddenly interrupted by quenching of the steel or rapid lowering of its temperature. We would find that any austenite that did not form pearlite would instead transform to **martensite,** yielding a mixed microstructure consisting of both martensite and pearlite. On quenching, the metastable FCC austenite attempts to reject the 0.8 wt% C held in solid solution because stable BCC α-Fe can only dissolve 0.01 wt% C. But the temperature drops so rapidly that carbon simply has no time to escape; rather it is trapped within the austenite, straining it to the point where a **diffusionless** or martensitic transformation occurs.

The martensite transformation is crystallographically interesting. Enclosed within the two FCC γ-Fe unit cells of lattice constant a is an incipient body-centered tetragonal cell, sketched in Fig. 9-5, whose lattice parameters are a' and c. Amid complex lattice shearing, involving atomic displacements of less than a lattice spacing, austenite distorts into an elongated body-centered tetragonal martensitic structure. This simplified model of the transformation illustrates how effortlessly the (111) plane in FCC Fe becomes the (011) plane in BCT Fe. Increasing carbon contents actually extend the c and shorten the a' dimensions. The microstructure of martensite reproduced in Fig. 9-3B reveals that it has a needlelike acicular morphology.

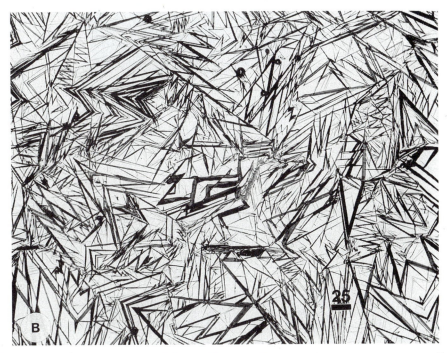

FIGURE 9-3 Microstructures of (A) bainite in 1060 steel (500×) and (B) martensite in a steel containing 1.8 wt% C (200×). Courtesy of G. F. Vander Voort, Carpenter Technology Corporation.

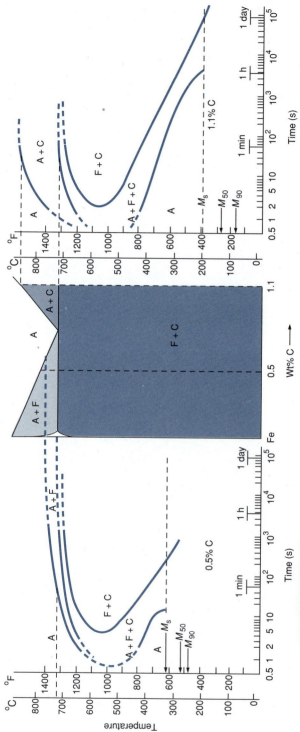

FIGURE 9-4 TTT curves and their relation to the Fe–Fe₃C phase diagram. *Left:* 1050 hypoeutectoid steel (0.50 wt% C, 0.91 wt% Mn). *Right:* 10110 hypereutectoid steel (1.10 wt% C, 0.30 wt% Mn).

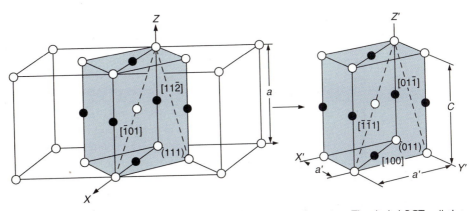

Simplified model for the crystallography of the martensite transformation. The shaded BCT cell that becomes martensite in the $X'Y'Z'$ coordinate system stems from the prior FCC cells in the XYZ coordinate system. Interstitial carbon sites are drawn as dark circles.

The important thing to remember is that the highly strained martensite is extremely hard and responsible for strengthening the steel. It is harder and stronger than the finest pearlitic structures described earlier. Carbon strongly influences the formation and hardness of martensite. Harder martensites are produced with higher carbon contents. And like many hard materials, quenched martensite is brittle, especially because it contains high levels of internal or residual stress. To relieve the latter it is recommended that freshly formed martensite be immediately annealed or **tempered.** Horror stories exist of quench-hardened dies loudly cracking, ruining the lunch of startled workers who planned to temper them after eating. Relief of the carbon supersaturation and stress occurs through outdiffusion of carbon to form carbide precipitates. The tempering process is accompanied by a slight, but tolerable reduction in hardness. Importantly, tempering toughens the steel considerably. Tempering temperatures of less than 500°C are typically employed for times on the order of minutes to a few hours depending on the steel and part size.

The martensite transformation has a number of other features relevant to its role in hardening steel. First, because it occurs by elastic shearing, the transformation essentially proceeds with the speed of sound. Further, the martensite transformation cannot be suppressed; it is triggered when the austenite temperature drops below the M_s (martensite start) temperature. Once the first bit of martensite forms, the driving force for further transformation is reduced because the structure is stabilized. Martensite and austenite coexist as the strain energy released by the former balances the chemical driving force to convert the latter. As the temperature drops, increasing amounts of martensite form until at the M_f (martensite finish) temperature, the transformation of austenite is complete. The M_s and M_f temperatures vary with the steel. To obtain 100% martensite it is necessary first to cool the steel so as to miss the nose of the

TTT curve; furthermore, the steel must be quenched below M_f. This eliminates the possibility that austenite, which can transform into brittle martensite if the service temperature drops sufficiently, will be retained.

EXAMPLE 9-1

Assume the cooling history is programmed to be *linear* on a TTT curve of eutectoid steel austenitized at 800°C. Estimate the critical cooling rate (dT/dt) at the time the knee in the TTT curve is just missed.

ANSWER   We must calculate dT/dt, but note that the abscissa is given in log t(s) units. As $d \log t = (1/2.30t) \, dt$, $dT/d \log t = 2.30t \, dT/dt$, or $dT/dt = (1/2.30t) \, dT/d \log t$. The cooling rate is time dependent. A straight line drawn starting at 800°C is tangent to the knee of the TTT curve (Fig. 9-2) at $t \sim 0.8$ seconds and $T \sim 530$°C. After substitution,

$$dT/dt = (1/2.30 \times 0.8)(530 - 800)/(\log 0.8 - \log 0.1) = -162\text{°C/s}.$$

In view of the difficulty in estimating t, this value must be regarded as approximate.

9.2.4. Continuous Cooling Transformation of Austenite

Aside from a few special treatments, steel is rarely **isothermally** transformed. Rather, it is cooled **continuously** from austenitizing temperatures to room temperature, and transforms incrementally in a complex way dependent on the cooling rate. The line drawn superimposed over the TTT curve of Fig. 9-2 represents such a typical cooling history. We would like to map the kinetics of austenite decomposition along this or other sloping cooling curves as we did previously along the horizontal isotherms; however, it is a more difficult experimental challenge to generate a *continuous cooling transformation* (CCT) curve than an *isothermal transformation* (IT or TTT) curve. Nevertheless, it is found that transformation during continuous cooling is delayed. This displaces the CCT curve down and to the right of the TTT curve as shown in Fig. 9-6 where the two behaviors are superimposed.

We shall make the important assumption that the more accessible TTT curves can be used to describe continuous cooling behavior. Avoiding the knee of a TTT curve ensures transformation of austenite to martensite. If it is desired to harden a part fully, then cooling rates throughout must be equal to or greater than the critical one (i.e., fall to the left of the knee). On the other hand, when a slower cooling-rate curve intercepts the upper portions of the TTT curve, transformation to pearlite or mixtures of pearlite and martensite can be expected. Although some judgment is required, TTT curves provide a

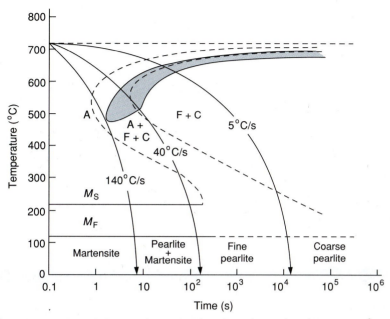

FIGURE 9-6   TTT and CCT (shaded) curves for eutectoid steel superimposed on the same set of temperature–time axes. Adapted from the *Atlas of Isothermal Transformation and Cooling Transformation Diagrams*, U.S. Steel Corporation, Pittsburgh (1959).

reasonable understanding of what to expect during practical heat treatment. The hardening response of steel due to different cooling rates and alloy compositions will now be addressed.

9.2.4.1. Different Cooling Rates

Consider a steel part of appreciable size and complex shape such as a large gear that is uniformly austenitized and then quenched. The thin teeth and surface necessarily cool more rapidly than the more massive interior portions of the gear. Therefore, a range of cooling rates can be expected throughout as schematically shown in Fig. 9-7A. Whereas martensite will form at the surface and near-surface regions, some transformation to pearlite in the thicker interior portions is likely. If the gear is to be hardened throughout, pearlite formation must be avoided. On the other hand, there are situations in which it is desirable to selectively harden only the teeth and surface of large gears, leaving the interior soft to better absorb impact loading. This is usually accomplished by electrical induction heating, where only the outer surface of the gear is heated to the austenite range by high-frequency (skin) currents. Before heat travels to the interior, the gear is quenched, creating the desired structure.

The response of a given part to different cooling ambients is not difficult to understand. Four common methods for cooling austenite, listed in order of

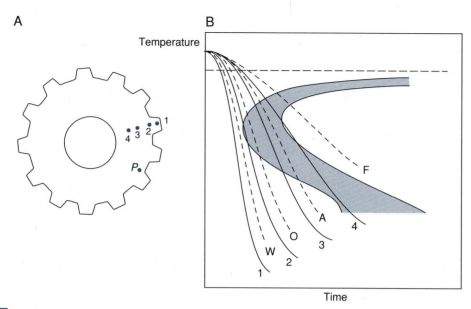

A

B

(A) Schematic cooling rates at locations 1, 2, 3, and 4 within a gear that is austenitized and quenched in a given medium, for example, oil. (B) Schematic cooling rates at point *P* on the gear due to water quenching (W), oil quenching (O), air cooling (A), and furnace cooling (F). In either case higher hardnesses are associated with more rapid cooling rates.

decreasing severity of quench rate, are water quenching, oil quenching, air cooling, and furnace cooling. In practice oil quenching is preferable to water quenching because it is not as prone to causing warpage and cracking. The effect of *different* quenching media on the hardening response at a *single* location in the gear is sketched in Fig. 9-7B. Whereas martensite will form during water and oil quenching, both air cooling and furnace cooling result in pearlite formation. The measured hardness also tracks the quench severity, with water quenching yielding the highest, and furnace cooling, the lowest values.

9.2.4.2. Effect of Alloying

Alloying has a dramatic influence on the response of a steel to heat treatment. It does this by altering the shape and relative position of the TTT curve on the time axis. The most important alloying element in steel is carbon and it exerts a strong effect in shifting the TTT curve to the right. Plain carbon steels that contain less than about 0.2 wt% C cannot be effectively hardened. For such compositions, parts thicker than a knife blade cannot be cooled fast enough to miss the TTT curve, which is shifted far to the left. At higher carbon levels, however, both the hardness of the martensite and the ease with which it forms increase.

Other common alloying elements in steel, for example, Mn, Ni, Si, Cr, Mo,

V, and W, also cause the TTT curve to translate to the right. In differing degrees and for different reasons these elements slow down the austenite-to-pearlite transformation. They are homogeneously distributed on substitutional sites within the austenite, but when this phase is no longer stable they partition to either the ferrite or the carbide phase. Manganese, nickel, and silicon segregate to the ferrite, whereas the other elements listed tend to form carbides. Pearlite formation relies on solid-state diffusion processes in the austenite to redistribute these atoms to the appropriate phase boundaries. But compared with carbon, which diffuses interstitially and hence rapidly, these substitutional atoms diffuse more slowly through the lattice, thus decelerating the overall transformation rate. Boron's interesting effect in inhibiting the austenite-to-pearlite transformation should be mentioned. Very small B additions have a profound effect in displacing the TTT curve to the right, a fact capitalized upon to conserve strategic metals. Apparently, boron segregates to and stabilizes the prior austenite grain boundaries, making them less accommodating sites for pearlite nucleation.

The size of the prior austenitic grains strongly influences the position of the TTT curve knee and the resulting critical cooling rate in carbon steels. A coarse-grained austenite provides few nucleation sites from which to launch the pearlite transformation, slowing it down. This effectively shifts the TTT curve to the right. On the other hand, low austenitizing temperatures and alloying elements like vanadium refine the austenite grain size, thereby accelerating the transformation. From what has been said it might be inferred that large grain size is desirable to slow the transformation of austenite. In fact, quite the opposite is true in practice. A fine grain size confers the toughness required of tool steels.

From just one illustrative TTT curve (Fig. 9-8), many of the benefits of alloying are immediately apparent. Relative to eutectoid steel, the TTT curve for 4340 steel is shifted toward longer times. The critical cooling rate required to miss the knee of the curve is considerably reduced, for example, by a factor of more than 10 relative to carbon steels. This means that it is possible to produce a martensitic structure even if the cooling rate is quite low. Where pearlite might be produced in eutectoid steel, martensite would form in the alloy steel. A corollary benefit is that less severe quenches can be used to produce martensite. What may be attainable only by water quenching a carbon steel could be easily realized by oil quenching an alloy steel. As already noted, the use of milder quenches in steel heat treatment is not only desirable, but imperative in alloy steels because they are susceptible to high residual stresses and quench cracking.

The advantage of being able to harden alloy steels during slow cooling is turned into a significant disadvantage during welding. Welding is a joining process in which the base metals experience temperature excursions ranging from above the melting point to room temperature. Steel surrounding the molten weld may *air harden* to brittle martensite that can crack in service if undetected. Those steels that are susceptible to this potential hazard should

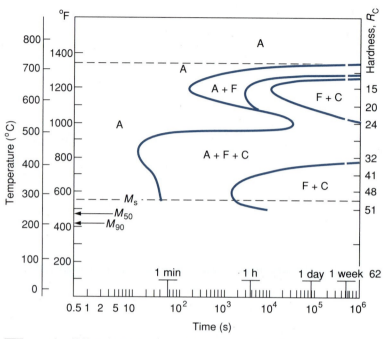

FIGURE 9-8 TTT curve for 4340 steel which contains 0.40 wt% C, 0.90 wt% Mn, 0.80 wt% Cr, 0.2 wt% Mo, and 1.83 wt% Ni. From the *Atlas of Isothermal Transformation and Cooling Transformation Diagrams*, U.S. Steel Corporation, Pittsburgh (1959).

be preheated prior to welding to ensure that the cooling rate in air is less than the critical one required to trigger martensite formation. This problem does not surface in commonly welded structural steels containing only 0.15 to 0.2 wt% C.

9.2.5. Hardenability

Hardenability is a measure of the **ability** to **harden** a steel to a given depth. It is not the same as hardness, which measures a material's ability to withstand penetration by a hard indenter. The factors that displace TTT curves to the right, inhibiting the transformation of austenite to pearlite, also increase the hardenability of the steel. These include increased carbon content, alloying elements, and large austenitic grain size. High hardenability implies ease in producing martensite at greater depths within a part and accomplishing it with less drastic cooling rates.

All of these associations have been distilled into a simple, universally accepted test—the Jominy end-quench hardenability test—that readily rates the hardenability of steels. All that is needed is a fixture with a hole in it to support a hot test bar and a source of water placed a fixed distance beneath it (Fig.

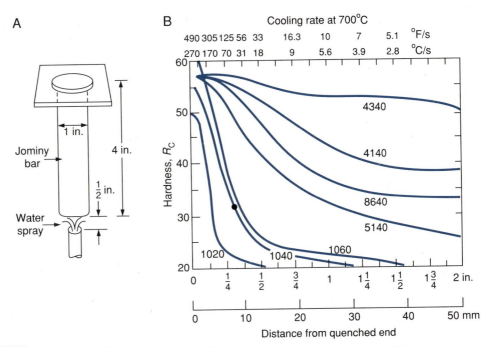

A

B

Cooling rate at 700°C

FIGURE 9-9 (A) Schematic of Jominy test apparatus. (B) Jominy hardness data for a number of different steels.

9-9A). A round Jominy bar of standard dimensions is austenized at an elevated temperature and then quickly transferred to the fixture, where it is end quenched in a specified way. After cooling, a flat is ground on the bar length and hardness measurements are recorded along it at regular intervals. A basic assumption of the test is that all steels cool at the *same* rate at corresponding positions along the bar. This is approximately true because of the common directional cooling and the fact that heat conduction, convection, and radiation properties of all heated steels are similar. (Stainless and other high-alloy steels with low thermal conductivities are exceptions.) Jominy test results for a number of steels are plotted in Fig. 9-9B with approximate cooling rates indicated on the upper abscissa. Because a given cooling rate is associated with a particular microstructure having a characteristic hardness, Jominy test data allow hardenability comparisons among different steels in parts of varied shape.

EXAMPLE 9-2

A company is bidding for a contract to produce 10,000 clevis blocks or chain adaptors used in aerial lifts (cherry pickers). The machine drawing for this steel component is indicated in Fig. 9-10. Specifications call for a hardness value of R_c 46 at the indicated point. Otherwise the choice of the material is left to the manufacturer, cost being the only consideration. The only steel on hand is 1040 and a quick decision as to which steel to use is required. If the only possibilities are those shown in Fig. 9-9B, how would you go about selecting a steel for this application?

ANSWER   First, a model of the clevis block should be machined from the 1040 steel. It should then be heat treated in the same manner as that anticipated for the production run. This is done, and suppose the measured hardness value is R_c 32. (Clearly 1040 steel will not do.) The hardness is then located on the Jominy curve for 1040 steel. But, the selected steel must have a hardness of R_c 46 or higher at the *same* Jominy bar distance. Of the steels displayed, 5140, 8640, 4140, and 4340 will all meet the hardness specifications after heat treatment. The most inexpensive steel should then be selected.

9.2.6. Additional Steel Hardening and Toughening Treatments

A number of special treatments aimed at strengthening, toughening, and improving the properties of steel have been devised over the years. Some of the interesting ones involve combinations of thermal transformations and plastic deformation and are known as **thermomechanical treatments** (TMTs). The

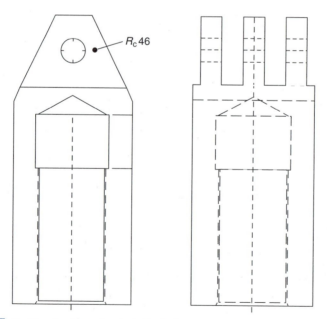

FIGURE 9-10 Machine drawings of a clevis block that must be heat-treated to a hardness value of R_c 46 at the indicated point.

latter are further distinguished according to whether deformation occurs (I) *before* austenite transforms (martensite forms in work-hardened austenite); (II) *during* austenite transformation (martensite forms in deformed metastable steels); or (III) *after* austenite transforms (thermal aging of transformation products). They all exploit the properties of TTT diagrams and brief consideration of them is a fitting way to end our discussion on heat treatment of steel. We start with two heat treatments that require no deformation.

9.2.6.1. Austempering

In austempering, steel is first quenched to the left of and below the nose of the TTT curve to a temperature at which bainite formation is feasible. Time is then allowed for isothermal transformation to bainite to proceed to completion (Fig. 9-11A). High hardness and toughness are the benefits of austempering. The process is not suited to steels that require long bainite transformation times.

9.2.6.2. Martempering (or Marquenching)

Martempering involves an interrupted quench of austenitized steel to reduce the overall cooling rate. Residual stress generation in the martensite and its potential for distortion and even cracking are minimized in this way. In martempering, steel is quenched into an oil or molten salt bath held just above or slightly below the M_s temperature. It is held there until a uniform temperature is attained, but less than the time required for bainite transformation to initiate (Fig. 9-11B). Gradual cooling below the M_f followed by standard tempering completes the process.

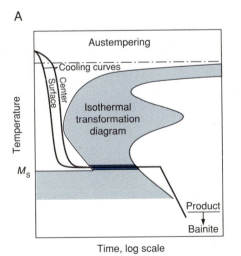
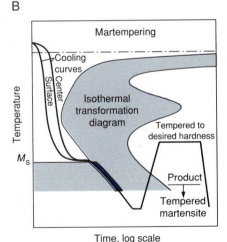

FIGURE 9-11 (A) Austempering and (B) Martempering treatments. From the *Atlas of Isothermal Transformation and Cooling Transformation Diagrams*, U.S. Steel Corporation, Pittsburgh (1959).

9.2.6.3. Ausforming (Ausforging)

An assortment of both high- and low-temperature TMT processes fall into the ausforming category and are depicted in Fig. 9-12. They feature mechanical deformation of both stable and unstable austenite with subsequent transformation to either ferrite and pearlite or martensite. Low- and medium-carbon steels are strengthened, whereas tool steels are toughened and made more wear resistant by such treatments. Perhaps the most widespread of these TMT processes are carried out in **high-strength low-alloy** (HSLA) steels, also known as microalloyed steels. These contain 0.05 to 0.2 wt% C and about 0.1 wt% of elements like Nb, Ti, and V. During hot rolling the austenite transforms to a fine-grained ferrite. As we shall see in Section 9.4.2, significant strengthening can occur through grain refinement. This plus greater corrosion resistance is the reason these microalloyed steels are successfully supplanting mild steel in many structural applications.

9.2.6.4. Transformation-Induced Plasticity (TRIP) Steels

TRIP steels have special compositions that stabilize austenite even at room temperature. In TMT (process II) the austenite is first deformed (rolled) to a 70 to 80% reduction, significantly raising the yield stress. The strain triggers

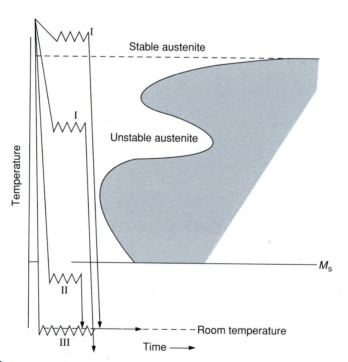

FIGURE 9-12   Schematic TTT curve showing classification of thermomechanical treatments during transformation of austenite. Included are TMT processes for HSLA (I), TRIP (II), and maraging steels (III). Adapted from M. A. Meyers and K. K. Chawla, *Mechanical Metallurgy*, Prentice–Hall, Englewood Cliffs, NJ (1984).

transformation of the martensite, which not only raises the strength but also, surprisingly, increases the toughness of the steel. Strengthening increases the local strain hardening rate so that neck formation and failure are inhibited (see Section 7.3.3). On the other hand, if a crack already exists, martensite transformation at the crack tip blunts its further propagation.

9.2.6.5. Maraging

Maraging steels are heavily alloyed (e.g., ~18 wt% Ni, 8 wt% Co, 5 wt% Mo, 0.5 wt% Ti) but contain no carbon. The lack of carbon makes for a relatively ductile martensite that can be mechanically processed with comparative ease. After martensite is aged or heated to low temperatures for prolonged times, precipitates form that appreciably strengthen the matrix. Aging or precipitation hardening treatments are commonly employed to strengthen nonferrous alloys and are discussed again in Section 9.5.2.

The great improvement in *both* fracture toughness and strength as a result of these two thermomechanical treatments is evident in Fig. 9-13.

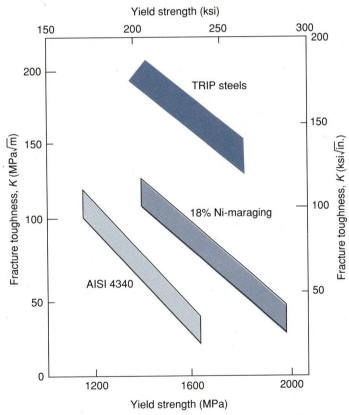

FIGURE 9-13 Fracture toughness versus yield strength for 4340, maraging, and TRIP steels. Adapted from V. F. Zackay, E. R. Parker, J. W. Morris, and G. Thomas, *Materials Science and Engineering* **16**, 201 (1974).

9.3. FERROUS AND NONFERROUS ALLOYS: PROPERTIES AND APPLICATIONS

The subject matter of this section could easily fill a book by itself. In fact, books have been written on specific alloy subcategories, for example, stainless, tool steels. To limit the treatment to manageable proportions, only representative iron (steel)-, aluminum-, and copper-base alloys will be given attention, and only to the extent of tabulating the mechanical properties and uses of just a few alloys. This is done in Tables 9-1, 9-2, and 9-3. It is hoped that these

TABLE 9-1   **MECHANICAL PROPERTIES AND APPLICATIONS OF A VARIETY OF STEELS**

Composition (wt%)	Condition	Strength (MPa)		Ductility (% elongation)	Applications
		Tensile	Yield		
Plain Carbon					
1020 (0.20C, 0.45Mn)	As rolled	448	331	36	Steel plate, structural sections
	Annealed	393	297	36	
1040 (0.40C, 0.75Mn)	As rolled	621	414	25	Shafts, studs
	Annealed	517	352	30	
	Hardened	800	593	20	
1080 (0.80C, 0.75Mn)	As rolled	967	586	12	Music wire, helical springs, chisels, die blocks
	Annealed	614	373	25	
	Hardened	1304	980	1	
Low Alloy					
4340 (0.40C, 0.90Mn, 0.80Cr, 0.2Mo, 1.83Ni)	Annealed	745	469	22	Truck, auto, and aircraft parts, landing gear
	Hardened	1725	1587	10	
5160 (0.60C, 0.80Cr, 0.90Mn)	Annealed	725	276	17	Automobile coil and leaf springs
	Hardened	2000	1773	9	
8650 (0.50C, 0.55Ni, 0.50Cr)	Annealed	710	386	22	Small machine axles
	Hardened	1725	1552	10	
Stainless					
430 (17Cr, 0.12C max) (ferritic)		550	375	25	Decorative trim, acid tanks, restaurant equipment
304 (19Cr, 9Ni) (austenitic)		579	290	55	Chemical and food processing equipment

Tool	Hardness (R_c)	Applications
T1 (18W, 4Cr, 1V, 0.75C)	60 at 260°C, 50 at 600°C	Lathe and milling machine tools, drills
M2 (6W, 5Mo, 4Cr, 2V, 0.9C)	60 at 260°C, 50 at 600°C	Taps, reamers, dies

tables will provide a comparative appreciation of these most abundantly produced metals. The following remarks are intended to heighten distinctions between and among these metals and provide perspectives on their use.

Iron-containing or ferrous alloys are produced in tonnages ten times those of aluminum- and copper-based alloys combined. The reasons are not hard to see. Iron is not only abundant, but easily and cheaply extracted. After alloying and fabrication, steels are significantly cheaper than either Al or Cu alloys. But the major reason for the use of iron and steel alloys is their superior strength. For structural applications (e.g., skyscrapers, bridges) as well as for sheet metal enclosures of all kinds (auto bodies, appliances, etc.) and inexpensive tools and wire, unalloyed plain carbon steels have an adequate combination of strength and ductility. For these purposes there is no necessity to harden steel further through alloying and additional heat treatment. In addition, a great deal of iron and steel is cast to meet assorted demands (engine blocks, machine bases, radiators, valves, etc.). Cast irons are not generally ductile, but are adequately strong in compression. In addition, malleable and ductile cast irons exist.

TABLE 9-2 **MECHANICAL PROPERTIES AND APPLICATIONS OF A VARIETY OF ALUMINUM ALLOYS**

Composition (wt%)	Condition	Strength (MPa)		Ductility (% elongation)	Applications
		Tensile	Yield		
Wrought Alloys					
1100 (99.0 min Al, 0.12Cu)	Annealed	89	24	25	Sheet metal
	Half-hard	124	97	4	
2024 (4.4Cu, 1.5Mg, 0.6Mn)	Annealed	220 (max)	97 (max)	12	Aircraft structural parts
	Heat-treated (T-6)[a]	500	390	5	
6061 (1.0Mg, 0.6Si, 0.27Cu, 0.2Cr)	Annealed	152 (max)	82 (max)	16	Truck and marine structures
	Heat-treated (T-6)	315	280	12	
7075 (5.6Zn, 2.5Mg, 1.6Cu, 0.23Cr)	Annealed	276 (max)	145 (max)	10	Aircraft and other structures
	Heat-treated (T-6)	570	500	8	
Cast Alloys					
356 (7Si, 0.3Mg)	Sand cast	207 (min)	138 (min)	3	Transmission and axle housings, truck wheels, pump housings
	Heat-treated (T-6)	229 (min)	152 (min)	3	
413 (12Si, 2Fe)	Die cast	297	145	2.5	Intricate die castings

[a] T-6 = solution treated, then aged.

TABLE 9-3	MECHANICAL PROPERTIES AND APPLICATIONS OF A VARIETY OF COPPER ALLOYS

Composition (wt%)	Condition	Strength (MPa)		Ductility (% elongation)	Applications
		Tensile	Yield		
Wrought Alloys					
OFHC[a] (99.99Cu)	Annealed	220	69	45	Bus conductors, vacuum
	Cold worked	345	310	6	seals, glass-to-metal seals, electrical wiring
Electrolytic Tough,	Annealed	220	69	45	Electrical wiring, auto
Pitch (0.04O)	Cold worked	345	310	6	radiators, gutters, roofing
Brass (70Cu,	Annealed	325	105	62	Radiators, lamp fixtures,
30Zn)	Cold worked	525	435	8	musical instruments, ammunition shells, locks, electrical sockets
Beryllium copper	Solution	410	190	60	Bellows, spring contacts,
(1.7Be, 0.20Co)	treated				diaphragms, fasteners, fuse clips, switch parts, welding equipment
	Precipitation hardened	1240	1070	4	
Casting Alloys					
Leaded bronze	As cast	255	117	30	Flanges, pipe fittings,
(5Sn, 5Pb, 5Zn)					plumbing goods, water pump impellers and housings, sculpture, ornamental fixtures
Aluminum bronze	As cast	586	242	18	Bearings, gears, corro-
(4Fe, 11Al)					sion-resistant vessels

[a] Oxygen-free, high-conductivity copper.

Steels are much harder than Al and Cu and, with higher melting points, are much more heat resistant. This earmarks the use of steels in all kinds of shaping and cutting tools not only for metals, but for polymers, wood, ceramics, fabrics, and so on, and for the critical manufacturing operations discussed in the previous chapter. In addition, there are a host of load-bearing components (e.g., gears, shafts) in static and rotating machinery, transportation vehicles, and earth-moving equipment that require hard wearing surfaces. For all of these purposes, medium- and high-alloy steels that can be hardened are essential. The sought after hardness and strength are attained but at the expense of lower ductility and some sacrifice in toughness. A comparison between plain carbon and alloy steels reveals the great strength benefits conferred by alloying and the quenching and tempering heat treatments they make possible. Among the most heavily alloyed iron-base alloys are the stainless and tool steels. Stainless steels come in ferritic, austenitic, and martensitic varieties that emphasize their structure. They combine processing capability with reasonable strength and high ductility in applications in which corrosion-free operation is essential. Tool steels are critical materials because they enable the important machining operations of turning, drilling, and milling to be realized.

Steels unfortunately have a high density and, with the exception of stainless varieties, tend to corrode in ambient environments. Furthermore, they are poor conductors of electricity and heat. In applications where strength is not a factor, alloys of aluminum and copper usually better address these property deficiencies and frequently become the materials of choice. In particular, aluminum has additional advantages in lightweight design for transportation needs. Otherwise, aluminum and copper alloys often compete; important examples are high electrical and thermal conductivity applications in wiring, cookware, and heat exchangers. Alloys of both metals are readily cast, Al for automobile engine components and housings and Cu for plumbing fixtures, pumps, valves, and marine components. They are also both readily mechanically formed and shaped by rolling, extrusion, and drawing into an assortment of products, for example, foil and sheet stock, tubing and pipes, architectural trim. Copper alloys are generally stronger and more ductile than aluminum alloys.

It is always a plus when additional strength can be imparted to nonferrous alloys without sacrificing other properties. A number of practical treatments for strengthening these metals are discussed later in the chapter. Even after such treatments nonferrous alloys do not generally approach the strength of steels. Strengthening Cu is not always an unmixed blessing, however. The necessary alloying for hardening simultaneously reduces the electrical conductivity. This is another example where property trade-offs must be optimally balanced.

Common to the above and other commercial metals are the cold working and annealing treatments they undergo. Because the former *strengthen* metals and the latter tend to *toughen* them, they are considered next.

9.4. MECHANICAL WORKING AND RECRYSTALLIZATION

9.4.1. Cold Working

The first metals shaped by humans (e.g., Au, Ag, Cu) were probably hammered cold into the desired form. Paul Revere, the American silversmith, carried on the tradition of hammering silver to create beautiful pieces that are preserved in many museums. Starting with a flat round sheet of silver he peened it uniformly in concentric circles, raising the metal into a bowl. Further hammering would tear the now less ductile silver so he annealed it. This softened the metal and restored his ability to cold work and shape it further until the next anneal was required, and so on. In this way narrow-necked decanters, candlesticks, goblets, and other items were fabricated. The beaten lead sheet sculpture shown in Fig. 9-14 was made using ball peen hammers but no anneals were required! Silver work hardens as it is *cold worked* at room temperature, where $T/T_M = 300/1234 = 0.24$. For lead the corresponding ratio of the working to melting temperatures is 0.50, a value that is suggestive of *hot working*. Thus, deformation of lead is accompanied by simultaneous softening.

| **FIGURE 9-14** | *The Prophet*, a hammered lead sheet sculpture by M. Ohring. The sculpture is 20 in. high. |

Lead does not work harden because atomic diffusion is appreciable during room temperature annealing. It is immaterial whether cold working occurs by hammering, cold rolling, extrusion, forging, or other processes; the microstructural change and property modification are similar if the plastic strains are comparable. Today all of the ductile metals in engineering use undergo cold working operations at one time or another to meet strength and dimensional specifications.

That working makes metals harder and stronger, but less ductile, has already been pointed out in Chapter 7 and in Section 8.3.4. Crystallographic distortion due to plastic deformation of grains, together with accompanying dislocation generation, interference, and repulsive interactions, accounts for the strengthening but reduced ductility and toughness. Our attention now focuses on the changes wrought by annealing.

9.4.2. Recrystallization

Annealing cold worked metals initiates a train of events, collectively known as **recrystallization,** that reverses the effects of the deformation. In particular, the prior softer but more ductile, unworked matrix is restored. Such a sequence of changes is followed metallographically in a relatively pure iron as a function of both temperature and time in Fig. 9-15. Final recrystallized grain sizes may even exceed those of the undeformed grains. Clearly, temperature has a greater influence than time in effecting microstructural change during recrystallization. The apparent structural reversibility during the working/annealing cycle gives it the appearance of a kind of phase transformation that can be described by the reaction

$$\text{metal (soft)} \xrightleftharpoons[\text{annealing}]{\text{cold work}} \text{metal (hard)}.$$

Strain energy, absorbed by the metal during deformation, is deployed in the stress fields that surround dislocations. This simultaneously strengthens the matrix and raises its free energy. Reduction of the dislocation density lowers the system free energy and this is precisely what happens during annealing and recrystallization. These processes generally require elevated temperatures and sufficient time for atoms to diffuse; only then can dislocations move, annihilate, climb, or disappear in large numbers. A key to understanding recrystallization behavior of different metals is the inverse relationship between melting point and rate of diffusion (see Section 6.3.2). This accounts for the fact that Ag must be heated to several hundred degrees centigrade to anneal it, whereas Pb recrystallizes at room temperature.

Much can be learned about recrystallization by annealing cold worked metals (typically deformed 50%) for a **constant time** (1 hour) at a series of **different temperatures.** Subsequent structural observations and property measurements (strength, elongation, electrical resistivity, etc.) have provided

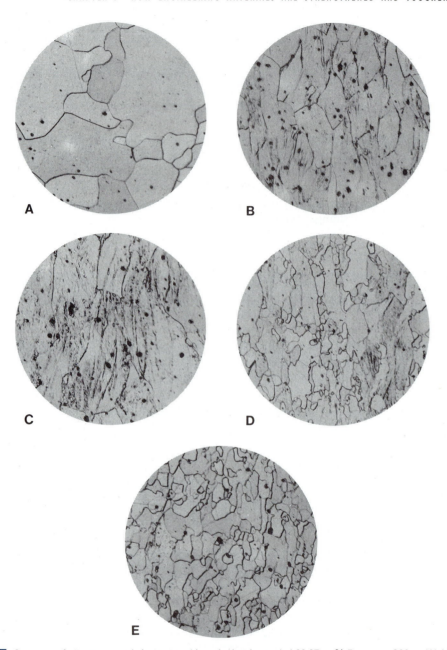

FIGURE 9-15 Sequence of microstructural change in cold worked and annealed 99.97 wt% Fe iron at 200×. (A) Hot rolled and annealed at 950°C for 2 hours. (B) Specimen A, but reduced 60% by cold rolling. (C) Specimen B, but annealed at 510°C for 1 hour. (D) Specimen B, but annealed at 510°C for 10 hours. (E) Specimen B, but annealed at 510°C for 100 hours. (F) Specimen B, but annealed at 675°C for 1 hour. (G) Specimen B, but annealed at 900°C for 1 hour.

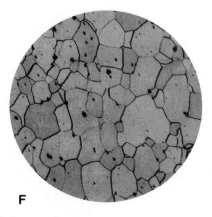
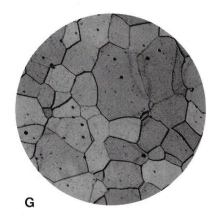

F   G

FIGURE 9-15 *(continued)*

a detailed portrait of resultant micro- and macroscopic change. Temperature-dependent changes in assorted mechanical properties following annealing are shown in Fig. 9-16. Of note are the countertrends in strength and ductility. A fractional strength or hardness drop of 50% arbitrarily defines the recrystallization temperature (T_R). It is customary to divide the response to annealing into three temperature regimes: **recovery, recrystallization,** and **grain growth.**

9.4.2.1. Recovery

Also known as primary recrystallization, the recovery stage is characterized by little drop in hardness and virtually no change in grain structure. Upon closer examination a redistribution of dislocations into vertical arrays of small misorientation angle, or so-called **subgrain boundaries,** occurs. The involved dislocation climb indicates that vacancies are mobile during recovery. Nevertheless, the density of dislocations is only slightly reduced, leaving barriers to their motion largely intact. As a result, the metal still retains much of its high-strength, low-ductility character. Notably, however, residual stress is largely eliminated.

9.4.2.2. Energetics of Recrystallization

When the cold worked matrix is heated to $\sim 0.3 T_M$ to $0.5 T_M$ (K) it recrystallizes. The major portion of the material response to annealing occurs during recrystallization. A large drop in strength is accompanied by a corresponding improvement in ductility. Actually, good compromises in strength and toughness can be struck in this temperature range, accounting for the fact that commercial annealing processes are conducted in this regime. The change in microstructure is dramatic (Fig. 9-15). At first, microscopic strain-free grains nucleate in local regions of the matrix that are particularly strained.

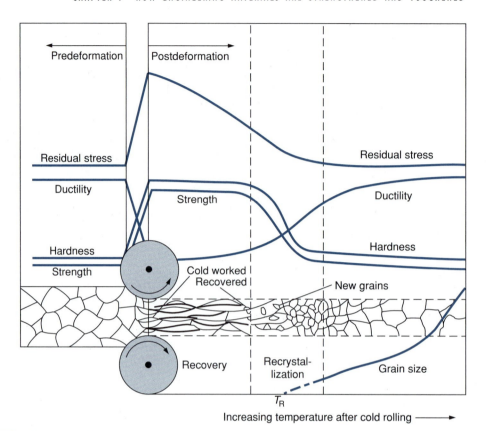

FIGURE 9-16   Schematic trends in structure and strength, hardness, and ductility properties of cold worked metals that are annealed at different temperatures for a fixed time.

These early changes can be understood with reference to nucleation models (see Section 6.4.1). Strain-free grains and the cold worked matrix represent the thermodynamically stable and unstable phases, respectively. They are separated in free energy by ΔE_S, the strain energy per unit volume. Nests and arrays of dislocations are now eliminated on a grand scale and the strain energy released provides the driving force to stabilize new grain boundaries having surface energy γ. The number of such strain-free nuclei is proportional to $\exp(-\Delta G_N^*/RT)$ (Eq. 6-20), with the energy barrier for nucleation (ΔG_N^*) related to $\sim\gamma^3/(\Delta E_S)^2$.

One consequence of these ideas is that a more heavily cold worked matrix will recrystallize at lower temperature. The reason is that ΔE_S is larger. This reduces ΔG_N^* and yields a higher recrystallized nucleus density at equivalent temperatures. Furthermore, a high nucleation density implies that the recrystallized grain size will be finer, as observed. More and more of the cold worked

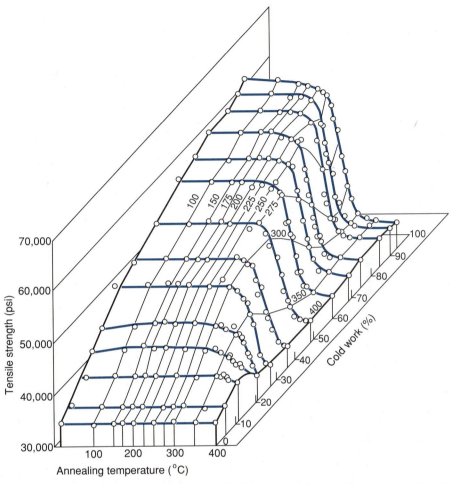

FIGURE 9-17 Tensile strength of oxygen-free high conductivity (OFHC) copper as a function of percentage cold work and annealing temperature. After M. V. Yokelson and M. Balicki, *Wire and Wire Products*, 1179 (1955).

matrix transforms in this way into new strain-free crystals which grow until they abut one another.

A useful representation (Fig. 9-17) connecting percentage cold work, tensile strength, and annealing temperature of pure copper illustrates the domain of interaction among these variables and how T_R varies with cold work.

9.4.2.3. Recrystallization Kinetics

Information on recrystallization times and the kinetics of softening is important when designing annealing heat treatments. Because we have already drawn the analogy to phase transformations, it is appropriate to extend the use of the Avrami equation (Eq. 6-22) to describe recrystallization kinetics. Therefore, the fractional amount of recrystallization (f_R) as a function of time (t) is given by

$$f_R = 1 - \exp(-Kt^n). \tag{9-1}$$

The temperature-dependent constant K is thermally activated [$K = K_0 \exp(-E_R/RT)$ with K_0 a constant], n is often equal to 1, and E_R is the activation energy for recrystallization. Fractional hardness decrease or fractional area of recrystallized grains is the way f_R is commonly measured experimentally. Recrystallization data for cold worked 3.25 wt% Si steel are displayed in Fig. 9-18 as a function of *variable time* at a number of *fixed temperatures*. Characteristic sigmoidal curves reflect the early incubation period for nucleation, followed by the subsequent growth and later impingement of recrystallized grains. By convention, the time required for transformation to proceed halfway, that is, $f_R = \frac{1}{2}$, is often taken as the recrystallization time $t_{1/2}$. After substitution in Eq. 9-1, with $n = 1$,

$$1/t_{1/2} = (K_0/\ln 2) \exp(-E_R/RT). \tag{9-2}$$

Lastly, T_R, the **recrystallization temperature,** is practically defined as that for which $f_R = \frac{1}{2}$ after a 1-hour anneal. Practice in applying these equations and definitions is given in the following illustrative problem.

EXAMPLE 9-3

Consider the recrystallization data of Fig. 9-18.
a. What is the activation energy for recrystallization?
b. What is T_R for this steel?

ANSWER   a. The times for which $f_R = \frac{1}{2}$ at the various annealing temperatures are read off the horizontal axis. Thus, $t_{1/2}$ (1000°C) = 0.0048 second, $t_{1/2}$ (911°C) = 0.051 second, ..., $t_{1/2}$ (600°C) = 1500 seconds, $t_{1/2}$ (550°C) = 20,000 seconds. A semilog plot of $t_{1/2}$ versus $T(K)^{-1}$ is shown in Fig. 9-19. The slope of the resulting Arrhenius curve ($-E_R/2.303R$) is 14,600. Therefore,

$$E_R = 14{,}600 \times 2.303 \times 8.31 = 279 \text{ kJ/mol or } 66.8 \text{ kcal/mol.}$$

This value is in excellent agreement with the activation energy for self-diffusion in Fe.

b. We seek the temperature at which half of the structure recrystallizes in 1 hour. The intersection of $f_R = \frac{1}{2}$ and $t = 3600$ seconds occurs at a roughly estimated recrystallization temperature of 580°C.

9.4.2.4. Grain Growth

After grains recrystallize they are stable, but their size can often be enlarged in a spectacular way. All that is required is to raise the temperature. As the

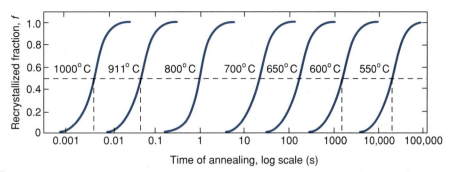

FIGURE 9-18 Recrystallization behavior of a 60% cold worked 3.25 wt% Si steel. Data from Research Center, U.S. Steel Corporation.

grains grow, the matrix softens significantly and becomes even more ductile. Morphological change remarkably parallels the enlargement of soap bubble grains with time. Grains coalesce as boundaries disappear in an effort to reduce the surface tension or area of the liquid soap film.

Earlier in Section 6.6.3 the thermodynamic and kinetic factors controlling grain growth were introduced. One aspect not treated, but important in alloy design, is the fact that foreign atoms reduce the grain size by exercising a drag on migrating grain boundaries. This means that the purer the metal, the larger

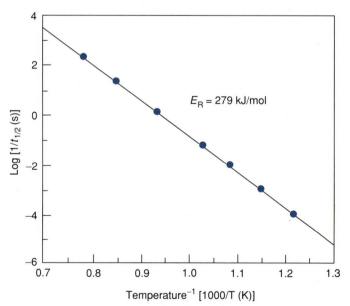

FIGURE 9-19 Arrhenius plot of log $1/t_{1/2}$ versus $1000/T$ (K) for the recrystallization of cold worked 3.25 wt% Si steel.

the grain size and the lower the value of T_R. Resistance to grain growth and thermal softening of pure metals can often be enhanced through small additions of specific alloying elements that segregate to grain boundaries. The binding between the alloying and host atoms immobilizies grain boundaries, making them effective barriers to dislocation pileups and annihilation processes. As a result, elevated temperature strength is conferred without terribly compromising properties (e.g., electrical conductivity) sensitive to alloying content.

9.4.2.5. Effect of Grain Size on Strength and Toughness

Grain size control is an extremely important engineering concern in the processing of all metals irrespective of purpose. The restricted dislocation motion associated with fine grain size increases strength prospects. In fact, a proposed relationship between yield strength (σ_0) and grain diameter or size (D_g),

$$\sigma_0 = \sigma_s + B D_g^{-1/2}, \tag{9-3}$$

known as the Hall–Petch equation, has been observed in many metals. In this equation σ_s is the yield stress of a single crystal and B is a constant. This relationship states that as the grain size becomes smaller the yield stress increases. But strength alone is not the only reason a fine and uniform grain size is universally desired.

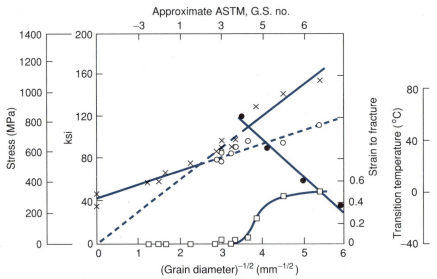

FIGURE 9-20 Dependence of yield strength, fracture strength, strain to fracture (all at −196 K), and transition temperature of a low-carbon steel as a function of grain size. ×, Fracture stress; ○, yield stress; □, strain to fracture; ●, transition temperature. Composite data drawn from *Relation of Properties to Microstructure*, American Society for Metals, Metals Park, OH (1954), and N. J. Petch, *Fracture*, Technology Press, MIT, Cambridge, MA and Wiley, New York, (1959).

Contrary to the rule enunciated at the end of Section 7.6.3, there is one practical way to increase both strength and toughness of a metal simultaneously and that is to reduce the grain size. Grain boundaries are the reason for this unmixed blessing. They are effective barriers that blunt the advance of cracks. Therefore, propagation can only occur by repeated crack initiation and tortuous change of direction at each grain boundary intersection. Such crack extension absorbs mechanical energy and effectively raises the matrix toughness. This is why the fracture stress rises with decreasing grain size in a Hall–Petch-like (Eq. 9-3) fashion. Several properties of mild steel that hinge on the grain size are depicted in Fig. 9-20. Interestingly, the ductile–brittle transition temperature also significantly drops through grain refinement, again in a Hall–Petch-like manner.

Achieving a fine and uniform grain size is usually a goal of primary processing operations, for example, casting, rolling, and annealing. Once a component is shaped, however, it is much easier to increase the grain size than to reduce it. The only way to accomplish the latter is through special solid-state phase transformations, for example, quenching and tempering of steel.

9.5. STRENGTHENING NONFERROUS METALS

9.5.1. Solid Solution Strengthening

Cold working and grain size reduction have been the only ways introduced thus far for strengthening both pure and alloyed nonferrous metals. Another method applicable to all metals is to simply dissolve other alloying elements in them. Because solute and solvent atoms differ in size and electronic character, local interactions with dislocation cores are inevitable. The size mismatch ($\Delta a/a$), defined as the ratio of the difference in the two atomic radii to the radius of the solvent atom, creates a strain whose stress field makes it harder for dislocations to move. Matrix lattice parameters change with alloying and so $\Delta a/a$ is a function of solute concentration, C. It is found that alloying raises the yield stress relative to that of a pure metal by an amount proportional to $\sim(\Delta a/a)^{3/2}C$.

9.5.2. Hardening by Precipitation

9.5.2.1. Introduction

Mechanical working is an effective way to harden nonferrous metals, but if tried on already shaped or formed parts like castings, they will distort or, perhaps, even break. Precipitation hardening treatments do have the capability of further strengthening shaped parts safely and are quite easy to implement. But not all alloy compositions are suitable candidates for these hardening heat treatments, and optimal ones must be selected. The first requirement is that the alloy system have a phase diagram (only binaries will be considered) with solid-state solubility limits that increase appreciably as the temperature is raised. Fortunately, many binary systems display this behavior, the classic example being the Al–Cu system shown in Fig. 9-21. Aluminum-rich alloys containing up to ~5 wt% Cu cross the sloped solidus phase boundary line and appear to be promising. If an alloy containing 4 wt% Cu were slowly cooled down to the ambient following high-temperature processing (forming or casting) in the α field, coarse θ (CuAl$_2$) phase precipitates would be rejected from the prior solid solution. The resulting two-phase structure is harder and stronger than pure Al, but not appreciably so.

More significant strengthening can be achieved if, instead of slow cooling from the α field, the alloy were quenched. Now Cu atoms are frozen into α solid solution and corresponding (metastable) θ phase formation is suppressed. This is a case of simple solid solution strengthening that was discussed briefly in the preceding section. Although normally a steppingstone in precipitation hardening treatments, significant strengthening can sometimes be achieved by quenching to retain a supersaturated solid solution. In Al–Mg alloys, for example, a quadrupling of the yield stress relative to annealed Al is attainable with a 5 wt% Mg addition. Nevertheless, it is usual to liberate the trapped solute from supersaturated solution by controlling the nucleation and growth rates of the rejected θ precipitates. An optimal precipitate phase size and dispersion are required for maximum strengthening and this is the issue we now address.

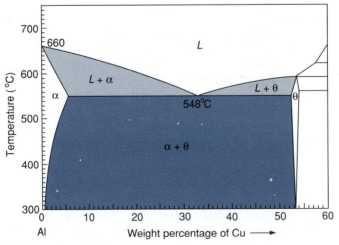

FIGURE 9-21   Equilibrium phase diagram of Al-rich portion of the Al–Cu system.

9.5.2.2. Precipitates, Dislocations, and Strength

Appreciation of the way in which precipitates interact with matrix atoms and dislocations is vital to understanding precipitation and other particle-based hardening mechanisms. If the particles are spaced very far apart they barely inhibit the motion of dislocations. The latter can easily balloon outward past the particles and either clone themselves via the Frank–Read mechanism or, perhaps, continue to glide on the slip plane until they encounter a more substantial barrier. But when the particles are closely spaced (Fig. 9-22) dislocations cannot bow sufficiently to either multiply or glide. The particles now behave as obstacles; more stress is required for further dislocation motion and the strain associated with it. Theory has shown that the shear stress (τ) required to expand a dislocation loop is inversely proportional to the interparticle spacing l; that is,

$$\tau = 2Gb/l, \tag{9-4}$$

where G is the shear modulus and $\mathbf{b}$ is Burgers vector. Substitution of typical values for G and $\mathbf{b}$ (0.2 nm) reveal that τ attains yield stress levels when l is approximately a few tens of nanometers.

The size and nature of the precipitate also influence its potential for strengthening. Large precipitates are not as effective as small ones because dislocation cores can best interact with objects of similar dimension. A kind of size *impedance match* is required between the interacting entities. Small precipitates, having just been rejected from the parent solid solution (Fig. 9-23A), are also likely to retain a strained bonding relationship to the lattice. A state of **coherency** then exists between these transitional or θ' precipitates and the matrix. This is the case modeled in Fig. 9-23B where the involved phases have different crystal structures. Surrounding such precipitates are so-called *Gunier–Preston zones* with stress fields that restrict dislocation motion; however, larger precipitates (Fig. 9-23C) are no longer coherent with the matrix and offer significantly less resistance to dislocation motion. These are the equilibrium θ phase precipitates.

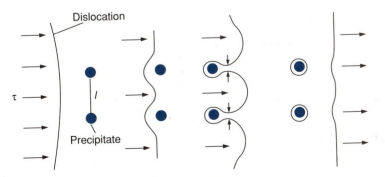

FIGURE 9-22   Model of dislocation interactions with precipitates.

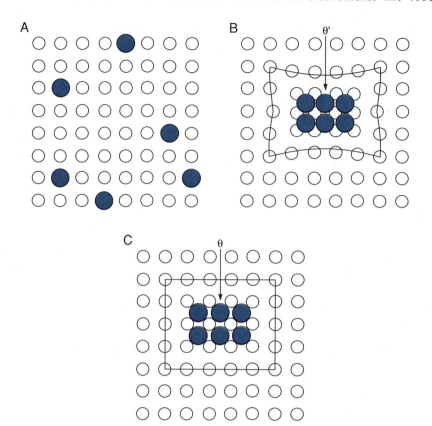

FIGURE 9-23   Depiction of coherent precipitate formation in a matrix. (A) Random solid solution. (B) Coherent precipitate. (C) Overaged precipitate.

9.5.2.3. Precipitation Hardening Treatment

Prerequisites for hardening—small, optimally dispersed precipitates—have been suggested by the previous two paragraphs. A recipe for hardening an Al–4 wt% Cu alloy therefore necessitates the following steps:

1. **Solution** heat-treat to solution temperature T_s (e.g., 540°C) in the α phase field.
2. **Quench** to a low temperature (e.g., 25°C) in the $\alpha + \theta$ field.
3. **Anneal** or age at temperature T_p (e.g., 150°C) for the time necessary to optimally form coherent θ' phase precipitates and develop maximum mechanical properties.

Aging is a generic term that describes time-dependent reactions and property changes in the solid state. Precipitation is an important aging reaction and the kinetics of the resultant strength and ductility changes are shown in Fig. 9-24 for a 6061 Al alloy. Care must be taken not to overage the precipitates for then

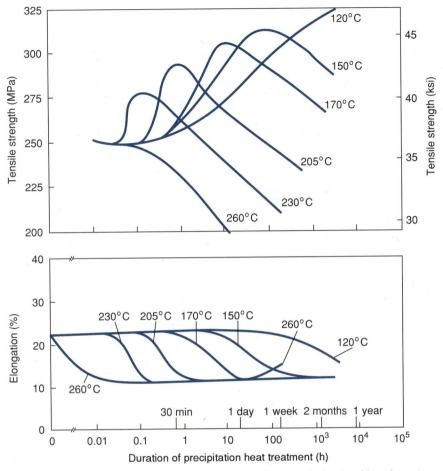

FIGURE 9-24 Tensile strength and elongation of 6061 aluminum alloy as a function of aging time. Note that an increase in strength is accompanied by a decrease in ductility. Formation of successive metastable coherent precipitates accounts for the rise in strength to the maximum value. Subsequent softening is associated with overaging and growth of equilibrium precipitates. From *Metals Handbook*, 9th ed., Vol. 4, American Society for Metals, Metals Park, OH (1981).

coherency is lost and the matrix actually begins to soften. The microstructure of coherent precipitates in the Al–3.8 wt% Cu precipitation hardening alloy is reproduced in Fig. 9-25.

An analogy between isothermal transformation of austenite and decomposition of the quenched Al α phase during precipitation hardening is worthy of comment. In fact, a TTT curve of the type drawn in Fig. 9-2 is appropriate for any isothermal, nucleation, and growth transformation. At aging (transformation) temperature T_p, α phase is stable until θ nucleates and the $\alpha + \theta$ field is entered. As in steel, aging at a lower temperature increases the driving force

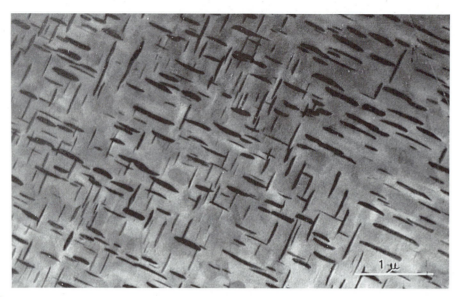

FIGURE 9-25   TEM image of an Al-3.8 wt% C alloy aged at 240°C for 2 hours showing θ' $CuAl_2$ precipitates. Specimen offered by S. Koda and taken by JEOL.

for transformation. There is a larger precipitate nucleation rate leading to a finer precipitate size. A harder microstructure results, but it takes a longer time for it to evolve.

9.5.2.4. Some Precipitation Hardening Alloys

There are commercial precipitation hardening alloys based on virtually all of the nonferrous metals. The Al–4 wt% Cu alloy was the first one developed, and it remains popular. Other commercially important Al precipitation hardening alloys contain zinc and magnesium either singly or in combination with other elements. Their properties are listed in Table 9-2.

Copper can undergo extraordinary hardening with about 2% beryllium. Tensile strengths approaching 200,000 psi (1.38 GPa), some three times that of very heavily cold worked pure copper, are possible. The resulting alloy is widely used for spring contacts. Other notable precipitation hardening Cu-base alloys contain Ti, Zr, and Cr. The latter are added in small amounts so as not to adversely lower the electrical conductivity.

The popular titanium alloy, Ti–6Al–4V, also precipitation hardens. It combines high tensile strength and good workability, qualities that are important in aircraft engine components.

9.5.3. Strengthening Superalloys

The general needs of the military and the particular demands of high-performance aircraft engines have done far more to accelerate the development of strengthened advanced metals and superalloys than any other set of applica-

| TABLE 9-4 | COMPOSITIONS AND PROPERTIES OF SELECTED SUPERALLOYS |

Alloy	Composition (wt%)	σ_0 (ksi at 25°C)	σ_{UTS} (ksi at 25°C)	E (Mpsi)	Elongation (%)
Hastelloy X	50Ni, 21Cr, 9Mo, 18Fe, Co + W ~ 2	55.2	108	29.8	57
Inconel 718	53Ni, 19Cr, 19Fe, 5Nb, 3Mo, 1Ti	185	198	29.0	28
Udimet 720	55Ni, 18Cr, 15Co, 3Mo, 5Ti, 2.5Al, 1.5W	177	225	33.0	16
Mar-M 509	54.5Co, 23.5Cr, 10.5Ni, 7.5W, 3.5Ta, Rem C, Zr, Ti	83	114	32.7	4
Haynes 25	54Co, 20Cr, 10Ni, 15W, 1Fe	67	146	33	64

Source: *Materials Engineering*, p. 38 (September 1989).

tions. This can be appreciated by once again referring to the components within the turbojet engine of Fig. 1-9 and recognizing that different needs are required of compressor and turbine hardware.

Compressor. Temperatures and pressures are moderate in this section of the engine and many metals could meet the strength requirements. Therefore, weight becomes a critical consideration, and their relatively low density is the reason titanium alloys are extensively used. Two phases are of interest in Ti alloys: high-temperature BCC β and low-temperature HCP α. By optimizing the proportion and morphology of these phases through assorted thermomechanical treatments, considerable improvement in creep and fatigue resistance has been achieved. High-temperature strength remains low, however, and this is why superalloys are needed.

Turbine. The harsh thermal, stress, and oxidation environment faced by components in the tail portion of the compressor, the combustion chamber, and the turbine have been satisfactorily countered only through the use of superalloys. Properties of several popular nickel- and cobalt-base superalloys are listed in Table 9-4. The nickel-base alloys derive their strengthening from the presence of blocklike Ni_3Al or γ' precipitates, distributed within the matrix of the γ phase, as shown in Fig. 9-26. Dislocations propagate easily enough through γ, a relatively ductile FCC austenite-like phase. But, upon encountering γ', dislocation motion is severely impeded. The reason stems from the ordered nature of this compound, which requires Al atoms to strictly occupy cube corners and Ni atoms to populate only face-centered sites of their FCC unit cell. Any dislocation disruption of this order would result in high-energy Ni–Ni or Al–Al pairings, and is strongly discouraged. With so many atomic compo-

FIGURE 9-26   Electron micrograph of the block structure of nickel–aluminum γ' phase in a nickel-base superalloy (84,000×). Courtesy of N. D. Pearson, United Technologies Research Center.

nents it is not surprising that a number of strengthening mechanisms interact synergistically to actually strengthen these alloys as the temperature increases. Like Ni_3Al, Ni_3Ti precipitates confer matrix hardening, whereas the carbides that are often present tend to stabilize grain boundaries. All of these particles are set within a tough γ solid solution matrix.

Their higher-temperature ($\sim 1100°C$ versus $\sim 1000°C$) strength earmarks cobalt rather than nickel-base superalloys for use in the combustion chamber and in neighboring stationary turbine vanes. Strengthening occurs by the presence of refractory metal (Mo and W) carbide particles which collect at grain boundaries, restraining them and the matrix from deforming under stress.

The remainder of the chapter is largely devoted to composites of which dispersion strengthened alloys are examples. It is to the very important subject of composites that we now turn.

9.6. MODELING COMPOSITE PROPERTIES

9.6.1. Introduction

About 30 years ago Dustin Hoffman, the hero of the film *The Graduate*, was given a one-word piece of sage advice that defined the future—"**plastics.**" Today the word would be "**composites.**" The development and wide-scale use of synthetic composites in recent years, particularly those containing fibers in a polymer matrix, have created a revolution in materials usage that continues to accelerate. Actually, the idea of using high-strength fibers to strengthen a cheap matrix is not new. The practice of embedding straw in mud bricks very likely predated the reference to the Hebrew slaves under the Pharaoh in the Bible (*Exodus* 5:18).

A working definition of a composite is a material that contains a physical mixture of two or more constituent phases that are chemically different and separated by a distinct interface. Two-phase mixtures like eutectics (Pb–Sn) and eutectoids (α-Fe–Fe_3C) would qualify as composites in a broad sense. So would the natural composites all around us, wood being a prime example. It consists of flexible cellulose fibers embedded in the stiffer amorphous lignin polymer matrix. But we are concerned with artificially synthesized composite structures, not those that form naturally. Composites generally exhibit a synergistic admixture of the attributes of the individual constituents such that properties of the resultant combination are improved.

In the hierarchy of composites **particulate composites** are the simplest. Next are the **fibrous composites,** which represent a more complex way of reinforcing the matrix. Lastly are **structural composites,** consisting of sandwiches or laminates of simpler composites. This section is devoted to the theoretical basis for the dependence of the elastic and plastic strength of composites on particle or fiber content. Section 9.7 is devoted to practical issues of composites, that is, their fabrication, properties, and applications. Throughout, we are specifically concerned with embedding high-strength, stiff, and often brittle particles or fibers in generally weaker, less stiff, but more ductile matrices.

9.6.2. Young's Modulus

9.6.2.1. Particle-Reinforced Composites

Consider a dispersion of isolated particles (p) in a matrix (m) as shown in Fig. 9-27A. We have already dealt with particulate composites in previous chapters (e.g., a cement–aggregate mix, WC in a Co matrix). The volume fraction of particles is V_p and they have a Young's modulus of E_p; the corresponding values for the matrix are V_m and E_m. What is Young's modulus of the composite? This is part of a broader question of estimating net properties, not only mechanical but electrical, thermal, and so on, of a composite material containing a dispersed phase. There are two simple ways of averaging properties of mixtures. One is a linear law of mixtures expressed by the weighted average of individual properties or

$$E_c = V_p E_p + V_m E_m, \tag{9-5}$$

where E_c is the modulus of the composite. An alternate possibility is to assume that the inverses of the moduli are additive:

$$E_c^{-1} = V_p E_p^{-1} + V_m E_m^{-1} \quad \text{or} \quad E_c = E_m E_p / (V_m E_p + V_p E_m). \tag{9-6}$$

These two approximations for E_c are plotted in Fig. 9-27B and represent limits to the maximum and minimum values of the modulus. Actual composite behavior falls between the two extremes.

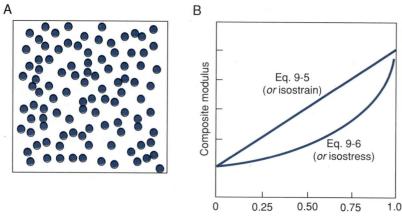

FIGURE 9-27   (A) Model of a composite containing a volume fraction V_p of dispersed particles. (B) Variation of E_c with V_p. The same behavior is exhibited by fibrous composites where isostrain and isostress moduli are defined. See Eqs. 9-8 and 9-10.

9.6.2.2. Fibrous Composites

These important composites consist of fibers, with a large length-to-diameter ratio, embedded in a matrix. For simplicity we consider long or continuous fibers that extend from one end to the other of a cube-shaped composite. With no loss in generality, the composite is equivalently visualized to consist of alternating parallel, rectangular layers of fiber and matrix as shown in Fig. 9-28A. It is a simple matter to calculate Young's modulus for the case where the composite is loaded *along the fiber direction* with force F_c. Both the fiber and the matrix are assumed to behave elastically with respective moduli E_f and E_m. Static equilibrium implies that F_c is equal to the sum of the forces carried by each constituent, that is, F_f and F_m. These forces, in turn, are given the product of the corresponding stresses, σ_f and σ_m, and the respective load-bearing areas of fiber and matrix. But, for a composite of unit volume (a unit cube), these areas are equal to the respective volume fractions, V_f and V_m, *in magnitude,* where $V_f + V_m = 1$. Deformation compatibility requires that fibers and matrix extend by the same strain e. Therefore, with the help of Hooke's law ($\sigma = Ee$),

$$1 \times \sigma_c = V_f\sigma_f + V_m\sigma_m \quad \text{and} \quad \sigma_c = V_fE_fe + V_mE_me. \quad (9\text{-}7)$$

But, Young's modulus for the composite is $E_c = \sigma_c/e$, with the result that

$$E_c = V_fE_f + V_mE_m. \quad (9\text{-}8)$$

This so-called **isostrain deformation** case essentially yields the same form of the elastic modulus given by Eq. 9-5.

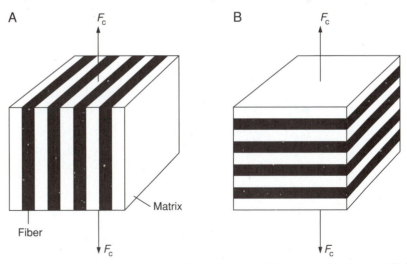

A F_c

B F_c

Matrix

Fiber

F_c

F_c

FIGURE 9-28 (A) Model of a fibrous composite with fibers aligned parallel to the loading direction. (B) Model of a fibrous composite with fibers aligned perpendicular to the loading direction.

Isostress deformation occurs when the fibers are oriented *perpendicular to the loading axis* (Fig. 9-28B). Both constituents now support the same load or stress σ, but elongate or strain different amounts because their elastic moduli differ. The total composite strain e_c is given by the sum of the individual strains (e_f and e_m), or

$$e_c = e_f + e_m = \sigma V_f/E_f + \sigma V_m/E_m. \qquad (9\text{-}9)$$

Again, Young's modulus for the composite is $E_c = \sigma/e_c$; therefore,

$$E_c = (V_f/E_f + V_m/E_m)^{-1} \quad \text{or} \quad E_c = E_m E_f/(V_m E_f + V_f E_m). \qquad (9\text{-}10)$$

Not surprisingly, this form of E_c is identical to that given in Eq. 9-6 for particulate composites. The isostrain case predicts upper-bound (maximum) moduli values, whereas lower-bound or minimum values are described by the isostress approximation. Randomly oriented two-phase mixtures will exhibit moduli that fall between these extremal values. Only in the case where the reinforcing phase is aligned along the loading axis is maximum elastic stiffness attained.

EXAMPLE 9-4

It is desired to make an aligned, continuous glass fiber composite with a ratio of elastic modulus (E_c) to density (ρ_c) of $E_c/\rho_c = 20$ GPa-m^3/Mg. (The reason for the modulus-to-density ratio will be apparent in Section 9.7.1.) For this purpose continuous glass fibers are distributed in a polyester resin matrix.
a. What volume fraction of fiber is required?
b. What fraction of the load is carried by fibers?
For the materials in question, $E_f = 69$ MGa, $\rho_f = 2.5$ Mg/m^3, $E_m = 3.0$ MGa, $\rho_m = 1.2$ Mg/m^3.

ANSWER   a. By Eq. 9-8, $E_c = V_f E_f + V_m E_m$. The definition of density is $\rho_c = M_c/v_c$, where ρ, M, and v are the indicated density, mass, and volume (not volume fraction), respectively, in the composite (c), fiber (f), and matrix (m). Therefore, $\rho_c = (M_f + M_m)/V_c$ and

$$E_c/\rho_c = (E_f V_f + E_m V_m)v_c/(M_f + M_m)$$
$$= (E_f V_f + E_m V_m)v_c/(v_f \rho_f + v_m \rho_m).$$

Because $V_f = v_f/v_c$ and $V_m = v_m/v_c$,

$$E_c/\rho_c = (E_f V_f + E_m V_m)/(V_f \rho_f + V_m \rho_m) = 20.$$

Letting $V_m = 1 - V_f$ and substituting values for the moduli results in the equation

$$69 V_f + 3.0(1 - V_f) = 20[2.5 V_f + 1.2(1 - V_f)].$$

Solving, $V_f = 0.525$.

b. The fraction of the total load carried by the fibers is $\sigma_f A_f / \sigma_c A_c$, where A_f and A_c are the fiber and composite cross-sectional areas. But, $A_f / A_c = V_f$, $\sigma_f = E_f e$, and $\sigma_c = E_c e$, so that

$$\sigma_f A_f / \sigma_c A_c = E_f e A_f / E_c e A_c = E_f V_f / (E_f V_f + E_m V_m).$$

After substitution,

$$69 \,(0.525)/(69 \times 0.525 + 3 \times 0.475) = 0.962.$$

Thus, with a little over half the composite volume, fibers support 96% of the total load.

9.6.3. Strength of Composites

9.6.3.1. Continuous Fibers

Until now our concern has been only with the *elastic* properties of continuous fiber composites. Beyond the elastic regime the stress–strain behavior of the individual constituents, as well as that of the composite, is schematically indicated in Fig. 9-29. The overall response is complex because the fibers only deform elastically until they fracture at the tensile stress [$\sigma_f(TS)$], whereas the matrix simultaneously deforms both elastically and plastically. As fibers have a far higher tensile strength than the matrix yield strength [$\sigma_m(YS)$] they continue to deform elastically while the matrix starts deforming plastically. With further load, fibers begin to fracture, and when they do they no longer support load; the stress drops but not abruptly because some fibers continue to remain intact. But after they break or are pulled out only the matrix is left to carry the load until the composite fractures.

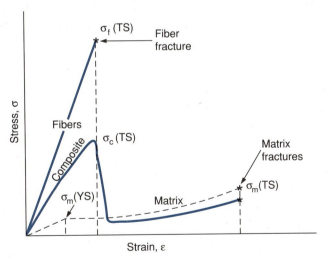

FIGURE 9-29 Stress–strain curve for a fibrous composite together with those of the fiber and matrix. The peak composite strength occurs when the fibers are on the verge of breaking. From M. F. Ashby and D. H. R. Jones, *Engineering Materials 2: An Introduction to Microstructures, Processing and Design*, Pergamon Press, Oxford (1986).

For design purposes, it is the peak or tensile stress of the composite $[\sigma_c(TS)]$ that is of interest. An estimate of the composite strength is

$$\sigma_c(TS) = \sigma_f(TS)V_f + \sigma_m(YS)V_m. \tag{9-11}$$

This expression reflects a simple weighted average of the fracture strength of the fibers and the yield strength of the matrix as these two stresses represent the upper load-bearing capacities. Once all of the fibers break the fracture strength of the matrix $[\sigma_m(TS)]$ limits the strength of the composite.

9.6.3.2. Effect of Fiber Length

It would be wrong to conclude from the discussion of composites to this point that only long continuous fibers that span the entire structure will do. If this were the case processing and forming difficulties would limit fibrous composites to very few applications. Fortunately, cut fibers that are convenient for molding and extrusion operations will impart substantial strengthening if they are longer than a critical length. To estimate this critical length let us consider Fig. 9-30, depicting a chopped fiber embedded in a matrix. Under a tensile load the axial force is transmitted to the fiber via the matrix along the cylindrical fiber–matrix interface. On a segment of fiber dx long, the force is given by $dF = \pi d\tau_m(s)dx$, where $\tau_m(s)$ is the interfacial *shear* stress and d is the fiber diameter. The force on the fiber segment differs depending on where it is located. If is is near the ends very little force is transferred to it from the matrix; near the center a longer length of matrix surrounds the fiber to transmit force. At a distance x from the end the cumulative transferred force is

$$F = \int_0^x \pi d\tau_m(s)dx = \pi d\tau_m(s)x. \tag{9-12}$$

The force that will break the fiber is $F_f = \pi d^2\sigma_f(TS)/4$. If these two expressions for force are equated, the critical distance $(x = x_c)$ is determined to be

$$x_c = d\sigma_f(TS)/4\tau_m(s). \tag{9-13}$$

Optimal fibers are thus predicted to be $2x_c$ in length. If they are shorter than this length they will not fail; neither will they reinforce the matrix or carry the maximum load they are capable of supporting. On the other hand, if fibers are longer than $2x_c$ the extra length is essentially wasted.

Again, it is the overall strength of the composite that is important. From Eq. 9-11,

$$\sigma_c(TS) = \tfrac{1}{2}\sigma_f(TS)V_f + \sigma_m(YS)V_m, \tag{9-14}$$

where the term $\tfrac{1}{2}\sigma_f(TS)$ is the *average* value of the fiber strength at $x = \tfrac{1}{2}x_c$. A comparison between Eqs. 9-11 and 9-14 shows that the resulting strength of the composite will be more than half that for continuous fibers only if the chopped fibers are aligned. If they are not aligned the strength is reduced accordingly.

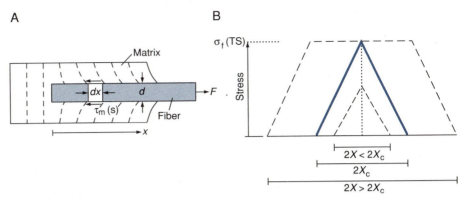

(A) Load transfer from the matrix to a discontinuous fiber. (B) Variation of stress with position in cut fibers that are less than, equal to, or longer than the critical length.

9.6.4. Toughness of Composites

Often, neither the polymer matrix nor the fibers used to reinforce it are tough materials. It is therefore not obvious why making a composite of the two (Fig. 9-31A) should improve toughness. But, surprisingly, this is precisely what happens. The reason can be seen in Fig. 9-31B where a crack has propagated *transverse* to the fiber axis. If the fibers are less than $2x_c$ in length they do not break. Rather, the crack opens as fibers are *pulled out* of the matrix, a process that requires work and increases the fracture surface area. Longer fibers pull out as well as crack, and energy is likewise expended. Cracks that travel *parallel* to the fibers have an easier time of it (Figs. 9-31C and D); they can propagate along the fiber–matrix interface in response to shear stresses or through the matrix under tension. This all underscores the importance of good interfacial adhesion between fiber and matrix. Other toughening mechanisms include deflection of cracks as they impinge on fibers and fiber bridging of cracks that restrain them from opening and advancing further.

Formulas that will not be derived here estimate the toughness as the work per unit area (W_c) of crack surface required to pull fibers from it by shearing the fiber–matrix bonds. When fibers of length L are shorter than $2x_c$ they cannot fracture and

$$W_c = \tau_m(s)V_f L^2/2d. \tag{9-15}$$

On the other hand, if the fibers are longer than $2x_c$ they will break instead of pull out. In such a case,

$$W_c = [\sigma_f(TS)]^2 V_f d/8\tau_m(s). \tag{9-16}$$

The first of these formulas emphasizes the importance of increasing the matrix shear strength in toughening the composite. In the second equation a high fiber strength is the route to toughening.

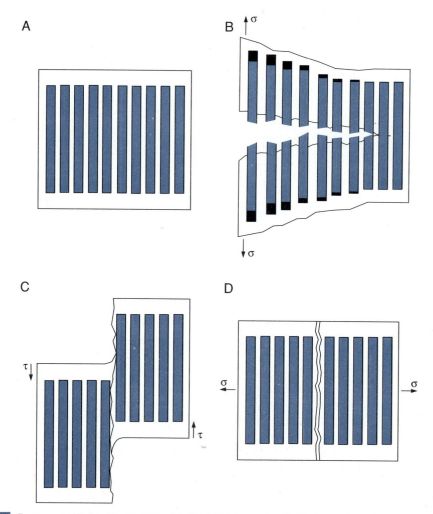

FIGURE 9-31 Fracture models in a fibrous composite. (A) Initial composite. (B) Under tensile pulling, fibers are pulled from matrix and also fracture. (C) Interfacial fiber/matrix failure under applied shear stresses. (D) Matrix fracture under tensile stresses applied normal to the fibers.

9.7. ENGINEERING COMPOSITES

9.7.1. Polymer Composites

9.7.1.1. Design

Representative mechanical properties of a number of important polymer materials were given in Table 7-4. Through reinforcement with the fibers listed in Table 9-5, the indicated commercially produced polymer composites result.

| TABLE 9-5 | PROPERTIES OF REINFORCING FIBERS, POLYMER MATRICES, AND COMPOSITES |

Material	Density (Mg/m^3)	E (GPa)	σ_{UTS} (GPa)	Fiber radius (μm)	Fracture toughness, K_{IC} (MPa-m$^{1/2}$)
Fiber					
E-glass	2.56	76	1.4–2.5	10	
Carbon					
High-modulus	1.75	390	2.2	8.0	
High-strength	1.95	250	2.7	8.0	
Kevlar	1.45	125	3.2	12	
Matrix					
Epoxy	1.2–1.4	2.1–5.5	0.063		
Polyester	1.1–1.4	1.3–4.5	0.060		
Composites					
CFRP (58% C in epoxy)	1.5	189	1.05		32–45
GFRP (50% glass in polyester)	2	48	1.24		42–60
KFRP (60% Kevlar in epoxy)	1.4	76	1.24		

After B. Derby, D. A. Hills, and C. Ruiz, *Materials for Engineering*, Longman Scientific and Technical, London (1992), and M. F. Ashby and D. R. H. Jones, *Engineering Properties 2: An Introduction to Microstructures, Processing and Design*, Pergamon Press, Oxford (1986).

That unreinforced polymers are considerably less stiff (modulus) and strong (yield and fracture strength) than other engineering materials (metals and ceramics) has been pointed out before. On the other hand, the glass, carbon, and aramid (Kevlar) fibers commonly used in composites have stiffness and strength properties comparable to those of engineering materials. The respective composites are widely known as GFRP, CFRP, and KFRP, where the last three letters refer to fiber-reinforced polymer. Not only are they strong and light, but they possess some toughness. Approximately 30% of the external surface area of the Boeing 767 airliner (Fig. 9-32) consists of such high-performance composites.

In designing structural components for automobiles and airplanes, important considerations are to enhance both stiffness and strength for a given *weight*. The concern for reduced weight is understandable because impressive energy savings can result. Therefore, in some applications material *density* is as important as the strength properties. Both material properties are combined in a quotient: the **specific modulus**, or E/ρ ratio, or the **specific strength**, σ_0/ρ. To appreciate the role of E/ρ in design consider the stress (σ) and displacement (δ) distributions in loaded struts, cantilever beams, and plates shown in Fig. 9-33. Formulas given for σ and δ are those normally derived in engineering

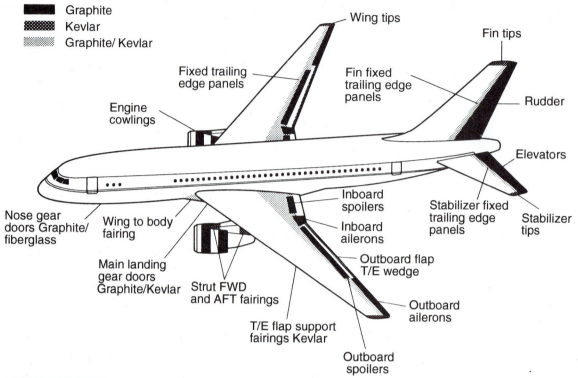

Graphite
Kevlar
Graphite/ Kevlar

Wing tips

Fin tips

Fixed trailing edge panels

Fin fixed trailing edge panels

Rudder

Engine cowlings

Elevators

Inboard spoilers

Stabilizer fixed trailing edge panels

Stabilizer tips

Nose gear doors Graphite/ fiberglass

Wing to body fairing

Inboard ailerons

Outboard flap T/E wedge

Main landing gear doors Graphite/Kevlar

Strut FWD and AFT fairings

Outboard ailerons

T/E flap support fairings Kevlar

Outboard spoilers

FIGURE 9-32   Use of high-performance polymer matrix composites on the Boeing 767 passenger jet. Courtesy of Boeing Corporation.

courses on statics or strength of materials. Struts, however, are easily enough analyzed by concepts already presented.

EXAMPLE 9-5

Compare the weights for optimum stiffness in high-strength steel, aluminum alloy, and fiberglass struts, beams, and plates. Assume the respective values for Young's modulus E and density ρ (GPa, Mg/m^3) in these materials are 207 and 7.8 (steel), 71 and 2.8 (aluminum), and 37.7 and 1.9 (GFRP or fiberglass).

ANSWER   For the geometries shown in Fig. 9-33, by eliminating t, the weights vary as $M = (Fl^2/\delta)[\rho/E]$ for struts; $M = 2(Fl^5/\delta)^{1/2}[\rho/E^{1/2}]$ for beams; $M = l^2(5Fw^2/32\delta)^{1/3}[\rho/E^{1/3}]$ for plates. Therefore, for the same load, structural dimensions, and resulting deflection, the materials selection criteria for struts, beams, and plates depend on minimizing ρ/E, $\rho/E^{1/2}$, and $\rho/E^{1/3}$, respectively.

Mode of loading	Optimum stiffness	Optimum strength
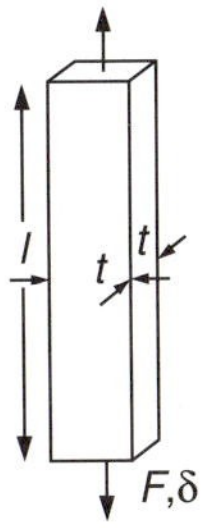 F,δ	$\delta = \dfrac{Fl}{Et^2}$ $M = \rho l t^2$ $= \left(\dfrac{l^2 F}{\delta}\right)\dfrac{\rho}{E}$ Maximize $\dfrac{E}{\rho}$	$\sigma_0 = \dfrac{F}{t^2}$ $M = \rho l t^2$ $= Fl\left(\dfrac{\rho}{\sigma_0}\right)$ Maximize $\dfrac{\sigma_0}{\rho}$
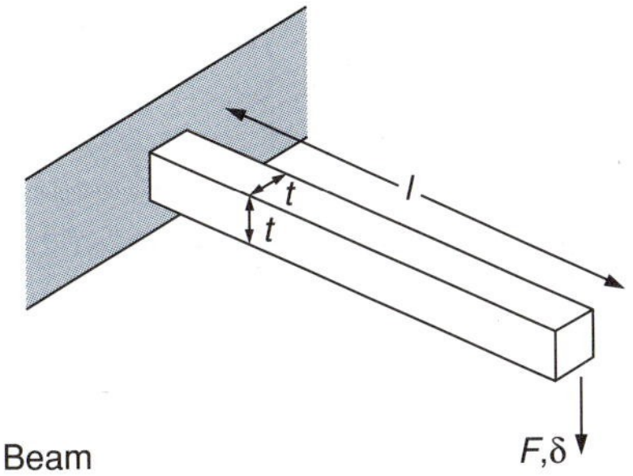 Beam F,δ	$\delta = \dfrac{4Fl^3}{Et^4}$ $M = \rho l t^2$ $= 2\left(\dfrac{Fl^5}{\delta}\right)^{1/2}\left(\dfrac{\rho}{E^{1/2}}\right)$ Maximize $\dfrac{E^{1/2}}{\rho}$	$\sigma_0 = \dfrac{6Fl}{t^3}$ $M = \rho l t^2$ $= l\,(6Fl)^{2/3}\left(\dfrac{\rho}{\sigma_0^{2/3}}\right)$ Maximize $\dfrac{\sigma_0^{2/3}}{\rho}$
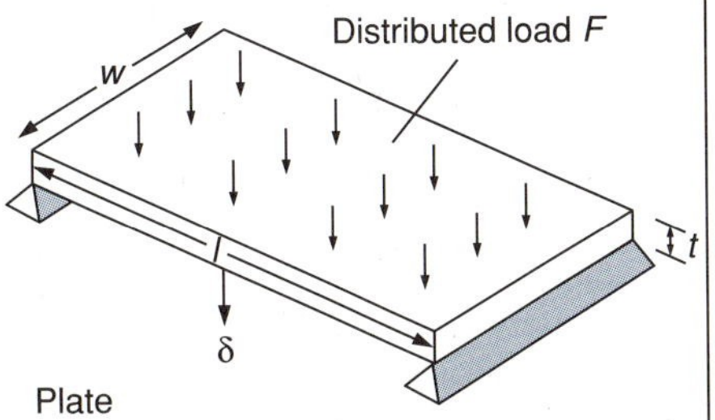 Plate	$\delta = \dfrac{5Fl^3}{32Ewt^3}$ $M = \rho l w t$ $= l^2\left(\dfrac{5Fw^2}{32\delta}\right)^{1/3}\left(\dfrac{\rho}{E^{1/3}}\right)$ Maximize $\dfrac{E^{1/3}}{\rho}$	$\sigma_0 = \dfrac{3Fl}{4wt^2}$ $M = \rho l w t$ $= \left(\dfrac{3Fl^3 w}{4}\right)^{1/2}\left(\dfrac{\rho}{\sigma_0^{1/2}}\right)$ Maximize $\dfrac{\sigma_0^{1/2}}{\rho}$

FIGURE 9-33   Strut, beam, and plate geometries and the combination of properties that maximize the stiffness and strength to weight. From M. F. Ashby and D. H. R. Jones, *Engineering Materials 2: An Introduction to Microstructures, Processing and Design*, Pergamon Press, Oxford (1986).

These minimize M. Maximum values for the *inverse* quantities are thus desirable and are given in the following table for the indicated materials.

		Steel	Aluminum	Fiberglass
Struts	E/ρ	27	25	20
Beams	$E^{1/2}/\rho$	1.8	3.0	3.2
Plates	$E^{1/3}/\rho$	0.76	1.5	1.8

These calculations show that both steel and aluminum struts are superior to those made of fiberglass insofar as optimal stiffness is concerned. For beams, however, aluminum is better than steel, which explains why it is used in airframes. In large load-bearing plates such as floor panels and flaps, fiberglass is superior to both metals.

The reason for combining strong fibers in a weak but cheap matrix is evident. Without the union neither of these materials would be as useful. High fiber and processing costs are a chief factor in limiting the extension of composite use to yet new applications.

9.7.1.2. Applications

Fiberglass (GFRP).   Produced in the largest quantities of any polymer matrix composite, fiberglass consists of glass fibers in assorted polymer matrices, polyesters being the most common. Newer versions are based on a nylon matrix and are generally stronger and more impact resistant. Glass fibers are readily drawn from the molten state; in addition to their high specific modulus (E/ρ) they are chemically inert. But first they must be coated by a thin polymer layer to prevent surface flaws and penetration by water. (Glass fiber for optical communications is also coated immediately after drawing for similar reasons.) To produce fiberglass composites it is then necessary to surround the fibers by the polymer matrix and then form the two together into useful shapes. Fiberglass applications include automotive and marine bodies, as well as walls and floorings in structures. Service temperatures do not extend much beyond 200°C when the matrix begins to flow.

Carbon and Other Fiber–Polymer Composites.   Carbon (graphite) fiber has a higher specific modulus than glass; it is also more resistant to elevated temperature and chemical exposure, but is considerably more costly. Therefore, CFRP composites are reserved for more demanding applications that can justify the added expense. In high-performance aircraft they account for weight savings of ~25% relative to metals. The widespread use of graphite-reinforced polymers in sporting equipment is well known to everyone. Tennis rackets, golf clubs, and sailboat masts are some of the applications. Less well known is their use in weaving equipment components. Their high stiffness and low weight make possible the design of very high speed looms.

Assorted polymers have also been impregnated with other fibers, notably Kevlar (KFRP), boron, and silicon carbide. They have found their way into assorted applications including sporting goods and aerospace, aircraft, and marine components.

9.7.1.3. Fabricating Polymer Composites

Numerous methods have been developed to fabricate fiber-reinforced polymer matrix composites as schematically indicated in Fig. 9-34. The first step

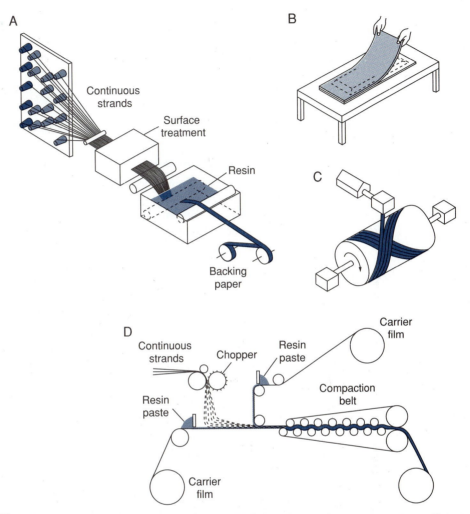

is to produce so-called prepreg tapes or sheets of the composite that will be subsequently consolidated into a laminated structure. To accomplish this, continuous fiber unwound from bobbins forms strands that are first coated with agents to promote good bonding to the matrix polymer (resin). These strands are then continuously immersed in resin baths that will become the matrix that surrounds the fibers (Fig. 9-34A). Sheets can be compression molded or hand laid in molds to make laminates (Fig. 9-34B). Tape prepregs can also be wound over large shapes and cured to make large lightweight containers

(Fig. 9-34C). By cofeeding and compacting chopped fiber and resin (Fig. 9-34D) sheet is produced. Shaped composite parts are commonly injection molded (Fig. 8-29C) by continuously adding chopped fiber (~1 cm long) along with the polymer feedstock. Controlling the fiber orientation is an important concern in this process.

9.7.2. Metal Matrix Composites

9.7.2.1. Particulate Composites

Although metal matrix composites (MMCs) have been of interest for several decades it is only recently that methods have been developed to incorporate large volume fractions (up to ~50%) of oxide, carbide, boride, and other particles (and fibers) into metal matrices. On the other hand, **dispersion-strengthened alloys** were the earliest MMCs and they contained only several volume percent particles, usually oxides. Even though both kinds of MMCs have significant advantages over the matrix metal, these extremes in particle content are capitalized upon in different applications.

By dispersing oxide particles in a ductile matrix (e.g., ThO_2 in Ni, Al_2O_3 in Al) the oxide retains its high hardness at elevated temperatures and helps restrain the softer, lower-melting-point metal from flowing. A significant improvement in resistance to thermal softening and creep can thus be achieved in dispersion-strengthened alloys. The principles governing precipitate size and distribution for optimal strengthening of precipitation hardening alloys (see Section 9.5.2) are also valid for dispersion-strengthened composites. Since *incoherent* interfaces develop between dispersant and matrix, coherency stresses cannot be relied upon. In their absence, submicrometer ($1~\mu m = 10^{-3}$ mm)-sized particles are essential for impeding dislocation motion, and the smaller the particles are, the greater their effectiveness.

In all MMCs the basic challenge is to disperse particles uniformly throughout the metal. The typical processing route has involved the powder metallurgy techniques of pressing and sintering a blended mixture of both metal and oxide powders, followed by extrusion. In this way pure copper is strengthened with alumina to produce a high-conductivity matrix that is resistant to softening up to ~900°C. Containing ~5 vol% Al_2O_3, this material is drawn into wire and used to support the very hot tungsten filament in incandescent lamps. Similar composites are used for electrical contacts. These products contain dispersions of oxides (e.g., CdO) in a metal matrix (e.g., Ag) that has a high electrical conductivity. Making and breaking contact in high-power electrical circuits is often accompanied by arcing and degradation of the metal through cratering, pitting, and erosion. The dispersant mechanically strengthens the contact metal against such wear and protects circuits from open circuiting.

Attempts to extend the creep resistance of nickel-base superalloys to yet higher temperatures spawned an interesting new batch processing method known as **mechanical alloying.** Particle dispersion is accomplished by high-energy ball milling of a mixture of metal and ceramic (e.g., Y_2O_3) powders. The process is so named because the brittle yttria particles are eventually so finely subdivided by the impacting steel balls that they alloy with the metal matrix on an atomic level! These dispersion-strengthened alloys show remarkable high-temperature strength after subsequent compaction, sintering, and extrusion.

Significantly higher levels of incorporated particulates generally make the resulting MMCs stiffer, stronger, lighter, and more resistant to wear and elevated temperature deformation. Recently, new solidification processing techniques have been developed to incorporate these higher particulate levels. One such method involves infiltration of liquid metal into porous ceramic preforms. As a result, a variety of aluminum, magnesium, and titanium MMCs containing SiC, Al_2O_3, BC, graphite, AlN, and other dispersants have either been produced commercially or are in development stages. The most widely used are Al–SiC MMCs, made using traditional Al–Cu and Al–Zn alloys as the matrices. In 1983 Toyota introduced the Al–SiC MMC piston for diesel engines and was followed by other manufacturers. Since then additional automotive and aerospace applications have emerged worldwide. Aluminum-base MMC applications include rocket motor components (Fig. 9-35), cylinder liners, control

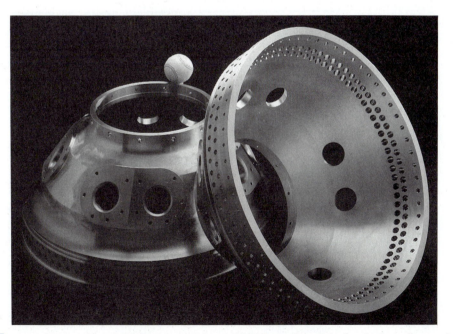

FIGURE 9-35 This rocket motor component is the world's largest MMC forging. It consists of an aluminum alloy matrix containing 25 vol% SiC particulates, and was fabricated by powder metallurgy techniques. Courtesy of Advanced Composite Materials Corporation.

arms, connecting rods, bicycle frames and wheels, tennis rackets, and structural materials for space platforms.

9.7.2.2. Fibrous Composites

The idea of strengthening metals via the *fibrous* composite route has also gained momentum in recent years. Carbon, boron, silicon carbide, and alumina, as well as metal fibers, have been incorporated in amounts of up to 50% by volume to stiffen Al, Mg, and Ti alloys for applications noted above. Fiber or whisker mats that are infiltrated with metals (squeeze casting) can be oriented, providing strength and stiffness advantages relative to particulate MMCs; however, the fiber surfaces must be specially treated to enhance metal wetting and interfacial adhesion, as well as limit undesirable reactions that degrade bonding.

Relative to the base metal, 50% increases in tensile strength and Young's modulus coupled with a similar decrease in density can be very roughly expected in both types of MMCs. Actual property enhancements are indicated in Table 9-6 and depend on the particular MMC and volume fraction of particles or fibers.

TABLE 9-6   **PROPERTIES OF SOME METAL AND CERAMIC MATRIX COMPOSITES AND THEIR REINFORCEMENTS**

Material	Density (Mg/m^3)	E (GPa)	σ_{UTS} (GPa)	Fiber radius (μm)
Fiber				
Silicon carbide				
Whisker	3.2	480	3	
Fiber	3.0	420	3.9	140
Alumina	3.2	100	1.0	3
Si$_3$N$_4$	3.2	380	2	
Boron	2.6	380	3.8	100–200
Composite				
2014–Al (20% Al$_2$O$_3$, particulate)		103	0.48	
2124–Al (30% SiC, particulate)		130	0.69	
6061–Al (continuous fiber–51% B)		231	1.42	—
6061–Al (discontinuous fiber–20% SiC)		115	0.48	—
6061–Al (no reinforcement)		69	0.31	—
Si$_3$N$_4$ (10% SiC whisker)			0.45	—
Al$_2$O$_3$ (10% SiC whisker)			0.45	—

9.8. CERAMICS AND HOW TO STRENGTHEN AND TOUGHEN THEM

9.8.1. Mechanical Properties of Ceramics

Mechanical properties of ceramics have been listed previously (e.g., Table 7-1) to invite comparisons with other materials. Table 9-7 is devoted solely to the mechanical properties of ceramics where their extraordinary attributes are worth repeating. Ceramics are seen to have very high elastic moduli and low densities (and, therefore, high E/ρ ratios), coupled with very high hardnesses and melting temperatures. Additional favorable properties in many applications include low thermal conductivity and thermal expansion coefficients. On the other hand, ceramics are very susceptible to **thermal shock.** This important property reflects the tendency of materials to fracture as a result of rapid temperature (T) change, usually during cooling. A quantitative, thermal shock figure of merit (S_T) can be defined and derived by combining several simple definitions; most have already been discussed elsewhere in the book:

1. Hooke's law $\qquad\qquad\qquad\qquad \sigma = Ee \qquad\qquad$ (Eq. 7-3)
2. Definition of strain $\qquad\qquad\qquad e = \Delta L/L \qquad$ (Eq. 7-2)
3. Definition of thermal expansion $\quad \Delta L/L = \alpha \Delta T \qquad$ (Eq. 2-19)
4. Definition of heat flow $\qquad\qquad Q = -\kappa \Delta T/\Delta x. \qquad$ (9-17)

TABLE 9-7 **PROPERTIES OF CERAMICS**[a]

Material	E (GPa)	ρ (Mg/m³)	T_M (K)	κ (W/m-K)	α (10^{-6}°C^{-1})	K_{IC} (MPa-m$^{1/2}$)	Modulus of rupture (MPa)
Diamond	1050	3.52		1100	1		
Tungsten carbide, WC	550	15.7		35	4	80	
Borides (Ti, Zr, Hf)	500	4.5		~30	7.8		
Silicon carbide, SiC	410	3.2	3110	84	5.3	3	200–500
Alumina, Al$_2$O$_3$	390	3.9	2323	~27	8.8	3.5	300–400
Beryllia, BeO	340	2.9	2570	~40	5	7.4	
Silicon nitride, Si$_3$N$_4$	310	3.2	2173	17	2.5	4	300–850
Magnesia, MgO	250	3.5	2620		10		
Zirconia, ZrO$_2$	200	5.5	2843	1.5	8	4–12	200–500
Mullite, 3Al$_2$O$_3 \cdot$ 2SiO$_2$	150	3.1	1850			4.6	
Titanium carbide, TiC	460	4.9	3067	34	8.3	0.5	
Titanium nitride, TiN	590	5.4	2950	30	9.3		
Sialons (Si–Al–N)	300	3.2		23	3.2	5	500–830
Silica, SiO$_2$	94	2.2	1980	2	0.55	1	
Granite	70	2.6			8		23
Glass	70	2.4	~1000	1	~6	1	50
Concrete	40	2.4		2	12	0.2	7

[a] Data were gathered from many sources. Property values are strongly dependent on composition and processing variables.

The last equation is the analog of Eq. 6-1 for conduction heat flow rather than diffusional mass flow phenomena. Each term in this equation was also discussed in Section 8.2.3. If S_T is arbitrarily defined as the stress developed normalized per unit flux of heat, per unit thickness of solid, then $S_T = \sigma/Q\,\Delta x$, or

$$S_T = E\alpha/\kappa. \tag{9-18}$$

The higher the value of S_T, the greater the induced stress; if the latter exceeds the tensile stress there is danger of fracture. One way to measure thermal shock resistance experimentally is to heat the material to successively higher temperatures, followed by a water quench. The higher the temperature without fracture, the greater the shock resistance. Amazingly, one can pour molten steel into a quartz crucible sitting on a cake of ice without breaking it (Fig. 9-36); however, do not try this for other crucible materials, like Al_2O_3.

A combination of Eqs. 7-2, 7-3, and 2-19 yields a widely used approximate formula to evaluate thermal stress: $\sigma = E\alpha\Delta T$. More exactly,

$$\sigma = E\alpha\Delta T/(1 - \nu), \tag{9-19}$$

where ν is the Poisson ratio. Note that thermal stresses develop only when the body is constrained or prevented from expanding or contracting.

EXAMPLE 9-6

a. Compare the thermal shock properties of SiO_2 and Al_2O_3.

b. Estimate the maximum temperature drop Shuttle Orbiter tiles made of either fused SiO_2 or Al_2O_3 can tolerate during reentry. Assume the fracture stress, σ_F (modulus of rupture), of SiO_2 is 100 MPa; for Al_2O_3, $\sigma_F = 300$ MPa.

ANSWER   a. From Tables 7-1 and 9-7, for SiO_2, $E = 94$ GPa, $\nu = 0.25$, $\alpha = 0.55 \times 10^{-6}\,°C^{-1}$, and $\kappa = 2$ W/m-K. For Al_2O_3, $E = 390$ GPa, $\nu = 0.26$, $\alpha = 8.8 \times 10^{-6}\,°C^{-1}$, and $\kappa = 27$ W/m-K. Using Eq. 9-18, a comparison of the two materials shows that

$$S_T(SiO_2) = (94\text{ GPa})(0.55 \times 10^{-6}\,°C^{-1})/(2\text{ W/m-K})$$
$$= 2.59 \times 10^{-5}\text{ GPa-m/W},$$
$$S_T(Al_2O_3) = (390\text{ GPa})(8.8 \times 10^{-6}\,°C^{-1})/(27\text{ W/m-K})$$
$$= 12.7 \times 10^{-5}\text{ GPa-m/W}.$$

b. Using Eq. 9-19, the tile temperature drop is estimated to be $\Delta T = \sigma(1 - \nu)/E\alpha$, where σ is taken as σ_F. After substitution,

$$\Delta T(SiO_2) = \frac{(100\text{ MPa})(1 - 0.25)}{(94\text{ GPa})(0.55 \times 10^{-6}\,°C^{-1})} = 1451°C.$$

Similarly,

$$\Delta T(Al_2O_3) = \frac{(300\text{ MPa})(1 - 0.26)}{(390\text{ GPa})(8.8 \times 10^{-6}\,°C^{-1})} = 64.7°C.$$

These results demonstrate the superiority of SiO_2 relative to Al_2O_3 when exposed to thermal shock conditions. The dominant role that the thermal expansion coefficient plays should be noted.

9.8.2. Toughening Ceramics

In addition to thermal shock, the intrinsically brittle nature of ceramics is reflected in very low values of fracture toughness. The driving force to create fuel-efficient, high-temperature turbine engines has spawned a worldwide research effort to enhance the fracture toughness of ceramics. A main strategy seeks to engineer the microstructure so that cracks have a more tortuous path to execute as they advance through the matrix. Introducing compressive stress into the ceramic surface, thereby offsetting the influence of harmful applied tensile stresses, is another approach. How these toughening mechanisms have been practically implemented will be addressed in turn.

9.8.2.1. Composites

Ceramics can be toughened by the same principle applied successfully in polymers, namely, through the use of composites. Very short, small-diameter fibers known as **whiskers** are incorporated into ceramic bodies up to a volume fraction of 25%. Because energy is expended in pulling whiskers out of the matrix and in bending or breaking them, crack propagation is blunted and the

FIGURE 9-36 Molten steel poured into a quartz crucible resting on a cake of ice. The resistance to thermal shock stems largely from the low thermal expansion of quartz. Courtesy of Corning Incorporated.

fracture toughness is effectively raised. In one application, cordierite glass–ceramic substrates, on which arrays of computer chips are mounted, have been toughened with 1-μm-diameter, 10-μm-long Si_3N_4 whiskers. In another, SiC fibers have toughened Al_2O_3 cutting tools used in machining superalloys. A number of reinforcements have been commercially available for some time including SiC, Al_2O_3, and boron fibers; typical fibers are 10 to 20 μm in diameter. Some of their properties and those of the ceramic matrices they strengthen are listed in Table 9-6. Prediction of composite properties follows the guidelines presented earlier for fiber-reinforced polymers.

Attaining the right bond strength between the fiber and the matrix is the key to the success of tougher ceramic composites. But this is easier said than done. Unfortunately, all too frequently improvement in mechanical properties of ceramics comes at the expense of ease in sintering. Porosity at the fiber–matrix interface causes a low bond strength and a lack of toughening. At the other extreme, too great a bond strength prevents fiber pullout and leads to fracture instead; again the result is low toughening.

9.8.2.2. Transformation Toughening

Pure zirconia (ZrO_2) undergoes a spontaneous tetragonal–monoclinic martensitic reaction at 1150°C; the associated ~9% volume expansion makes fabrication processes difficult because of stress-induced cracking. Zirconia can be toughened, however, through a 3.6 wt% Y_2O_3 addition. Different amounts of CaO and MgO will also work. In such alloys at about 1100°C, tetragonal ZrO_2 is stable, but this phase can be retained at room temperature during relatively rapid cooling. The phase diagram of Fig. 9-37A shows that the stable phase is a monoclinic ZrO_2-rich solid solution, which coexists with this now unstable tetragonal phase in metastable equilibrium. Transformation from the tetragonal to the monoclinic phase is sluggish but is speeded up by high-temperature aging or triggered by cracks present in the matrix. An ~3% volume *increase* due to atomic reorganization accompanies the transformation. In the process, a compressive stress field develops around transformed particles. The compressive stresses seal cracks shut and, in blunting their further advance, toughen the matrix as schematically shown in Fig. 9-37B. Fracture toughness values approaching 10 MPa-m$^{1/2}$, which are several times higher than those for other ceramics, have been measured in these partially stabilized zirconia (PZT) materials. This is the reason they have been called *ceramic steels*.

9.8.2.3. Surface Toughening

Introduction of desirable residual compressive stresses in ceramic surfaces can be achieved chemically. Indiffusion of metal ions larger than those of the matrix stuffs them into smaller vacant holes, compressing the diffused surface layers. Ion exchange of this type is extensively practiced in toughening glasses. In such materials the normally destructive applied tensile stresses are effectively reduced to safe levels by the subtractive effect of the residual compressive stresses.

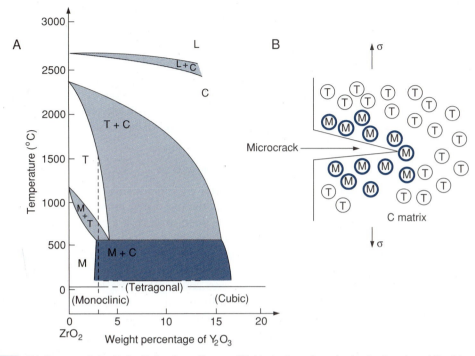

FIGURE 9-37 (A) Portion of the ZrO_2–Y_2O_3 phase diagram. (B) Mechanism of toughening in zirconia stabilized by yttria. The stress-induced transformation of tetragonal (T) ZrO_2 grains to the monoclinic (M) structure expands the volume at the crack tip, squeezing it shut.

An allied method borrowed from semiconductor technology, known as *ion implantation,* has been employed on an experimental basis to toughen ceramic surfaces by the same principle. As the name implies very energetic ions (~50 keV) of the desired metals or nonmetals are propelled by large electric fields and implanted ~1 μm beneath the surface.

9.8.2.4. Grain Size Control

Controlling the grain size distribution can be effective in toughening ceramics. In noncubic materials like Al_2O_3 and TiO_2, misorientation and roughening of grain boundaries are promoted by the anisotropic crystals. Effectively larger intergranular fracture surfaces are created in this way.

9.8.3. High-Performance Ceramic Composites

9.8.3.1. Hard-Layered Composites

The marriage of hard coatings (<10 μm thick) to underlying substrates has been driven by the need to complement the desirable properties of each. For example, many structural substrate materials do not have the necessary hardness and wear resistance, or high-temperature stability in oxidizing environments. On the other hand, ceramic coatings of TiC, TiN, Al_2O_3, ZrO_2, and other compounds do possess these properties, but do not qualify as structural materials because they cannot be easily fabricated in bulk form. Coatings of the first three compounds on steels and sintered WC have served to appreciably extend the life of cutting tools. These layered composites are created by the chemical vapor deposition techniques discussed in Section 5.2.2. Impressive gains in wear resistance have also been recorded on ball bearing components coated with TiC.

9.8.3.2. Carbon–Carbon Composites

We conclude the chapter by introducing some of the most extraordinary composite materials yet conceived. Their manufacture starts off in an unusual way. Carbon or graphite fibers or yarns in the form of monofilaments and twisted multifilaments are woven, braided, or knitted into clothlike preforms. These are placed into suitable molds or retention containers in which the void spaces are filled with carbonaceous tars and resins. In alternative chemical vapor deposition processes, carbon-bearing gases are the source of the infiltrated carbon deposit. Multiple heating (pyrolysis) and carbon infiltrating cycles dispel the volatile species and increase the preform density to the desired value. Finally, the component is entirely converted into graphite by heating to the extraordinarily high graphitizing temperatures of 2200 to 3000°C. The product is finally coated with SiC, which inhibits oxidation.

The resulting carbon–carbon composites are tougher and more resistant to thermal shock than ceramics, but they are also softer and must not be abraded. High cost limits their use to critical components such as rocket motor nozzle liners and aircraft afterburners.

9.9. PERSPECTIVE AND CONCLUSION

It is the fate of each material to have a combination of desirable as well as undesirable properties. The driving force to continually improve them is to capture, retain, and extend their commercial usage. For mechanical properties this poses different challenges depending on the material involved. Ductile materials like metals are already tough and also moderately stiff and hard; the primary task is to harden, strengthen, and protect them from environmental damage. At the other extreme, ceramics are very stiff, hard, and strong; they must be toughened. Between these extremes are polymeric materials which are neither stiff, hard, strong, nor tough; however, they are light, cheap, and easily shaped. High stiffness, strength, and toughness then become the generic goals that we set for all materials. How closely properties converge to these universal ideals depends on the particular material in question. In general, remarkable advances have been and continue to be made in all material classes.

9.9.1. Metals

Metals start off with perhaps the best initial complement of mechanical properties, and that is why they have been the materials of choice in structural and machine applications for so long. *Cold working, alloying, quenching–tempering* (for steels), and *precipitation hardening* treatments (for nonferrous alloys) are the traditional ways to strengthen and harden metals. In all of these cases there is some sacrifice in ductility and toughness. Alloy design and treatments then seek to strike the optimal compromise between yield strength and fracture toughness, two key properties that usually vary inversely with one another. Disrupting the lattice perfection and reducing the size of structural features in the matrix constitute the strategy employed to increase the strength of metals. The reason it works is that dislocation movement is impeded. But, restricted dislocation motion also limits the development of crack-tip plastic zones that are effective in blunting crack propagation. The result is reduced fracture toughness. To meet the demands of harsh, high-temperature applications, special refractory metals (e.g., superalloys) containing desirable phases, precipitates, and dispersants have been developed. Not to be outdone by other material competitors, particulate and fiber-reinforced metal matrix composites of aluminum and magnesium are enjoying increasing use.

9.9.2. Polymers

For a myriad of non-load-bearing applications (e.g., packaging, insulation, assorted molded consumer goods, toys), unstrengthened polymers will do just fine. But for more demanding engineering applications, particularly in transportation where low weight is critical for speed or energy conservation, polymer composites have filled a vital need. The fiber-reinforced polymer matrices have been simultaneously stiffened, strengthened, and toughened by dispersing stiff, strong filaments of glass, carbon, or polymer within them. Resulting fiber/matrix admixtures have yielded enhanced composite elastic moduli and toughness. The former is due directly to the presence of the fibers which carry the majority of the load; the latter occurs because the difficulty in pulling fibers out arrests crack extension. Continuous aligned fibers confer maximum benefits, but cut fibers are advantageously used provided they exceed a critical length. Without their incorporation into composites both the fibers and the polymer matrices would have more limited engineering use.

9.9.3. Ceramics

Blessed with high stiffness, strength, and temperature capabilities, ceramics have two fatal attributes: low fracture toughness and high susceptibility to thermal shock. These shortcomings have been addressed by a number of different strategies including the use of fiber composites, transformation toughening (in ZrO_2), and residual surface compressive stresses. Despite improvements, ceramics still largely lack sufficient toughness to be included in most fracture-safe structural designs. But, in the form of filaments and coatings, ceramics have transferred their desirable properties to *other engineering materials* in creating bulk and surface composites.

A broad comparison of the many materials discussed in this chapter is made in Fig. 9-38 based on specific strength and use temperature criteria. The following items are worth noting:

1. The higher the service temperature, the fewer the number of suitable materials, and the lower the specific strength.
2. At any given use temperature, polymer or ceramic composites exceed the specific strength of both MMCs and metals.
3. Material costs scale directly with specific strength and temperature of use.

Strengthening and toughening are increasingly a matter of processing compatible combinations of different material classes to capitalize on the best properties each constituent has to offer.

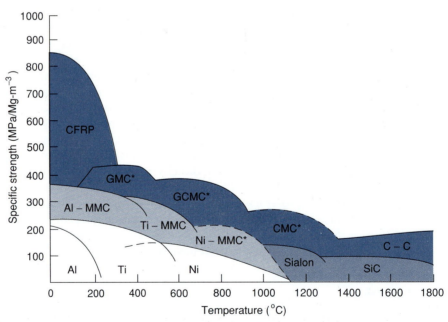

FIGURE 9-38   Specific strength of materials as a function of temperature. CFRP, carbon fiber-reinforced polymer; GMC, glass matrix composites; GCMC, glass–ceramic matrix composites; CMC, ceramic matrix composites; MMC, metal matrix composites; C–C carbon–carbon composites; sialon contains Si, Al, O, and N. Research materials are denoted by stars. After R. J. Jeal, *Metals and Materials* **4**, 539 (1988).

Additional Reading

M. F. Ashby and D. R. H. Jones, *Engineering Materials 2: An Introduction to Microstructures, Processing and Design,* Pergamon Press, Oxford (1986).

J. A. Charles and F. A. A. Crane, *Selection and Use of Engineering Materials,* 2nd ed., Butterworths, London (1989).

W. F. Smith, *Structure and Properties of Engineering Alloys,* McGraw–Hill, New York (1981).

G. Weidmann, P. Lewis, and N. Reid (Eds.), *Materials in Action Series: Structural Materials,* Butterworths, London (1990).

QUESTIONS AND PROBLEMS

9-1. Devise heat treatments for 1080 steel that would effect the following conversions:

   a. Pearlite to martensite
   b. Martensite to coarse pearlite
   c. Martensite to fine pearlite
   d. Pearlite to bainite
   e. Bainite to pearlite
   f. Martensite to bainite
   g. Spherodite to pearlite

Refer your treatments to the TTT curve.

9-2. Design heat treatments for 1080 steel that would effect the following conversions:

   a. 100% pearlite to a mixture of 50% pearlite and 50% martensite
   b. Mixture of 75% pearlite and 25% martensite to 100% tempered martensite
   c. Spherodite to a mixture of 50% pearlite and 50% martensite
   d. Austenite to a mixture of 50% martensite and 50% tempered martensite

Refer your treatments to the TTT curve.

9-3. Since time immemorial blacksmiths have forged stone carving tools starting with essentially pure iron, containing perhaps 0.05 wt% carbon. Assuming you have such round bar blanks and a blacksmith's hearth, detail the steps you would follow to forge a flat-edged stone carving chisel. Note that chisels must be very hard and tough at the cutting edge, but soft where grasped by the hand to absorb mechanical impact and prevent a ringing sensation.

9-4. It has been noted several times in this book that heat conduction is analogous to mass diffusion. Suppose a cylindrical Jominy bar, initially at 800°C, is suddenly quenched at one end to 20°C. In snapshot fashion, schematically sketch temperature–distance profiles that evolve in the bar at a sequence of increasing times during cooling.

9-5. An austenitized Jominy bar is end quenched employing oil rather than water. How would this change the resulting hardness–distance profile of the Jominy curve?

9-6. During viewing of microstructures along a water-quenched Jominy bar of 1080 steel, the following phases were observed starting from the quenched end: martensite; martensite plus pearlite; fine pearlite; coarse pearlite. At what distances along the 4-in. Jominy bar would each of these structures be observed?

9-7. An approximate relationship between impact strength (E_{IS}) and tensile strength

(σ_{UTS}) of quenched and tempered steels is $E_{IS} = 100 - \frac{5}{12}\sigma_{UTS}$, where the units of E_{IS} and σ_{UTS} are ft-lb and ksi, respectively.

a. Graph this relationship to scale.

b. In rough terms where do the following structures fall on this curve for a given steel: quenched martensite, tempered martensite, spheroidite?

c. Suppose you wish to select an optimal set of mechanical properties that maximizes the product of E_{IS} and σ_{UTS}. What combination of property values is required? (*Hint:* Inscribe the largest rectangular area that fits beneath the E_{IS}–σ_{UTS} curve.)

9-8. Vanadium, a common alloying element in tool steels, refines the grain size by reducing the prior austenitic grain size. Consider a steel with and without V. Jominy bars of both steels are austenitized at high temperatures and end quenched. Sketch the expected microstructures along the bar for each steel. (*Note:* Pearlite nucleates at austenite grain boundaries.)

9-9. When viewed in an optical microscope fresh martensite appears light and tempered martensite appears dark after proper metallographic preparation. Based on this information design a set of heat treatments that would enable the M_S temperature to be determined *metallographically*.

9-10. The heat treatments noted below are carried out in 4340 steel. In each case what phases will be present at the end of the treatment?

a. Austenitize at 750°C; quench to 0°C in 1 second.

b. Austenitize at 750°C; rapid quench to 650°C; isothermally anneal for 1000 seconds; quench to 0°C in 1 second.

c. Austenitize at 750°C; very rapidly quench to 250°C; heat to 350°C; isothermally anneal for 500 seconds; quench to 0°C in 1 second.

d. Austenitize at 800°C; rapidly quench to 550°C; isothermally anneal for 1000 seconds; heat to 750°C; quench to 0°C in 1 second.

e. Austenitize at 800°C; rapidly quench to 350°C; isothermally anneal for 400 seconds; heat to 550°C for 700 seconds; quench to 0°C in 1 second.

9-11. Why does the hardness of martensite, pearlite, and spherodite increase with carbon content?

9-12. a. Measurements of the lattice constants of iron–carbon martensites reveal that

$$a' \text{ (nm)} = 0.293 + 0.012 \times (\text{wt\% C} - 0.6 \text{ wt\% C}),$$
$$c \text{ (nm)} = 0.285 - 0.002 \times (\text{wt\% C} - 0.6 \text{ wt\% C}),$$

for carbon contents between 0.6 and 1.6 wt%. As the carbon content increases, does the volume available to each carbon atom increase or decrease?

b. Is your answer consistent with the effect of carbon content on martensite hardness?

9-13. A gear made of 5140 steel has a hardness of R_c 30 when quenched in oil. What probable hardness would a 4340 steel gear of identical dimensions exhibit if heat-treated in the same way?

9-14. To capitalize on weight savings a steel is quenched and tempered, strengthening it from 1100 MPa, where K_{1C} is 76 MPa-m$^{1/2}$, to 2000 MPa, where $K_{1C} =$ 26 MPa-m$^{1/2}$. What change in critical flaw size occurs for a design stress of 800 MPa?

9-15. Select an optimal set of mechanical properties that maximizes the product of fracture toughness and yield strength for the AISI 4340 steel in Fig. 9-13. Over the property range shown what combination of toughness and strength values is required? (*Hint:* Inscribe the largest rectangle that fits below the curve.)

9-16. Refer to the Al–Si phase diagram in Fig. 15-21:
   a. Suggest an alloy composition that might be amenable to precipitation hardening.
   b. What sequence of heat treatments would you recommend?
   c. It is found that the amount of precipitation hardening achieved in this system is very small compared with that in Al–Cu. Suggest a possible reason why.

9-17. Design an aging heat treatment (temperature and time) for a 6061 aluminum alloy that would yield a tensile strength of 275 MPa coupled with an elongation of 20%. (The treatment should be economical to perform.)

9-18. Magnesium-rich Mg–Al alloys are used as sacrificial anodes to protect ship hulls against corrosion (see Section 10.2.3). If these cast alloys are to be mechanically strengthened, suggest a possible way to do it.

9-19. Some 20,000 lb of an aluminum alloy containing 4 wt% Cu and 1 wt% Mg is used in floor beams of a transport plane. As a weight saving measure the same volume of an aluminum–lithium alloy containing 3 wt% Li and 1 wt% Cu is suggested as a replacement. What weight saving will occur? Assume weighted averages of density.

9-20. From Fig. 9-17 determine the following for pure copper:
   a. Recrystallization temperature for 95% cold work
   b. Recrystallization temperature for 20% cold work
   c. Percentage deformation required to raise the tensile strength of annealed Cu to 50,000 psi
   d. Maximum tensile strength for a 90% deformation
   e. Tensile strength after a 60% deformation and 1-hour anneal at 225°C
   f. Annealing temperature required to produce a tensile strength of 47,000 psi after an 85% deformation at room temperature

9-21. Make a plot of T_R for copper as a function of percentage cold work.

9-22. Distinguish recovery, recrystallization, and grain growth stages in annealed cold worked metals in terms of their effects on:
   a. Microstructure
   b. Strength or hardness
   c. Defect structure

9-23. Complete recrystallization of a 74% cold worked, 99% pure aluminum alloy requires 1000 hours at 232°C. At 363°C recrystallization is complete in 0.01

hour. At what temperature will Al completely recrystallize in an hour if the process is thermally activated?

9-24. What shear stress must be applied to move dislocations past 0.5-μm-diameter dispersed Al_2O_3 particles spaced 50 nm apart in an aluminum matrix?

9-25. a. What is the modulus of elasticity of pure copper dispersion-strengthened with 5 vol% Al_2O_3 particles?

b. If the oxide particle diameter is 50 nm what approximate dispersion-strengthening increment can be expected over the 10,000-psi yield stress for Cu?

9-26. In Fig. 9-20 both the yield (σ_0) and fracture (σ_F) strengths are fit to linear equations that express the $D_g^{-1/2}$ dependence, that is,

$$\sigma_0 = \sigma_s + BD_g^{-1/2} \quad \text{and} \quad \sigma_F = \sigma_{sF} + B_F D_g^{-1/2}.$$

a. Evaluate constants σ_s and B for yield and σ_{sF} and B_F for fracture stress data.

b. By equating the expressions for σ_0 and σ_F determine the critical grain size where there is a transition in mechanical behavior.

c. Schematically sketch the stress–strain curve for grain sizes smaller and larger than the critical one.

d. For what grain diameter will low-carbon steel have a transition temperature of $-40°C$?

9-27. A composite consists of 40% by volume continuous E-glass fibers aligned in an epoxy matrix. If the elastic moduli of the glass and epoxy are 76 and 2.7 GPa:

a. Calculate the composite modulus parallel to the fiber axis.

b. What is the composite modulus perpendicular to the fiber axis?

9-28. A composite consists of 35 vol% continuous aligned Kevlar fiber in epoxy whose elastic modulus is 2.3 GPa.

a. Calculate the composite modulus ($E_\parallel$) parallel to the fiber axis.

b. What enhancement in ($E_\parallel$) occurs if the elastic modulus of the matrix is increased to 3.7 GPa?

c. What increase in the percentage fiber volume of the original composite will yield the same $E_\parallel$ enhancement as that in part b?

9-29. Consider a polymer with elastic modulus E_p that contains two different continuous fibers with moduli E_1 and E_2. Derive an expression for Young's modulus (E_c) that would be measured in the following composites if the corresponding fiber volume fractions are V_1 and V_2. (Assume V_1 and V_2 are relatively small.)

a. A composite where both fibers are aligned parallel to the axis of measurement

b. A composite where both fibers are aligned perpendicular to the axis of measurement

9-30. It is desired that the fibers support 90% of the load in a composite containing aligned, continuous high-modulus carbon fibers embedded in an epoxy matrix. What volume fraction of fiber will be required? (For epoxy assume $\rho = 1.3$ Mg/m^3 and $E = 3$ GPa.)

9-31. Young's modulus for a composite containing 60 vol% continuous aligned fiber is 44.5 GPa when tested along the fiber axis but only 7.4 GPa when tested normal to the fiber axis. Determine Young's modulus for the fiber and for the matrix.

9-32. A composite containing 25 vol% aligned continuous carbon fibers in an epoxy matrix fails at a tensile stress of 700 MPa. If the fiber tensile strength is 2.7 GPa what is the matrix yield strength?

9-33. In an aligned fiberglass composite the tensile strength of the 10-μm-diameter fibers is 2.5 GPa, the matrix yield strength is 6 MPa, and the interfacial shear strength is 0.1 GPa.
   a. What is the optimal cut fiber length?
   b. If chopped glass fiber of optimal length is used, what volume fraction should be added to the matrix to produce a composite tensile strength of 1 GPa?

9-34. An optical fiber essentially consists of a 125-μm-diameter glass filament (see Section 13.6.2) surrounded by a 250-μm-diameter urethane–acrylate polymer.
   a. Why must the glass filament be coated?
   b. Young's modulus for the polymer is 10 MPa and that for the glass fiber is 80 GPa. If an elastic load is applied to the composite, what fraction is carried by the fiber?
   c. To determine the polymer–glass interfacial bond strength a 1-cm length of composite is embedded in a bracing medium and a force is applied to the glass. If a 2.5-kg force is required to extract the filament, what is the bond strength?

9-35. a. What is the value of E/ρ for a composite containing 35 vol% continuous aligned Kevlar fibers ($E = 125$ GPa, $\rho = 1.45$ Mg/m^3) in an epoxy matrix ($E = 3.2$ GPa, $\rho = 1.3$ Mg/m^3)?
   b. High-strength carbon fibers are substituted for Kevlar in this composite. What volume fraction of continuous aligned carbon fibers will yield the same E/ρ value?

9-36. A single cylindrical fiber of diameter $2r$ is embedded into a matrix to a depth of d as suggested by Fig. 9-30A. The ratio of the fiber tensile strength σ_f to the maximum interfacial shear stress τ_m between the fiber and matrix is 10. What fiber aspect ratio or value of d/r is required so that the fiber breaks before it is pulled out?

9-37. Three candidate materials for poles used in pole vaulting competitions are listed below:

Material	Density (g/cm^3)	Young's modulus (GPa)	Price ($/lb)
Al alloy	2.80	74	1.3
Glass fiber/epoxy	2.01	54	2.2
Carbon fiber/epoxy	1.68	140	67.0

For equivalent pole dimensions which material would you select? Why?

9-38. If the yield stresses for steel, aluminum, and titanium alloy materials are 1000, 500, and 900 MPa, respectively, compare the weights for optimal strength design of struts, beams, and plates.

9-39. a. Steel rails 30 m long are laid down at a temperature of 25°C. What should the gap between the rail ends be if they are to just touch when the temperature reaches 40°C?

b. What will the gap length be at $-10°C$?

c. What stress will develop in the rails if the temperature rises to 42°C?

(*Note:* For steel $\alpha = 11 \times 10^{-6}/°C$.)

9-40. A strip of copper with a thermal expansion coefficient (α) of $17 \times 10^{-6} \, °C^{-1}$ is clad to a stainless-steel strip of the same dimensions with $\alpha = 9 \times 10^{-6} \, °C^{-1}$. When heated this bimetallic composite strip bends. In which direction does it bend and why?

9-41. Order the following materials in increasing ability to withstand thermal shock: BeO, SiC, ZrO_2, TiN, Si_3N_4.

9-42. Mention three ways to toughen ceramics. To what extent do toughened ceramics approach the toughness of metals?

9-43. Values of the fracture stress in ceramic materials are subject to a wide statistical variation unlike the tensile strength of metals, which shows little spread. Why?

9-44. Approximately 7 mol% Y_2O_3 toughens ZrO_2 in a stabilized cubic form that can be successfully fabricated without cracking. What weights of ZrO_2 and Y_2O_3 are required to make 100 g of this composition?

10 DEGRADATION AND FAILURE OF STRUCTURAL MATERIALS

10.1. INTRODUCTION

Engineering responsibility does not simply end with the manufacture or construction of a component, system, or structure. Should the product degrade or fail in service there can be potentially severe implications with respect to human safety and financial loss. This chapter is concerned with the reasons that **structural** materials degrade or fail in service and some of the practical methods employed to ameliorate these damaging effects. In a similar vein, the last chapter of this book deals with many of the same concerns, but in **electronic** materials and devices. Collectively, both chapters deal with the serious issues of product quality, assurance, and reliability and, in a broader sense, with national competitiveness. The cost to American industry and society of materials degradation and failure has been variably estimated to range between a few tens and as much as a hundred billion dollars or more annually. A figure at the upper limit may not seem incredible when costs of downtime, lost production, equipment replacement, hospitalization or health care for injured parties, insurance and legal fees, and other items are taken into consideration.

It is appropriate to start by defining what we mean by **degradation** and **failure**. When a component or system ceases to meet intended performance specifications we may broadly say that it has failed. A component does not necessarily have to break into pieces for it to fail, although that is certainly the most immediate and unambiguous sign of failure. Quality-control practices weed out components that **fail** inspection; however, we are not concerned with these, but rather with service failures. Wear in metal forming dies that results in parts that no longer maintain tolerances means that the tools have failed to meet their intended purpose. Similarly, turbine blades that stretch beyond designed dimensions during operation have essentially failed. In these examples each component has technically "failed" but not broken, and may even be repairable so that it can be returned to service.

In this chapter, however, *failure* signifies mechanical fracture or breakage. All too frequently catastrophic fracture occurs suddenly and without warning. After postmortem analysis of the fractured components or structure it often happens that the mechanical design and stress analysis were not at fault. Rather, the wrong material was chosen, it was not processed or fabricated properly for the job, or the subtle progressive damage to it was undetected. Nothing will concentrate an engineer's mind more than involvement in a failure of some consequence and its analysis. This is, perhaps, the most compelling reason for engineers of all types to understand materials.

Degradation denotes the progressive loss in ability of a component, system, or structure to meet performance specifications. Corrosion and wear are two important mechanisms of degradation that are discussed in this chapter. Often one learns to live with such degradation if there is no danger of imminent failure. An example is corrosion damage to an automobile fender. Wear of varied items such as tools, machine parts, and auto components is also frequently tolerated up to a point. But all too often, degradation may ultimately lead to catastrophic failure (fracture) if the progressive damage is not recognized and halted in time.

For each of the topics discussed in this chapter the approach taken stresses the causes and manifestations of damage phenomena, the materials that are most vulnerable, and the practical strategies for minimizing degradation and failure. The important issue of time dependence of damage is addressed where possible. Representative case histories of degradation and failure phenomena illustrate the breadth of the concerns.

10.2. CORROSION

10.2.1. Elementary Aspects of Electrochemistry

10.2.1.1. Standard Electrodes

Corrosion may be broadly defined as an unintentional attack on a material through chemical reaction with a surrounding medium. According to this definition ceramics, concrete, and polymers also corrode but our discussion here is limited to metals. And of the metals, corrosion damage to iron and steel far exceeds that to other metals. As we shall see the typical corrosion scenario involves the equivalent of two somehow different metal electrodes, exposed to an aqueous ionic electrolyte and connected by a conductor of electrons; in the process metal is lost from one electrode, signifying that corrosion has occurred.

Much about aqueous corrosion can be learned by studying what happens when metals are placed in contact with solutions (electrolytes) containing either ions of the same metal or hydrogen ions. Since electrical currents flow in electrochemical systems we must simultaneously consider reactions at two electrodes, the anode and cathode, as shown in Fig. 10-1. To physically order the tendency of metals (M°) to become ions (M^{n+}) they are placed in a 1 M solution of the metal ions contained within a so-called half-cell. Through loss of n electrons (n = ion valence), the reversible reaction

$$M^\circ \rightleftarrows M^{n+}(1\ M) + ne^- \tag{10-1}$$

occurs. Metal undergoing this reaction in the forward direction are said to *oxidize*; in the reverse direction, the metal ions are *reduced* to metal. The other half of the system is a standard or reference cell. It consists of a chemically unreactive platinum electrode immersed in a 1.0 M acid (e.g., HCl) solution where the electrode reaction involving reduction of H^+ to H_2 can be written as

$$2H^+(1\ M) + 2e^- \rightleftarrows H_2 \qquad (1\ \text{atm}). \tag{10-2}$$

When the two half-cells are connected through a porous membrane that does not let the solutions mix, a voltage or electromotive force, E°, develops across

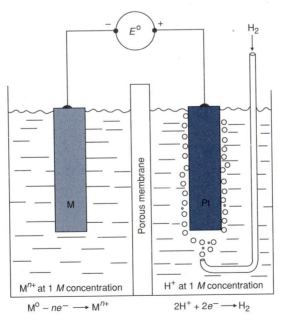

M^{n+} at 1 M concentration   H$^+$ at 1 M concentration

$$M^\circ - ne^- \longrightarrow M^{n+} \qquad\qquad 2H^+ + 2e^- \longrightarrow H_2$$

FIGURE 10-1   Electrolytic cell composed of two half-cells. On the left, metal electrode M$^\circ$ is immersed in a 1 M solution of M^{n+} ions. The standard hydrogen electrode on the right is saturated with H$_2$ and E° is the standard half-cell potential.

the electrodes of this galvanic cell. This experiment is repeated for a series of metals to obtain the well-known **standard electromotive force series,** and measured values of the **standard electrode potentials** are listed in Table 10-1. In the convention adopted here the **oxidation** potential polarities are tabulated. (Note that some authors use **reduction** potentials of opposite sign.)

A number of items are worth noting about the information in this table:

1. All values refer to 1 *M* solutions at 1 atm and 25°C.

2. The electrode potential values ($E°$) are given for the case where *no current flows*. Under this condition the electrode reaction is reversible.

3. Metals that have a positive electrode potential behave as cathodes relative to hydrogen. Ions of such metals in solution would tend to be reduced to metal. The higher the metals are situated in the table, the more pronounced is their tendency to remain cathodes and be unreactive in solution.

4. Metals that have a negative electrode potential behave as anodes relative to hydrogen and would tend to oxidize by dissolving in solution. The lower the metals are situated in the table, the more pronounced is their tendency to become anodic and corrode.

TABLE 10-1   **STANDARD ELECTRODE POTENTIALS AT 25°C** [a]

Electrode reaction	Electrode potential, $E°$(V), relative to standard hydrogen
$Au \longrightarrow Au^{3+} + 3e^-$	+1.498
$2H_2O \longrightarrow O_2 + 4H^+ + 4e^-$	+1.229
$Pt \longrightarrow Pt^{2+} + 2e^-$	+1.200
$Ag \longrightarrow Ag^+ + e^-$	+0.799
$2Hg \longrightarrow Hg_2^{2+} + 2e^-$	+0.788
$Fe^{2+} \longrightarrow Fe^{3+} + e^-$	+0.771
$4(OH)^- \longrightarrow O_2 + 2H_2O + 4e^-$	+0.401
$Cu \longrightarrow Cu^{2+} + 2e^-$	+0.337
$Sn^{2+} \longrightarrow Sn^{4+} + 2e^-$	+0.150
$H_2 \longrightarrow 2H^+ + 2e^-$	0.000
$Pb \longrightarrow Pb^{2+} + 2e^-$	−0.126
$Sn \longrightarrow Sn^{2+} + 2e^-$	−0.136
$Ni \longrightarrow Ni^{2+} + 2e^-$	−0.250
$Co \longrightarrow Co^{2+} + 2e^-$	−0.277
$Cd \longrightarrow Cd^{2+} + 2e^-$	−0.403
$Fe \longrightarrow Fe^{2+} + 2e^-$	−0.440
$Cr \longrightarrow Cr^{3+} + 3e^-$	−0.744
$Zn \longrightarrow Zn^{2+} + 2e^-$	−0.763
$Al \longrightarrow Al^{3+} + 3e^-$	−1.662
$Mg \longrightarrow Mg^{2+} + 2e^-$	−2.363
$Na \longrightarrow Na^+ + e^-$	−2.714

[a] Cathodic behavior or reduction occurs more readily the *higher* the electrode reaction is in the table. Anodic behavior or corrosion occurs more readily the *lower* the electrode reaction is in the table.

5. The similarity between the electromotive force series and the Ellingham diagram (see Fig. 5-2), which rates the tendency of metals to oxidize, should be recognized. As we shall see, oxidation of a metal in an aqueous solution or in gaseous oxygen requires both electronic plus ionic motion in liquid and solid (oxide) electrolytes, respectively.

There is nothing that prevents us from considering half-cell combinations between any two metals. Thus, the different metals can either lie above or below hydrogen, or one can lie above and the other below. As an example of the latter consider the combination of half-cells composed of copper and iron:

$$Fe^o \rightleftharpoons Fe^{2+}(1\ M) + 2e^-, \qquad E^o(Fe) = -0.440\ V; \qquad (10\text{-}3a)$$
$$Cu^o \rightleftharpoons Cu^{2+}(1\ M) + 2e^-, \qquad E^o(Cu) = +0.337\ V. \qquad (10\text{-}3b)$$

The overall reaction is obtained by adding (subtracting) the half-reactions according to the rules of chemistry, which yields

$$Fe^o + Cu^{2+}(1\ M) \rightleftharpoons Cu^o + Fe^{2+}(1\ M),$$
$$E^o_{cell} = -0.440 - (+0.337) = -0.777\ V. \qquad (10\text{-}4)$$

It is the sign and magnitude of the net cell potential, $E^o_{cell} = E^o(Fe) - E^o(Cu)$, that have important implications with respect to the tendency of metals to corrode. The negative sign tells us that the reaction will go spontaneously as written. In this galvanic cell Fe metal oxidizes and is the anode, whereas Cu is the cathode. The magnitude of the cell potential may be viewed as the *driving force* for current flow in the circuit. And we shall see that if current flows, corrosion occurs. In the language used it appears there might be a connection between E^o_{cell} and free energy change (ΔG^o) because both are additive in chemical equations and serve as driving forces for change. In fact, an important result from the thermodynamics of electrochemical systems connects the two quantities as

$$\Delta G^o = nE^o_{cell}F, \qquad (10\text{-}5)$$

where ΔG^o is the standard free energy of the cell reaction per mole, n is the number of electrons transferred, and F is the Faraday constant ($F = 96,500$ C/mol $= 96,500$ A-s/mol). In the case of the reaction given by Eq. 10-4,

$$\Delta G^o = 2(-0.777\ V) \times (96,500\ C/mol) = -150\ kJ, \text{ or } -35.8\ kcal/mol.$$

Note that in the convention adopted here a negative free energy change, implying reaction as written, is associated with a negative cell potential.

10.2.1.2. Nonstandard Electrodes

In actual corrosion situations the electrolytes are virtually never standard 1 M solutions but much more dilute. Fewer ions to promote the reverse reaction means a more negative electrode potential. Changes in the half-cell as well as full cell potentials, *relative* to the standard electrode potentials, can be expected in such cases. From Eq. 5-4 we recall that the chemical equilibrium constant is related to the free energy change; therefore, for an arbitrary half-cell reaction we may write with the aid of Eq. 10-5,

$$[M^{n+}]/[M^{o}] = \exp\{-[\Delta G^{o} - \Delta G]/RT\} = \exp\{-n(E^{o} - E)F/RT\}.$$
(10-6)

Customarily, the state of the pure metal is defined such that $[M^{o}] = 1$. For a 1 M electrolyte, $[M^{n+}] = 1$; therefore, satisfaction of Eq. 10-6 requires that $E = E^{o}$ and $\Delta G = \Delta G^{o}$. These are precisely the conditions that define the standard electrode. But, for other than 1 M electrolytes, the half-cell electrode potential changes from E^{o} to E. Correspondingly, the nonstandard free energy (ΔG) is associated with this altered value of E. When the full (galvanic) cell reaction between anode A and cathode C (i.e., $A^{o} + C^{n+} \rightarrow A^{n+} + C^{o}$) is considered,

$$([A^{n+}] \cdot [C^{o}])/([A^{o}] \cdot [C^{n+}]) = \exp\{-n(E^{o}_{cell} - E_{cell})\, F/RT\}.$$
(10-7)

Equations 10-6 and 10-7 are forms of the Nernst equation which relate the *individual* electrode and *full cell* potentials to temperature and electrolyte concentration. A more familiar form for the half-cell, taking $[M^{o}] = 1$, is

$$E = E^{o} + (RT/nF)\ln[M^{n+}] = E^{o} + (0.059/n)\log[M^{n+}] \quad \text{V (25°C)}.$$
(10-8)

Similarly, for the full cell potential,

$$E_{cell} = E^{o}_{cell} + (0.059/n)\log([A^{n+}]/[C^{n+}]) \quad \text{V (25°C)}.$$
(10-9)

EXAMPLE 10-1

What cell potential would arise in a cell containing an Fe electrode in a 0.001 M $FeCl_2$ solution and a Cu electrode in a 1.5 M $CuCl_2$ solution?

ANSWER   Employing Eq. 10-9, we know that $E^{o}_{cell} = -0.777$ V, $[Fe^{2+}] = 10^{-3}$, $[Cu^{2+}] = 1.5$, and $n = 2$. After substitution,

$$E_{cell} = -0.777 + (0.059/2) \times \log(10^{-3}/1.5) = -0.871 \quad \text{V}.$$

10.2.2. Faraday's Law: Corrosion Rate

So far very little has been said of corrosion. The electrochemical cells of the preceding section were in equilibrium and no current was drawn. If current is allowed to flow then they behave as batteries that can power electrical circuits. But in the process of doing useful work the cell deteriorates or *corrodes*. In applications like propelling torpedos, metals at the extreme ends of the standard electromotive force series are employed in batteries. The intent is to quickly draw as much current as possible to power the propulsion motors. In this case rapid corrosion is a desideratum! In other battery applications long operation at reasonable potentials governs the choice of electrodes and electrolytes. Often an additional requirement is the provision for reversing any "corrosion" through recharging.

Once current flow occurs an estimate of the extent of metal loss or corrosion is given by **Farraday's law** of electrolysis. The latter states that *for every 96,500 C of charge [current I (A) × time t (s)], one gram equivalent weight [atomic weight M (g/mol)/valence n] of metal dissolves at the anode.* An identical amount of reaction (e.g., plating of metal, hydrogen evolution, etc.) occurs at the cathode. The weight loss, w (in grams), is given by

$$w = ItM/nF, \tag{10-10}$$

and w/t may be viewed as a corrosion rate. If corrosion occurs uniformly over a given area, then weight loss can be easily converted to an equivalent reduction in metal thickness. Common units for corrosion rates are g/m^2-year, μm/year, and in./year (ipy).

EXAMPLE 10-2

a. What is the open-circuit potential of the cell in Fig. 10-2 if both anode and cathode electrolytes are 1 M?

b. If 1 mA of current flows how long will it take for 0.5 mg of the anode to dissolve?

c. If the total anode area is 100 cm^2 what is the metal thickness loss if corrosion is uniform?

d. Under what conditions will the potential reverse?

ANSWER   a. Because Fe lies below Sn in Table 10-1 we may assume it is the anode. By a reaction similar to Eq. 10-4, Fe$^\circ$ + Sn^{2+} = Fe^{2+} + Sn$^\circ$, and

$$E^\circ_{cell} = E^\circ(Fe) - E^\circ(Sn) = -0.440 - (-0.136) = -0.304 \text{ V}.$$

b. This question is an application of Faraday's law. For Fe, $M = 55.9$ and $n = 2$. Substitution into Eq. 10-10 yields $t = nFw/IM = (2 \times 96,500 \times 0.0005)/(0.001 \times 55.9) = 1730$ seconds, or 28.8 minutes.

c. The loss of Fe is 0.0005 g/100 cm^2 = 5 × 10^{-6} g/cm^2. Dividing by the density of Fe (7.86 g/cm^3), the thickness of Fe lost is 5 × 10^{-6}/7.86 = 0.636 × 10^{-6} cm = 6.36 nm.

d. The potential reverses when $E_{cell} = 0$. If Sn now becomes anodic or active then the above reaction and cell potential reverse. From Eq. 10-9, $E_{cell} = 0 = -0.304 + (0.059/2) \times \log([Fe^{2+}]/[Sn^{2+}])$. Solving, $([Sn^{2+}]/[Fe^{2+}]) = 4.95 \times 10^{-11}$.

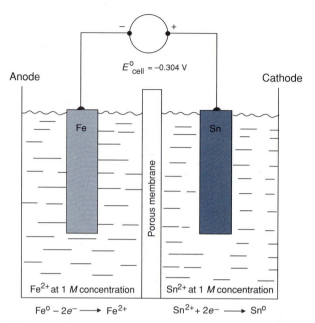

FIGURE 10-2 Cell potential developed by addition of the half-cell reactions for iron and tin electrodes immersed in 1 M aqueous solutions of their respective salts (e.g., $FeCl_2$ and $SnCl_2$).

This example illustrates that Sn is normally cathodic to Fe and this accounts for its use in "tin can" containers; however, in contact with certain foods Sn reacts chemically to form complex (organic) ions such that the concentration of Sn^{2+} is reduced dramatically. When this happens the potential markedly shifts in the anodic direction and Sn rather than Fe may corrode.

10.2.3. Manifestations of Corrosion

It has been already noted that all corrosion phenomena require metal anodes and cathodes connected by an electron conductor (metal) and contacted by an electrolytic medium, such that an internal short-circuit current flows. As we shall see these requirements are physically manifested in many varied ways so that it is often not easy to recognize all of the disguises corrosion displays. These ominous faces of corrosion are responsible for a large toll of material damage when not designed against. It is instructive, therefore, to enumerate some of the common types of corrosion, first as model textbook cells and then in the form they assume in actual practice. We start with galvanic action.

10.2.3.1. Galvanic (Two-Metal) Corrosion

Galvanic (two-metal) corrosion involves two different metals and can be ideally studied in the electrochemical cells of the previous section. An important subcategory of galvanic corrosion occurs when the electrode areas differ. In the equivalent electric circuit of a corrosion system it is necessary that the same total **current** flows through anodic and cathodic portions. If this happens the **current densities** (current/area) and corrosion rates will vary inversely with anode size. Small anodes in contact with large cathodes will thus corrode more severely than large anodes in contact with small cathodes. For example, small steel (iron) rivets that are used to clamp large plates of copper will corrode through readily in a saltwater environment. On the other hand, copper rivets clamping large steel plates would survive much longer, from a corrosion standpoint, because the iron anode current density is significantly reduced. These effects are illustrated in Fig. 10-3.

The preferential corrosion of small anodes in contact with large cathodes is capitalized on in protecting normally anodic steel. Underground steel pipes and ship hull steels are protected against corrosion in this way. In the case of ships more strongly anodic zinc or magnesium alloy plates are bonded to the much larger hull, which now assumes a cathodic character. The Zn and Mg become **sacrificial** anodes, which corrode, but effectively protect the steel (Fig.

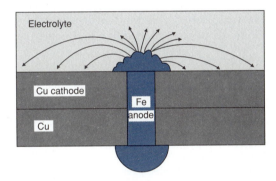

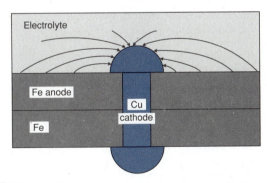

FIGURE 10-3 Effect of electrode area on corrosion. Small-area Fe (anode) rivets joining large-area Cu plates (*above*) will corrode more readily than small Cu rivets joining large Fe plates (*below*). Ionic current flow lines in surrounding electrolyte are shown.

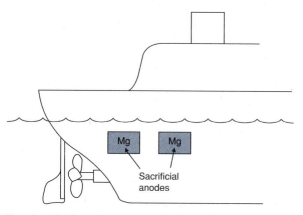

FIGURE 10-4 Use of sacrificial anodes to protect steel ship hulls.

10-4). Submerged piping is also commonly protected by impressing voltages that make them cathodic.

Since sacrificial anodes are placed below the seawater line, this is an opportune time to introduce Table 10-2 ("The Galvanic Series in Sea Water"). Like the electromotive force series, this table orders metals with respect to their anodic and cathodic tendencies. But unlike the former case, the electrolyte is

TABLE 10-2 **GALVANIC SERIES IN SEAWATER**

↑	Gold
Increasingly	Graphite
cathodic	Titanium
(protected)	Silver
	18 Cr–8Ni stainless steel (passive)
	11–30 wt% Cr stainless steel (passive)
	Silver solder
	Monel (70 wt% Ni, 30 wt% Cu)
	Bronzes (Cu–Sn)
	Copper
	Brasses (Cu–Zn)
	Nickel
	Tin
	Lead
	18Cr–8Ni stainless steel (active)
	Steel or iron
	Cast iron
	2024 aluminum
Increasingly	Cadmium
anodic	Aluminum
(corroded)	Zinc
↓	Magnesium

seawater rather than 1 M laboratory solutions; also, the corrosion behavior of alloys important in naval and maritime applications is included. There are a number of surprises in this table. For example, Zn and Al have switched places relative to their positions in Table 10-1, as have Ni and Sn. Because the tendency of metals to corrode under service conditions is strongly influenced by the environment, the galvanic series is usually more appropriate in engineering design than the electromotive series.

10.2.3.2. Single-Metal Corrosion

Caused by the Metal. Until now we have dealt exclusively with two-metal (galvanic) corrosion situations. In practice, however, there are probably more cases of corrosion involving a single metal. Paradoxically, the single metal contains both anodes and cathodes. The fundamental reason for this can be tied to nonuniform composition and mechanical processing. This establishes nonuniform distributions of alloying elements, precipitates, inclusions, and so on, as well as structural heterogeneities represented by grain boundaries and dislocations. Free energy gradients develop on a local scale, with some atoms more energetic than others or, equivalently, some surfaces more anodic than others. Outwardly single, homogeneous, and even pure metals are anything but from a corrosion standpoint.

As a simple example, consider a zinc metal rod dipped in dilute HCl acid (Fig. 10-5A). As everyone knows Zn dissolves while H_2 gas simultaneously

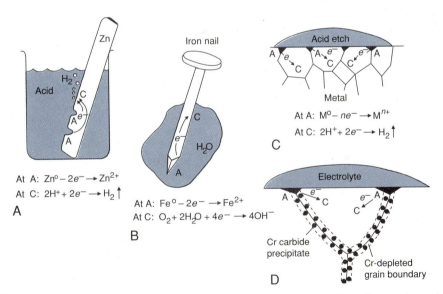

At A: $Zn^0 - 2e^- \rightarrow Zn^{2+}$
At C: $2H^+ + 2e^- \rightarrow H_2 \uparrow$

A

At A: $Fe^0 - 2e^- \rightarrow Fe^{2+}$
At C: $O_2 + 2H_2O + 4e^- \rightarrow 4OH^-$

B

At A: $M^0 - ne^- \rightarrow M^{n+}$
At C: $2H^+ + 2e^- \rightarrow H_2 \uparrow$

C

D

FIGURE 10-5 Corrosion action within a single metal. (A) Local anodes and cathodes develop on Zn dipped in acid. (B) Plastically deformed regions of nail are anodic relative to undeformed areas. (C) Preferential corrosion at anodic grain boundaries relative to cathodic bulk grains. (D) Grain boundaries depleted of Cr corrode relative to the stainless-steel matrix.

bubbles off. Closely separated areas of the same rod are different electrically; the acid exploits these differences and activates anodic and cathodic regions. At anodes, Zn^{2+} ions are released, whereas at cathodes, H_2 is discharged. Another example involves preferential corrosion of cold worked areas relative to undeformed regions of a nail (Fig. 10-5B). Atoms at the head and point are more energetic than those in the unstressed shank, so the former regions become anodic to the latter.

A not unrelated form of attack, known as **stress corrosion**, affects some metals that are simultaneously stressed in tension while immersed in a corrosive medium. Apparently, metal atoms at crack tips are made even more reactive by tensile stresses. The synergism between corrosion and crack extension accelerates failure of the metal. Only specific environments for particular metals are effective in inducing stress corrosion cracking. Stainless steel in chloride solutions and brass in ammonia environments are the classic examples of metals that suffer such attack.

Grain boundaries are a favored and important region for corrosion attack. Atoms are more energetic there than in grain interiors, making grain boundaries anodic relative to the bulk metal. Such preferential corrosion is the mechanism responsible for the controlled etching of metals that delineates grain boundaries for metallographic observation (Fig. 10-5C). A far less desirable form of grain boundary corrosion sometimes occurs in certain stainless steels that contain greater than ~0.03 wt% carbon. When heated and then slowly cooled through the temperature range 500 to 800°C, as during welding, chromium carbides (Cr_6C_{23}) tend to nucleate and precipitate at grain boundaries (Fig. 10-5D). This locally depletes the surrounding region below the 12 wt% Cr level that conferred the "stainless" quality to iron in the first place. The steel is now "sensitized" and susceptible to **intergranular corrosion**. Stainless steels can be spared this problem by lowering the carbon content or by tying it up in the less harmful form of titanium or niobium carbides. Rapid cooling of the steel from an elevated temperature carbide (re)solution treatment will also help.

Caused by the Electrolyte.   Single metals can be perfectly homogeneous and free of defects and still corrode, because of inhomogeneous distributions of electrolytes that effectively divide the same metal into anodic and cathodic regions. To study the factors that influence corrosion in such cases let us consider the so-called **concentration cell** of Fig. 10-6, in which iron is used for both electrodes. There is a lower concentration of Fe^{2+} ions in the left half-cell than in the right half-cell. According to the Law of Mass Action there will be a greater tendency for Fe° to ionize and form Fe^{2+} in the more dilute solution than in the more concentrated one. Thus, we expect the left-hand electrode to become anodic, a result borne out by considering the equation

$$E_{cell} = (0.059/n)\log([A^{n+}]/[C^{n+}]). \qquad (10\text{-}11)$$

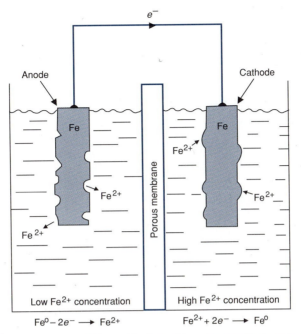

e^-

Anode

Cathode

Fe

Fe

Fe^{2+}

Porous membrane

Fe^{2+}

Fe^{2+}

Fe^{2+}

Fe^{2+}

Low Fe^{2+} concentration

High Fe^{2+} concentration

$Fe^0 - 2e^- \longrightarrow Fe^{2+}$ $Fe^{2+} + 2e^- \longrightarrow Fe^0$

FIGURE 10-6 Concentration cell. Identical Fe electrodes are immersed in electrolytes containing different concentrations of Fe^{2+}.

This equation is simply Eq. 10-9 modified by taking $E_{cell}^o = 0$. A negative value of E_{cell} results when $[A^{n+}] < [C^{n+}]$, and this is consistent with the cell concentrations chosen and the sign convention adopted.

A practical example of corrosion stemming from the same metal exposed to two different local ionic concentrations occurs when disks or propellers rotate in an electrolyte (Fig. 10-7). At the outer radius the high linear velocity of the fluid efficiently disperses any buildup of ions. This region therefore becomes anodic with respect to the cathodic metal closer to the shaft, where the liquid is less disturbed and the ionic concentration is higher. As a result outer regions suffer corrosion. Rotating impellers also frequently undergo **cavitation** damage which, though not strictly caused by corrosion, nevertheless often results in severe metal loss (e.g., Fig. 10-8). Cavitation is caused by the copious generation of minute bubbles at the interface between rapidly flowing liquids and solid surfaces. The rapidly collapsing bubbles impart surprisingly large stresses to the surface that erode it away.

Another important variant of the concentration cell, but this time consisting of different oxygen levels, is exemplified in Fig. 10-9. Contrary to prior expectations the iron electrode that is *least* exposed to O_2 becomes the anode. In this case the anode reaction is still

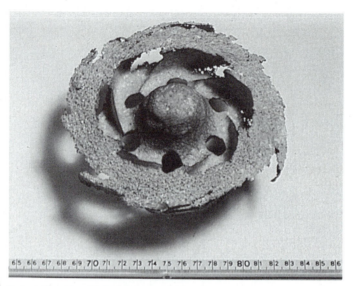

FIGURE 10-7   Bronze (Cu–10 wt% Sn) pump impeller exposed to 1% sulfamic acid at 30°C, suffering acid corrosion failure in a few hours. From E. D. D. During, *Corrosion Atlas: A Collection of Illustrated Case Histories,* Elsevier, Amsterdam (1988).

FIGURE 10-8   Stainless-steel pump impeller exhibiting cavitation pitting. The environment was skimmed milk at 70°C. From E. D. D. During, *Corrosion Atlas: A Collection of Illustrated Case Histories,* Elsevier, Amsterdam (1988).

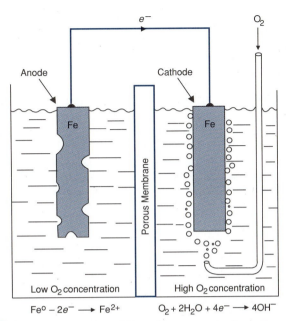

FIGURE 10-9 Oxygen concentration cell. Corrosion occurs at the electrode exposed to the *lower* oxygen concentration.

$$Fe^o \rightleftarrows Fe^{2+} + 2e^-, \qquad (10\text{-}3a)$$

but, at the oxygen-rich cathode, hydroxyl ions are produced according to the reaction

$$O_2 + 2H_2O + 4e^- \rightleftarrows 4OH^{-1}. \qquad (10\text{-}12)$$

Although the anode from which O_2 was excluded appears as bright as ever, the cathode rusts by forming a layer of $Fe(OH)_3$. The combination of Fe^{2+} and OH^- yields $Fe(OH)_2$, and with additional O_2, the oxidation to the Fe^{3+} state occurs through the reaction

$$4Fe(OH)_2 + O_2 + 2H_2O \rightleftarrows 4Fe(OH)_3 \qquad (\text{rust}). \qquad (10\text{-}13)$$

This is the reason for the widespread practice of deaeration or removal of oxygen from boiler and industrial waters. Another implication is corrosion *below* the water line in submerged steel piers.

A related corrosion problem occurs when steel surfaces are covered with dirt, scale, or other pieces of steel, for example, a washer, riveted or welded sections, that trap electrolyte in between (Fig. 10-10). Such **crevice corrosion** is caused by **oxygen starvation**, and those regions so deprived become anodic and corrode relative to the cathodic regions exposed to air. An area effect is also at play here. In the oxygen-starved crevices there is a small anode in contact with a very large cathode.

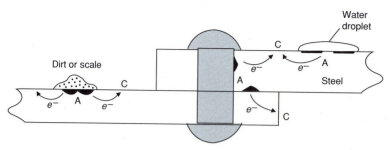

FIGURE 10-10 Examples of crevice corrosion in riveted steel plates. In the anodic (A) crevices the reaction is $Fe^0 = Fe^{+2} + 2e^-$, whereas the reaction $O_2 + 2H_2O + 4e^- = 4OH^{-1}$ occurs on the cathodic (C) steel surfaces. Similar reactions occur under and near the water droplet.

10.2.4. Corrosion Case Histories

In practice, corrosion does not always assume the classic forms just presented. A number of case histories illustrated in Fig. 10-11 attribute degradation to the harmful influence of certain chemical species (e.g., chlorine, hydrogen, and ammonia) and stress.

FIGURE 10-11 Case histories of corrosion and degradation. (A) Chloride pitting attack of a stainless-steel gas cooler pipe. (B) Uniform corrosion of aluminum roofing in an atmosphere contaminated by alkali halides. (C) Swelling and cracking in an ethylene–propylene rubber flange seal as a result of exposure to HCl, chloroform, and carbon tetrachloride. (D) Formation of cavities during cold hydrogen attack of unalloyed steel. The environment was water with absorbed carbon dioxide. (For part E, see color plate.) (F) Transgranular stress corrosion of aluminum brass (Cu–2 wt% Al) in steam containing traces of ammonia. From E. D. D. During, *Corrosion Atlas: A Collection of Illustrated Case Histories*, Elsevier, Amsterdam (1988).

B

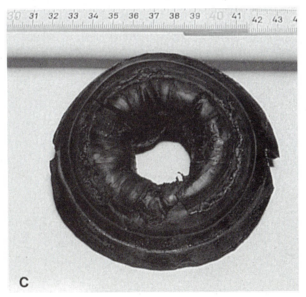

C

FIGURE 10-11 *(continued)*

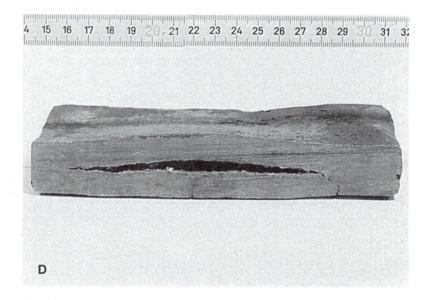

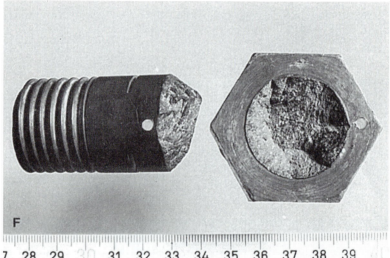

FIGURE 10-11 (continued)

10.2.4.1. Chlorine

Chlorine in salt form is not only a relatively abundant element in marine and industrial environments, but a chemically reactive one. Stainless steels and aluminum alloys are particularly susceptible to chloride attack, as seen in Figs.

10-11A and B. Through reaction of salt and water enough chloride ion is released on metal surfaces to cause either localized **pitting corrosion** in the case of stainless steel or uniform attack in the case of aluminum. Examples of similar chloride attack of aluminum and gold employed in integrated circuits are discussed in Section 15.5.5 together with appropriate chemical reactions. Metals are not the only materials that suffer from the presence of chlorine. Absorption of HCl caused the rubber gasket shown in Fig. 10-11C to swell and crack.

10.2.4.2. Hydrogen

Hydrogen is another very harmful element in virtually all of the major metals. In fact, great efforts are made to degas steel, aluminum, and copper alloy melts prior to mechanical forming operations. Despite this, hydrogen enters metals from ubiquitous corrosive environments and can cause blisters and cavities in extreme cases (Fig. 10-11D). Hydrogen probably dissolves in the metal in atomic form, and the eventual formation of H_2 is accompanied by development of internal stresses and microcrack nucleation. The actual mechanisms remain elusive as the evidence evaporates when the cracks open. In the presence of external stress, hydrogen can severely embrittle steels, a subject addressed again in Section 10.5.4.

10.2.4.3. Ammonia

Environments containing ammonia gas and solutions of ammonium ions are common in industry. Brasses are susceptible to ammonia corrosion particularly in the presence of stress. Although the progressive buildup of blue-colored corrosion products containing the complex $Cu(NH_3)_4^{2+}$ ion occurred in the valve (Fig. 10-11E), sudden brittle fracture occurred in the bolt (Fig. 10-11F). Stress accelerated the latter, and crack propagation occurred across grains (transgranular) with the assistance of corrodent penetration.

10.2.5. Polarization and Corrosion Kinetics

10.2.5.1. Polarization Phenomena

Some metals become **passive** or resistant to corrosion in a given electrolyte at times, but dissolve rapidly in it under other conditions. Why? Inhibitors added to electrolytes are effective means of retarding corrosion. How do they work? A closer look at electrode reaction rates is the clue to understanding these and other important electrochemical phenomena. Reversible reactions at single electrodes, like that described by Eq. 10-1, occur at the same rate in either direction. Metal atoms can be oxidized to ions and, conversely, ions can be discharged as metal during reduction. As charge is transferred between electrode and electrolyte in both cases, we may imagine that electron current densities (A/cm^2) $i(\rightarrow)$ and $i(\leftarrow)$ respectively flow in the indicated reaction directions. At equilibrium the electrode potential is E° and no net current flows, that is, $i(\longrightarrow) = i(\longleftarrow) = i_{ex}$, where i_{ex} is known as the **exchange current density**.

When a net current density, i_{net}, is drawn from the electrode surface, a bias is introduced. Now the currents do not balance and $i(\longrightarrow) \neq i(\longleftarrow)$. The difference is i_{net}. The equilibrium electrode potential E° splits into two different values depending on current: one for each electrode reaction as schematically shown in Fig. 10-12. Each electrode is said to be **polarized**, and the change in potential is known as the **polarization** (or overvoltage). *The polarization of the anode is always positive and that of the cathode is always negative.* Reaction rate theory introduced in Section 6-6 can explain these effects if it is recognized that i is proportional to corrosion rate $\dot{R}_c$. The overvoltage η (in volts) is a kind of driving force that can either accelerate or inhibit the electrode reaction rate. Lowering the effective activation energy enhances the anode reaction rate; similarly, the cathode reaction is suppressed because the effective activation energy is increased:

Anode: $\dot{R}_c \sim \exp(-qE^\circ + nq\eta)/RT$ or $i(\longrightarrow) \sim \exp(nq\eta/RT)$
$$(10\text{-}14\text{a})$$

Cathode: $\dot{R}_c \sim \exp(-qE^\circ - nq\eta)/RT$ or $i(\longleftarrow) \sim \exp(-nq\eta/RT).$
$$(10\text{-}14\text{b})$$

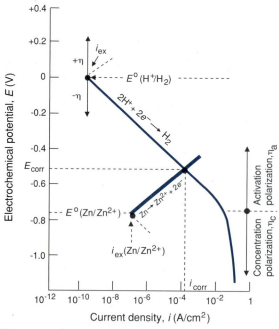

FIGURE 10-12 Polarization effects at zinc (anode) and hydrogen (cathode) electrodes.

These equations are commonly recast into the form

$$\eta = c_1 \pm c_2 \log i \quad \text{or} \quad E = c_3 \pm c_4 \log i, \qquad (10\text{-}15)$$

where c_1, c_2, c_3, and c_4 are constants.

The polarization behavior displayed as lines in Fig. 10-12 generally follows the trends suggested by Eq. 10-15. Two additive contributions to polarization can often be discerned in these plots: **activation** polarization, η_a, and **concentration** polarization, η_c.

Activation Polarization.   As the name implies, activation polarization occurs because mass transfer at the electrode–electrolyte interface is limited by the *activation* energy barrier of a slow reaction step. Thus, to reduce hydrogen ions to hydrogen gas at a zinc cathode, via the reaction $2H^+ + 2e^- \longrightarrow H_2$, the following sequential steps must be successfully negotiated:

1. Hydrogen ions must migrate through the electrolyte.
2. Reduction to neutral hydrogen atoms (H°) must occur at the Zn surface.
3. Two hydrogen atoms must combine to form a hydrogen gas molecule.
4. Hydrogen bubbles must nucleate and be evolved.

The slowest of these steps will control the electrode reaction.

Concentration Polarization.   Slow diffusion of ions through the electrolyte is the cause of concentration polarization, which is frequently observed at cathodes. In the case of hydrogen, for example, discharge at the electrode–electrolyte interface may be rapid enough, but the slow step involves the difficulty in replenishing the H^+ ions consumed. An electrolyte layer depleted of H^+ ions builds up at the cathode interface, slowing electrolytic action. High discharge rates and low H^+ concentrations are conducive to this type of polarization.

When two half-cells are joined together to form a battery, polarization has the undesirable effect of robbing the battery of energy by reducing the output voltage. On the other hand, polarization is very beneficial when it comes to corrosion; without it our metal structures would rapidly decompose into rubbles of corrosion products.

Corrosion reactions can now readily be understood in terms of combinations of polarization curves for the two involved electrode reactions. In Fig. 10-13 the corrosion of iron, by any of the mechanisms and forms discussed earlier, is schematically represented. Oxidation of Fe to Fe^{2+} occurs simultaneously with the reduction of H_2O to OH^- ions. This only happens at *one* specific cell operating point characterized by corrosion potential E_{corr} and corrosion current density i_{corr}. If we wish to reduce the rate of corrosion it is clear that lowering the value of i_{corr} is a strategy that will work.

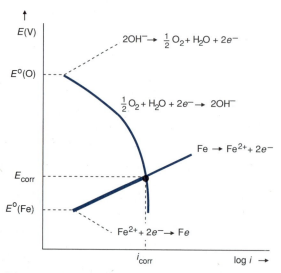

$$2OH^- \rightarrow \tfrac{1}{2}O_2 + H_2O + 2e^-$$

$$\tfrac{1}{2}O_2 + H_2O + 2e^- \rightarrow 2OH^-$$

$$Fe \rightarrow Fe^{2+} + 2e^-$$

$$Fe^{2+} + 2e^- \rightarrow Fe$$

FIGURE 10-13　Schematic polarization diagram representing corrosion of Fe in water containing oxygen. The cathode reaction involves oxygen reduction.

10.2.5.2. Inhibitors and Passivation

Adding **inhibitors** is practiced widely to slow corrosion reactions. Anodic inhibitors (e.g., CrO_4^{2-}, NO_2^-) reduce i_{corr} by raising the anode polarization curve in Fig. 10-14A; alternatively, the more common cathodic inhibitors (e.g., zinc salts, phosphates) lower the cathode polarization curve in Fig. 10-14B. The cooling system of cars is protected by inhibitor formulations typically containing silicates, borates, molybdates, nitrates, and nitrites that are added to the glycol antifreeze.

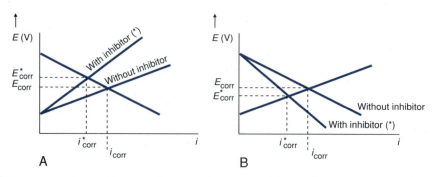

FIGURE 10-14　Effect of inhibitors on the polarization curve: (A) Anodic inhibitors. (B) Cathodic inhibitors.

Certain metals exhibit interesting polarization effects when the electrode potential is increased to more noble (cathodic) values. At low potentials the electrode is **active** as shown in Fig. 10-15, behaving in the manner previously described. When the so-called **passivation potential** (E_p) is exceeded, i_{corr} drops by several orders of magnitude and the metal becomes passive. A thin surface film, usually an oxide of the base metal, forms and protects it from further attack. Passivation is induced by applying anodic currents to the surface because a positive polarization develops in this way. Dipping iron into a strong oxidizing agent like HNO_3 will, surprisingly, passivate it and halt corrosion. But if the protective film is destroyed, or the electrode potential raised still more, a sudden increase in current occurs. The metal becomes active again. In the latter case the metal enters the so-called *transpassive region* of behavior. Stainless steel as well as aluminum, nickel, and titanium alloys also exhibit passivity.

10.2.6. Preventing Corrosion through Design

This chapter has focused attention on the role of materials and their selection when combating corrosion in different environments. But such considerations will go for naught if the structural design is not made correspondingly corrosion safe. A number of simple examples illustrating both *recommended* and *not recommended* design practice are shown in Fig. 10-16 without comment. It is left as a challenge to indicate reasons for the poor design.

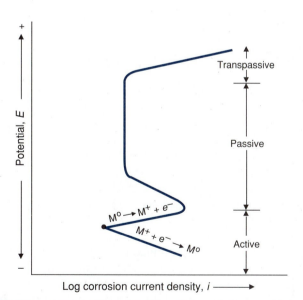

FIGURE 10-15 Polarization curve illustrating active, passive, and transpassive behaviors. Passivity is characterized by low corrosion current densities.

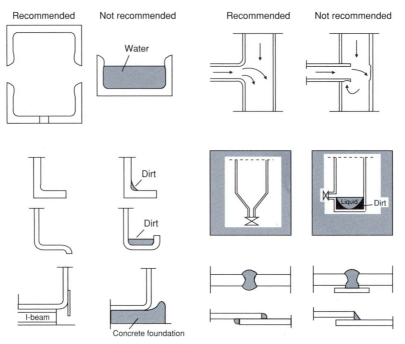

FIGURE 10-16   Recommended and not recommended structural design features from the standpoint of corrosion. From E. Mattson, *Basic Corrosion Technology for Scientists and Engineers,* Wiley (Halsted Press), Chichester (1989).

10.3. GASEOUS OXIDATION

Negative free energy changes for metal oxide formation (see Fig. 5-2) make it abundantly clear that a universal response of metal surfaces exposed to an oxygen-bearing atmosphere is to oxidize. Although oxide formation occurs during **wet** corrosion processes, the oxides considered in this section are the products of **dry** oxidation in elevated-temperature oxygen ambients. The oxidation product may be a thin adherent film that protects the underlying metal from further attack, or it may be a thicker porous layer that flakes off and offers no protection. In a broad sense both types of oxides grow by similar mechanisms involving mass transport of metal and oxygen ions.

Consider the formation of an oxide layer in an oxygen gas ambient as shown in Fig. 10-17. Two simultaneous ion transport processes occur during oxidation. At the metal–oxide interface neutral metal atoms lose electrons and become ions that migrate through the oxide to the oxygen–oxide interface. The released electrons also travel through the oxide and serve to reduce oxygen molecules to oxygen ions at the surface. These oxidation–reduction reactions are simply

$$M^\circ \rightarrow M^{n+} + ne^- \qquad \text{(at metal–oxide interface)} \qquad (10\text{-}16a)$$

and

$$\tfrac{1}{2}O_2^\circ + 2e^- \rightarrow O^{2-} \qquad \text{(at oxygen–oxide interface)}. \qquad (10\text{-}16b)$$

Ionic mass transfer through a solid oxide (electrolyte) is very much slower than through a liquid electrolyte. This is why elevated temperatures are required to generate oxide layers of appreciable thickness. If the metal cations migrate more rapidly than oxygen anions the oxide will grow at the outer surface. Oxides of Fe, Cu, Cr, and Co grow in this way. On the other hand, oxide forms at the metal–oxide interface when oxygen migrates more rapidly than metal. This is the case in oxides of Ti, Zr, and Si. The necessity for electron motion also has an important bearing on oxide growth. In good insulators like SiO_2 and Al_2O_3 electron mobility is low so that the oxidation–reduction reaction is inhibited. As a result very thin protective native oxides, often less than ~ 3 nm thick, form, which then stop growing.

The model developed for the oxidation of silicon (see Section 6.2.3), together with Eqs. 6-9 and 6-10, in particular, is also applicable to metals. Both parabolic growth under diffusion-controlled conditions and linear oxide growth when interfacial reactions limit oxidation are frequently observed; however, not all oxidation processes fit these transport mechanisms and other growth rate laws have been experimentally observed in specific temperature and oxygen concentration ranges. For example, cubic growth ($x_0^3 = C_1t + C_2$, with C_1 and C_2 constant, and t the time) and logarithmic growth ($x_o = C_3 \ln[C_4t + C_5]$, with C_3, C_4, and C_5 constant) rates for oxidation are just two of several that are observed.

The physical integrity of the oxide coating is the key to whether the underlying metal will suffer further degradation or not. If the oxide is thin and dense, then it can generally be tolerated. But, if it is porous and continues to spall

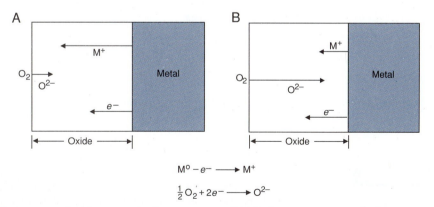

FIGURE 10-17 Oxidation of solids as a result of ionic transport. (A) Metal migrates more rapidly than oxygen in the oxide. (B) Oxygen migrates more rapidly than metal in the oxide.

off, the exposed metal surface will suffer further deterioration. Whether the oxide is dense or porous can frequently be related to the ratio of the oxide volume produced to the metal volume consumed when it oxidizes. The quotient, known as the **Pilling–Bedworth ratio** (PBR), is given by

$$\text{oxide volume/metal volume} = (M_o/a\rho_o)/(M_m/\rho_m) = M_o\rho_m/aM_m\rho_o, \tag{10-17}$$

where M and ρ are the molecular weight and density, respectively, of the oxide (o) and metal (m), and a is the number of metal atoms per oxide molecule, M_aO_b. If the ratio is less than 1 then lack of compatibility with the metal may cause the oxide to split, much like dried wood, and afford little protection to the metal underneath. Oxygen has direct access to the metal surface where chemical reaction will readily cause the oxide to grow linearly in time. Oxides of the alkali metals Li, Na, and K are of this type. On the other hand, if the PBR is between 1 and 2 the oxide will tend to be coherent with the metal and be protective. Further growth can occur only by ionic diffusion in this case. But, if the PBR is too large (i.e., >2) the oxide may be subjected to large compressive stresses causing it to buckle and flake off.

In addition to the PBR, another important consideration in assessing oxide (or nitride, carbide, sulfide, etc.) coating integrity is the difference in thermal

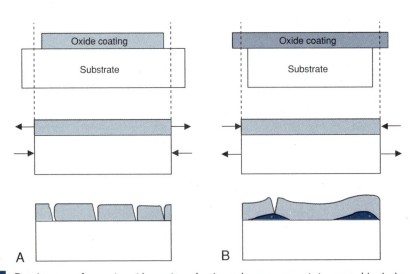

FIGURE 10-18 Development of stress in oxide coatings. At elevated temperatures it is assumed both the coating and metal substrate are unstressed. (A) If $\alpha_o > \alpha_m$, the oxide contracts more than the substrate and it will be residually stressed in tension. Cracking of the oxide may occur in this case. (B) If $\alpha_m > \alpha_o$, the greater contraction of the metal forces the oxide into compression and possible delamination.

expansion coefficient between it (α_o) and the substrate metal (α_m). Assume that metal and oxide are unstressed when the latter forms at elevated temperatures. If $\alpha_o > \alpha_m$, then the layer will contract more than the metal on cooling to ambient temperature. But compatibility requires that the oxide and metal have a common length. A compromise is struck and the layer is placed in tension with the substrate in compression as shown in Fig. 10-18A. If, however, the reverse is true of the expansion coefficients, the stresses reverse. The latter is generally the case for many practical oxides (Fig. 10-18B). If the residual stresses exceed the fracture stress, either cracking or buckling of the oxide may occur.

EXAMPLE 10-3

a. Assume that iron can oxidize to either FeO, Fe_2O_3, or Fe_3O_4 at the same temperature in different oxygen atmospheres. What is the value of the PBR for each oxide?

b. When Fe is exposed to O_2 at 800°C what oxides form and in what sequence?

ANSWER The following table includes relevant data and calculated PBR values.

	M	$\rho(g/cm^3)$	a	PBR
Fe	55.85	7.86	—	—
FeO	71.85	5.7	1	1.77
Fe_2O_3	159.7	5.24	2	2.14
Fe_3O_4	231.5	5.18	3	2.10

As a sample calculation, substitution in Eq. 10-17 yields PBR $(Fe_2O_3) = (159.7)(7.86)/2(55.85)(5.24) = 2.14$. Similarly for the other PBR values.

b. Reference to the Fe–O phase diagram (Fig. 10-19) reveals that Fe_2O_3 is the most oxygen-rich oxide, but it is successively reduced to Fe_3O_4 and finally to FeO by the availability of Fe. This same sequence of oxides is observed during the development of scale when steel is hot rolled. At the atmosphere–scale surface the oxide is Fe_2O_3. Next comes Fe_3O_4 and at the Fe interface the stable oxide is FeO or wüstite. It is typically found that FeO constitutes about 85% of the scale thickness, Fe_3O_4 10 to 15%, and Fe_2O_3 only 0.5 to 2%.

What has been said about oxidation applies equally well to sulfidation of metals in SO_2 or H_2S ambients. Metal sulfides are particularly deleterious because of their low melting temperatures. Liquid sulfide films tend to wet grain boundaries, making metals susceptible to intergranular cracking or hot shortness during elevated temperature processing (see Section 8.3.5).

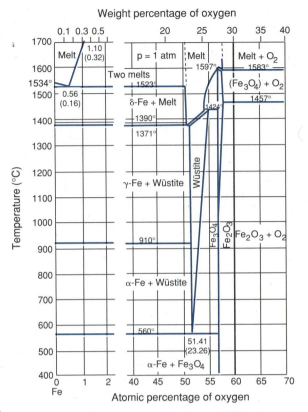

Portion of the equilibrium phase diagram of the Fe–O system.

10.4. WEAR

10.4.1. Introduction

The subject of wear is concerned with interacting material surfaces that are in frictional contact and in relative motion. Wear is one component of a relatively new and interdisciplinary field of scientific and engineering study known as **tribology**. Derived from the Greek word *tribos* meaning "rubbing," tribology is concerned with adhesion, hardness, friction, wear, erosion, lubrication, and related phenomena. Both friction and wear effects are not only limited to engineering systems. They arise in the movement of animal joints and lead to arthritis and degradation of bone–socket interfaces, necessitating replacement prostheses like that shown in Fig. 1-10. But we also rely on controlling friction when we walk or participate in sports such as rock climbing, ice skating, and skiing.

Since the dawn of civilization humans have struggled to overcome friction when moving heavy loads. Consider the Egyptian bas-relief of Fig. 10-20. It dates to 1880 BC and illustrates the use of rollers and sledges to facilitate transport of a large statue, estimated to exert a force of 6×10^5 N, along a wooden track. Closer examination reveals that a man, perhaps one of the first lubrication engineers, is pouring a liquid (oil?) in the path of motion. Assuming that each of the 172 famished slaves pulls with a force of 444 N (100 lb) the coefficient of friction (f), as defined in physics courses, is $f = (172 \times 444)/(6 \times 10^5) = 0.13$. The four-millennia evolution of methods to support and move loads from sliding to rolling friction, from single-contact to multiple-contact wheels, has been accompanied by a corresponding decrease in the coefficient of friction.

Like corrosion, wear degradation involves two different materials and mass loss from usually one of them. But, whereas mass transfer phenomena during corrosion are due to chemical driving forces, wear is produced by interfacial mechanical forces. Wear is of crucial importance in machine components such as ball bearings and computer disk heads where high relative velocities between the contacting surfaces (i.e., ball and race, head and disk) occur. Heads that fly over but very close to the surface of computer disks normally do not contact them except during starting and stopping of the drives. Occasionally, the contact is such that a "crash" occurs, with attendant loss of stored information. In this case as well as in bearings much damage can be produced by very little mass loss.

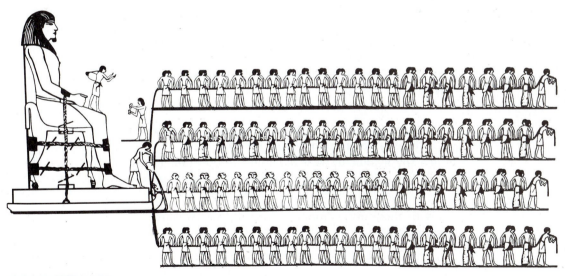

FIGURE 10-20   Egyptian bas-relief dating to 1880 BC, illustrating transport of an Egyptian colossus. From J. Halling, *Principles of Tribology*, Macmillan Press, London (1978).

From an engineering standpoint three separate categories of behavior can be distinguished based on the relative magnitudes of friction and wear.

1. *Friction and wear are both low.* This is the case in bearings, gears, cams, and slideways.
2. *Friction is high but wear is low.* This combination is desired in devices that use friction to transmit power such as clutches, belt drives, and tires.
3. *Friction is low and wear of one body is high.* This is the situation that prevails in material removal processes such as machining, cutting, drilling, and grinding. In these operations the tools also suffer wear and must be periodically sharpened.

10.4.2. Types of Wear

A number of wear mechanisms have been identified and these are discussed in turn.

10.4.2.1. Adhesive Wear

Material surfaces that appear smooth on a macroscopic level actually consist of a distribution of microscopic asperities or peaks. When bodies containing two such surfaces are brought together contact occurs at relatively few asperities. Under loading the stress builds up at the asperities, and above the yield stress they deform and tend to weld together. Continued tangential sliding causes the welds to shear. If shear occurs at the interface between asperities then there is no wear. More commonly, however, shear occurs away from the interface because the interfacial bond strength exceeds the cohesive strength of one (or both) of the contacting bodies. In this case, material is transferred from one surface to the other, usually from the softer to the harder body. With further rubbing the transferred material becomes detached to form loose wear particles. In severe cases of adhesive wear, smearing, galling, and seizure of the surfaces may occur.

Three facts of adhesive wear are universally recognized:

1. The volume (V) of wear material is proportional to the distance (L) over which relative sliding occurs.
2. The volume of wear material is proportional to the applied load (F).
3. The volume of wear material is inversely proportional to the hardness (H) or yield stress of the *softer* material.

With these observations combined, the Archard expression for adhesive wear emerges as

$$V = KFL/3H, \qquad (10\text{-}18)$$

where K is dimensionless constant known as the wear coefficient. Values of K listed in Table 10-3 are typically less than 0.001. K for steel surfaces decreases

TABLE 10-3	APPROXIMATE VALUES OF THE WEAR COEFFICIENT, *K*, IN AIR

Unlubricated

Mild steel on mild steel	10^{-2}–10^{-3}
60–40 brass on hardened tool steel	10^{-3}
Hardened tool steel on hardened tool steel	10^{-4}
Polytetrafluoroethylene (PTFE) on tool steel	10^{-5}
Tungsten carbide on mild steel	10^{-6}

Lubricated

52100 steel on 52100 steel	10^{-7}–10^{-10}
Aluminum bronze on hardened steel	10^{-8}
Hardened steel on hardened steel	10^{-9}

From S. Kalpakjian, *Manufacturing Processes for Engineering Materials*, 2nd ed., Addison–Wesley, Reading MA (1991).

by many orders of magnitude in the presence of lubricants relative to the unlubricated case. The reason stems from the development of interfacial incompressible lubricant (fluid) films that keep the bearing surfaces from contacting.

EXAMPLE 10-4

In a 2-cm-diameter ball bearing composed of 52100 steel balls and races, how many revolutions would be required to produce a wear volume of 10^{-6} cm^3 under an applied load of 50 kg, if the hardness is 1000 kg/mm^2? Assume metal loss occurs by adhesive wear.

ANSWER   From Table 10-3 the value for *K* is chosen to be 10^{-9}. The relative distance *L* traveled along the race is $2\pi(10)N$, where *N* is the number of bearing revolutions. For a wear volume of $V = 10^{-6}$ cm^3 or 10^{-3} mm^3, Eq. 10-18 yields $10^{-3} = (10^{-9})(50)[2\pi(10) N]/3(1000)$. Solving, $N = 9.55 \times 10^5$.

10.4.2.2. Abrasive Wear

Abrasive wear covers two different situations in practice. Both involve the plowing or gouging out of the softer material by a harder surface. In the first case a rough, hard surface rubs against a softer material as exemplified by the action of a file or an emery paper on a soft metal. This type of wear has been eliminated in modern machinery by minimizing surface roughness. In the second type of abrasive wear, airborne dust, grit, and the products of adhesive or corrosive wear manage to find their way to the interface between the contacting materials. Sealing and filtration are the only practical precaution against the ingress of airborne abrasive particles. The volume of material removed by abrasive wear is given by a formula similar to Eq. 10-18.

10.4.2.3. Fatigue Wear

Fatigue wear results when repeated loading and unloading of the contacting surfaces occur. Unlike adhesive and abrasive wear in sliding contact, which are operative from the moment rubbing commences, surfaces that are lubricated do not immediately suffer such wear. This is the case in well-designed ball and roller (containing cylindrical bearing surfaces) bearings. But these bearings are prone to rolling contact fatigue, which manifests itself after long times if the amplitude of the reversed stresses is large enough. Wear initiates at metallic inclusions, pores, and microcracks, and eventually results in dislodged particles. The testing of large numbers of roller bearings has revealed that their life is inversely proportional to F^3.

10.4.2.4. Corrosive Wear

Corrosive wear requries both corrosion and rubbing. In the corrosive atmosphere, either liquid or gaseous, the corrosion products that form at the contacting surfaces are poorly adherent and may act as abrasive particles to accelerate wear. In the case of oxidation wear, however, protective oxides that are removed may readily regrow, limiting the extent of damage.

10.4.2.5. Fretting Wear

Fretting wear may be viewed as a type of fatigue wear that occurs under conditions of high-frequency vibratory motion of small amplitude (in the range $1-200 \mu$m). Many sequential damage processes occur during fretting including breakup of protective films, adhesion and transfer of material, oxidation of metal wear particles, and nucleation of small surface cracks.

10.4.2.6. Delamination Wear

Delamination wear takes the form of regular detachments of platelike particles from wearing surfaces due to the influence of high tangential (friction) forces in the surface contact zone. During cyclic loading the cracks that develop propagate parallel to the surface at a depth governed by the stress, material properties, and coefficient of friction.

10.4.2.7. Erosion

When a stream of abrasive solid particles is directed at a surface it erodes at a rate that is strongly dependent on the impingement angle ϕ as shown in Fig. 10-21. Furthermore, the wear rate exhibits opposing trends depending on whether ductile (metal) or brittle (ceramic) surfaces are involved. Low impingement angles stretch ductile materials in tension, causing them to fracture, whereas normal impact peening compresses the surface, strengthening it. Conversely, brittle materials suffer more erosion with increasing impingement angle.

Erosion degradation is common in earth moving, mining, and agricultural equipment handling rock, particulates, and slurries. It is also a troublesome

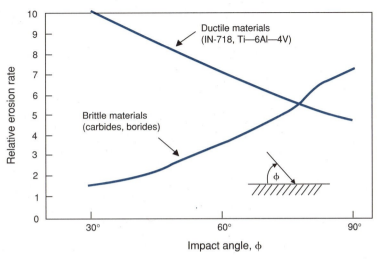

FIGURE 10-21 Relative erosion rates in ductile and brittle materials as a function of particle impact angle, ϕ.

form of wear in gas turbine engines that ingest airborne particulates. Erosion also occurs during electrical arcing and leads to pitting degradation of electrical contacts. When arcing occurs between the copper commutator and brushes of a motor, both metal and brush loss cause the contact to deteriorate. Spark erosion is not always deleterious, however. It is capitalized upon in spark cutting or electrical-discharge machining, a technique employed to cut hard metals through controlled incremental erosion.

10.4.3. Wear in Cutting Tools

Wear of cutting tools has a major impact on the costs sustained by the metal removal industries and figures prominently in the price of machined products. An instant in the high-speed cutting action of a tool against a workpiece is frozen in the depiction of Fig. 10-22A. Severe local stresses are generated to continually shear and fracture the metal forming the chip. The friction generated between the chip and the tool can result in local temperatures approaching 1000°C or more depending on tool and workpiece materials, cutting speeds and feed rates, extent of lubrication, and so on (Fig. 10-22B). Wear craters and material loss, accelerated by abrasive particles, thermal softening, and oxidation and chemical reactions, occur on the tool (rake) face, side or flank face, and nose (Fig. 10-22C). The tool eventually fails and must be either sharpened or replaced because the work dimensions and surface finish no longer meet specifications. In 1907, F. W. Taylor formulated a widely used relationship between steel tool life (T_t) and machining velocity (V_t), namely, $V_t T_t^n = C$, where C and n are constants.

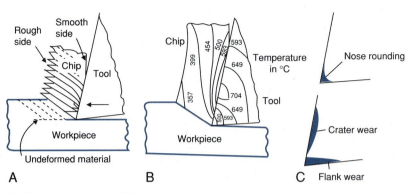

FIGURE 10-22   (A) Cutting action of a tool bit against a metal workpiece in a two-dimensional machining operation. (B) Typical temperature profiles within tool and chip. (C) Tool degradation caused by nose rounding (above), and flank and crater wear (below).

The life of a tool is directly proportional to the product of its fracture toughness (K_{1C}) and hardness. Although the two properties vary inversely with one another, the hundredfold tool life improvement in the past century (see Fig. 1-2) has been realized largely by increasing hardness. Today TiC, TiN, and Al_2O_3 coatings (~5–10 μm thick), used singly or in combination, are the chief means of combating tool wear.

10.5. FRACTURE OF ENGINEERING MATERIALS

This section and succeeding ones on fatigue and creep are of considerable practical importance. The theoretical basis for fracture analysis is the Griffith crack theory, which was presented in Section 7.6.3. Here the discussion takes a practical turn to examine actual manifestations and morphologies of failure and, where possible, the operative mechanisms.

10.5.1. Fracture of Glass Fibers

It is fitting to start with fracture in glass fibers because they were originally studied by Griffith. More importantly, optical fibers composed of ~125-μm-diameter, coaxial silica glass filaments that can be tens of kilometers long have revolutionized communications systems (see Section 13.6). Data on the strength dependence of these fibers as a function of crack size are presented in Fig. 10-23. To withstand high service stresses it is essential that fibers be defect free or, if not, have very small surface flaws. In addition, glass generally exhibits the phenomenon of static fatigue and glass fiber is no exception. This means that if fibers are found to have a proof stress σ_p during a test time t_p, then

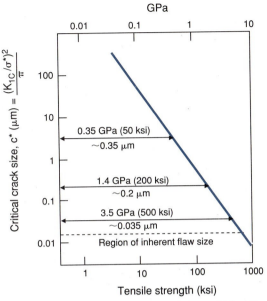

FIGURE 10-23 Critical crack size dependence on strength of optical (glass) fibers. Courtesy of G. Bubel, AT&T Bell Laboratories.

they will safely support an applied stress σ ($\sigma < \sigma_p$) until failure at time t_f, according to the relationship

$$(\sigma/\sigma_p)^n = t_p/t_f,$$ (10-19)

where n is a constant known as the slow crack growth exponent. Static fatigue data for optical fibers are given in Fig. 10-24. Large values of n are desirable because relatively small decreases in strength will occur with time.

EXAMPLES 10-5

   a. How well does Griffith crack theory describe the results displayed in Fig. 10-23?
   b. What is the value of K_{1C} for silica employed in optical fiber?
   c. Using Fig. 10-24 what is the value of the slow crack growth exponent for a fiber that was found to have a proof stress of 207 MN/m²?

ANSWERS   a. Griffith crack theory (Eq. 7-27) suggests that c^* varies as σ^{-2}. Therefore, a plot of log c^* versus log σ should have a slope of -2. Careful measurement with a ruler demonstrates that for every two-orders-of-magnitude drop in c^*, there is very nearly a one-order-of-magnitude increase in σ. Therefore, the experimental data strongly support the Griffith dependence.

b. As $K_{1C} = \sigma(\pi c^*)^{1/2}$, simple substitution and calculation using three pairs of indicated data yield the following:

σ(MPa)	c^* (m)	K_{1C}(MPa-m$^{1-2}$)
350	3.5×10^{-7}	0.367
1400	2.0×10^{-7}	1.11
3500	3.5×10^{-8}	1.16

For fused silica the accepted value is 0.79 MPa-m$^{1/2}$.

c. If logs of both sides of Eq. 10-19 are taken and terms rearranged, $\log t_f = -n \log \sigma + \log t_p + n \log \sigma_p$. The last two terms are constants, and a plot of $\log t_f$ versus $\log \sigma$ has a slope of $-n$. Calculation of the slope yields $n = -(12 - 0)/(\log 53 - \log 130) = 30.8$.

Note that higher proof stress fiber has a smaller n value.

10.5.2. Fracture of Glass

The brittle behavior of glasses and ceramics makes them exemplary illustrative materials from the standpoint of fracture mechanics theory. But theory does not always prepare us for practical fractures. Consider the fracture surface of the glass plate shown in Fig. 10-25. The very flat semicircular region, commonly called the **mirror area**, is characteristic of low-speed fracture in glass parts. It is usually oriented perpendicular to the direction of maximum tensile stress. Furthermore, the mirror area is often associated with a recogniz-

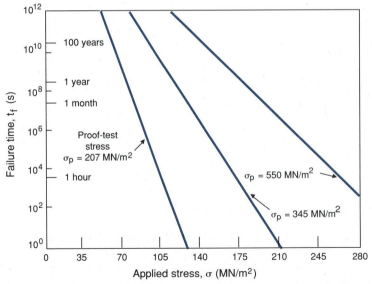

FIGURE 10-24 Static fatigue life of optical (glass) fibers having different proof stresses. From S. E. Miller and A. G. Chynoweth (Eds.), *Optical Fiber Communications*, Academic Press, New York (1979).

able defect or discontinuity in the surface of the glass. Other features commonly observed are so-called **Wallner lines,** which are also associated with low-speed fractures. These lines appear as tiny ridges or ripples in the fracture surface where the crack seems to have stopped, waited for the stress to realign, and then started again. Wallner lines are always concave toward the crack initiation point which can then be located by backtracking along the ripples. Lastly are the **hackle marks,** which are characteristic of high-speed fractures. When cracks travel at speeds higher than 5000 ft/s, these hackle marks are visible along the direction of propagation as tiny fractures in the newly formed surface. They are perpendicular to both mirror and Wallner lines (if present). A magnifying

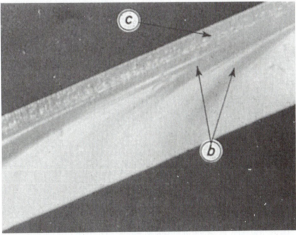

FIGURE 10-25   Fracture of glass plate showing mirror area ⓐ, Wallner lines ⓑ, and hackle marks ⓒ. From R. Matheson, *Why Glass Parts Fail,* General Electric.

glass is frequently sufficient to reveal these telltale signs. Fractures in glass seldom travel at a constant velocity but speed up, slow down, and change direction before stopping.

10.5.3. Ductile Fracture of Metals

Metals fracture in both ductile and brittle modes and sometimes in a combined ductile–brittle manner, as schematically indicated in Fig. 10-26A. Although these are relative terms, ductile fracture is accompanied by large percentage elongation and extensive plastic deformation in the region of the crack tip. Tensile tests provide an unambiguous way to relate fracture appearance to the nature of loading. Ductile metals neck and display the common **cup–cone** fracture morphology shown in Fig. 7-5. The process of ductile fracture may be broadly viewed in terms of the sequential microvoid nucleation and growth processes, schematically indicated in Fig. 10-26B. Shrouded in mystery, the earliest stage of fracture spawns isolated microscopic cavities. These nucleate heterogeneously at inclusions, second-phase particles, and probably at grain boundary junctions. Microvoid coalescence then yields a flat elliptical crack that spreads outward toward the periphery of the neck. Finally, an overloaded outer ring of material is all that is left to connect the specimen halves, and it fails by shear. Further examination reveals equiaxed or spherical **dimples** on the flat crater bottom loaded in tension and elongated ellipsoidal dimples on the shear lips oriented at 45° (Fig. 10-27). In high-purity FCC and BCC metals

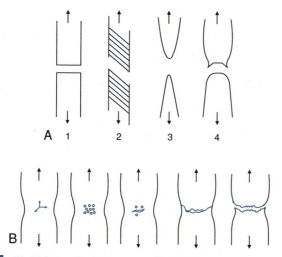

FIGURE 10-26   (A) (1) Brittle fracture in single and polycrystals. (2) Shear fracture in ductile single crystals. (3) Ideal ductile fracture in polycrystals. (4) Ductile cup–cone fracture in polycrystals. (B) Stages in the development of a cup-cone fracture caused by void coalescence. From *Radex Rundschau*, **3/4** (1978).

free of inclusions, however, necking to ~ 100% area reduction (i.e., to a point) is possible.

10.5.4. Brittle Fracture of Metals

Brittle fracture is the most feared of all. It often occurs under static loading without advance warning of impending catastrophe as, for example, in the Liberty Ship fractures and numerous recorded failures of very large liquid (e.g., molasses, oil) storage containers made of steel. Brittle fractures in metals propagate rapidly and in their wake leave relatively flat surfaces behind that can often be fitted together. At low magnification, arrow-shaped chevron markings are frequently evident on the fracture surface and these often conveniently point back to the point of failure initiation. Chevron patterns actually represent a transition morphology between brittle **cleavage** fracture at low temperatures and ductile shear fracture at higher temperatures, as the ship plate fractures in Fig. 10-28 demonstrate.

Higher magnification reveals that **transgranular** cracking often occurs during brittle fracture, especially at low temperatures. In such cases the crack propa-

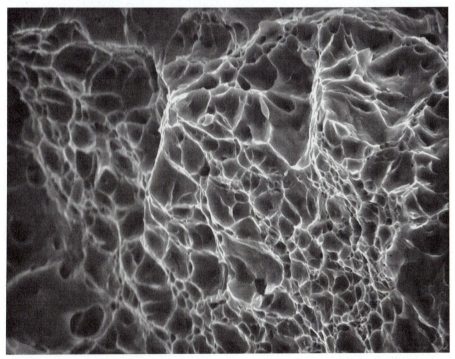

FIGURE 10-27 SEM image of dimples in the ductile tensile fracture of a stainless steel (3920×). Courtesy of G. F. Vander Voort, Carpenter Technology Corporation.

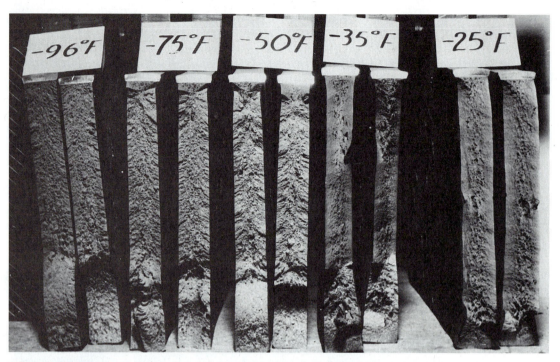

FIGURE 10-28   Chevron features in the fracture surface of ship plate steels tested at various temperatures. At −50°F the chevrons are most distinctive. Courtesy of G. F. Vander Voort, Carpenter Technology Corporation.

gates *across* and through the grain interiors. The mechanism by which transgranular fracture occurs is cleavage, a process that occurs preferentially along specific crystallographic planes. Under stress it is easier for planes to separate through bond breaking than to deform by slip. Cleavage is observed in BCC as well as HCP metals and is quite common in ionic and covalent materials. Characteristics of such fractures include flat facets, like those seen in a 2.5% Si–Fe alloy that fractured at −195°C (Fig. 10-29A). River patterns are also common on the cleaved surfaces. Low temperatures and high strain rates are conducive to transgranular fracture. Fatigue cracks often propagate in a transgranular mode.

When cracks propagate *along* grain boundaries then we speak of **intergranular** fracture. Such failures (Fig. 10-29B) reveal clearly outlined grains that appear to be deeply etched or stand out in relief like rock candy. Grain boundary decohesion between contiguous grains, caused by the presence of harmful segregated impurities and second-phase particles or by penetration of aqueous corrodents, is a common feature. Intergranular embrittlement is also promoted by elevated temperature creep and stress corrosion. Such damage is generally made more visible by polishing flat metallographic sections that contain the

FIGURE 10-29 (A) SEM image of *transgranular* fracture in Fe–2.5 wt% Si tested at −195°C (194×). (B) SEM image of high-temperature *intergranular* fracture in a nickel-base superalloy (49×). Courtesy of G. F. Vander Voort, Carpenter Technology Corporation.

intergranular cracks (Fig. 10-30). Auger spectroscopy (see Section 2.3.2) has often detected the presence of elements such as sulfur and phosphorus in postmortem failure analyses of embrittled steel grain boundaries.

Brittle fracture attributed to hydrogen has been accepted as one of the leading causes of failure in high-strength steels. Known as **hydrogen embrittlement**, the cracks that nucleate where excess (atomic) hydrogen accumulates propagate in an intergranular fashion. Although it may be incorporated during the melting of steel, hydrogen more often penetrates from aqueous and corrosive environments. Hydrogen is apparently drawn to incipient or preexistent grain boundary cracks, stabilizing them against the tendency of the surrounding metal to blunt their advance. Softer, tougher steels are generally more resistant to hydrogen embrittlement.

10.5.5. Failure and Fracture of Polymers

Fracture of thermosetting polymers is much like that of other brittle solids. In the presence of surface flaws or sharp notches, critical levels of stress act to sever covalent bonds and cause fracture. But fracture of thermoplastic poly-

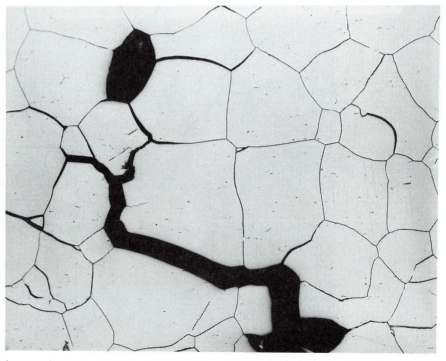

FIGURE 10-30   Intergranular stress cracking in a nickel-base alloy exposed to a sulfur-containing atmosphere at 1120°C (97×). Courtesy of G. F. Vander Voort, Carpenter Technology Corporation.

mers is different and can be brittle, ductile, or some admixture of the two, in nature. It all depends on the service temperature, T, relative to the glass transition temperature, T_G. When $T < T_G$, brittle fracture occurs, whereas more ductile fracture modes are evident when $T > T_G$. As noted in Chapter 7, tensile (ductile) stretching is viscoelastic in nature, and during nonuniform plastic deformation, the interesting phenomenon of **crazing** occurs. Crazing is a kind of tortuous plastic flow that produces a large number of discontinuities, actually voids, that are the seeds of the eventual failure. Crazes, like that shown in Fig. 10-31, have a number of characteristics:

1. They appear to be surface cracks, but are not in reality.
2. The plane of the craze is normal to the tensile stress direction.
3. They reflect light in a manner similar to that of cracks.

As the localized tangled web of thin polymer fibrils in crazes entraps voids, crazes possess a kind of sponginess. Optimists would say that crazes are a source of toughness. The fact that rubber toughening of commercial thermoplastics depends on the creation and stabilization of large craze fields supports this view. Pessimists, on the other hand, view crazes as the onset of brittle fracture. Surprisingly, the work per unit area required to cause fracture in amorphous thermoplastic polymers is many times larger than that in silica glasses and ceramics. Apparently, it takes more energy to align molecules within crazes than to break covalent bonds.

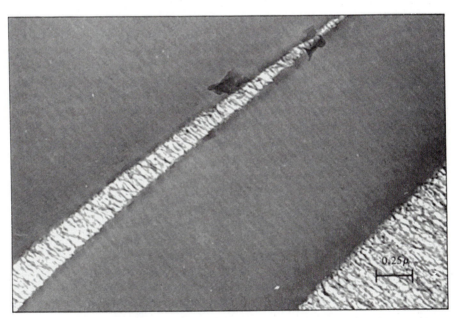

0.25μ

FIGURE 10-31 Transmission electron micrograph of a craze in poly(phenylene oxide). Photo taken by R. P. Kambour, GE Corporate Research and Development.

10.6. ELEVATED TEMPERATURE CREEP DEGRADATION AND FAILURE

Several defect mechanisms for thermal softening and elevated temperature degradation under stress, that is, creep, are depicted in Fig. 10-32. Irrespective of material class, the plastic strain rate in each of the mechanisms varies with stress and temperature according to Eq. 7-29, which is reproduced here:

$$\dot{\varepsilon} = A\sigma^m \exp(-E_c/RT).$$

This equation applies strictly to steady-state creep. But, with little loss in generality, an equation of the sort

$$1/t_R = B\sigma^{m^*} \exp(-E_{CR}/RT) \qquad (10\text{-}20)$$

is expected to also predict creep rupture failure times (t_R). A new set of constants—B, m^*, and E_{CR}, the effective activation energy for creep rup-

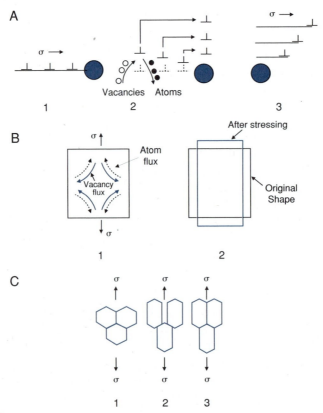

FIGURE 10-32   Microscopic mechanisms of creep: (A) Dislocation climb. (B) Herring–Nabarro creep. (C) Coble creep.

ture—hold. These can, in principle, be determined from accelerated creep rupture testing.

10.6.1. Dislocation Creep

In crystalline metals and ceramics, plastic flow ceases whenever dislocation motion is impeded. Dislocation barriers and impediments such as precipitates, dissolved atoms, and other dislocations that are so effective at low temperatures are not nearly as formidable when the temperature is raised. One reason is due to **dislocation climb**, a mechanism illustrated in Fig. 10-32A. In this process the dislocation sheds its bottom line of atoms by a counter vacancy diffusion flux into the core. The dislocation is incrementally lifted one lattice spacing as a result. Through repeated climb, dislocations vault over obstructions to their motion and, once again, can glide freely on new slip planes. The limiting step is a single diffusive atomic jump, and that is why E_c is often the same as the activation energy for lattice diffusion. And because stress strongly affects dislocation motion, **power law creep** with $m = \sim 5$ to 7 is predicted. Dislocation creep was apparently the operative mechanism in Example 7-8.

10.6.2. Diffusional Creep

Diffusional creep is also diffusion controlled but occurs at low stress levels and does not require dislocations. One version shown in Fig. 10-32B envisions axial stretching of a single crystal under load by diffusion of *vacancies* from the axis to the sides and a counter diffusional current of *atoms*. This so-called *Herring–Nabarro creep* involves diffusion across the crystal, and the resulting strain rate is linearly dependent on stress ($m = 1$).

At lower temperatures and for finer-grained matrices, so-called *Coble creep* is operative. It occurs via low-activation-energy grain boundary paths (Fig. 10-32C), and again the value of m is equal to unity. Stress-driven diffusional creep can generally be understood as a combination of grain elongation, separation, and grain boundary sliding. Creep may also be thought of as an application of the Nernst–Einstein relationship (Eq. 6-29), where velocity is replaced by the strain rate, and force by stress.

10.6.3. Creep Failure

Observations of creep failure often reveal the dominant role grain boundaries play in the phenomenon. Intergranular fracture, shown in Fig. 10-29B, is a leading manifestation of creep damage. Failure is the outgrowth of a number of interrelated processes, for example, cavitation or void nucleation, their growth and linkage at grain boundaries, and viscous sliding of grains relative to adjacent ones. The latter mechanism has been demonstrated by noting offsets in lines scribed across boundaries between grains that have creeped. Employing single-crystal turbine blades to minimize these deleterious effects of grain boundaries during high-temperature and stress service is a popular strategy; superalloy, metal matrix, ceramic matrix, and carbon–carbon composites are alternative options in addressing general creep problems.

10.7. FATIGUE

10.7.1. Morphology and Mechanism of Fatigue

The fracture surface of shafts that have suffered fatigue is distinctive and not hard to recognize. Such a fracture is shown in Fig. 10-33 and consists of three main regions. The first is associated with the fracture initiation site, where local stress concentration (point C) nucleates a tiny circular crack. Next, most of the sectional area is covered with **beachmarks** that spread in concentric circular or semicircular ridges away from the crack nucleus. Beachmarks denote the instantaneous positions of the advancing crack and give the fracture surface a clamshell appearance. Finally, only a small area of the cross section at the shaft keyway remains to support the load, and it ruptures by a ductile overload fracture.

Machined grooves, threads, and deep fillets also serve as crack nuclei. Even dislocation models of crack nucleation, that occur without benefit of a surface flaw, have been proposed. One such mechanism involves the alternate inward and outward glide of dislocations with stress until slip band *intrusions* and *extrusions* develop on the surface. An exploded view of such surface defects is shown in Fig. 7-9C. Once nucleated at an intrusion, the incipient crack grows slowly (perhaps only 10^{-12} cm per cycle on average) across several grains via slip planes. This is *stage I* fatigue propagation, and results in a small, flat, featureless crack that acts as the springboard for further damage.

In *stage II*, a well-defined crack now begins to propagate at a much faster rate ($\sim 10^{-4}$ cm per cycle) perpendicular to the direction of maximum tensile stress. Advance of the crack proceeds by repetitive sharpening and blunting of the crack tip as tension changes to compression. Each stress cycle leaves a **striation** behind that is easily visible in the scanning electron microscope (Fig. 10-34). In forensic investigations these are often analyzed to determine the crack origin and propagation rate. There may be hundreds or even thousands of microscopic striations in a single macroscopic beachmark.

10.7.2. Fracture Mechanics and Fatigue

Observation of discrete striations has supported the fracture mechanics approach to fatigue that was briefly introduced in Section 7.7.4. We now amplify on the predictive capability inherent in this approach. It should be recalled that through experiment the instantaneous crack length c is measured as a function of N, the number of stress cycles. Therefore dc/dN can be

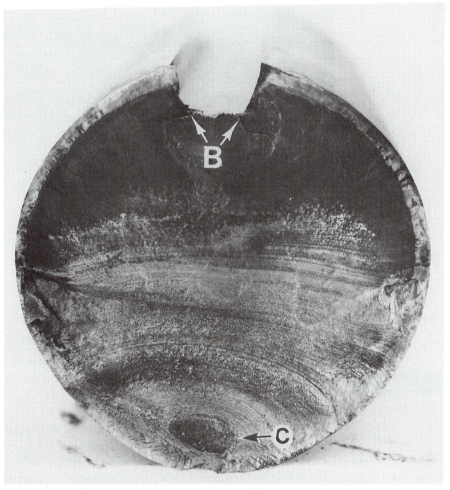

FIGURE 10-33 Fracture surface of a $4\frac{1}{4}$-in.-diameter, 4320 steel drive shaft that has undergone fatigue failure. Courtesy of G. F. Vander Voort, Carpenter Technology Corporation.

extracted as a result. Fatigue–crack growth behavior for a number of metals and nonmetals is reproduced in Fig. 10-35. Despite the limited data it is evident that crack growth rates in polymers and ceramics typically exceed those in metals by several orders of magnitude for the same value of ΔK. The data are consistent with a crack extension that depends on some power of the applied stress. Hence, a frequently employed expression is

$$dc/dN = B(\Delta K)^m = B[\Delta\sigma(\pi c)^{1/2}]^m, \qquad (10\text{-}21)$$

where B and m are constants derived directly from experiment. Physically, the crack growth rate is necessarily proportional to the amplitude of the applied

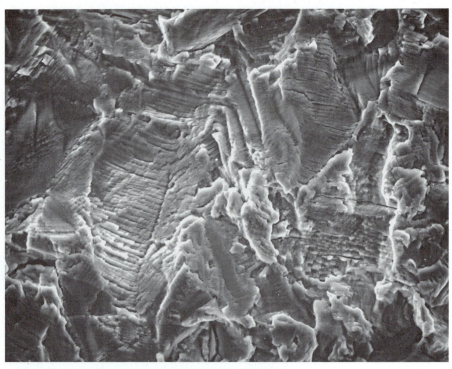

FIGURE 10-34   Fatigue striations in a nickel-base superalloy viewed in the SEM at 2000×. Courtesy of G. F. Vander Voort, Carpenter Technology Corporation.

peak-to-peak stress, $\Delta\sigma$. Explicitly, $\Delta\sigma$ is equal to the difference between the maximum and minimum applied stress and, through Eq. 7-28, is proportional to ΔK. Corrections for cracks of different geometry alter the constants slightly. This differential equation then directly links the crack size to applied stress and material constants. Below the indicated stress levels, cracks do not grow; above them crack growth no longer obeys Eq. 10-21.

If the crack size is initially c_0, the number of cycles, N_f, required to extend it to a final length c_f is easily obtained by first separating variables c and N:

$$\int_0^{N_f} dN = \int_{c_o}^{c_f} \frac{dc}{B[\Delta\sigma(\pi c)^{1/2}]^m}.$$

Direct integration and substitution of limits yield

$$N_f = \frac{2[c_o^{-(m-2)/2} - c_f^{-(m-2)/2}]}{(m-2)\,B\,\pi^{m/2}\,(\Delta\sigma)^m} \qquad (m \neq 2). \qquad (10\text{-}22)$$

Application of this equation will be made subsequently in the failure analysis detailed in Example 10-6.

Two additional equations have proven useful in predicting both low and

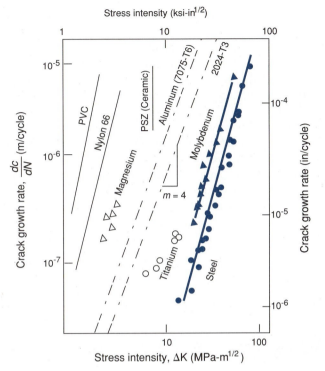

FIGURE 10-35 Fatigue crack propagation in various metals and nonmetals. These data support the power law relationship between dc/dN and ΔK. The data are taken from various sources.

high cycle fatigue life or number of cycles to failure of *uncracked* components. For *low cycle* fatigue, the Coffin–Manson law states that

$$N_f^{0.5} (\Delta \varepsilon_P) = C_1 \qquad (C_1 = \text{constant}), \qquad (10\text{-}23)$$

where $\Delta \varepsilon_P$ is the plastic strain range. This formula applies when the maximum applied stress exceeds the yield stress. On the other hand, when applied stresses do not exceed the yield stress, *high-cycle* fatigue failures are described by Basquin's law:

$$N_f^{\alpha} (\Delta \sigma) = C_2 \qquad (C_2 = \text{constant}). \qquad (10\text{-}24)$$

Constant α lies between ~ 0.06 and 0.12 for many materials. Both equations are not unlike the Taylor formula governing tool life given earlier (i.e., $V_t T_t^n = C$).

10.7.3. Combating Fatigue

Combating fatigue damage is fairly straightforward once it is recognized that the following factors affect fatigue strength adversely:

1. *Stress concentrators.*   Keyways on shafts, holes, abrupt changes in cross sections, sharp corners, and other stress raisers are design features that should be avoided in components subjected to repetitive loading. Stress concentrators mean operation at higher stress levels on the *S–N* curve (Fig. 7-38). This in turn reduces the number of stress cycles to failure.

2. *Rough surfaces.*   Crack nucleation is facilitated at microcrevices and grooves on rough surfaces. Grinding, honing, and polishing of surfaces remove these sources of potential cracks.

3. *Corrosion and environmental attack.*   Localized chemical attack will always reduce fatigue life. Metal removed from exposed grains and grain boundaries leaves a generally rough surface, pits, and corrosion products behind that serve as incipient cracks.

4. *Residual tensile stresses.*   To produce equivalent fatigue damage, less additional load is required when the surface is stressed in residual tension relative to the unstressed state. Therefore, by introducing desirable residual compressive stresses in surface layers by peening, higher levels of applied tension can be tolerated. As a result, fatigue strength and life can be increased.

5. *Low-Strength Surfaces.*   Strengthening surfaces can enhance fatigue resistance; carburization and deposition of hard coatings are ways to achieve this.

10.8. FRACTURE CASE HISTORY

Any component or system failure of consequence, where fracture is involved, will invariably be the subject of one or more engineering reports as to cause. All too often monetary and legal considerations result in conclusions and opinions by "engineering experts" that are biased, contradictory, and inaccurate. It is therefore important to conduct failure analyses that are supported by sound scientific and engineering principles. Furthermore, high ethical standards must be adhered to. The following failure analysis of a fractured helicopter float cross tube illustrates the method.

During a reportedly normal landing, the float cross tube on a helicopter fractured. The helicopter was involved in transporting supplies during oil exploration on Cooke Island in Alaska where the temperature was −18°C. There was no personal injury, but because two similar helicopters suffered cross tube failures, this was a matter of serious concern. A schematic of the helicopter indicating the position of failure and the assumed loading conditions is shown in Fig. 10-36.

The float tube is made of 7075-T6, a precipitation hardening aluminum–zinc alloy. Visual examination of the fractured tube cross section revealed an elliptical flaw (Fig. 10-37), measuring 0.191 cm deep and 0.465 cm along the surface, that was apparently the origin of failure. This conclusion was backed by the observation of fatigue-like striations in the scanning electron microscope. Further examination of the tube clearly showed the presence of shallow surface cracks from processing (0.0013 cm deep) distributed along its entire length.

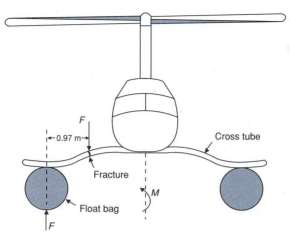

FIGURE 10-36 Schematic of helicopter float tube failure and assumed applied loading. From D. W. Hoeppner, in *Case Studies in Fracture Mechanics* (T. P. Rich and D. J. Cartwright, Eds.), Army Materials and Mechanics Research Center, Watertown, MA (1977).

Hypothesized as stemming from stress–corrosion processes, it was the probable fate of one such incipient crack to grow to critical dimensions and cause this failure.

Stress analysis was performed assuming the loading shown in the free body diagram of Fig. 10-36. The float tube is acted upon by the aircraft load F (assumed to be 2030 lb = 9013 N) and a float bag reaction of the same amount. In static equilibrium these forces are balanced by moment M ($M = Fl$, where l is the moment arm = 0.97 m). Elementary analysis (from statics or strength of materials courses) shows that the average float tube stress σ_a is given by

$$\sigma_a = M/\pi r^2 w = (9013 \text{ N} \times 0.97 \text{ m})/\pi(0.0361 \text{ m})^2 \, (7.95 \times 10^{-3} \, m)$$
$$= 2.69 \times 10^8 \text{ Pa} \, (= 39{,}000 \text{ psi}),$$

where quantities r and w are the tube radius (0.0361 m) and wall thickness (7.95×10^{-3} m), respectively. The actual stress is probably somewhat higher because of impact loading during landing, an effect that conservatively raises the calculated stress by a factor of 1.5. Therefore, a representative average stress adjacent to the flaw is assumed to be 269 MPa $\times$ 1.5 = 403 MPa (58,400 psi).

The last issue to be addressed is whether the observed crack is critical from the standpoint of fracture mechanics. Using the simple surface flaw fracture model, the stress intensity K is given by (Eq. 7-28)

$$K = 0.98\sigma_a \, (\pi c)^{1/2}, \tag{10-25}$$

where the factor 0.98 accounts for the flaw shape and dimensions. If c is taken to be 0.00191 m, K is evaluated to be 30.6 MPa-m$^{1/2}$. This value falls within the accepted range of K_{1C} (i.e., 25–31 MPa-m$^{1/2}$) for 7075-T6.

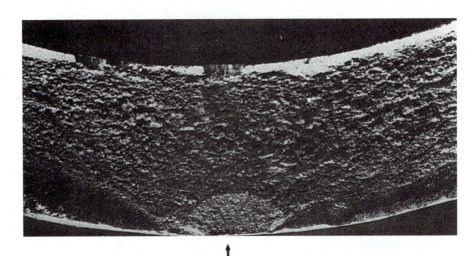

FIGURE 10-37 Photograph of the float tube cross section showing the elliptical surface fatigue crack that probably initiated the fracture. From D. W. Hoeppner, in *Case Studies in Fracture Mechanics* (T. P. Rich and D. J. Cartwright, Eds.), Army Materials and Mechanics Research Center, Watertown, MA (1977).

The conclusion, therefore, is that the crack appears to be of critical size and would undergo rapid fracture. Furthermore, even smaller cracks may be critical under stress levels present in the cross tube.

EXAMPLE 10-6

Making reasonable assumptions, estimate the number of fatigue stress cycles or landings the helicopter could sustain prior to fracture if the initial flaw size were 0.000013 m deep.

ANSWER   For 7075-T6, Fig. 10-35 indicates Eq. 10-21 can be approximately expressed by $dc/dN = 0.7 \times 10^{-10} (\Delta K)^4$ m/cycle. Before Eq. 10-22 can be used, c_f must be known. The observed crack depth is approximately $c_f = 0.00191$ m. As an alternative estimate, the maximum critical crack length prior to fast fracture is $c_f = 1/\pi (K_{1C}/\sigma_o)^2$. From Table 9-2, $\sigma_o = 500$ MPa. Therefore, $c_f = 1/\pi [28$ MPa-m$^{1/2}/500$ MPa$]^2 = 0.00100$ m. All terms have been accounted for except $\Delta\sigma$, which is taken to be 403 MPa. After substitution in Eq. 10-22,

$$N_f = 2(0.000013^{-1} - 0.00191^{-1})/[(2)(0.7 \times 10^{-10}) \pi^2 (403)^4].$$

Evaluation yields $N_f = 4190$ cycles or landings. Note that this calculation strongly depends on the value selected for $\Delta\sigma$. (Using $c_f = 0.0010$ m does not substantially change this result.)

It is not known how many prior landings the helicopter made, but it was in service a total of 1619 hours.

10.9. PERSPECTIVE AND CONCLUSION

Materials generally "fail" because of compositional or structural change brought about by chemical and mechanical driving forces. (Other forces or stimuli can also serve as agents of degradation and failure, for example, radiation, light, electric fields, and some of these are discussed in connection with electronic materials in Chapter 15.)

The potential for aqueous metallic corrosion exists whenever metals acting as cathodes and anodes are surrounded by electrolyte and connected electronically so as to complete an electrical circuit. In such circumstances, **voltages** arise at the electrode–electrolyte interfaces that generate corrosion **currents**. The electrode potentials and corrosion currents are necessarily governed by *Ohm's law*, whereas the link to the amount of chemical reaction or electrode dissolution is made via *Faraday's law*. Similar considerations apply to elevated-temperature oxidation of metals where solid oxide electrolytes are involved. One of the interfaces bounding the oxide essentially behaves as the anode (e.g., the metal–oxide); the cathode is the other interface (e.g., the oxide–ambient). Simple volume considerations roughly predict whether the oxide formed is protective or not.

What makes corrosion so insidious is the many disguises and configurations it assumes. More obvious are the cases of galvanic corrosion because then two different metals are involved. Both the electromotive force and galvanic series can then serve as guides to assess the prospects for corrosion. The more problematical corrosion situations arise when a single metal is involved. Regions of the metal that behave as effective anodes and cathodes can depend on (1) local differences in the metal alone (due to stress and composition variations, structural singularities—grain boundaries, precipitates, etc.); (2) the electrolyte alone (due to different local ionic or oxygen concentrations); or (3) some combination of the two. For example, irons and steels corrode when they are inhomogeneously deformed (e.g., car fenders, nails) or differentially exposed to oxygen, water, or ions (e.g., oxygen starvation, acidic or basic solutions). Strategies to combat corrosion include restricting the use of galvanic couples, keeping the sizes of anodes large and cathodes small, protecting surfaces prone to corrosion (with paint, or Zn, Sn, Cr, etc.), using sacrificial anodes, impressing voltages to counter corrosion potentials, and using inhibitors and passivators.

The mechanical degradation and failure (fracture) mechanisms treated in the remainder of the chapter stem either from forces that act locally at the interface between materials in contact or from forces that are applied at a distance and stress the body throughout. Wear is an example of the first case; creep, fatigue, and fracture exemplify the second category. Although the details are different, at a fundamental level, fracture mechanics is applicable to both of these mechanically functional failures. While fracture in bulk solids is the result of tensile stresses, surfaces undergoing wear are nominally stressed in compression. But parasitic tensile stresses are also operative and they cause particle detachment. Because of the tendency of compressive stresses to seal cracks, wear usually leads not to catastrophic failure, but rather to surface degradation.

The central equation of fracture mechanics, $K_{1C} = \sigma (\pi c)^{1/2}$, very succinctly summarizes the roles that materials **processing, properties, and engineering design** play in influencing susceptibility to fracture. *Processing* or manufacturing is impossible without creating defects of size c in the product. Material *properties* are reflected in the fracture toughness K_{1C}, and *design* stresses are given by σ. When the right-hand side of the equation just exceeds the left-hand side, an unsafe situation exists. Fracture in glass, ceramics, and metals can be profitably modeled by this equation, not only under static, but also cyclic loading conditions. In particular, a basis for predicting fatigue failure times is an outcome of fracture mechanics theory. Raising the fracture toughness, a subject addressed in other chapters, remains a primary materials goal.

Additional Reading

C. R. Brooks and A. Choudhury, *Metallurgical Failure Analysis,* McGraw–Hill, New York (1993).

M. G. Fontana and N. D. Greene, *Corrosion Engineering,* 2nd ed., McGraw–Hill, New York (1978).

D. R. H. Jones, *Engineering Materials 3. Materials Failure Analysis: Case Studies and Design Implications,* Pergamon Press, Oxford (1993).

E. Mattson, *Basic Corrosion Technology for Scientists and Engineers,* Wiley (Halsted Press), Chichester (1989).

Metals Handbook, 8th and 9th eds., Vols. 10 and 11, American Society for Metals, Metals Park, OH (1975 and 1986).

E. Rabinowicz, *Friction and Wear of Materials,* Wiley, New York (1965).

QUESTIONS AND PROBLEMS

10-1. The famous wrought iron pillar of Delhi, India, has not rusted in 1700 years of exposure to the ambient. In addition to iron the 7.2-m-high pillar contains 0.15 wt% carbon and 0.25 wt% phosphorus, and has a magnetic coating that is 50 to 600 μm thick. Provide possible reasons for the rust-free state of preservation.

10-2. A galvanic cell consists of a magnesium electrode immersed in a 0.1 M $MgCl_2$ electrolyte and a nickel electrode immersed in a 0.005 M $NiSO_4$ electrolyte.

These two half-cells are connected through a porous plug. What is the cell electromotive force (emf)?

10-3. Hot water heaters contain magnesium sacrificial anodes. A 0.4-kg anode lasts 15 years before it is consumed. What (continuous) current must have flowed to produce this amount of corrosion?

10-4. It is common to attach brass valves onto iron pipes.
a. Which metal is likely to be anodic?
b. Why do such apparent galvanic couples tend to survive without extensive degradation?

10-5. A current density of 1.1 $\mu A/cm^2$ is measured to flow when iron corrodes in a dilute salt electrolyte. How much loss of iron, in thousandths of an inch, will occur in a year?

10-6. The silver anodes that are used in electroplating silverware corrode and transfer the silver through the electrolyte to deposit on the cathodes (forks, spoons, etc.). How long would it take a kilogram of silver to dissolve or corrode in this way if a current of 100 A flowed through the plating tank?

10-7. Explain the following observations concerning corrosion:
a. Corrosion is more intense on steel pilings somewhat below the water line.
b. A rubber band stretched tightly around a stainless-steel plate submerged in a salt solution will slowly but readily cut through it much like a saw blade.
c. Corrosion is often observed near dents in a car fender.
d. Stainless-steel knives sometimes exhibit pitting in service.
e. Corrosion often occurs in the vicinity of very dilute electrolytes.
f. Two-phase alloys corrode more readily than single-phase alloys.

10-8. Localized pitting corrosion is observed to occur on an aluminum plate that is 1.2 mm thick. The pits are 0.2 mm in diameter and appear to extend into the plate without much widening. A hole appeared after 50 days.
a. What single pit current density was operative in this case?
b. Similar damage developed over an area of 100 cm^2 with a pit density of 25 pits/cm^2. What is the total corrosion current?

10-9. A metal M corrodes in an acid solution and both oxidation and reduction reactions are controlled by activation polarization. The electrode potentials and exchange current densities for the metal and hydrogen electrodes are

$$E_o = -0.6 \text{ V}, \quad i_{ex} = 10^{-7} \text{ A/cm}^2 \qquad \text{for M/M}^{2+}$$

and

$$E_o = 0 \text{ V}, \qquad i_{ex} = 10^{-10} \text{ A/cm}^2 \qquad \text{for H/H}^+.$$

Further, the metal oxidation potential rises by 0.1 V for every 10-fold increase in current density; similarly, the hydrogen reduction potential falls by 0.13 V for a 10-fold increase in current density. What is the value of the corrosion potential? What is the magnitude of the corrosion current?

10-10. In seawater which of the following metals corrodes preferentially: (1) Titanium or steel? (2) Brass or stainless steel? (3) Copper or nickel? (4) Monel or brass? (5) Silver or gold?

10-11. Which of the two half-cells will be anodic when connected via a salt bridge
1. $Zn/0.5\ M\ ZnCl_2$ or $Ni/0.005\ M\ NiCl_2$?
2. $Cu/0.15\ M\ CuSO_4$ or $Fe/0.005\ M\ FeSO_4$?
3. $Mg/1\ M\ MgSO_4$ or $Al/0.01\ M\ AlCl_3$?
4. $Ag/0.1\ M\ AgNO_3$ or $Cd/0.05\ M\ CdCl_2$?

10-12. In a laboratory experiment six different alloys (A, B, C, D, E, F) were tested to determine their tendency to corrode. Pairs of metals were immersed in a beaker containing a 3% salt solution. A digital voltmeter displayed the following potentials between the electrodes with the indicated polarities. (Alloys at the column tops are connected to the positive meter terminal, and alloys at the left side of the rows are connected to the negative terminal.)

	A	B	C	D	E	F
A	0	+0.163	−0.130	−0.390	+0.130	+0.033
B	×	0	−0.325	−0.650	−0.013	−0.098
C	×	×	0	−0.243	+0.163	+0.039
D	×	×	×	0	+0.487	+0.260
E	×	×	×	×	0	−0.013
F	×	×	×	×	×	0

Order the metals from the most anodic to the most cathodic.

10-13. Distinguish between the following terms:
a. Activation polarization and concentration polarization
b. Ohm's law and Faraday's law
c. Electromotive force series and galvanic series
d. Anodic inhibitor and cathodic inhibitor
e. Passive and active electrodes
f. Standard and nonstandard electrodes

10-14. Illustrate through an example the following types of corrosion:
a. Uniform corrosion attack   b. Pitting corrosion
c. Grain boundary corrosion   d. Stress corrosion cracking
e. Crevice corrosion   f. Corrosion fatigue

10-15. Steel screws used as fasteners on aluminum siding surprisingly underwent severe corrosion even though iron is normally cathodic to aluminum. Provide a possible explanation for this occurrence.

10-16. Of the half million bridges in the United States, 200,000 are deficient and, on average, 150 to 200 suffer partial or total collapse each year. Corrosion of both steel and concrete is responsible for many failures. Suggest some details of actual damage mechanisms. The March 1993 issue of *Scientific American* discusses this important problem.

10-17. The corrosion current density in a dental amalgam (an alloy containing mercury, here Ag–Hg) filling is $1\ \mu A/cm^2$.
a. How many univalent ions are released per year if the filling surface area is $0.12\ cm^2$?
b. What is the metal thickness loss if the density and the molecular weight of the filling are $10\ g/cm^3$ and 160 amu, respectively?

10-18. Prove that the ratio of the volume of oxide produced by oxidation to that of the metal consumed by oxidation = PBR = $M_o\rho_m/aM_m\rho_o$ (Eq. 10-17).

10-19. Predict the relative protective nature of Al_2O_3 and AlN coatings on Al and TiO_2 and TiN coatings on Ti. Assume the following densities: ρ (Al_2O_3) = 3.97 g/cm^3, ρ (AlN) = 3.26 g/cm^3, ρ (TiO_2) = 4.26 g/cm^3, ρ (TiN) = 5.22 g/cm^3.

10-20. According to the Pilling–Bedworth ratio which of the following oxide coatings are expected to be protective: (a) UO_2 on U? (b) U_3O_8 on U? (c) ThO_2 on Th? Note that the densities are ρ (UO_2) = 11.0 g/cm^3, ρ (U_3O_8) = 8.30 g/cm^3, and ρ (ThO_2) = 9.86 g/cm^3.

10-21. According to the Pilling–Bedworth ratio which of the following grown films are expected to protect the underlying semiconductor? (a) SiO_2 on Si? (b) GeO_2 on Ge? (c) Si_3N_4 on Si? Note that ρ (Si) = 2.33 g/cm^3, ρ (SiO_2) = 2.27 g/cm^3, and ρ (Ge) = 5.32 g/cm^3.

10-22. For a precision bearing to lose no more than 10^{-11} cm^3 when its lubricated hardened steel surfaces contact each other for a total length of a mile, what is the maximum load that should be applied to it? Assume a hardness of 1100 kg/mm^2.

10-23. A tungsten carbide tool slides over an unlubricated mild steel surface that has a hardness of 1000 MPa. What distance of travel between the two contacting surfaces is required for a volume loss of 0.0001 cm^3 by adhesive wear if a load of 50 kg is applied?

10-24. Comment on the relative magnitudes of friction and wear that are desired between the following contacting surfaces:
1. A diamond stylus and the grooves of a plastic phonograph record
2. Ball and race contact in ball bearings
3. A pickup head flying over (and contacting) the surface of a computer disk
4. A high-speed steel drill and the aluminum plate being drilled
5. Brake shoe and drum in a car
6. Copper and copper in an electrical switch
7. Boot sole and rock during mountain climbing
8. Ski and snow

10-25. When glass is loaded in a vacuum instead of the ambient, static fatigue is largely suppressed. Suggest a reason why.

10-26. Surfaces of thermally toughened (tempered) windshield glass have a residual compressive stress while the interior is stressed in tension. When the thickness of the compressed layers is thin, surface damage due to particle impact or deep scratching has led to catastrophic fracture of the glass. Why?

10-27. Improvement in a turbine blade alloy composition enables it to operate at a stress of 25,000 psi instead of under the former conditions of 22,000 psi maintained at 1100°C. Alloying did not change the 150 kJ/mol activation energy for creep, nor the m = 5.7 stress exponent value. What blade operating temperatures are now possible if the steady-state creep strain rate is to remain unchanged?

10-28. Assume the same stress develops in WC and Al_2O_3 tool bits during identical machining processes. As crack formation is a cause of tool bit failure what is the ratio of the critical flaw dimensions in Al_2O_3 that can be tolerated relative to WC?

10-29. The fracture toughness of a given steel is 60 MPa-m$^{1/2}$ and its yield strength is given by σ_o (MPa) = $1400 - 4T$, where T is the temperature in degrees Kelvin. Surface cracks measuring 0.001 m are detected. Under these conditions determine the temperature at which there may be a ductile–brittle transition in this steel.

10-30. The quenched and aged Ti–6Al–4V alloy used in the Atlas missile has yield points of 229 ksi at $-320°F$, 165 ksi at $-70°F$, and 120 ksi at $70°F$. Suppose small cracks 0.02 in. long were discovered on cryogenic storage containers made from this alloy whose fracture toughness is 50 ksi-in.$^{1/2}$. Estimate the ductile-to-brittle fracture transition temperature of this alloy. Would it be safe to expose these containers to liquid nitrogen temperatures (77 K)?

10-31. A stainless steel with a yield stress of 350 MPa is used in Dewars that hold liquid nitrogen. Due to poor design the steel is constrained from contracting and sustains a 150°C temperature difference when filled with refrigerant. If the coefficient of thermal expansion is 9×10^{-6} °C^{-1}, what thermal stress is produced? Fatigue failure becomes a potential problem when the thermal stress exceeds half the yield stress. Is there reason to worry about thermally induced fatigue damage?

10-32. A highly strained component of the steel whose low cycle fatigue properties are given in Example 7-9 fails in service after 70 strain reversals.
   a. What practical treatment would you recommend to extend life?
   b. If another application failure occurs after 10^6 reversals, what treatment would you recommend to prolong fatigue life in this case?

10-33. A person weighing 91 kg (200 lb) broke a leg and a 316 stainless-steel bone pin was screwed into both halves of the broken bone to help them mend together. The pin failed by fatigue, undergoing an estimated 10^6 stress cycles prior to fracture. Calculate the operative stress range on the bone pin given the following information: For 316 stainless steel $B = 6.22 \times 10^{-21}$ and $m = 3.38$. The fracture cross-sectional area is measured to be 0.613 cm^2. $c_i = 0.0254$ cm (the size of surface scratches) and c_f is measured to be 0.122 cm. Is the calculated stress reasonable? Explain

10-34. It is suggested that incandescent light bulb life is governed by low cycle thermal fatigue. If the thermal strain is induced by a 2000°C rise or drop in temperature every time the tungsten filament is turned on or off, predict the number of cycles to failure. Assume C_1 in Eq. 10-23 is 0.2.

10-35. An aluminum aircraft alloy was tested under cyclic loading with $\Delta\sigma = 250$ MPa and fatigue failure occurred at 2×10^5 cycles. If failure occurred in 10^7 cycles when $\Delta\sigma = 190$ MPa, estimate how many stress cycles can be sustained when $\Delta\sigma = 155$ MPa.

ELECTRICAL PROPERTIES OF METALS, INSULATORS, AND DIELECTRICS

11.1. INTRODUCTION TO ELECTRICAL CONDUCTION IN SOLIDS

In this chapter and the next our attention turns to the electrical properties of materials. These are the properties of concern in electrical power generation and transmission facilities, consumer electronics, and computer and telecommunications equipment. Much of the book's subject matter to this point, particularly the mechanical properties, ultimately derives from the implications of crystal or molecular structures, that is, their perfection or lack thereof, their phase shapes, sizes and distributions, and the way atoms move around within these **physical structures.** In contrast, electrical as well as optical and magnetic properties of materials are concerned with the **electronic structure** of atoms and the response of electrons to electromagnetic fields.

Simple aspects of the theory of electronic structure of the different classes of solids were presented in Chapter 2. Notably energy bands and their occupation by electrons was introduced. It was suggested that metals were distinct from insulators, and that semiconductors occupied a middle ground between the two. Simple band diagram representations in Fig. 11-1 graphically distinguish these three types of solids. Semiconductors have a relatively small energy gap, E_g, of ~1 eV separating the nearly full valence and almost empty conduction bands. Insulators have a larger energy gap (5–10 eV) between the even fuller valence band and virtually unoccupied conduction band. The size of the energy gap may be viewed as a barrier to electrical conduction and that is why

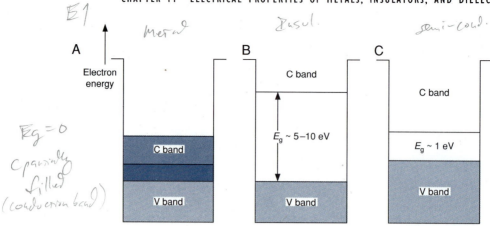

E1

Metal   *Insul.*   *semi-cond.*

Eg = 0
C partially
filled
(conduction band)

FIGURE 11-1   Comparison of electron band diagrams for (A) metals, (B) insulators, and (C) semiconductors.

insulators are poor conductors. Conversely, metals have no effective energy gap because valence and conduction bands either overlap (Fig. 11-1A) or the conduction band is only partially filled. Therefore, they are good conductors. This chapter focuses on these two extremes of conduction behavior.

Irrespective of material class involved, or its physical state (gas, liquid, or solid), an electric current I (amps or A) flows when a concentration of carriers n (number/m^3) with charge q (coulombs or C) moves with velocity v (m/s) past a given reference plane in response to an applied electric field $\mathscr{E}$ (V/m). The magnitude of the current density J (A/m^2), or current per unit area, that flows is then expressed by the simple relationship

$$J = nqv. \tag{11-1}$$

For small electric fields the carrier velocity is proportional to $\mathscr{E}$ so that

$$v = \mu\mathscr{E}. \tag{11-2}$$

The proportionality constant μ is known as the mobility (m^2/V-s). Substitution yields

$$J = nq\mu\mathscr{E}. \tag{11-3}$$

We may write this as $J = \sigma\mathscr{E}$, where σ (ohm-m)$^{-1}$ is the electrical conductivity with a value of $nq\mu$. The resistivity ρ (ohm-m) of a material is the inverse of its conductivity so that

$$\rho = 1/\sigma = (nq\mu)^{-1}. \tag{11-4}$$

Combining the last two equations, we obtain J (A/m^2) = (1/ρ) (ohm-m)$^{-1}$ $\mathscr{E}$ (V/m), whose units can be reduced to the more recognizable form, amps = volts/ohm, or **Ohm's law.** Although resistivity is a property of the material, the resistance of a specific circuit element depends on its geometry and dimen-

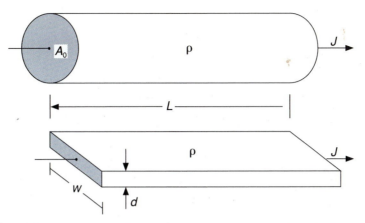

FIGURE 11-2 *Top:* Electrical resistance of a conductor with resistivity ρ, length L, and uniform cross-sectional area A_0 is $\rho L/A_0$. *Bottom:* Electrical resistance of plate or film conductor is $\rho L/dw$.

sions (Fig. 11-2). For a conductor of uniform cross-sectional area A_0 and length L, the resistance is $R = \rho L/A_0$.

Materials can be reliably distinguished on the basis of both the magnitude of the resistivity and the rate at which it changes with temperature. The typical resistivity range of a number of metals, semiconductors, and insulators is graphically depicted in Fig. 11-3. Of note is the huge ~24-order-of-magnitude difference in resistivity between the best (metals) and worst (insulators) conductors, respectively. There is no other material property that spans so large a range of magnitude. Physically, metals can be further distinguished from all

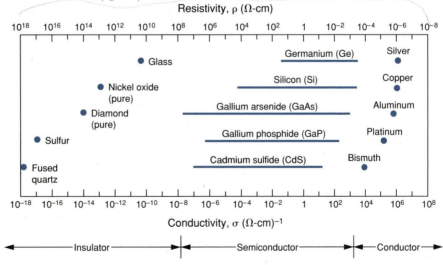

FIGURE 11-3 Typical electrical resistivity range for insulators, semiconductors, and conductors. From S. M. Sze, *Semiconductor Devices: Physics and Technology*, Wiley, New York (1985). Copyright © 1985 by AT&T; reprinted by permission.

other classes of materials because they become poorer conductors as the temperature is raised, implying a positive resistivity temperature dependence (i.e., $d\rho/dT > 0$). Semiconductors and insulators, however, become better conductors ($d\rho/dT < 0$). Even here there are occasional exceptions to the rule over small temperature excursions [e.g., doped semiconductors in the so-called exhaustion range (see Section 12.2.3)].

Descriptions of electrical conductivity are concerned with the magnitude and attributes of the materials constants in Eq. 11-4. Corollary issues revolve about how n and μ vary as a function of temperature, composition, defect structure, and electric field. An alternative complementary approach to understanding electrical properties is via electron band diagram considerations. We adopt this approach at times but also make extensive use of charged carrier dynamics to provide a balanced portrait of electrical conduction and related phenomena.

In addition to high resistivity, some insulators possess important dielectric properties that are exploited for use in capacitors. Electrons and ionic charge in dielectrics tend to oscillate in concert with the applied ac electric field frequency; this is distinct from the migration and net displacement carriers undergo in a dc field. The implications of this in assorted applications are addressed later.

11.2. ELECTRONS IN METALS

11.2.1. Free Electrons Revisited

The description of electronic structure in Chapter 2 suggests that metals contain a typical distribution of atomic core levels topped by a densely packed continuum of conduction or free electron levels. Their energies were derived assuming the electrons were confined to a well from which they could never escape. A more realistic picture of a metal was suggested in Section 2.4.3.2. Instead of walls extending to infinite energy there is now a finite well that confines the free electrons as shown in Fig. 11-4. The energy possessed by electrons in the highest occupied level is known as the **Fermi energy**, E_F. But electrons at energy E_F still have to acquire the **work function energy**, $q\Phi$, a type of electrostatic energy barrier, before they are totally free of the metal. Typical values of $q\Phi$ range between 2 and 5 eV.

Electron energy levels within the metal are slightly altered relative to those in the infinite well but just as densely packed. To handle such an unwieldy number of *discrete* electrons and energy levels the obvious strategy is to define *continuous* distributions. Sums that are difficult to add are then converted to functions that are frequently easier to integrate. The details of how this is done

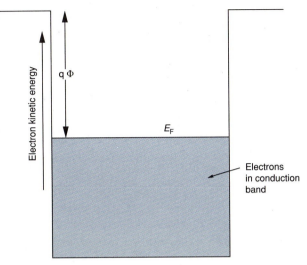

Electron kinetic energy

$q\Phi$

E_F

Electrons in conduction band

FIGURE 11-4   Free electron model of a metal having Fermi energy E_F and work function energy $q\Phi$.

are beyond the scope of the book; however, two widely used functions facilitate the statistical accounting of the conduction electrons, and are useful in interpreting the electrical properties of metals as well as semiconductors.

11.2.1.1. Fermi–Dirac Function

The first is the Fermi–Dirac electron distribution function, which has the form

$$F(E) = 1/\{\exp[(E - E_\text{F})/kT] + 1\}. \tag{11-5}$$

It specifies the probability, $F(E)$, that an electron level having energy E is occupied. At $T = 0$ K electrons fill all of the available states up to E_F. Simple substitution, where kT has the usual meaning, shows that if the electron level E is less than E_F, then $F(E) = 1$ at $T = 0$ K [because $\exp(-\infty) = 0$]. Similarly, if the electron energies lie above E_F, then $F(E) = 0$ (because $\exp \infty = \infty$). This step function-like behavior is graphically sketched in Fig. 11-5. If the temperature is raised above 0 K, electrons at the Fermi level can be excited to higher unoccupied levels. This frees electrons that are lower in the distribution to rise. It turns out that $F(E)$ rises for energies above E_F, and falls (antisymmetrically) for energies below E_F. The distribution spreads a bit but is still a pretty sharp step. At high energy, $\exp[(E - E_\text{F})/kT] > 1$, and $F(E) \sim \exp[-(E - E_\text{F})/kT]$. But this is the dependence of the **Boltzmann distribution**. In both metals and semiconductors it is often frequently possible to replace the Fermi–Dirac by the Boltzmann distribution function (see Eq. 5-13).

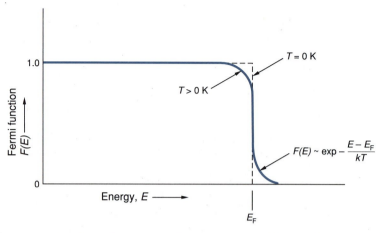

FIGURE 11-5   Variation of the Fermi–Dirac function, $F(E)$, with energy, E. Shown is the behavior of $F(E)$ at $T = 0$ K and $T > 0$ K.

11.2.1.2. Density of States

The second function is the density of states $N(E)$, and it specifies the total number of states that lie between energy E and energy $E + dE$. We have already noted in Chapter 2 that as E increases, the number of states also increases; but the great degeneracy at large quantum numbers serves to limit the magnitude of energy reached. Mathematically, $N(E)$ has a simple parabolic dependence on energy given by

$$N(E) = \tfrac{1}{4}\pi V(8m_e/h^2)^{3/2}\,E^{1/2}, \tag{11-6}$$

where m_e is the electron mass, h is Planck's constant, and V is the volume of the metal. The product of $N(E)F(E)$, the electron distribution function, is significant because it accounts for the statistical summation of states in a physically acceptable manner. First, electron states $[N(E)]$ must exist, and then the probability $[F(E)]$ that they are occupied must be considered. Accounting for both factors, as shown in Fig. 11-6, is a common way of describing the conduction electron distribution. We can combine Eqs. 11-5 and 11-6 to determine the value of E_F for metals at 0 K by considering the integral

$$N_e = 2\int_0^{E_F} N(E)F(E)\,dE. \tag{11-7}$$

This expression simply states that the total number of electrons in the metal (N_e) must be distributed over filled states, each containing two electrons (spin up, spin down). At 0 K this integral is easy to evaluate because $F(E) = 1$, and the highest occupied energy is E_F; we must simply integrate $E^{1/2}$. Therefore, solving for the upper limit,

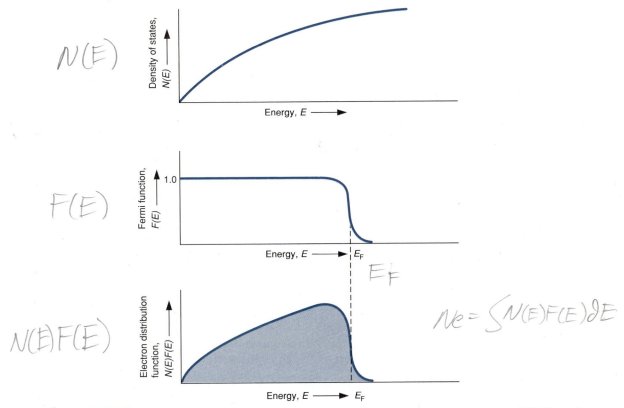

N(E)

F(E)

N(E)F(E)

E_F

$N_e = \int N(E)F(E)\, \partial E$

FIGURE 11-6 Variation of the product of the density of electron states and Fermi–Dirac function, $N(E)F(E)$, with energy, E.

$$E_F = \frac{h^2}{8m_e}\left(\frac{3N_e}{\pi V}\right)^{2/3}.$$ (11-8)

EXAMPLE 11-1

a. Calculate E_F for copper.

b. What is the probability that electron states are occupied in Cu at 25°C if the states have energies of $E = E_F$, $E = E_F + 0.1$ eV, and $E = E_F - 0.1$ eV?

ANSWER a. Calculation of E_F requires first knowing N_e/V, or the number of electrons contained within 1 m³ of Cu. The density of Cu is 8.92 Mg/m³ and its atomic weight is 63.5 g/mol or 63.5×10^{-6} Mg/mol. Therefore, there are

$$[(6.02 \times 10^{23} \text{ atoms/mol}) \times (8.92 \text{ Mg/m}^3)]/63.5 \times 10^{-6} \text{ Mg/mol}$$
$$= 8.46 \times 10^{28} \text{ atoms/m}^3,$$

and because there is one conduction electron per atom, $N_e/V = 8.46 \times 10^{28}$ electrons/m^3. After substitution into Eq. 11-8,

$$E_F = \left[\frac{(6.62 \times 10^{-34})^2}{8 \times 9.1 \times 10^{-31}}\right]\left(\frac{3 \times 8.46 \times 10^{28}}{\pi}\right)^{2/3}$$
$$= 1.12 \times 10^{-18}\,J = 7.01\,eV.$$

b. At $E = E_F$, direct substitution into Eq. 11-5 yields $F(E_F) = 1/(\exp 0 + 1) = \frac{1}{2}$. This problem suggests a definition of the Fermi energy as that energy for which $F(E)$ is one half. For $E = E_F + 0.1$ eV,

$$F = [\exp(0.1/8.63 \times 10^{-5} \times 298) + 1]^{-1} = 0.020.$$

For $E = E_F - 0.1$ eV,

$$F = [\exp(-0.1/8.63 \times 10^{-5} \times 298) + 1]^{-1} = 0.980.$$

Therefore, F changes by 96% over a 0.2 V change about E_F.

11.2.2. Electron Emission

In several important applications the objective is to eject free (as well as core) electrons from metals into the surrounding vacuum. There are at least four important ways to liberate electrons from the attractive pull of the positive ions and they are schematically differentiated in Fig. 11-7.

11.2.2.1. Thermionic Emission

If a metal is heated to sufficiently high temperatures, the energy of some conduction electrons will rise high enough on the Fermi distribution to extend beyond the work function (Fig. 11-7A). At this point they have essentially left the metal. If means are provided to draw off the electrons, a current flows. This thermionic current density (J_T) is given by the Richardson–Dushman equation

$$J_T = AT^2\exp(-q\Phi/kT)\quad A/m^2, \tag{11-9}$$

where A is a constant whose theoretical value is 1.20×10^6 A/m^2-K^2. Thermionic emission is thus thermally activated with an activation energy equal to the work function. Note that in the remainder of the book activation energies will usually be given in eV/atom units. Before the age of solid-state devices, vacuum tubes were employed in electronics and these relied on thermionic emission from heated filaments. This phenomenon is also commonly exploited in electron beam melting and welding processes, as well as in X-ray generation and electron optics applications.

11.2.2.2. Photoemission

In this case electrons are emitted from metals when they are illuminated by radiation whose energy $(h\nu)$ exceeds the work function. The physics

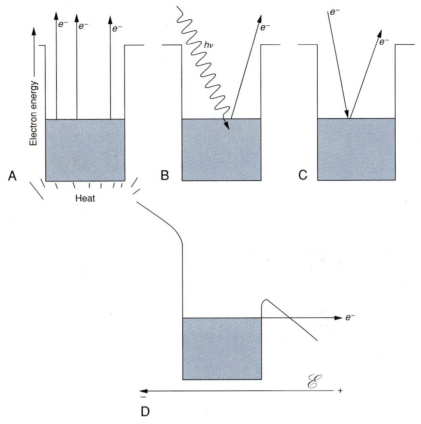

Four electron emission processes operative in metals. (A) Thermionic emission (heat in–electron out). (B) Photoemission (photon in–electron out). (C) Secondary electron emission (electron in–secondary electron out). (D) Tunneling (electron emission in very high electric field).

involved is described by the well-known Einstein equation for the photoelectric effect

$$\tfrac{1}{2}m_e v^2 = h\nu - q\Phi, \qquad (11\text{-}10)$$

where the term on the left is the **maximum** kinetic energy achieved by emitted electrons of velocity v. Electrons at the Fermi level are emitted with maximum energy whereas those lower in the distribution leave with less energy (Fig. 11-7B). An important device that capitalizes on the photoelectric effect is the photomultiplier tube. Incoming light causes photoelectron emission from a first cathode. The emitted electrons then impinge at the next dynode electrode, where still more (secondary) electrons are emitted. Successive emission at other dynodes causes electron multiplication and a large output current signal. In this way even a single photon can be detected.

11.2.2.3. Secondary Electron Emission

When high-energy charged particles such as electrons or ions strike metal surfaces, both the atomic core and the free electrons contained within a subsurface region are excited (Fig. 11-7C). The deposited energy is dissipated by deexcitation processes that may be manifested in photon, X-ray, and Auger electron emission (see Section 2.3) as well as the ejection of low-energy secondary electrons. These have energies of a few to tens of electron volts and emanate from only the uppermost atomic layers near the surface. Scanning electron microscope images are produced by capturing secondary electrons emitted in response to the impinging high-energy (30-keV) electron beam.

11.2.2.4. Field Emission

All of the emission processes considered above rely on electrons gaining sufficient energy to rise *vertically* above an effective surface barrier. In field emission electrons actually leave the metal without increasing their energy. What is required is a very large applied electric field that is normally produced by raising a metal surface to a high negative electrical potential. This creates an asymmetrically distorted and slanted potential well (Fig. 11-7D), so that instead of a high thick uniform barrier we now have a high thin triangular barrier. Rather than climb over the energy barrier, electrons burrow or **tunnel** through it! Surprisingly, electrons need no thermal stimulation to accomplish tunneling, an effect that has quantum-mechanical origins. In effect, there is a probability that waves associated with the well electrons can exist outside the barrier provided there are vacant electron states available for emitted electrons to occupy. Field emission and tunneling phenomena occur in scanning tunneling, field ion, and (certain) scanning electron microscopes. Essential to all of these microscopy techniques is the electron emission that occurs from specially prepared emitters.

Devices based on electron emission processes are usually configured with dual electrodes—the cathode emitter and anode—enclosed within a vacuum envelope. There are important materials choices, electrode spacings and geometries, and operating conditions that must be optimized for efficient device function. For example, low-work function materials are required for thermionic emission and field emission processes. In addition, high-melting-point metals are essential to reach the temperatures needed for sufficiently large thermionic currents. Table 11-1 lists workfunctions and related material properties of various cathode materials. With the exception of field emitters, the cathodes, or surfaces from which electron emission occurs, are generally planar, operate in relatively low electric fields, and are not spaced particularly close to the anode. Field emission, on the other hand, requires high electric fields, $\mathscr{E}$, and these can be generated by applying modest voltages, V, to metal needlelike tips thinned to a radius of curvature, r, of 100 nm or less. (Recall that $\mathscr{E} \approx V/r$.) Electron tunneling phenomena are not limited to field emission applica-

| TABLE 11-1 | CHARACTERISTICS OF THERMIONIC EMITTERS | | | |

Material	T_M (K)	Operating temperature (K)	Work function (eV)	Constant A (10^6 A/m²-K²)
Metals				
W	3683	2500	4.5	0.60–0.75
Ta	3271	2300	4.2	0.4–0.6
Mo	2873	2100	4.2	0.55
Th	2123	1500	3.4	0.60
Ba	983	800	2.5	0.60
Cs	303	293	1.8	1.6
Coated Cathodes				
W + Th	—	1900	2.6	0.03
W + Ba	—	1000	1.56	0.015
W + Cs	—	1100	1.36	0.032
Ni/BaO + SrO	—	1100	1.0	10^{-4}–10^{-5}

After R. M. Rose, L. A. Shepard, and J. Wulff, *The Structure and Properties of Materials*, Vol. IV, Wiley, New York (1966).

tions but occur at metal–semiconductor contacts and in superconducting devices. In fact, any time positively and negatively charged conductors separated by an insulator are brought to within ~5 nm or less of each other, there is a probability that tunneling can occur.

EXAMPLE 11-2

a. A tungsten thermionic emitter coated with ThO_2 has a work function of 2.5 eV. How much more current will there be from such an emitter relative to pure W ($q\Phi = 4.5$ eV) if both are heated to 2400 K?

b. Assuming 100% efficiency in photoelectric conversion, what is the minimum incident photon beam power density required to produce a photoelectron current equal to the thermionically emitted current from W in part a? What photon wavelength is required?

ANSWER   a. Using Eq. 11-9 the ratio of the currents from the two emitters is

$$J_T(\text{W–ThO}_2)/J_T(\text{W}) = [\exp(-q\Phi_{\text{W–ThO}_2}/kT)]/[\exp(-q\Phi_\text{W}/kT)].$$

After substitution,

$$\frac{J_T(\text{W–ThO}_2)}{J_T(\text{W})} = \frac{[\exp(-2.5/8.63 \times 10^{-5} \times 2400)]}{[\exp(-4.5/8.63 \times 10^{-5} \times 2400)]} = 1.56 \times 10^4.$$

The implications of lowering workfunctions are that cathode surface areas and their temperatures can be proportionately reduced while the same emission current is maintained.

b. The thermionic emission current density from W is (Eq. 11-9)

$$J_T(W) = 1.20 \times 10^6 \times (2400)^2[\exp(-4.5/8.63 \times 10^{-5} \times 2400)] = 2530 \text{ A/m}^2$$
$$= 2530 \text{ C/s-m}^2 \text{ or } 2530/1.60 \times 10^{-19} = \underline{1.58 \times 10^{22} \text{ electrons/s-m}^2.}$$

The minimum photon energy required to produce 1 photoelectron is 4.5 eV, or 7.2×10^{-19} J. Therefore,

$$(1.58 \times 10^{22} \text{ electrons/s-m}^2) \times (7.2 \times 10^{-19} \text{ J/electron}) = 1.14 \times 10^4 \text{ J/s-m}^2,$$
$$\text{or } \underline{1.14 \times 10^4 \text{ W/m}^2}$$

is the required power density.

The wavelength of photons required to just eject photoelectrons is given by Eq. 2-7. Substituting, $\lambda = 1.24/4.5 = 0.276 \ \mu\text{m}$. This radiation occurs in the ultraviolet.

11.2.3. Electron Energy Gaps in Solids

The free electron model used to this point has explicitly assumed a constant lattice potential or zero electric field everywhere. But this is difficult to accept in a lattice of periodically placed ion cores. The pull on electrons should differ depending on whether they are close to ion centers or in between neighboring nuclei. In general, then, the electron potentials and forces they give rise to are periodic in space, reflecting the crystallographic arrangement of atoms. The specific mathematical form of the periodic potential for electrons is complex, however.

To see the consequences of a periodic lattice potential let us return to Eqs. 2-12 and 2-13. The electron kinetic energy ($E = p^2/2m_e$) is reproduced here as

$$E = \frac{h^2 n_x^2}{8 m_e L^2} = \frac{h^2 k^2}{8\pi^2 m_e}, \tag{11-11}$$

and it is clear by comparison that k, defined as the **wavenumber** (wavevector in three dimensions), has substituted for $n_x\pi/L$. A plot of E versus k is shown in Fig. 11-8A, and describes a parabola that is characteristic of free electrons in a **non**periodic lattice. We also recall that electron waves propagating to and fro within a well of length L have wavelengths $\lambda = 2L/n_x$ or, equivalently, $\lambda = 2\pi/k$. (Note that these definitions and equations are consistent with the de Broglie relationship.)

Next, consider electron wavelengths that are very short and even comparable to distances between atoms. If the lattice spacing is a, then the *smallest* possible wavelength that could propagate in the lattice is $\lambda = 2a$. In this case the electron wave amplitude will be zero at any given atom and at each of its flanking atoms. When it is recalled that Bragg's law can be written as $\lambda = 2a \sin \theta$, then for normal incidence where $\theta = 90°$, the condition $\lambda = 2a$ holds. During

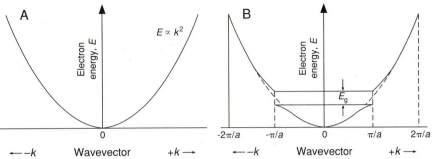

FIGURE 11-8 Conduction electron energy, *E*, versus wavevector, *k*. (A) Free electron parabola. (B) In periodic lattice energy gaps (E_g) appear at $k = \pm\pi/a$, $\pm 2\pi/a$, ... etc.

X-ray diffraction we know that when the Bragg requirement is met, X rays are ejected from the crystal. Similarly, during electron diffraction, a parallel phenomenon, satisfaction of the Bragg condition means that electrons of the critical wavelength are also expelled from the crystal. Therefore, on the *E*-versus-*k* curve peculiar effects can be expected when $\lambda = 2a$ or, equivalently, when $k = \pi/a$.

Ejected or expelled electrons mean there are no states that will accommodate them. Thus at the critical value of *k*, corresponding to wave propagation in the +*x* direction, there is a **gap** in the allowable electron energy levels. The same forbidden energy gap (E_g) exists for electrons traveling in the −*x* direction. Account is taken of these effects in Fig. 11-8B. Note that electrons of long wavelength (*k* small) are barely aware of the periodic structure they are propagating through; in this regime the parabolic *E*-versus-*k* behavior is hardly altered. Only when electron wavelengths become comparable to lattice dimensions is *E*-versus-*k* behavior appreciably changed. Electron states confined between $k = -\pi/a$ and $k = +\pi/a$ are said to belong to the first **Brillouin zone**. Just as there are higher orders of diffraction, so we can imagine *k* extending from π/a and $-\pi/a$ to $+2\pi/a$ and $-2\pi/a$, respectively. In between are states corresponding to the second Brillouin zone. For a simple cubic lattice of *N* unit cells, each zone contains *N* states or 2*N* electrons.

Although slightly more cumbersome to deal with, it is not conceptually difficult to picture electron propagation in two or three dimensions. For example, in two dimensions,

$$E = \frac{h^2(k_x^2 + k_y^2)}{8\pi^2 m_e}. \tag{11-12}$$

In k_x, k_y space, constant energy contours are circular because $k_x^2 + k_y^2 = \text{const.}$ is the equation of a circle. All occupied electron states are contained within the circle; states of the highest energy lie on the circumference and correspond to the Fermi energy level. In three dimensions the Fermi surface encloses a sphere of occupied states (see Fig. 11-10).

Despite the impression given by the above development, energy gaps do *not* effectively exist in the band structure of metals. To see why, we must view *E*-versus-*k* curves in three electron propagation directions, and not only in the *x* direction. Forbidden energy gaps in one crystallographic direction are offset by allowed states in other directions of electron propagation, so that a continuous distribution of levels is always available for occupation.

In the case of the monovalent alkali metals like Na and K, half of the states in a zone are occupied by conduction electrons while the other unoccupied half are easily accessible upon electron excitation to higher levels. For multivalent metals the situation is more complex because there is zone overlap. Thus, divalent metals will have just enough electrons to fill the first Brillouin zone, but overlap with the second zone causes its states also to fill. A whole new group of energy levels is thus made available for potential occupation. The situation is quite different in semiconductors and insulators where true forbidden energy gaps do exist. What effect this has on electrical conduction in these materials will be explored subsequently.

11.3. ELECTRON SCATTERING AND RESISTIVITY OF METALS

11.3.1. Electron Scattering in Metals

Let us consider a classical model for the influence an applied electric field, $\mathscr{E}$, has on the motion of a conduction electron. Newton's law of motion ($F = m_e a$) for an electron driven by an electric field force is given by

$$q\mathscr{E} = m_e \frac{dv}{dt}, \tag{11-13}$$

where v, m_e, and q are the electron velocity, mass, and charge, respectively; The force, F, is associated with $q\mathscr{E}$, while dv/dt is the electron acceleration a. By integrating, $v = q\mathscr{E}t/m_e$ is obtained. But this solution for v is not physically meaningful because it increases without bound in time. Rather, we expect that sooner or later the electron will collide with some obstacle (e.g., vibrating atoms, impurity atoms, defects) in the lattice, scatter, and then start to drift in the field direction until it scatters again, and so on. Just before scattering the electron velocity reaches a maximum value, only to be reduced after scattering to a minimum value because the electron trajectory reverses. If we assume that the time between collisions is 2τ, where τ is the so-called relaxation time, the idealized time-dependent history of electron motion is shown in Fig. 11-9. In actuality the time between collisions varies statistically about 2τ, as do the terminal velocities. Halfway between the velocity extremes an average drift velocity (v_D) can be defined. A more physically appealing expression for the now bounded electron velocity is $v_D = q\tau\mathscr{E}/m_e$. When v_D is substituted for v

in Eqs. 11-1 to 11-4, the resultant conductivity is

$$\sigma = nq^2\tau/m_e. \tag{11-14}$$

It is instructive to calculate the mean distance Λ between collisions in a metal because the longer it is, the greater the electrical conductivity. Clearly, $\Lambda = 2\tau v_F$, where v_F is the velocity at which electrons at the Fermi level travel. In actuality, it is only Fermi level electrons that can move freely, for only they have empty states to move into. Lower lying electrons are surrounded by filled states of immediately higher or lower energy. We may assume that the Fermi kinetic energy, E_F, is given by $\frac{1}{2}m_e v_F^2$, so that $v_F = (2E_F/m_e)^{1/2}$. Combining all of these contributing terms yields

$$\Lambda = 2\tau v_F = 2(m_e\sigma/nq^2) \times (2E_F/m_e)^{1/2}. \tag{11-15}$$

Importantly, Λ is proportional to the conductivity. In the case of copper, substitution of $\sigma = 5.8 \times 10^7$ (Ω–m)$^{-1}$ (at 25°C), $m_e = 9.1 \times 10^{-31}$ kg, n (from Example 11-1) = 8.46×10^{28} electrons/m^3, $q = 1.6 \times 10^{-19}$ C, and E_F (from Example 11-1) = 1.12×10^{-18} J yields $\Lambda = 7.6 \times 10^{-8}$ m, or 76 nm. This huge mean free path between electron collisions, which is more than 200 times larger than the lattice parameter of Cu, is hard to rationalize in terms of classical physics; and besides, at very low temperatures Λ might be 1000 times larger! Clearly such behavior indicates that free electron motion in metals (and semiconductors) cannot be explained by Newtonian mechanics. Rather, quantum effects are at work here.

The quantum-mechanical picture of electron scattering is very complex and only a hint of it is suggested in Fig. 11-10. Within the indicated Brillouin zone is a partially filled band of occupied states that is enclosed by the circular Fermi surface. As already noted, only electrons with energies around E_F can move.

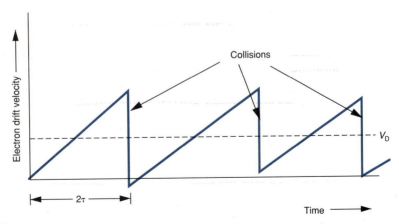

FIGURE 11-9 Representation of classic electron drift velocity with time in metals. After scattering events, the electron reaccelerates.

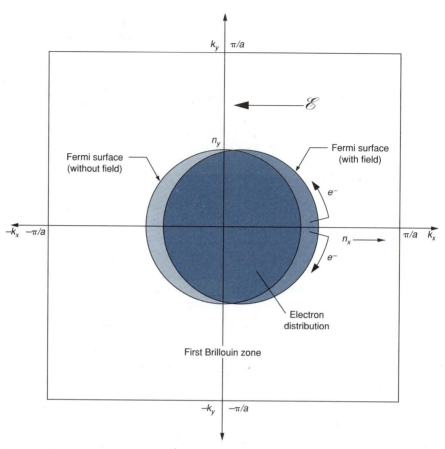

FIGURE 11-10   Motion of electrons in a metal subjected to an external electric field. Only electrons at the Fermi level are capable of occupying new states.

With no applied electric field, just as many electrons move in the positive x as in the negative x direction, occupying states with larger $+n_x$ or $-n_x$ quantum numbers. The same applies to electron motion in the y direction so the overall effect is that no **net** current flows. Application of an electric field provides the driving force that displaces the Fermi distribution toward the positive electrode. Electrons are scattered more frequently at the advancing edge of the Fermi distribution than at the trailing edge. As they scatter, the values of k reverse sign. Electrons thus peel off the front edge in an ablative-like motion and are transported to the back edge of the distribution.

11.3.2. Resistivity Contributions: Matthiessen's Rule

If we could attach ourselves to an electron while it speeds through a relatively pure metal at very low temperatures, we would note that it scatters where there are disruptions in the periodic lattice potential. Thus, chance encounters with impurity atoms, point defects (vacancies, interstitials), line defects (dislocations of all types and configurations), and surface defects (grain boundaries, stacking faults) would result in scattering. Interestingly, if records were kept of the mean free path, or distance Λ between scattering for each type of scatterer, we would note that the average values would differ. There would be an average value of Λ_i for impurities that differed from Λ_v for vacancies, or from Λ_b for grain boundaries, and so on.

As the temperature is raised the atoms are thermally agitated and the effect this has on electrons dwarfs all of the other scattering contributions. Lattice vibrations and the larger effective area or cross section that atoms now present to electrons are the source of this thermal scattering. To roughly estimate the temperature dependence of scattering, consider vibrating atoms to behave like masses on springs (see Section 5.3.4). The average kinetic energy is the same as their average potential energy, and the latter is given by $\frac{1}{2}k_s r^2$, where k_s is the spring constant and r is the vibrational amplitude. From kinetic theory an association can be made between the total mechanical energy of an oscillating atom ($\sim k_s r^2$) and its thermal energy ($\sim kT$). Therefore, r^2 is proportional to temperature. As the mean distance between thermal scattering events (Λ_T) is inversely proportional to the scattering cross section or area (πr^2), we conclude that Λ_T varies as T^{-1}. Meanwhile, scattering from impurities and defects is not appreciably different from scattering at very low temperatures.

Whenever we have a number of freely competing independent events, those that occur most frequently dominate the overall behavior. When translated to scattering events, the average overall mean free path in the system is governed by the mechanism with the smallest Λ, or

$$\Lambda^{-1} = \Lambda_T^{-1} + \Lambda_i^{-1} + \Lambda_d^{-1}. \tag{11-16}$$

In this formula, Λ_d is the collective mean free path for defect scattering. Since the electrical conductivity is proportional to Λ, and the resistivity (ρ) to Λ^{-1},

$$\rho_{total} = \rho_T + \rho_i + \rho_d. \tag{11-17}$$

This equation is a statement of **Matthiessen's rule** and indicates that the various contributions to the resistivity of metals are independently additive. The graphic representation of Fig. 11-11 reinforces the additive nature of Matthiessen's rule at any given temperature.

11.3.3. Effect of Temperature, Composition, and Defects

Unlike all other materials, metals become more resistive as the temperature is raised. The linear slope of ρ versus T is essentially related to the temperature coefficient of resistivity, α, an important material property that is usually defined by the equation

$$\rho_T = \rho_0(1 + \alpha\Delta T). \tag{11-18}$$

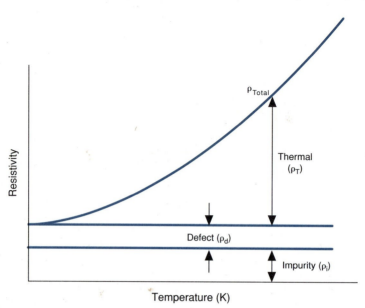

FIGURE 11-11
Representative temperature dependence of the electrical resistivity of a metal. Thermal, impurity, and defect contributions are independently additive.

The quantity ρ_0 is the resistivity at 20°C and ΔT is the temperature difference relative to 20°C. Values for ρ_0 and α are given in Table 11-2 for a number of metals. At elevated temperatures α is constant, whereas at cryogenic temperatures the resistivity is nonlinear and varies as $\sim T^5$. The almost unchallenged use of copper and aluminum for common electrical wiring applications is due to their low resistivities. For electrical heating applications, however, high resistivities are essential. The following example illustrates the material properties required and considerations involved in designing electrical heating elements.

EXAMPLE 11-3

A 1000-W toaster is required to operate on 110 V. It is decided to employ nichrome windings which can tolerate operating temperatures of 1300 K. What length of winding is required and what diameter wire should be used?

ANSWER   Let l be the length and d the diameter of the required wire. The resistance R of such a wire is $R = \rho_0[1 + \alpha\Delta T] \times (l/\pi d^2/4)]\Omega$. For nichrome at 1300 K, $R = 1.1 \times 10^{-6}[1 + 0.0004(1027 - 20)] \, 4 \times l/\pi d^2 \Omega$. Evaluation yields $R = 1.96 \times 10^{-6} \times l/d^2 \, \Omega$. The power drawn is $P_e = V^2/R$, or $P_e = (110)^2 \, d^2/1.96 \times 10^{-6} \times l = 6.17 \times 10^9 \, d^2/l$ W. We have one equation and two unknowns, d and l.

$R = \dfrac{\rho l}{A}$

If it is assumed that all of the electrical power is dissipated by radiation, the Stefan–Boltzmann law (see Section 5.3.4.2) can be used to connect the radiated power, P_R, to the absolute temperature of the heating element. Therefore,

$$P_R = \varepsilon_R \sigma_R (T^4 - T_0^4) A_0, \qquad (11\text{-}19)$$

where A_0 is the radiating surface area, πdl. The emissivity of the surface is ε_R and assumed to be 0.5. σ_R, the Stefan–Boltzmann constant, is equal to 5.67×10^{-8} W/m^2-K^4, and T_0 is the ambient temperature, ~298 K. Substitution yields $P_R = 0.5 \times 5.67 \times 10^{-8}(1300^4 - 298^4)\pi d \times l = 2.54 \times 10^5 \, d \times l$ W. Equating the input electrical power and the output radiated power to 1000 W results in

$$6.17 \times 10^9 \, d^2/l = 1000 \text{ W, and } 2.54 \times 10^5 \, d \times l = 1000 \text{ W.}$$

Solving these two simultaneous equations yields $d = 8.61 \times 10^{-4}$ m, or 0.0861 cm, and $l = 4.57$ m.

Alloying elements invariably raise the resistivity of the metal to which they are added. This is different from the behavior in semiconductors and insulators, which become better conductors when alloyed. The contribution selected elements make to the resistivity of copper is shown in Fig. 11-12. One important commercial product for electrical applications is oxygen-free high conductivity (OFHC) copper. It has a purity of 99.995% Cu. Copper refiners make great efforts to remove elements like phosphorus and iron, which have a deleterious

TABLE 11-2 ELECTRICAL RESISTIVITY OF METALS[a]

Metal	Resistivity at 20°C, ρ (10^{-9} Ω-m)	Temperature coefficient of resistivity, (C^{-1})
Aluminum (annealed)	28.3	0.0039
Copper (annealed standard)	17.2	0.0039
Brass (65 wt% Cu–35 wt% Zn)	70	0.002
Gold	24.4	0.0034
Silver	15.9	0.0041
Contact alloy (85 wt% Ag–15 wt% Cd)	50	0.004
Platinum (99.99%)	106	0.0039
Iron (99.99%+)	97.1	0.0065
Steel (wire)	107–200	0.006–0.0036
Silicon iron (4 wt% Si)	59	0.002
Stainless steel (18 wt% Cr, 8 wt% Ni)	73	0.00094
Nickel (99.95%+ Co)	68.4	0.0069
Nichrome (80 wt% Ni–20 wt% Cr)	1100	0.0004
Invar (65 wt% Fe, 35 wt% Ni)	81	0.0014
Tungsten	55	0.0045

[a] Data were taken from several sources.

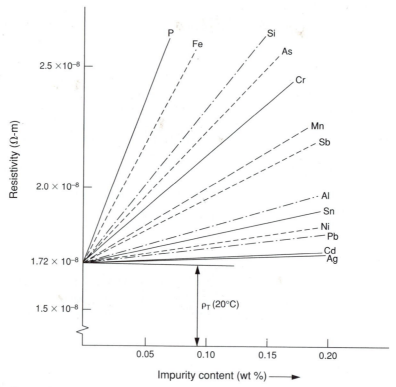

FIGURE 11-12   Change of electrical resistivity of copper as a function of alloying element and content.

effect on the conductivity of OHFC. On the other hand, it is frequently desirable to combine high conductivity with high strength or resistance to elevated-temperature softening. These almost mutually exclusive properties can sometimes be realized through alloying with elements that do not degrade conductivity appreciably, but allow for some strengthening. Solid solution strengthening with small silver additions and precipitation hardening with zirconium are two ways in which copper has been modified to meet such demands.

A common method for assessing the purity and defect content of "pure" metals is through the measurement of the residual resistivity ratio (RRR). The latter is defined as the ratio of the resistivity at 298 K to that at liquid helium temperature (4.2 K). Thus,

$$RRR = \rho_{total}(298)/\rho_{total}(4.2) = \sim \rho_T/(\rho_i + \rho_d), \qquad (11\text{-}20)$$

using Eq. 11-17. At room temperature the thermal contribution is much greater than the impurity and defect contributions to the resistivity so that the latter can be neglected in comparison. At low temperatures, however, thermal scattering is reduced to the point where it contributes much less to the resistivity than impurity and defect scattering. For highly purified metals that are annealed to eliminate defects, RRR becomes quite large and can reach values in the thousands. The value of RRR is a sensitive indicator of impurities and is sometimes demanded by purchasers of pure metals as a measure of quality.

11.4. THERMAL CONDUCTIVITY OF MATERIALS

Conduction of heat parallels the conduction of charge in metals. Because the same free electrons that contribute to electrical currents also transport energy (heat) their role in heat flow is expected to be important. The phenomenon of heat flow was defined in Sections 8.2.3 and 9.8.1, where **thermal conductivity** (κ in units of W/m-K) is the material constant of concern; it plays the same role in heat transport that diffusivity (D) does in mass transport. In addition to electron waves, lattice vibration waves or **phonons** also transport heat. During heat flow, the respective mean free paths between scattering events for electrons and phonons are Λ_e and Λ_p.

Heat conduction in solids can be profitably modeled as if the process occurred in a gas (Fig. 11-13). Consider a molecule of mass M, with heat capacity c (J/kg), moving in the x direction through a region where there is a temperature drop δT. The thermal energy it transports (δU) is equal to $\delta U = Mc\delta T$. But molecules move in both directions. If a **net** number of n molecules per unit volume move in the x direction with velocity v, then the total thermal energy

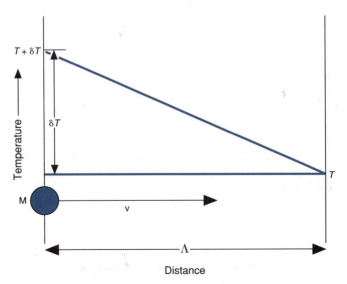

| **FIGURE 11-13** | Model for electron motion in a temperature gradient.

conducted per unit area per unit time is $Q = Mcnv\delta T$. The geometry suggests that $\delta T = -(dT/dx)\Lambda$, where Λ is the appropriate mean free path. Therefore,

$$Q = -\tfrac{1}{3}Mcnv\Lambda \frac{dT}{dx}. \qquad (11\text{-}21)$$

In this formula the negative sign physically accounts for heat flow counter to the sign of the temperature gradient. Because molecules travel in three coordinate directions the factor of $\tfrac{1}{3}$ selects only those moving in the x direction. Comparison between Eqs. 9-17 and 11-21 reveals that the thermal conductivity is defined by $\kappa = \tfrac{1}{3}Mcnv\Lambda$. The direct dependence on Λ, which may be taken as either Λ_e or Λ_p, or some combination of the two, should be noted.

In good metallic heat conductors both the velocity and mean free path for electrons exceed the corresponding values for phonons by a factor of 10 to 100. On the other hand, the heat capacity for phonons exceeds that of free electrons by a similar factor. [The reason has to do with the number of heat carriers involved. All of the vibrating atoms (phonons) absorb heat, but only a small fraction of electrons at the Fermi level can do so.] All in all then, the electron contribution to the thermal conductivity is typically 10 to 100 times that for phonons. Metals with high electrical conductivity are also expected to be good thermal conductors because the number of electrons (n) is large. Factors that reduce the electrical conductivity such as alloying will also adversely affect thermal conductivity. Alloying elements reduce both Λ_e and Λ_p, and values of κ can be lowered into the range associated with thermal insulators. An example is austenitic stainless steel (18 wt% Cr, 8 wt% Ni), which is a relatively poor heat conductor.

The closeness of the two properties in metals is reflected in a relationship known as the Weidemann–Franz law,

$$\kappa = L\sigma T = LT/\rho, \qquad (11\text{-}22)$$

where L is a constant whose value is theoretically predicted to be 2.44×10^{-8} W-Ω/K^2. Values of thermal conductivity for common metals are listed in Table 11-3 and span the range ~15 to 430 W/m-K at 300 K. When combined with the resistivities from Table 11-2, the values of L obtained generally agree with the predicted one.

Insulators have very few electrons so heat is transferred via phonons. But they are not as efficient as electrons because they are readily scattered by lattice defects. Glasses and amorphous ceramics are extreme in this regard because of their high defect content; they have lower thermal conductivities than crystalline ceramics. In general, κ for this class of materials is low, typically ranging from ~2 to 50 W/m-K (Table 11-3). When phonons have long mean free paths due to few lattice imperfections, however, the thermal conductivity can be surprisingly large. For example, κ for diamond is four times that of copper, accounting for its use as a heat sink in advanced laser devices. At elevated temperatures, κ falls because Λ_p decreases. But, at still higher temperatures,

| **TABLE 11-3** | **THERMAL CONDUCTIVITY OF MATERIALS AT 300 K**[a] |

Thermal conductivity (W/m-K)

Metals		$[L (10^{-8})]$[b]	Ceramics		Polymers	
Aluminum	240	[2.26]	Soda glass	1	Polyethelene	0.4
Copper	397	[2.28]	Borosilicate glass	1	Polypropylene	0.2
Brasses	121	[2.82]	Dense Al_2O_3	25.6	Polystyrene	0.1–0.15
Gold	315	[2.56]	Silicon carbide	84	PVC	0.15
Silver	427	[2.26]	Silicon nitride	17	PMMA	0.2
Iron	78	[2.52]	Zirconia	1.5	Epoxies	0.2–0.5
Mild steel	60		Sialons	20–25	Elastomers	~0.15
Low-alloy steels	40					
High-alloy steels	12–30					
Stainless steel	15					
Nickel	89	[2.03]				
Superalloys	11					
Titanium	22					

[a] Data were taken from several sources.
[b] Weidemann–Franz ratios for several metals are listed, in units of W-Ω/K^2.

there is an anomalous increase in κ caused by radiation heat transfer. Porosity in ceramics is beneficial in developing insulating properties for fire brick applications because heat transfer through pores filled with still air is very low.

11.5. SUPERCONDUCTIVITY

11.5.1. Physical Characteristics

The phenomenon of superconductivity was discovered in 1911 by Kamerlingh Onnes when he observed that the electrical resistance of mercury vanished below 4.15 K. Instead of the expected resistivity–temperature dependence for **normal** metals, an anomalous drop in conductivity for the superconductor occurred. Actually, the resistivity was not exactly 0 Ω-cm, but rather estimated not to exceed 10^{-20} Ω-cm. This is some 14 orders of magnitude below that of normal metals. The promise of low-Joule heating losses in electric power and transmission systems accounts for the intense interest in superconductors.

In addition to zero resistivity, superconductivity is defined by the **Meissner effect.** If we imagine a magnet moving toward the surface of a superconductor, Faraday's law predicts that an electromotive force will be generated and establish a current flow in the latter. In a superconductor these currents produce a magnetic field that *exactly cancels* the field of the first magnet. Therefore, no magnetic field penetrates the bulk of the superconductor. As the distance between the two decreases, the repulsive forces will rise and may exceed the pull of gravity. In such a case the magnet will float or levitate above the superconductor (Fig. 11-14). This is visual proof of magnetic flux exclusion or diamagnetism, that is, the Meissner effect. If, however, the magnetic field rises above a critical value, superconductivity is destroyed. Magnetic flux penetrates everywhere and the levitating magnet drops.

Of the several material properties that influence the viability of superconductors in engineering applications, there are four key ones:

1. *Critical Temperature* (T_c).   Above T_c superconductivity is extinguished and the material goes normal. Liquid helium (4.2 K) cooling systems are not cheap and therefore a high T_c value is imperative. Progress in raising T_c over the years (until about 1986) had been slow, as seen in Fig. 11-15.

2. *Critical magnetic field* (H_c).   Superconductivity disappears when the magnetic field (more correctly, magnetic flux density) rises above H_c; therefore, high values of the latter are desired.

3. *Critical current density* (J_c).   A superconductor cannot carry a current *density* in excess of J_c without going normal; the high magnetic fields the currents produce are the reason. If each of the three critical properties is plotted on the coordinate axes as shown in Fig. 11-16, the enveloping surface, in phase diagram-like fashion, distinguishes between superconducting and normal states. It is obviously desirable to extend this surface outward as far as possible in all three variables.

4. *Fabricability*.   The ability to fabricate the superconductor into required conductor shapes and lengths is essential.

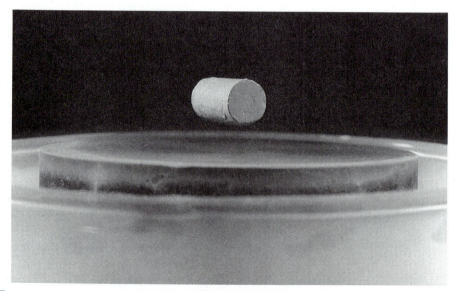

FIGURE 11-14   Levitation of a cylindrical magnet above a high-temperature $YBa_2Cu_3O_7$ superconductor disk cooled to liquid nitrogen (77 K) temperatures. Courtesy of GE Corporate Research and Development.

11.5.2. Materials

11.5.2.1. Metals

Since 1911 superconductivity has been found to occur in some 26 metallic elements and in hundreds, or perhaps thousands, of alloys and compounds; some of these are listed in Table 11-4 together with the key properties of interest. There are two kinds of superconductors: type I (or soft) and type II (or hard).

Type I superconductors. Except for niobium and vanadium the remaining elements are type I superconductors. In them the superconducting transition

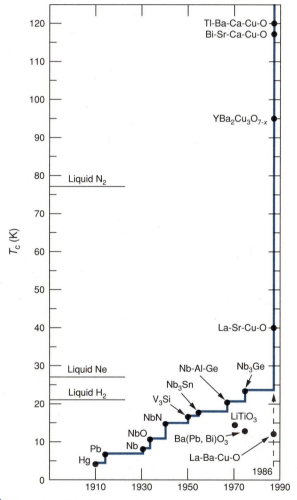

FIGURE 11-15 Chronological development of superconductor materials and corresponding T_c value. From D. R. Clarke, *Advanced Ceramic Materials 2*, Vol. 3B, p. 288 (1988).

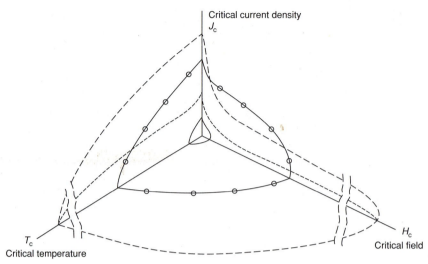

Critical temperature–magnetic field–current density space for assorted superconductor materials. The superconducting states lie between the surface contours and the origin. —, Type I superconductors, for example, Hg and Pb. ○—○, Type II superconductor alloys and compounds, for example, NbTi and Nb₃Sn. ———, Oxide superconductors, for example, YBaCuO and BiSrCaCuO thin films. - - -, Oxide superconductors, for example, bulk YBaCuO and BiSrCaCuO. From M. Wood, in *The Science of New Materials* (A. Briggs, Ed.), Blackwell, Oxford (1992).

TABLE 11-4   **PROPERTIES OF SUPERCONDUCTING MATERIALS**

Element	T_c (K)	B_c (T)[c]	Alloy or compound	T_c (K)	B_c (T)	J (A/m²)
Al	1.19	0.0099				
In	3.41	0.029	Nb–Ti	8	12	3×10^9
La	5.9	0.100	Nb–Zr	11	11	3×10^9
Nb	9.2	0.20,[a] 0.30[b]	Nb₃Sn	18.3	26	3×10^{10}
Pb	7.18	0.080	Nb₃Ge	22.5	38	
Re	1.70	0.020	V₃Si	17	25	
Sn	3.72	0.031	NbN	17.3	47	
Ta	4.48	0.083	YBa₂Cu₃O₇	93		10^{10}
Tc	8.22	—	BiSrCaCuO	107		
V	5.13	0.13,[a] 0.70[b]	TlBaCaCuO	120		

Note: Values are for critical magnetic flux density B_c, rather than critical magnetic field H_c, and are measured at 0 K. $[B_c(T) = 4\pi \times 10^{-7} H_c(A/m).]$

[a] $B_c(l)$.

[b] $B_c(u)$.

[c] Units of B_c are Tesla (T).

is abrupt, and magnetic flux penetrates only for fields larger than H_c. Values for both T_c and H_c are typically low in these materials.

Type II superconductors. Type II superconductors have two critical values for H_c, a lower value $H_c(l)$ and an upper value $H_c(u)$. For magnetic fields below $H_c(l)$ these materials behave like type I superconductors: there is total flux exclusion. But for magnetic fields above $H_c(l)$ but below $H_c(u)$, there is some flux penetration and a mixed superconducting state exists. Magnetic fields larger than $H_c(u)$ will make the material go normal. Importantly, type II superconductors such as Nb_3Sn, Nb–Ti, Nb–Zr, and Nb_3Ge have relatively high T_c and very large H_c values. This has earmarked them for use in transmission lines and magnet wire and made them favored candidates for emerging applications. Unfortunately, many of the high-T_c metal alloys and compounds are not ductile. The coextrusion process for producing NbTi wires, described in Section 8.3.5, indicates what is involved in fabricating usable shapes.

11.5.2.2. Ceramics

Seventy-five years after the initial discovery, superconductivity was found to occur in special ceramic materials (e.g., $YBa_2Cu_3O_7$, whose perovskite structure is shown in Fig. 4-24). Resistivity data of the type shown in Fig. 11-17 were recorded, and T_c values of ~90 K were attained. These results electrified the world because cheap refrigeration systems based on liquid nitrogen (77 K) created hopes for wide-scale use of these new materials. Intense research to

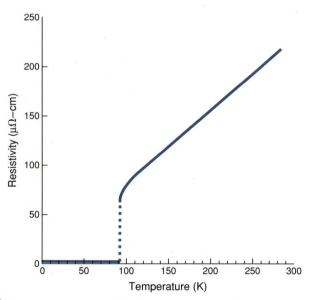

FIGURE 11-17 Electrical resistance of a $YBa_2Cu_3O_7$ superconducting film as a function of temperature. Courtesy of Y. I.-Qun Li and B. Gallois.

uncover yet new superconductors with higher T_c values is under way. As of this writing, oxides containing thallium have T_c values as high as 125 K. Despite the initial euphoria, applications based on these superconducting materials have been relatively slow in emerging. Candidate materials with otherwise favorable electrical and temperature characteristics have proven difficult to fabricate. This is particularly true of ceramic superconductors, as they are intrinsically brittle.

Lastly, the chain compound $(SN)_x$ is the first example of an inorganic superconducting polymer. With a T_c of only 0.3 K it is likely to remain a scientific curiosity.

11.5.3. Applications

Superconductors are obvious choices in applications requiring high current densities, very large magnetic fields, and devices that must be energy efficient. Low-energy-loss, superconducting power transmission lines are always desirable but only practical in special cases where the necessary refrigeration systems are cost effective. Powerful electromagnets consisting of dc powered coils of multiturn superconducting wire are commercially used in both magnetic resonance spectrometers and medical imaging systems. Type II superconductors (e.g., NbTi, Nb_3Sn) are exclusively used for these purposes because of their high H_c values. The greatest potential for superconductor usage, however, appears to be in transportation systems. Strong **magnetic** fields that **levitate** trains (**MAGLEV**) are a viable way to achieve high-speed rail transportation without friction and hence minimum energy expenditure. An air-cushioned, low-noise ride for the traveler and few environmental problems add to the attractive features. Levitated, linear electric motor-propelled trains have been built in Japan and experimental models have achieved speeds of 500 km per hour. It should be noted that these systems employ low-temperature superconductors for the required magnetic fields used in suspending the train cars. Conventional magnets, which eliminate the need for cryogenic refrigeration, are currently part of the design strategy adopted by U.S. efforts in this area.

A couple of superconductor applications are illustrated in Fig. 11-18.

11.6. CONDUCTION BEHAVIOR IN INSULATING SOLIDS

11.6.1. Charge Carriers and Mobility

Glasses, ceramics, and polymers are the materials that spring to mind when we think of insulators. What distinguishes them electrically from conductors and semiconductors? Even among a given material class there are better and poorer insulators whose resistance can differ from one another by many orders of magnitude. Why? These and other questions are addressed here. A good

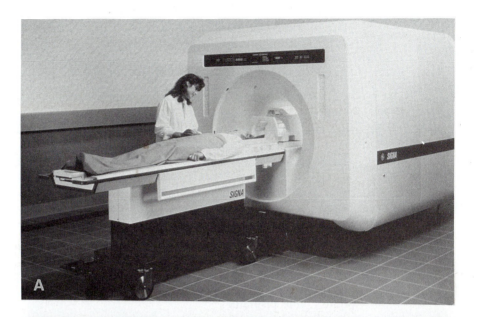

FIGURE 11-18 Applications of superconductors. (A) Magnetic resonance imaging system. Courtesy of GE Medical Systems; (B) A 13-ft-long rotor that operates at 4 K is being lowered into an experimental superconducting electric generator. In tests, 20,600 kW of electricity, or twice the power of conventional generators of comparable size, was produced. Courtesy of GE Corporate Research and Development.

place to start is Eq. 11-4, which defines the conductivity as $\sigma = nq\mu$. Each of the terms is important and deserves scrutiny.

11.6.1.1. The Charge Carrier

Current carriers in insulators can be negatively charged electrons, positively charged holes (the absence of electrons), ions of either positive or negative sign, and point defects (vacancies) of either charge sign. In all cases the charge magnitude is the electronic charge q. Directed motion of any of these entities past a reference plane will contribute to the current. Because the energy gap is large (5–10 eV) in these materials, very few electrons are promoted from the valence to the conduction band, where they can conduct electricity. Therefore, the major insulating solids normally have few conduction electrons. This contrasts with metals and semiconductors with their immensely larger mobile electron (and hole) populations. Usually, ions and defects are the charge carriers in glasses and ceramics. But at elevated temperatures, electrons that are excited across the energy gap also contribute to conduction processes. Electrons, interestingly, dominate conduction in thin-film amorphous oxide insulators.

11.6.1.2. Carrier Concentrations

At low temperatures the majority of the carriers in glasses, ceramics, and polymers are the impurity atoms (ions). These are, for example, cations (e.g., Na^+, Ca^{2+}) that are intentionally added to increase the fluidity of glass, ions used in sintering aids for ceramics, and compounds employed in processing polymers. Collectively, they are the **extrinsic** sources of carriers that are present at concentration levels of n_{ex}. Contrary to their effect in metals, such impurities always raise the conductivity of insulators and semiconductors. In addition, there are **intrinsic** carriers. The Schottky defects (vacancies) that were introduced in Section 3.5 are examples of such carriers in ionic solids. Their concentration, n_{in}, hinges solely on temperature according to a Boltzmann dependence, $n_{in} \sim \exp(-E_{in}/kT)$ (see Section 5-8), where E_{in} is the energy required to make an intrinsic carrier. Note that the latter could be electrons, in which case E_{in} would be the energy gap (or, more correctly, half the energy gap). The total carrier concentration is the sum of the two, or $n = n_{in} + n_{ex}$.

11.6.1.3. Carrier Mobility

Now that the source, magnitude, and concentration of carriers have been identified, all that remains is to get them to move in an electric field. But this is easier said than done. First, there has to be a place for the ions to go, or states that electrons can occupy. Second, carriers generally have to surmount an energy barrier (E_M) by a kind of diffusive hopping to arrive at these destinations. Therefore, the mobility will entail a Boltzmann factor: $\mu \sim \exp(-E_M/kT)$. The previously developed Nernst–Einstein equation is another way to visualize mobility. As $\mu = v/\mathscr{E}$, the use of Eq. 6-29 yields $\mu = qD/kT$ (note $F = q\mathscr{E}$).

A combination of factors then yields

$$\sigma = A(n_{in} + n_{ex})(q/kT) \exp(-E_M/kT), \qquad (11\text{-}23)$$

where A is a constant. Examples that illustrate the above model of ionic conduction are depicted in Fig. 11-19 for the Schottky and Frenkel mechanisms.

Silicate glasses have a relatively large number of ionic carriers as well as sites for them to move into. A high conductivity results, making them poor insulators. They become better insulators if the silica in the network is replaced with Al_2O_3 or B_2O_3, and the Na^+ ions with Ba^{2+} or Pb^{2+}. The former would reduce the number of bridging oxygens about which cations congregate, and the latter are large ions that move with difficulty. Crystalline ceramics have both few carriers and few sites to accommodate them. This is a consequence of their strong bonds, which mean high activation energies for defect creation and motion. They make excellent insulators, as do pure polymers, which have many sites available but few carriers to fill them. Elevated impurity contents

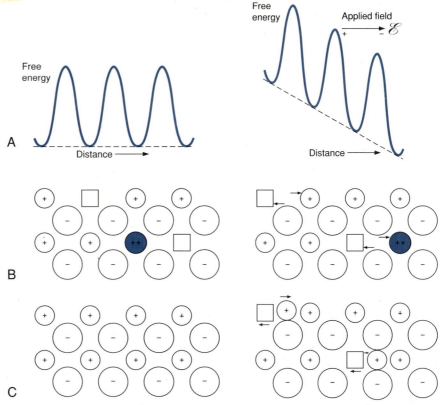

FIGURE 11-19 Electrical conduction in ionic solids. *Left:* (A) Lattice potential for positive charge carriers with no applied field. (B) Lattice containing a divalent impurity and associated vacancy. (C) Perfect lattice *Right:* (A) Lattice potential for positive charge carriers in an applied field. (B) Schottky conduction involving exchange of (negatively charged) vacancies with positive ions. (C) Frenkel conduction involving interstitial motion of ions.

can alter these characteristics dramatically. In fact, semiconducting polymers such as polyacetylene with conductivities of 1.5×10^7 $(\Omega\text{-m})^{-1}$ have been synthesized. In these materials certain electrons along the polymer chain become delocalized.

11.6.2. Sensing Oxygen

Oxygen sensor applications include monitoring the oxygen level in automotive fuel mixtures and exhausts and determining the oxygen content in steel melts. The reason for considering them here is that the high-temperature conductivity of the ceramic $ZrO_2–Y_2O_3$ is capitalized on in these devices. This material is the solid ionic electrolyte in a voltaic cell shown schematically in Fig. 11-20. Platinum electrodes are exposed to different oxygen pressures and the cell essentially transports oxygen from the high- to low-pressure side until the pressures equilibrate. The basic reversible reaction is

$$O_2 \text{ (atmosphere)} + 4e^- \rightleftarrows 2O^{2-} \text{ (electrolyte).} \qquad (11\text{-}24)$$

Borrowing the Nernst relationship from the preceding chapter (Eq. 10-8), we have

$$E_{cell} = E^0_{cell} + (RT/nF) \ln(P_1/P_2) \qquad V, \qquad (11\text{-}25)$$

where P_1 and P_2 are the low and high oxygen pressures involved. Since P_2 is normally the ambient oxygen pressure, which is constant, the cell potential is proportional to $\sim\ln P_1$. In automotive applications P_1 ranges between $\sim10^{-2}$ atm for slightly lean air–fuel mixtures and $\sim10^{-20}$ atm for fuel-rich compositions. The sensitivity of the sensor is roughly 50 mV for a 10-fold change in

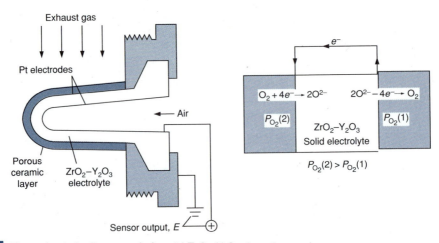

FIGURE 11-20   Electrochemical cell composed of a solid $ZrO_2–Y_2O_3$ electrolyte used to sense oxygen.

O_2 level at ~500°C. Therefore, an output voltage range of about 0.9 V that varies somewhat with temperature can be expected.

The basic materials problem is to design the solid electrolyte. Desirable properties include the ability to readily partake in reversible oxygen–oxygen ion reactions at elevated temperatures, high conductivity, rapid transport of O^{2-} ions, and mechanical stability during thermal cycling. The weakly bonded, open matrix of zirconia allows for oxygen exchange at the electrodes, while introduction of oxygen vacancies makes it more conducting. For the latter purpose, Y_2O_3 works well because a single molecule substitutes for two ZrO_2, leaving one oxygen vacancy behind. Besides, there is an added dividend because yttria stabilizes the zirconia and toughens it against thermal shock (see Section 9.8.2).

11.7. DIELECTRIC PHENOMENA

11.7.1. Definitions

Dielectrics are materials possessing high electrical resistivities. A good dielectric is therefore a good insulator but the reverse is by no means true. Electrically, dielectrics are called upon to perform functions in ac circuits other than simple isolation of components. It is the electric dipole (positive and negative charges separated by a distance) structure they possess intrinsically, or assume in direct and alternating electric fields, that is the key to understanding their nature and use. A capacitor consisting of two parallel conducting plates of area A, separated a distance L by a vacuum space, is a good place to start. The ratio of the charge, Q (coulombs, C), stored on the plates to the voltage, V, across them is defined as the capacitance, C (farads, F):

$$C = Q/V. \tag{11-26}$$

However, the ability to store charge depends on the geometry of the plates and the filling between them (Fig. 11-21A). For a vacuum capacitor,

$$C = \varepsilon_o A/L, \tag{11-27}$$

where ε_o is the permittivity of free space ($\varepsilon_o = 8.85 \times 10^{-12}$ F/m). If a dielectric is now inserted, keeping the voltage constant, it is observed that more charge (Q') can be stored on the plates. This effect can be interpreted as signifying a higher capacitance which can be accounted for in Eq. 11-27 by substituting ε for ε_o, where ε is the permittivity of the dielectric. Dielectric properties of different materials are compared via the *relative* permittivity or **dielectric constant,** defined as $\varepsilon_r = \varepsilon/\varepsilon_o$. Values of the dielectric constant for assorted materials are given in Table 11-5 together with other relevant dielectric properties introduced later.

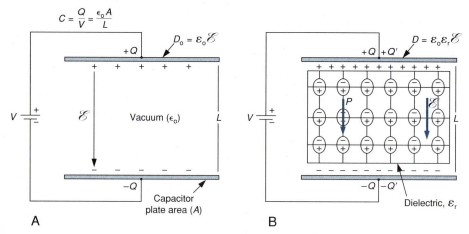

$$C = \frac{Q}{V} = \frac{\epsilon_0 A}{L}$$

A

B

(A) Capacitor plates with an intervening vacuum space. (B) Capacitor filled with a dielectric. In this case more charge is stored on the plates for the same voltage.

DIELECTRIC PROPERTIES OF SELECTED MATERIALS (AT 300K)[a]

Material	Dielectric constant (at 10^6 cycles)	Dielectric strength (kV/mm)	Loss (tan δ)
Air	1	3.0	
Alumina	8.8	1.6–6.3	0.0002–0.01
Cellulose acetate	3.5–5.5	9.8–11.8	
Glass (soda-lime)	7.0–7.6	1.2–5.9	0.004–0.011
Mica	4.5–7.5	2.0–8.7	0.0015–0.002
Nylon	3.0–3.5	18.5	0.03
Phenol–formaldehyde resin (no filler)	4.5–5.0	11.8–15.8	0.015–0.03
Polyester resin	3.1	17.1	0.023–0.052
Polyethylene	2.3	18.1	0.0005
Polystyrene	2.4–2.6	19.7–27.6	0.0001–0.0004
Porcelain (high-voltage)	6–7	9.8–15.7	0.003–0.02
Pyrex glass		500	
Rubber (butyl)	2.3	23.6–35.4	0.06
Silica thin films (gate oxide)	3.8	700	0.0002
Silicone molding compound (glass-filled)	3.7	7.3	0.0017
Steatite	5.5–7.5	7.9–15.8	0.0002–0.004
Styrene (shock-resistant)	2.4–3.8	11.8–23.6	0.0004–0.02
Titanates (Ba, Sr, Mg, Pb)	15–12,000	2.0–11.8	0.0001–0.02
Titanium dioxide	14–110	3.9–8.3	0.0002–0.005
Urea–formaldehyde	6.4–6.9	11.8–15.8	0.028–0.032
Vinyl chloride (unfilled)	3.5–4.5	31.5–39.4	0.09–0.10
Zircon porcelain	8–9	9.8–13.8	0.0006–0.002

[a] Data were taken from various sources.

11.7.2. Polarization

11.7.2.1. Macroscopic Polarization

Internal dipole moment alignment, or **polarization** in the presence of the electric field, is the reason dielectrics enhance capacitance. Polarization can arise, for example, because positive and negative ions or charge centers within molecules and atoms separate and assume the direction of the applied field. Although all of the dipoles within the dielectric align, it is those dipoles immediately facing the plates that effectively have uncompensated charges, as seen in Fig. 11-21B. This polarization charge on the dielectric surface attracts additional opposite charge from the battery. An increased charge density on the capacitor plates is the result.

How the polarization charge density, P, is related to the plate charge densities, with and without the dielectric, is developed in basic physics courses. From electrostatics we know that in a given electric field ($\mathscr{E} = V/d$ in V/m), the surface charge density, D, on capacitors filled with vacuum and dielectric are

$$D_o = \varepsilon_o \mathscr{E} \quad \text{C/m}^2 \tag{11.28a}$$

and

$$D = \varepsilon_o \varepsilon_r \mathscr{E}, \tag{11.28b}$$

respectively. The difference between the two is defined as the polarization,

$$P = D - D_o = \varepsilon_o(\varepsilon_r - 1)\mathscr{E}, \tag{11.29}$$

a quantity that increases linearly with the magnitude of the electric field.

11.7.2.2. Microscopic Polarization

As already noted, dipoles within the dielectric are responsible for the polarization. There are two kinds of dipoles in materials—those that are **induced** and those that are **permanent**—and both cause polarization or charge separation. A measure of the latter is the dipole moment, μ_d, defined as $\mu_d = Qd$, where Q is the magnitude of the charge and d is the distance separating the pair of opposite charges. The connection between the macroscopic polarization and microscopic dipole moments is given by

$$P = \left(\sum \mu_d\right)/V_o = \left(\sum Qd\right)/V_o, \tag{11-30}$$

where the sum over the number of moments per unit volume (V_o) of solid is carried out vectorially.

Dielectrics are good insulators so that charge is not easily transported in them. This does not prevent the charge within atoms and molecules from responding to applied electric fields, however, by parting slightly from one another or even reorienting. When this happens the dielectric is polarized.

Three basic types of polarization that contribute to the total magnitude of P in a material have been identified:

1. *Electronic.*   At any instant, atomic electron charge clouds are displaced toward the positive end of an applied electric field while the positively charged nucleus is attracted to the negative end (Fig. 11-22A). If the field reverses direction, the atomic charge reorients in concert but the magnitude of μ_d is preserved. If, however, the electric field is turned off, the polarization vanishes; clearly, *electronic polarization* is an *induced* effect. It can persist to extremely high electric field frequencies because electron standing waves within atoms have very high natural frequencies ($\sim 10^{16}$ Hz). The electronic polarization (P_e) is proportional to the magnitude of $\mathscr{E}$ in small electric fields, and can follow its oscillations at optical frequencies. Because of this, there is a very important connection between dielectric and optical properties. It will not be proved here but the *refractive index* is equal to the square root of the *dielectric constant*:

$$n_r = \varepsilon_r^{1/2}. \qquad (11.31)$$

The importance of dielectric materials in optical applications is more fully drawn in Chapter 13.

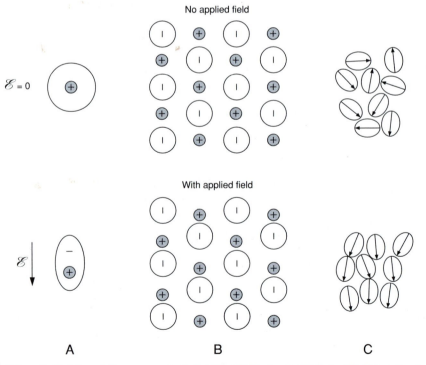

No applied field

With applied field

$\mathscr{E} = 0$

$\mathscr{E}$

A   B   C

FIGURE 11-22   Schematic depiction of the sources of polarizability. (A) Electronic. (B) Ionic. (C) Orientational.

2. *Ionic.* Many dielectrics are ionic materials. Application of alternating electric fields deform the cation and anion bonds, *inducing* net dipole moments (Fig. 11-22B) and an ionic polarization given by P_i. Again, these dipoles can follow alternating electric fields but the ions are too heavy to oscillate in phase with them much beyond frequencies of about 10^{13} Hz.

3. *Orientational.* Dielectrics often possess microscopic *permanent* dipole moments. Examples include water and transformer oils, where one end of the molecule is effectively positive while the other end is negative (Fig. 11-22C). Application of electric fields causes these molecular dipoles to orient; but they are also buffeted by the surrounding thermal vibrations which tend to randomize their orientations. The orientational polarization (P_o) is larger the greater the electric field and the lower the temperature.

By summing the individual contributions, the total polarization is obtained as

$$P = P_e + P_i + P_o. \tag{11-32}$$

11.7.3. Frequency Response, Dielectric Loss, and Breakdown

11.7.3.1. Frequency Response

The dielectric constant as a function of frequency is schematically depicted in Fig. 11-23 for a typical dielectric. All three polarization mechanisms are operative at the lowest frequencies and, therefore, ε_r is high. As the frequency is raised, one polarization mechanism after another is frozen out. The first to

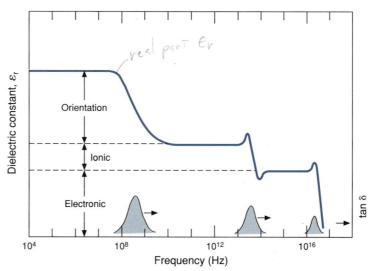

real part ε_r

FIGURE 11-23 Typical frequency response of the dielectric constant of solids. Electrons can respond to alternating electric fields below $\sim 10^{16}$ Hz. Ions and molecular dipoles respond to frequencies less than $\sim 10^{13}$ Hz and $\sim 10^8$ Hz, respectively. Dielectric loss (shaded) peaks occur at the indicated frequencies.

stop contributing to ε_r is the orientational component, then the ionic, and lastly the electronic.

11.7.3.2. Dielectric Loss

Both dielectric loss and breakdown are undesirable characteristics to which all dielectric materials are susceptible. Although it is possible to live with dielectric loss through proper electrical design, dielectric breakdown causes a catastrophic failure of the material. Dielectric loss can be understood in electrical engineering terms. In ideal capacitors it is well known that the ac current *leads* the voltage by 90°. But real capacitors have a resistive component that make them *lossy* so they dissipate some of the applied ac energy as Joule heat. This slightly reduces the lead angle by δ degrees. The quantity tan δ, plotted schematically in Fig. 11-23, is a measure of the ratio of the electrical energy lost per cycle to the maximum energy stored. Values for tan δ are given in Table 11-5.

It is instructive to consider the analogy to a mechanical system, where stress and strain are like electric field and polarization, respectively. If the system is elastically stressed in tension, unloaded, then elastically compressed, and finally unloaded, no energy is dissipated in the cycle because strain is in phase with the stress. Similarly, if the polarization is in phase with the ac driving field, as at very low frequencies, the response is "elastic" and there is no energy loss. But usually the dipole oscillations are not in phase with the field; polarization at high frequencies may cause changes in energy level populations, and orienting polar molecules suffer a kind of friction when rotating relative to neighbors. By these and other mechanisms energy is ultimately dissipated as heat. The dipoles can almost follow the field and the imperfect coupling of electrical energy makes for lossy motion. A similar thing happens in vibrating solids where the energy loss is caused by so-called **internal friction.** Energy dissipation in such cases is often attributed to dislocation and point defect motion, diffusional processes, and viscous flow.

11.7.3.3. Dielectric Breakdown

Application of a very high electric field to an insulator can set in motion a train of events that rapidly leads to an irreversible breakdown. The breakdown field (in V/cm or kV/mm) is a common measure of the dielectric strength of the material, as is tensile strength for mechanical behavior. Typical values are included in Table 11-5. Manifestations of dielectric breakdown include melted and even vaporized material, holes, and craters. Often other electrical components that rely on the high-voltage isolation dielectrics provide also fail. Dielectric breakdown can be made visible in large glass (or polymer) specimens by exposing them to gamma ray (ionizing) radiation and then mechanically stressing the surface at a point. The dramatic lightning-like discharge, captured in

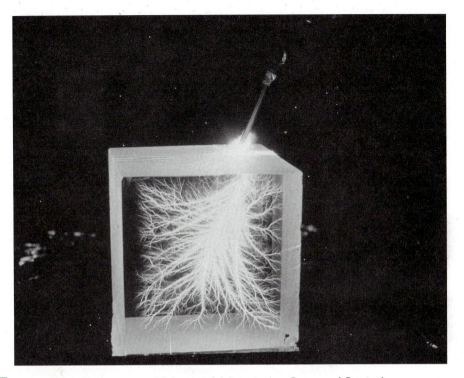

FIGURE 11-24 Lichtenberg discharge pattern of dielectric breakdown in glass. Courtesy of Corning Inc.

Fig. 11-24, initiated when the stored radiation-induced charge was set in motion by the large built-in electric field.

Several mechanisms for dielectric breakdown have been suggested. In one, a small number of carriers in the conduction band are accelerated in the field and these collide with atoms, ionizing them. This releases more carriers for further impact ionization and the current rapidly builds. But the resultant Joule heating causes insulators to become better conductors that can pass more current. The process feeds on itself until a thermal runaway causes local failure. Shaping insulators in special ways can effectively reduce their tendency to break down or conduct at high voltages. For example, the familiar convoluted or ribbed surface profile on large cone-shaped, ceramic insulators, used to isolate high-voltage transformers and transmission lines, is not a decorative feature. Rain water is less likely to wet the contour undersides, thus minimizing the hazard of surface conduction short-circuiting of the bulk insulator.

Dielectric breakdown is an important reliability problem in the very thin oxides employed in microelectronic devices. This subject is dealt with again in the last chapter of the book.

11.8. DIELECTRIC MATERIALS AND APPLICATIONS

11.8.1. Traditional Dielectrics

Thus far we have focused on ceramic, polymer, and glass dielectrics used in capacitors and for high-voltage insulation in assorted conventional electrical applications. There are situations requiring either low- or high-dielectric-constant materials. In high-speed pulse circuits and transmission lines, for example, low capacitance is essential because the RC time constant (R = resistance) governs the rate of buildup and decay of signals. Low values of ε_r are then sought in components like printed circuit boards to minimize the possibility of signal overlap. On the other hand, a large value of ε_r is needed to store large amounts of charge in physically small capacitors. Except for dielectric heating applications, low dielectric loss is usually desired. High dielectric breakdown strength is essential in all applications. Often breakdown is the overriding concern. Table 11-5 is devoted, for the most part, to conventional dielectric materials. But there are other dielectrics that have emerged to fill new needs, and they are discussed next.

11.8.2. Ferroelectrics

Ferroelectric materials have *permanent* electric dipoles and therefore exhibit a **spontaneous polarization,** that is, polarization even without an applied electric field. They are analogous to the more common ferromagnetic materials that contain permanent *magnetic dipoles* and are therefore *spontaneously magnetized*. This will be appreciated once Chapter 14 is completed. In fact, the adoption of the *ferro* prefix points to other similarities (like the hysteresis loop, see Section 14.4.1) even though ferroelectrics rarely, if ever, contain iron. The structure and properties of ferroelectrics are the keys to understanding their role in transducer and sensor applications.

One of the most important ferroelectrics is $BaTiO_3$, whose crystal structure at elevated temperatures is like that of an ideal perovskite shown in Fig. 4-22. It consists of octahedral cages of O^{2-} ions joined at corners. Within each octahedron is a Ti^{4+} ion while a Ba^{2+} ion occupies the cube center. A closer look reveals that the structure is symmetrical or has a center of symmetry at high temperatures. But below the critical **Curie temperature** of 125°C a structural transformation occurs. The cubic structure becomes unstable and undergoes a tetragonal distortion. Because the cations and anions shift in opposite directions, the centroids of positive and negative ionic charge no longer coincide. As a result the crystal develops a *net* polarization. The ion positions deviate very slightly from those in the cubic cell as indicated in a new view of the unit cell in Fig. 11-25. The difference is enough to develop a net dipole per unit cell as the following calculation shows.

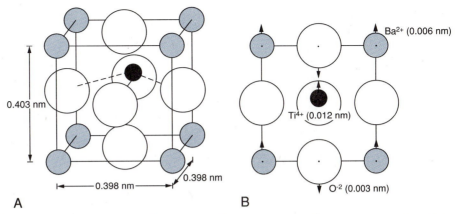

A   B

FIGURE 11-25 Permanent dipole structure of $BaTiO_3$ below the Curie temperature. (A) The cubic structure has a slight tetragonal distortion. (B) Directions and magnitudes of the ion displacements relative to the midplane oxygen ions (e.g., Ti^{4+} moves up 0.012 nm, O^{2-} moves down 0.003 nm, and Ba^{2+} moves up 0.006 nm).

EXAMPLE 11-4

a. Calculate the polarization in a unit cell of $BaTiO_3$ if the ion positions are located as shown in Fig. 11-25.

b. If the relative dielectric constant is 4300, what is the effective internal electric field corresponding to this polarization?

ANSWER a. Assuming all calculations of displaced ions are made relative to the midplane oxygen ions, there are one Ti^{4+}, two O^{2-}, and eight Ba^{2+} ions to consider.

1. The Ti^{4+} ion is displaced by $d = 0.012$ nm. $Q = 4 \times (1.6 \times 10^{-19}C)$. Therefore,

$$Qd = 4 \times (1.6 \times 10^{-19}) \times (0.012 \times 10^{-9}) = +7.68 \times 10^{-30} \quad \text{C-m.}$$

2. The top and bottom O^{2-} ions are each shared by two cells and count only as one ion. Therefore,

$$Qd = -2 \times (1.6 \times 10^{-19}) \times (-0.003 \times 10^{-9}) = +0.96 \times 10^{-30} \quad \text{C-m.}$$

3. The eight Ba^{2+} ions are shared by eight cells and count as one ion. Therefore,

$$Qd = +2 \times (1.6 \times 10^{-19}) \times (0.006 \times 10^{-9}) = +1.92 \times 10^{-30} \quad \text{C-m.}$$

The total polarization is $P = \left(\sum Qd \right)/V_o$, where

$$\sum Qd = (7.68 + 0.96 + 1.92) \times 10^{-30} = 10.6 \times 10^{-30} \quad \text{C-m}$$
$$V_o = (0.403 \times 0.398 \times 0.398) \times 10^{-27} \quad m^3 = 0.0638 \times 10^{-27} \quad m^3.$$

$P = \epsilon_0 (\epsilon_r - 1) E$

Therefore, $P = (10.6 \times 10^{-30})/(0.0638 \times 10^{-27}) = 0.166$ $C\text{-}m^{-2}$. The experimentally measured polarization is 0.26 $C\text{-}m^{-2}$.

b. From Eq. 11-29,

$\mathscr{E} = P/\varepsilon_0(\varepsilon_r - 1) = 0.166/8.85 \times 10^{-12}(4300 - 1) = 4.36 \times 10^6$ V/m.

The above calculation suggests that one might sustain an electrical shock by touching a crystal of $BaTiO_3$. This does not happen, however. To lower the large external electrostatic energy that totally aligned electric dipoles would create, the matrix decomposes into an array of grainlike **domains.** In each domain all of the unit cell dipoles are aligned and the polarization is ~0.26 $C\text{-}m^{-2}$. But the polarization vector differs in neighboring domains, and when averaged over many domains throughout the volume, the *net* value of P vanishes. When such a material is polarized in electric fields, favorably oriented domains grow at the expense of others, and now the net value of P does *not* vanish. What is so special about ferroelectrics is that their polarization can be *reversed* by changing the field direction. And when ferroelectrics are subjected to cycles of reversed electric fields, the polarizations respond in concert and trace out a **hysteresis loop,** as shown in Fig. 11-26. Importantly, polycrystalline sintered powder compacts display these effects and are commercially employed in devices; expensive single crystals are not necessary.

We can understand the processing and use of these materials from this figure. Starting with randomly oriented domains (1) an electric field applied at elevated temperatures produces a net macroscopic polarization; domain growth (2) and alignment (3) near saturation are facilitated in this so-called **poling** process. Then the temperature is lowered with the field in place. This locks in a remnant polarization (4) of magnitude P_r, even when the field is removed. In state 4 the ferroelectric exhibits electromechanical effects when electric fields are applied, as discussed below. On the other hand, state 1 allows for normal capacitor behavior (in low reversible, electric fields). Parallel domain and hysteresis phenomena occur in ferromagnetic materials (see Chapter 14). On this basis, ferroelectrics can be understood by interchanging the analogous quantities—magnetic and electric fields, and magnetization and polarization.

Properties of a number of important ferroelectric materials are listed in Table 11-6. Anomalously large dielectric constants are noteworthy because they enable capacitors and microwave components (e.g., waveguides, couplers, and filters) to be miniaturized. The latter are frequently used in communications, navigation, and various types of radar.

11.8.3. Piezoelectric Materials and Devices

All ferroelectric materials are **piezoelectric** but the reverse is not true; single crystal quartz is an example of a piezoelectric material that is not ferroelectric. When subjected to a stress, piezoelectric crystals develop an electrical polarity. Compressive stresses will cause charge to flow through a measuring circuit in

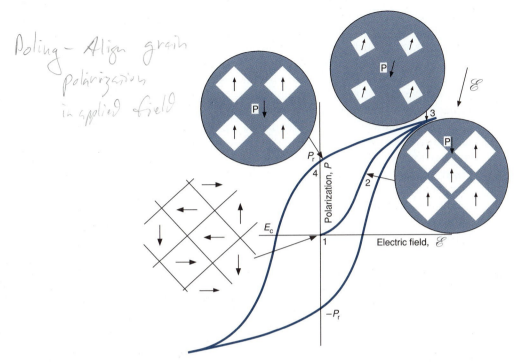

Poling – Align grain polarization in applied field

FIGURE 11-26	Typical hysteresis loop in ferroelectric materials. As the dc electric field is cycled to a maximum value with one polarity, decreased to zero, and then similarly increased and returned to zero with the opposite polarity, the net polarization traces the hysteresis loop shown. The corresponding variation in the domain structure is indicated, with the matrix polarized in the field direction shaded.

		PROPERTIES OF SOME PIEZOELECTRIC MATERIALS
TABLE 11-6		

Material	Density (Mg/m^3)	Dielectric constant	Curie temperature (°C)	Saturation polarization (C/m^2)	Coupling coefficient, K
Quartz	2.65	4.6	575		
Li_2SO_4	2.06	10.3			
$BaTiO_3$	5.7	1900	130	0.26	0.38
$PbTiO_3$	7.12	43	494		
PZT-4[a]	7.6	1300	320	~0.5	0.56
PZT-5[a]	7.7	1700	365	~0.5	0.66
$LiNbO_3$	4.64	29	1210	0.74	0.035
Rochelle salt		5000	24		

[a] Lead zirconate titanate.

Sources: *Handbook of Tables for Applied Engineering Science,* 2nd ed., CRC Press, Boca Raton, FL (1973), and A. J. Moulson and J. M. Herbert, *Electroceramics,* Chapman and Hall, London (1990).

one direction; when tensile stresses are applied, the direction of current flow reverses. The stress either shortens or lengthens the dipole length, altering the polarization and surface charge in the process. Conversely, piezoelectric crystals also exhibit inverse or **electrostrictive** effects. An applied electric field forces the dipole spacing to extend or shrink, depending on the orientation of the field with respect to the polarization direction. In this case the crystal undergoes either a tensile or a compressive strain. These coupled mechanical and electrical effects, schematically depicted in Fig. 11-27, form the basis of a number of commercial devices known as **transducers** and **actuators.** The former are sensing devices that convert various stimuli (stress, pressure, light, heat, etc.) into electric signals; the latter provide displacement of electromechanical devices (e.g., linear motors, hydraulic cylinders) as an output.

Underwater detection of submarines by sonar is an instructive engineering application because both the direct and the inverse piezoelectric effects are involved. As illustrated in Fig. 11-28, ac electric power fed into a piezoelectric transmitter generates a vibratory mechanical motion that launches sound waves into the water. Waves that hit remote submerged objects are reflected back, radar style, and are detected by an array of hydrophones. Piezoelectric transducers in the latter detect the pressure waves of the echo and convert them into electric signals. Ultrasonic imaging of fetuses in the womb and detection of flaws in metals are related examples that rely on piezoelectrics for mechanical–electrical conversion. Phonograph cartridges and actuators used to translate specimens in the scanning tunneling microscope (see Section 3.4.2) are two of the many additional applications of these materials.

To efficiently generate strong signals and detect weak ones, piezoelectric materials must have a large value of the **coupling coefficient,** K. It turns out that K^2 is equal to the ratio of mechanical energy output to electrical energy

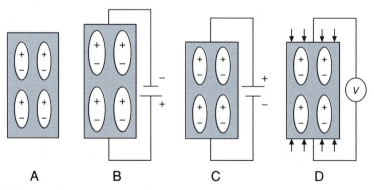

A   B   C   D

Schematic of electrostriction and piezoelectric effects. (A) No applied stress or electric field. (B) With applied electric field, specimen elongates. (C) With reverse electric field, specimen shortens. (D) Applied stress induces polarization and external voltage.

FIGURE 11-28 Use of piezoelectric materials in underwater surveillance of a submarine.

input. As the effects are reversible, K^2 is also the ratio of electrical energy output to mechanical energy input. Ceramics known as **PZT**s have the highest values of K. These materials are essentially combinations of PbO, ZrO_2, and TiO_2, have perovskite structures, and may be thought of as binary $PbZrO_3$–$PbTiO_3$ alloys. For the 50–50 alloy, K is the largest, having a value of ~ 0.6. The special processing noted earlier is necessary to develop optimal properties. Related materials like $LiNbO_3$ are also piezoelectric and, additionally, exhibit important acousto-optical effects.

11.8.4. Liquid Crystals

The chapter ends with a discussion of dielectric liquid crystal display (LCD) materials. These interesting fluids consist of rodlike organic polymers (e.g., polypeptides, cyanobiphenyl compounds) that can be aligned in electric fields. It is the ordering that makes their dielectric and optical properties anisotropic, and this behavior is exploited in LCDs used in such products as wrist watches and electronic calculators. In fact, thin-film transistor/liquid crystal displays will be the computer monitors of the 1990s; they have already found applications in lap-top computers and in hand-held televisions.

Three major types of liquid crystals are schematically indicated in Fig. 11-29:

1. **Nematic** liquid crystals are the most liquidlike because they are the least ordered. Nevertheless, the rods are reasonably parallel and translate relative to one another as well as rotate about their long axis. In electric fields, the

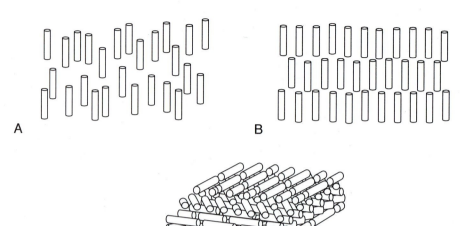

A   B

C

FIGURE 11-29   Liquid crystal materials. (A) Nematic. (B) Smectic. (C) Cholesteric.

rods can line up either parallel to (homeotropic alignment) or perpendicular to (heterotropic alignment) the field axis.

2. There are several types of **smectic** liquid crystals. A common configuration shown in Fig. 11-29B is similar to that of the nematic case except that the smectic is layered.

3. The **cholesteric** phase resembles the nematic on a very local scale. Further examination shows that over broader dimensions the layers rotate relative to one another, giving the structure the appearance of a spiral staircase. The pitch varies in different materials and can be either right or left handed.

The most common LCD devices are based on twisted nematic liquid crystals, and a cell employing them is shown in Fig. 11-30A. At each end of the cell is a transparent electrode (a film of indium–tin oxide) and an optical polarizer plate. Some 10 μm of liquid crystal fluid occupies the space between the electrodes. Each electrode plate is grooved and serves to conveniently mechanically align the liquid crystal molecules, that is, parallel to the plane of optical polarization at the top and perpendicular to it at the bottom. The twist of the molecules causes a similar twist in the electric and optical polarizations from top to bottom. Light that enters the cell is plane polarized at the top, rotated 90° in the liquid, and exits through the bottom polarizer. In a strong electric field ($\sim 5 \times 10^5$ V/m produced by ~ 5 V) the molecules align perpendicular to the electrodes (Fig. 11-30B). Now the uniaxial electric polarization is incapable of rotating the polarization of the light, so that it cannot pass through the lower polarizer. In this way light and dark contrast is achieved.

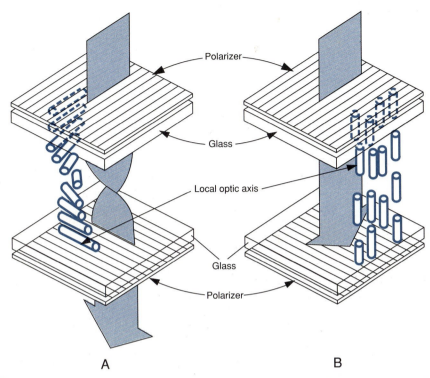

FIGURE 11-30 Display system employing nematic liquid crystals. (A) Without electric field, light is transmitted. (B) With applied electric field, light is blocked.

11.9. PERSPECTIVE AND CONCLUSION

Opposite extremes in the electrical behavior of materials were exposed in this chapter. Electrical conductivity distinctions between metals and nonmetals essentially hinge on the nature of the electron distributions in atoms. In ionic and covalent solids only very fixed numbers of electrons are tolerated by atoms; local excesses or deficiencies in their number are resisted. An excess electron flows to an atom with one too few and vice versa; free electrons are very scarce. Electron currents largely counterbalance one another so the net current is vanishingly small. In metals, however, electron numbers about atoms are not so rigidly controlled. Enough free electrons are always available to smooth local depletions and accumulations; electrons fed into a wire at one end will, therefore, be readily removed in equal number at the other end.

Metals are the best conductors of electricity and, therefore, have the lowest resistivities ($\rho \sim 10^{-8}$–10^{-6} Ω-m). **Insulators,** comprising ceramics, glasses, and polymers, have the highest resistivities ($\rho \sim 10^{9}$–10^{18} Ω-m); conduction by ions, not electrons, is often the chief mechanism for charge transport. As the temperature is raised, metals become poorer conductors and insulators become better conductors. Both intentionally added alloying elements and unintentional impurities lower the conductivity of metals, but make insulators better conductors. Dislocations, vacancies, and grain boundaries degrade the conductivity of metals but make insulators more conductive. The effect of these variables on resistivity in metals is largely additive according to Matthiessen's rule. This essentially means that the increase in resistivity due to a small concentration of another metal in solid solution is independent of temperature. For insulators, the effects of temperature and impurities are multiplicative. In particular, the resistivity falls exponentially with temperature according to a Boltzmann factor; in contrast, a more modest linear rise in resistance with temperature is observed in metals. The resistivity can, therefore, easily change by orders of magnitude in insulators, but much less than this in the case of metals.

Because electrons move so easily within them, metals cannot support electric fields inside or across their bounding surfaces. On the other hand, **dielectric** materials have very few mobile charge carriers. In the process of supporting electric fields, local charge separation into dipoles occurs and the dielectric becomes polarized. This effect occurs at all electric field frequencies, but the source of the polarization changes. At the lowest frequencies, atomic and ionic charges respond and net dipoles are induced where none existed before; permanent molecular dipoles can also reorient in phase with the exciting field. But, as the driving electric field frequency increases, the contributions to the polarization vanish one by one, as their ability to oscillate in phase diminishes. Finally, only atomic electrons are left to follow the highest (optical) frequencies. At this point dielectrics are important for their optical properties.

Despite their vast differences, metals and nonmetals occasionally exhibit surprisingly similar electrical behavior. Work functions of metals and ceramics are alike in magnitude, and comparable thermionic (as well as secondary) electron emission can occur from them. Polymers have been prepared that have higher electrical conductivities (on a weight basis) than copper. And of course there are the new high-T_c ceramic superconductors. Making nonmetals more conductive should be viewed in the context of imparting complementary attributes that extend the range of materials utilization. This has been done successfully with mechanical properties and is a general goal of materials science.

Additional Reading

N. Braithwaite and G. Weaver (Eds.), *Materials in Action Series: Electronic Materials,* Butterworths, London (1990).

R. M. Rose, L. A. Shepard, and J. Wulff, *The Structure and Properties of Materials,* Vol. IV, Wiley, New York (1966).

L. Solymar and D. Walsh, *Lectures on the Electrical Properties of Materials,* 4th ed., Oxford University Press, Oxford (1988).

C. A. Wert and R. M. Thomson, *Physics of Solids,* 2nd ed., McGraw–Hill, New York, (1970).

11-1. Distinguish between the following pairs of terms:
   a. Fermi–Dirac distribution function and Maxwell–Boltzmann distribution function
   b. Fermi–Dirac distribution function and density of states
   c. Thermionic emission and photoemission
   d. Fermi energy and work function
   e. Field and secondary electron emission
   f. Brillouin zone and energy band

11-2. The *average* energy of free electrons in a metal at 0 K is $3/5E_F$. Show this result by integrating $EF(E)N(E)$ over the appropriate energy range and normalizing the result.

11-3. A metal has a Fermi energy of 4.5 eV. At what temperature is there a 1% probability that states with energy of 5.0 eV will be occupied?

11-4. An insulator has a band gap of 6.2 eV. What is the probability of valence-to-conduction band transitions at 1250°C?

11-5. What is the longest wavelength that will cause photoemission of electrons from the following tungsten-coated cathodes: W–Th, W–Ba, and W–Cs?

11-6. The thermionic emission current from electron beam guns is often used to weld, melt, or evaporate materials cleanly in a vacuum environment. To efficiently evaporate aluminum to coat potato chip bags, a power of 2 kW is required to heat the Al charge. The power supply delivers a 10-kV potential between the cathode (emitter) of the electron beam gun filament and ground (Al).
   a. For a tantalum emitter what minimum surface area of filament would be required?
   b. For a thoriated tungsten emitter what minimum surface area of filament would be required?

11-7. A cable consists of three strands of copper wire each 1 m long and 250 μm in diameter. What is the room-temperature resistance of the cable?

11-8. Approximately how many collisions do electrons make each second with lattice scattering sites in pure silver at 25°C?

11-9. Short, narrow thin-film aluminum stripes are used as interconnections in integrated circuits as shown in Fig. 1-14C. Design rules call for 0.5-μm-thick $\times$ 0.5-μm-wide conductors that carry a current density of no more than 1×10^9 A/m^2. For a conductor with a resistivity of 2.9×10^{-8} Ω-m that is 0.0001 m long, what is the voltage across the stripe when the maximum current flows?

11-10. A 120-V source is placed directly across a tungsten filament 10 m in length and 50 μm in diameter.
   a. What current is passed initially?
   b. What is the power in watts?
   c. If the filament were perfectly insulated and heated, approximately what

temperature would it reach if the initial current flowed for 1 second? Assume the heat capacity of W is 25 J/mol.

d. What would the resistance be at this temperature?

11-11. "On October 21, 1879, Thomas Edison's famous (tungsten filament) lamp attained an incredible life of 40 hours. The entire electrical industry has as its foundation this single invention—the incandescent lamp" (J. W. Guyon, GEC (1979).

   a. Comment on the second sentence.

   b. How many different materials and processing methods can you identify in a common light bulb.

11-12. The cubic composites shown in Fig. 9-28 are made up of slabs with electrical resistivities ρ_A and ρ_B. In each composite the volume fraction of A is V_A, and that of B, V_B. What resistance would be measured in the direction of F_c in each case?

11-13. What typical metals and electrical properties (e.g., high, low, intermediate resistivity, temperature coefficient of resistivity) would you recommend for the following applications?

   a. Electrical furnace windings

   b. Electrical power transmission lines

   c. Electrical resistance standards

11-14. Consult an appropriate reference to find out how metal resistance thermometers and strain gauges operate. In each case what material properties are capitalized upon?

11-15. Copper and aluminum cables compete for electrical wiring applications. Consider two wires, one of Cu and the other of Al, having the same resistance. What is the ratio of the weights of these wires if they have the same length?

11-16. Using the data of Fig. 11-12, construct resistivity-versus-temperature plots for pure copper, Cu alloyed with 0.10 wt% Cr, and Cu alloyed with 0.10 wt% Fe. (The plot should extend from -100 to $100°C$.)

11-17. A vacuum-melted oxygen-free copper has measured resistivities of 1.72×10^{-8} Ω-m at 298 K and 5.97×10^{-11} Ω-m at 4.2 K.

   a. What is the residual resistivity ratio?

   b. Estimate a value for RRR at 77 K, the temperature at which liquid nitrogen boils.

11-18. At 298 K lead has a resistivity of 20.6×10^{-8} Ω-m, whereas at 4.2 K sensitive instruments indicate that the resistivity does not exceed 10^{-25} Ω-m.

   a. What is the apparent residual resistivity ratio?

   b. It is known that for the purest metals, RRR values do not range much beyond 100,000. Is the measurement on Pb therefore incorrect? Explain.

11-19. A need arises for more conductive polymer parts. Suggest a possible way to fabricate them, employing common materials and processing methods.

11-20. a. The heated surfaces of stainless-steel cookware are clad with copper. Why?

   b. Magnesium is a better conductor of heat than magnesium oxide. Why?

   c. In metals the electrical and thermal conductivities are correlated, but in ceramic oxides and polymers they are not. Why not?

   d. A vacuum is an excellent thermal insulator. Why?

11-21. For superconductors below the critical temperature, T_c, the critical magnetic flux varies with temperature T as $B_c = B_0(1 - T^2/T_c^2)$, where B_0 is the critical flux at 0 K.

    a. Through a plot of B_c versus T, pictorially indicate the variation between the variables.

    b. What is the value of B_c (in Tesla) at $T = 10$ K for Nb_3Sn?

    c. For a Nb_3Sn superconducting magnet that must deliver flux densities of 23 T (Tesla), below what temperature must the magnet operate to still be superconducting?

11-22. What are some reasons that poor insulators may not make good dielectrics for capacitor applications.

11-23. A parallel plate capacitor with electrodes measuring 0.1×0.1 m, spaced 10 μm apart, is filled with a dielectric with a relative permittivity of 4.2.

    a. What is the capacitance?

    b. In an electric field of 10,000 V/m, what is the surface charge density on the plates?

    c. What is the polarization?

11-24. Distinguish between the following pairs of terms:

    a. Dielectric loss and dielectric breakdown

    b. Electronic and ionic polarizability

    c. Ferroelectric and ferromagnet

    d. Dielectric constant and refractive index

    e. Nematic and smectic liquid crystals

    f. Sensor and transducer

11-25. The capacitance of capacitors containing PVC dielectric dropped by a factor of 2 as the frequency was raised from 60 to 10^7 Hz. With polyethylene dielectric there was no corresponding capacitance change. Account for the difference in behavior.

11-26. The relative dielectric constant for inorganic glasses and polymers increases with temperature. Why? Above the glass transition temperature the increase is particularly noticeable. Why?

11-27. In diamond only electronic polarizability is present. If the index of refraction of diamond is 2.42, what is the relative dielectric constant?

11-28. The problem of making capacitances of reasonable magnitude reduces to stacking a large area of thin dielectric material, with metal plate contacts, into a small volume. In one type of capacitor a polymer ribbon is coated with evaporated (offset) metal contacts on either side, and then pleated like computer paper. The result is a capacitor in the shape of a rectangular prism. Suppose it is desired to make a 0.01-μF capacitor using a 50-μm-thick polystyrene dielectric sheet with a folded area of 1 cm². In addition the metal thickness per dielectric layer is 2 μm. How high will the capacitor be?

11-29. With reference to the previous problem suggest possible ways to reduce the capacitor volume.

11-30. A 100-nm-thick SiO_2 capacitor dielectric suffered breakdown within 10^{-4} seconds when an electric field of 1.4×10^9 V/m was applied. A filament 1 μm² in cross-sectional area was observed to short the electrodes and showed definite signs of melting. If SiO_2 melts at 1700°C and the average heat capacity is 1 J/g-°C, estimate the magnitude of the current that flowed. The density of SiO_2 is 2.27 g/cm^3 and the heat of fusion is 15.1 kJ/mol (5.70×10^2 J/cm^3).

11-31. A capacitor operating at 10^6 Hz has a 0.1-mm-thick polyethylene dielectric spacer. To improve capacitor reliability the dielectric was replaced by a 0.01-mm-thick SiO_2 spacer. For the same applied voltage V_0 what change in surface charge density will occur?

11-32. Estimate the velocity of ionic diffusion of Co^{2+} in CoO at 1400°C in an applied electric field of 1000 V/m.

11-33. In piezoelectric materials the fractional change in polarization is linearly proportional to the fractional change in length or strain, that is, $\underline{\Delta P/P = \Delta l/l}$. Young's modulus for a certain piezoelectric material is equal to 80 GPa and a compressive stress of 700 MPa is applied. $P_a = P_{ascal} = N/m^2$
   a. Is the polarization increased or decreased?
   b. What is the percentage change of the polarization?

11-34. Consider a piezoelectric capacitor shaped like a slab 1 mm thick and 3 mm^2 in area. Property values for the piezoelectric include $P = 40 C/m^2$, $\varepsilon_r = 75$, and E (Young's modulus) = 74 GPa. What force must be applied to the surface to create an electric field of 3 kV/mm that will enable an electric spark to be generated in air? Suggest a use that capitalizes on this effect.

12
SEMICONDUCTOR MATERIALS AND DEVICES: SCIENCE AND TECHNOLOGY

12.1. INTRODUCTION

Semiconductors are a unique class of materials that have transformed society and technology in truly revolutionary ways. A reasonably large number of different semiconductor materials are used in assorted electronic and electro-optical devices. They are broadly divided into elemental and compound semi-conductors. The relevant portion of the Periodic Table for these materials is indicated in Fig. 12-1. Elemental semiconductors stem from column IVA and notably include silicon, the most important of all semiconductors. In addition, there is germanium, the first semiconductor to be widely exploited in devices. Today, Ge has been virtually entirely supplanted by Si. Carbon in the form of diamond is also semiconducting. Diamond's semiconducting properties have not yet been capitalized upon, but its remarkably high thermal conductivity has been used to draw heat away from semiconductor lasers, enabling them to operate more reliably.

Compound semiconductors are composed of elements drawn almost entirely from four columns of the Periodic Table, with two residing on one side next to column IVA and two adjacent to it on the other side. Columns IIB and IIIA, containing metals with nominal valences of 2 and 3, respectively, thus combine with the nonmetals of columns VA and VIA that have respective valences of 5 and 6. Important elemental and binary semiconductor materials are listed in Table 12-1, together with a complement of physical and electrical properties

II B	III A	IV A	V A	VI A
	B Boron	C Carbon	N Nitrogen	
	Al Aluminum	Si Silicon	P Phosphorus	S Sulfur
Zn Zinc	Ga Gallium	Ge Germanium	As Arsenic	Se Selenium
Cd Cadmium	In Indium	Sn Tin	Sb Antimony	Te Tellurium
Hg Mercury		Pb Lead	Bi Bismuth	

FIGURE 12-1   Elements of the Periodic Table that play an important role in semiconductor technology.

whose meanings will become clearer as the chapter develops. It is interesting to note that the compounds are either of the III–V type (e.g., GaAs, InP) or the II–VI type (e.g., CdTe, ZnS). Thus, eight electrons are available (3 + 5 = 8 or 2 + 6 = 8) to be shared by the atoms. Four covalent bonds per atom are generally formed but some compounds have considerable ionic character. There are, of course, no I–VII semiconductor compounds. Instead, common alkali-halides form.

TABLE 12-1   **SEMICONDUCTOR PROPERTIES**

Material	Lattice parameter (nm)	Melting point (K)	Energy gap (eV at 25°C)	Electron mobility (cm²/V-s)	Hole mobility (cm²/V-s)
Diamond	0.3560	−4300	5.4	1,800	1400
Si	0.5431	1685	1.12 I[a]	1,450	450
Ge	0.5657	1231	0.68 I	3,600	1900
ZnS	0.5409	3200	3.54 D	120	5
ZnSe	0.5669	1790	2.58 D	530	28
ZnTe	0.6101	1568	2.26 D	530	100
CdTe	0.6477	1365	1.44 D	1,050	100
HgTe	0.6460	943	−0.15	25,000	350
CdS	0.5832	1750	2.42 D	340	50
AlAs	0.5661	1870	2.16 I	1,200	420
AlSb	0.6136	1330	1.60 I	200	420
GaP	0.5451	1750	2.26 I	110	75
GaAs	0.5653	1510	1.43 D	8,500	400
GaSb	0.6095	980	0.67 D	5,000	850
InP	0.5869	1338	1.27 D	4,600	150
InAs	0.6068	1215	0.36 D	30,000	460
InSb	0.6479	796	0.165 D	80,000	1250

[a] I refers to indirect band gap; D refers to direct band gap (see Section 13.4.1.).
Sources: S. M. Sze, *Semiconductor Devices: Physics and Technology*, Wiley, New York (1985), and B. R. Pamplin, in *Handbook of Chemistry and Physics* (R. C. Weast, Ed.), CRC Press, Boca Raton, FL (1980).

Irrespective of whether they are elemental or compound, all semiconductors share a number of common attributes:

1. The electrical resistivity falls between the 10^{-8} to 10^{-6} Ω-m value associated with metals and the $\sim 10^{8}$ Ω-m or greater value for insulators (see Fig. 11-3).

2. Small additions of certain alloying elements or dopants can cause enormous enhancements in electrical conductivity. (In Si, for example, addition of only one part per billion of phosphorus reduces the resistivity from 2200 Ω-m in very pure Si to about 2 Ω-m at room temperature.)

3. Semiconductors become better conductors as the temperature is raised.

4. Unlike metals, but like dielectrics, semiconductors can support internal electric fields, and these play a large role in electronic device behavior.

5. Interesting electrical and optical effects can occur at junctions between different semiconductor materials.

6. Single-crystal rather than polycrystal semiconductor materials are required for the overwhelming majority of device applications. Increasingly, single-crystal semiconductors must be fabricated in thin-film form.

7. These materials are generally hard and brittle.

8. Semiconductors are expensive materials on a bulk basis because applications require very pure and frequently scarce raw materials; however, as minuscule amounts are used, processing costs largely determine the price of devices. The added value that processing gives these materials is enormous.

This chapter has been reserved for exploring electrical phenomena in homogeneous semiconductors, and how electronic devices are fabricated and operate. The focus is primarily, but not exclusively, on silicon. Compound semiconductor electro-optical applications involving photon emission and absorption phenomena are considered in the next chapter.

12.2. CARRIERS AND CONDUCTION IN HOMOGENEOUS SEMICONDUCTORS

12.2.1. Bonds and Bands

12.2.1.1. Intrinsic Semiconductors

Aspects of the *electronic* and *crystallographic* structures of semiconductors were presented in Chapters 2, 3, and 11. A useful representation of pure or **intrinsic** silicon combining features of both structures is shown in Fig. 12-2. The Si lattice is flattened into a plane schematically preserving the character of the tetrahedral covalent bonding (see Fig. 3-6). Each of the four **bonds** contains two (hybridized) valence electrons (see Section 2.4.4) shared by neighboring Si atoms. At very low temperatures all of the bonds have their complete complement of electrons and so we say that the valence **band** is filled. As the

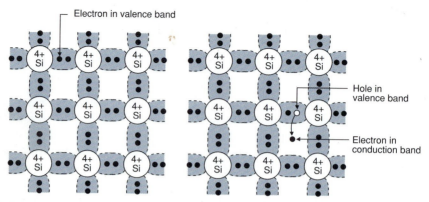

FIGURE 12-2   Simple two-dimensional model of the valence electrons in a semiconductor matrix. Creation of intrinsic electron–hole pairs is depicted. Electrons populate the conduction band, leaving holes in the valence band.

temperature is raised, however, thermal stimulation breaks some of these valence bonds and the **electrons** are now promoted to the conduction band. Simultaneously, **holes** (or the absence of electrons) are left in the valence bond network; we speak of hole creation in the valence band.

The energy band diagram of Fig. 12-3 concisely depicts all of the energy states available for electron occupation. At 0 K, valence electrons populate states in the valence band up to a maximum energy level, E_V. From the top of the valence band to the bottom of the conduction band (E_C), there is an

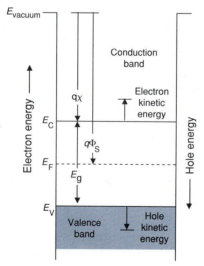

FIGURE 12-3   Band diagram of pure silicon.

energy gap free of states where electrons are denied existence. Above E_C, there is a continuum of electron states but they are all vacant. As noted above, thermal generation of **electron–hole pairs** results in equal numbers of intrinsic conduction band electrons and valence band holes. Although associated with a single broken bond, the electron and hole are independent carriers that will have separate histories. In the conduction band electrons behave very much like the free or conduction electrons in a metal (see Section 11.2). They too are like electrons in a well. Their energies are approximately given by Eq. 11-11, and the electron distribution is characterized by the Fermi–Dirac function (Eq. 11-5) and density of states (Eq. 11-6). Most importantly, such electrons are free to conduct electricity. Similarly, there are hole states and energies in the valence band. Holes are very much like electrons because they are also carriers of electricity. Holes, however, do not move as rapidly as electrons; because they are positively charged they travel counter to electrons in the same electric field. Increasing energy for them is equivalent to decreasing energy for electrons, and vice versa.

The magnitude of the energy gap, E_g, is a key characteristic of the semiconductor and values of it are entered in Table 12-1. For Si, $E_g = 1.12$ eV, an energy that is typically sufficient to cause atomic diffusion in solids.

Two other energies play an important role in semiconductors and we will use them subsequently. The first is the **Fermi energy**, E_F, which was defined in Section 11.2.1. In an intrinsic semiconductor, E_F lies precisely in the middle of the energy gap. The reason can be gleaned from the definition of E_F offered in Eq. 11-5 as that energy state for which the Fermi–Dirac function, $F(E)$, or probability of electron occupation, is equal to 1/2. At 0 K, states at the top of the valence band are occupied, $F(E_V) = 1$, whereas those at the bottom of the conduction band are empty, $F(E_C) = 0$. Thus, E_F ought to be somewhere in between. In addition, for every electron that is excited to the bottom of the conduction band as the temperature is raised, a **hole** is created at the top of the valence band. The depopulation of valence states is balanced by the occupation of conduction states, so again E_F falls midway between their respective energies. The other important energy is the semiconductor **work function**, $q\Phi_S$. As in metals the work function is the energy required to elevate an electron from E_F to a level just outside the semiconductor known as the vacuum level. When electrons enter a semiconductor from the vacuum and fall to level E_C (not E_F) they lose an energy $q\chi$ known as the **electron affinity**. Note that χ, like Φ_S, has units of volts, and that multiplying by the charge q yields electron-volt or energy units.

12.2.1.2. Extrinsic Semiconductors

Semiconductors become useful in devices only when selected impurities or dopants are introduced. If we consider our original intrinsic Si lattice and add group VA elements (e.g., P, As, Sb), it acquires an n-type character, meaning that excess electrons are available for conduction. Figure 12-4A illustrates the

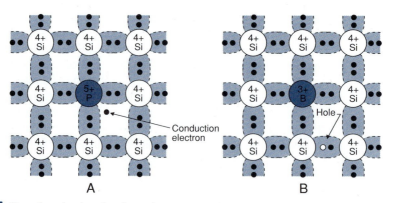

FIGURE 12-4   Two dimensional semiconductor lattice representation of (A) n-type doping and (B) p-type doping.

consequences of n-type doping. The pentavalent impurity atom substitutes for Si. After fulfilling all of silicon's covalent bonding needs an extra electron remains, and it orbits the dopant nucleus much like the electron around the hydrogen proton. With a small amount of thermal excitation this electron detaches and enters the conduction band; in the process the dopant is ionized, acquiring a positive charge of $1+$. The implications of n-type dopant incorporation can be visualized in the corresponding band diagram of Fig. 12-5A, where important group VA element ionization energies, E_d, are seen to lie only about 0.05 eV below E_C. Because n-type dopant elements **donate** electrons to the conduction band they are known as **donors**. The extrinsic concentration of electrons excited into the conduction band from ionized donors, n_{ex}, is approximately given by

$$n_{ex} = N_d \exp[-(E_C - E_d)/kT] \quad \text{electrons/cm}^3, \qquad (12\text{-}1)$$

where N_d is the dopant concentration and kT has the usual meaning. Note that the Boltzmann factor describes this excitation process, where $E_C - E_d$ may be viewed as an activation energy. At low temperatures, n_{ex} will be ~ 0; at very high temperatures, $(E_C - E_d)/kT \sim 0$, and $n_{ex} = N_d$. Practically speaking, usually enough thermal energy is available to largely ionize donors in Si and GaAs at 300 K.

In an entirely analogous fashion, elements from column IIIA can substitute for Si in the lattice and create a p-type semiconductor possessing an excess of holes. The reason is due to the trivalent nature of the dopant acceptor atoms. Neutral acceptors are one electron shy of completing their covalent bonding requirements. They **accept** this electron from the Si valence band, leaving a hole behind, and this is the reason for calling such dopants **acceptors**. The corresponding physical and electronic structures are depicted in a schematic way in Figs. 12-4B and 12-5B, respectively. Acceptor states are usually located less than 0.1 eV above E_V and, like donor states, are easily ionized thermally.

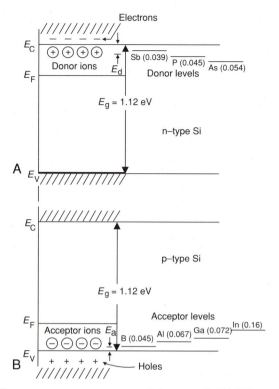

Band diagram representations of dopants in Si. (A) *Left:* n-type dopants, *Right:* Donor levels for Sb, P, and As. (B) *Left:* p-type dopants, *Right:* Acceptor levels for B, Al, Ga, and In.

The ionized acceptor atom then acquires a negative charge of $1-$. A formula similar to Eq. 12-1, with $E_a - E_V$ replacing $E_C - E_d$, governs the concentration of valence band holes of extrinsic origin, where E_a is the acceptor energy level.

Note that prior forbidden states in the energy gap have now become accessible to dopant electron levels. The doping of silicon is also accompanied by changes in the position of the Fermi level, as schematically shown in Fig. 12-5. An upward displacement of E_F toward E_C in n-type material is due to the now increased probability of electron occupation of higher energy states. Therefore, the energy for which $F(E) = \frac{1}{2}$ is proportional to the number of easily ionized donors and rises above the intrinsic or midgap value of E_g. For similar reasons E_F drops toward E_V in proportion to the number of ionized acceptors present. Most donor levels in GaAs are closer to the band edge than those in Si; for example for S, Se, and Sn, $E_d = 0.006$ eV. Acceptor levels for Be and Mg are 0.028 eV above the valence band edge.

In summary, the conduction and valence bands derive their charge carriers from two sources. Thermal generation is the first and it spontaneously yields equal numbers of intrinsic electrons and holes. Ionized donor or acceptor dopant atoms are the second source for either electrons or holes. We can only exercise reasonable control over extrinsic carriers through doping practice.

12.2.2. Carrier Concentrations

When proper account is taken of the density of states and the Fermi–Dirac function through Eq. 11-7, the temperature dependence of the electron concentration in the conduction band is approximately given by

$$n = 4.84 \times 10^{15} \, T^{3/2} \, \exp[-(E_g - E_F)/kT] \quad \text{electrons/cm}^3. \qquad (12\text{-}2)$$

Similary, the hole concentration (p) in the valence band is given by

$$p = 4.84 \times 10^{15} \, T^{3/2} \, \exp(-E_F/kT) \quad \text{holes/cm}^3. \qquad (12\text{-}3)$$

These formulas apply to both intrinsically and extrinsically generated carriers and have a number of important consequences.

1. For an intrinsic semiconductor, $n = p = n_i$, where n_i is the concentration of intrinsic electrons (or holes). Equating n and p in Eqs. 12-2 and 12-3 yields $E_g - E_F = E_F$. Therefore, $E_F = E_g/2$. This proves our previous contention that the Fermi energy lies in the middle of the energy gap. For Si at $T = 300$ K, $n_i = 1.0 \times 10^{10}$ cm^{-3}. (Actually a more exact value is $n_i = 1.45 \times 10^{10}$ electrons/cm^3.) The intrinsic carrier densities in Ge, Si, and GaAs are plotted as a function of $1/T$ in Fig. 12-6. The slopes of the displayed Arrhenius plots essentially yield respective values of $E_g/2$.

2. The product of n and p, where n and p are different (extrinsic), is equal to $2.34 \times 10^{31} \, T^3 \exp(-E_g/kT)$. But this, in turn, is also equal to n_i^2. Therefore,

$$n \cdot p = n_i^2. \qquad (12\text{-}4)$$

Equation 12-4 expresses the fact that a law of mass action governs the equilibrium electron and hole concentrations in semiconductors. The major carriers are those present in greatest profusion. Minority carriers are the other carrier species. In n-type material, electrons are the major and holes the minority carriers; the situation is reversed in p-type material.

EXAMPLE 12-1

Silicon is doped with phosphorus atoms. If 10^{16} donors per cm^3 are ionized at 300 K, determine (a) the carrier concentrations, (b) the temperature to which intrinsic Si would have to be heated to produce an equivalent number of electrons, (c) the location of the Fermi level, (d) the band diagram, and (e) the actual doping level.

ANSWER   a. The concentration of electrons is equal to 10^{16} electrons/cm^3. Electrons are the majority carrier in this case. The hole concentration is equal to $p = n_i^2/n$. After substitution, $p = (1.45 \times 10^{10})^2/10^{16} = 2.10 \times 10^4$ holes/cm^3. Holes with a concentration a trillionfold lower than that of electrons are the minority carriers.

b. From Fig. 12-6 it is apparent that when $1000/T = 1.505$, the intrinsic carrier density is 10^{16}/cm^3. Therefore, $T = 1000/1.505 = 664$ K, or 391°C.

c. Using Eq. 12-3, $E_F = -kT \ln[p/4.84 \times 10^{15} \ T^{3/2}]$. Substitution yields $E_F = -(8.63 \times 10^{-5})(300) \ln[(2.10 \times 10^4/4.84 \times 10^{15})(300^{3/2})] = 0.90$ eV.

d. As shown in Fig. 12-7, E_F lies 0.22 eV below the conduction band edge.

e. By Eq. 12-1, $N_d = n_{ex} \exp[(E_C - E_d)/kT]$. Substituting $N_d = 10^{16} \exp[0.045/(8.63 \times 10^{-5})(300)] = 5.69 \times 10^{16}$/cm^3.

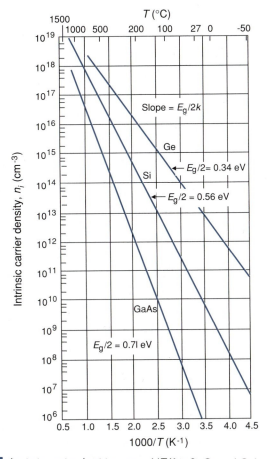

FIGURE 12-6   Intrinsic carrier densities versus $1/T$ K in Si, Ge, and GaAs.

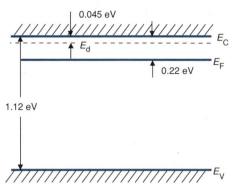

FIGURE 12-7   Location of Fermi level in Example 12-1.

12.2.3. Electrical Conduction in Semiconductors

Let us return to our flattened lattice model of Si and assume there are electrons in the conduction band as well as holes in the valence band. An electric field of indicated polarity is applied as shown in Fig. 12-8A. The electrons drift in zigzag fashion toward the positive electrode. They migrate within spaces surrounding lattice sites that may be imagined to constitute the conduction band. Holes, meanwhile, continuously exchange places with valence bond electrons, all the while executing a net motion toward the negative electrode (Fig. 12-8B). Through an arbitrary plane normal to the axis of current flow there are negatively charged electrons moving to the right with velocity v_n and positively charged holes migrating to the left with velocity v_p. An extension of Eq. 11-1 must now consider two carriers and yields

$$J = nqv_n + pq_pv_p, \tag{12-5}$$

where q_p is the hole charge and J is the current density. Physically, both currents **add** in the same direction because charges of opposite sign are compensated by oppositely directed velocities; that is, two negatives make a positive. Furthermore, Eq. 11-3 suggests that

$$J = \{nq\mu_n + pq_p\mu_p\}\mathscr{E}, \tag{12-6}$$

where $\mathscr{E}$ is the electric field. In parallel with Eq. 11-4 the conductivity σ is given by

$$\sigma = nq\mu_n + pq_p\mu_p = |q|(n\mu_n + p\mu_p), \tag{12-7}$$

where $|q| = |q_p|$. There are now two different mobilities, μ_n and μ_p, to contend with. Figure 12-9 reveals that electron mobilities are higher than hole mobilities, and that electron mobilities in GaAs exceed those in Si. The first of these facts has led to the preferred development of n-type devices controlled by the faster electron motion; the second has made GaAs attractive for high-speed device applications.

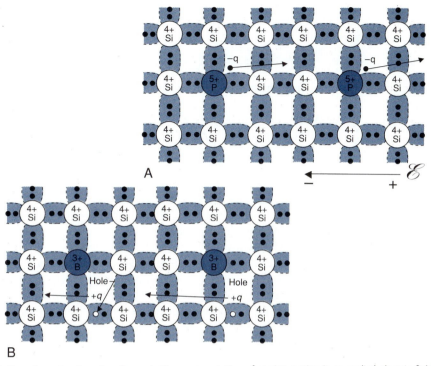

A

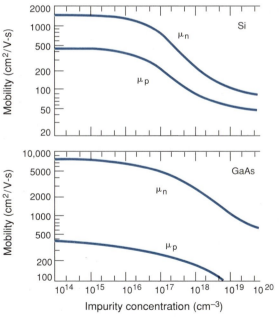

B

FIGURE 12-8 Two-dimensional semiconductor lattice representation of carrier motion in an applied electric field.

FIGURE 12-9 Electron and hole mobilities in Si and GaAs as a function of impurity concentration. From S. M. Sze, *Semiconductor Devices: Physics and Technology,* Wiley, New York (1985). Copyright © 1985 by AT&T; reprinted by permission.

EXAMPLE 12-2

What is the electrical conductivity of the material in Example 12-1?

ANSWER For electrons, $n = 10^{16}$ cm^{-3} and $\mu_n = 1200$ cm^2/V-s (from Fig. 12-9). For holes, $p = 2.10 \times 10^4$ cm^{-3} and $\mu_p = 400$ cm^2/V-s. Therefore, by Eq. 12-7, $\sigma = (1.6 \times 10^{-19})(10^{16} \times 1200 + 2.10 \times 10^4 \times 400) = 1.92$ (Ω-cm)$^{-1}$.

A very handy summary of the dependence of resistivity (the reciprocal of conductivity) on doping level is given in Fig. 12-10. In the illustrative problem just completed a conductivity value of 1.92 (Ω-cm)$^{-1}$ corresponds to a resistivity of 1/1.92 (Ω-cm)$^{-1}$, or 0.52 Ω-cm, in good agreement with the 10^{16} cm^{-3} doping level given.

An **electron** energy band diagram view of conduction that complements the carrier transport model considered above is shown in Fig. 12-11. The applied electric field tilts the bands. Electrons that stream toward the positive electrode scatter along the way and lower their energies. Contrariwise, holes float upward in the field because they reduce their energy in this way. Later, we shall see that devices depend on carrier motion in response to *internal* electric fields at semiconductor junctions.

The temperature dependence of conductivity is an issue of interest. Both the carrier concentrations and the mobilities vary, the former in the manner shown

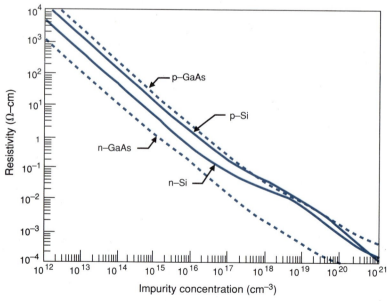

FIGURE 12-10 Electrical resistivity versus impurity concentration in Si (—) and GaAs (- - -) at 300 K. From S. M. Sze, *Semiconductor Devices: Physics and Technology,* Wiley, New York (1985). Copyright © 1985 by AT&T; reprinted by permission.

in Fig. 12-6 and the latter as indicated by Fig. 12-12. As in metals, electrons and holes in semiconductors undergo thermal (lattice) and impurity scattering processes that contribute to the electrical resistivity. The former varies as $T^{-3/2}$ and the latter as $T^{3/2}$. Stronger temperature dependencies are exhibited in semiconductors than for the equivalent processes in metals (e.g., $\sim T^{-1}$ for thermal scattering and temperature independent for impurity scattering). In addition, mobilities are also influenced by the doping level. Therefore, involved temperature-, and composition-dependent carrier concentrations and mobilities all conspire to complicate the electrical conductivity response of semiconductor materials.

At elevated temperatures intrinsic conduction dominates. The intrinsic number of thermally generated carriers exceeds the extrinsic number of carriers introduced by doping. As the weak $T^{-3/2}$ mobility temperature dependence is totally swamped by the strong $\exp(-E_g/2kT)$ temperature dependence of the intrinsic carrier concentration (Eq. 12-2), the conductivity follows the latter exponential behavior. At lower temperatures ionized donors and acceptors keep the carrier concentrations roughly constant over a wide temperature span known as the **exhaustion range.** Combined with a weakly varying mobility, the conductivity is reasonably constant. At much lower temperatures donors do not ionize and the conductivity precipitously drops. An indication of the complex conduction behavior in doped Si as a function of temperature is shown in Fig. 12-13.

Fortunately, in Si the conductivity is broadly constant in the neighborhood of room temperature for the doping levels required in devices. Germanium, however, becomes intrinsic near room temperature, leading to unacceptably large changes in conductivity. Annoying fadeout of sound in early Ge-based transistor radios played in the hot sun led to their demise and to the adoption of a Si-based device technology.

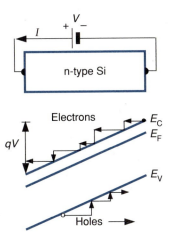

FIGURE 12-11 Energy band model of carrier motion in semiconductors subjected to an electric field.

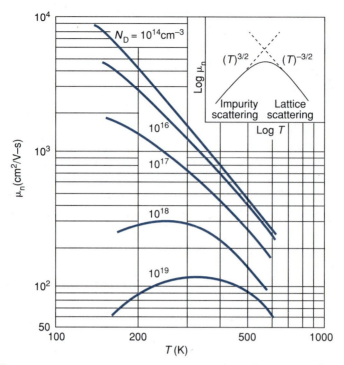

FIGURE 12-12 Carrier mobility versus temperature in Si. From S. M. Sze, *Semiconductor Devices: Physics and Technology,* Wiley, New York (1985). Copyright © 1985 by AT&T; reprinted by permission.

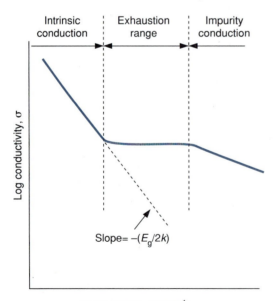

FIGURE 12-13 Typical variation of electrical conductivity with temperature in semiconductors.

12.2.4. The Hall Effect

How can we determine whether a semiconductor is n or p type? They look alike and a simple measurement of resistance will not distinguish them. To our rescue comes the Hall effect, a phenomenon exhibited by conductors carrying current when placed in a magnetic field. We know that current-carrying conductors in a magnetic field experience forces because this is how motors work. In physics courses it was shown that a current of electrons moving with velocity v_n, in a field of magnetic flux, B, perpendicular to it, experiences a Lorentz force given by $F_L = -qv_nB$. The negative sign is due to electrons whose charge is $-q$. A hole current of positive sign would be deflected in the opposite direction, and this difference is the basis of the measurement.

Consider now an n-type semiconductor crystal where the electric field ($\mathscr{E}_y$) is applied in the y direction and the magnetic field in the z direction (B_z), as shown in Fig. 12-14. Electrons flowing in the y direction experience a Lorentz force in the x direction (right-hand rule). They pile up at the front and back surfaces of the crystal (if they do not leak out the side walls) and establish an electric (Hall) field $\mathscr{E}_H$ that discourages further electron motion. An equilibrium is struck and the Hall field then balances the force due to the magnetic field. Physically, $q\mathscr{E}_H = -qv_nB_z$. But qv_n may be written in terms of the current density J_y in the y direction as $qv_n = J_y/n$ from Eq. 12-5, where n is the carrier concentration. Therefore, the Hall field is given by

$$\mathscr{E}_H = -J_yB_z/nq = R_HJ_yB_z. \qquad (12\text{-}8)$$

R_H is equal to $-1/nq$ and is known as the Hall coefficient. For electrons, R_H is negative and for holes it is positive. Through measurement of the Hall voltage

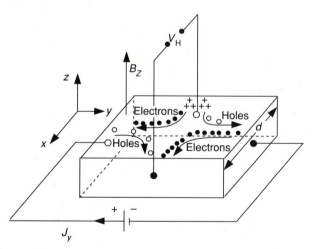

FIGURE 12-14 Hall effect geometry in a semiconductor crystal. Electric and magnetic fields are applied in y and z directions, respectively, while Hall field develops along x direction.

($V_H = d\mathscr{E}_H$) developed across the crystal of thickness d, it is possible to determine its dopant concentration and the semiconductor type (i.e., n or p). Although the polarity of V_H yields the latter, its magnitude provides the former. The Hall effect also occurs in metals but its magnitude is very small.

Commercial gaussmeters that measure the strength of magnetic fields capitalize on the Hall effect. For this application we note that for electrons $V_H = -dJ_y B_z / nq$. But because $J_y = nq\mu_n \mathscr{E}_y$ (Eq. 12-6), $V_H = -\mu_n B_z \mathscr{E}_y d$; the latter formula provides a way to determine μ_n as all other terms are known. Thus, for given applied electric and magnetic fields, the largest signal voltage arises from high-mobility semiconductors. That is why InSb is often used as the Hall effect probe.

EXAMPLE 12-3

A portable n-type InSb Hall probe gaussmeter is designed to operate at 3 V (applied along the crystal). What voltage output can be expected for a magnetic field of 1 Gauss?

ANSWER Using $V_H = \mu_n B_z \mathscr{E}_y d$, it is given that $\mu_n = 8$ m^2/V-s (Table 12-1), $B_z = 1$ Gauss or 10^{-4} Wb/m^2, and $\mathscr{E}_y d = 3$ V. Therefore, $V_H = 8 \times 10^{-4} \times 3 = 2.4 \times 10^{-3}$ V.

see notes for figure

12.3. PHENOMENA AT SEMICONDUCTOR JUNCTIONS

12.3.1. Contact Potential Between Metals

Virtually all of the important solid-state devices exploit electrical phenomena that occur at junctions between semiconductors and other semiconductors, metals, and insulators. Rather than semiconductors it is easiest to start off with two metals, A and B, with work functions $E_F(A)$ and $E_F(B)$, that sandwich an insulator (e.g., vacuum) in between. The band diagrams of the isolated metals are shown in Fig. 12-15A together with their respective work functions, $q\Phi(A)$ and $q\Phi(B)$. It is arbitrarily assumed that $E_F(A) < E_F(B)$ and therefore $q\Phi(A) > q\Phi(B)$. A conducting wire connects the metals in Fig. 12-15B, providing a conduit for electrons. Whenever electronic materials contact each other an important law of electrochemical (i.e., thermodynamic) equilibrium applies: *the Fermi energies must be equilibrated throughout.* This happens by transfer of electrons from B to the lower unfilled levels of A until the level of the electron sea is the same in both metals. But A becomes negatively charged and B positively charged in the process, favoring a subsequent reversal of charge transfer. A dynamic equilibrium is rapidly established, resulting in no further motion of electrons in either direction. An electric potential known as the

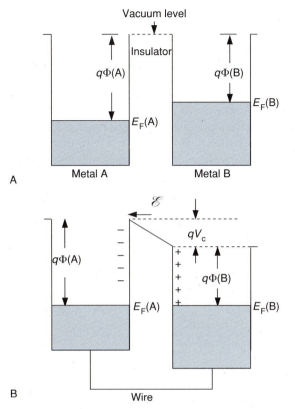

FIGURE 12-15 Band diagram illustration of contact potential between two metals. (A) Before contact. (B) After contact electron transfer between metals establishes an electric field in the vacuum space.

contact potential (qV_C) develops between the metals of magnitude equal to the Fermi energy difference or, alternatively, to the work function difference. Thus,

$$qV_c = E_F(B) - E_F(A) \quad \text{or} \quad qV_c = q\Phi(A) - q\Phi(B). \quad (12\text{-}9)$$

An electric field is established in the insulator space between the metals as evidenced by the slanted line on the band diagram. (Recall that the gradient of the potential is the field.) Although a battery appears to be present, energy cannot be extracted from a system at thermodynamic equilibrium. If it were possible, a perpetual motion machine could be constructed.

12.3.2. Semiconductor Junctions

12.3.2.1. No Applied Field

What happens at a p–n junction is crucial to understanding the behavior of diode and transistor devices. The treatment of junction behavior and devices will stress discussions and pictures but hardly involve any mathematical modeling. Structural implications of joining together isolated n-type and p-type semiconductors at a common junction are shown in Fig. 12-16A. Prior to junction formation, the n-type material consists of a uniform gas of mobile conduction electrons that permeates a distribution of positively charged immobile or fixed ionized donor atoms. The negative electron charge balances the fixed positive charge while the vast majority of matrix atoms are electrically neutral. On the p-type side a complementary state of affairs exists. There, a uniform distribution of mobile holes populate the valence bonds that surround the negatively charged immobile or fixed ionized acceptors. These two charge densities balance within a matrix that is overwhelmingly electrically neutral.

This idyllic state is rudely interrupted at the instant the junction is formed. A burst of current flows as electrons and holes diffuse toward each other and recombine on both sides of the junction. Quite simply, recombination occurs when electrons in the conduction band fill holes in the valence band. As a result, a kind of "no carrier's land" or so-called **space charge region** is created.

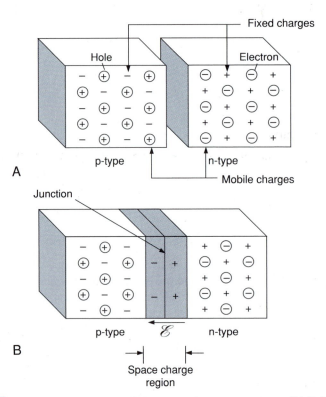

FIGURE 12-16   Structural representation of the semiconductor p–n junction. (A) Before contact of p- and n-type semiconductors. (B) After contact is made a space charge or depletion region develops, establishing an electric field.

It is also known as a **depletion region,** because both the mobile electrons and holes are depleted within it. In the process of electron–hole annihilation through recombination, the n-type semiconductor acquires an overall positive charge, and the p-type semiconductor a net negative charge. These overall positive/negative charge distributions create an internal or **built-in electric field** that discourages further carrier motion across the insulating space charge layer (Fig. 12-16B).

Electron band diagrams are a good way to visualize what is happening at a p–n junction. A few rules must be obeyed in constructing band diagrams of two (or more) materials in contact.

1. The Fermi level must be at the same level on both sides of the junction if there is no externally applied field.
2. Far from the junction the semiconductors are unchanged and must therefore exhibit the band structure of homogeneous materials. Merely bringing semiconductors together does not change their individual Fermi energy levels, work functions, or energy band gaps.
3. Only in the junction region, where built-in electric fields exist, are the bands bent or curved.
4. A potential energy step, qV_0, due to a built-in (contact) potential, V_0, develops at the junction. It is equal in magnitude to the difference in Fermi energies, $E_F(n) - E_F(p)$, or, equivalently, to the difference in work functions, $q\Phi(p) - q\Phi(n)$. (See Eq. 12-9 and Fig. 12-15.)
5. Externally applied electric potentials will displace the relative positions of E_F and the band edges by amounts over and above those produced by rules 1 to 4.

On the band diagram in Fig. 12-17A the original built-in potential V_0 at the junction is physically represented by a step. The construction requires that the sloping conduction and valence band lines smoothly connect the band edges of the homogeneous semiconductors on either side of the junction.

12.3.2.2. In an Applied Field

Figures 12-18A and 12-17B illustrate the results when a **reverse bias** voltage V is applied across the p–n junction. Reverse bias means applying the **negative** terminal of a battery to the **p** side and the **positive** terminal to the **n** side. Respective majority carriers (electrons in n-type material and holes in p-type material) are pulled away from the junction, depleting and widening the space charge region even further. The resulting band diagram shows a junction step larger than that in the unbiased case, indicating a larger electric field. Far from the junction the Fermi levels are no longer equal, indicating that equilibrium no longer exists. Importantly, virtually no current flows across the junction.

Figures 12-18B and 12-17C illustrate what happens when a **forward bias** voltage is applied across the p–n junction. Forward bias means applying the **positive** battery terminal to the **p** side and the **negative** terminal to the **n** side.

Handwritten annotations (top):

QUESTION: $V_{turn-on} \simeq 0.6 - 0.7V$

$V_{turn-on} \neq V_0$ (Built in)

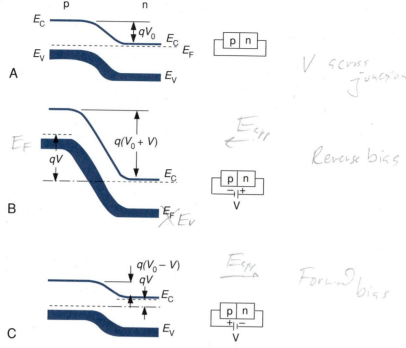

Handwritten annotations (right of figure):

V across Junction

$E_{S/I}$ ← Reverse bias

$E_{S/I}$ → Forward bias

Energy band diagram representation of a semiconductor p–n junction. (A) With no applied bias. (B) Under reverse bias. Both the width of the depletion region and the magnitude of the internal electric field increase. (C) Under forward bias. Both the width of the depletion region and the magnitude of the internal electric field decrease.

Handwritten annotation (below caption):

RB - Increased Barrier   Prob $\propto e^{qV/kT}$

The potential step, electric field, and width of the depletion region across the junction are all reduced simultaneously. Majority electrons from the n side are drawn to the p side, and majority holes from the p side to the n side where recombination can efficiently occur. Any carriers lost to recombination are replenished by the external battery so that a continuous current flows. The larger the extent of forward bias, the larger the current flow. We have just shown that the p–n junction behaves like a rectifier; current flows only for a positive voltage but is cut off for a negative voltage.

12.3.3. Thermoelectric Effects

12.3.3.1. Seebeck Effect

Both metals and semiconductors exhibit interesting thermoelectric effects whose practical exploitation requires junctions. But unlike the types we have been discussing, soldered, welded, or pressed junctions will work; bulk rather than interfacial properties of materials are involved. There are two thermoelec-

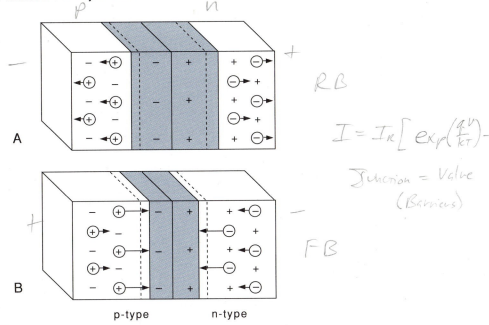

FIGURE 12-18 Structural representation of the semiconductor p–n junction under bias. (A) Junction under reverse bias. Electrons and holes separate as depletion region widens. (B) Junction under forward bias. Electrons and holes recombine as depletion region narrows. −, Fixed negative charges; +, fixed positive charges; ⊖, mobile electrons; ⊕ mobile holes.

tric effects of engineering importance, and they are schematically depicted in Fig. 12-19. As the basis for the operation of **thermocouple** temperature sensors, the **Seebeck** effect of Fig. 12-19A is more familiar. Quite simply, if conductors A and B are connected so as to form two junctions, and they are maintained at different temperatures, a measurable electromotive force (emf) develops as shown. To see why, consider a homogeneous metal bar whose ends are at different temperatures, as shown in Fig. 12-20. More electrons are promoted to the conduction band at the hotter end than at the cooler end, as reflected in different Fermi function behavior at each end. This causes electrons to diffuse toward the cold end and charge it negatively. A potential known as the *Seebeck voltage* develops along the bar that discourages further electron flow. In equilibrium, the Seebeck potential balances the thermally induced driving force for electron flow so that the electron concentration gradient along the bar vanishes.

If materials A and B were the same, the sum of Seebeck voltages from the two B segments would necessarily equal the Seebeck voltage from leg A; the emf detected by a voltmeter would be zero. If, however, B is different from A, a net emf equal to the difference in Seebeck potentials for the two materials develops. Its magnitude varies with the difference in junction temperatures, and frequently in a linear fashion over a limited temperature range. Values for

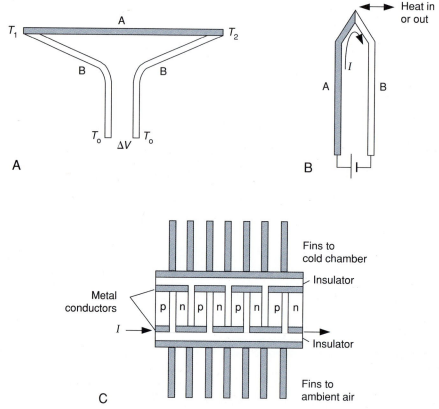

FIGURE 12-19   Thermoelectric effects. (A) Thermocouple arrangement illustrating the Seebeck effect. (B) Peltier effect. (C) Thermoelectric refrigerator.

the Seebeck voltage (α_A or α_B) produced by a 1°C change in temperature are entered in Table 12-2. Of note is the fact that the Seebeck voltage is 1000-fold larger in semiconductors than in metals.

In designing a useful thermocouple we must consider a number of properties in addition to the Seebeck potential output. For high-temperature applications the materials pairs must be resistant to degradation (e.g., due to oxidation, interdiffusion, reactions) that may cause voltage output instabilities. In addition, thermocouple materials must usually be fabricated in long lengths and be sufficiently flexible to be positioned within furnaces. For all of these reasons thermocouples are invariably composed of metal wires. The associated emf detection equipment more than compensates for low sensor output emfs. Popular thermocouples and their maximum usable temperatures include: copper–constantan (60% Cu–40% Ni), 300°C; chromel (90% Ni–10% Cr) and alumel (94% Ni–2% Al–3% Mn–1% Si), 1200°C; platinum–rhodium (13% Rh), 1500°C; and tungsten–rhenium, greater than 1500°C.

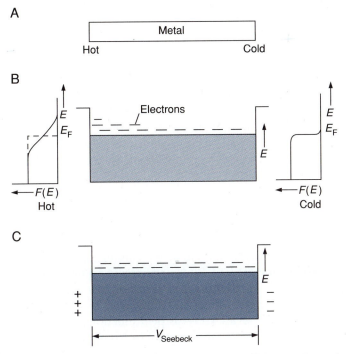

FIGURE 12-20   (A) A metal bar with ends at different temperatures. (B) Nonequilibrium free electron distribution at either end. (C) A Seebeck voltage develops when electrons equilibrate.

TABLE 12-2   **THERMOELECTRIC MATERIALS**[a]

Material	Temperature (°C)	Seebeck coefficient α (μV/K)	Figure of merit, Z ($K^{-1} \times 10^{-3}$)
Thermoelement			
Al	100	−0.20	
Cu	100	+3.98	
W	100	+3.68	
ZnSb	200	+220	
Ge	700	−210	
$Bi_2Te(Se)_3$	100	−210	
TiO_2	725	−200	
Couples			
Chromel–constantan			0.1
Sb–Bi			0.18
ZnSb–constantan			0.5
PbTe(p)–PbTe(n)			1.3
Bi_2Te_3(p)–Bi_2Te_3(n)			2.0

[a] Data were taken from various sources.

12.3.3.2. Peltier Effect

When current I flows through a junction of dissimilar conductors, heat can be either absorbed or liberated, depending on the current polarity (Fig. 12-19B). This phenomenon, known as the *Peltier effect*, has been exploited in small commercial thermoelectric refrigerators, as well as in the inverse device, the thermoelectric generator. To design useful devices and systems, materials with large Peltier coefficients are required; the latter scale directly with Seebeck coefficients. In addition, the materials must have a high electrical conductivity (σ) to minimize Joule heating and a low thermal conductivity (κ) to reduce heat transfer losses. A figure of merit, Z, combines these factors as

$$Z = \alpha_{AB}^2/\{(\kappa/\sigma)_A^{1/2} + (\kappa/\sigma)_B^{1/2}\}^2, \tag{12-10}$$

where α_{AB} is defined as the Seebeck coefficient difference, $\alpha_{AB} = \alpha_A - \alpha_B$. Readers will recognize that for metals, κ/σ is approximately constant at a given temperature due to the Wiedemann–Franz relationship (Eq. 11-22). Therefore, Z varies as α_{AB}^2 in metals. Because of their large Seebeck coefficients, semiconductors have Z values a factor of 10 or more times greater than those for metals and are the only practical choices for reasonably efficient refrigerator or generator operation. One configuration of a thermoelectric refrigerator consisting of a series array of p–n junctions is shown in Fig. 12-19C. With passage of current, heat is essentially pumped from junctions near the region to be cooled to the junctions at the ambient-air heat sink.

12.4. DIODES AND TRANSISTORS

12.4.1. Semiconductor Diode

The rectifying behavior of p–n junctions is exploited in devices known as diodes. Electrical engineers always represent devices in terms of their voltage–current responses or characteristics. In semiconductor diodes the characteristics are simple and displayed in Fig. 12-21. For positive applied voltages the current appears to rise exponentially. This can be physically appreciated by noting that the potential step at the junction is very much like an energy barrier to charge motion. Its magnitude depends on the applied voltage, V, and therefore, it is very likely that the current I is proportional to an appropriate Boltzmann factor, that is, $\sim \exp(qV/kT)$. In this expression the value of kT at room temperature is equal to 0.025 eV. Bias voltages are generally larger than 0.025 V so that $|qV/kT| \gg 1$. Under forward bias, $V > 0$ and $I \sim \exp(qV/kT)$ is large, as observed. The junction behavior under both bias directions is expressed by the single diode equation

$$I = I_R[\exp(qV/kT) - 1], \tag{12-11}$$

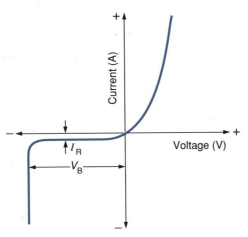

FIGURE 12-21 Current–voltage characteristics of a diode.

where I_R is a constant known as the reverse current. For reverse bias, V is negative and $\exp(-qV/kT)$ is very small if V is sizable. A small, constant, negative reverse current, $I = -I_R$, flows then. In short, the diode equation expresses the existence of a barrier to current flow in one direction and its absence in the other.

At specific large reverse voltages, V_B, a phenomenon not unlike breakdown in dielectrics occurs. Electrons and holes are accelerated to sufficiently large velocities and kinetic energies, so that collisions with the covalent valence electrons cause the latter to be excited into the conduction band. This process repeats and feeds on itself, producing a large avalanche current. Zener diodes exploit this effect and are used to regulate circuit voltages to desired values. Generally, high doping levels mean greater conduction and lower Zener breakdown voltages; alternatively, low doping levels yield high-breakdown-voltage devices.

12.4.2. Metal–Semiconductor Diode

Metal–semiconductor combinations are interesting because this is the way to make electrical contact to devices. Surprisingly, this structure may exhibit rectifying properties. Intentionally fabricated devices with this structure are known as *Schottky diodes*. The reason for their behavior is seen in the band diagrams of Fig. 12-22. Comparison with Fig. 12-17A reveals a similar band structure, at least on the n-type semiconductor side. In the unbiased state the built-in potential step and bent band features resemble aspects of the semiconductor diode. When either positive or negative bias is applied, the bands shift in comparable ways. Therefore, rectifying behavior can be expected in the metal–semiconductor junction.

Normally, ohmic behavior with low contact resistance is desired because current then flows equally and unimpeded in either direction. Unintentional rectification at a semiconductor contact must be eliminated because it interferes with proper circuit functioning. What metal–semiconductor combinations will lead to ohmic rather than rectifying contacts? This is an important and vexing materials issue that surfaces every time new semiconductor materials and devices emerge. As a matter of fact it is not always easy to make good contacts to Si and GaAs devices. Materials like Au–Ge–Ni annealed from 400 to 600°C are required to make low-resistance contacts to GaAs, underscoring the art involved in making good contacts.

The barrier height, Ψ_B (Fig. 12-22), is clearly the obstacle to electron motion at contacts. Electrons from the semiconductor have to climb a wide hill to get into the metal, and vice versa for metallic electrons to pass into the semiconductor. Lowering Ψ_B would help but this is easier said than done; however, by heavily doping the semiconductor the depletion width can be reduced greatly. This makes it possible for electrons to burrow or **tunnel** through the barrier rather than surmount it. Heavy doping is thus a practical way to reduce the metal–silicon contact resistance.

12.4.3. Transistors

12.4.3.1. Bipolar Transistor

When p–n junctions are connected back to back, one forward biased and the other reverse biased, a device known as a **bipolar junction transistor** emerges. The p–n–p transistor shown in Fig. 12-23 is considered simply because the direction of majority carrier (hole) motion coincides with that of the current flow. There is also an n–p–n version of this device, which is easier to fabricate and more commonly used. By reversing the bias voltages, carrier

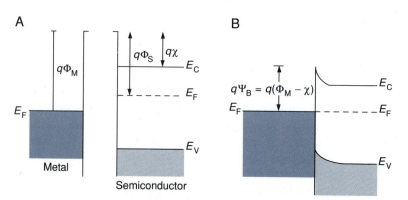

FIGURE 12-22   Band diagram representations of the metal–semiconductor contact. (A) Isolated metal and n-type semiconductor. (B) Metal and semiconductor form a contact that has diode characteristics.

roles, and current direction, the discussion applies to this device as well. In transistors three semiconductor regions can be distinguished—the emitter, base, and collector—and their respective band diagrams are shown. When unbiased, the Fermi levels line up throughout the entire device (Fig. 12-23A). During transistor operation one p–n junction (the left one) is forward biased and the other (on the right) is reverse biased. The bands bend and move up or down as shown (Fig. 12-23B) according to the rules established earlier for individual junctions. In operation, the heavily doped emitter injects majority (hole) carriers into the lightly doped base region. These holes would recombine with electrons in the base of a normal diode. But, because the base is made very thin (e.g., 1 μm thick), the holes diffuse right through it before they have a chance to recombine. Those holes that traverse the base–collector interface are now efficiently swept to the collector contact by the electric field in the junction. Equivalently, holes float uphill on the band diagram, which is their way of reducing energy.

In well-designed transistors the collector current, I_C, is only slightly smaller than the emitter current, I_E. The difference or base current, I_B, is typically 1% of I_E or I_C. A small variation in the emitter–base voltage, V_{E-B}, produces an exponential increase in the injected emitter current by virtue of Eq. 12-11, as well as a much smaller rise in I_B. Therefore, large changes in the collector current effectively result from small variations in the base current. Current gain or amplification is the chief attribute of transistors; it is defined by the ratio I_C/I_B, and typically assumes values greater than 100. Transistor action essentially occurs because the emitter–base voltage in one part of the device

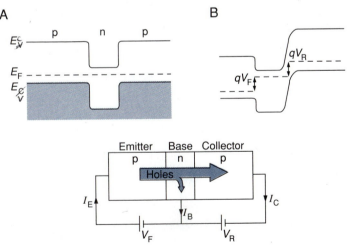

FIGURE 12-23 (A) Band diagram of an unbiased p–n–p transistor. (B) Band diagram of the biased p–n–p transistor. (C) The p–n–p transistor with bias voltages and hole currents schematically indicated. Minority electrons flow counter to the hole current.

controls the collector current in another part of the device. Specific doping levels, bias voltages, and base dimensions are required for optimal transistor performance.

12.4.3.2. Planar Bipolar Transistor

Discrete or single bipolar transistors of the type just discussed were the first to reach the market. Later, the concept of integrating many bipolar transistors on a single wafer meant a new processing philosophy. Rather than fabricating semiconductor junctions "side to side" as in Fig. 12-23, a "top-to-bottom" thin-film technology had to be adopted. By sequentially doping through the wafer thickness, the n–p–n (or p–n–p) junction depths and the base thickness, in particular, can be very well controlled. The problem is how to make contact to the emitter, base, and collector. Obviously, contacts can be made only on the top surface of the wafer and novel methods were developed to accomplish this, as shown in Fig. 12-24. Electrons injected from the emitter cross the narrow base region vertically down into a highly conducting, heavily doped *buried layer* that serves as a horizontal extension of the collector. They flow

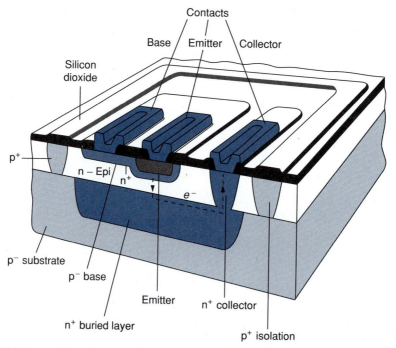

FIGURE 12-24 Schematic of the planar n–p–n bipolar transistor. Electron path from the emitter–base–collector is dashed. Smaller hole currents flow in the opposite direction. The n-Epi is a lightly doped n-type epitaxial Si layer (see Sections 5.2.2 and 12.5.2.2). From W. E. Beadle, J. C. C. Tsai, and R. D. Plummer, *Quick Reference Manual for Silicon Integrated Circuit Technology*, (1985). Published by John Wiley and Sons. Copyright © 1985 by AT&T; reprinted by permission.

laterally through the buried layer and then veer up to the collector contact at the top surface. Provision must also be made to isolate (insulate) the transistor from other nearby circuit elements and devices. Processing details to achieve these ends are discussed in Section 12.6.

The operating principle of the planar bipolar transistor is the same as that for the discrete bipolar transistor. Because of their very high operational speeds, these devices are widely used in computer circuits.

12.4.3.3. MOS Field Effect Transistor

The MOS field effect transistor derives its name from the critical metal-oxide-semiconductor structure of the device and from the electric field that is placed transverse to it. A typical device structure is shown in Fig. 12-25 and consists of three electrically contacted regions; two are the **source** and **drain** and the other is the **gate**. Source and drain regions are doped alike (either n or p) and separated by Si, doped oppositely (either p or n). The latter sits directly under a very thin gate (SiO_2) oxide covered with a gate electrode. This oxide is the dielectric that fills a capacitor whose plates consist of the underlying semiconductor and overlying gate electrode (usually a heavily doped, conducting polycrystalline Si layer). Voltage required to switch this device on or off is applied across this MOS capacitor.

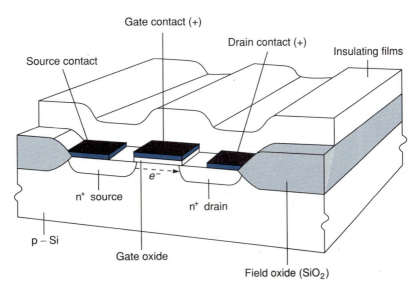

FIGURE 12-25 Schematic of the n-channel enhancement-mode MOS transistor. Source electrons (dashed) travel to the drain region when the applied gate voltage inverts the p-doped Si beneath the very thin gate oxide. Thin films of SiO_2, phosphorus–silica glass, and silicon nitride are deposited to respectively insulate, smooth or planarize the surface, and hermetically enclose the device. From W. E. Beadle, J. C. C. Tsai, and R. D. Plummer, *Quick Reference Manual for Silicon Integrated Circuit Technology*, (1985). Published by John Wiley and Sons. Copyright © 1985 by AT&T; reprinted by permission.

Let us consider the operation of the depicted n-channel MOS (NMOS) **enhancement-mode transistor**. The drain is made positive with respect to the source but no current flows horizontally initially because of the reverse-biased junctions. A positive voltage is now applied to the gate electrode. The majority holes are repelled from the oxide interface, depleting the Si underneath it of these carriers. Still no current flows. With higher positive gate voltages, electrons are drawn to the interface and the underlying Si begins to acquire an n-like character; that is, it is being inverted. At a critical **threshold voltage** a point is reached where a thin *n-type channel* spans the source and drain. Formation of the short-circuit conducting path triggers electron flow through the channel from source to drain. The device is turned on going from 0 to 1, or from nonconducting to conducting state. Switching a signal rather than amplifying it is of concern here.

Another type of transistor device that is turned on initially, but off during operation, can be imagined easily enough. The **depletion-mode transistor** starts with a similar source–drain structure, but the area under the gate is now doped to create a permanent n-type channel. By applying a negative voltage to the gate electrode, holes are drawn to the SiO_2 interface and invert the channel to p-type. At a critical voltage the electron path is pinched and the current flow stops; the device switches from 1 to 0.

Just as there are the NMOS devices described above, there are p-channel MOS (PMOS) versions of these field effect transistors. Collectively these devices are the basic building blocks of logic circuits, memory devices, and microprocessors. A popular combination of NMOS and PMOS transistors is known as CMOS. These devices are commonly employed in logic applications because they dissipate very little power and tolerate a wide variation in power supply voltages. They consist of enhancement-mode NMOS and PMOS transistors fabricated on a single substrate and tied to a common gate.

12.4.3.4. GaAs Transistors

Although electrons travel about five times faster in GaAs than in Si, hole motion is much slower. This means that p-type GaAs devices, for example, p-channel transistors, are not practical; therefore, most of the attention is given to the more efficient n-channel technology. GaAs metal–semiconductor field effect transistors employ different materials (e.g., no SiO_2) but operate on similar principles. Known as metal–semiconductor field effect transistors (MESFETs), they are faster and dissipate less power than comparable Si devices. There are four basic types of GaAs integrated circuits, based on depletion-mode (D-MESFET), enhancement-mode (E-MESFET), high-electron-mobility (HEMT), and heterojunction bipolar (HJBT) transistors. Hailed by proponents as the "digital integrated circuits of the future," advances in Si technology have thus far conspired to perpetually confine GaAs and other III–V semiconductors to a runner-up status. This is not the situation in optoelectronic devices, a subject treated in the next chapter.

12.5. MATERIALS ISSUES IN PROCESSING SEMICONDUCTOR DEVICES

12.5.1. Scope

It is much easier to draw schematics of devices and circuit diagrams on paper than to realize them in physical terms. The latter requires that the individual character and behavior of different electrical materials be creatively merged to create p–n junctions or MOS structures. Further, it is necessary to contact and interconnect them to other components of the circuit and to provide for the necessary electrical insulation. Not just any materials will do. Semiconductors must be single crystals; be free of dislocations; have the right doping level; be easy to prepare, shape, and process; be compatible with other materials and components; not degrade over time; and so on. In a similar vein different specific attributes are required of metal contacts and interconnections, insulation, and materials used to package the completed electronic products. These issues are at the heart of a very broad interdisciplinary effort on the part of all engineers to design semiconductor devices and circuits, process and fabricate them, package them in systems, and ensure their reliable functioning. In this and the next section the processing and fabrication steps involved in making devices and integrated circuits are introduced.

12.5.2. Growing Single Crystals

12.5.2.1. Bulk Crystals

Silicon. Preparing single crystals is the first and in some ways the key step in semiconductor technology. By *single crystals* we not only mean bulk or thick crystals that are grown from the melt (see Section 6.4.3) and converted into wafers for further processing. Usually, single-crystal semiconductor films less than 1 μm thick must be grown on suitable bulk single-crystal substrates; techniques for their production will be discussed subsequently. Bulk single crystals of silicon are grown by melting a charge in a silica crucible and inserting a small cooled single-crystal seed into it, as shown schematically in Fig. 6-15. The cylindrical crystal that solidifies is simultaneously pulled up and rotated in such a way as to extend its length and maintain a constant diameter.

By neglecting heat radiation from both the melt and the crystal, it is easy to make a simple heat balance at the interface between the two phases (Fig. 12-26). During growth, heat generated in the melt flows into the cooled crystal. Heat flow (in units of J/m²-s) originating from the liquid stems from two sources: heat conduction of amount $\kappa_L(dT/dx)_L$ (see Eq. 9-17) and latent heat evolution of magnitude $\rho H v$. The sum of these two contributions, $\kappa_L(dT/dx)_L + \rho H v$, must be equal to the heat conducted away into the crystal, or $\kappa_S(dT/dx)_s$. In these expressions L and S refer to the liquid and solid, κ is the thermal conductivity (W/m-K), and dT/dx is the temperature gradient.

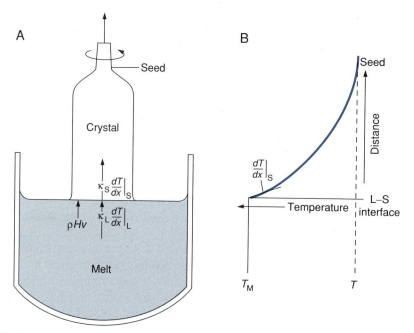

Heat and mass transport effects during Czochralski single-crystal growth. (A) Heat fluxes at the crystal–melt interface. (B) Vertical temperature profile in the melt and crystal.

Latent heat of fusion, H (J/kg), liberated during solidification contributes to the heat balance in direct proportion to the rate of melt transformation or to the velocity of crystal growth, v (m/s). Because of the way terms are defined, the density, ρ (kg/m³), must also be included. Physically, $\kappa_S (dT/dx)_S \gg \kappa_L (dT/dx)_L$ (see Fig. 12-26), and therefore the maximum velocity at which crystals can be pulled from the melt is given by

$$v = (\kappa_S/\rho H)(dT/dx)_S. \qquad (12\text{-}12)$$

Note that higher crystal growth rates are promoted by large temperature gradients in the solid. Production requirements attempt to maximize v, but not at the expense of crystal quality. Following growth, the crystals are shaped round by grinding and then sawed and polished to produce wafers. Presently, commercial integrated circuit production lines process 15- and 20-cm-diameter Si wafers.

Composition variations and growth defects are primary concerns that limit crystal quality. Look at almost any phase diagram (e.g., Si–Ge in Fig. 5-14) and note that the liquidus and solidus lines drop for small solute additions to silicon. If we start with a melt of composition C_o, in the two-phase liquid plus solid field, the liquid will have equilibrium composition C_L and the solid, composition C_S. But, C_L is always greater than C_S in this case. Thus at the

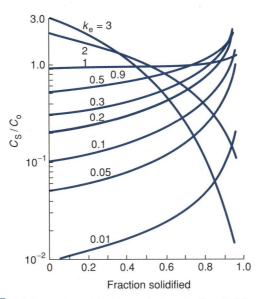

FIGURE 12-27   Axial impurity profiles in crystals directionally pulled from melts with different k_e values.

crystal–melt growth interface the solid is always purer than the liquid. The **equilibrium segregation coefficient**, k_e, defined as C_S/C_L, is a measure of this solute partitioning. It is less than unity for important dopants and impurities in Si [e.g., $k_e(B) = 0.8$, $k_e(P) = 0.35$, $k_e(As) = 0.3$, and $k_e(O) = 0.25$].

As the first solid freezes a small amount of solute is rejected to the liquid ($k_e < 1$) and the crystal is purer than the initial melt composition. Because the solid crystal is cooler than the melt, solute atoms do not diffuse very far from where they were incorporated in the solid during solidification; however, complete homogenization occurs in the melt. This same solute partitioning is repeated incrementally until the entire crystal is pulled. A variable axial distribution of solute develops along the grown crystal. Purer than the initial melt of concentration C_o near the seed, the crystal is less pure at the other end. Axial dopant concentration (C_S) profiles in the crystal are shown in Fig. 12-27, plotted as a function of fraction solidified, for different values of k_e. Meanwhile, the melt is the sink that collects all of the rejected solute. In addition to this there are radial distributions of solute caused by the complex convective swirling of the melt at the crystal interface.

More serious in many ways is oxygen incorporation into Si crystals. Slight dissolution of the quartz crucible is the source of oxygen which reacts to form tiny precipitates of SiO_2. These are sites for the nucleation of dislocation and stacking fault defects, which occasionally propagate into crystal layers where devices are being fabricated. When this happens, device yields suffer and reliability problems arise.

Compound Semiconductors.   Single crystals of GaAs and InP are important starting points in the manufacture of assorted high-speed electronic and optical communications devices. Crystal growth poses a greater challenge in these materials than in Si. The reason is that Ga and In have very low melting points, whereas As and P have higher melting points; in addition the latter have much higher vapor pressures than the group III metals. Pairs of elements cannot be simply weighed out and melted together because evaporation of the group V elements would destroy the critical 1 : 1 stoichiometry that must be maintained.

One way to overcome this problem is to use a two-zone horizontal furnace, schematically indicated in Fig. 12-28. In the case of GaAs, liquid Ga is heated slightly above the melting point (1238°C) of this compound, whereas As is heated to ~610°C. This creates a temperature gradient in the system. At the respective operating temperatures the decomposition pressure of GaAs and vapor pressure of As are both approximately 1 atm. A slight overpressure of As causes vapor transport to the Ga, resulting in the formation of GaAs at a seed crystal. During crystal growth, the temperature in the high-temperature zone is gradually reduced, translating the gradient to the right and causing incremental solid GaAs growth all the while. When done carefully, large crystals of GaAs can be grown by this so-called Bridgeman technique. InP is harder to grow because compound decomposition pressures of ~40 atm have to be contained. Like silicon, these crystals are sliced and polished after growth.

12.5.2.2. Thin-Film Semiconductor Crystals

As all of the active devices in integrated circuits are confined to the upper few micrometers of the wafer, thin single-crystal films are critical. There are a number of ways to grow such films of Si and compound semiconductors.

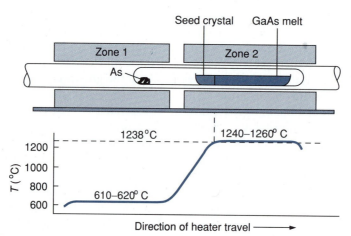

FIGURE 12-28   Horizontal Bridgeman technique to grow GaAs single crystals in a two-zone furnace. From S. M. Sze, *Semiconductor Devices: Physics and Technology*, Wiley, New York (1985). Copyright © 1985 by AT&T; reprinted by permission.

Chemical vapor deposition (CVD) is one that was introduced in Section 5.2.2 as means of depositing Si **epitaxial** films on a Si wafer substrate. One may well ask why an extra few micrometers of **homoepitaxial** Si is needed, since all it appears to do is thicken the wafer. Actually, epitaxial Si films are purer and frequently more defect free than the substrate wafer; furthermore, they can be doped independently of it. Epitaxial Si dramatically increased the yield of early thin-film transistors and is currently used in the processing of bipolar and field effect transistors.

Again, relative to Si the deposition of epitaxial compound semiconductor films poses a significantly greater challenge in materials synthesis. First there is the need to critically control the composition, of not only binary compounds (e.g., GaAs) but often ternary (e.g., $Al_x Ga_{1-x} As$, $1 > x > 0$) and even quaternary alloys as well. The magnitude of the energy band gap is very much dependent on composition and the exploitation of this fact has led to families of lasers, light-emitting diodes, and detectors tuned to operate at specific wavelengths. This subject is dealt with more fully in the next chapter. Second, the film and substrate must have very closely matched lattice constants, a. Otherwise the interface will contain dislocation defects that impair charge transport or seriously degrade light emission and absorption processes. Lattice mismatch or misfit is defined as

$$f = (a_s - a_f)/a_f, \tag{12-13}$$

where subscripts f and s refer to film and substrate values. For high-quality devices, misfit values of less than 0.002 are required. What happens when films are deposited on substrates is schematically indicated in Fig. 12-29. If the

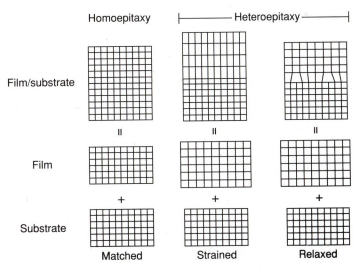

FIGURE 12-29 Atomic arrangements in the film and substrate during homoepitaxy, strained film heteroepitaxy (no interfacial defects), and, heteroepitaxy. Large lattice misfits give rise to interfacial dislocations.

film and substrate have different compositions we speak of **heteroepitaxy** and heterojunctions are created between them. Semiconductor laser structures may consist of several such junctions composed of differently doped materials. In such cases, f is small and the interfaces are defect free. The straining of interatomic bonds is not sufficient to nucleate dislocations. But for larger misfits, the only way to relax accumulated bond strain energy is to generate undesirable misfit dislocations at the interface.

In Section 5.5.2 the basis of the still commercially important liquid phase epitaxy (LPE) method was described. Demand for high-performance devices has necessitated new methods with greater control of cleanliness, purity, stoichiometry, and film thickness. There are two important ways of depositing epitaxial semiconductor films from the vapor phase. They involve either chemical or physical methods to first place the involved atoms in the vapor phase. These then condense in a sequentially ordered, atom-by-atom fashion on the receiving substrate. In chemical vapor deposition, precursor gases containing the group III and V elements are fed into a heated reactor (Fig. 12-30). They react to form the solid compound that deposits, while the by-product gases exit the system. For example, GaAs films can be synthesized by reacting trimethylgallium (a gas) and arsine (AsH_3) at 700°C and ~0.001 atm:

$$(CH_3)_3Ga + AsH_3 \Leftrightarrow GaAs + 3CH_4. \tag{12-14}$$

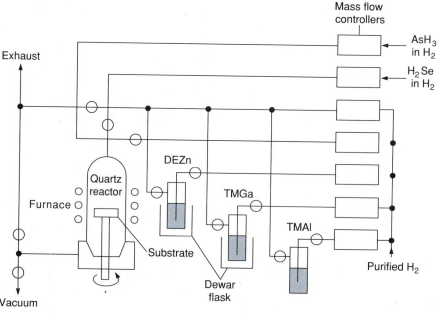

FIGURE 12-30 CVD reactor employed to grow epitaxial films of compound semiconductors. TMGa, trimethylgallium; TMAl, trimethylaluminum; DEZn, diethylzinc.

The popular molecular beam epitaxy (MBE) method depends on a very highly controlled simultaneous thermal evaporation of the involved atoms. These issue as beams from different heated sources, under extraordinarily clean, high-vacuum conditions. A modern multichamber MBE system and schematic of the deposition chamber are shown in Figs. 12-31A and B, respectively. Temperatures and resulting vapor pressures (Fig. 5-6) are selected to enable optimal single-crystal film deposition rates. By MBE methods, a 1-μm-thick

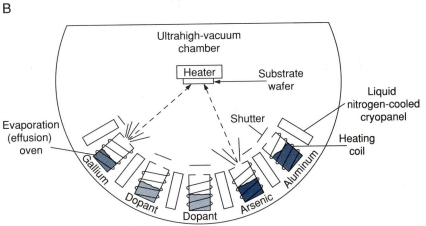

FIGURE 12-31 (A) Photograph of a multichamber MBE system. Courtesy of Riber Division. (B) Schematic of the MBE deposition geometry showing evaporation sources arrayed around the substrate.

film is typically grown in an hour, whereas CVD methods take about a minute to deposit the same film thickness.

Critical to ensuring single-crystal formation is a combination of a low film nucleation rate and a high film growth rate. If many film nuclei form on the substrate, a worthless polycrystalline deposit results. A low gas supersaturation and deposition rate will create few nuclei. On the other hand, high substrate temperatures enhance diffusion rates and facilitate atomic incorporation into lattice sites. This promotes the early lateral extension of the single-crystal film and growth perfection as it thickens.

EXAMPLE 12-4

A dream of semiconductor technology is to integrate faster and optically active GaAs devices with the massive signal processing capability of Si devices, all on a single robust Si wafer. For this purpose GaAs films would have to be epitaxially grown on Si. Comment on this possibility and how the interface is likely to be deformed.

ANSWER   The lattice misfit between these semiconductors is $f = [a(\text{Si}) - a(\text{GaAs})]/a(\text{GaAs})$. Substituting a values from Table 12-1 yields $f = [0.5431 - 0.5653]/0.5653 = -0.0393$. This calculated misfit is much too large to avoid interfacial dislocation formation.

If, along the interface, atoms are forced to assume some average lattice parameter, the latter will be larger than that for Si and smaller than that for GaAs. Therefore, to reach a compromise, GaAs would be compressed and Si would be stretched in tension.

GaAs films have been deposited on Si. Devices fabricated in the GaAs have performed poorly. The use of transition layers with graded lattice parameters is one approach to eliminating interfacial defects.

12.5.3. Doping

12.5.3.1. Diffusion

Doping semiconductors to the desired type and to the required concentration is carried out both *during* and *after* bulk single-crystal growth. Diffusion is one of the methods used to accomplish it after crystal growth. For example, starting with a p-type Si wafer, a diode can be created by diffusing an n-type dopant into the surface at an elevated temperature. Diffusion theory was addressed at length in Chapter 6 and perhaps should be reviewed at this point. Its application to junction formation is illustrated in Example 6-2. When rather deep junction depths are involved and lateral device features are not critical, diffusional doping methods are employed. For this purpose the silicon surface is masked with a patterned SiO_2 film that is effectively impervious to the penetration of dopant. Instead diffusion occurs where bare Si is exposed, as shown in Fig. 12-32A. A reliable set of diffusivity data (Fig. 6-11B) is required for precise doping profiles. But for high-performance devices with shallow junction depths, ion implantation is the preferred method of doping.

12.5.3.2. Ion Implantation

Ion implantation made possible the operation of digital watches and pocket calculators by controlling the doping level under the gate oxide. This lowered the threshold or critical voltage required to switch field effect transistors on or off. Unlike diffusion, where the surface dopant concentration is higher than in the crystal interior, the reverse is true in ion implantation. How can this be if the surface is the entry to the interior? The answer is that ion implantation

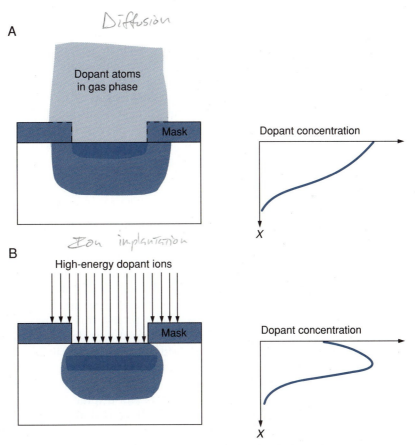

| FIGURE 12-32 | Dopant profiles produced by: (A) Thermal diffusion. In this diffused profile the dopant concentration peaks at the surface. (B) Ion implantation. The gaussian-shaped, ion-implanted dopant profile peaks beneath the surface. |

relies on propelling dopants into the crystal as though shot out of a cannon. It does this by first stripping electrons from gaseous dopant atoms, making ions of them. By applying voltages of ~50 to 100 kV, these ions are accelerated to kinetic energies large enough to embed them into Si. Possessing energies that are thousands of times larger than the binding energy of lattice atoms, the dopant ions wreak havoc in passing through the solid. At first a lightning-like ion current excites the atomic electrons and ionizes atoms, absorbing energy in the process. When the dopants lose enough energy they slow sufficiently and then begin to collide violently with Si nuclei, knocking some off lattice sites. Each dopant ion executes a different odyssey and comes to rest at a different depth from the surface. The ion range is statistically distributed. Lighter boron atoms typically penetrate 300 nm, whereas heavier phosphorus atoms only travel 100 nm at 100 keV. Ion-implanted dopant profiles have a gaussian shape and are compared with diffusional profiles in Fig. 12-32.

Advantages of ion implantation include selective and controllable doping to high or low levels, vacuum cleanliness, room-temperature processing (unlike thermal diffusion), and the ability to create novel depth profiles (by sequential implantations). Disadvantages include high equipment cost and postimplantation annealing to restore crystalline perfection to the ion-damaged matrix. Ion penetration in Si can fortunately be blocked by interposing thin films of SiO_2 or photoresist materials (discussed next). This makes it possible to selectively dope geometrically patterned regions, provided an appropriate stencil mask is applied to the wafer surface. Lithography is the method used to accomplish patterning and it is described next.

12.5.4. Lithography

Perhaps the most critical technology in microelectronics is lithography. It enables processing (e.g., diffusion, ion implantation, deposited film removal or etching) to occur in only selected patterned areas of a wafer surface. Lithography has obvious parallels to photography, but instead of producing flat images on paper, a three-dimensional relief topography is created in materials like Si, SiO_2, silicon nitride, aluminum, and photoresist. The process can be readily understood by studying Fig. 12-33. Briefly, it is desired to transfer the pattern shown on the mask to the SiO_2-coated Si wafer. (Section 6.2.3 discusses the oxidation of Si.) A photoresist film is spun on the surface. As the name implies, *photoresists* are sensitive to light and **resistant** to chemical attack by liquid or gaseous etchants. Some are also sensitive to electrons and X rays. They consist of specially compounded polymers and display two broad types of behavior. When exposed to ultraviolet light the crosslinked polymer chains of **positive resists** undergo scission or cutting, rendering them more soluble in a developer; unexposed areas resist chemical attack. Under the same stimulus, **negative resist** molecules polymerize while unexposed regions are soluble and dissolve in the developer. Next, the exposed SiO_2 film is removed by etching and finally, the

Lithography

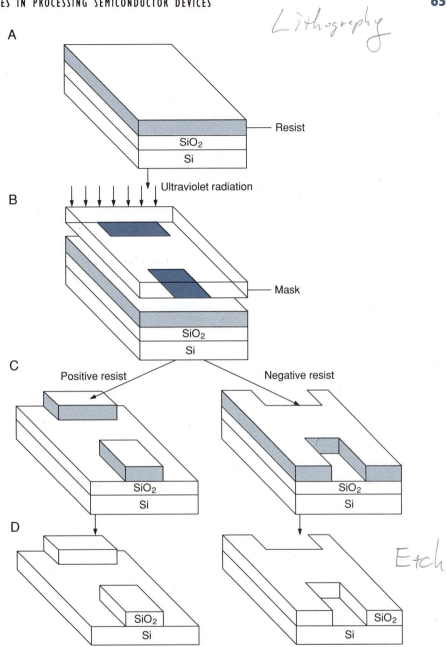

Etch

FIGURE 12-33   Schematic of the lithography process to define patterned oxide regions on the wafer surface. A photoresist layer is first applied to an oxidized Si wafer. Ultraviolet light passing through a mask exposes the photoresist to define the desired pattern geometry. In a positive resist, the mask pattern is defined in the underlying SiO$_2$ layer after resist removal and etching. In a negative resist, the region surrounding the mask pattern is defined.

photoresist is stripped away. This leaves the bare Si surface geometrically patterned by an SiO_2 mask. This structure may now be ion implanted, for example, to dope Si, or coated with metal to make a contact. In the latter case, another lithography step would be required to pattern the metal and etch it away from undesired locations.

As we shall see later, the process just described must be repeated at five or more different depths or levels in the Si wafer as actual integrated circuits unfold. What is absolutely critical is to align successive levels so that the correct registry (e.g., contact metal over a contact hole) is maintained simultaneously over millions of devices on large wafers. The present state of the art in this mind-boggling technology employs lithography techiques capable of creating ~ 0.5-μm circuit features. This is targeted to shrink below 0.3 μm at the beginning of the 21st century, in the inexorable drive for higher device packing densities. Submicrometer circuit features are shown in Fig. 1-14C and in Fig. 15-4.

12.5.5. Deposition of Conducting and Insulating Films

12.5.5.1. Conducting Films

Evaporation.   There is a need to **contact** and **interconnect** all of the doped semiconductor electrodes (e.g., emitter, base, collector in bipolar transistors or source, drain and gate in field effect transistors) to other devices on the chip. Assorted metals and alloys are employed for these purposes. Many contact materials have been tried and used, including Al and Al alloys, W, PtSi, Pd_2Si, and $TiSi_2$; interconnection lines between contacts, on the other hand, are invariably made of Al or Al–Cu alloys. Just why these metals are used as *metallizations* and not others is due to the following desirable material properties:

1. High electrical conductivity
2. Low contact resistance to Si
3. No tendency to react chemically with or interdiffuse in Si
4. Resistance to corrosion or environmental degradation
5. Electromigration resistance (*Electromigration* is the transport of metal atoms in conductors carrying large current densities; see Section 15.5.3.)
6. Ease of deposition
7. Compatibility with other materials and steps (e.g., lithography, etching) in the fabrication process

The physical vapor deposition (PVD) techniques of **evaporation** and **sputtering** are the means employed to deposit metal films. Sequential atom-by-atom deposition results in film nucleation and subsequent growth. But unlike epitaxial deposition, both methods yield polycrystalline films with submicrometer grain sizes, partly because the substrates are often amorphous (e.g., SiO_2). Evaporation is the simpler of these techniques and involves heating the source metal until it evaporates at appreciable rates. The process is carried out in a chamber (Fig. 12-34) operating at vacuum pressures that are typically 10^{-10} atm. Wafer substrates, carefully positioned relative to the evaporation source, are rotated to yield uniform metal coverage over the stepped terrain of the processed surface. Evaporation rates are directly proportional to the product of the vapor pressure of the element(s) in question (obtained from Fig. 5-6) and the size and shape of the source. The deposited film thickness is inversely proportional to the square of the source–substrate distance. A formula derived from the kinetic theory of gases can be used to calculate the arrival rate, R_e, of evaporated atoms at a substrate located directly above the source and a distance L from it. Presented without proof, it is given as

$$R_e = \frac{2.67 \times 10^{25}\, P_{atm}}{(MT)^{1/2} \pi L^2} \quad \text{atoms/cm}^2\text{-s}. \tag{12-15}$$

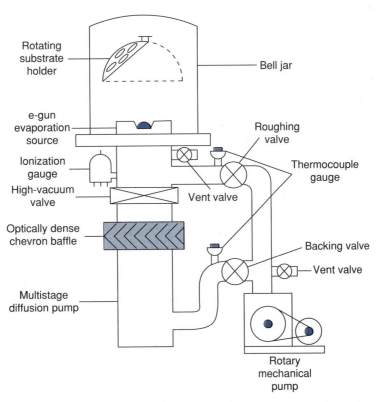

| **FIGURE 12-34** | Schematic of evaporation system. Evaporated atoms from the heated source condense on the rotating Si wafer substrates. |

In this formula, M is the evaporant atomic weight, and T and A are the source temperature (K) and surface area, respectively. The quantity P_{atm} is the vapor pressure of the evaporant in atmospheres.

EXAMPLE 12-5

What film thickness of Al will deposit on a substrate located 25 cm above an Al source heated to 1500 K? The source area is 1 cm^2 and the deposition time is 1 minute.

ANSWER   From Fig. 5-6 the vapor pressure of Al is $\sim 2 \times 10^{-5}$ atm. For Al, $M = 27$. Substitution in Eq. 12-15 yields

$$R_e = (2.67 \times 10^{25})(2 \times 10^{-5})/[27 \times 1500]^{1/2}(\pi (25)^2)$$
$$= 1.35 \times 10^{15} \quad \text{atoms/cm}^2\text{-s.}$$

In 60 seconds, 8.11×10^{16} Al atoms will deposit on each square centimeter of substrate. Al, an FCC metal, has four atoms per unit cell and a lattice parameter of 0.405 nm. Therefore, there are $8.11 \times 10^{16}/4 = 2.03 \times 10^{16}$ unit cells lying above each square centimeter of substrate area. If n is the number of cells stacked vertically, $n \times 1 \text{ cm}^2/(4.05 \times 10^{-8} \text{ cm})^2 = 2.03 \times 10^{16}$ unit cells. Solving, $n = 33.3$ and the film thickness is $33.3 \times 0.405 = 13.5$ nm.

Typical Al stripes in integrated circuits are ~ 1 μm thick and 1 μm wide in cross section and are deposited at rates of 1 μm/min. Other products like plastic sheet and mirrors are also metallized.

Sputtering.   In this thin-film deposition technique the metal or alloy to be deposited is fashioned into a target plate (cathode) electrode and placed parallel to the substrate or anode within a chamber. An inert gas (Ar) at low pressure (~ 0.01 atm) is introduced, and a potential difference of a few thousand volts dc is established between the electrodes. Under these conditions an electrical **glow discharge** is sustained between electrodes, much like what occurs in the neon sign tube. The gas atoms become ionized and, together with electrons and neutral atoms, create a plasma. Energetic positive ions are drawn to the negative cathode, where they impact and dislodge atoms. These then fly through the sputtering system (Fig. 12-35), deposit on the substrate, and build the film. Many variants of sputtering exist, including ac methods to deposit insulating films. Low film deposition rates generally hamper sputtering processes. A great advantage of sputtering, however, is the maintenance of stoichiometry in deposited *alloy* films. During evaporation of alloys, on the other hand, the constituent elements evaporate more or less independently. Deposit and source compositions are different and stoichiometry is not maintained. Sputtering is used to

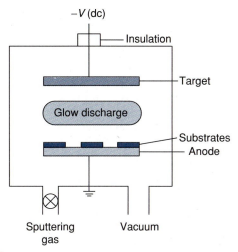

FIGURE 12-35   Schematic of sputtering system. Ionized sputtering gas ions impact the cathode target surface, ejecting metal atoms which deposit on the substrate.

deposit metal films in other technologies as well. Deposition of magnetic alloy coatings on disks for data storage applications is a growing application.

12.5.5.2. Insulating Films

As all devices and interconnections support voltages and carry current while sitting on a conducting Si wafer, strategically placed insulators are needed to isolate them electrically. Otherwise, the chip components would short-circuit and a meltdown would occur when powered. There is also a need to passivate microelectronic chips after they are fabricated, and silicon nitride is often used for this purpose. It is an effective barrier against moisture and alkali ion (Na^+) penetration. The *growth* of thin SiO_2 films by thermal oxidation of Si has already been considered. These very high quality insulator films are used as gates in field effect transistors. In the next generation of devices they will be considerably thinner than 10 nm and must be immune from defects and dielectric breakdown phenomena. Aside from this critical (grown) oxide, an assortment of other insulator films are *deposited*. They include materials like SiO_2, silicon nitride, and phosphosilicate glass. All of them are amorphous in structure and typically ~1 μm thick. Although they can all be deposited by evaporation or sputtering, CVD methods are employed in microelectronic processing. Reactors employed are similar to those used to oxidize Si (Fig. 6-7) or deposit epitaxial films. Typical reactions with precursor gases (on the left) and products (on the right) include, for SiO_2,

$$Si(OC_2H_5)_4 \longrightarrow SiO_2 + \text{by-products} \qquad \text{at } 700°C;$$

for silicon nitride,

$$4NH_3 + 3SiCl_2H_2 \longrightarrow Si_3N_4 + 6HCl + 6H_2 \qquad \text{at } \sim750°C,$$
$$SiH_4 + NH_3 \longrightarrow SiNH + 3H_2 \qquad \text{at } 300°C;$$

and for phosphosilicate glass,

$$SiH_4 + 4PH_3 + 6O_2 \longrightarrow SiO_2-2P_2O_5 + 8H_2 \qquad \text{at } 450°C.$$

It is found that the composition, structure, and properties of such CVD films vary widely with deposition conditions. Silicon nitride is a case in point. At very high temperatures dense, crystalline Si_3H_4 is formed. At lower temperatures considerable hydrogen is incorporated into films, particularly if the deposition is done with the assist of a plasma discharge. Such films are amorphous and less dense, but sufficiently insulating and passivating. In general, sequential film growth and deposition processing steps are carried out at successively lower temperatures so as not to alter the prior processed and doped materials.

12.6. FABRICATION OF INTEGRATED CIRCUIT TRANSISTORS

The chapter closes with a couple of flowcharts illustrating the integration of process steps in the fabrication of Si microelectronic transistors.

12.6.1. Bipolar Transistor

1. The entire surface of the initial lightly doped p-type Si wafer having either a (100) or a (111) orientation is oxidized. Regions are then lithographically defined where the SiO_2 is etched away to the bare Si surface (Fig. 12-36).

2. A buried layer of heavily doped n-type Si is created. Arsenic dopant is normally used for this purpose and it is either diffused or ion implanted into the wafer. The Si surface is reoxidized.

3. SiO_2 is entirely removed and an n-type epitaxial Si layer (epi-layer) is deposited by CVD methods. Next the surface is reoxidized.

4. An isolation lithography step defines the rectangular regions, which will each contain one transistor. A p-type dopant is diffused into the perimeter network. Within the wafer, p–n junctions are created and serve to electrically isolate each of the rectangles when reverse biased. (A moat of insulating SiO_2 serves this function in modern devices.) Again the surface is oxidized.

5. A small rectangular region is defined by lithography to create the p-type base window. Ion-implanted or diffused boron is used to dope the base. The Si is reoxidized.

6. Lithography is employed to open two additional windows simultaneously, one for the emitter and one for the collector. Note that the emitter region is nested within that of the base. A high level of phosphorus dopant is then

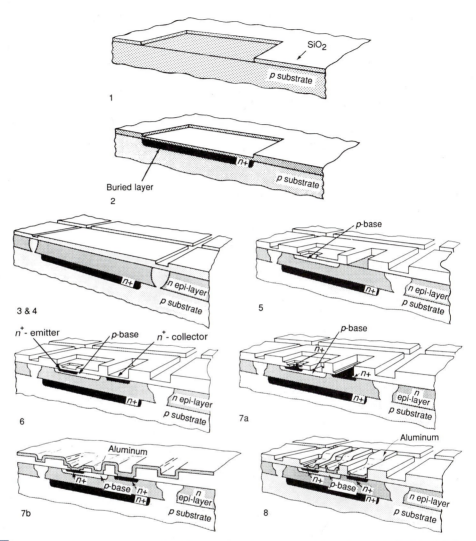

FIGURE 12-36 Sequence of processing steps used to fabricate a bipolar transistor. Adapted from W. C. Till and J. T. Luxon, *Integrated Circuits: Materials, Devices, and Fabrication*, Prentice–Hall, Englewood Cliffs, NJ (1982).

simultaneously introduced into the emitter and collector by thermal diffusion or implantation. Once again, SiO_2 covers all.

7. A contact mask and lithography step is required to etch away the oxide at the contact windows (a). Then, aluminum metal is evaporated or sputtered over the entire surface (b).

8. A metallization pattern is applied by photolithography. Unlike other steps where material is deposited, here the metal is *etched* from regions where it is not wanted. The interconnection stripe geometry is produced as a result.

This completes the process, aside from making contacts to pads, bonding leads, and encapsulation. Finished chips or dies, whether bipolar or MOS treated next, are enclosed within the familiar packages shown in Fig. 12-37 and are ready for insertion into circuit boards.

12.6.2. NMOS Field Effect Transistor

1. A lightly doped p-type (100) oriented Si wafer is the starting point. The wafer is first thermally oxidized (~50 nm thick) and then covered with a deposited Si_3N_4 layer (100 nm thick). Photoresist is applied and patterned (Fig. 12-38).

2. The photoresist serves as the mask that shields the eventual transistor region against the implantation of boron into the surrounding area. Boron serves to limit the channel length (Chanstop) and electrically isolate it from neighboring transistors.

3. The Si_3N_4 layer not covered by photoresist is etched away and the resist is also removed. A layer of SiO_2 is thermally grown everywhere except in the Si_3N_4-protected areas. This 0.5- to 1-μm-thick **field oxide** film isolates the individual transistors from one another. The remaining Si_3N_4 and underlying SiO_2 films are both stripped, and the bare Si surface is oxidized to produce a ~20-nm **gate oxide** across the channel region.

4. A heavily doped polycrystalline Si layer is deposited by CVD methods.

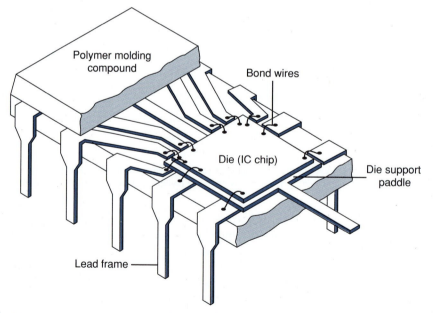

FIGURE 12-37 Chip or die encapsulated within a dual-in-line (DIP) package.

The **polysilicon** is lithographically patterned and removed everywhere but in a small central region, where it will become the gate electrode.

5. This gate serves as the mask for the ion implantation of an n-type dopant (usually As) that creates both the source and drain regions. In this way the source and drain are symmetrically located or **self-aligned** with respect to the gate.

6. A phosphorus-doped oxide glass is then deposited and heated to make it flow viscously to **planarize** or smooth the overall topography of the structure. This glass also provides additional insulation. Contact windows are then defined lithographically and etched to expose the semiconductor surface.

7. The last step is metallization. Metal is evaporated over the entire structure. A final lithography step is required to delineate the conductor stripe network.

It is instructive to compare the two processes for fabricating transistors. In the bipolar case there were seven film formation processes, six lithography operations, and four doping steps. Fabrication of the NMOS transistor is

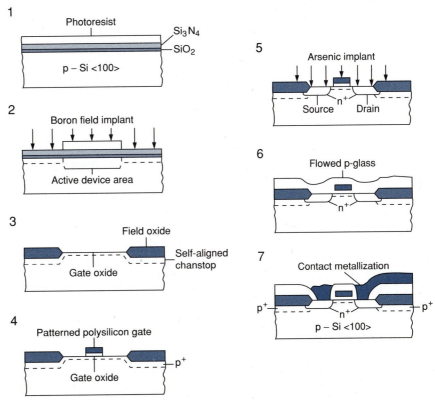

FIGURE 12-38 Sequence of processing steps used to fabricate an NMOS field effect transistor. Adapted from S. M. Sze, *Semiconductor Devices: Physics and Technology*, Wiley, New York (1985). Copyright © 1985 by AT&T; reprinted by permission.

simpler, requiring seven film formation processes, four lithography operations, and three doping or ion implantation steps. The saving of two lithographic operations and one doping step is significant.

12.7. PERSPECTIVE AND CONCLUSION

Never in history has a class of materials so profoundly impacted our lives, in so short a time, as semiconductors. These remarkable materials are drawn largely from column IV and the two flanking columns on either side of the Periodic Table. They are composed of elements that had little commercial use prior to 1950, with the exception of Si in the form of ferrosilicon for steel making. As the name implies the key property capitalized upon is *semi*conductivity. These materials do not quite have the high electrical conductivity of metals, but with doping there are sufficient charge carriers that move fast enough for device applications. On the other hand, semiconductors are more conductive than insulators and, like dielectrics, possess the important attribute of supporting electric fields, especially across junctions. It is the combination of a relatively large number of mobile electron and hole charge carriers, contained within a dielectric-like matrix, that is so unusual. This enables interesting asymmetrical charge transport to occur across junctions of dissimilar semiconductors (or semiconductors and metals) by applying small voltages. It is the important basis for the functioning of many devices. Unlike all other engineering materials, semiconductors must be single crystals and have extreme levels of purity combined with carefully controlled doping levels; otherwise, their wonderful attributes are significantly diminished.

Charge carriers in semiconductors stem from two sources. **Intrinsic** carriers are generated when the temperature is high enough; **extrinsic** carriers arise when dopants are introduced. The situation is entirely analogous to that for ionic solids, where Schottky defects (a cation vacancy–anion vacancy pair) are like intrinsically produced electron–hole pairs. Also, cation or anion vacancies, introduced by impurity ions of valences different from those of the matrix, are like majority electrons or holes produced when penta- and trivalent dopants are added. And aside from the large differences in magnitude, the conductivity behavior as a function of temperature is also similar.

Unlike dielectrics there are few applications for homogeneous semiconductors; Hall effect devices are an exception. Rather, both n and p doped semiconductor regions that contact each other at junctions are required. At the single p–n junction of a diode, electric signals are rectified, whereas n–p–n or p–n–p transistor structures allow signals to be amplified. The reliable fabrication of these and other devices in both Si and GaAs has been one of the miracles of modern manufacturing technology. All of the major classes of engineering materials are brought together, albeit in minute amounts: *semiconductors* for active devices, *metals* to contact and interconnect them, *insulators* to electrically isolate devices and circuit elements, and *polymers* to encapsulate and package finished integrated circuit chips. The devices are present in great profusion (as many as a few million per chip), so that lateral circuit feature sizes are now less than 1 μm. In addition, layers of conducting and insulating films—the former less than 1 μm thick, the latter as thin as 0.01 μm—define the depths involved. Unit materials processing steps include diffusional and ion implantation doping, growth of epitaxial semiconductors, and deposition of metals and insulators. These processes are limited laterally by lithographic patterning of features, and critically defined through the depth by control of time–temperature–composition variables.

One legacy of integrated circuit technology has been the dramatically improved understanding of the behavior of *all* materials at the atomic level. Materials and device characterization tools and techniques largely developed for this technology, for example, high-resolution electron microscopy (both TEM and SEM), coupled with an impressive array of chemical analytical instruments, have enriched every scientific and engineering discipline.

Additional Reading

N. Braithwaite and G. Weaver, *Materials in Action Series: Electronic Materials,* Butterworths, London (1990).
C. R. M. Grovenor, *Microelectronic Materials,* Adam Hilger, Bristol (1989).
R. C. Jaeger, *Introduction to Microelectronic Fabrication,* Vol. 5, Addison–Wesley, Reading, MA (1988).
J. W. Mayer and S. S. Lau, *Electronic Materials Science: For Integrated Circuits in Si and GaAs,* Macmillan, New York (1990).
W. S. Ruska, *Microelectronic Processing,* McGraw–Hill, New York (1987).
S. M. Sze, *Semiconductor Devices: Physics and Technology,* Wiley, New York (1985).

QUESTIONS AND PROBLEMS

12-1. What is the conductivity of intrinsic Si at 300 K?

12-2. For the same carrier concentrations, order the following n-type semiconductor materials with respect to their room-temperature conductivities. Assume $n \gg p$.
   a. Ge   d. CdTe
   b. InP   e. ZnS
   c. GaP

12-3. A milligram of arsenic is added to 10 kg of silicon.
   a. How many unit cells of Si are associated with an As atom?
   b. Estimate the electrical resistivity of this semiconductor.

12-4. Silicon is doped to a level of 10^{18} boron atoms/cm^3 at 350 K.
   a. Determine the carrier concentrations.
   b. Determine the location of the Fermi level.
   c. What fraction of the total current is carried by holes in this semiconductor?

12-5. a. Explain why the Fermi level for n-type Si falls toward the middle of the energy gap as the temperature rises.

b. Similarly, explain why the Fermi level for p-type Si rises toward the middle of the energy gap as the temperature increases.

12-6. a. What fraction of the total current is carried by electrons in intrinsic silicon?

b. What fraction of the total current is carried by holes in intrinsic GaAs?

12-7. One Si wafer has a concentration of 2×10^{18} antimony dopant atoms/cm^3, and a second Si wafer contains 4×10^{18} gallium atoms/cm^3.

a. What is the majority carrier in each wafer?

b. At what temperature will both wafers contain the same number of ionized carriers?

12-8. Consider a Si-based n-type field effect transistor with a 2-μm channel length. If 5 V is applied between source and drain, how long will it take electrons to traverse the channel? What is a rough upper limit to the frequency response of a circuit employing transistors of this type?

12-9. At the melting point of Si, how many electrons are excited from the valence band to the conduction band? What fraction of the total number of valence electrons (that are responsible for the covalent bonding) does this correspond to?

12-10. A common way to stack the semiconductor junctions in thermoelectric refrigerators is through a cascade structure. In this arrangement, one bank of thermocouples is used to provide a cold sink for hot junctions of a second bank of couples. The latter, in turn, provides a still colder sink for the third stage, and so on.

a. Sketch such a cascade-type thermoelectric refrigerator.

b. What is the advantage of this structure relative to that of Fig. 12-19C?

12-11. Thermistors are devices composed of mixtures of transition metal oxides. Because their resistivity changes in a thermally activated way like semiconductors, they are used to sense temperature. For such a thermistor the resistivity at 100 K is 10^{10} Ω-m; at 500 K the resistivity is 10 Ω-m.

a. Suggest a formula that describes the resistivity as a function of temperature.

b. It is desired to use this device to sense changes in temperature in the vicinity of 30°C. What is the smallest temperature change that can be detected if the resistance can be measured to 0.01%?

12-12. Is the mobility of electrons higher in pure copper or in pure silicon? Explain.

12-13. Contrast the effect on electrical conductivity of doping pure Si with 0.01 wt% Al with that of alloying pure Al with 0.01 wt% Si. What is the approximate increase or decrease in electrical conductivity in each case?

12-14. A current of 1 A flows through the p–n junction of a diode that is forward biased with 0.30 V. What reverse current will flow at 298 K when the polarity is reversed?

12-15. In a common base transistor circuit, the dependence of collector current, I_C, on emitter–base voltage, V_{E-B}, is $I_C = A \exp(B V_{E-B})$. If $I_C = 10$ mA when

V_{E-B} is 5 mV, and $I_C = 100$ mA when V_{E-B} is 25 mV, predict the collector current when V_{E-B} is 50 mV.

12-16. Sketch the band diagram across an n–p–n transistor in both the unpowered and powered states.

12-17. How would you expect increasing temperature to affect the operation of a p–n junction diode?

12-18. The maximum solubility of boron in silicon is 600×10^{24} atoms/m³.
  a. What atomic percentage does this correspond to?
  b. What weight percentage does this correspond to?
  c. What is the maximum electrical conductivity the B-doped Si can attain?
  d. What are the implications of dopant solubility on device processing and properties?

12-19. The maximum solubility of phosphorus in silicon is 1200×10^{24} atoms/m³. A surface source with one-tenth the maximum P solubility is used to dope Si containing 1×10^{16} boron atoms/cm³ in fabricating a solar cell. At what depth will the junction depth be located after a 15-minute diffusion where $D = 1.2 \times 10^{-12}$ cm²/s?

12-20. Find the electron and hole concentrations in silicon at 300 K for (a) Wafer A doped to a level of 1×10^{16} boron atoms/cm³ and (b) Wafer B containing two dopants: boron at a level of 1×10^{17} atoms/cm³ and phosphorus at a level of 4×10^{16} atoms/cm³. Determine the resistivity of (c) wafer A and (d) wafer B.

12-21. Given an unknown semiconductor specimen, what electrical measurements would you conduct to obtain a value for the mobility of majority carriers?

12-22. In very pure germanium the mobility is 50 m²/V-s at 4 K. Estimate the mobility at 300 K.

12-23. a. Refer to the Ge–Si phase diagram and determine the equilibrium segregation coefficient of Ge in Si at 1400, 1350, and 1300°C. Is k_e constant?
  b. Determine the equilibrium segregation coefficient of Si in Ge at 1050, 1000, and 950°C.
  c. What phase diagram features are required for k_e to be constant as a function of temperature?

12-24. Explain what would happen to the contour of a single crystal if, during Czochralski growth, the melt temperature were to suddenly rise because of a faulty temperature controller. What would happen if the melt temperature were to drop momentarily?

12-25. In Fig. 12-27 an analytical expression for $C_S/C_o = k_e(1 - f)^{k_e-1}$ where C_o is the initial concentration and f is the fraction of melt solidified.
  a. Write an analytical expression for C_L/C_o.
  b. A silicon crystal containing the same initial overall concentrations of boron and phosphorus is grown. At what value of f will the concentration of B equal that of P?

12-26. A 50-kg silicon melt initially contains 10^{16} phosphorus atoms/cm^3.
   a. What is the weight of phosphorus in the melt?
   b. What is the composition of the crystal that first solidifies onto the seed?
   c. If the bottom half of the grown crystal is discarded what is the *average* doping level in the remaining crystal?
   (*Note*: See previous problem.)

12-27. Calculate the value of the misfit and describe the state of stress in the following two single-crystal film–substrate combinations:
   a. An AlAs film layer on a GaAs substrate
   b. A CdS film layer on an InP substrate
   c. A ZnSe film on a GaAs substrate

12-28. a. Explain how ion implantation can be used to make a p–n junction.
   b. How would you produce a subsurface dopant concentration profile that is roughly trapezoidal in shape?
   c. An ion-implanted dopant distribution, gaussian in shape, is heated to an elevated temperature. Sketch the resulting dopant concentration profile.

12-29. Ion implantation has been used in nonsemiconductor doping applications to modify surfaces of mechanically functional components. What components would you recommend for such a treatment and what ions would you implant?

12-30. Semiconductor technology has helped to spin off new high-tech materials industries, processes, and equipment. Give one example of each.

13
OPTICAL PROPERTIES OF MATERIALS

13.1. INTRODUCTION

The 20th century, and particularly the latter part of it, can truly be called the Age of Optical Materials. This appellation is warranted in view of the ubiquitous role of optical materials and devices in some of the most important arenas of human activity, for example, fibers for optical communications; lasers and imaging displays for medical applications; optical coatings and devices (solar cells) for energy conservation; lenses, filters, and optically active materials for observing, detecting, displaying, and recording images in microscopy, photography, media, and information applications. We certainly have come a very long way when we consider that human involvement with optical materials and properties, prior to the past few centuries, was limited largely to aesthetic purposes, for example, mirrors, glazes for pottery, colored glass, pigments for paints, jewelry. But this does not even begin to accurately convey the dramatic progress that has been made. An excellent illustration of this can be seen in Fig. 13-1, where the improvement in the optical transparency of glass with time is depicted. Until this century one could barely see through window glass. Even up to 1970 only ~10% of the light incident on a block of optical glass a meter thick would be transmitted through it, as can be appreciated if you try to view a pane of glass sideways. Today, the specially purified glass used in optical fibers would allow 99.998% of the light through. This is equivalent to 98% transmission through a block of optical fiber quality glass 1 km thick!

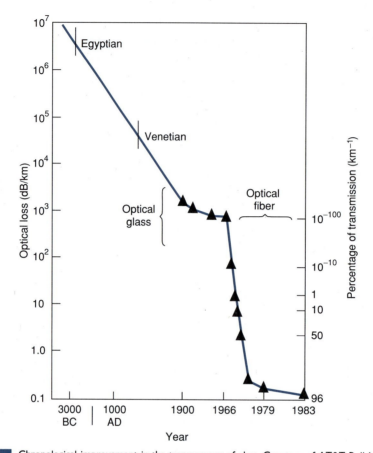

FIGURE 13-1 Chronological improvement in the transparency of glass. Courtesy of AT&T Bell Laboratories.

Optical materials can be broadly divided into two categories: passive and active. The active category includes those materials that exhibit special optical properties in response to electrical, magnetic, mechanical, optical, thermal, and other stimuli and interactions. Examples include electro-optical devices such as lasers, light-emitting diodes, and photodiodes, as well as magneto-optical recording media, luminescent materials, polymer photoresists, and liquid crystal displays. Passive optical materials encompass everything else, including inactive, active materials. For example, silicon is active in a solar cell but passive when used in optical coatings. In this chapter examples of both passive and active optical properties are treated.

All classes of engineering materials—metals, ceramics, semiconductors, and polymers—have representatives with interesting optical properties. In many of the modern optical applications, thin-film materials play a critical role and an introduction to this subject is therefore offered.

The unifying concept that embraces all optical properties is the interaction of electromagnetic radiation with the electrons of the material. On this basis, optical properties are interpretable in terms of the electronic structure as influenced by atomic structure and bonding. Therefore, aspects of both quantum mechanics and electromagnetic wave theory are required to account for a fundamental understanding of optical phenomena. For the engineering applications we are concerned with, however, simple equations distilled from elementary physics are all that is necessary, and these will be offered without proof.

13.2. INTERACTION OF LIGHT WITH SOLIDS

13.2.1. General Considerations

The regions of the electromagnetic spectrum we are concerned with in this chapter range from the ultraviolet (UV), through the visible, and into the infrared (IR). These optical regimes are depicted in Fig. 13-2 together with the defining wavelengths and equivalent photon energies. The connection between these two quantities is given by the second formula (Eq. 2-2) in this book,

$$E = h\nu = hc/\lambda \qquad \text{or} \qquad E(\text{eV}) = 1.24/\lambda(\mu\text{m}), \tag{13-1}$$

where E is photon energy, ν is frequency, λ is wavelength, c is the speed of light, and h is Planck's constant. As valence and conduction electron energies in solids typically range between 0.1 to 10 eV, the correspondence to photon energies in the IR, visible, and UV is evident. Core electrons possessing energies of thousands of electron volts interact with X-rays, as noted in Section 2.3.1, and their effects are not discussed here. When photon and electron energies

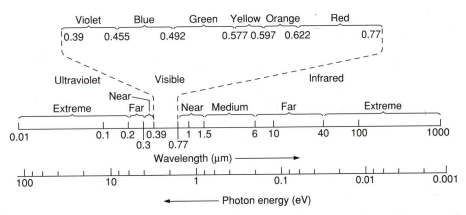

FIGURE 13-2 Photon wavelengths, colors, and energies associated with the infrared, visible, and ultraviolet regimes of the electromagnetic spectrum. From S. M. Sze, *Semiconductor Devices: Physics and Technology*, Wiley, (1985). Copyright © 1985 by AT&T; reprinted by permission.

are reasonably close in magnitude we speak of resonance and expect absorption. Light incident on a solid with intensity I_o is then attenuated as it passes through it because of photon–electron interactions. At a distance x, the intensity I (commonly in units of power density, e.g., W/m^2) is given by

$$I = I_o \exp(-\alpha x), \tag{13-2}$$

where α is the absorption coefficient (in units of m^{-1} or cm^{-1}). The **absorption** that occurs is wavelength dependent and related to an important dimensionless optical constant of the material known as the **index of absorption** (k); that is, $\alpha = 4\pi k/\lambda$.

Although opaque materials like metals and semiconductors strongly absorb light (in the visible), transparent materials like silica glass, ceramic, and polymer dielectrics allow light to readily penetrate so that **transmission** is high. In previous physics study it was learned that the **index of refraction** (n_r) is the relevant material constant that describes light transmission through glass and lenses. By definition, $n_r = c/v$, where v is the velocity of light in the medium. Light incident at an arbitrary angle ϕ_1 is refracted or bent as it crosses the interface between optically different materials as shown in Fig. 13-3. Then Snell's law holds; that is,

$$n_r(1)/n_r(2) = \sin \phi_2/\sin \phi_1 = v_2/v_1, \tag{13-3}$$

where $n_r(1)$ and $n_r(2)$ are the refractive indices in adjacent media 1 and 2. Note also that n_r varies with wavelength of light, a phenomenon known as

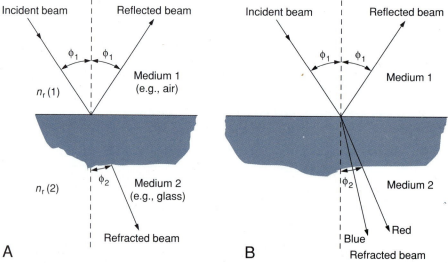

FIGURE 13-3 Optical effects at the interface between media possessing different refractive indices. (A) Depiction of Snell's law. Light travels faster in low-index medium. (B) Dispersion of light. Red light travels slightly faster than light of shorter wavelength, and is bent or dispersed less.

dispersion. The higher value of n_r for blue light means it travels more slowly than red light and bends more, as shown.

All materials can be described by a complex index of refraction that combines the attributes of both n_r and k. As already suggested, $k > n_r$ in opaque solids; conversely, $n_r > k$ in transparent solids. Incident light is not simply absorbed or transmitted, however. A fraction R is reflected and a fraction S is scattered from the incident surface. Including the fraction absorbed, A, and the fraction transmitted, T, conservation of energy requires that the sum of these contributions be

$$A + T + R + S = 1. \tag{13-4}$$

In the field of optics the reflected fraction is an important quantity and electromagnetic theory shows that for light passing through air ($n_r = 1$), impinging normally on a **transparent** material of index n_r, and $k = 0$,

$$R = [(n_r - 1)^2/(n_r + 1)^2]. \tag{13-5}$$

For glass with $n_r = 1.5$, $R = 0.5^2/2.5^2 = 0.04$. Thus, 4% of the light is immediately reflected from the surface. Even this small amount is an intolerable loss in many optical applications, and as we shall see later, it is reduced through the use of antireflection coatings. If, however, light impinges normally on an **absorbing** material,

$$R = [(n_r - 1)^2 + k^2]/[(n_r + 1)^2 + k^2]. \tag{13-6}$$

A couple of practical examples will indicate some implications of a number of the equations in this section.

EXAMPLE 13-1

a. The optical constants of silver metal at a wavelength of 0.55 μm (550 nm) are $n_r = 0.055$ and $k = 3.32$. What is the fraction of light reflected (i.e., reflectance) from silver at this wavelength?

b. How thin does silver have to be to transmit 50% of the light incident on it?

ANSWER a. Silver is an absorbing material so Eq. 13-6 applies. Substitution yields $R = [(0.055 - 1)^2 + 3.32^2]/[(0.055 + 1)^2 + 3.32^2] = 0.982$. The high reflectivity of silver earmarks its use for mirrors.

b. From Eq. 13-2, $x = -\alpha^{-1} \ln(I/I_o)$, or $x = -(4\pi k/\lambda)^{-1} \ln(I/I_o)$. Substituting, $x = -(4\pi \times 3.32/5.5 \times 10^{-5})^{-1} \ln(0.5) = 9.14 \times 10^{-7}$ cm, or 9.14 nm. Half-silvered mirrors that reflect as well as transmit light in assorted optical applications are, therefore, composed of a thin film deposited on a glass substrate. Such films are prepared by physical vapor deposition processes such as evaporation (see Section 12.5.5.1).

13.2.2. Interaction of Light with Electrons in Solids

13.2.2.1. Conducting Solids

In conducting solids like metals and semiconductors, a great number of electrons are available to interact with electromagnetic radiation. For example, metals contain a large density of empty, closely spaced electron states above the Fermi level. Incident photons over a broad wavelength range are readily absorbed by conduction band electrons. These excited electrons occupy higher energy levels where they may undergo collisions with lattice atoms. The extra energy is dissipated through lattice vibrations that heat the metal, and we speak of absorption. Alternately, if the probability of colliding with an ion is small, the electron will emit a photon as it drops back to a lower energy level. This is the origin of the strongly reflected beam exhibited by metals in the visible and infrared region. When the time it takes for electrons to become excited and deexcited is short compared with the period of the incident electromagnetic wave, high reflectivity can be anticipated. Physically, the electrons are able to respond quickly enough to establish an electrical shield against the (more slowly) oscillating electric field associated with the incident wave, preventing it from penetrating. These conditions are well obeyed for metals in the infrared.

Of the nearly 80 metallic elements only gold and copper (and alloys of these metals) are significantly colored. All the other elements are white or gray, although in portions of the visible spectrum some metals have a slight tinge of color. Aluminum and chromium have a faint bluish, and nickel a weak yellowish, coloration. The overwhelming majority of metals reflect all portions of the visible spectrum and thus appear white. Gold and copper preferentially absorb in the green region, and these metals respectively assume the color of the reflected yellow and red light. These effects are illustrated in Fig. 13-4,

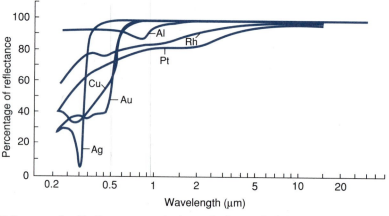

FIGURE 13-4   Reflectance of gold, silver, copper, aluminum, rhodium, and platinum mirror coatings as a function of wavelength.

where the reflectivity of a number of metals is displayed as a function of incident wavelength. A noteworthy feature of the optical response is the decreased reflectivity in the ultraviolet and, in particular, the abrupt absorption edge exhibited in Au, Cu, and Ag. The absorption now is not of the free carrier, **intraband** type discussed previously, but is instead due to **interband** electron transitions (e.g., a 3*d* to 4*s* transition in Cu).

In semiconductors it is somewhat startling to note such a strong variation in the absorption coefficient with wavelength as that indicated in Fig. 13-5. The several orders of magnitude change in α at the wavelength corresponding to the energy band gap has important implications in semiconductor optoelectronic devices. By now we are all familiar with the important interband transition that occurs in semiconductors when electron–hole pairs are created. If the incident wavelength has a higher energy associated with it than the energy band gap, electrons can successfully negotiate this interband transition. The

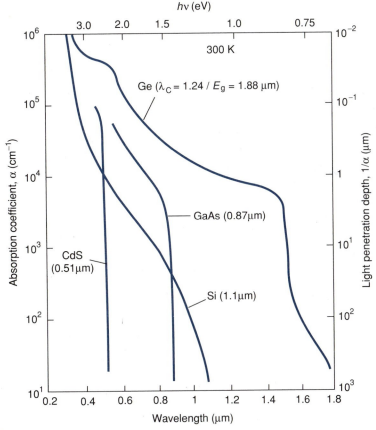

FIGURE 13-5   Optical absorption coefficients in semiconductor materials. From S. M. Sze, *Semiconductor Devices: Physics and Technology*, Wiley, (1985). Copyright © 1985 by AT&T; reprinted by permission.

critical wavelength, λ_c, required is given by Eq. 13-1, namely, λ_c (μm) = 1.24/ E_g (eV). In the case of GaAs, where E_g = 1.43 eV, λ_c = 0.87 μm. Photons of shorter wavelength (with higher energy than E_g) are then strongly absorbed, generating free electrons that occupy excited levels in the continuum of conduction band states. Semiconductors resemble metals then and are highly reflective. Photons with longer wavelengths than λ_c (with lower energy than E_g) are not absorbed. In this (infrared) range semiconductors are transparent because mechanisms no longer exist to excite transitions across the band gap.

13.2.2.2. Insulating Solids

Materials like silica glass, ceramics, and polymers contain no free electrons. Therefore, electromagnetic radiation interacts in a fundamentally different way with such solids than with conductors. The differences are related to distinctions between the electrical properties of insulators (dielectrics) and conducting solids (see Chapter 11). After all, both electrical and optical phenomena in dielectrics, as well as conductors, stem from applied ac electric fields ($\mathscr{E}$) and the response of charge to it. When the frequency is low, dc being an extreme case, $\mathscr{E}$ fields can be applied via surface contacts from an electric power supply. But when the frequency is extremely high, as in light, the $\mathscr{E}$ field component of electromagnetic radiation interacts with the material in a contactless way. In either case the same kinds of polarization phenomena occur. Electrons and ions oscillate to and fro, while molecular dipoles try to reorient in an attempt to respond in concert with the rapidly alternating electric field. We speak of the **dielectric constant**, ε_r, as the material property that reflects the extent of the polarization when the frequency is low. Similarly, when the frequency is high, only electronic polarizability contributes, and the appropriate constant is called the **index of**

| TABLE 13-1 | **REFRACTIVE INDICES OF VARIOUS MATERIALS**[a] |

Material	Refractive Index	Material	Refractive Index (Average)
Glasses		Ceramics	
Silica glass	1.46	CaF_2	1.434
Soda-lime glass	1.51	MgO	1.74
Borosilicate glass	1.47	Al_2O_3	1.76
Flint optical glass	1.6–1.7	Quartz (SiO_2)	1.55
		Rutile (TiO_2)	2.71
Polymers		Litharge (PbO)	2.61
Polyethylene	1.51	Calcite ($CaCO_3$)	1.65
Polypropylene	1.49	$LiNbO_3$	2.31
Polystyrene	1.60	PbS	3.91
Polymethyl methacrylate	1.49	Diamond	2.42

[a] Data were taken from several sources.

refraction (n_r). The underlying continuity in the electrical and optical properties is expressed in the connection between ε_r and n_r at high frequency:

$$\varepsilon_r = n_r^2. \tag{13-7}$$

A few aspects of n_r are worth noting while perusing Table 13-1, which contains values of n_r for different materials. As might be expected, n_r increases with the number of charges or dipoles per unit volume that interact with the electric field. A composite portrait of a high-n_r material is one that has high mass density, containing atoms with high atomic number, large ionic polarizability, and a large degree of covalent bonding. In SiO_2, for example, in order of increasing density, $n_{r,glass} = 1.46$, $n_{r,tridymite} = 1.47$, $n_{r,cristobalite} = 1.49$, and $n_{r,quartz} = 1.55$. Also, in order of increasing covalent bonding, $n_{r,ZnCl_2} = 1.68$, $n_{r,ZnO} = 2.08$, $n_{r,ZnS} = 2.37$, $n_{r,ZnSe} = 2.57$, and $n_{r,ZnTe} = 3.56$.

13.3. APPLICATIONS OF THE OPTICAL PROPERTIES OF METALS AND DIELECTRICS

13.3.1. Metals

13.3.1.1. Mirrors

In addition to reflections from pools of water, highly polished metals have served as mirrors since antiquity. Until relatively recently, mirrors have necessarily been small because of the weight, expense, and difficulty of polishing bulk quantities of noble metals—Au, Ag, Cu, and their alloys. Nowadays thin metal films or coatings (typically less than 1000 nm thick) of the appropriate metals are deposited on flat glass or curved substrates for this purpose. High reflectivity in the visible is of course the chief property required of metals for mirror applications, and Fig. 13-4 provides such information for the most widely used materials. In commonly employed front surface mirrors, resistance to mechanical scratching (of Au, Ag, Cu, Al), oxidation (of Al), and tarnishing (of Ag) are additional concerns. Despite its relatively low reflectivity, rhodium, a metal that is more expensive than gold, has found application in telescope mirrors, in optical reflectivity standards, and in mirrors for medical purposes. The reasons are the high hardness and environmental stability of this metal.

13.3.1.2. Carat Gold Alloys

The color of gold alloys is interesting scientifically and important in jewelry applications. In the caratage scale, pure gold is assigned a value of 24 carat (kt). Other gold alloys are based on proportional weight ratios; 10-carat gold contains 10/24 parts by weight of gold, and so on. Gold is most frequently alloyed with silver and copper to create colored gold jewelry alloys. As indicated

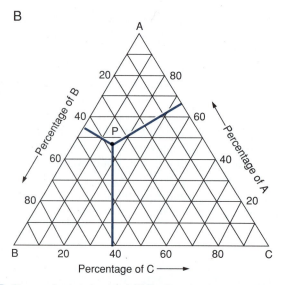

B

FIGURE 13-6 (For part A, see color plate.) (B) Equilateral triangle representation of ternary A–B–C alloy compositions. Based on the geometric theorem mentioned in the text, the alloy of composition P contains 47.5% A, 37.5% B, and 15% C.

by Fig. 13-6A, a variety of colors ranging from red to yellow to green to white with many shades in between are available in this ternary alloy system.

Although we have not formally discussed three-component alloys in this book, their properties are often represented on equilateral triangles. At each apex is a pure component, for example, 100% Au. Since the base opposite it represents 0% Au, it also corresponds to the collection of Ag–Cu binary alloys. In crossing from the base to the apex, the percentage of Au linearly increases from 0 to 100%. Each point within the triangle corresponds to a different ternary composition, which is easily determined by first dropping altitudes to the three bases. A line through the point parallel to any base yields the composition of that particular component. Similarly for the other two elements. This method of representing composition is based on the geometric theorem that the sum of the altitudes from any point within an equilateral triangle is constant (Fig. 13-6B).

The equilateral triangle format is a natural way to represent ternary phase equilibria at a given temperature or isotherm. By stacking isotherms vertically on a temperature axis, a three-dimensional phase diagram that takes the shape of a triangular prism emerges. Its internal space is subdivided by phase boundaries into complex fields that can admit mixtures of up to four phases. Binary phase diagrams like those in Chapter 5 occupy each prism face.

EXAMPLE 13-2

Assume you are given the task of designing the following alloys for the cheapest cost. (Assume the price of Au is \$400/oz; Ag, \$5/oz; and Cu, \$0.1/oz.)
a. A red 8-kt gold alloy
b. A green–yellow 18-kt gold alloy.
How much of each element should be added and what is the alloy cost per ounce?

ANSWER Because Au costs 80 times more than Ag and 4000 times more than Cu, the objective is to minimize the weight of Au. The possible compositions that yield 8- and 18-kt alloys fall on the horizontal lines representing 33.3 wt% Au ($\frac{8}{24} \times 100$) and 75 wt% Au ($\frac{18}{24} \times 100$), respectively.

a. The cheapest red gold alloy would contain about 33.3 wt% Au, 66.7 wt% Cu, and no Ag. The cost would be indistinguishable from the price of Au. Of course, if the color, a subjective matter, is not the right shade of red, up to ~20 wt% Ag should be substituted instead of Cu.

b. The cheapest green–yellow 18-kt gold alloy would contain 75 wt% Au, ~20 wt% Ag, and ~5 wt% Cu. The material cost of such an alloy is $\frac{75}{100}$ (\$400) $+ \frac{20}{100}$ (\$5) $+ \frac{5}{100}$ (\$0.1) = \$301.05/oz.

At least three different mechanisms have been identified as causing color changes in metals during alloying:

1. A *change in the wavelength* associated with the absorption edge (e.g., Ag–Au alloys)
2. A *change in the relative reflectivity* without a shift in the absorption edge (e.g., Au–Pd alloys)
3. *Formation of compound phases* having an entirely different spectral reflectivity

An interesting example of case 3 occurs in the gold–aluminum alloys, where fine-diameter Au wire is used to bond to Al contact pads of integrated circuits. The reaction between the two metals produces the intermetallic compound $AuAl_2$, which is purple. Viewed with alarm as a potential reliability problem in electrical contacts, the compound is known as "purple plague" (see Section 15.5.4). When cast in rings this purple gold makes for a remarkably attractive jewelry alloy.

13.3.2. Dielectric Optical Coatings

13.3.2.1. Antireflection Coatings

To appreciate the importance of reducing the reflectivity at optical surfaces, consider the following example.

EXAMPLE 13-3

An optical system consists of 20 air–glass interfaces at lenses, prisms, beam-splitters, and so on. The glass employed has an index of refraction of $n_r = 1.50$.
a. Neglecting absorption, what is the transmission of the system?
b. If the reflectivity at each interface is reduced to 0.01 what is the transmission?

ANSWER   a. At interface 1, the reflectivity is R_1 and the transmission is $1 - R_1$. At interface 2, the transmission is $(1 - R_1)(1 - R_2)$. Therefore, for 20 interfaces the transmission is $T = (1 - R_1)(1 - R_2)(1 - R_3) \cdots (1 - R_{20})$. Because the same glass is used $R_1 = R_2 = R_3 = \cdots = R_{20} = R$ and $T = (1 - R)^{20}$. If $n_r = 1.5$, $R = 0.04$ from Eq. 13-5. Substitution yields $T = (1 - 0.04)^{20} = 0.442$. Thus, less than half of the light incident is transmitted through the system.
b. For $R = 0.01$, $T = (1 - 0.01)^{20} = 0.818$. The transmission is almost doubled, representing a significant increase in optical performance.

Use of antireflection (AR) optical coatings is a practical method for reducing reflectivity at glass surfaces. A common example is the plum–purple-colored AR coating on camera lenses. They are also used in solar cells and on ophthalmic, microscope, telescope, and binocular lenses as well. The wave interference effect shown in Fig. 13-7 is responsible for antireflection properties. Although grossly exaggerated in the figure, imagine that light impinges normally on a planar surface containing a transparent AR coating of thickness d. Rays that bounce off the top and bottom (interfacial) coating surfaces will be **out of phase** if the light path difference, $2d$, is equal to an integral number of **half-wavelengths**, that is, $2d = m\lambda/2$, where m is an integer. Due to destructive interference the incident light is not reflected. When $m = 1$ we speak of the quarter-wave AR coating with thickness $\lambda/4$. For light of 0.6 μm, a 0.15-μm-thick coating is required. Films this thin are usually thermally evaporated.

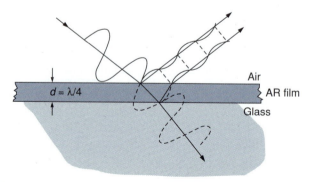

Air
$d = \lambda/4$ AR film
Glass

FIGURE 13-7   Interference of light from the top and bottom surfaces of a quarter-wave antireflection coating. From R. W. Ditchburn, *Light*, 3rd ed., Vol 1, Academic Press, London (1976).

Properties of the most common optical coating materials are displayed graphically in Fig. 13-8, where typical refractive indices and ranges of transparency are indicated. Theory shows that an AR coating with an index of refraction equal to $n_r^{1/2}$ is optimal because it yields the lowest reflectivity. Thus, for a glass lens with $n_r = 1.5$, an AR film with an index of refraction equal to $1.5^{1/2} = 1.22$ should be chosen, other things being equal. The fact that n_r for MgF_2 is reasonably close to the optimal value is one reason for its widespread use.

13.3.2.2. Other Optical Coatings

Although AR coatings account for the majority of all optical coatings they are not the only ones of value. Figure 13-9 provides an indication of the varied

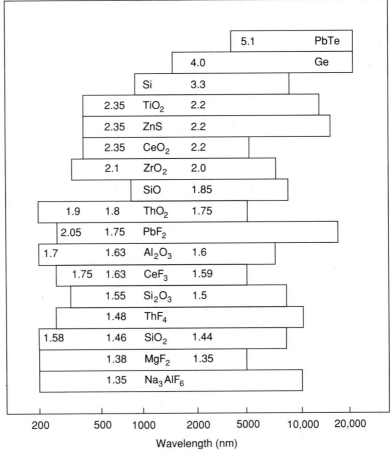

FIGURE 13-8   Spectral region of high transparency in dielectric films. Indices of refraction are given at the indicated wavelength. After G. Hass and E. Ritter, *Journal of Vacuum Science and Technology* **4**, 71(1967).

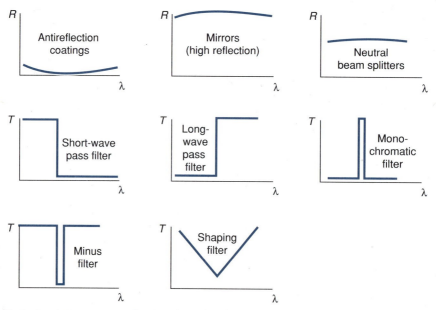

FIGURE 13-9   Desired optical response as a function of wavelength in assorted applications. From H. Pulker, *Coatings on Glass*, Elsevier, Amsterdam (1984).

spectral response of reflectance and transmission of coatings used in assorted applications. Interestingly, it is possible to create dielectric coatings that are highly reflecting and behave as mirrors. Instead of n_r values *smaller* than those of the glass they coat as in AR coatings, mirrors arise with (usually multilayer) films possessing n_r values *larger* than that of the glass substrate. Also for an enhanced reflected beam, there must be constructive interference between the two rays leaving the coating surfaces. This occurs when the path difference ($2d$) is equal to an integral number of **full** wavelengths. Therefore, $\lambda/2$ coatings are required. Dielectric mirrors are employed in high-performance optical systems such as lasers. Unlike metal mirrors which absorb some light, dielectric mirrors depend on interference effects, so that there is very little absorption.

Light filters that transmit in some regions of the spectrum but not in others have been employed in slide and movie projectors, energy-saving devices, and assorted photography applications. Filtering of visible light from the heating effects of IR radiation is important in projectors in order not to damage photographic emulsions. Thin optical films and coatings also help to conserve energy. For solar water heaters high-thermal-absorption coatings enable efficient capture of the sun's heat with little loss from reradiation. In another application, optically transparent window panes coated with a so-called *heat mirror* reflect heat back into the home in winter.

13.4. ELECTRO-OPTICAL PHENOMENA AND DEVICES

13.4.1. Introduction to Optical Effects in Semiconductors

Unlike the passive optical components treated to this point, electro-optical (also called optoelectronic) phenomena rely on the active coupling between electrical and optical effects. Semiconductors and, to a lesser extent, ferroelectrics are the primary materials in which these effects are large enough to be capitalized upon in practical devices. The three basic radiative transitions can be simply understood by referring to Fig. 13-10 depicting the valence and conduction bands of the involved semiconductor or active electro-optical medium. Either photons or electronic signals can trigger these electro-optical responses.

1. The process of **absorption** of an incident photon excites an electron from the valence to the conduction band, generating mobile carriers or an electric signal. Thereafter, electronic circuitry is activated and the signal can be amplified as in a **photodiode detector** or be used to deliver external power as in a **solar cell**.

2. The reverse process of photon **emission** occurs during deexcitation of electrons from the conduction to valence band. In **light-emitting diodes** (LEDs), for example, the emission is triggered by an electric current that passes through the junction causing electron–hole recombination and a corresponding *spontaneous* emission of monochromatic light. The light can now be viewed directly in a display, made to pass through lenses of an optical system, or enter an optical fiber. Similar excitation and deexcitation processes were considered in Section 2.2.4 for electrons undergoing transitions between atomic energy levels E_1 and E_2.

3. The last case in Fig. 13-10 has features of the previous two. Through absorption of light, electrons ascend from E_1 to E_2 as shown in Fig. 2–3. But,

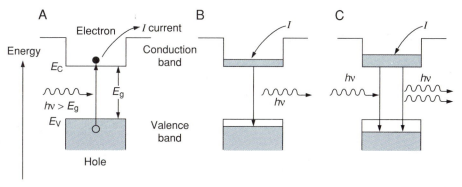

FIGURE 13-10 Three basic electron transition processes between valence and conduction bands of semiconductors: (A) Absorption. (B) Spontaneous emission. (C) Stimulated emission.

as predicted by Einstein in 1917, if a photon with energy $E_2 - E_1$ impinges on such an excited atom, it can be **stimulated** to emit a photon. This photon has the same energy and phase as the original one. If instead of allowing the light to exit the medium it is made to bounce back and forth in phase, while impinging on other atoms in excited states, the light intensity will build. This is the basis of the laser effect. An electric current initiates the usual spontaneous emission of monochromatic light from the junction in semiconductor **laser diode** devices; however, the light is captured and reflected between parallel end face mirrors in the device, stimulating further emission. The optical amplitude builds and light of high intensity is created. Everything happens so fast that the laser output (pulse) is a nearly instantaneous response to the original input electric signal (pulse). Like case 2, the laser light enters an optical system for further processing.

Unfortunately, not all semiconducting materials, including silicon and germanium, are capable of exhibiting large electro-optical effects that are necessary in lasers and light-emitting diodes. Rather, we must rely on compound semiconductors to make these devices because of quantum effects related to the nature of the transition between conduction and valence band levels. When the transition occurs in a direct way there is a vertical line connecting electron states in the two bands, as shown in Fig. 13-11. It was shown in Fig. 11-8 that one way to represent the band structure was through an E-versus-k plot. But because k is essentially the electron momentum it means that momentum, which is the same in both bands, is conserved in a direct transition. Light emission and absorption processes are efficient under such conditions. Materials like GaAs, InAs, and InP are examples of **direct semiconductors**.

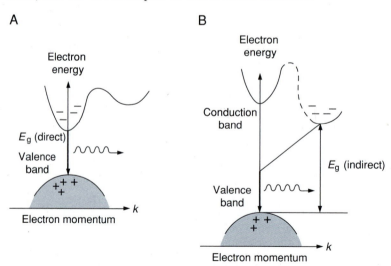

FIGURE 13-11   Direct and indirect energy band transitions in semiconductors. From M. Ohring, *The Materials Science of Thin Films*, Academic Press, Boston (1992).

In **indirect semiconductors**, on the other hand, the transition is slanted so that momentum is not conserved. Instead, some momentum is transferred to the crystal atoms, resulting in lattice heating and energy loss. This makes light absorption or emission processes far less efficient in indirect compared with direct semiconductors. In addition to Si and Ge, AlAs and GaP are indirect semiconductors.

13.4.2. Variable-Energy Band Gaps

Electro-optical devices necessarily operate at a fixed wavelength that is determined by the magnitude of the energy gap of the semiconductor. But what if we want a device to operate at a certain wavelength or color, and there is no material available with the necessary value of E_g? Fortunately, it is often possible to design and synthesize semiconductor alloys that will have the requisite energy gap value. To assist in the process an important representation of energy band gaps and lattice parameters (a) in compound semiconductors is shown in Fig. 13-12. Lines connecting individual compounds (e.g., GaAs and AlAs) represent ternary semiconductor compositions (e.g., $Al_xGa_{1-x}As$, where x is the atomic fraction of Al). For many systems (e.g., GaAs–AlAs), the resulting value of E_g for the ternary is the weighted average of the binary E_g values. As ternary alloy semiconductor films are deposited on semiconductor

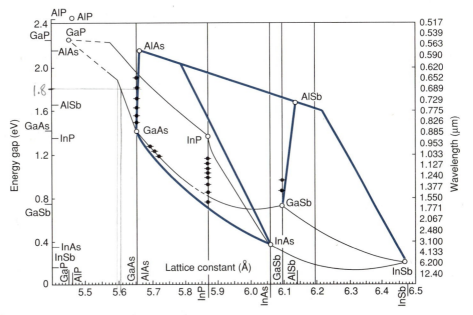

FIGURE 13.12 Energy band gaps and corrresponding lattice constants for important compound semiconductor materials. Lines connecting two compound semiconductors characterize ternary alloy behavior. Solid lines indicate direct semiconductors; dashed lines, indirect semiconductors.

substrates (e.g., a GaAs wafer) a good lattice match (see Section 12.5.5.2) between the two is essential. Therefore, information on lattice spacings is also provided in Fig. 13-12, enabling a sense of the lattice misfit to be gauged. Vegard's law, suggesting that the ternary lattice parameter can be obtained by the weighted average of the binary lattice parameters, is generally a good assumption. Thus, linear interpolation between *binary* compound energy gaps and lattice parameters often yields reasonably accurate values of E_g and a, respectively, for ternary alloys.

EXAMPLE 13-4

Binary GaAs and InAs compounds can be alloyed to produce $Ga_x In_{1-x} As$ ternary solid solution semiconductor alloys, where x is the atomic fraction of Ga and $1 - x$ is the atomic fraction of In.

a. What value of x is required if high-quality films are to be lattice matched to InP substrates?

b. What would be the operating wavelength of a light-emitting diode made from this ternary alloy?

ANSWER   a. To be lattice matched to InP, the ternary film must have the same lattice parameter as InP, or 0.5869 nm (see Table 12-1). Therefore, interpolating between the lattice constants for GaAs ($a = 0.5653$ nm) and InAs ($a = 0.6068$ nm),

$$1 - x = (0.5869 - 0.5653)/(0.6068 - 0.5653) = 0.52, \text{ and } x = 0.48.$$

A film composition of $Ga_{0.48} In_{0.52} As$ is required.

b. The operating value of E_g can be read off Fig. 13-12 as the intersection of the GaAs–InAs curve with the vertical line representing the lattice parameter of InP. As a result $E_g = 0.7$ eV, corresponding to an IR wavelength of 1.77 μm.

13.4.3. Solar Cells

The promise of clean energy through direct conversion of sunlight into electricity has made solar cells a favorite component of national energy generation strategy. We already know that light incident on a bulk semiconductor creates electrons in the conduction band and holes in the valence band. This also happens in a diode if the p–n junction region (Fig. 13-13A) is close enough to the surface so that photons can penetrate to it. Once the junction is illuminated, carriers that were immobile in the space charge region of the prior nonilluminated or dark junction now move. Electrons lose energy by occupying lower conduction band levels in the n-type material; similarly, holes lose energy by floating upward to fill valence band levels in p-type material.

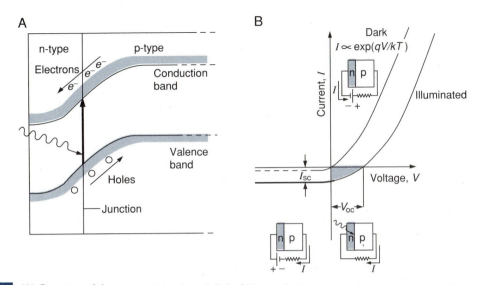

FIGURE 13-13 (A) Operation of the p–n junction solar cell. Light falling on the junction generates electron–hole pairs. Electrons separate to the n region, holes to the p region. The resulting current flow powers an external circuit. (B) Circuit polarities and current–voltage characteristics of a solar cell. Power is delivered externally in shaded fourth quadrant.

Many of these carriers will promptly recombine. But some may separate further to produce the equivalent of positive and negative battery electrodes, or a **photovoltaic** effect. The illuminated junction then acts like a photon-activated electrical pump generating a current in the diode (cell) that can power an externally connected circuit.

To appreciate how a solar cell operates it is instructive to review the current (I)–voltage (V) characteristics of the diode (see Section 12.4.1). These are displayed in Fig. 13-13B where the dark characteristics in the forward (first quadrant)- and reverse (third quadrant)-biased cells are the same as those for a diode (see Fig. 12-21). In particular, the direction of the current delivered from the battery to the diode should be noted. When the external battery is removed and light shines on the device, power is delivered to the circuit. Simultaneously, the I–V characteristics shift so that there is no longer an operating state where $V = 0$ and $I = 0$. Instead, within the fourth quadrant there is an open-circuit voltage, V_{oc} (when $I = 0$), and a short-circuit current, I_{sc}, when $V = 0$.

Solar cells have been fabricated from a host of different semiconductor materials but only silicon cells have been commercialized on a large scale. Interestingly, both polycrystalline and even amorphous Si cells can be purchased. The efficiency of energy conversion is about 12% for crystalline cells, meaning that of the 1 kW/m^2 of solar power density falling on earth, 120 W of electric power would be generated from 1 m^2 of cell area.

13.4.4. Luminescence and Color

13.4.4.1. Definitions

In Section 5.3.4 it was noted that visible light is emitted from heated solids, a phenomenon known as **incandescence**. Light emission by other methods of stimluation is known as **luminescence**. There are basically three important types of luminescence depending on the means of exciting electrons to higher energy levels.

1. *Photoluminescence.*   Light emission from solids through *photon* excitation is known as photoluminescence. The fluorescent light bulb capitalizes on this phenomenon. Ultraviolet light from a mercury discharge impinges on phosphors (e.g., doped CaF_2) that coat the tube interior, and these in turn exhibit photoluminescence by emitting visible light of longer wavelength.

2. *Cathodoluminescence.*   *Electron* bombardment resulting in light emission is known as cathodoluminescence, a phenomenon relied upon in cathode-ray tubes.

3. *Electroluminescence.*   Light emission generated by electric currents and fields is known as electroluminescence.

Luminescent materials are further divided depending on the time difference between absorption of the excitation energy and emission of the light. When these successive processes are very rapid and occur within 10^{-8} second, the effect is known as **fluorescence**. Emission that takes longer ($>10^{-8}$ second) is called **phosphorescence**. Most commercial phosphors exhibit delayed emission of milliseconds.

13.4.4.2. Phosphors and Their Behavior

A material that successfully luminesces is called a **phosphor**, a word derived from the element phosphorus, which glows in the dark as it oxidizes. Phosphors are basically large-band-gap semiconductors. One objective for display applications (e.g., TV color monitors) is to get them to emit colored light in desired regions of the visible spectrum. To see what is required to do this, let us consider the band diagram for a luminescent material shown in Fig. 13-14. Excitation with absorbed light or electrons of energy E_g or greater causes the usual valence-to-conduction band transition. Subsequent interband photoemission may occur as in case A, but we know that the emitted light has a wavelength longer than that corresponding to E_g. For pure ZnS, a popular phosphor material, E_g is 3.54 eV, corresponding to a wavelength of 0.350 μm in the UV. This means that the downward transition to the valence band must originate from a level below that of the conduction band edge; that is, the level must lie somewhere in the gap, if visible light of lower energy is to be emitted (case B).

Clearly, to make ZnS a useful phosphor, interlevel spacings less than 3.54 eV have to be created. We have already learned two potential strategies for

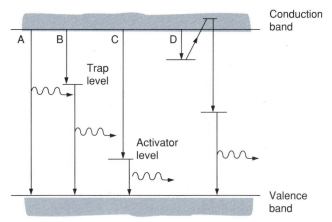

FIGURE 13-14 Energy band diagram illustrating photon emission in a luminescent material like ZnS. (A) Photon emission with full band gap energy. (B, C) Photon emission with less than full band gap energy because of either trap or activator level transitions. (D) Complex photon emission process.

accomplishing this. The first involves alloying it with a semiconductor having a smaller energy gap. For ZnS the compound CdS (E_g = 2.42 eV) has been successfully employed to reduce the width of the energy gap to correspond more closely to optical wavelengths. By these means emission anywhere from red to blue has been obtained in ZnS–CdS alloys. The second approach is simply to dope ZnS with donor and acceptor atoms that will directly create levels throughout the energy gap. Much experimentation on choice of dopant and amounts to add is required to obtain useful phosphors. In the case of ZnS, for example, Ag additions of ~0.01 at.% produce blue light emission (~0.48 μm), whereas Cu gives a greener light (~0.54 μm). If, however, 0.005 at.% Ag is added to a 50–50 ZnS–CdS alloy, yellow light emission is obtained.

The Ag and Cu dopants added to ZnS phosphors are known as **activators** and they create levels that lie above the valence band (Fig. 13-14C). Other levels lying beneath the conduction band behave as electron **traps**. Electrons initially excited to the conduction band can execute a variety of odysseys during deexcitation. For example, they can alternately populate conduction band and trap levels and then descend to recombine with vacant activator states accompanied by light emission (Fig. 13-14D). The greater the number of electrons participating in the process, the greater the emission intensity. There are many variations on the theme of recombination, and the longer it takes to occur, the greater the *persistence* of the phosphor. Low electron mobility, combined with enough trapping obstructions, makes electrons survive longer prior to recombination and extend the persistence. A simple exponential decay formula is generally used to describe the kinetics of the decay process,

$$I(t) = I_o \exp(-t/\tau), \tag{13-8}$$

where I_o is the initial luminescent intensity, and $I(t)$, the intensity after time t. The time constant, τ, a measure of the persistence, is a property of the phosphor; it depends on the trap density as well as the energy difference, E_t, between trap levels and the conduction band edge; that is, $1/\tau(s^{-1}) \sim \exp(E_t/kT)$. Because of the Boltzmann factor, emission persists for long times at low temperatures.

Color television screens use phosphors in an interesting way. The picture tube contains three electron guns, each producing a separate beam that is focused on only one set of phosphors (Fig. 13-15). Thus, one beam strikes only phosphors emitting red light, a second beam only blue light phosphors, and a third only green light phosphors. The inside of the glass envelope contains a broad periodic array of the three types of phosphor dots, each precisely located. Close to the phosphor screen is a shadow mask, a sheet steel shield perforated by very closely spaced holes (~0.3 mm in size) precisely aligned over the phosphor dots. (A 19-in. tube would have ~440,000 holes.) As the electron beams raster across the screen they pass through the holes and the three colors emitted are simultaneously merged. Almost 16,000 lines are scanned per second, and the phosphor persistence has to be controlled so that successive scans do not blur the picture.

13.4.4.3 Color of Materials

The subject of color in luminescent materials has been dealt with at length. But other materials, like glasses and many dielectrics, inorganic compounds,

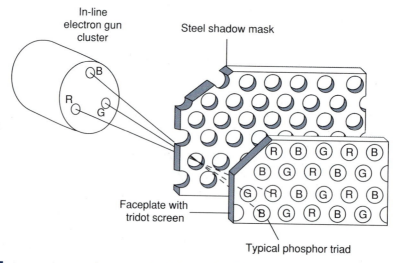

FIGURE 13-15   Section of a color TV picture tube shadow mask and its placement relative to the underlying red (R), green (G), and blue (B) phosphors, and the electron beam guns that activate them.

and semiconductors that are not luminescent, are also colored. Why? Actually, the similarities between luminescent and other colored materials are greater than the differences; in the latter, the color does not persist. We all know that isolated atoms emit visible light when properly excited (e.g., sodium yellow light, 0.589 μm), and that the involved levels split and broaden into bands when the solid is formed. For electrical applications they are called the valence and conduction band, but for optical properties the upper band is sometimes referred to as the *absorption* band. Suppose the energies associated with the absorption band correspond to blue light ranging in wavelength from 0.46 to 0.55 μm, as shown in Fig. 13-16. If white light were incident, the blue component would be absorbed. The solid appears yellow–orange–red depending on the level(s) from which the light is finally reradiated. This situation arises in CdS with a band gap of $E_g = 2.42$ eV corresponding to the blue portion of the spectrum. The nonabsorbed, reemitted light has an energy of ~1.8 to 2.2 eV, corresponding to the yellow-orange color that characterizes its appearance.

Silica glasses are colored by incorporating transition metal ions during melting in amounts ranging from 0.1 to ~5%. Thus, V^{4+} as well as colloidal Au and Cu colors glass red; Fe^{2+} and Co^{2+}, blue; Cu^{2+}, blue-green; Fe^{3+}, yellow green; Mn^{3+}, purple; and so on. The ions substitute for matrix atoms and disturb the electronic structure of the glass by creating new levels in the band gap. Specific wavelengths are now preferentially absorbed and emitted at these impurity levels, giving rise to a limitless collection of optical transitions.

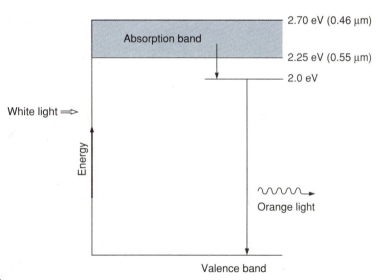

FIGURE 13-16   Schematic of band structure in colored materials.

13.5. LASERS

13.5.1. Introduction

The laser, one of the most important inventions of the 20th century, has played a very important role in numerous scientific and technological endeavors. High-resolution spectroscopy, audio and video recording, surveying and range finding, surgery, optical communications, cutting, drilling, welding, and supermarket checkout scanners suggest a wide range of applications. Light from ordinary light sources can perform some of these functions. But they cannot do it as well as lasers. The reason is that light from conventional light sources is **incoherent**. Light rays emanate from independent excited atoms and the emission is uncoordinated. As a result the waves are out of phase with one another and travel in random directions. Importantly, such light has low intensity. In physics it was learned that sine waves (e.g., $A \sin kt$, t = time, k = Const) of amplitude A have an intensity I given by A^2. If we consider a light beam containing a large number (N) of out-of-phase waves of the same frequency, the total incoherent intensity is the sum of the individual intensities:

$$I_{\text{incoherent}} = A_1^2 + A_2^2 + A_3^2 + \ldots + A_N^2. \tag{13-9}$$

Now consider a **coherent** light source, where all of the waves are in phase. The amplitudes are added first and then squared, yielding the result:

$$I_{\text{coherent}} = (A_1 + A_2 + A_3 + \ldots + A_N)^2. \tag{13-10}$$

If $A_1 = A_2 = \ldots A_i = A$, then $I_{\text{incoherent}} = NA^2$ and $I_{\text{coherent}} = N^2 A^2$. The coherent intensity is vastly larger than the incoherent intensity as substitution of, say, $N = 10^{18}$ will easily show. Lasers behave like Eq. 13-10 and ordinary light sources like Eq. 13-9.

13.5.2. Stimulated Emission

Let us try to see what conditions must be fulfilled for **stimulated emission** and laser (*l*ight *a*mplification by *s*timulated *e*mission of *r*adiation) action to occur. Competing with stimulated emission is **spontaneous emission** (not to be confused with stimulated emission) and **absorption**. If we consider the two levels in Fig. 2-3, then a consequence of the Boltzmann distribution is that the instantaneous population (n) of electrons distributed between them is

$$n_2/n_1 = \exp[-(E_2 - E_1)/kT]. \tag{13-11}$$

Because $(E_2 - E_1)/kT > 0$ there are more electrons in the ground state than in the excited state (i.e., $n_1 > n_2$), and this is the usual state of affairs. As we shall see, lasers require the reverse situation, or a **population inversion**, where $n_2 > n_1$.

For electrons located in level E_2, the rate of *spontaneous* emission is proportional to their population and the rate at which they descend, or $\rho_{21}n_2$, where ρ_{21} is the emission rate constant. The rate of *absorption* for electrons in level E_1 is proportional to the product of the number of electrons in the lower level, the impinging photon energy density (Ψ), and the absorption rate constant ν_{12}, or $n_1\Psi\nu_{12}$. Lastly, the rate of *stimulated* emission is proportional to the number of interactions between photons and excited electrons that lead to transitions to the ground state. Mathematically, this is expressed by the product of n_2, the same value of Ψ, and rate constant ν_{21}, that is, $n_2\Psi\nu_{21}$.

In lasers we want to maximize the rate of stimulated emission relative to the other competing processes. Two ratios are instructive in this regard:

1. Stimulated emission rate/spontaneous emission rate $= n_2\Psi\nu_{21}/\rho_{21}n_2$ $= \Psi\nu_{21}/\rho_{21}$.
2. Stimulated emission rate/absorption rate $= n_2\Psi\nu_{21}/n_1\Psi\nu_{12} = n_2\nu_{21}/n_1\nu_{12}$.

The first condition implies that a strong photon field must be present; the second that an electron population inversion, $n_2/n_1 > 1$, is necessary. Both must be effected simultaneously for laser action and this is not easy to do. Nevertheless, there are a reasonably large number of lasing systems, but relatively few of them have been exploited as commercial lasers.

13.5.3. Laser Systems

Scientific aspects of lasers are embodied in characteristic energy level diagrams for the lasing medium. Promising systems must then be engineered to allow the desired transitions to occur. Consideration of a few varied laser systems illustrates how these issues are addressed.

13.5.3.1. He–Ne Laser

A schematic of the popular helium–neon gas laser is shown in Fig. 13-17 together with the energy level diagram it capitalizes on. The coincidence in energy between the $2s$ state of He and the $3s$ state of Ne is critical to the behavior of this laser. A couple of kilovolts (dc) is applied to a $90:10$ He–Ne gas mixture maintained at about 3×10^{-3} atm pressure. The discharge ionizes the He, which efficiently transfers its energy to the Ne. The "matched energy" level ensures that the majority of the Ne atoms will be in this *metastable 3s* level. Without He, there is a vanishingly small probability that this will happen. Fortunately, electrons survive a bit longer ($\sim 6 \times 10^{-6}$ second) in this metastable state than usual, before relaxing to lower levels. A population inversion relative to the $2p$ state is achieved. The characteristic red light (632.8 nm) from the $3s$-to-$2p$ transition is emitted and bounces back and forth between the end mirrors, building the intensity of the photon field. This stimulates further emission from other Ne atoms and finally an intense beam issues from the laser.

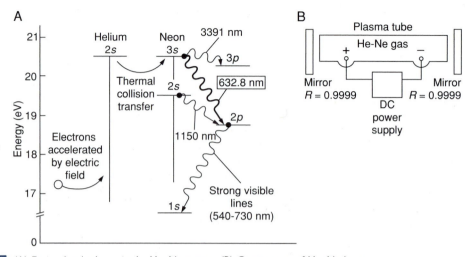

FIGURE 13-17 (A) Energy level scheme in the He–Ne system. (B) Components of He–Ne laser.

13.5.3.2. Ruby Laser

The ruby laser uses a ruby rod, which in reality is a single crystal of Al_2O_3 doped substitutionally with ~0.05% Cr ions. The laser and its electron energy levels are schematically represented in Fig. 13-18. Surrounding the laser rod is a xenon flash lamp capable of emitting a very intense (incoherent) light that

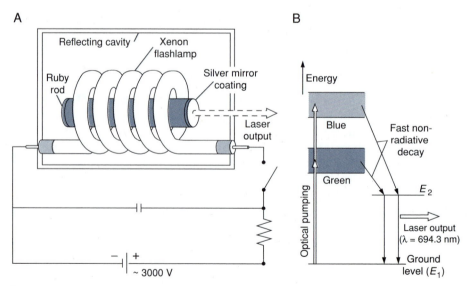

FIGURE 13-18 (A) Schematic of the ruby laser. (B) Energy level transition scheme in the ruby laser.

"pumps" or excites electrons into the blue and green absorption bands of ruby. After a rapid nonradiative transition to the E_2 excited state of Cr^{3+}, a population inversion is created. With deexcitation, the characteristic (coherent) red laser light is emitted at a wavelength of 694.3 nm. The sequence of events during laser action within the crystal bounded by end mirrors is described schematically in Fig. 13-19.

13.5.3.3. Semiconductor Laser

The semiconductor laser is a special kind of diode containing very heavily doped n- and p-type regions. In these devices direct band gap compound

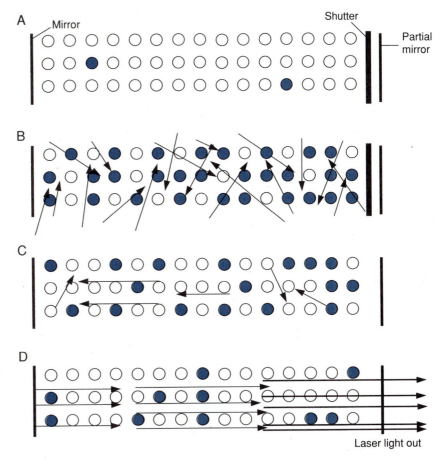

FIGURE 13-19 Schematic operation of the ruby laser. (A) Equilibrium state. (B) Pump light excites Cr atoms, causing population inversion and emission of photons. Limited stimulated emission with shutter closed. (C) With shutter open, photons traveling parallel to laser axis are reflected back and forth, stimulating a photon cascade. (D) The coherent photon beam builds in amplitude and exits the partially silvered mirror as a pulse of laser light. ◉, Excited Cr atom; ○, Cr atom in ground state.

semiconductors are essential for efficient light production, and GaAs and InP base materials are almost universally employed. In Fig. 13-20A the equilibrium band diagram for an unbiased homojunction GaAs laser is shown. The surprising fact that the Fermi levels in the n and p regions lie within the conduction and valence bands, respectively, is a consequence of the high doping levels. It also makes lasing action possible. To see why, consider that the diode is sufficiently forward biased (Fig. 13-20B). The junction width narrows in such a way that the electron-rich conduction band of n-GaAs overlaps the hole-rich valence band of p-GaAs; this creates the necessary population inversion of electrons. As large currents flow, electrons and holes recombine and light is emitted. If mirrors are placed on the device end faces, light is confined to the optical cavity at the junction where it coherently builds in intensity until light emission occurs.

Several semiconductor laser structures are shown in Fig. 13-21. The simplest consists of a homojunction composed of n- and p-doped GaAs regions, and light is emitted in a plane across the entire laser width (A). More advanced lasers operate at lower power levels and also confine the light laterally. These so-called double-heterojunction (DH) lasers (B, C) have a more complex layered semiconductor structure consisting of junctions between doped GaAs and $Al_xGa_{1-x}As$ films. Semiconductor lasers are minuscule compared with all other lasers. This can be appreciated by referring to the stunning TEM cross section of the InP-based laser shown in Fig. 3-38B.

13.5.4. Laser Applications

Laser applications can be roughly divided according to the power level required. The different types of lasers are listed in Table 13-2 together with some pertinent characteristics. For traditional monochromatic light generation,

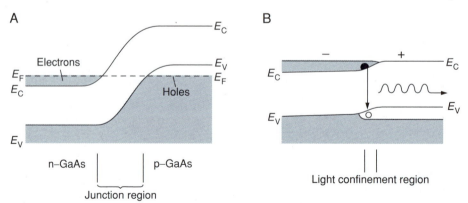

FIGURE 13-20   Semiconductor laser operation. (A) Band diagram of n- and p-doped GaAs homojunction laser in equilibrium. (B) Band diagram for forward-biased diode. Population inversion occurs in the junction where light is emitted.

detection, and processing using lenses, mirrors, prisms, diffraction gratings, and so on, low-power He–Ne and semiconductor lasers are sufficient. Examples of such usage include spectroscopy, optical communications, bar code scanners, video and audio recording systems (e.g., compact disks). In the latter case the audio signal to be recorded modulates the output of a laser beam that burns the information, in the form of little craters, into a plastic film on the disk. On playback, low-power laser light reflected from the disk provides the optical signal that is converted back to sound. The process shown in Fig. 13-22 is analogous to phonograph recording except that no (noisy) contact is made.

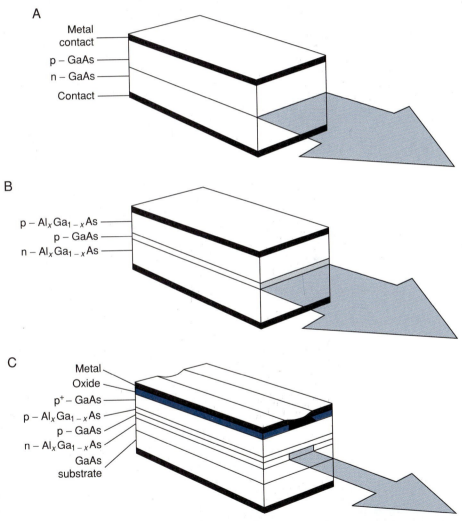

FIGURE 13-21 (A) Schematic of homojunction laser. (B) Double-heterojunction (DH) laser. (C) DH laser with stripe geometry. From S. M. Sze, *Semiconductor Devices: Physics and Technology*, Wiley, New York. (1985). Copyright © 1985 by AT&T; reprinted by permission.

TABLE 13-2   **OPERATING PARAMETERS FOR SEVERAL COMMON LASERS**

Type	Wavelength	Power/Energy	Type of Output	Beam Diameter
Helium–neon	632.8 nm	0.1–50 mW	cw	0.5–2 mm
Ruby (Cr:Al$_2$O$_3$)	694.3 nm	0.03–100 J	Pulsed	1.5 mm–2.5 cm
Carbon dioxide	10.6 μm	3–100 W	cw	3–4 mm
Nd : YAG (solid)	1.064 μm	0.04–600 W	cw	0.75–6 mm
Nd–glass (solid)	1.06 μm	0.15–100 J (per pulse)	Pulsed	3 mm–2.5 cm
Argon ion	488 and 515 nm	5 mW–20 W	cw	0.7–2 mm
Dye (liquid)	400–900 nm (tunable)	20–800 mW	cw/pulsed	0.4–0.6 mm
GaAs (semiconductor diode)	780–900 nm	1–40 mW	cw/pulsed	Diverges rapidly

From F. L. Pedrotti and L. S. Pedrotti, *Introduction to Optics,* 2nd ed., Prentice–Hall, Englewood Cliffs (1993).

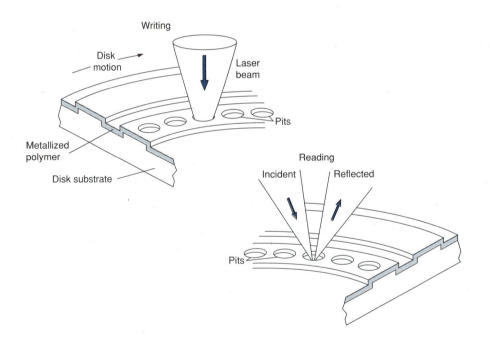

FIGURE 13-22   Schematic of the optical recording and playback process. During recording, information is stored in the form of pits, which are burned into the polymer disk surface by a laser beam. Upon playback, a weak laser beam scattered from the disk surface is detected, enabling the information to be retrieved.

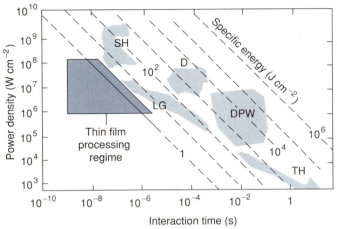

FIGURE 13-23 Laser processing regimes illustrating the relationships between power density, interaction time, and energy density. D, drilling; SH, shock hardening; LG, laser glazing; DPW, deep-penetration welding; TH, transformation hardening. The regime used in the surface modification of thin films is shown. From M. Ohring, *The Materials Science of Thin Films*, Academic Press, Boston (1992).

High-power laser applications are more spectacular because, like ray guns in science fiction, the beam vaporizes, melts, or heats the surface it impinges on. Industrial applications include scribing, welding, cutting, and hardening operations. Laser processing regimes in assorted applications are shown in Fig. 13-23, where the required power densities are depicted as a function of interaction or treatment times. The latter are important because although some lasers provide light continuously (cw or continuous wave), others yield pulses of light. Selection of a laser for a particular processing application involves consideration of light absorption by the workpiece, energy required, and processing time.

13.6. OPTICAL COMMUNICATIONS

13.6.1. Introduction

The earliest optical communications systems probably involved sending smoke signals (fire modulated by waving blankets) that were easily visible across mountains, where the messages were detected. A source and detector consisting of two paper or metal cups connected by a string constitute another communications system familiar to children. By replacing the string with a special glass fiber that can be tens of miles long, and the cups by a laser at the transmitting end and a photon detector at the receiver end, we have the outline of a modern optical communications system. The latter, shown schematically in Fig. 13-24, has auxiliary electronic circuitry to convert sound or video signals

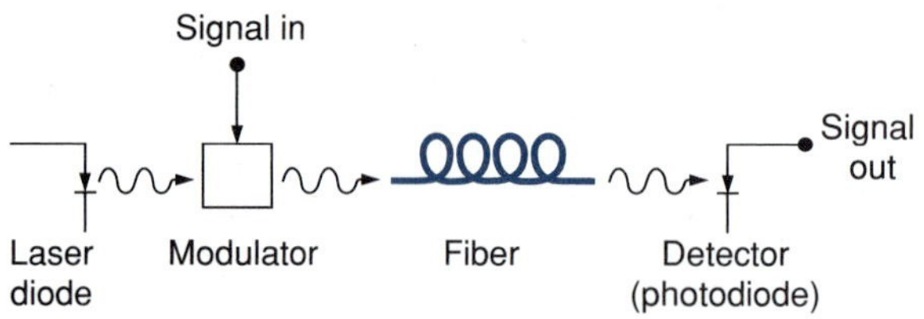

FIGURE 13-24 Schematic of an optical communications system.

into a train of electronic pulses that modulates or encodes the light wave carrier with the information. A series of on–off coded light pulses enter the fiber and, at the opposite end, are detected and converted back to sound or video. For long-distance communications systems, intermediate repeater stations are required to boost the attenuated light intensity to required levels.

To appreciate the advantages of optical communications we note that the human voice requires a frequency range from 200 to 4000 Hz. It makes no difference whether the 3800-Hz band width is transmitted in this frequency range, or in the ranges 10,200 to 14,000 Hz, 1,000,200 to 1,004,000 Hz, and so on. Clearly, the higher the frequency of the carrier wave, the greater the potential number of separate conversations or channels of 3800 Hz each. (Important electronic techniques known as multiplexing enable many channels to be synchronized in time so that each is separate and messages are not garbled.) Although communications in the radio and microwave region of the spectrum have frequency bands in the range of $\sim 10^6$ and $\sim 10^{10}$ Hz, respectively, optical frequencies of $\sim 10^{15}$ Hz allow vastly more conversations or information to be transmitted. The "information highway" is based on this fact.

Optical communications systems have the following additional advantages: high capacity, small size, low weight, quality transmission (minimal crosstalk), low cost, and generally good security. In the development and improvement of optical communications technology, materials have played a critical role, none more, perhaps, than the glass fiber, to which we now turn our attention.

13.6.2. The Fiber

13.6.2.1. Light Transmission

Optical communications require two important attributes of the fiber, namely, that light rays be solely confined to it and that long-distance transmission occur with minimum loss in light intensity. The optical fiber waveguide (Fig. 13-25A) consists of the active inner cylindrical fiber core (8.5 μm in diameter) and the surrounding cladding (~ 125 μm in outer diameter), whose respective indices of refraction are $n_r(f)$ and $n_r(c)$, where $n_r(f) > n_r(c)$. Whereas the cladding is silica, the core is doped with GeO_2 (germania), which raises $n_r(f)$ about a percent above $n_r(c)$. Consider light ray 1 in the high-index core

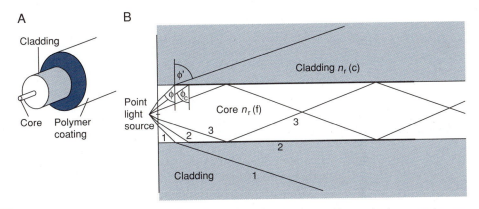

FIGURE 13-25 (A) Structure of single-mode optical fiber. The diameters of the core, cladding, and polymer coating are 8.5, 125, and 250 μm, respectively. (B) Refraction and propagation of light in an optical fiber.

that impinges on the cladding interface at angle ϕ, as in Fig. 13-25B. It will be refracted according to Snell's law,

$$n_r(f) \sin \phi = n_r(c) \sin \phi', \qquad (13\text{-}12)$$

where ϕ' is the angle of refraction. But the angle ϕ is such that light enters the cladding. To confine the light solely to the core, ϕ' must be greater than 90°. Therefore, when light is launched at an angle of incidence that is greater than a critical value ϕ_c [given by $\sin \phi_c = n_r(c)/n_r(f)$, i.e., when $\phi' = 90°$], light is no longer lost to the cladding. At angle ϕ_c light is *internally reflected* and confined to the core–cladding interface (ray 2 in Fig. 13-25B). For angles greater than ϕ_c, light is totally confined to the core (e.g., ray 3) and propagates in zigzag fashion down the fiber.

13.6.2.2. Modes of Transmission

Consider two light pulses launched at one end of the fiber. One propagates precisely down the core axis. The second undergoes a zigzag motion down the fiber, reflecting off the core–cladding interface as it goes. A remote detector will sense the first light pulse sooner than the second pulse because the latter travels a longer distance. The time difference may be very short indeed, but large enough to cause errors in the transmission of information. It is therefore desirable to have only **single-mode** transmission where light travels down the fiber core axis without interference by other modes of transmission. Small core diameters ensure single-mode operation and virtually all fiber used for communications is of this type.

13.6.2.3. Loss

Light intensity in fibers declines exponentially with distance, according to Eq. 13-2. It is customary to express this loss in terms of decibels (dB), where the attenuation (in dB/km) is defined as

$$-\text{loss(dB/km)} = (1/L)\,10\log(I/I_o). \tag{13-13}$$

In this definition I and I_o are the light intensity at the detector and source, respectively, and L is the distance between them in kilometers.

EXAMPLE 13-5

a. Calculate values of the intensity reduction per kilometer for light transmitted through Egyptian and Phoenician glass using information contained in Fig. 13-1. What are the corresponding values of the absorption coefficient for each glass?

b. What is the absorption coefficient for 1983 optical fiber?

c. Suppose light signals in fibers have to be reamplified at repeated stations when I/I_o falls to 0.3. How far apart should repeater stations be placed?

ANSWER   a. From the definition of Eq. 13-13, $\log(I/I_o) = -L(\text{loss in dB/km})/10$ and $L = 1$ km. For Egyptian glass, the optical loss is approximately -6×10^6 dB/km (note minus sign for loss), so that $\log(I/I_o) = -1 \times 6 \times 10^6/10$, or $I/I_o = 10^{-600,000}$, or $\exp(-1.38 \times 10^6)$. From Eq. 13-2, $1.38 \times 10^6 = \alpha x$. For $x = 1000$ m, $\alpha = 1.38 \times 10^3$ m^{-1}.

A similar calculation for Phoenician glass, where the optical loss is approximately -10^5 dB/km, yields $I/I_o = 10^{-10,000} = \sim\exp(-23,000)$, and $\alpha = 23$ m^{-1}.

b. For 1983 optical fiber, the loss is approximately -0.11 dB/km, and $\log(I/I_o) = -0.11/10$. Therefore, $I/I_o = 10^{-0.011} = \exp(-0.025)$. As $\alpha x = 0.025$, for $x = 1000$ m, $\alpha = 0.000025$ m^{-1}.

c. From Eq. 13-13, $L = -10[\log(I/I_o)]/(\text{loss in dB/km})$, and after substitution, $L = -10(\log 0.3)/0.11 = 47.5$ km.

13.6.2.4. Absorption

In optical fibers the absorption coefficient is a sum of two contributions that cause the *bulk* and *intrinsic* losses as a function of wavelength, shown in Fig. 13-26. Bulk absorption has been attributed to light-absorbing impurities (e.g., Fe^{2+} and OH^- ions). On the other hand, scattering of light by imperfections (e.g., air bubbles, scratches) and both density and compositional fluctuations are sources of intrinsic loss. The latter contribution, known as Rayleigh scattering, varies as λ^{-4}. Because this term is large at small wavelengths, UV light and visible light are not used in optical communications. Instead, the loss is minimum at ~1.3 to 1.5 μm, and this is the reason for light transmission in this IR wavelength window.

13.6.2.5. Fiber Fabrication

The first step in producing fiber is to make a preform. A modified chemical vapor deposition (MCVD) technique, schematically shown in Fig. 13-27, is

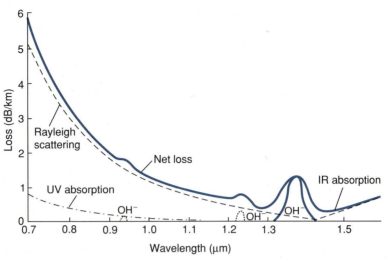

Optical loss as a function of wavelength in silica fibers. Courtesy of AT&T Bell Laboratories.

widely used for this purpose. Very pure precursor gases of $SiCl_4$ and $GeCl_4$ are mixed with oxygen, transported down a rotating silica tube, and reacted at ~1600°C. As a result of oxidation, small solid "soot" particles of SiO_2 doped with GeO_2 are produced. The cotton candy-like soot deposits on the tube walls, and when enough has collected it is densified by heating to ~2000°C. At this very high temperature, surface tension and external pressure promote the sintering that causes the tube to collapse radially until the center hole is eliminated. The dopant profile is carefully controlled during deposition to ensure that the subsequently drawn fiber will have a larger index of refraction at the core than in the cladding. Phosphorus and fluorine are also introduced to facilitate processing and assist in achieving the required index of refraction profile across the preform.

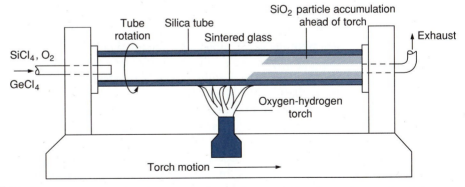

Modified chemical vapor deposition process for producing optical preforms. Glass working lathes are employed to grip and rotate the silica tube, as well as translate the torch.

Next, the ~2 cm diameter, ~1 m long preforms are drawn to fiber in a process that remotely resembles wire drawing of metals. High draw towers photographed in Fig. 13-28 produce miles of fiber from a single preform. The preform is heated again, this time to greater than 2000°C, and necked down. Fiber of uniform diameter is then drawn from the melt in a way that faithfully reproduces the radial index profile in much shrunken dimensions. Immediately after the fiber cools sufficiently, its pristine surface is coated with a protective polymer coating. The intent is to prevent mechanical reliability problems posed by defects and cracks (see Sections 7.6.3 and 10.5.1). A schematic of the drawing and coating processes is shown in Fig. 13-29.

13.6.2.6. Erbium-Doped Fiber

Installing and maintaining undersea optical communications systems is an expensive proposition. If electronic components in repeaters fail, cables must be hoisted to surface ships and repair costs can run into millions of dollars. The lasers used to boost attenuated light signals have traditionally posed a reliability concern. That is why alternate methods of optical amplification have been actively explored. A recent exciting development involves doping short lengths of the glass core with erbium (Er^{3+}) ions. By doing this, an Er laser (light amplifier) is effectively created and integrally incorporated within the fiber. The laser can be reliably pumped with 0.98-μm light and emits light at 1.5 μm, a wavelength where optical fiber losses are minimal (see Fig. 13-26). Future optical communications systems are targeted to contain this cheaper and more reliable technology for boosting the intensity of light pulses.

13.6.3. Sources and Detectors

Laser sources and photodetectors must be fabricated to operate at the same wavelength as that of the fiber transmission. Let us try to design and select some of the materials required to function at 1.5 μm. But first we must seek a substrate on which to grow the devices. As the optical fiber must be coupled to these devices without loss of light, a substrate that is *transparent* to 1.5-μm photons is desirable. This means that the energy gap must be larger than 0.83 eV (by Eq. 13-1, $E_g = 1.24/1.5 = 0.83$ eV). Reference to Fig. 13-12 shows that of the readily available substrates (GaAs, InP, GaP), only InP can form alloys with the desired E_g. If devices are to be grown on InP they must also be lattice matched to it. For example, we have shown in Example 13-4 that the $Ga_{0.48}In_{0.52}As$ is lattice matched to InP, but E_g is only 0.70 eV. The obvious way to raise E_g and remain lattice matched to InP is to alloy $Ga_{0.48}In_{0.52}As$ with InP. Therefore, **quaternary** alloys of precise composition containing Ga, As, In, and P are needed and such materials have been used in the manufacture of these high-performance device applications.

FIGURE 13-28 Photograph of optical fiber draw tower. Courtesy of F. DiMarcello, AT&T Bell Laboratories.

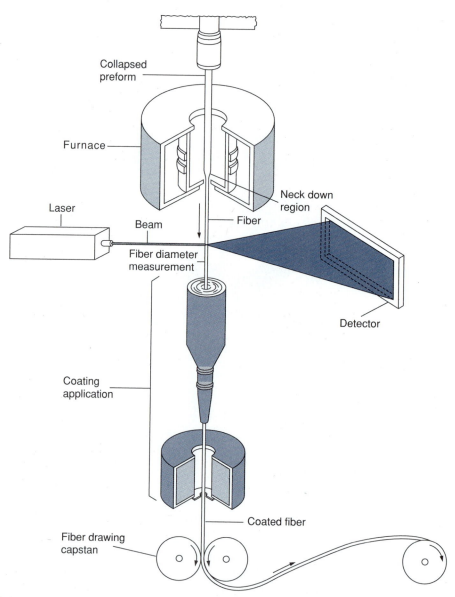

FIGURE 13-29   Schematic of the fiber drawing process. From D. H. Smithgall and D. L. Myers, *The Western Electric Engineer*, Winter (1980).

13.7. MISCELLANEOUS OPTICAL PROPERTIES AND EFFECTS

A number of issues that relate to the optical properties of glasses, ceramics, and polymers remain to be addressed. One has to do with light reflection from their surfaces and transmission through their interiors. In the case of reflection, incident light is either scattered **specularly** (Fig. 13-30A), as from a mirror; directionally, as from a smooth surface (Fig. 13-30B); or diffusely in all directions as from a very rough surface (Fig. 13-30C). Case B has characteristics of cases A and C, so that the reflection varies as a function of polar angle with an intensity given by the envelope of the vector lengths. Collectively, **gloss** is the term given to the quality of surface reflectance.

Transmission of light through these materials depends strongly on inhomogeneities in the refractive index. The latter arise from differences in density between amorphous and crystalline regions, as well as from distributions of filler particles, pigments, voids, cracks, and so on. Scattering from such particles and defects reduces the **clarity** of the material. In addition, light that enters and is forward scattered through small angles contributes to the **haze** of the body. Both kinds of reflection and transmission phenomena may be viewed as perturbations on effects displayed by ideal homogeneous optical materials with perfectly smooth surfaces.

Polymers have been somewhat neglected in our discussion of optical properties. Although they behave optically in much the same way as inorganic dielectrics, they exhibit some special properties and have applications that are worthy of note. In polymers, absorption of optical radiation in the IR occurs by excitation of bond and chain vibrations; however, in the visible and UV, electrons are stimulated to rearrange within atoms and molecules. On the whole, very few pure polymers absorb visible radiation and they are therefore

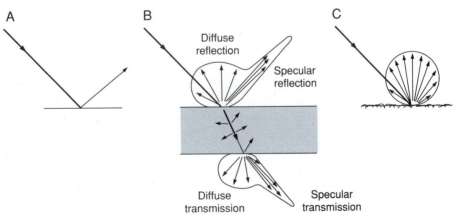

FIGURE 13-30 Modes of light reflection, transmission, and scattering from surfaces. (A) Specular scattering. (B) Directional scattering. (C) Diffuse scattering.

colorless. Absorption at different wavelengths in the IR is, however, plentiful and varied. This information is put to good use in analytical chemistry, where the complex IR absorption spectra serve to fingerprint the types of organic molecular units that are present.

A number of polymers have interesting optical properties that have important commercial uses. Photoresist materials are examples that were discussed in Section 12.5.4. Their critical role in lithographic patterning of integrated circuit features hinges on the unusual effect light has in inducing chemical changes, that is, polymerization or chain scission. The latter effect, which breaks bonds, is, incidentally, one reason why polymers degrade. Optical radiation (UV) as well as ionizing radiation (X rays, γ rays) severs bonds, leaving the chain fragments at the mercy of oxidation reactions. Polymer embrittlement and discoloration are the results of such exposure.

Another application involves the alteration of optical properties under applied stress. Structurally, isotropic materials are also optically isotropic, and there is only one refractive index. In certain noncubic crystals and anisotropic materials, however, the refractive index may assume different values in different

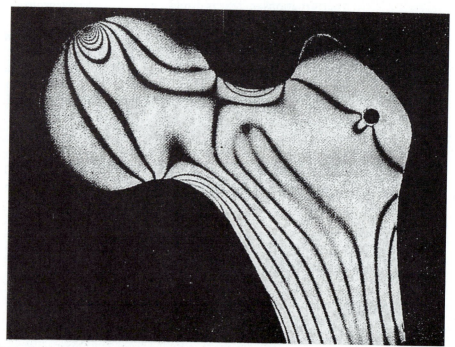

FIGURE 13-31 Photograph illustrating photoelastic effect in a stressed polymer in order to simulate bone behavior. From A. W. Miles and K. E. Tanner (Eds.), *Strain Measurement in Biomechanics*, Chapman and Hall, London (1992).

directions, and the material is said to be doubly refracting or **birefringent**. Surprisingly, amorphous polymers that are normally optically isotropic become birefringent as the molecules are aligned under stress. The same thing happens in flowing polymer melts. A method for conducting photoelastic stress analysis in complex mechanical parts and structures is then suggested by the stress birefringence phenomenon. The component or structure is fashioned in polymer (e.g., epoxies, polyesters, polymethyl methacrylate), then loaded and viewed in transmitted light between crossed polarizers. Two components of the refracted ray travel at different speeds, producing interference effects and fringes that are readily visible, as shown in Fig. 13-31. Regions of stress concentration are qualitatively apparent and can be quantitatively calculated through further analysis.

13.8. PERSPECTIVE AND CONCLUSION

Optical properties, perhaps more than any other property, are a triuimph of the quantum theory of atoms and solids. The electric field component of high-frequency electromagnetic light waves interacts strongly with the electrons of the solid. But electrons in the atoms of solids conform to a quantum order, and therefore, restrictions in their interactions with light are expected. All interactions can essentially be reduced to the simple picture containing a combination of the following features: a photon entering or leaving, and a pair of discrete electron levels (1 and 2), one of which is vacant, separated by energy $E_2 - E_1$. Electrons at E_1 energized by light absorption occupy a new energy E_2 ($E_2 > E_1$). Alternately, electrons that were previously excited to high energy levels (e.g., E_2) have a natural tendency to retreat to lower energy levels (E_1), emitting a photon of wavelength λ to conserve energy. The ubiquitous Bohr formula, $E_2 - E_1 = hc/\lambda$ (Eq. 13-1), precisely governs both types of events, which are played out in countless ways within materials. Sometimes, $E_2 - E_1$ represents a difference in atomic levels, as in the He–Ne laser. But more often it represents **interband** transitions, for example, the energy gap between the valence and conduction bands in semiconductors or between the absorption band and ground state in colored glasses. Transitions within a single band (**intraband**) also occur between closely spaced E_1 and E_2 levels, as in metals. In all cases of interest in this chapter, the energy difference corresponds to wavelengths extending from the UV, through the visible, to the IR; core level transitions normally involve the more energetic X-ray wavelengths. The variation of optical properties like color, reflectivity, and transparency in different materials can be broadly understood in terms of their electronic structure. Importantly, the above concepts provide guidelines for designing devices that produce, capture, or transmit light efficiently.

Each of the classes of materials exhibits important optical properties that have been harnessed for useful purposes. The high density of conduction electrons in metals is responsible for their high reflectivity, which is capitalized on in mirrors and in applications where aesthetic appearance is a consideration. At the other extreme (in optical as with other properties) are the insulators and dielectrics composed of polymers, oxides, ionic materials, and so on. They contain very few unbound electrons that can interact with impinging light, but many electric dipoles that either exist or can be potentially induced. Polarization alters the speed of light in these solids, but there is very little absorption, contrary to the behavior in metals. As a result they are largely transparent and uncolored. Thin-film optical coatings to reduce reflection from underlying substrates are an important application of these materials. Coloration of non-metallic materials occurs when enough electron levels are positioned within the energy gap to undergo transitions that yield visible light. Introduction of metal cations not only may impart color but also may slow the speed of optical transitions so that the material luminesces under suitable stimuli.

Semiconductors are the most exciting optical materials because of the active electro-optical devices that are possible. Unlike insulators and dielectrics, the electrons that undergo optical transitions are sufficiently numerous and mobile to also be directly coupled to external electrical circuits. The Bohr formula still applies, with E_2 and E_1 the conduction and valence band edges, respectively. **Absorption** of light resulting in elevation of electrons from E_1 to E_2 is the energy conversion mechanism in solar cell p–n junctions. **Spontaneous emission** of light achieved by passing forward currents across p–n junctions of light-emitting diodes causes electrons to descend from E_2 to E_1. And if the emitted light is trapped within the cavity of specially doped and structured diodes so that a large in-phase photon field builds, **stimulated emission** of light can occur, as in lasers. Importantly, energy band gaps in compound semiconductors can be engineered through alloying to correspond to specific photon wavelengths for particular electro-optical applications. Defect-free, thin single-crystal semiconductor films are required for these devices, necessitating lattice matched structures.

Optical communications systems couple very high performance laser light sources and photon detectors to fibers of the most transparent glass ever made. Together with semiconductor electronics they have fashioned the integrated Computer–Communications–Information Age.

Additional Reading

P. Bhattacharya, *Semiconductor Optoelectronic Devices,* Prentice–Hall, Englewood Cliffs, NJ (1994).

J. P. Powers, *An Introduction to Fiber Optic Systems,* Irwin, Homewood, IL (1993).

S. M. Sze, *Semiconductor Devices: Physics and Technology,* Wiley, New York (1985).

A. Yariv, *Introduction to Optical Electronics,* 2nd ed., Holt, Rinehart and Winston, New York (1976).

QUESTIONS AND PROBLEMS

13-1. By means of approximate percentage transmission-versus-wavelength curves distinguish the following materials:
   a. A transparent colorless solid
   b. A transparent blue solid
   c. An opaque solid
   d. A translucent colorless solid

13.2. Consider a glass slide with an index of refraction of $n_r = 1.48$ and with light passing through it incident on the glass–air interface.
   a. What is the speed of light in the glass?
   b. At what angle of incidence (measured from the normal) will light undergo total internal reflection in the glass?
   c. Can light undergo total internal reflection when traveling from air into the glass?

13.3. Six percent of the light normally incident on a nonabsorbing solid is reflected back. What is the index of refraction of the material?

13.4. Light is incident on the surface of a parallel plate of material whose absorption coefficient is 10^3 (cm^{-1}) and index of refraction is 2.0. If 0.01 of the incident intensity is transmitted, how thick is the material?

13.5. Sapphire, which is essentially pure Al_2O_3, has no absorption bands over the visible spectral range and is therefore colorless. When about 1% Cr^{3+} ions are added in the form of Cr_2O_3, there are two absorption bands: a deep one near 0.4 μm and a shallow one around 0.6 μm. Roughly sketch the transmission spectrum of chromium-doped alumina in the visible range. What is the color of this material?

13-6. The optical constants of the colored metals copper and gold are tabulated at the indicated wavelengths:

Metal	0.50 μm		0.95 μm	
	n_r	k	n_r	k
Copper	0.88	2.42	0.13	6.22
Gold	0.84	1.84	0.19	6.10

In each case calculate the fraction of incident light reflected from the surface, or the reflectance. Compare your values of R with those plotted in Fig. 13-4.

13-7. What are the optical and metallurgical advantages and disadvantages of using gold, silver, aluminum, and rhodium metals for mirror applications?

13-8. Which of the optical coating materials displayed in Fig. 13-8 could be used to make a one-layer AR coating that is to operate at 0.50 μm on a glass lens having an index of refraction of 1.52?

13-9. A germanium crystal is illuminated with monochromatic light of 0.5-eV energy. What is the value of the reflectance at the incident surface?

13-10. What types of oxides added to silica melts will tend to increase the index of refraction of the resulting glass? What scientific principles underscore the influence of specific metal ions?

13-11. Specifications for the design of an optical system containing 17 glass–air interfaces call for total minimum transmittances of $T = 0.72$ at $\lambda = 0.45$ μm, $T = 0.91$ at $\lambda = 0.60$ μm, and $T = 0.75$ at $\lambda = 0.77$ μm. The same AR coating will be used on each optical glass substrate of refractive index 1.52. What is the required reflectance of this AR coating at each wavelength?

13-12. A certain material absorbs light strongly at a wavelength of 160 nm and nowhere else, being transparent up to 5 μm. Is this material a metal or a nonmetal? Why?

13-13. The relaxation time for a zinc-based phosphor bombarded with an electron beam is 3×10^{-3} second.
   a. How long will it take the light to decay to 20% of the initial light intensity?
   b. If 15% of the intensity is to persist between successive beam scans on a television screen, how many frames per second must be displayed?

13-14. Phosphorescent materials have long persistence times. Ultraviolet light incident on such a phosphorescent material results in reemitted light after the UV is turned off. If the reemitted light decays to half the original intensity in 5 minutes, how long will it take for the intensity to decline to one-tenth of its original value?

13-15. Solar cells are made from both silicon and GaAs. Cells fabricated from GaAs can be much thinner than Si cells, thus conserving precious gallium and arsenic resources. Why are thinner GaAs cells possible?

13-16. a. At a wavelength of 0.4 μm what is the ratio of the depths to which 98% of the incident light will penetrate silicon relative to germanium?
   b. What is the corresponding ratio of depths at a wavelength of 0.8 μm?
   c. Such light causes Si and Ge to become more conductive. Why?

13-17. By consulting Table 12-1 determine the wavelength of light that is associated with the energy band gap of Si, GaP, and ZnSe.

13-18. Compound semiconductors containing binary, ternary, and quaternary combinations of Ga, In, As, and P are, by far, the easiest to grow and fabricate into high-quality electro-optical devices.
   a. On this basis explain why infrared devices commonly employ these elements.
   b. Explain why blue light-emitting devices are not common.
   c. Similarly, explain why ultraviolet light-emitting semiconductor devices are rare.

13-19. There has been a great deal of interest in developing a blue light semiconductor laser operating in the range of 2.6 eV. By consulting Table 12-1 select a possible candidate material and suggest a substrate that is closely lattice matched to it.

13-20. The junction depth of a silicon solar cell lies 200 nm below the surface. What fraction of 550-nm light incident on the device will penetrate to the junction? (Assume no light reflection at the surface.)

13-21. How would the answer to the previous problem change if the solar cell were fabricated from GaAs?

13-22. A measure of the electric power delivered by a solar cell is the fourth quadrant area enclosed by the I–V characteristics. Specifically, the power is reported as the area of the largest rectangle that can be inscribed in this quadrant. The cell "fill factor," a quantity related to efficiency, is the ratio of this maximum power rectangle to the product of V_{oc} and I_{sc}. Suppose the I–V characteristic in the fourth quadrant is a straight line joining V_{oc} and I_{sc}. What are the values of the maximum cell power and fill factor?

13-23. In making LEDs that emit 1.8-eV red light, direct band gap GaAs is used as the substrate. A series of graded alloys is first grown epitaxially to smooth out interfacial defects caused by lattice mismatch. The required direct band gap alloy film of $GaAs_yP_{1-y}$ is then deposited.

    a. In GaAs–GaP alloys what value of y corresponds to the transition from direct to indirect band gap behavior?

    b. What value of y is needed to produce the desired red light?

    c. What wavelength light will be emitted from the LED?

13-24. Yellow, orange, and green LEDs, surprisingly, use *indirect* band gap gallium phosphide substrates onto which $GaAs_yP_{1-y}$ alloy films are deposited. (Although not as efficient as direct band LEDs, they are used.)

    a. What weight percentages of Ga, As, and P are required for an LED operating at 2.00 eV?

    b. What band gap energy is associated with a $GaAs_{0.1}P_{0.9}$ semiconductor?

13-25. Comment on the purity or monochromatic character of the light emitted during an electron transition in an atom compared with that in an extended solid source of many atoms.

13-26. a. Consider a 100-W bulb viewed at a distance of 1 m through a 1-mm-diameter aperture. What radiant power falls on the eye?

    b. A 5-mW He–Ne laser emits a 1-mm-diameter beam that does not spread appreciably. What laser light power falls on the eye when viewed at 1 m? (From the answer it is clear why we should never look at a laser beam directly.)

13-27. For an optical recording application it is desired to vaporize some 100,000 hemispherical pits per second in a polymer having a density of 1.5 g/cm^3. About 2 kJ/g is required to vaporize this polymer. If each pit is assumed to have a diameter of 5 μm, what laser power is required for this application?

13-28. One way to make a short planar waveguide is to dip a glass slide (containing ~12% Na) into a $AgNO_3$ melt for a period of time. The waveguide produced (at each surface) consists of the unaltered glass slide and air regions sandwiching the altered slide surface (core) layer in between. Explain what optical property the core layer must have.

13-29. A planar optical waveguide is modeled by a stack of three glass slides in perfect contact. The two outer (cladding) slides of refractive index $n_r = 1.46$ sandwich a (core) slide of refractive index $n_r = 1.48$ in between. Surrounding the waveguide is air.

    a. For total internal reflection to occur in the core, determine the angle of incidence, ϕ (measured from the normal), for light incident on the core–cladding interface originating from the core side.

b. Sketch, by means of a ray diagram, what happens when the angle of incidence is slightly greater or slightly smaller than the critical one.

c. If light rays impinge on the core–cladding interface at 85° (with respect to the normal), how many times will they bounce off the interface if the glass slides are 75 mm long and 1 mm thick.

13-30. a. Silica optical fibers for communications purposes require 1.3- to 1.5-μm light for most efficient operation. Why?

b. Optical communications applications have largely driven the development of optoelectronic devices operating in the infrared region of the spectrum. Explain why.

13.31. Repeater stations for underwater cables containing optical fibers are spaced 45 km apart. The fiber loss is known to be −0.2 dB/km. What is the ratio of the light intensity leaving a repeater to that entering it?

14 MAGNETIC PROPERTIES OF MATERIALS

14.1. INTRODUCTION

Although the Chinese exploited lodestone (Fe_3O_4) for four millennia in compasses to aid navigation, the fundamental nature of this and other magnetic materials only began to be understood in this century. Discovering the roles composition, structure, and processing play in influencing magnetic properties is the key to developing better magnets; these coupled materials science and engineering activities persist to the present day. Judging by Figs. 14-1 and 14-2, progress in the last century has been impressive. Advances in the strength (magnetic, not mechanical) of permanent or **hard** ferromagnetic materials (Fig. 14-1) allow for greater efficiency and miniaturization of such products as motors, loudspeakers, tool holders, and door catches.

Similar advances have occurred in another class of magnetic materials that are employed on a large scale in generators, transformers, inductors, and motors for electric power generation, transmission, storage, and conversion applications, respectively. These important **soft** ferromagnetic materials perform the critical function of shaping and concentrating magnetic flux in alternating current equipment, a task that must be accomplished with minimum energy loss. The almost two-orders-of-magnitude decline in this loss (the so-called core loss of Fig. 14-2), since the beginnings of the electric power industry, has been responsible for huge energy savings. For example, annual 60-Hz core energy losses in the United States are currently estimated to be about 10^{14} W-h,

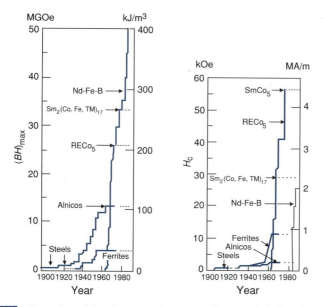

FIGURE 14-1 Chronology of the advances made in raising the strength (*left*) and coercive field, H_c (*right*), of permanent magnets. The quantity $(BH)_{max}$, a figure of merit for the strength of permanent magnets, is defined in Section 14.5.3. From National Materials Advisory Board Report NMAB-426 on Magnetic Materials, 1985.

a magnitude equal to the output of several large utility power-generating plants.

All materials display magnetic effects that can basically be divided into two broad categories. We have already alluded to the most important class of magnetic materials—the ferromagnets. These are spontaneously magnetized, which means they retain their magnetic state even in the absence of a magnetic field. Ferromagnets exhibit very large effects that are immediately obvious. On the other hand, magnetic effects are small in paramagnetic and diamagnetic materials; the latter develop measurable magnetic properties only in the presence of large magnetic fields. Even so, the *induced* effects are weak and generally necessitate sensitive instruments for detection. Ferromagnets typically display a millionfold or more greater response to applied magnetic fields than para- and diamagnets and, for this reason, are the only magnetic materials employed commercially. They, therefore, receive the overwhelming share of attention in this chapter. Although not directly related to the magnetic materials we are concerned with, mention must be made of the high-field superconducting *electromagnets*. Used in medical magnetic resonance imaging (MRI) systems and magnetically levitated (MAGLEV) trains, these and other applications were noted in Section 11.5.3.

The phenomenon of magnetism was first understood by physicists, who recognized its connection to electricity and incorporated it into the elegant

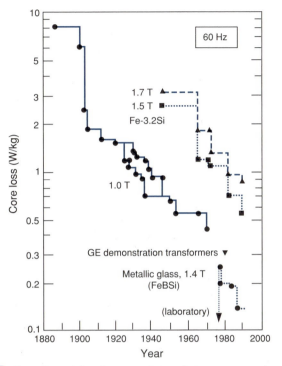

FIGURE 14-2 Chronology of the advances made in reducing core energy losses in soft magnetic materials used for line frequency transformers. From F. E. Luborsky, in *Proceedings of NATO Advanced Study Institute on Glasses* (A. F. Wright and J. Dupuy, Eds.), Martinus Nijhoff, The Hague (1985).

framework of **macroscopic** electromagnetic theory. Thus, the chapter starts with some brief comments on elementary aspects of this subject as it impinges on magnetic materials. Next, an extended **microscopic** focus on specific materials attempts to explain the role of atomic electrons, magnetically active atoms, composition, structure, and processing treatments in producing magnetic effects, in both zero and applied magnetic fields.

This chapter concludes with a discussion of applications other than those noted above, that is, computer memories and information storage.

14.2. MACROSCOPIC INTERACTION BETWEEN MAGNETIC FIELDS AND MATERIALS

The ability to attract steel and redistribute iron fillings in a characteristic pattern is familiar evidence that a magnetic field (H) is established in space by a bar magnet (Fig. 14-3A). A magnetic field is distributed geometrically in an identical fashion in a solenoid, a long wire closely wound in a helix, carrying direct current as shown in Fig. 14-3B. And when an iron bar fills the air

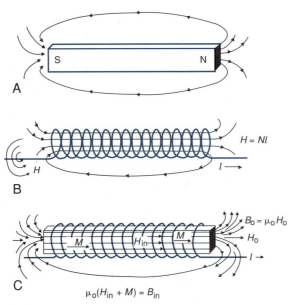

FIGURE 14-3 (A) Magnetic field surrounding a bar magnet. (B) Magnetic field surrounding an air-filled solenoid. (C) Magnetic field surrounding a solenoid filled with an iron core.

space of the solenoid core (Fig. 14-3C), the external magnetic field strength is dramatically amplified but exhibits the same spatial character. If one did not know a solenoid was there it would be natural to attribute its influence on nearby magnets, pieces of steel, or iron filings, to the presence of an equivalent bar magnet. An inverse effect, where magnets (in motion) or changing magnetic fields induce voltages and currents in wires, is known as *Faraday's law* and is capitalized upon in generating electric power. From these (and other observations) the strong equivalence between *currents* and *magnetic fields* has emerged as a cornerstone of physics. Because all phenomena, even extending down to the atomic level, are electromagnetic in origin, these fundamental interactions are extremely important.

The implications of Fig. 14-3 can be summarized in a few formulas that are simple enough to write but not as easy to grasp. Magnetic units (SI throughout) are a bit of an annoyance and take some getting used to if one is not familiar with them. We start with a hollow solenoid powered with a current I (amperes). A **magnetic field** H, a vector (with units of ampere-turns/m or just A/m), is produced having a magnitude given by

$$H = NI \quad \text{A/m}, \tag{14-1}$$

where N is the number of turns per meter. Magnetic lines of force or flux concentrated in the solenoid core flair out in the surrounding medium, creating a **magnetic induction** (vector) field B, which has units of Webers per square meter (Wb/m^2) or Tesla (T). Gauss (G) is a widely used measure of B in cgs-emu units, and 10^4 G = 1 Wb/m^2 = 1 T. The relationship between B and H *in vacuo* is simply $B = \mu_o H$, where μ_o is the permeability of free space, with a magnitude equal to $4\pi \times 10^{-7}$ Wb/A-m or, alternately, henry/m.

What happens to B when a material is inserted into the core depends on what the material is. In general, however, the relationship

$$B = \mu_o H + \mu_o M \qquad (14\text{-}2)$$

holds, where M, a new vector quantity known as the **magnetization,** arises. For the material-filled solenoid, B is the same or continuous inside (in) the core as well as outside (o) in the vacuum; thus, $B_{in} = B_o = B$. But H is *not* the same inside and outside because of the presence of M; that is, $\mu_o(H_{in} + M) = B = \mu_o H_o$.

Although their origins will be discussed later, a small magnetic dipole moment (an atomic bar magnet) can be thought of as the equivalent magnetic response of each atom in most, but not all, solids. Each of these dipoles has a magnitude μ_m that is a product of the magnetic pole strength (m_p) and distance of pole separation d: $\mu_m = m_p\,d$. The magnetization is defined to be a vector sum of all these moments on a per unit volume (V) basis and can be written as

$$M = (\Sigma \mu_m)/V \quad \text{A/m.} \qquad (14\text{-}3)$$

If $V = A_o d$, where A_o is the cross-sectional area of the body, then $M = (\Sigma m_p)/A_o$. When all the moments are aligned, the magnetization reaches a maximum or saturation value M_s. Increasing the magnetic field further causes B to rise further, but not M_s (Eq. 14-2).

In **ferromagnetic** materials like Fe, Ni, Co, and their alloys, the individual permanent atomic magnetic moments all point in a common direction over vast stretches of lattice terrain. And furthermore, even a small H field is enough stimulus to cause these atomic moments to align in the field direction. As a result, M will be large, greatly increasing the magnitude of B. **Paramagnetic** materials also have permanent moments but they are randomly oriented so that the net magnitude of M is very small. It is also small in **diamagnetic** materials, but in this case the magnetism is induced where none existed before. These two types of magnetic materials are briefly discussed again in Section 14.3.2.

Mathematically, the important equation

$$B = \mu H \qquad (14\text{-}4)$$

connects B and H in materials, where μ is the **permeability** in units of Wb/A-m or henry/m. Permeability is a measure of the ease with which magnetic flux penetrates a material. For ferromagnets it is a function of H. In Fig. 14-4 we see that M is linearly proportional to H if the magnetic field is small. Therefore,

$$M = \chi H, \tag{14-5}$$

where the constant of proportionality, χ, is known as the **magnetic susceptibility,** a dimensionless quantity. A combination of Eqs. 14-2, 14-4, and 14-5 yields

$$B/H = \mu = \mu_0(1 + \chi), \tag{14-6}$$

illustrating the connection between the permeability and the susceptibility.

Values of χ in para- and diamagnetic materials are slightly positive and negative, respectively, by about a part per million. For example, the paramagnets Al, Pt, Ta, and Ti have room-temperature susceptibilities of $+0.6$, $+1.1$, $+0.87$, and $+1.25$ (10^{-6}), respectively; similarly, respective susceptibility values for the diamagnets Be, Cu, Ge, and Ag are -1.0, -0.086, -0.12, and -0.20 (10^{-6}). In ferromagnetic materials, χ is usually many millions of times larger. And because it varies in a complex way with H, so does μ. Another quantity frequently quoted when comparing magnetic materials is the **relative permeability,** μ_r. Defined as μ/μ_0, μ_r is dimensionless and equal to $1 + \chi$. The slope of the $B-H$ response at small H values is a practical measure of μ_r.

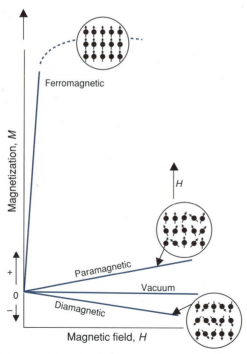

FIGURE 14-4 Magnetization versus magnetic field for paramagnets, diamagnets, and ferromagnets. The magnetic susceptibility is positive in paramagnets and ferromagnets, and negative in diamagnets.

14.3. ATOMIC BASIS OF MAGNETISM

14.3.1. Atomic Magnetic Moments

Electrons circulating around atomic nuclei provide the same connection between magnetic effects and electric currents on an atomic scale that solenoids do on a macroscopic scale. The analogy is shown in Fig. 14-5. One atomic electron orbit of radius r is like a single-turn solenoid or loop of current. Therefore,

$$B = \mu_o H = \mu_o NI. \tag{14-7}$$

For a magnet with no applied field, $H = 0$ and Eq. 14-2 yields

$$B = \mu_o M. \tag{14-8}$$

Therefore, $M = NI$. But from earlier considerations, M for single atomic dipole is given by $M = \mu_M/dA_o$. By combining the above equations,

$$\mu_m = IA_o, \tag{14-9}$$

where the one-turn solenoid length may be taken to be d and $N = 1/d$. This important result shows that *a loop of electron current is equivalent to a magnetic moment.*

Furthermore, the current or charge per unit time is $q/(2\pi r/v)$, where q and v are the electronic charge and velocity, respectively, and $2\pi r$ is the circular perimeter of the current loop of radius r. Therefore, from Eq. 14-9, $\mu_m = [q/(2\pi r/v)]\pi r^2 = qvr/2$. After numerator and denominator are multiplied by the electronic mass, m_e, it is clear that μ_m is proportional to $m_e vr$ or the **angular**

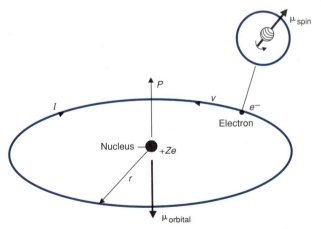

FIGURE 14-5 Origin of the magnetic moment in atoms due to an electron orbiting the nucleus ($\mu_{orbital}$) and to electron spin about its axis (μ_{spin}).

momentum of the orbiting electron. Thus, *atomic magnetic moments are directly proportional to the angular momentum of electrons*. The preceding two italicized statements summarize in a nutshell the atomic basis for magnetism. Not much more will be said about angular momentum except that it is quantized. In fact, the familiar quantum number (see Section 2.2.3) designations for atoms—l, m, and s—are all directly related to quantized angular momenta of specific electrons. The first two have to do with **orbital** and the last with **spin** angular momentum. It is the **spin** angular momentum with quantum number s ($= \pm\frac{1}{2}$) that plays an important role in ferromagnetism.

In a single atom the total magnetic moment is the vector sum of the orbital and spin components. But each component, in turn, is the vector sum of the contributions from all of the individual electrons. It is often the case, however, that the spin contribution dominates the magnetic properties of the atom. Quantum theory has shown that the magnetic moment due to one electron spin has a magnitude equal to the Bohr magneton, μ_B (1 $\mu_B = 9.27 \times 10^{-24}$ A-m^2).

14.3.2. Paramagnetism and Diamagnetism

Let us calculate the magnetic moment of a sodium atom. Of the two $1s$ electrons, one spin vector is up and the other is down, to satisfy the Pauli principle; the vector sum vanishes and there is no magnetic moment contribution from these electrons. Similarly, there is no *net* spin from the two $2s$ and the six $2p$ electrons because of the even numbers involved. But there is one remaining $3s$ electron whose spin is uncompensated. Therefore, atoms of sodium and other alkali metals have a net magnetic moment of 1 μ_B. Even though these metals are **paramagnetic** and possess permanent magnetic moments, the vector directions are in disarray, pointing every which way, as shown in Fig. 14-6A. The net magnetization, or vector sum over all atoms, is expected to be small because of the random ordering of moments. Application of a strong magnetic field helps to align them in the field direction, raising M (Figs. 14-6B,C). Low temperatures will prevent thermal energy from randomizing the moment directions and thus effectively promote higher M values in the presence of H. Theoretical considerations show that the effect of H and T on M can be combined as the single factor H/T, or the quotient of the variables.

Compared with ferro- and paramagnets, the response of **diamagnetic** materials (e.g., Au, Bi, and Si) is strange indeed. There are no permanent moments initially and when an H field is applied, magnetic moments are **induced** in atoms. But they oddly point in a direction *opposite* to H, as though the material wants to expel the magnetic flux. Recall that this phenomenon was also characteristic of the behavior of superconductors in magnetic fields (see Section 11.5.1). Actually, diamagnetism is a manifestation of Lenz' law applied to

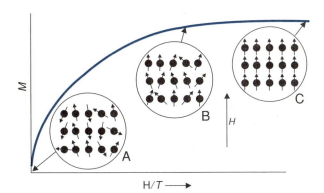

FIGURE 14-6 Magnetization of alkali metal paramagnets as a function of H/T. The progressive alignment of magnetic moments from **A** to **C** occurs through combinations of increased magnetic field and reduced temperature.

current loops. If an external magnetic field is impressed on such a loop, a current flows and establishes a magnetic field that counteracts the influence of the imposed field. As Lenz' law is perfectly general, all materials display diamagnetic effects, but may not exhibit a negative susceptibility because of overriding paramagnetic or ferromagnetic contributions. The behavior of diamagnetic materials in magnetic fields is schematically indicated in Fig. 14-4.

14.3.3. Ferromagnetism

14.3.3.1. Occurrence

Of all the magnetic "isms," ferromagnetism is, by far, the most important and the one given the most attention in theory and practice. Like paramagnets, ferromagnetic materials have permanent magnetic moments; but unlike them, moments of neighboring atoms align in a single direction, so that a **spontaneous net magnetization** exists over large regions of the material. Importantly, all of this happens in the absence of an external magnetic field. In this section the origin of ferromagnetism and its temperature dependence are addressed. Issues dealing with domains and the important behavior of ferromagnets in magnetic fields are deferred to the next section.

Concerning counting of the $3d$ shell electron spins in the transition metals and some of their important ions, Fig. 14-7 reveals a twist that should be noted. Hund's rule states that in unfilled shells the spins take on the *maximum* value permissible. This means that all five of the $3d$ electrons in Mn^{2+} have spins oriented in the same direction. Proceeding in this manner, it is evident that each of the three ferromagnetic metals Fe, Co, and Ni has a net number of unpaired spins; that is, Fe has 4, Co has 3, and Ni has 2. Moments of $4\mu_B$, $3\mu_B$, and $2\mu_B$, respectively, would thus be predicted for these atoms.

Ion	Number of electrons	Electronic structure — 3d shell					4s	Magnetic moment (Bohr magnetons)
Cr^{2+}, Mn^{3+}	22	↑	↑	↑	↑	(empty)		4
Fe^{3+}, Mn^{2+}	23	↑	↑	↑	↑	↑		5
Fe^{2+}	24	↑↓	↑	↑	↑	↑		4
Co^{2+}	25	↑↓	↑↓	↑	↑	↑		3
Ni^{2+}	26	↑↓	↑↓	↑↓	↑	↑		2
Cu^{2+}	27	↑↓	↑↓	↑↓	↑↓	↑		1
Zn^{2+}, Cu^+	28	↑↓	↑↓	↑↓	↑↓	↑↓		0
Atom								
Fe	26	↑↓	↑	↑	↑	↑	↑↓	4
Co	27	↑↓	↑↓	↑	↑	↑	↑↓	3
Ni	28	↑↓	↑↓	↑↓	↑	↑	↑↓	2

FIGURE 14-7 Electronic structure and net magnetic moment of selected ions and atoms in the transition metal series.

EXAMPLE 14-1

a. What is the saturation magnetization M_s and corresponding value of B in nickel, assuming that unpaired $3d$ electrons are the source of magnetic effects? Nickel has a face-centered cubic structure with $a = 0.352$ nm.

b. *Metallic* nickel has a saturation magnetization of $0.6\mu_B$ per atom, which decreases when copper is alloyed with it. How much Cu is predicted to cause ferromagnetism to vanish in Ni–Cu alloys?

ANSWER a. By Eq. 14-3, $M = (\Sigma \mu_m)/V$ A/m. Within an FCC unit cell there are four atoms of Ni, each of which carries a magnetic moment of $2\mu_B$. Therefore,

$$M_s = \frac{[(4 \text{ atoms/cell}) \times (2\mu_B/\text{atom}) \times (9.27 \times 10^{-24} \text{ A-m}^2/\mu_B)]}{(0.352 \times 10^{-9} \text{ m})^3/\text{cell}}$$
$$= 1.70 \times 10^6 \text{ A/m}.$$

If Eq. 14-8 is extended, $B_s = \mu_o M_s$, where corresponding values of B_s and M_s are substituted. Thus, $B_s = 4\pi \times 10^{-7}$ Wb/A-m $\times 1.70 \times 10^6$ A/m $= 2.14$ Wb/m^2, or 2.14 T. (As noted above, M_s for Ni *metal* is actually only $0.6\mu_B$ per atom. Therefore, $M_s = (0.6/2)1.70 \times 10^6 = 5.1 \times 10^5$ A/m. Similarly, $B_s = (0.6/2)2.14 = 0.64$ T.)

b. The fractional amount of vacant space in the $3d$ shell of Ni is 0.6 (or the same as M_s). It is assumed that all of the $4s$ electrons in Cu fill the $3d$ shell upon alloying. For a 60 at.% Cu addition, the d shell will be completely filled and there will no longer be a net magnetic moment.

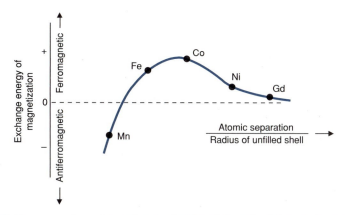

Magnetic exchange interaction as a function of the ratio of the atomic separation to the radius of the 3d electron orbit. On this basis Fe, Ni, Co, and Gd are ferromagnetic and Mn is not.

All of the other $3d$ metals shown in Fig. 14-7 and many metals with $4d$ and $5d$ electrons also have unpaired spins. Why are they also not ferromagnetic? The answer has to do with the strength of the magnetic coupling between spins belonging to adjacent atoms. The energy that acts to align spin moments is known as **exchange energy**, E_{Ex}. Quantum mechanical in origin, exchange energies are complex quantities. They may be positive or negative depending on whether spins line up parallel or antiparallel, respectively. In ferromagnetism, E_{Ex} is positive, whereas in **antiferromagnetic** materials like MnO, Mn, and Cr, E_{Ex} is negative; in antiferromagnets, spin directions alternate up and down in adjacent atoms. The correlation in Fig. 14-8 provides a clue to what is required for ferromagnetism to occur in metals, namely, that the ratio of atomic separation distance to the radius of the $3d$ shell be larger than a certain amount ($\sim$1.5). Under such conditions, E_{Ex} is positive.

Ferromagnetism is a rather common phenomenon and has been observed in a large number of metal alloys, many ceramics (e.g., ferrites, garnets), ionic solids (e.g., $CrBr_3$), a few semiconductors (e.g., MnSb), as well as in amorphous and crystalline matrices of these materials. Furthermore, there are optically opaque as well as transparent ferromagnets.

14.3.3.2. Temperature Dependence

Ferromagnetic materials are spontaneously magnetized but this does not mean that every last moment is oriented in the same direction. Certainly the overwhelming majority are, and at 0 K all moments are theoretically predicted to be perfectly aligned. In this case the magnetization saturates or reaches its maximum value, M_s ($T = 0$ K). Above 0 K some of the spins orient in random directions, a trend that accelerates as the temperature rises higher and higher. Finally, at the **Curie temperature** (T_c), the magnetization vanishes.

The temperature dependence of the magnetization is depicted in Fig. 14-9 and can be qualitatively understood in thermodynamic terms. What is at play here is the eternal tug between internal energy and entropy. Exchange energy is minimized when all of the moments align cooperatively. But the alignment is offset by an entropy increase due to thermal energy absorption by the magnetic dipoles. This causes some of them to break rank from a common spin direction and assume random orientations. The randomizing effects are more pronounced near T_c, where, like falling dominoes, the collective spin ordering collapses rapidly. Above T_c the material behaves like a paramagnet.

14.3.4. Ferrimagnetism

If one did not know better it would be very easy to mistake **ferrimagnets** for ferromagnets. Like the latter, ferrimagnets exhibit large magnetic effects and have a magnetization that abruptly disappears at the Curie temperature. There are magnetically soft and hard varieties which undergo hysteresis effects and domain behavior similar to that described in Section 14.4. What makes ferrimagnets different is the origin of the magnetism. In ferrimagnets, magnetism arises from magnetic ions located within different sublattices that may have different magnetizations and spin orientations. In addition, they are ceramic materials and therefore possess high electrical resistivity. This is a property that makes them useful for things other than magnets to stick on refrigerators. There are two important classes of ferrimagnetic materials and they include ferrites and garnets.

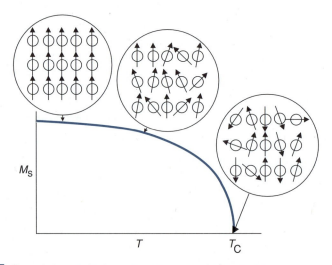

FIGURE 14-9   Saturation magnetization versus temperature in ferromagnets.

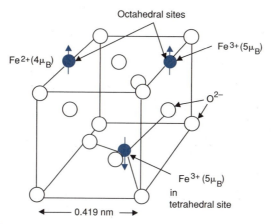

Octahedral sites

$Fe^{2+}(4\mu_B)$

$Fe^{3+}(5\mu_B)$

O^{2-}

$Fe^{3+}(5\mu_B)$
in
tetrahedral site

0.419 nm

FIGURE 14-10 The (inverse) spinel structure of ferrite of the type $MeFe_2O_4$, emphasizing the magnetization of ions in octahedral and tetrahedral sites within one octant of the normal unit cell.

14.3.4.1. Ferrites

The most well known ferrite is the oxide magnetite, or Fe_3O_4, also known as lodestone. This oxide has an *inverted* **spinel** structure generally described as an FCC lattice of O^{2-} ions that form cages surrounding interstitial sites. (Spinels are also viewed as having an admixture of rocksalt and zinc blende packing.) Figure 14-10 shows how the structure is built up. In the indicated cubic cell, one-eighth the size of the normal unit cell, there are eight tetrahedral sites. One of these is occupied by an Fe^{3+} ion with a spin pointing downward. Two of four octahedral sites in the cell are also occupied, one by an Fe^{3+} ion and one by an Fe^{2+} ion, both with upward spin. Thus, the compound can be written as $Fe^{2+}Fe_2^{3+}O_4$ in this octant cell, or $(Fe^{2+}Fe_2^{3+}O_4)_8$ in the normal cell.

EXAMPLE 14-2

What is the saturation magnetization of $Fe^{2+}Fe_2^{3+}O_4$ if the cube dimension in Fig. 14-10 is 0.419 nm?

ANSWER Referring to Fig. 14-7, we are now interested in the ionic states Fe^{2+} and Fe^{3+}. The indicated summation of moments is tabulated as follows.

Ion	Number	Site	Spin direction	Ion moment, μ_B	Total moment, μ_B
Fe^{2+}	1	Octahedral	↑	4	4
Fe^{3+}	1	Octahedral	↑	5	5
Fe^{3+}	1	Tetrahedral	↓	5	−5

Overall there is a moment imbalance of $4\mu_B$. The value of M_s is therefore $M_s = (\Sigma M)/V = 4\mu_B \times (9.27 \times 10^{-24} \text{ A-m}^2/\mu_B)/(0.419 \times 10^{-9} \text{ m})^3 = 5.04 \times 10^5$ A/m.

This compares with a measured value of 5.3×10^5 A/m.

Other magnetic ferrites have inverted spinel structures in which the magnetic moment may be viewed as deriving solely from the divalent cation. The Fe^{3+} ions on tetrahedral and octahedral sites couple antiferromagnetically, resulting in a cancellation of the moment from this source. A rich variety of ferrite alloys having the general formula $Me^{2+}Fe_2^{3+}O_4$, where Me^{2+} is a divalent metal ion that substitutes for Fe^{2+} (e.g., Ni^{2+}, Co^{2+}, Zn^{2+}), is possible. For example, in $Co^{2+}Fe_2^{3+}O_4$, a mole of CoO is alloyed with one of Fe_2O_3. The subunit cell magnetization depends only on the Co^{2+} ion moment of $3\mu_B$. Because of moment cancellation in adjacent sublattices, the saturation magnetization of ferrites is generally less than that of metal magnets.

Hexagonal ferrites are an important class of magnetic ferrites that have a complex hexagonal structure similar to that of the mineral magnetoplumbite. For our purposes these ferrites have the formula $MO \cdot 6Fe_2O_3$ or $MFe_{12}O_{19}$, where M is Ba, Sr, or Pb. Per formula unit, the 12 Fe^{3+} ions are arranged on several different sites as follows: 2 Fe^{3+} ions are in tetrahedral sites with spin up; 2 Fe^{3+} ions are in octahedral sites with spin up; 7 Fe^{3+} ions are in octahedral sites with spin down; 1 Fe^{3+} ion is in fivefold coordination with spin down. In summing, there are four uncompensated Fe^{3+} ion moments each of magnitude $5\mu_B$. Therefore, the net formula moment is $4 \times 5\mu_B = 20\mu_B$. The hexagonal ferrites are used to make permanent magnets and are addressed again in Section 14.5.3.2.

14.3.4.2. Garnets

If instead of alloying a divalent oxide with Fe_2O_3 as in ferrites, a trivalent oxide M_2O_3 were used, garnets with a formula $3M_2O_3 \cdot 5Fe_2O_3$, or equivalently $M_3Fe_5O_{12}$, are produced. These materials are structurally similar to Al_2Mg_3-Si_3O_{12}, the formula for natural garnet gemstones. In magnetic garnets, M is usually yttrium (Y) or another rare-earth metal. In the most important technical garnet, yttrium–iron–garnet or YIG ($Y_3Fe_5O_{12}$), the resulting magnetization is due to two Fe^{3+} ions or $2 \times 5\mu_B = 10\mu_B$. Applications of these interesting magnetic materials are discussed in Section 14.5.4.3.

14.4. THE MAGNETIZATION PROCESS: MAGNETIC DOMAINS

14.4.1. The Hysteresis Loop

Ferromagnetic (and ferrimagnetic) materials that are spontaneously magnetized are not necessarily **magnets**. Iron nails and steel chisels do not generally have the ability to attract steel paper clips. (This apparent paradox is addressed when domains are considered next in Section 14.4.2.) Both the nails and chisel have a zero *net* magnetization, but they can be magnetized or made into magnets through exposure to magnetic fields. One way to make the steel chisel a magnet is to place it within a multiturn solenoid powered with a direct current; this is how electromagnets work. A measure of the magnetic induction (B) produced in the steel chisel core is the axial force it exerts on a standard external piece of iron. As the current increases, the $B = 0$, $H = 0$ state changes and follows the arrows as shown in Fig. 14-11. With high enough current, B practically levels off at the saturation value B_s in the first quadrant. At this point the chisel behaves like a magnet and exerts a maximum pull on the iron. If the current (and H) is reduced to zero, the magnetic states follow a path back different from that originally traversed until the value B_r, known as the **remanent induction,** is reached. The chisel has become magnetized in the absence of an H field. To explore the magnetization process further the current polarity is reversed. Magnetic states in the second quadrant are accessed until the steel is demagnetized ($B = 0$) at a magnetic field of $-H_c$, known as the **coercive field.** The chisel exerts no force on the iron at this point. Further increase of reverse current causes a saturation of induction in the opposite direction, that is, $-B_s$. Finally as the (reverse) current is first reduced to zero, changed back to the initial polarity, and then increased, the magnetic path proceeds from the third quadrant, through the fourth, and back into the first quadrant to complete the

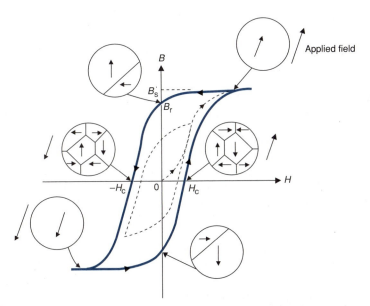

FIGURE 14-11 *B–H* hysteresis curve in ferromagnets. Corresponding magnetic domain (see Section 14.4.2) motion shows that magnetization effects differ depending on whether magnetic fields are increasing or decreasing. Note that ferromagnetic domain behavior parallels that of ferroelectric domains in Fig. 11-26.

loop. Further current cycling reproduces this so-called *B–H* or **hysteresis loop** almost indefinitely. If magnetic saturation does not occur, smaller hysteresis loops are traced as shown.

It is clear that for a given *H* there can be many different *B* values, depending on the magnetization history. For example, none of the three paths traversed through the first quadrant of the hysteresis curve superimpose. The fact that the paths are not *reversible* is a certain sign that energy is being lost in the magnetization process. We have seen analogous behavior before in lossy dielectrics and in plastic stress–strain cycling. In magnetic materials the area enclosed by the hysteresis loop is equivalent to the energy lost or dissipated in one cycle. Just what causes the loss will be explored when magnetic domain behavior is discussed in the next section.

There are ac methods to experimentally trace the envelope of the loop, as well as to record the *initial* stage of the magnetization process. The initial slope of the *B–H* variation is the permeability (Eq.14-4), a magnetic property of considerable interest in the varied soft magnetic material applications. To obtain μ, the material is fashioned into a toroid and converted into an inductor by wrapping many turns of wire about it. By measuring the ac current through it, and the voltage across it, the initial magnetization curve for a nickel zinc ferrite was obtained as shown in Fig. 14-12.

14.4.2. Magnetic Domains

14.4.2.1. Origin of Domains: Magnetostatic Energy

We now return to the issue raised above: How is it possible for a collection of atoms that are spontaneously magnetized in the same direction to be demag-

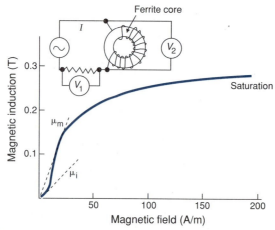

FIGURE 14-12   Magnetization curve for Ni–Zn ferrite obtained using the indicated ac circuit. Both the initial (μ_i) and maximum (μ_m) permeabilities are derived from the indicated slopes of the *B–H* curve. Voltages V_1 and V_2 are respectively proportional to *H* and *B*.

netized macroscopically? If the material were not demagnetized it is clear that, like a bar magnet, an external magnetic field would surround it (Fig. 14-13A). But we know that this field is capable of doing work. It, therefore, has a spatial **magnetostatic energy** density (E_M in units of J/m^3) associated with it, much like the electrostatic energy density emanating from point charges. A measure of E_M is the density of magnetic flux lines in free space. Importantly, E_M can be reduced by a simple internal rearrangement of moments resulting in side-by-side oppositely oriented regions or **magnetic domains,** as shown in Fig. 14-13B. Saturation magnetization (M_s) is reached in each domain but their vectors are antiparallel. In accord with the natural tendency to reduce energy, the domain structure undergoes further change (Figs. 14-13C,D) until there is no external flux. Closure domains oriented with M_s perpendicular to the original ones are responsible for this. Magnetic domains in an iron single crystal have been made visible by the Bitter technique, which employs tiny Fe particles to decorate domain boundaries (Fig. 14-13E). In situations where very little magnetic flux emerges externally, the material is essentially demagnetized. The magnetic domain structure is generally independent of the polycrystalline grain structure.

14.4.2.2. Domain Configurations

In addition to Fig. 14-13, other domain geometries are possible. What rules govern domain sizes, shapes, and configurations? Just as with the common grain structure of crystalline solids, complex energy considerations and compro-

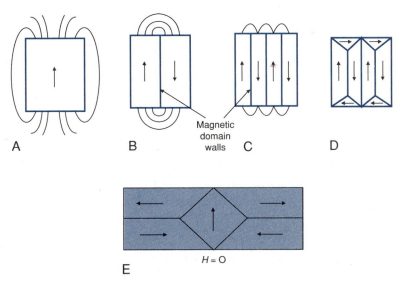

FIGURE 14-13 Ferromagnetic domain structure accompanying the reduction in magnetostatic energy. (A) Single domain. (B) Two domains of opposite magnetization. (C) Four domains of alternating magnetization. (D) Closure domains minimize magnetostatic energy. (E) Domain pattern in a single crystal of iron when $H = 0$. Vertical and horizontal lines lie in $\langle 100 \rangle$ directions.

mises are at work. Prominent among these are the roles of exchange energy, **magnetocrystalline anisotropy energy,** and **domain wall energy** in addition to magnetostatic energy. For example, minimizing magnetostatic energy essentially means the breakup of a single domain, as we have just seen. But having domains whose moments point antiparallel and perpendicular to the original orientation raises issues of competing energy reduction mechanisms. If two neighboring, oppositely magnetized domains are considered, as in Fig. 14-14A, the transition region or domain boundary must necessarily consist of continuously rotated or tilted spin moments. Minimum exchange energy requires that the angle between adjacent spin vectors be as small as possible, however. This means extensive widening of the domain boundary to accommodate the full 180° rotation. But magnetocrystalline anisotropy energy (E_A) now enters the picture, and E_A is reduced when the domain boundary is narrow or when adjacent spin vector angles are large. The equilibrium width of the domain wall (d_D) is a compromise between these opposing trends, as shown in Fig. 14-14B, and is typically 100 nm wide.

Magnetocrystalline anisotropy energy arises because the magnetization differs depending on crystallographic direction. If a single crystal of iron is magnetized along each of its three most important directions, M_s is found to be largest along [100], least along [111], and between these extremes along [110], as depicted in Fig. 14-15. A small magnetic field will cause magnetic saturation along [100] but much larger fields are required along [111]. Thus, [100] is the easy axis of magnetization, and [111], the hard axis; for complex reasons the spins of Fe atoms prefer to lie in the [100] direction. The energy to make M lie along a particular axis relative to the easy axis is defined as E_A, a quantity that has a graphical interpretation. Just as the integrated product of stress and

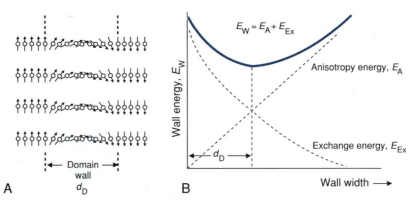

A   B

FIGURE 14-14 (A) Change of the spin vector from one domain to another across a domain boundary. Spins align in parallel orientation to minimize exchange energy, but in parallel as well as antiparallel *crystallographic* orientations to reduce magnetocrystalline anisotropy energy. (B) The equilibrium domain wall width is determined by minimizing the sum of exchange and anisotropy energies. After C. R. Barrett, W. D. Nix, and A. S. Tetelman, *The Principles of Engineering Materials*, Prentice–Hall, Englewood Cliffs, NJ (1973).

strain is the mechanical strain energy density (J/m³), so too the product of M (or B) and H is E_A (J/m³). In Fig. 14-15, E_A is the shaded area representing the energy required to magnetize Fe in the easy direction. Domains in Fe are preferably polarized along [100] or, equivalently, [010] and [001], as suggested in Fig. 14-13. Across vertical or horizontal domain walls, the M vector turns from an easy axis, through hard magnetization directions, and then into the antiparallel easy direction again. Narrow domain walls clearly reduce E_A. A structure with too few domains is not stable because, although both E_A and E_{EX} are reduced, E_M is increased. Similarly, if there are too many domains the reverse is true. Equilibrium lies somewhere in between.

14.4.2.3. Types of Domain Walls

Three different types of walls or boundaries separating adjacent domains have been identified in ferromagnets, and these are shown in Fig. 14-16. The first is the 180° **Bloch** wall. If the M vector lies parallel to the book page within both domains, then M twists *out* of and back *into* the page across the Bloch wall. There is also a 180° tilt rotation of the M vector in the case of the **Ne'el** wall (Fig. 14-16B). But in contrast to the Bloch wall, the rotation of the M vector occurs *in* the plane of the page. Domain walls oriented at 45° in Fig. 14-13D lie in [110] directions and represent a 90° tilt. The last wall type is known as the **crosstie** and it consists of tapered Bloch wall lines jutting out in both directions from a Ne'el wall spine. Bloch and Ne'el walls are the common ones in bulk ferromagnets, whereas crosstie walls are frequently seen in thin

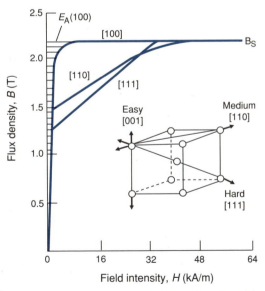

FIGURE 14-15   Magnetization curves for iron in [100], [110], and [111] directions. The shaded area is a measure of the magnetocrystalline anisotropy energy.

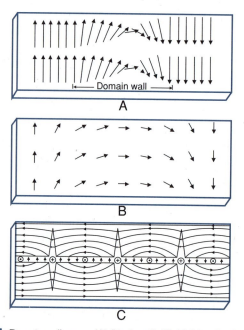

FIGURE 14-16   Domain wall types: (A) Bloch wall. (B) Ne'el wall. (C) Crosstie wall.

ferromagnetic films. Each of the domain walls has an associated energy (E_W) with a typical magnitude of $\sim 10^{-3}$ J/m². This compares with grain boundary surface energies of ~ 1 J/m² in metals. Depending on material and geometry, the value of E_W can vary.

14.4.2.4. Domain–Magnetic Field Interactions

Until now we have considered domains in the absence of external magnetic fields. But we know that when H is applied, a net magnetization is produced and, therefore, some alteration in the domain structure must occur. Actually, domains assume a lively mobility during three distinct regimes of the magnetization process.

1. *Reversible domain wall motion* is the first thing that happens when H is increased slightly. In the unmagnetized matrix, defects, impurity atoms, precipitates, and so on serve initially to anchor domain walls, and small values of H do not appear to disrupt these interactions; there is no hysteresis on cycling.

2. When H is increased, *irreversible domain wall translation* occurs, and if the field is removed, hysteresis does occur. Domains favorably oriented with respect to H grow at the expense of unfavorably oriented ones. The magnetic field is not strong enough in polycrystalline materials to drive the magnetization beyond easy directions. Unlike the relatively slow movement of boundaries characterizing grain growth during annealing of metals, domain walls translate

at great speeds. The difference, of course, is due to thermally activated atomic motion in the former case and collective, coordinated spin reorientation in the latter case. Walls become pinned at new sites, and a residual magnetization is left when H is removed. During this second stage of magnetization, hysteresis loops of smaller area than the saturated one are generated during field cycling.

3. Lastly, in very high magnetic fields, *domain rotation* occurs to turn the magnetization forcibly to line up with the H field direction. This may mean rotation from easy to hard magnetization directions until B_s is eventually achieved. These dynamical domain effects, schematically illustrated in Fig. 14-11, dissipate energy and are responsible for hysteresis loop energy loss.

When H is reversed the sequence just described is played backward, but not too faithfully because of hysteresis effects. The demagnetizing field produces reverse domains that encounter a new set of obstacles, preventing the original path from being retraced. Two of the important states derived from the hysteresis curve—saturation induction (B_s) and coercive force (H_c)—are key properties of the magnetic materials to which we now turn our attention.

14.5. FERROMAGNETIC MATERIALS AND APPLICATIONS

14.5.1. Hard and Soft Magnets

Magnetic materials are broadly separated into two categories: soft and hard. The division is based on the value of H_c and characterized by distinct hysteresis curves schematically shown in Fig. 14-17.

Hard magnetic materials are *hard* to magnetize and *hard* to demagnetize. High values of both B_s and H_c characterize these materials and make them excellent permanent magnets. The former establishes the high induction field needed to strongly attract ferromagnetic materials; the latter prevents magnetic effects from being influenced or even destroyed by nearby magnets or external fields.

Soft magnetic materials, on the other hand, are *easy* to magnetize and *easy* to demagnetize. This enables them to reverse magnetization readily in response to alternating electric fields where they are required to concentrate magnetic flux in transformers and inductances. Small coercivities enable high-permeability, soft magnetic materials to fulfill these functions with little energy loss. In addition to thin hysteresis loops, high values of B_s are also desirable to minimize the size of the soft magnetic material required. A number of the key properties of important hard and soft magnetic materials are entered in Tables 14-1 and 14-2.

14.5.2. Creating a High Coercive Field

Knowing the type of hysteresis curve required for an engineering application is one thing; creating a hard magnetic material that has the desired coercive

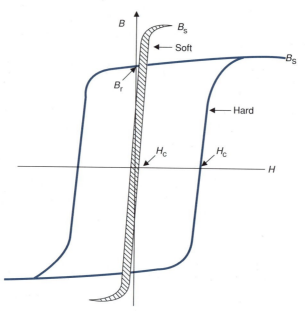

FIGURE 14-17 Difference between hysteresis curves for soft and hard magnetic materials. Typically, $H_c < 40$ A/m in good soft magnets, and $H_c > 40$ kA/m in good hard magnets.

force is another matter. What general principles are there to guide us in designing magnetic properties? In the case of the hard magnets, we require a large value of H_c. With $H = H_c$ when $B = 0$, Eq. 14-2 yields

$$0 = \mu_o(H_c + M) \qquad (14\text{-}10)$$

TABLE 14-1   **PROPERTIES OF HARD MAGNETIC MATERIALS**[a]

Material and composition (wt%)	Remanence, B (T)	Coercivity, H_c (kA/m)	$(BH)_{max}$ (kJ/m³)	Curie temperature, T_C (°C)
Carbon steel (0.9C, 1Mn)	0.95	4	1.6	~768
Cunife (20Fe, 20Ni, 60Cu)	0.54	44	12	410
Alnico V (50Fe, 14Ni, 25Co, 8Al, 3Cu)	0.76	123	36	887
Samarium cobalt (SmCo₅)	0.9	600	140	727
Nd₂Fe₁₄B	1.1	900	220	347
Barium ferrite (BaO–6Fe₂O₃)	0.4	264	28	447

[a] Data were taken from various sources.

| **TABLE 14-2** | **PROPERTIES OF SOFT MAGNETIC MATERIALS**[a] | | | |

Material and composition (wt%)	Maximum relative permeability	Saturation induction, B_s (T)	Coercive field (A/m)	Resistivity ($\mu\Omega-m$)
Commercial iron ingot (99.95Fe)	5,000	2.2	80	0.1
Silicon–iron, oriented (97Fe, 3Si)	15,000	2.01	12	0.47
Permalloy 20Fe (79Ni, 21Fe)	10^5	1.1	4	0.2
Supermalloy (79Ni, 15Fe, 5Mo)	10^6	0.80	0.16	0.6
Ferroxcube A (MnZn ferrite)	1400	0.25	0.8	10^6
Amorphous Fe–B–Si	$>10^5$	1.6		10^3

[a] Data were taken from several sources.

or $H_c \sim |M|$. An upper magnitude for M, and therefore H_c, is the saturation magnetization value (M_s) that single domains, by definition, exhibit. Fine particles that are ~100 nm or so in size cannot geometrically sustain more than one domain because the domain wall thickness is typically the same size. Consider now the magnetization of such ellipsoidal or elongated single-domain particles, where the easy direction lies parallel to the long axis. If a strong reverse magnetic field is applied, the magnetization cannot change by domain wall motion because there are no domain walls. Instead, the magnetization rigidly rotates through the hard direction and snaps into the antiparallel easy orientation. But such M rotation is possible only in large fields, that is, a high H_c value. Therefore, a strategy to produce high-coercive-field magnets is to consolidate single-domain powders. This is in fact done in rare-earth magnets and in magnetic tape, as noted below.

Raising H_c by subdividing a matrix into small regions that reorient with difficulty can be accomplished by means other than the use of powders. Reverse domains sometimes, but not always, have difficulty in nucleating at grain boundaries or in heavily deformed or strained ferromagnetic matrices. Fine distributions of magnetic phases as a result of transformation are also often effective in enhancing H_c.

14.5.3. Hard Magnetic Materials

14.5.3.1. *BH* Product

Good hard magnetic materials, like good soft magnetic materials, require a high-saturation induction; but unlike them, they require a high coercivity also. Both of these properties are combined in a single quantity, the maximum

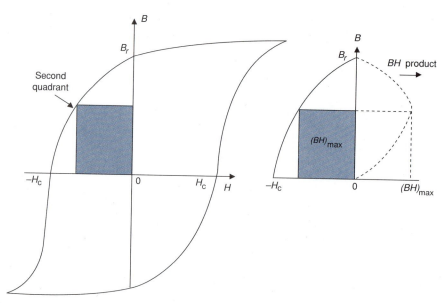

FIGURE 14-18 Method for obtaining $(BH)_{max}$ from the hysteresis curve. $(BH)_{max}$ is the product of the B and H lengths defining the largest rectangular area that can be inscribed within the second quadrant of the B–H loop.

BH product, or $(BH)_{max}$, which is used as a convenient figure of merit for the strength of permanent magnet materials. $(BH)_{max}$ is derived from the demagnetization portion, or second quadrant of the B–H curve, as shown in Fig. 14-18. The products of pairs of B and H values from the second quadrant of the hysteresis loop are replotted on the right as B versus BH, from which $(BH)_{max}$ is readily obtained. This is the way the values reported in Table 14-2, and noted in Fig. 14-1, were obtained. Because it is derived from the hysteresis curve, the BH product is also known as the energy (density) product with units of J/m³-cycle.

EXAMPLE 14-3

A certain rare-earth alloy magnet has a demagnetization curve that is parabolic in shape and can be expressed by the formula $B = 1.1 - 4.4 \times 10^{-6}H^2$, where B is in T, and H is in kA/m.
   a. What is H_c?
   b. What is $(BH)_{max}$?

ANSWERS   a. H_c is the value of H for which $B = 0$. Therefore, $0 = 1.1 - 4.4 \times 10^{-6}H_c^2$ and $H_c = [(1.1/4.4) \times 10^6]^{1/2} = -500$ kA/m.
   b. The desired $(BH)_{max}$ can be obtained by setting the derivative of BH with respect to H to zero. As $BH = 1.1H - 4.4 \times 10^{-6}H^3$, $d(BH)/dH = 1.1 -$

$3 \times 4.4 \times 10^{-6}H^2$. Therefore, $H = H_{max} = [(1.1/13.2) \times 10^6]^{1/2} = -289$ kA/m. At this value of H_{max}, $B_{max} = 1.1 - 4.4 \times 10^{-6}H^2_{max} = 1.1 - 4.4 \times 10^{-6}$ $(289)^2 = 0.732$ T. Finally, $(BH)_{max} = B_{max}H_{max} = 0.732 \times 289 = 212$ kA-T/m $= 212,000$ Wb-A/m^3, or 212 kJ/m^3.

14.5.3.2. Materials

Alnico. One of the most widely produced permanent magnet materials derives its name from the *al*uminum, *ni*ckel, and *co*balt additions to the iron base metal. Popular since the 1930s, these alloys have been continuously improved, and applications based on them have expanded in concert. The Alnicos are brittle and therefore shaped by casting or powder metallurgy methods. Because they are precipitation hardening alloys, the first step involves solution treatment at 1300°C. Upon cooling to 800°C a two-phase decomposition into α and α' occurs. The former is the weakly magnetic matrix phase that contains the precipitated α' phase distributed through it. Rich in Fe and Co, α' has a high magnetization, and its distribution and needlelike shape (Fig. 14-19) are largely responsible for Alnico's high coercivity. Solidification in externally applied magnetic fields aligns precipitates *anisotropically* and increases the aspect (length-to-width) ratio of α', raising H_c in the process.

Alnico development is an excellent example of the integration of processing, structure, and properties to improve performance.

FIGURE 14-19   Structure of Alnico. TEM image taken by Y. Iwama and M. Inagaki. Courtesy of JEOL USA, Inc.

Rare Earth Magnets.   The most important rare-earth magnet is $SmCo_5$. In single-phase form, this alloy has an energy product four to six times that of the Alnicos and coercivities that are 5 times higher. This means greater miniaturization of magnets used in motors, surgically implanted pumps and valves, and electronic wristwatch movements. Magnetism in these alloys is derived microscopically from unpaired $4f$ electrons of Sm; on a macroscopic basis the coercivity arises because of domain pinning at grain boundaries and surfaces. Sm–Co alloys are fabricated by powder metallurgy methods where, interestingly, powders are aligned during pressing in a magnetic field. Sintering is done at low temperatures to prevent grain growth. There are other similar Sm–Co alloys containing copper or iron that are precipitation hardenable and can be heat treated to yield extraordinarily high $(BH)_{max}$ products.

One of the significant advances in permanent magnet materials was the development of neodymium–iron–boron (NdFeB) in 1984. These magnets have $(BH)_{max}$ products of ~ 300 kJ/m^3, the highest yet reported. Produced by both rapid solidification and powder metallurgy processing, this magnetic material is expected to find applications requiring low weight and volume, for example, automobile starting motors. High cost is a disadvantage of rare-earth magnets.

Ferrites.   It may come as somewhat of a surprise that hard ferrites account for slightly more than 60% of the total global dollar value (1.8 billion dollars in 1988) and more than 97% of the total weight (250,000 tons) of permanent magnets produced. These magnets are based largely on barium and strontium ferrite with compositions of $BaO \cdot 6Fe_2O_3$ and $SrO \cdot 6Fe_2O_3$, respectively. They are made by pressing and sintering powders whose grain sizes are large compared with the domain size. Thus, the magnetization is believed to occur by domain nucleation and motion, rather than by the rotation of single-domain particles. Hard ferrites have low $(BH)_{max}$ products compared with other permanent magnets. Nevertheless, their low cost, moderately high coercivity, and low density are responsible for widespread use in motors, generators, loudspeakers, relays, door latches, and toys.

14.5.4. Soft Magnetic Materials

The key requirements for soft magnetic materials are high permeability and low coercivity. For the former it is helpful to align the easy axis of magnetization parallel to the field direction. This eliminates time spent in rotating misoriented domains, enhancing permeability as a result. A practical way to achieve this in iron–silicon alloy sheet is to develop the proper texture or crystallographic orientations as noted below. Low-coercivity fields mean elimination of impurities, inclusions, and grain boundaries that impede the motion of domain walls. How some of these broad design guidelines are implemented in practice is discussed in the context of several important soft magnetic materials.

14.5.4.1. Iron–Silicon

Electrical-grade steel alloys for the generation and distribution of electrical energy (i.e., transformers) are the magnetic metals produced in the largest tonnage. Alloy improvement over the years is a testament to the persistence of materials science and metallurgical engineering methods, as the following brief chronology illustrates. Prior to 1900, mild steels were used for the above purposes. In that year the benefit of silicon additions to iron was discovered. It was found that 3 wt% Si in Fe not only increased the permeability, but simultaneously reduced the coercive force relative to unalloyed iron. Furthermore, the rise in electrical resistivity was an added bonus. High resistance is important because alternating magnetic fields induce corresponding **eddy** currents (I_e) by Faraday's law. Eddy current flow in magnetic metals produces undesirable Joule heating energy losses of amount $I_e^2 R = V^2/R$. Through Ohm's law, high core resistance (R) lowers Joule heating losses for the same operating voltage V. Thus, reduction of the core losses (a sum of Joule and hysteresis losses) outweighed the slight reduction in saturation induction resulting from Si additions. Polycrystalline Fe–Si sheet materials soon replaced the conventional materials and enjoyed wide use for the next three decades. By the use of laminations rather than bulk metal cores, eddy current paths were interrupted, further lowering core losses.

The randomly oriented grain structures were far from ideal, however, because magnetic saturation was reached only by applying fields well above H_c. This limited useful magnetic inductions to $B \sim 1$ Wb/m^2. But it was known that single crystals of Fe could reach inductions greater than ~ 2 Wb/m^2 along preferred directions (see Fig. 14-15). An ideal transformer would therefore consist of single-crystal sheets oriented along [100] directions. Important progress toward approximating this goal was made in 1935 with certain cold working and annealing steps that produced a crystallographic **texture** characterized by [100] directions in (110) planes lying parallel to the rolling plane. This cube-on-edge texture in these grain-oriented materials can be understood by reference to Fig. 14-20. Later progress in lowering hysteresis losses was made by reducing the carbon content through hydrogen annealing, by adjusting the grain size, by using insulating coatings to separate magnetic layers, and by introducing tensile stresses in the sheet. Roughening the sheet by laser scribing to disrupt eddy current paths is the latest beneficial treatment. Through incremental improvement over a century, core losses dropped from approximately 8 to 0.4 W/kg (Fig. 14-2).

14.5.4.2. Ferrites

The story of *soft* ferrite development dates to the mid-1930s, when significant advances in magnetic materials began to unfold. Over the years important compositions that were developed included (MnZn)Fe$_2$O$_4$, (MnCu)Fe$_2$O$_4$, and (NiZn)Fe$_2$O$_4$. In addition, processing improvements in the next four decades led to a 400-fold reduction in core loss and a corresponding shrinkage in size of components made from ferrites (e.g., transformer cores).

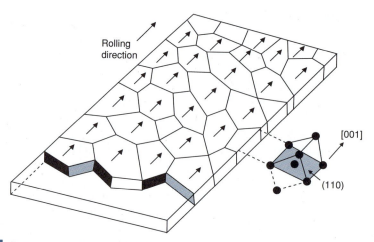

FIGURE 14-20   Illustration of preferred (110) [001] texture in rolled polycrystalline (BCC) Fe–3% Si transformer sheet.

Ferrite products are produced by powder pressing and sintering methods and one of the common shapes produced is the toroid. When wound with wire, toroids make efficient transformer cores, and because they are electrically insulating there is very little eddy current loss, even when operated at microwave frequencies. This coupled with high-saturation magnetization has enabled ferrites to find applications in television and radio components such as line transformers, deflection coils, tuners, and rod antennas, as well as in assorted small to medium power supplies. The use of low-coercivity, square-hysteresis-loop ferrites for computer memory applications capitalized on the 1 and 0 states produced by positive and negative magnetizations. These were generated by either clockwise or counterclockwise current flows in the cores. "Now just a memory," billions of ferrite cores typically less than 0.5 mm in diameter were produced annually for computer use, until semiconductor memory technology replaced them around 1970. The cores were periodically arrayed in a woven framework and the centers were threaded by three sets of wires (word, bit, and sense lines) that provided the currents and magnetic fields necessary to generate and detect signals.

14.5.4.3. Garnets

Yttrium–iron–garnet (YIG) and derivatives based on it (e.g., gadolinium gallium garnet) display a plethora of interesting magnetic phenomena of interest to both scientists and engineers. In fact, the role of YIG in providing fundamental knowledge about magnetism in nonmetallic materials has been likened to that of fruit flies in genetic studies! Importantly, garnets exhibit magneto-optic as well as magneto-acoustic effects. Because garnets have lower M_s values than ferrites, they have not competed with them in traditional soft magnet applications; instead, they have found use in magnetic bubble memory and assorted microwave devices.

FIGURE 14-21 Serpentine domain pattern in yttrium–iron–garnet viewed by transmitted polarized light. The magnetization is perpendicular to the plane of the page, e.g., up for dark domains and down for light domains.

The magneto-optic effect is interesting and deserves further comment. All magnetic materials considered until now do not transmit light. In bulk form, YIG is black, but thin films (about several micrometers thick) are transparent to visible light. In most magnetic thin films the magnetization (M) vector lies *parallel* to the film plane. But in garnet films, M is *perpendicular* to the film plane and points either up or down. When plane-polarized light passes through it parallel to the direction of M (up), its plane of polarization is rotated a few degrees clockwise; the reverse is true when M is oriented down. This is the magneto-optic **Faraday effect.** By use of an analyzer plate, light and dark contrast is obtained from the magnetized domains, as illustrated in Fig. 14-21.

Magnetic bubble memory devices present an interesting application of magnetic garnet thin films. If such a film is imagined in the plane of the page, then with the application of suitable H fields, the serpentine domains shrink into tiny cylindrical magnetic domains or bubbles, ~0.5 μm in diameter. In actual devices an array of conductors and permalloy thin films, in the patterned shape of chevrons, I and T bars, are deposited over the bubble film to guide and detect bubble motion. (Permalloy is an important soft magnetic material containing 79 at.% Ni and 21 at.% Fe.) The bubbles can then be generated, moved, switched, counted, and annihilated in a very animated way in response to the driving fields, as suggested by Fig. 14-22. Bubble memory devices with capacities in excess of a megabit are available. Some of their characteristics include high storage densities ($\sim 10^9$ bits/cm^2), absence of mechanical wear during operation, nonvolatile memory, and a wide temperature range of read–write memory.

14.5.4.4. Metallic Glasses

These relatively recent metal alloys typically contain one or more transition metals (e.g., Fe, Ni, Co) and one or more metalloids (e.g., B, Si, C). Melts quenched at extremely high rates, as described in Section 5.6.4, are continuously cast to yield ribbon ($\sim 50 \, \mu$m thick) that is thinner than a sheet of paper. Based on their composition, structure, and processing, one might not expect them to be useful magnetic materials. Nevertheless, several alloys (e.g., 79at.%Fe-13at.%B-8at.%Si) are blessed with all of the properties desired of soft magnetic materials such as high-saturation induction, low coercivity, high permeability, and low core loss from dc to several megahertz. Their low anisotropy and lack of grain boundaries mean there is little to prevent reverse domains that nucleate from readily spreading. Potentially attractive as a replacement for electrical steels, their lower ductility, higher cost, and potential instability during aging are issues of concern.

14.5.5. Magnetic Recording

The needs of both professional and consumer audio, video, and computer tapes and disks are currently met by an assortment of magnetic materials in

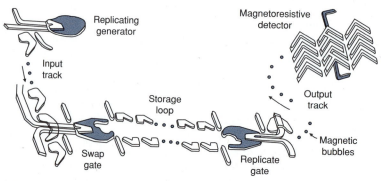

FIGURE 14-22　Schematic arrangement of a complete magnetic bubble memory device. From D. K. Rose, D. J. Silverman, and H. A. Washburn, in *VLSI Electronics: Microstructure Science,* (N. Einspruch, Ed.), Vol. 4, Academic Press, New York (1982).

both particle and thin-film form. But, the insatiable appetite for information storage continues to demand ever-increasing recording densities at lower prices. In magnetic recording systems, information is digitally stored as 1 and 0 states, corresponding to positive and negative magnetization levels. To better appreciate the material requirements necessary to cheaply and reliably store and read large densities of information at high speeds, it is helpful to understand the recording process.

Time-dependent electrical input signals are converted into spatial **magnetic** patterns (**writing**) when the **storage medium** translates relative to a recording head, as schematically shown in Fig. 14-23. The medium is either a **magnetic** tape or a flat disk, whereas the head is a gapped *soft* ferrite with windings around the core portion. Input electrical signals are instantaneously converted into magnetic flux that is concentrated in the head. Fringe flux flaring out in the narrow gap intercepts the nearby magnetic medium, magnetizing it 1 or 0 over small regions; the smaller the region, the higher the storage density. As soft magnetic materials of high permeability are easily magnetized, they **might** at first glance appear to be ideal for media applications. But the magnetization patterns must be permanently stored for later playback, and the moment **the** medium moves beyond the head, they would vanish. Instead, the medium **must** possess both high permeability and high coercivity. The latter prevents the magnetization from being erased or altered when exposed to external magnetic fields.

During playback (**reading**) the magnetized medium is run past the head. Now moving magnetic fields, originating from the medium, readily induce voltage signals (by Faraday's law) in the stationary head pickup coils. Electrical amplification regenerates the sound, picture, or information recorded or stored in the first place.

From the foregoing it is apparent that magnetic recording systems require opposite but complementary magnetic properties, that is, soft magnetic materials for the recording and playback heads and hard magnetic materials for the

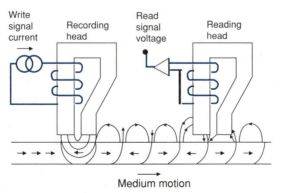

FIGURE 14-23 Schematic of magnetic recording and playback processes. Reproduced with permission, from the *Annual Review of Materials Science*, Volume 15, © 1985, by Annual Reviews Inc.

storage media. Traditionally, high-permeability soft MnZn ferrites and alloys like permalloy have been used for heads. Tape and disk media have been dominated by the use of very small ($<0.5 \times 0.06 \ \mu$m) elongated particles of γ-Fe_2O_3, γ-Fe_2O_3(Co), or CrO_2 that are essentially single domains. The high coercivity in these high-aspect-ratio domains is due to **shape anisotropy,** or the difficulty of reversing magnetization in them. They are deployed in a passive binder on flexible polymer tapes or on more rigid metal disks.

The increasing use of thin magnetic film media has been driven by the desire to reduce the size of personal computers and portable video recording and playback systems. Thin films enable higher recording densities to be achieved. The reason is the combined efficiency of 100% packing of magnetic material in films, compared with 20 to 40% in particulate media, and the generally higher magnetization possible with cobalt alloy films.

In addition to the recording systems emphasized above, there are additional uses for magnetic materials in audio, video, VCR, computer, and other equipment. Motors and head phones or speakers are examples. From the contents of this chapter it should not be too difficult to prescribe material requirements for them.

14.6. PERSPECTIVE AND CONCLUSION

Magnetism is the only material property treated in this book, other than X-ray or Auger electron spectroscopy (Chapter 2) that depends on the nature of core rather than valence electrons. But, unlike the spectra that are dependent on the *quantized energies* of core electrons, it is the *quantized angular momenta* of these same electrons that are important in magnetism. Electrons orbiting nuclei or spinning about their own axes possess angular momentum. The latter produce magnetic moments in much the same way that circulating currents do in a solenoid. Of the two, electron spin is the major source of magnetic effects in atoms. In the common ferromagnetic metals, for example, it is the unfilled $3d$ electron shell (or $3d$ band in solids) that has been identified as the source of ferromagnetism. It is also the source of magnetic ions and ferrimagnetism in ceramic oxides.

Once it is established that atoms or ions have a net magnetic moment, the next level of magnetic hierarchy is dependent on their interactions. For this purpose atoms are essentially replaced by small bar magnet vectors, and statistical vector addition provides a model for the type of magnetic behavior exhibited. When the vectors are strongly and spontaneously aligned over large distances, we speak of **ferromagnetism**; if randomly oriented, we have **paramagnetism**; when they alternate in direction so that neighboring vectors are always antiparallel to one another, **antiferromagnetism** is displayed; and if moments in adjacent sublattices are of different magnitude and sign, but vectorially add to yield a net (spontaneous) magnetization, the phenomenon is **ferrimagnetism**. Only the first and last of these effects have commercial significance. Magnetic domains within these materials have a size and geometric distribution that depend on the material and a host of energies (i.e., magnetostatic, exchange, anisotropy, domain wall). A compromise is struck among them to minimize the overall energy of the system. When exposed to magnetic fields, domain walls undergo a spirited motion that instantaneously reduces the energy at each stage of magnetization. First there is expansion in the size of favorably oriented domains and shrinkage of unfavorably oriented ones. Urged by high enough magnetic fields, the domain magnetization vector rotates until saturation occurs in the field direction.

Engineering applications of magnetism are intimately tied to the hysteresis behavior displayed. These fall into two categories depending on whether soft or hard magnets are involved. Soft magnetic materials are virtually always coupled to ac electrical circuits, where signals and currents alternately magnetize and demagnetize them, as in transformers. High permeability and low coercivity are required to achieve saturation with small signals or H fields. If $H = 0$, the induction does not persist, but quickly vanishes. As long as operation is at low frequencies (e.g., 60 Hz) the hysteresis energy loss is tolerable and metal alloys are suitable. But at audio and higher frequencies, it is imperative to reduce core losses. High-resistivity soft ferrites are then called for.

Hard magnetic materials find applications where it is important that the state of magnetization persist for long times. Permanent magnets are the most widely recognized use for hard magnets. Another extremely important application involves information storage on audio and video disks as well as on computer hard drives and diskettes. The high coercive field is relied upon to preserve the magnetization in these cases.

Development of commercial soft and hard magnetic materials has been a history of unrelenting progress in alloy development and processing treatments. Starting from essentially zero a century ago, the economic impact due to all magnetic materials is now estimated to be ~1.5% of the U.S. gross national product.

Additional Reading

F. N. Bradley, *Materials for Magnetic Functions,* Hayden, New York (1971).

F. Brailsford, *Physical Principles of Magnetism,* Van Nostrand, London (1966).

N. Braithwaite and G. Weaver, *Materials in Action Series: Electronic Materials,* Butterworths, London (1990).

A Goldman, *Modern Ferrite Technology,* Van Nostrand Rheinhold, New York (1990).

QUESTIONS AND PROBLEMS

14-1. It is required to design a solenoid that will develop a magnetic field of 10 kA/m in vacuum when powered with 1 A. The solenoid is 0.3 m in length and 2 cm in diameter.

a. How many turns of wire are required?

b. If the solenoid is wound with 0.5-mm-diameter copper wire, what dc voltage is required to power it?

14-2. Application of a magnetic field of 1.720×10^5 A/m causes a magnetic induction of 0.2162 T in a material. Calculate its permeability and susceptibility. What sort of magnetic material is it?

14-3. A current of 2.5 A flows through a 0.25-m-long, 1000-turn solenoid contained in a vacuum chamber. When placed in a pure oxygen environment the magnetic induction exhibits an increase of 1.04×10^{-8} W/m². What is the magnetic susceptibility of oxygen?

14-4. Explain the physical difference between magnetization, M, and induction, B.

14-5. Consider three spheres of different materials but the same radius. One is a ferromagnet, the second is a paramagnet, and the third is a diamagnet. Each is placed in a uniform magnetic field of the same magnitude.

a. Roughly sketch the lines of magnetic flux through each sphere. Explain your answer.

b. A fourth sphere is superconducting. Sketch the lines of flux in this case.

14-6. Two ferromagnetic metal bars have identical shapes, dimensions, and appearance. One is a magnet (permanently magnetized) but the other is not. Devise a way to determine which one is the magnet without the use of instruments.

14-7. Schematically sketch the hysteresis loop for a ferromagnet at a temperature close to 0 K, just below the Curie temperature, and just above the Curie temperature.

14-8. Provide reasons for, or examples that illustrate, the following statements:

a. All magnets are ferromagnetic materials, but not all ferromagnetic materials are magnets.

b. Not all elements with incomplete $3d$ bands are ferromagnetic.

c. Ferromagnetic domain boundaries do not coincide with grain boundaries.

d. Mechanical hardness in a magnetic metal promotes magnetic hardness; similarly, mechanical softness promotes magnetic softness.

14-9. The $B-H$ curve for a hard ferromagnet can be described by Fig. 14-24A.

a. What is the total hysteresis energy loss per cycle?

b. What is the $(BH)_{max}$ product?

14-10. The $B-H$ curve for a hard ferromagnet outlines the parallelogram of Fig. 14-24B.

a. What is the total hysteresis energy loss per cycle?

b. What is the $(BH)_{max}$ product?

14-11. Mention two ways in which soft and hard magnets differ.

14-12. a. Sketch the $B-H$ response for Ti and Ge on the same scale as NiZn ferrite shown in Fig. 14-12.

b. What is the maximum permeability of NiZn ferrite?

14-13. Magnetic shielding materials protect electrical instruments and systems from interference by external magnetic fields, for example, the earth's magnetic field.

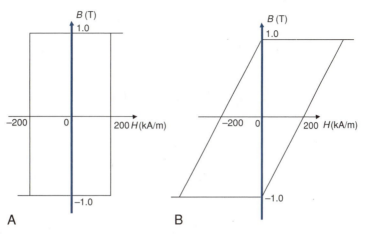

FIGURE 14-24 B–H curves for use in problems 14-9 and 14-10.

    a. Which of the following types of materials are suitable for magnetic shielding applications: paramagnetic, diamagnetic, soft ferromagnetic, hard ferromagnetic?

    b. What specific property is important for magnetic shielding?

14-14. For a permanent magnet material the hysteresis loop second quadrant BH product varies parabolically with B as

$$BH(\text{J/m}^3) = -2 \times 10^5 B^2 + 2 \times 10^5 B \qquad \text{(units of } B \text{ are T).}$$

    a. Plot BH versus B.

    b. What is the maximum value of BH or $(BH)_{\text{max}}$?

    c. What is the value of H at $(BH)_{\text{max}}$?

14-15. When is the electrical resistance of a ferromagnet an issue of concern in applications? What materials are available to address required electrical resistance needs?

14-16. Ferromagnets exhibit *magnetostrictive* effects that are the analog of electrostrictive effects discussed in Section 11.8.3.

    a. Suggest a potential application that capitalizes on magnetostrictive effects.

    b. Explain how hum in transformer laminations can be explained by magnetostriction.

14-17. A soft ferrite core weighing 20 g has rectangular hysteresis characteristics with $B_s = 0.4$ T and $H_c = 0.7$ A/m. The ferrite density is 5 g/cm^3 and the heat capacity is 0.85 J/g-°C.

    a. Estimate the temperature rise after one magnetizing cycle if the process is carried out adiabatically, that is, with no heat loss.

    b. At 60-Hz operation how long would it take for the magnet temperature to rise from 25 to 420°C, the Curie temperature?

    c. How will the magnetization vary above the Curie temperature?

14-18. Predict the value of the saturation magnetization of body-centered cubic iron metal at room temperature.

14-19. The ferrite $NiFe_2O_4$ is a derivative of magnetite, with Ni^{2+} substituting for Fe^{2+}.
   a. If the lattice constant is unchanged, what is M_s for this ferrite?
   b. Suppose that Zn^{2+} is substituted for half of the Ni^{2+} so that the ferrite formula is now $Ni_{0.5}Zn_{0.5}Fe_2O_4$. What is M_s for this ferrite?

14-20. Yttrium–iron–garnet ($Y_3Fe_5O_{12}$) can be written in the form $\{Y^{3+}\}_3$ $[Fe^{3+}]_2(Fe^{3+})_3O_{12}^{2-}$ to clarify the ionic valences. In a program to synthesize new garnets, suggest three new compositions that might be tried.

14-21. Computer memories require stable binary states—1 or 0, or $+M$ and $-M$—for operation. Electrical methods must be provided to write in a 1 or a 0 and read out a 1 or a 0. Explain how magnetic bars plus wires (solenoids, etc.) could be configured to work as memory elements. What material hysteresis loop would be ideal for this application?

14-22. Distinguish between hysteresis losses and eddy current losses in soft ferromagnets.
   a. When do the former dominate the latter?
   b. When do the latter dominate the former?
   c. Mention an application that capitalizes on eddy current losses.

14-23. An Alnico magnet bar whose coercive field is 120 kA/m is contained within a 0.2-m-long, 2000-turn air solenoid far from either end. The axis of the bar is oriented parallel to the solenoid axis. Approximately what current will cause the Alnico magnet to demagnetize and essentially go on to reverse its polarity?

14-24. The magnetic anisotropy energy, E_A, can be computed from data plotted in Fig. 14-15. If E_A is defined as $\int_0^{B_{max}} H\,dB$, estimate its value for the [111] direction in Fe. Does the [110] or [111] direction appear to have the higher value of E_A?

14-25. Distinguish among the following energies:
   a. Magnetostatic
   b. Exchange
   c. Magnetocrystalline
   d. Domain wall

14-26. a. What magnetic properties are required of computer diskettes?
   b. What magnetic properties are required of reading and writing heads?
   c. Suggest ways to increase magnetic storage capacity.

14-27. Why do high-coercive-field magnets have low initial permeabilities?

14-28. How would you make a flexible magnet that would conformally hug a sheet steel surface?

14-29. For the following applications state whether high or low values of B_s and H_c are required.
   a. Strong electromagnet   b. Strong permanent magnet
   c. Magnetic tape for VCR   d. Magnetic bubble memory
   e. Loudspeaker   f. Transformer core
   g. Compass needle   h. VCR tape head

15

FAILURE AND RELIABILITY OF ELECTRONIC MATERIALS AND DEVICES

15.1. INTRODUCTION

We dealt with the subject of failure and degradation of materials in mechanically functional components and structures in Chapter 10. No less important are failures in electronic materials, components, and devices. In August 1993, fate provided a spectacular example. The failure in the Mars Observer spacecraft that doomed the 1 billion dollar NASA project to study this planet was first attributed to transistor failure, but later to mechanical malfunction.

The purpose of this chapter is to introduce (1) the subject of failure modes and mechanisms in electronic materials and devices, (2) simple mathematical techniques for handling failure data, and (3) methods for predicting failure times. Each of these topics has been treated extensively in books and only highlights of this broad interdisciplinary subject are presented in this chapter. Some aspects were addressed in Chapter 10. In comparing these chapters, we see that there are probably more issues of common concern than issues of singular applicability to either mechanical or electrical systems. Both mechanical and electrical applications are vitally concerned with the question of **reliability**. This word is familiar in other contexts, but it will be redefined here as the **probability** of operating a product for a given **time** under specified **conditions** without **failure**. Reliability reflects the performance of products over time in order to assess their future dependability and trustworthiness. Other similarities between failures in electrical and mechanical systems include experimental methods and techniques of failure analysis and the common statistical evaluation of failure test data. Even some failure mechanisms like corrosion are shared. Distinctions between them largely stem from differences in the nature of the materials and devices involved and the particular operative failure mechanisms. Reliability concerns in electronics also evolved differently, and it is instructive to start by presenting a sense of this history.

15.2. RELIABILITY IN ELECTRONICS: PAST, PRESENT, AND FUTURE

15.2.1. A Brief History

In few applications are the implications of **un**reliability of electronic (as well as mechanical) equipment as critical as they are for military operations. Although World War I reflected the importance of chemistry, all of our subsequent wars, including the Gulf War, clearly capitalized on advances in electronics to control the machinery of warfare. Early military electronics, however, was far from reliable. For example, during World War II it was found that 50% of all stored airborne electronics became unserviceable prior to use. The Air Force reported a 20-hour maximum period of failure-free operation on bomber electronics. Similarly, the Army reported high truck and power plant mortalities, and the Navy did not have a dependable torpedo until 1943. Horror stories persisted (and still do) about the large fraction of military electronics that did not operate successfully when required. Up until the 1970s the problem of unreliability was met by increasing spare parts inventories. But this created the new problems of logistics and big business associated with the growth of military logistics commands. The realization sank in that the luxury of massive duplication was too expensive, and drove the military to quantify reliability goals. To reduce the logistics monster and increase the operational time of equipment, a several-orders-of-magnitude enhancement in product reliability was required.

The period from the late 1940s to the 1970s witnessed two major revolutions in electronics. First, solid-state transistors replaced vacuum tubes; later, discrete solid-state devices were superseded by integrated circuits (ICs). Early transistors were not appreciably more reliable than their vacuum tube predecessors, but power and weight savings were impressive. Although discrete transistors were advantageously deployed in a wide variety of communications, transportation, audiovisual, sensing, and measurement applications, information processing required far greater densities of devices. These were provided by ICs. But increased device packing densities raised a variety of new reliability problems associated not only with devices, but with the rapidly increasing number of contacts, interconnections, and solder joints.

15.2.2. Some Trends

It is instructive to quantitatively sketch the time evolution of IC production and characteristics to intelligently assess what the future may bring in terms of reliability issues. Starting in the early 1960s the IC market was based primarily on bipolar transistors. Since the mid-1970s, however, ICs composed of MOS transistors have prevailed because of the advantages of device miniaturization, high yield, and low power dissipation. Then in successive waves, IC generations based on these transistors arose, flowered, and were superseded. They were characterized by small-scale integration (**SSI**), medium-scale integration (**MSI**), and large-scale integration (**LSI**). The present very large scale integration (**VLSI**) era, characterized by ~10^6 devices per chip, will in time give way to ultralarge-scale integration (**ULSI**). By the year 2000, digital MOS ICs will dominate a market projected to be almost 100 billion dollars; likewise, the broader market for electronic equipment is expected to climb at a rate in excess of 10% a year to a trillion dollars.

To accommodate the burgeoning demand for IC chips, the Si wafer from which they are derived has steadily increased in diameter from 0.75 in. (1.91 cm) in 1953 to 8 in. (20.3 cm) in 1995. This means that more than 1000 chips measuring 0.2×0.2 in.^{2} can be produced from a single wafer. And 12-in. (30.5-cm) diameter wafers are projected to be the industry standard early in the 21st Century. Larger wafers make great demands on processing technology but ultimately reduce manufacturing costs. Even so, real estate on the surface of an IC is expensive, and this accounts for the continual shrinkage in size of both bipolar and MOS transistor features, as shown in Fig. 15-1. The transistor count on MOS logic and microprocessor chips as a function of time is plotted in Fig. 15-2. An almost steady doubling of the device density every 2 years from 1970 to 1990 means that by the end of the century, a staggering 100 million or more devices per chip may be anticipated. Associated with the increased device density are shrinking features. Specifically, Fig. 15-3 reveals that the MOS channel length and gate oxide thickness have each declined by roughly an order of magnitude in the last two decades. Similarly, conductor stripe linewidths have shrunk, and will continue to do so, as demonstrated dramatically in Fig. 15-4. By the year 2000, reliability engineers will be confronted with a 0.2-μm (1 μm = 10^{-6} m) channel length, a less than 0.2-μm metal linewidth, and a 5-nm gate oxide thickness. The latter corresponds to only about 50 atoms in a row.

What implication these trends have for future reliability is not altogether certain. Although past experience is a useful initial guide, history has shown that there will be surprises.

15.2.3. Some Definitions

Each of the boldface words used in the second paragraph of this chapter to define reliability is significant and deserves further commentary. Under **probability** we include the sampling of products for testing, the statistical

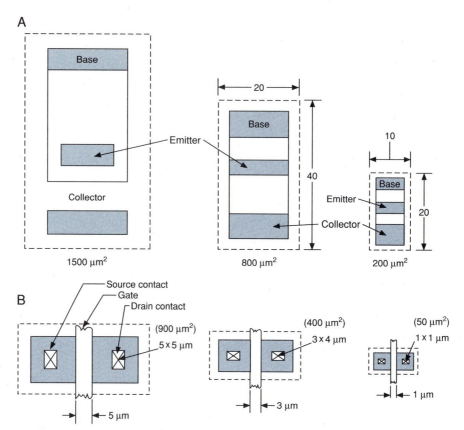

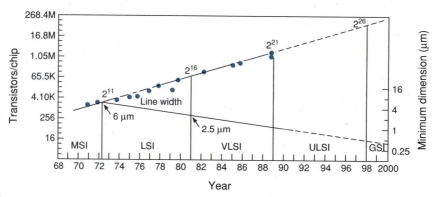

| **FIGURE 15-1** | Reduction in the surface dimensions of transistors. (A) Bipolar transistor (top view). From D. Rice, *Electronics* **53**, 137 (1979). (B) MOS transistor (top view). From E. C. Douglas, *Solid State Technology* **24**, 65 (1981). |

| **FIGURE 15-2** | Number of transistors in MOS microprocessors and custom logic as a function of the year of announcement. The transistor density per chip reflects a doubling of device packing every 2 years. After A. Reisman, *Proceedings IEEE* **71**, 550, © 1983 IEEE. |

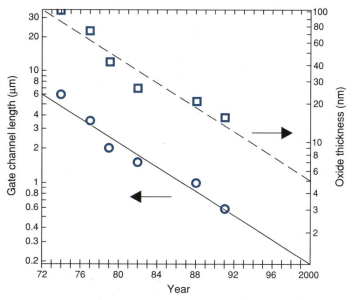

FIGURE 15-3   Shrinkage of gate channel length (○) and oxide thickness (□) as a function of time. After D. L. Crook, *Proceedings IEEE Reliability Physics Symposium,* © 1990 IEEE.

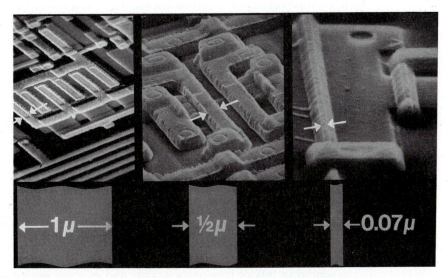

FIGURE 15-4   Aluminum stripe linewidths. *Left:* 1-μm stripes were introduced in approximately 1985. *Center:* 0.5-μm stripes were introduced in approximately 1990. *Right:* 0.07-μm stripes will be introduced beyond the year 2000. Courtesy of P. Chaudhari, IBM Corporation.

analysis of failure test results, and the projection of future behavior. Failure distributions enable some level of confidence to be achieved in predicting future behavior without exact knowledge of the underlying physical mechanism of degradation. Even though it is not feasible to predict the lifetime of an individual electronic component, an average probable lifetime can frequently be estimated.

The **time** dependence of reliability is implied by the definition, and therefore the time variable appears in all of the failure distribution functions defined subsequently. Acceptable times for reliable operation may be measured in decades for civilian communications equipment; but for guided missiles, only minutes are of concern. The time dependence of degradation of some physical parameter such as voltage, current, or device dimension is a cardinal concern.

Specification of the **conditions** of product testing or operation is essential in predicting reliability. For example, elevating the temperature is the universal way to accelerate failure. It, therefore, makes a big difference whether the test temperature is 25, −25 or 125°C. Similarly, other specifications might be the level of voltage or humidity, during testing or use.

Lastly, the question of what is meant by **failure** must be addressed. Does it mean performance degradation? If so, by how much? Does it mean that a specification is not being met? If so, whose specification? Despite these uncertainties, it is generally accepted that product failure occurs when the required function is not being performed. A key quantity that is a quantitative measure of reliability is the failure rate. This is the rate at which a device or component can be expected to fail under known use conditions. Such information is important to both manufacturers and purchasers of electronic devices. With it, engineers can usually predict the reliability that can be expected from devices in operating systems. Failure rates in semiconductor devices are generally determined by device design and issues related to processing, that is, the number of in-line process control inspections used, their levels of rejection, and the extent of postprocess screening or weeding out of weak products.

15.3. MATHEMATICS OF FAILURE AND RELIABILITY

15.3.1. Basic Statistics

A useful skill for a reliability engineer is the ability to convert a mass (mess?) of failure data into a form amenable to analysis. As an example, consider a company that makes electrical fuses for trucks that are rated at 5 A. (Fuses fail by melting when current-induced Joule heat cannot be effectively removed.) Because fuses protect other circuitry from electrical overload failure, it is vital that they function properly and not operate above the rated value. Fuses are produced in lots of tens of thousands and it is not practical to test them all and reject those that fail (or open) either above, say, 5.5 A or below 4.5 A. Either circuit damage or less than optimal operation may be the outcome of

TABLE 15-1	CURRENT (A) REQUIRED TO CAUSE FAILURE IN FUSES								
4.64	4.95	5.25	5.21	4.90	4.67	4.97	4.92	4.87	5.11
4.98	4.93	4.72	5.07	4.80	4.98	4.66	4.43	4.78	4.53
4.73	5.37	4.81	5.19	4.77	4.79	5.08	5.07	4.65	5.39
5.21	5.11	5.15	5.28	5.20	4.73	5.32	4.79	5.10	4.94
5.06	4.69	5.14	4.83	4.78	4.72	5.21	5.02	4.89	5.19
5.04	5.04	4.78	4.96	4.94	5.24	5.22	5.00	4.60	4.88
5.03	5.05	4.94	5.02	4.43	4.91	4.84	4.75	4.88	4.79
5.46	5.12	5.12	4.85	5.05	5.26	5.01	4.64	4.86	4.73
5.01	4.94	5.02	5.16	4.88	5.10	4.80	5.10	5.20	5.11
4.77	4.58	5.18	5.03	5.10	4.67	5.21	4.73	4.88	4.80

Data and problem from P. A. Tobias and D. C. Trindade, *Applied Reliability,* Van Nostrand–Reinhold, New York (1986).

operating currents beyond these extremes. To determine that fuses operate within specifications without excessive variability, a random population of fuses is chosen and tested by increasing the current until they fail or open circuit. The failure currents measured for 100 fuses are entered in Table 15-1, and it is clear that although values cluster about 5 A, there is a spread that warrants further statistical analysis.

To proceed it is convenient first to create a failure frequency distribution, as indicated in Table 15-2. For this purpose the range represented by minimum and maximum current values is arbitrarily subdivided into a number of cells (11 in this example) and the number of fuses destroyed in each current interval noted. The results are best represented in the form of a histogram shown in Fig. 15-5, where the ordinate is the number (or percentage) of failures. As the

TABLE 15-2	FREQUENCY TABLE FOR FUSE FAILURE DATA

Interval current range (A)	Number of failures in interval
4.395–4.495	2
4.495–4.595	2
4.595–4.695	8
4.695–4.795	15
4.795–4.895	14
4.895–4.995	13
4.995–5.095	16
5.095–5.195	15
5.195–5.295	11
5.295–5.395	3
5.395–5.495	1
	100

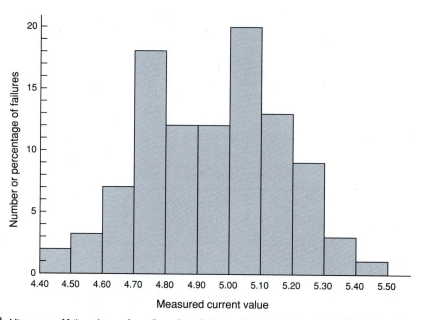

FIGURE 15-5   Histogram of failure data on fuses. From P. A. Tobias and D. C. Trindade, *Applied Reliability*, Van Nostrand Reinhold, New York (1986).

histogram is approximately symmetrical about 5 A with only a ±5% spread, the distribution of failures is deemed acceptable.

No discussion about statistics would be complete without introduction of the **normal** or **gaussian** distribution. The **probability distribution function** (PDF) that defines it is $f(x)$, and it has the familiar mathematical form

$$f(x) = [1/\sigma(2\pi)^{1/2}] \exp[-(x - \mu)^2/2\sigma^2], \tag{15-1}$$

where the quantities σ and μ are defined in the bell-shaped curve of unit area (Fig. 15-6A). It is evident that μ is the mean value of the distribution and σ is a measure of the spread of values about the mean. This function could be applied to the fuse current data, where x represents the current, $\mu = 5$ A, and σ (left to the reader as an exercise) can be evaluated by curve fitting.

Equation 15-1 is widely employed in industry to monitor the extent of process and production quality control. For example, "Six sigma in everything we do by 1992" was a battle cry used at the Motorola Corporation to increase competitiveness by enhancing the quality of products. Whereas 68.26% of the area under the normal distribution or product distribution curve falls within 1σ ($-\sigma$ to $+\sigma$), and 99.7% within 3σ, 99.9999998% falls within 6σ (-6σ to $+6\sigma$). Manufacturing processes, however, experience shifts in the mean as great as $\pm1.5\sigma$ in practice because equipment, operators, and environmental conditions are never constant. Six sigma quality is therefore defined as the number of defects that occur when such shifts happen (Fig. 15-6B). This, in effect means tolerating no more than 3.4 defective parts in one million (3.4 ppm). Of course, achieving six sigma quality is easier said than done but it remains a vital challenge for industry. The mathematically inclined can calculate what seven sigma implies!

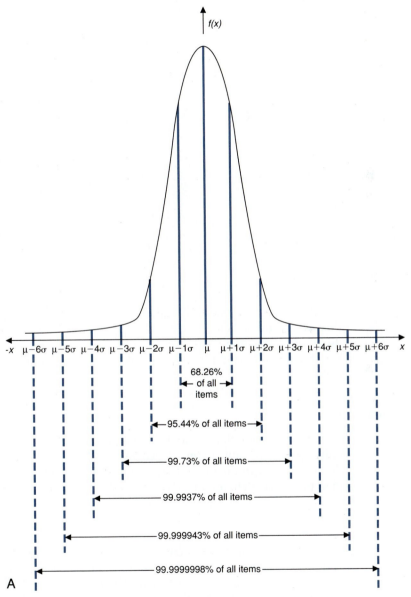

FIGURE 15-6 (A) Gaussian distribution curve. (B) Six sigma capability. After P. E. Fieler, *IEEE Tutorial Notes, International Reliability Physics Symposium,* © 1990 IEEE.

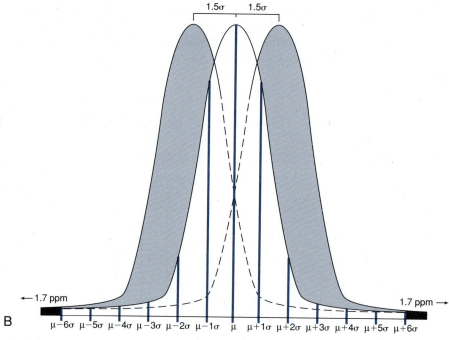

FIGURE 15-6   Continued.

It is common to plot failure data in another way. If the number of failures that occur at less than or equal to a given value is noted, then a **cumulative distribution function** (CDF) can be defined. Such cumulative distribution numbers for the indicated fuse current values are listed in Table 15-3 and plotted in Fig. 15-7.

15.3.2. Failure Rates

The fuse current data treated above were essentially time independent. Reliability, however, implies time dependence of failure, with time reckoned from the instant the product is put into service. To include time (t), we first define PDF to be $f(t)$. Therefore, CDF, defined by $F(t)$, is given by the integral

$$F(t) = \int_{-\infty}^{t} f(z)\, dz, \tag{15-2}$$

where the lower time limit may be taken to be $-\infty$ for mathematical purposes, and z is a dummy variable of integration. For a given population, $F(t)$ is the fraction that fails up to time t. A reliability (or survival) function, $R(t)$, can then be defined to indicate the surviving fraction where $F(t) + R(t) = 1$. The **failure rate** (also known as the **hazard rate**), $h(t)$, is defined as the number of

TABLE 15-3	CUMULATIVE FREQUENCY FUNCTION FOR FAILURE

Upper value in interval	Cumulative number of failures
4.495	2
4.595	4
4.695	12
4.795	27
4.895	41
4.995	54
5.095	70
5.195	85
5.295	96
5.395	99
5.495	100

products (e.g., devices) that failed between t and $t + \Delta t$, per time increment Δt, as a fraction of those that survived to time t. We write this complex definition of the failure rate as

$$h(t) = [F(t + \Delta t) - F(t)]/\Delta t [1 - F(t)]$$
$$= f(t)/[1 - F(t)] = f(t)/R(t), \qquad (15\text{-}3)$$

employing the definition of $F(t)$ and its derivative.

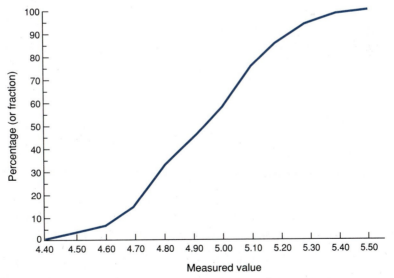

FIGURE 15-7	Cumulative distribution function for the fuse data. From P. A. Tobias and D. C. Trindade, *Applied Reliability*, Van Nostrand Reinhold, New York (1986).

A good way to appreciate how failure rates are calculated is through a couple of examples.

Suppose that a population of 525,477 solid-state devices is heated to an elevated temperature and the number of device failures as a function of time is recorded in the accompanying table. (Devices are frequently "burned in" prior to use to weed out the weak ones through temperature acceleration of failure rates.) Determine the failure rates, $h(t)$, at the indicated times.

Time (h)	Number failed	Cumulative failures	$h(t)$	Number of device hours	FITs
1	6253	6253	0.0119	525,477 × 1	1.19×10^7
2	1034	7287	0.00199	519,224 × 2	9.96×10^5
3	617	7904	0.00119	518,190 × 3	3.97×10^5
4	419	8323	0.000810	517,573 × 4	2.02×10^5
5	502	8825	0.000971	517,154 × 5	1.94×10^5
6	401	9226	0.000777	516,652 × 6	1.29×10^5
7	297	9523	0.000577	516,252 × 7	8.22×10^4
8	214	9737	0.000415	515,954 × 8	5.18×10^4
9	206	9943	0.000400	515,740 × 9	4.44×10^4
10	193	10136	0.000375	515,534 × 10	3.74×10^4

Sample calculation at 5 hours: $h(t) = 502/(525,477 - 8323) = 0.000971$.

Because the failure rates are very low, a new numerically larger unit, the **FIT**, has been defined and is widely used in the microelectronics industry. The FIT, which stands for *failure in time*, equals the number of failures in 10^9 device-hours. For example, 1 FIT is 1 failure in 10^7 devices after 100 hours of operation or 1 failure in 10^6 devices after 1000 hours, and so on. Employing this definition, failure rates in terms of FITs were generated and the values are displayed in the last column of the above table. After 5 hours, for example, $h(t) = [502/(517,154 \times 5)] \times 10^9 = 1.94 \times 10^5$ FITs.

15.3.3. Exponential, Lognormal, and Weibull Distributions

We have already introduced the normal distribution function whose PDF was given by Eq. 15-1. Three additional probability distribution functions are widely used to characterize failure and reliability in material systems. Only the exponential distribution is dealt with through subsequent illustration, that is, in Example 15-2. Perhaps with the exception of the lognormal, the mathematical forms of all of these distribution functions are not difficult and are presented for completeness.

1. *Exponential*

$$f(t) = \lambda \exp(-\lambda t), \qquad \lambda \text{ is a constant} \tag{15-4a}$$
$$F(t) = 1 - \exp(-\lambda t) \tag{15-4b}$$

2. *Lognormal*

$$f(t) = \frac{1}{t\sigma\sqrt{2\pi}} \exp\left\{ -\frac{(\ln t - \mu)^2}{2\sigma^2} \right\} \tag{15-5a}$$

$$F(t) = \Phi[\sigma^{-1}\ln(t/\mu)], \qquad \text{where } \Phi(z) = 1 + \tfrac{1}{2}\text{Erfc}(z/2^{1/2}) \tag{15-5b}$$

The complementary error function (Erfc) is related to the integral of the gaussian function, and was introduced in Chapter 6 in connection with the solution to diffusion problems. Comparison with Eq. 15-1 reveals that the lognormal distribution is largely derived from the normal distribution by substituting $\ln t$ for t. Furthermore, μ has the physical significance of representing the median time (or time when 50% of the distribution will fail, i.e., $\mu = t_{50}$). As an example, failure data on semiconductor lasers are plotted in lognormal fashion in Fig. 15-8.

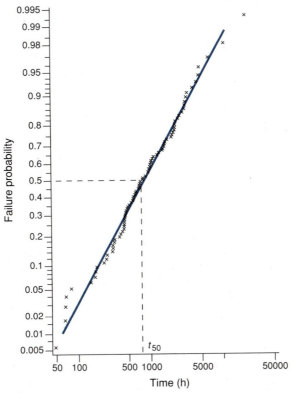

FIGURE 15-8 Lognormal distribution plot for failures of GaAs–AlGaAs lasers aged at 70°C. From F. R. Nash, *Estimating Device Reliability: Assessment of Credibility*, Kluwer Academic, Boston (1993).

3. Weibull

$$f(t) = \left[\frac{m}{t} \left(\frac{t}{c} \right)^m \right] \exp(-t/c)^m \qquad (m, c \text{ are constants} > 0) \qquad (15\text{-}6a)$$

$$F(t) = 1 - \exp(-t/c)^m. \qquad (15\text{-}6b)$$

Time dependent dielectric breakdown effects in SiO_2 are plotted in Weibull form in Fig. 15-9.

Which distribution to use for a given set of data is an important question in reliability modeling. Lognormal and Weibull distributions outwardly fit most sets of data equally well, as shown in Figs. 15-10A,B for the case of lightbulb wearout failures. Closer examination reveals that the Weibull distribution provides a better fit to short-time failures, whereas the lognormal plot is better at predicting longer lifetimes.

Lognormal distributions tend to apply when degradation occurs over time because of diffusion effects, corrosion processes, and chemical reactions. Weibull distributions appear to be applicable in cases where the weak link, or the first of many flaws, propagates to failure. Dielectric breakdown, capacitor failures, and fracture in ceramics are typically described by Weibull distributions. Readers with excellent memories will recall that the time dependence of $F(t)$ in the Weibull distribution is mathematically identical in form to that of

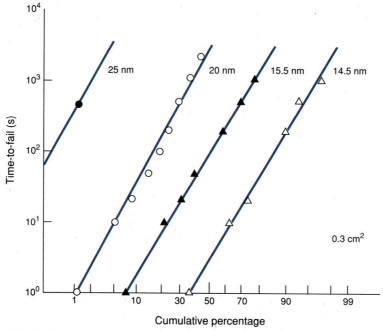

FIGURE 15-9 Weibull distribution of dielectric breakdown of SiO_2 films as a function of film thickness at room temperature. Stress field = 8 MV/cm. Courtesy of A.M-R Lin, AT&T Bell Laboratories.

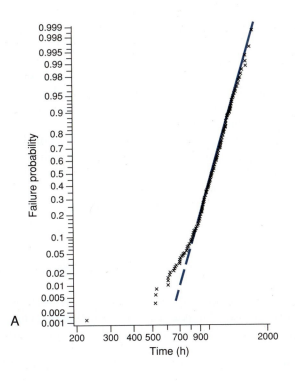

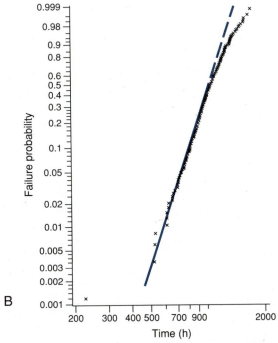

FIGURE 15-10 (A) Lognormal failure probability plot of lightbulb lifetimes. (B) Weibull failure probability plot of lightbulb lifetimes. From F. R. Nash, *Estimating Device Reliability: Assessment of Credibility*, Kluwer Academic, Boston (1993).

the fractional amount of transformation, f, in the Avrami equation (Eq. 6-22). Is this a coincidence or is there some underlying physical connection? Though independently derived, the similarity suggests that degradation, in some instances, may involve time-dependent **nucleation** of damage and subsequent **growth** until failure occurs.

Plotting failure data that fit lognormal and Weibull distributions is facilitated with the use of special lognormal or Weibull graph paper because linear plots result. It is when predictions must be projected beyond the range of recorded failure data that choice of the distribution function becomes important. For these purposes a more advanced knowledge of statistics and probability theory is required.

EXAMPLE 15-2

Suppose the lifetime population of devices is described by the exponential distribution whose CDF is $F(t) = 1 - \exp(-0.0001t)$.

a. What is the probability that a new device will fail after 1000 hours? After 5000 hours? Between these times?

b. What is the failure rate at 1000 hours? After 5000 hours?

ANSWER   a. By substitution, $F(t) = 1 - \exp[-0.0001(1000)] = 0.095$. Similarly, at 5000 hours, $F(t) = 0.39$. The probability of device failure between 1000 and 5000 hours is $F(5000) - F(1000) = 0.39 - 0.095 = 0.30$.

b.

$f(t) = dF(t)/dt = 0.0001 \exp(-0.0001t).$

$h(t) = f(t)/1 - F(t) = 0.0001 \exp(-0.0001t)/\exp(-0.0001t) = 0.0001.$

For this distribution the failure rate is seen to be constant or independent of time.

15.3.4. The Bathtub Curve

The bathtub curve, named for its shape and shown in Fig. 15-11, is perhaps the most famous graphical representation in the field of reliability. Plotted is the failure rate, $h(t)$, versus time. The resulting curve describes not only the behavior of engineering components, but also the lifetimes of human populations.

1. The first part of the curve is known as the *early failure* or "infant mortality" period. It is characterized by a *decreasing failure rate*. During this period the weak or marginally functional members of the population fail. The widely employed practice of screening out obviously defective components, as well as weak ones with a high potential for failure, is based on this portion of the curve. In screening processes, products must survive the ordeal of some sort

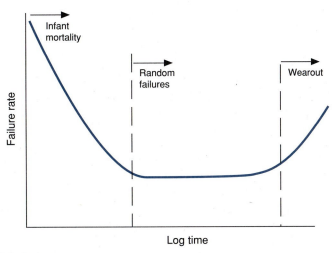

FIGURE 15-11 Bathtub curve.

of initial stressing (e.g., burn-in at high temperature, application of electrical overstress, temperature cycling). Particulates and contamination during the processing of IC chips are leading causes of early failures.

2. Next is a long, roughly flat portion known as the **intrinsic failure** period. Failures occur randomly in this region and the *failure rate is approximately constant*. Most of the useful life of a component is spent here, so that much reliability testing is conducted to determine values of $h(t)$ in this region.

3. Finally, there is the **wearout** failure regime. Here, components degrade at an accelerated pace so the *failure rate increases* in this region. One of the important goals in processing and manufacturing is to extend the life of components prior to the onset of wearout. Remarkable strides have been made in this direction. Actually, many devices do not readily fail, but rather become obsolete as a result of design changes and introduction of new technology.

It should be noted that although experimental failure data often reflect the entire shape of the bathtub curve, this is not true of the mathematical distribution functions. These distributions fit only one or, at most, two regions of the curve reasonably well.

15.4. FAILURE MECHANISMS

15.4.1. Definitions and Categories

We now turn our attention to the subject of actual physical failure of devices, a concern that occupies the remainder of the chapter. At the outset a distinction between failure **mode** and failure **mechanism** should be appreciated, even though the terms are sometimes used interchangeably. A failure mode is the recognizable electrical effect that indicates a failure has occurred. Thus a short or open circuit and an electrically open device input are symptomatic of electrical failure. Each mode could, in principle, be caused by one or more different failure mechanisms, however. Failure mechanisms are the specific microscopic physical, chemical, metallurgical, and environmental phenomena or processes that are directly responsible for device degradation or malfunction. For example, open circuiting could occur because of corrosion in an interconnect, a wire bond that detached from the bonding pad, or because excessive current initiated a fuse-like failure. High electric field dielectric breakdown in gate oxides is another example of a failure mechanism.

Semiconductor device failure mechanisms have been categorized into three main types: intrinsic failures, extrinsic failures, and electrical stress (in-circuit) failures.

1. **Intrinsic** failure mechanisms are operative within the silicon chip or die. The sources that cause such failures may be defects in the silicon wafer (e.g., dislocations, stacking faults) or defects introduced during subsequent processing. Included are failure mechanisms in grown or deposited insulators, metallizations, contacts to devices, and so on. Such failures are associated with the "front end" or chip processing stage of manufacturing.

2. **Extrinsic** failures are identified with the interconnection and packaging stages of manufacturing (i.e., the "back end").

3. **Electrical stress** failures are event dependent, and caused largely by over-stressing and excessive static charge generated during use by the consumer. They are often related to poor initial design of circuits or equipment, and to mishandling of components.

15.4.2. Failure Locations on the Integrated Circuit

The failure mechanisms are briefly surveyed below in terms of the critical regions of the IC affected. A more extensive treatment of several of these mechanisms is offered in Section 15.5. Some failures are exposed during testing of processed chips and cause rejects that reduce the **yield** of shippable products. Others pose true reliability problems that will surface in time during service.

15.4.2.1. Substrate

The most obvious failure in the silicon substrate is brittle cracking induced by scribing and separating dies, or by excessive stress produced in wire bonding and packaging operations. Thermal expansion differences and the stresses they produce are often the cause of such failure. Other sources of substrate failures include crystallographic defects (dislocations, stacking faults, oxygen precipitates), in both the Si wafer and subsequently grown epitaxial layers. When defects laden with impurity precipitates are located at device junctions abnormally large leakage currents may flow. Large crystal defects can create low-resistance paths or "pipes," which short-circuit junctions in bipolar transistors. Another class of substrate problems has been attributed to diffusion and ion implantation effects. Included are assorted anomalous diffusion phenomena associated with specific dopant/crystal defect combinations or metal impurity (e.g., Au, Fe, Cu)/precipitate reactions in the wafer. Gross processing defects (e.g., regions which are either poorly defined lithographically, over or under-etched, or which possess residual photoresist, contaminants) are additional sources of difficulty. A final type of failure is associated with radiation (alpha particles or gamma rays), which produces current pulses that result in so-called "soft errors." Stored 1's in the memory are changed to 0's, and vice versa as a result of radiation damage. Trace amounts of radioactive elements (e.g., Pb) in the Si are frequently the culprits.

15.4.2.2. Oxide

Oxide-related failures stem from a number of sources including processing (e.g., cracks due to thermal stresses, pinholes due to photoresist and mask defects, contaminants at the substrate interface, nonuniformity in oxide thickness due to mask misalignment). There are also extrinsic and intrinsic defect sources (e.g., fixed ionic and electronic charge near the Si/SiO_2 interface, mobile alkali ions, i.e., Na^+, K^+, and bulk trapped electronic charge) that may cause time-dependent dielectric breakdown. An example of such failure is shown in Fig. 15-12, where the pinhole craters and melted oxide reveal that thermal runaway has occurred (see Section 11.7.3.3). Short of breakdown, excessive oxide charge causes instabilities in the electrical characteristics of MOS devices.

15.4.2.3. Metallization

Metallization failures include failures associated with contacts and interconnections. Solid-state diffusional reaction between Si and Al at a contact (Fig. 15-13) is an example of the former that is discussed at length in Section 15.5.4. Interconnecting metal stripes are subject to thinning, corrosion, and electromigration, a mass transport-induced type of damage that leads to void formation and open circuiting. Recently a troublesome failure mechanism known as stress voiding has surfaced in submicrometer-wide Al interconnections. It is apparently caused by residual stress in the metal film and is manifested by slitlike cracks at grain boundaries that eventually cause open circuits. Even unpowered chips are prone to such degradation. Electromigration failures are discussed again in Section 15.5.3.

15.4.2.4. Bonding

Interconnection bond failures occur at the interface between the integrated circuit chip (die) metallization and the external conducting wire leads to the outside world. Bonding failures are attributed to low fatigue and fracture

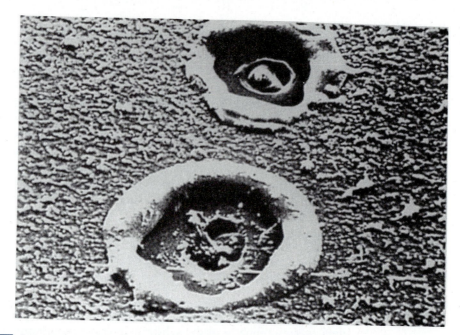

FIGURE 15-12 Scanning electron microscope image of breakdown in gates of MOS devices. The oxide thickness is 40 nm and the crater diameter is ~3 μm. From D. R. Wolters and J. J. van der Schoot, *Philips Journal of Research* **40**, 115 (1985).

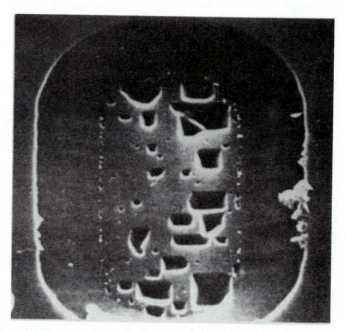

FIGURE 15-13 Alloy penetration pits formed by interdiffusion of aluminum and silicon. The contact reaction occurred during sintering at 400°C for 1 hour. From C. M. Bailey and D. H. Hensler, AT&T Bell Laboratories.

strength of the bond joint due to improper bonding pressure and temperature, contamination by impurities (e.g., metal, oxide, residual photoresist, and etchant) and the formation of intermetallic compounds. The latter, discussed further in Section 15.5.4, are frequently brittle, surrounded by voids, and subject the joint to stress. A collection of bonding failures is shown in Fig. 15-14.

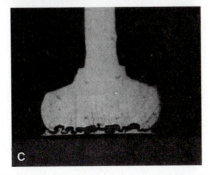

FIGURE 15-14 (A) SEM image of bond wire fracture during elevated temperature cycling. Courtesy of S. J. Kelsall, Texas Instruments. (B) SEM image of bond lifted from substrate contact pad. Courtesy of T. M. Moore, Texas Instruments. (C) Compound and void formation at the interface between a gold–aluminum bond. From R. J. Gale, *22nd Annual Proceedings Reliability Physics Symposium* (1984).

15.4.2.5. Packaging

Packaging failures are identified with the encapsulants that seal chips within packages and with the attachment system employed to connect them to printed circuit boards. Injection molding of the polymer encapsulants not infrequently results in lifted or broken bonds (Fig. 15-14B), and trapped air pockets. Poor package hermiticity also leads to ingress of metallic impurities and ionic species (Na^+, K^+, Cl^-), which, in combination with water, accelerate interconnect corrosion. Trapped moisture is also the cause of "popcorn" failure of plastic packages shown schematically in Fig. 15-15. Water vapor pressure buildup as a result of heat generated during subsequent soldering caused package cracking.

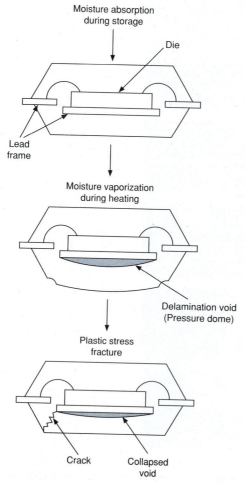

FIGURE 15-15   Sequence of "popcorn" cracking when soldering an IC package containing previously absorbed moisture. After R. C. Blish and J. T. McCullen, Intel Corporation.

Rather than being encased in a plastic package one at a time, many dies are now soldered directly to large ceramic (e.g., Al_2O_3) or even Si substrates to form multichip modules (MCMs).

Soldering, the universally employed means of joining packages or chips to printed circuit boards or MCMs, is an important source of reliability problems. These include poor wetting and adhesion, mechanical deformation and failure of joints, current-induced filament formation, and loss of joint integrity caused by vibration of the electrical enclosures. Solder problems often arise from the Joule heat liberated in IC chips. For example, the Pentium microprocessor chip generates approximately 15 watts in a rather small volume. Thermal mismatch stresses that arise from temperature and thermal expansion differences between the chip and the substrate often cause the weak, low melting solders (e.g., Pb–Sn) to creep, fatigue, and even fracture.

Some of the more prevalent failure mechanisms, and the probable factors that activate them, are presented in Table 15-4.

15.4.3. Electrical Overstress and Electrostatic Discharge

Two additional failure mechanisms of great importance are **electrical overstress (EOS)** and **electrostatic discharge (ESD)**. Charge accumulation and flow

TABLE 15-4   **TIME-DEPENDENT FAILURE MECHANISMS IN SILICON DEVICES**

Device location	Failure mechanism	Relevant factors[a]	Accelerating variables	Activation energy, E (eV)
Silicon dioxide and Si–SiO_2 interface	Surface charge accumulation	Mobile ions, V, T	T	1.0–1.05 (depends on ion density)
	Dielectric breakdown	$\mathscr{E}$, T	$\mathscr{E}$, T	0.2–1.0
	Charge injection	$\mathscr{E}$, T	$\mathscr{E}$, T	1.3 (slow trapping)
		Si/SiO_2 charge		1.0 (hot electron injection)
Metallization	Electromigration	T, J, gradients of T and J, grain size	T, J	0.5–1.2
	Corrosion (chemical, galvanic)	Contamination, H, V, T	H, V, T	0.3–1.1 (strong H effect)
	Contact degradation	T, metals, impurities	T	
Bonds and mechanical interfaces	Intermetallic growth	T, impurities, bond strength, stress	T	1.0–1.05 in Al–Au
	Fatigue	Bond strength, temperature cycling	T, thermal stresses	
IC package	Lack of hermiticity, leaks	Pressure differential, atmosphere	Pressure, H	

[a] V is voltage, $\mathscr{E}$ is electric field, T is temperature, J is current density, and H is humidity.
From W. J. Bertram, in *VLSI Technology* (S. M. Sze, Ed.), 2nd ed. McGraw–Hill, New York (1988).

are basic causes of EOS and ESD, but the source of the charge is generally different in each mechanism. In EOS large currents often flow through devices because of excessive applied fields that arise from poor initial circuit design, mishandling, or voltage pulses. Joule heating and accompanying local hot spots at junctions are some of the features associated with EOS (see Fig. 15-16).

In ESD, static charge buildup as a result of triboelectric (rubbing) effects can generate potentials ranging from tens of volts up to 30 kV! In a humid environment charge leaks away and the tribopotentials are 10 to 100 times smaller. The ordering of triboelectric effects in different materials is illustrated in Table 15-5. When two materials in the series are rubbed together, the one that is higher in the series acquires a positive potential relative to the lower one. Resulting discharge currents cause indiscriminate damage, i.e., failures are not specific to any particular component or device. Oxides, interconnections, contacts, and device junctions are all vulnerable.

15.4.4. Failure Frequency

Over the years failure mechanisms have been ranked according to frequency of occurrence. As expected, such rankings depend on the type of device or circuit involved, the use to which these are put, the operating conditions, and,

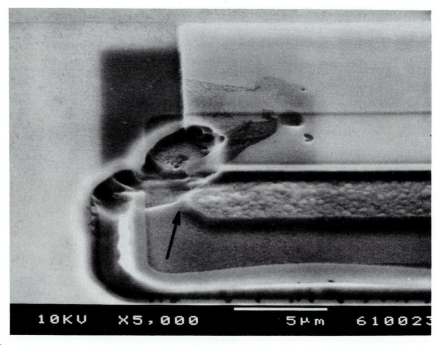

FIGURE 15-16 Scanning electron microscope image of damage to protection diode device during electrical overstress testing. Courtesy of M. C. Jon, AT&T Bell Laboratories.

importantly, the specific manufacturer. A very rough guide to the incidence of failures (Table 15-6) suggests that oxide, metallization, interconnection, packaging, and EOS/ESD malfunctions are the most troublesome. Similar failure frequencies arise in every generation of devices; if lessons are not learned then we are doomed to repeat mistakes. But lessons are hard to learn when percentages reported for the same failure mechanism sometimes differ among investigators and with the test method employed.

| **TABLE 15-5** | **ELECTROSTATIC TRIBOELECTRIC SERIES** |

Most positive (+)
    Air
    Human skin
    Asbestos
    Fur (rabbit)
    Glass
    Mica
    Human hair
    Nylon
    Wool
    Silk
    Aluminum
    Paper
    Cotton
    Steel
    Wood
    Sealing wax
    Hard rubber
    Nickel, copper
    Brass, silver
    Gold, platinum
    Acetate fiber (rayon)
    Polyester (Mylar)
    Celluloid
    Polystyrene (styrofoam)
    Polyurethane (foam)
    Polyethylene
    Polypropylene
    Polyvinyl chloride
    Silicon
    Teflon
    Silicone rubber
Most negative (−)

From R. Y. Moss, Caution—Electrostatic discharge at work, *IEEE Transactions on Components, Hybrids and Manufacture Technology 5*, 512, © 1982 IEEE.

| TABLE 15-6 | **PERCENTAGE FREQUENCY OF FAILURE IN DIFFERENT IC CIRCUITS** |

Failure allocation	USAF[a] TTL	BT MOS	Bell TTL	Bell CMOS	Bell MOS	RAC LSI	Telettra CMOS
Substrate	4	—	—	—	—	1	1
Oxide	27	10	20	1	75	11	25
Metallization	4	63	30	34	—	24	17
Interconnections	9	1	37	5	7	8	2
Package	15	—	—	—	—	3	1
EOS/ESD	24	—	4	60	17	11	43
Not identified	17	26	9	—	1	42	11

[a] USAF, U.S. Air Force; RAC, Rome Air Command; BT, British Telecom; Bell, AT&T Bell Laboratories.

15.5. SPECIFIC EXAMPLES OF FAILURE MECHANISMS

15.5.1. Dielectric Breakdown

The very thin gate oxides (now approaching ~10 nm in thickness) of field effect transistors have long been a reliability concern. Mechanisms of dielectric breakdown are complex; a simple model for the progression of events was presented in Section 11.7.3.3. Much failure testing of the breakdown in high-quality SiO_2 films has revealed an intrinsic breakdown voltage of about 1.1×10^7 V/cm, as shown in Fig. 15-17. This corresponds to a voltage drop of 0.11 V across a typical ion 0.1 nm in size. The energy associated with this potential is small compared with that required to cause valence-to-conduction band transitions.

From a reliability standpoint, time-dependent dielectric breakdown failures are particularly troublesome. After applied voltages are sustained for a period, breakdown occurs. This type of delayed failure is reminiscent of the phenomenon of static fatigue in glass (see Section 10.5.1). The results of dielectric breakdown testing, depicted in the Weibull plot of Fig. 15-9, reveal that thin oxides present a greater failure risk. Although not shown, the mean time to failure (MTF) is also thermally activated and, when combined with the displayed voltage dependence, the useful equation

$$MTF = K_1[\exp(0.33 \text{ eV}/kT)]\exp(-2.47 \text{ V}) \qquad (15\text{-}7)$$

emerges. The constant K_1 is inversely proportional to electrode area.

15.5.2. Hot Electron Transport

Another SiO_2 gate-related problem actually initiates in the underlying silicon of field effect transistors. To achieve high speed, the channel length of n-MOS

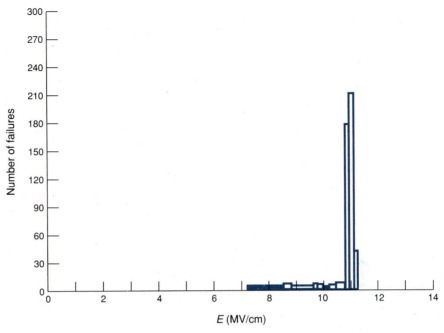

FIGURE 15-17 Histogram of the frequency of dielectric breakdown events in gate oxides as a function of voltage. From D. R. Wolters and J. J. van der Schoot, *Philips Journal of Research* **40**, 115 (1985).

devices is made as small as possible, resulting in very high electric fields. Electrons flowing across the channel between source and drain are accelerated to the point where they acquire energies in excess of normal levels for free electrons in Si; they become **hot electrons**. Upon scattering from Si atoms, some electrons acquire a transverse trajectory. They are injected upward into the oxide near the drain where they are trapped, as schematically indicated in Fig. 15-18. Hot electrons can produce transistor instabilities, which, interestingly, are larger at lower rather than at higher temperatures. Better design of channel dimensions and use of lower doping levels are helpful in eliminating hot electron effects.

15.5.3. Electromigration

Imagine an electrical wire carrying current I. As long as I is not too large, the conducting wire will last indefinitely. But when, as we have seen earlier, the current approaches a critical value, the probability increases that it will fail like a fuse. Most metal wires will melt by Joule heating when the **current density**, J, or current per unit area, A_0 ($J = I/A_0$), reaches a value of about 10^4 A/cm^2. In early integrated circuits the aluminum thin-film interconnections (~ 5 μm wide and ~ 1 μm thick) failed all too frequently by open circuiting,

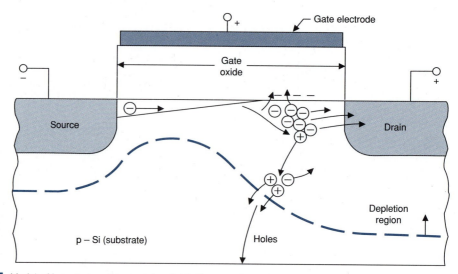

FIGURE 15-18   Model of hot electron transport in a field effect transistor.

posing a serious reliability problem. It was a case of million dollar computers protecting 10 cent fuses by failing first! The source of failure was puzzling until it was realized that J was often $\sim 10^6$ A/cm². Immediate fuselike failure did not occur because the very thin metal interconnections were essentially bonded to the much more massive Si wafer, which conducted the damaging Joule heat away. Now, some 30 years later, about half of a typical IC chip surface area is crisscrossed by a myriad of microscopic interconnections that resemble a city roadmap. Despite design changes suggesting a maximum of $J = \sim 10^5$ A/cm², similar high-current density degradation still poses a reliability concern.

Closer examination of failed conducting stripes reveals an assortment of damage manifestations such as voids, hill-like (hillock) growths, whiskers, thinned sections, and grain boundary cracks, some of which are shown in Fig. 15-19. The collection of mass transport phenomena leading to morphological degradation is known as **electromigration** because **electric**al current drives atom **migration**. At an atomic level the electrons streaming to the positive contact (anode) are believed to impart momentum to atoms, as illustrated in Fig. 15-20A. Under electron impact, atoms at the lattice saddlepoint configuration are pushed over the energy barrier, causing them to exchange places with vacancies in a directed manner. This basic current-assisted diffusion step is then repeated over and over again, usually within grain boundaries. Eventually, macroscopic depletion of atoms (i.e., voids) occurs in some regions with a corresponding accumulation of matter (i.e., hillocks) elsewhere (Fig. 15-20B). Thinned regions pass higher current densities and become hotter, accelerating electrotransport even more; failure by open circuiting then becomes inevitable.

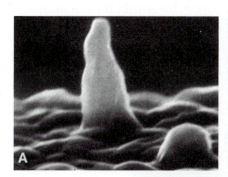
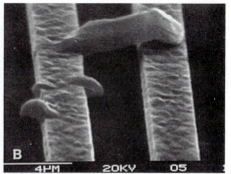
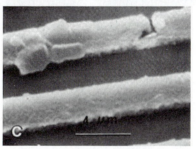

FIGURE 15-19 Manifestations of electromigration damage in Al thin films viewed in the SEM. (A) Hillock growth. Courtesy of L. Berenbaum, IBM Corporation. (B) Whiskers bridging two conductors. Courtesy of R. Knoell. (C) Mass depletion and stripe cracking. Courtesy of S. Vaidya, AT&T Bell Laboratories.

Analysis of extensive accelerated test data accumulated in powered interconnections has revealed a general relationship between mean time to failure and J, given by

$$(\text{MTF})^{-1} = K_2[\exp(-E_e/kT)]J^n. \tag{15-8}$$

In this widely used formula E_e is the activation energy for electromigration damage and it has a value close to that for grain boundary diffusion. In addition, K_2 is a constant, and so is n with a typical value between 2 and 3. Small additions of copper to the universally employed aluminum metallizations increase resistance to electromigration by impeding transport of aluminum atoms and raising the value of E_e. Electromigration failure times are usually modeled by lognormal distributions.

EXAMPLE 15-3

During accelerated testing at 150°C, conducting stripes failed in 100 hours when passing 3×10^6 A/cm². It is desired that the lifetime of the same stripes be 2 years when carrying 4×10^5 A/cm². If the activation energy for electromigration is 0.6 eV and $n = 2$, what is the maximum operating temperature?

ANSWER   Using Eq. 15-8 the following ratio is obtained, where a refers to 2 years and b to 100 hours, and $T = 150°C$ (423 K):

$$\frac{[\mathrm{MTF}(a)]^{-1}}{[\mathrm{MTF}(b)]^{-1}} = \frac{[\exp(-E_e/kT_a)]\, J_a^n}{[\exp(-E_e/kT_b)]\, J_b^n}.$$

In 2 years there are $2 \times 24 \times 365 = 17{,}520$ hours. Substitution yields

$$\frac{[17{,}520]^{-1}}{[100]^{-1}} = \frac{[\exp(-0.6/8.63 \times 10^{-5}\, T_a)]\,(4 \times 10^5)^2}{[\exp(-0.6/8.63 \times 10^{-5} \times 423)]\,(3 \times 10^6)^2}.$$

Solving, $\exp(-0.6/8.63 \times 10^{-5}\, T_a) = 2.34 \times 10^{-8}$ and, therefore, $T_a = 395$ K or 122°C.

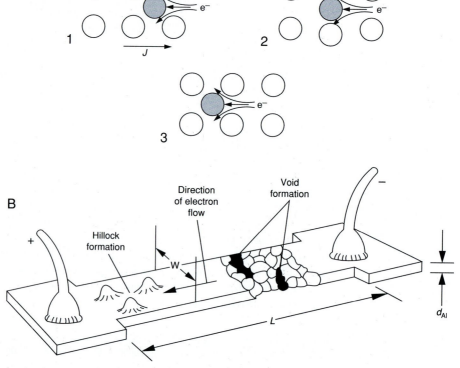

FIGURE 15-20   (A) Atomic model of electromigration involving electron momentum transfer to metal ion cores during current flow. (B) Model of void and hillock damage along a powered polycrystalline conductor. Typical stripe dimensions are $L = 10\ \mu\mathrm{m}$, $w = 1\ \mu\mathrm{m}$, and $d_{\mathrm{Al}} = 1\ \mu\mathrm{m}$.

15.5.4. Interdiffusion at Contacts

Many reliability problems stem from interdiffusion effects at the interface between two dissimilar materials (e.g., solder joint, semiconductor contact). To better understand the nature of the reaction let us consider the Al–Si system, an important contact combination. The phase diagram for Al–Si (Fig. 15-21) indicates a eutectic reaction; some solid solubility of Si in Al that increases as the temperature is raised, reaching ~0.3 wt% at 400°C; and, importantly, *no compounds*. Imagine a diffusion couple composed of equal volumes of Si and Al bonded across a planar interface that contains a row of minute, inert markers

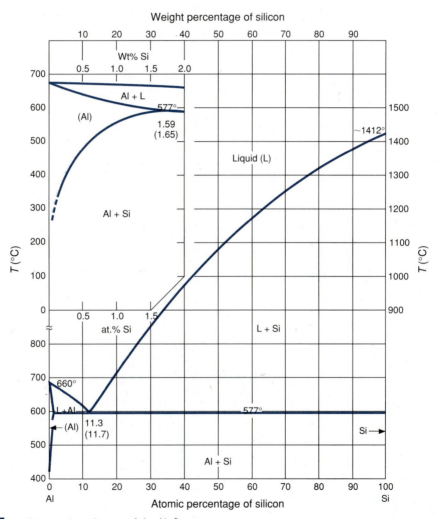

FIGURE 15-21 Equilibrium phase diagram of the Al–Si system.

(Fig. 15-22A). There is no reason to believe that Si atoms exchange with vacancies at the same rate that Al atoms do. Suppose Al–vacancy exchanges are more prevalent during diffusion. Therefore, in a given time less Si is transported to the right than Al to the left (Figs. 15-22B,C). But, if the couple suffers no volume change the marker has effectively shifted to the right (Fig. 15-22D). This phenomenon is known as the **Kirkendall effect**, and it has been observed in many binary bulk metal systems. Not only is there a chance that short circuit atomic transport will occur, but unequal vacancy diffusion fluxes tend to promote local excess vacancy concentrations. Vacancy condensation and void formation are thus possibilities. In fact, there tends to be a vacancy buildup on the Al side balanced by a corresponding depletion on the Si side. The resulting

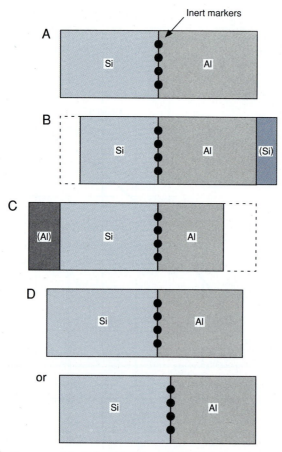

| **FIGURE 15-22** | Illustration of the Kirkendall effect in an Al–Si diffusion couple. (A) Initial geometry with inert, nondiffusing markers at the interface between Al and Si. (B) Transport of Si relative to the marker. (C) Transport of Al relative to the marker. (D) Final position of the marker. |

interdiffusion and potential porosity pose significant reliability problems, as illustrated by the next two examples.

Prior to the emergence of the VLSI era, the simple Al–Si contact was the usual way to make electrical connection to Si devices. In Fig. 15-23, a common mechanism of contact degradation is depicted. To reduce the thin layer of native SiO$_2$ on Si, the contact is heated to 400°C. During this step, Si readily diffuses into Al to satisfy its solubility requirements (Fig. 15-21), whereas Al, correspondingly, countermigrates into the Si. But very small amounts of Al dissolve in Si so that the excess precipitates in the form of conducting filaments that can "spike" or short-circuit shallow junctions (see Fig. 15-13). The solution

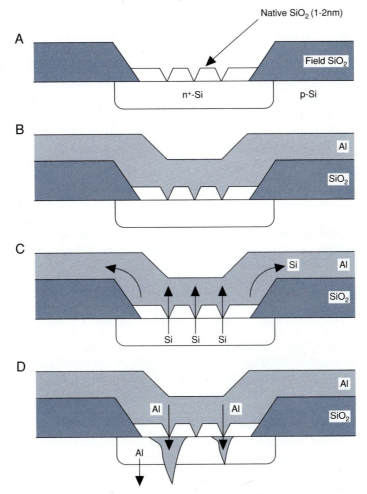

FIGURE 15-23 Schematic sequence of Al–Si interdiffusion at a contact resulting in junction spiking.

to this troublesome problem is simple, namely, to use an Al metallization saturated with Si. In such a case the driving force for interdiffusion of the two elements vanishes. About 1wt% Si is sufficient to eliminate junction spiking without adversely affecting the high conductivity of Al.

As a second example, consider degradation caused by the **purple plague**. This colorful appellation refers to the purple $AuAl_2$ (see Section 13.3.1) intermetallic phase formed when Al and Au react. The compound was known for at least 70 years before it surfaced as a reliability concern in bonding lead wires to microelectronic circuits. Common bonding configurations susceptible to purple plague formation involve fine Au wires or wedges bonded to thin films of Al on Si, and Au wire or balls bonded under heat and pressure (thermal compression) to Al films. At elevated temperatures and long times such bonds all too frequently fail by lifting, creating an open circuit (Fig. 15-14C). Bonds often lift as a result of annular microcrack formation in the compound reaction zone and local loss of Al. What is so alarming about the presence of $AuAl_2$ is the Kirkendall porosity that accompanies it. Although $AuAl_2$ alone has an acceptable electrical conductivity, its presence correlates with bond embrittlement and lack of strength.

As both phenomena discussed in this section are diffusion controlled, the rates of degradation are thermally activated and depend on temperature in the usual Arrhenius fashion.

15.5.5. Corrosion

Corrosion is a common reliability problem in IC devices because all of the necessary ingredients—metals, high electric fields, ionic contaminants (e.g., Cl^-), and moisture—are simultaneously present. In the case of aluminum metallizations, the trouble begins with ionization of the hydrated oxide due to Cl^- according to the reaction

$$Al(OH)_3 + Cl^- = Al(OH)_2Cl + OH^-. \qquad (15\text{-}9a)$$

This yields the water-soluble salt $Al(OH)_2Cl$, and once dissolution of the surface aluminum oxide occurs, the underlying Al combines directly with Cl^- to form $Al(Cl)_4^-$, especially when the Al is biased positively. Further reaction with water regenerates Cl^- to continue the corrosion process by the reaction

$$Al(Cl)_4^- + 3H_2O = Al(OH)_3 + 3H^+ + 4Cl^-. \qquad (15\text{-}9b)$$

Gold is not noted for its tendency to degrade, but electrolytic corrosion has been observed in certain alloy contact systems (e.g., Ti–Pd–Au, Ni–Cr–Au, and Ti–W–Au). The most likely reactions are

$$Au + 4Cl^- = AuCl_4^- + 3e^- \qquad \text{(at the anode)}, \qquad (15\text{-}10a)$$
$$AuCl_4^- + 3e^- = Au + 4Cl^- \qquad \text{(at the cathode)}. \qquad (15\text{-}10b)$$

Dissolution of Au at the anode is followed by migration of $AuCl_4^-$ to the cathode, resulting in dendritelike, Au metal bridges that short neighboring conductors.

As another example of corrosion, consider the packages in which IC chips are encased. The different metals employed in the lead frame, aluminum bond pad, solder, and lead wire are the source of corrosion in improperly encapsulated chips. Moisture and Cl^- ions can penetrate the epoxy-filled packages by creeping along the lead frame metal–polymer interfaces, as shown in Fig. 15-24. In addition, harmful ionic contaminants may be initially present in the polymer packaging materials and solder fluxes used.

15.5.6. Failure of Incandescent Lamps

As a change of pace, the book ends with a discussion of the failure of incandescent lightbulbs. Tungsten filament failures in lightbulbs are probably the most common and annoying reliability problem of an electrical nature that we normally deal with; happily, they are easily and cheaply dealt with. The "hot spot" theory of failure maintains that the temperature of the filament is higher in a small region because of some local inhomogeneity in the tungsten. Wire constrictions or variations in resistivity or emissivity can constitute inhomogeneities. With time, the temperature difference (ΔT) between the hot spot and the rest of the filament, at temperature T, increases. Preferential evaporation of W not only darkens the inner surface of the bulb, but thins the filament, making the hot spot hotter, resulting in yet greater evaporation. In bootstrap

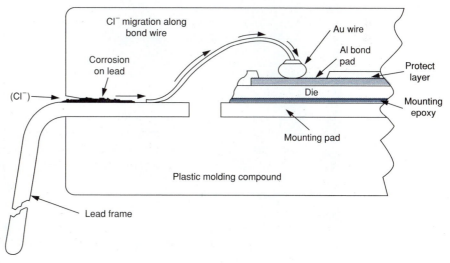

FIGURE 15-24 Chloride corrosion of gold wire at a bond in a dual-in-line plastic IC package. From L. Gallance and M. Rosenfield, *RCA Review* **45**, 256 (1984).

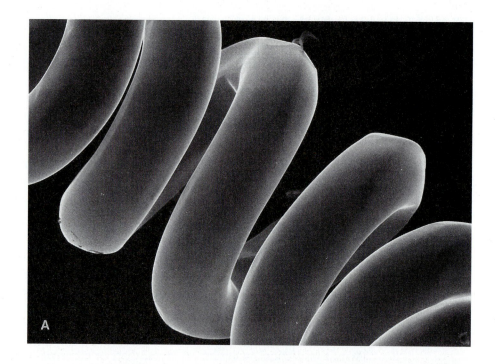

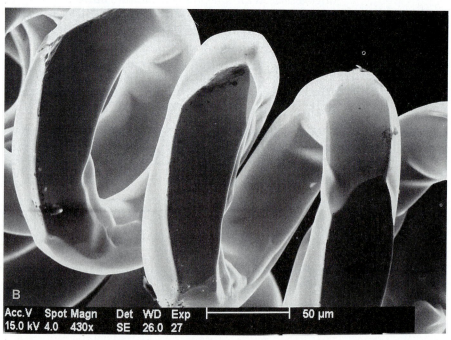

Acc.V Spot Magn   Det WD Exp
15.0 kV 4.0 430x   SE 26.0 27   50 µm

FIGURE 15-25 (A) SEM image of unheated 50-µm-diameter tungsten lamp filament. (B) Crystallographic faceting of heating filament that failed. Courtesy of R. Anderhalt, Philips Electronic Instruments Company, a Division of Philips Electronics North America Corporation.

fashion the filament eventually melts or fractures. In this model, applicable to filaments heated *in vacuo*, the mean life of the bulb (MTF) is inversely proportional to the vapor pressure of tungsten (P_W): MTF $\sim (P_W)^{-1}$. Section 5.3.1 reminds us that $P_W = P_0 \exp(-\Delta H_{vap}/RT)$, where ΔH_{vap} is the heat of vaporization for tungsten, and P_0 is a constant.

Common lightbulbs are filled with argon. The main parameter affecting bulb life is still the tungsten vapor pressure, but Ar alters the transport of W. Tungsten atoms no longer fly off in linear paths, but may swirl around by diffusion or convection in the Ar ambient. Bulb dimensions and P_W both play important roles in influencing MTF. Filaments are also coiled. When straight wires are coiled, the *effective* surface area responsible for mass loss is reduced by a factor of about 2. Although this may be expected to enhance life proportionately, other W transport mechanisms are operative. A high concentration gradient of tungsten in the coil promotes gas-phase diffusion, while severe temperature gradients foster surface diffusion along the filament. The latter is responsible for the undesirable crystallographic faceting of filament grains shown in Fig. 15-25.

Tungsten halogen lamps emit considerably more light than ordinary incandescent bulbs because the filament is operated at higher temperatures. The reason they do not fail is the chemical vapor transport of the evaporated W back to the filament. Small amounts of iodine and bromine added to the argon enables the formation of volatile WI_6 and WBr_6 compounds, which circulate and regenerate W on decomposition.

15.6. PERSPECTIVE AND CONCLUSION

It is only fitting that a book for engineering students end with subject matter dealing with the failure and reliability of electronic materials and devices. Our future ability to compete and standard of living depend largely on the quality as well as reliability of electronic materials and devices in communications, information processing, robotics and manufacturing, sensing and control, audiovisual, and other equipment. But more immediately, failure and reliability constitute the bottom line—the residual concerns that remain after design, material selection based on structure–property relationships, and processing or manufacturing are completed and the product is in service. Unlike immediately apparent failure manifestations in mechanically functional materials, electron microscopes are often needed to detect damage sites in electrically functional materials and devices. The latter provide a rich interactive arena in which semiconductors, metals, polymers, and ceramics all have significant roles, displaying assorted electrical, mechanical, and thermal degradation phenomena.

In a fundamental sense, time-dependent degradation phenomena stem from the motion of electrons, ions, and atoms initially situated at "normal" or harmless sites to locations where they alter the material properties for the worse. For example, **electron** transport is involved in dielectric breakdown, electrical overloading, and electrostatic discharge effects. **Ion** migration occurs during corrosion phenomena, penetration of moisture into electrical devices, and instabilities in gate oxides. **Atom** movements can cause harmful compositional and morphological changes during interdiffusion and compound formation between metals, during metal/semiconductor reactions at contacts, and during electromigration damage in interconnections. It is the size of features and devices that sets limits on the amount of charge and material transport that can be tolerated before reliability is compromised. And because these dimensions are already uncomfortably small compared with material transport distances, future shrinkage of device features will pose serious reliability issues or even upper bounds to device packing densities.

Almost all of the above material transport effects are **accelerated** by increasing the temperature. We should therefore appreciate the key equation governing degradation rates and failure,

$$(\text{MTF})^{-1} \approx \exp(-E/kT), \tag{15-11}$$

where MTF is the mean time to failure and E is the activation energy for the specific damage mechanism (see Table 15-4). In addition, driving forces for material transport and degradation include concentration gradients, mechanical stresses, electric fields, temperature gradients, humidity, and other forces. When sufficient failure time data are available, statistical treatment of them is essential to predict future behavior.

There is a sense of *déjà vu* with respect to reliability; similar failures arise in every generation of devices. Forecasting future failure trends in new ICs is sometimes risky because field reliability data are usually available only when the technology they refer to is already obsolete. Furthermore, advancing technology alters the failure allocation picture, sometimes in unexpected ways.

Additional Reading

E. A. Amerasekera and D. S. Campbell, *Failure Mechanisms in Semiconductor Devices,* Wiley, Chicester (1987).

Metals Handbook, Failure Analysis and Prevention, Vol. 11, 9th ed., American Society for Metals, Materials Park, OH (1986).

F. R. Nash, *Estimating Device Reliability: Assessment of Credibility,* Kluwer Academic, Boston (1993).

A. G. Sabnis, *VLSI Electronics Microstructure Science.* Vol. 22. *VLSI Reliability*, Academic Press, San Diego (1990).

QUESTIONS AND PROBLEMS

15-1. In shrinking the area dimensions of field effect transistors (FET) from 6000 to 50 μm^2, what increase in the number of devices is possible on an IC chip measuring 1 cm $\times$ 1 cm? Assume that FETs cover the entire chip.

15-2. a. If a current density of no more than 1×10^5 A/cm^2 is allowed in the year 2000, estimate the maximum current that will flow through conducting stripes 0.3 μm thick and 0.5 μm wide.
   b. If the dielectric breakdown strength of 1.1×10^7 V/cm is not to be exceeded, estimate the maximum gate voltage that will be tolerated in the year 2000.

15-3. "Degradation of materials stems from the movement of atoms and charge from harmless to harmful sites." Provide examples that support this statement.

15-4. Distinguish among exponential, lognormal, and Weibull distributions.

15-5. Derive an expression for the hazard rate for a Weibull distribution of failures.

15-6. Schematically sketch the bathtub life curve for a product if the infant mortality portion is modeled by a Weibull distribution, and the random failure period is described by an exponential distribution. The curve should be plotted as the logarithm of the hazard rate versus the logarithm of the time.

15-7. From what you know about error functions (see Chapter 6) and gaussians (see Chapter 15) prove that 68.26% of the area under the normal distribution curve lies within $\pm\sigma$.

15-8. What is the significance of six sigma quality?

15-9. The life distribution of a population of devices is given by $F(t) = 1 - \exp(-0.00007t^2)$, with t in hours.
   a. What kind of distribution is this?
   b. What is the probability that a new device will fail after 6 hours?
   c. What is the failure rate at 6 hours? After 20 hours?

15-10. What are the similarities between mechanically and electrically functional materials with regard to reliability issues? What differences are there?

15-11. a. The bathtub curve describes the lifetimes of human populations. Indicate where on this curve the various causes of death from childhood to old age (e.g., birth defects, heart disease, AIDS, cancer, Alzheimer's) belong.
   b. What effect will scaling IC chips to smaller dimensions have on the shape of the bathtub curve.

15-12. Voltage breakdown testing results on 200 capacitors yielded the indicated frequency table of failures.

Interval (MV/cm)	Frequency of failures (%)	Interval (MV/cm)	Frequency of failures (%)
0–0.5	5	4.5–5.0	6
0.5–1.0	0	5.0–5.5	1
1.0–1.5	2	5.5–6.0	5
1.5–2.0	3	6.0–6.5	8
2.0–2.5	0	6.5–7.0	12
2.5–3.0	4	7.0–7.5	25
3.0–3.5	4	7.5–8.0	13
3.5–4.0	3	8.0–8.5	6
4.0–4.5	3		100

   a. How many capacitors failed in the voltage range from 6.5 to 7.0 V?
   b. Plot the devices failed (percentage) versus the breakdown voltage (MV/cm).

   c. Calculate the cumulative distribution function for the capacitor failure data and plot the CDF data versus the breakdown voltage (MV/cm).

15-13. Suppose the failure behavior shown in Fig. 15-8 is for a group of potential lasers that are to be used in repeater stations of transoceanic light wave communications systems. Comment on their suitability for such an application if the activation energy for laser failure is 1.0 eV.

15-14. The lifetime of SiO_2 films is 200 hours during accelerated dielectric breakdown testing at 120°C and 10 V. What lifetime can be expected at 25°C and an operating voltage of 8 V?

15-15. An engineer suggests that by changing the temperature of 15.5-nm-thick SiO_2 films exposed to 8 MV/cm, they will last as long as 20-nm-thick films operating at 25°C under the same electric field. What temperature is required? Is it a practical suggestion?

15-16. Distinguish between the following pairs of terms:
   a. Electrical overstress and electrostatic discharge
   b. Purple plague and junction spiking
   c. Electromigration and diffusion
   d. Intrinsic failures and extrinsic failures
   e. Conduction electrons and hot electrons

15-17. Explain the following observations regarding electromigration:
   a. Conducting stripes with grains that vary widely in size and shape are more prone to damage than stripes containing equiaxed grains of uniform size.
   b. Conducting stripes that have a bamboo grain structure, with grain boundaries oriented normal to the stripe axis, exhibit very long lifetimes.
   c. Single-crystal metal stripes virtually never fail during electromigration.

15-18. During electromigration of a metal conducting stripe, the mean time to failure is determined to be $(MTF)^{-1} = K_2 [\exp(-E_e/kT)] J^n$, where $E_e = 1.0$ eV and $n = 3.4$.
   a. If MTF = 340 days at 100°C and $J = 2 \times 10^6$ A/cm², calculate the value of K_2.
   b. How long would conductors survive at 45°C and $J = 1 \times 10^6$ A/cm² if $E_e = 0.8$ eV?

15-19. It is decided to increase the average MTF for conductors of the previous problem by strategies that raise E_e and reduce n.
   a. Suggest a way to increase E_e.
   b. For operation at 45°C and $J = 1 \times 10^6$ A/cm², is it better to increase E_e by 10% or reduce n by 50%?
   Hint: Calculate $(MTF)^{-1} d(MTF)/dE_e$ and $(MTF)^{-1} d(MTF)/dn$.

15-20. What kinds of degradation and failure mechanisms are expected in the materials that make electrical connections between a silicon die and the metal lead frame of an IC package?

15-21. Compare the severity of corrosion damage in microelectronic and mechanically functional applications (see Chapter 10) when normalized to the same corrosion current.

15-22. Consider a diffusion couple composed of equal volumes of gold and aluminum with inert markers at the interface.
a. If the diffusivity of Al exceeds that of Au, sketch the final position of the markers after diffusion.
b. Where are voids expected to form in the diffusion zone?

15-23. It is known that the purple plague intermetallic compound $AuAl_2$ grows parabolically in time. The compound thickness x varies with time t according to $x^2 = A[\exp(-E_c/kT)]t$, where A is a constant, E_c is the activation energy for growth, and kT has the usual meaning. Limited thin-film testing has shown that $x = 410$ nm after 2 hours at 200°C and $x = 210$ nm after 2 hours at 175°C. How thick a compound layer can be expected at a contact between the metals during service at 35°C for 2500 hours?

15-24. The acceleration factor (AF) is defined as the ratio of the degradation rate at an elevated temperature to that at a low temperature.
a. Write an expression for AF if the degradation activation energy is E and the two temperatures are T_1 and T_2 ($T_2 > T_1$).
b. What is AF for corrosion ($E = 0.7$ eV) between $T = 25$ and 55°C?
c. What is AF for electromigration ($E = 0.5$ eV) between 25 and 100°C?
d. What is AF for electromigration ($E = 1.2$ eV) between 25 and 100°C?

15-25. Humidity is a factor that accelerates degradation in electronic components. Which materials are susceptible and what mechanisms are operative in causing damage?

15-26. What differences, if any, would you expect in the filament failure mechanisms of automobile tail lightbulbs compared with household lightbulbs? The former are powered by 12 V dc and the latter by 120 V ac.

15-27. Lamp filament failure rates are directly proportional to the vapor pressure of tungsten. What decrease in the mean time to failure can be expected for a 2% increase in the operating temperature of 2100°C? The heat of vaporization of W is 770 kJ/mol.

15-28. Electric fuse materials suffer catastrophic failure through melting and vaporization when the rated current is exceeded. What materials selection factors must be considered in designing a fuse.

15-29. Everyone, at one time or another, has struggled with bad electrical contacts in connections or in mechanical switches. List possible physical and chemical phenomena between contacting metals that could lead to contact degradation and loss of electrical continuity.

15-30. Mechanical stress effects and failures similar to those described in Chapter 10, but on a much smaller scale, occur in IC silicon dies, metal contacts and wires, solder, and plastic packages. Provide actual (or potential) examples of such degradation and failure.

15-31. Quite often two variables are operative and simultaneously accelerate failure of electronic materials. Mention at least two instances where this is so.

15-32. Because of limited testing, failure rates of a given device were reported to be 25,010 FITs at 130°C and 3207 FITs at 90°C. In addition, the uncertainty or error as to what constitutes actual failure is estimated to be $\pm 10\%$ of the number of FITs reported.

a. Calculate the apparent activation energy for failure based on the given information.

b. What is the error in activation energy due to the uncertainty in identifying the number of FITs?

APPENDIX A

PROPERTIES OF SELECTED ELEMENTS (AT 20°C)

Element	Symbol	Atomic number	Atomic weight (amu)	Crystal structure[a]	Lattice constant(s) [a(nm), c(nm)]	Atomic radius (nm)	Ionic radius (nm)	Valence	Density (g/cm³)
Aluminum	Al	13	26.98	FCC	0.4050	0.143	0.057	3+	2.70
Argon	Ar	18	39.95						
Arsenic	As	33	74.92	Rhomb		0.125	0.04	5+	5.72
Barium	Ba	56	137.33	BCC	0.5019	0.217	0.136	2+	3.5
Beryllium	Be	4	9.012	HCP	0.229, 0.358	0.114	0.035	2+	1.85
Boron	B	5	10.81	Rhomb		0.097	0.023	3+	2.34
Bromine	Br	35	79.90			0.119	0.196	1−	
Cadmium	Cd	48	112.41	HCP	0.298, 0.562	0.148	0.103	2+	8.65
Calcium	Ca	20	40.08	FCC	0.5582	0.197	0.106	2+	1.55
Carbon (gr)	C	6	12.011	Hex	0.246, 0.671	0.077	~0.016	4+	2.25
Cesium	Cs	55	132.91	BCC	0.6080	0.263	0.165	1+	1.87
Chlorine	Cl	17	35.45			0.107	0.181	1−	
Chromium	Cr	24	52.00	BCC	0.2885	0.125	0.064	3+	7.19

(continues)

APPENDIX A *(continued)*

Element	Symbol	Atomic number	Atomic weight (amu)	Crystal structure[a]	Lattice constant(s) [a(nm), c(nm)]	Atomic radius (nm)	Ionic radius (nm)	Valence	Density (g/cm³)
Cobalt	Co	27	58.93	HCP	0.251, 0.407	0.125	0.082	2+	8.85
Copper	Cu	29	63.55	FCC	0.3615	0.128	0.096	1+	8.94
Fluorine	F	9	19.00				0.133	1−	
Gallium	Ga	31	69.72	Ortho		0.135	0.062	3+	5.91
Germanium	Ge	32	72.59	Dia cub	0.5658	0.122	0.044	4+	5.32
Gold	Au	79	196.97	FCC	0.4079	0.144	0.137	1+	19.3
Helium	He	2	4.003						
Hydrogen	H	1	1.008				0.154	1+	
Iodine	I	53	126.91	Ortho		0.136	0.220	1−	
Iron	Fe	26	55.85	BCC	0.2866	0.124	0.087	2+	7.87
Lead	Pb	82	207.2	FCC	0.4950	0.175	0.132	2+	11.3
Lithium	Li	3	6.94	BCC	0.3509	0.152	0.078	1+	0.53
Magnesium	Mg	12	24.31	HCP	0.321, 0.521	0.160	0.078	2+	1.74
Manganese	Mn	25	54.94	Cubic	0.8914	0.112	0.091	2+	7.43
Mercury	Hg	80	200.59			0.150	0.112	2+	14.2
Molybdenum	Mo	42	95.94	BCC	0.3147	0.136	0.068	4+	10.2
Neon	Ne	10	20.18						
Nickel	Ni	28	58.69	FCC	0.3524	0.125	0.078	2+	8.9
Niobium	Nb	41	92.91	BCC	0.3307	0.143	0.069	4+	8.6
Nitrogen	N	7	14.007			0.071	0.015	5+	
Oxygen	O	8	16.00			0.060	0.132	2−	
Phosphorus	P	15	30.97	Ortho		0.109	0.035	5+	1.83
Platinum	Pt	78	195.08	FCC	0.3924	0.139	0.052	2+	21.4
Potassium	K	19	39.10	BCC	0.5344	0.231	0.133	1+	0.86
Silicon	Si	14	28.09	Dia cub	0.5431	0.117	0.039	4+	2.33
Silver	Ag	47	107.87	FCC	0.4086	0.144	0.113	1+	10.5
Sodium	Na	11	22.99	BCC	0.4291	0.186	0.098	1+	0.97
Sulfur	S	16	32.06	Ortho		0.104	0.174	2−	2.07
Tin	Sn	50	118.69	Tetra		0.158	0.074	4+	7.30
Titanium	Ti	22	47.88	HCP	0.295, 0.468	0.147	0.064	4+	4.51
Tungsten	W	74	183.85	BCC	0.3165	0.137	0.068	4+	19.3
Uranium	U	92	238.03	Ortho		0.138	0.105	4+	19.0
Vanadium	V	23	50.94	BCC	0.3028	0.132	0.061	4+	6.1
Zinc	Zn	30	65.39	HCP	0.266, 0.495	0.133	0.083	2+	7.13
Zirconium	Zr	40	91.22	HCP	0.323, 0.515	0.158	0.087	4+	6.49

[a] FCC, Face-centered cubic; Rhomb, rhombohedral; BCC, body-centered cubic; HCP, hexagonal close-packed; Hex, hexagonal; Ortho, orthorhombic; Dia cub, diamond cubic; Tetra, tetragonal.

APPENDIX B

VALUES OF SELECTED PHYSICAL CONSTANTS

Constant	Symbol	Value
Absolute zero of temperature		$-273.2°C$
Avogadro's number	N_A	6.023×10^{23}/mol
Boltzmann's constant	k	1.381×10^{-23} J/K
		8.63×10^{-5} eV/K
Bohr magneton	μ_B	9.274×10^{-24} A/m^2
Electronic charge	q	1.602×10^{-19} C
Electron mass	m_e	9.110×10^{-31} kg
Faraday's constant	F	9.649×10^4 C/mol
Gas constant	R	1.987 cal/mol-K
		8.314 J/mol-K
Gravitational acceleration	g	9.806 m/s
Permeability of vacuum	μ_0	$4\pi \times 10^{-7}$ Wb/A-m
Permittivity of vacuum	ε_0	8.854×10^{-12} F/m
Planck's constant	h	6.626×10^{-34} J-s
Velocity of light in vacuum	c	2.998×10^8 m/s

APPENDIX C

Length

$1 \text{ m} = 10^{10} \text{ Å}$	$1 \text{ Å} = 10^{-10} \text{ m} = 10^{-8} \text{ cm}$	$1 \text{ ft} = 0.305 \text{ m}$
$1 \text{ m} = 10^{9} \text{ nm}$	$1 \text{ nm} = 10^{-9} \text{ m}$	$1 \text{ in.} = 2.54 \text{ cm}$
$1 \text{ m} = 10^{6} \text{ } \mu\text{m}$	$1 \text{ } \mu\text{m} = 10^{-6} \text{ m} = 10,000 \text{ Å}$	$1 \text{ m} = 3.28 \text{ ft}$
$1 \text{ m} = 10^{3} \text{ mm}$	$1 \text{ mm} = 10^{-3} \text{ m}$	$1 \text{ cm} = 0.394 \text{ in.}$
$1 \text{ m} = 10^{2} \text{ cm}$	$1 \text{ cm} = 10^{-2} \text{ m}$	

Volume

$1 \text{ m}^3 = 10^6 \text{ cm}^3$	$1 \text{ cm}^3 = 10^{-6} \text{ m}^3$
$1 \text{ m}^3 = 35.3 \text{ ft}^3$	$1 \text{ ft}^3 = 0.0283 \text{ m}^3$

Mass

$1 \text{ kg} = 10^3 \text{ g}$	$1 \text{ g} = 10^{-3} \text{ kg}$
$1 \text{ kg} = 2.205 \text{ lb}_\text{m}$	$1 \text{ lb}_\text{m} = 0.454 \text{ kg}$

Density

$1 \text{ g/cm}^3 = 1 \text{ Mg/m}^3$	$1 \text{ Mg/m}^3 = 1 \text{ g/cm}^3$
$1 \text{ g/cm}^3 = 62.4 \text{ lb}_\text{m}/\text{ft}^3$	$1 \text{ lb}_\text{m}/\text{ft}^3 = 1.602 \times 10^{-2} \text{ g/cm}^3$

(continues)

793

Force

1 N = 0.1020 kg$_f$	1 kg$_f$ = 9.808 N
1 N = 0.2248 lb$_f$	1 lb$_f$ = 4.448 N
1 N = 10^5 dynes	1 dyne = 10^{-5} N

Stress

1 Pa = 1 N/m^2	1 N/m^2 = 1 Pa
1 MPa = 145 psi	1 psi = 6.90×10^{-3} MPa
1 kg/mm^2 = 1422 psi	1 psi = 7.03×10^{-4} kg/mm^2
1 Pa = 10 dynes/cm^2	1 dyne/cm^2 = 0.10 Pa

Fracture Toughness

1 MPa-m$^{1/2}$ = 910 psi-in.$^{1/2}$ 1 psi-in.$^{1/2}$ = 1.10×10^{-3} MPa-m$^{1/2}$

Energy

1 J = 0.239 cal	1 cal = 4.184 J	1 eV/atom = 23,060 cal/mol
1 J = 10^7 ergs	1 erg = 10^{-7} J	1 J = 0.738 ft-lb$_f$
1 J = 6.24×10^{18} eV	1 eV = 1.602×10^{-19} J	1 J = 1 V-A
1 J = 9.48×10^{-4} Btu	1 Btu = 1054 J	

1 kWh = 3413 BTU

Power

1 W = 1 J/s = 0.239 cal/s	1 Btu/h = 0.293 W
1 W = 1 V-A/s	1 hp = 746 W
1 W = 3.414 Btu/h	1 hp = 550 ft-lb$_f$/s

Temperature

T(K) = 273 + T(°C)	T(°C) = T(K) − 273
T(°C) = 5/9[T(°F) − 32]	T(°F) = 9/5[T(°C)] + 32

Viscosity

1 Pa-s = 10 P   1 P = 0.1 Pa-s

Magnetic Induction (flux density)

1 T (Tesla) = 1 Wb/m^2 = 10^4 G   1 G = 10^{-4} T

Magnetic Field

1 A-turn/m = $4\pi \times 10^{-3}$ Oe   1 Oe = 79.6 A-turn/m

ANSWERS TO SELECTED PROBLEMS

Chapter 1

1-4. ~2.8%

1-5. 173 square miles

1-13. a. 4.95 years

b. 3.5%

1-19. ~124 times

Chapter 2

2-1. a. 4.41×10^{23}

b. 2.0×10^{-7}

2-3. 867 g Ni, 133 g Al

2-5. 1.47×10^{16} photons/s

2-7. a. 4.29×10^{-24} N-s

b. 4.29×10^{-24} N-s

c. 40.8 m/s

2-9. 4906 eV, 0.253 nm

2-11. a. Yes for Cr, no for Mo

b. No for Cu, yes for Cr

2-12. Mo

2-16. 42 at% Cu–58 at% Ti.

2-18. a. $r = r_0 = (2B/A)^{1/6}$

b. $B/A = 1.22 \times 10^{-58}$

c. $-\frac{1}{2}A/r_0^6$

2-20. 352 kJ

2-22. a. LiF

b. $U_0(\text{LiF}) = -572$ kJ, $U_0(\text{NaBr}) = -416$ kJ

2-24. a. $+1 \times 10^{-11}$ N, -0.6×10^{-11} N

b. $+68.7 \times 10^{-11}$ N, -109×10^{-11} N

2-27. a. 1.27×10^{-24} N-s

b. 3.32×10^{-32} N-s

c. 1.27×10^{-25} N-s

2-30. 2.41×10^{-21} J or 0.015 eV

2-31. 1.66×10^{-10} m

Chapter 3

3-1. a. 0.263 nm

b. 0.180 nm

3-3.   0.388

3-5.   0.907, 4.02×10^7 atoms/cm

3-7.   0.0329, or 3.29%

3-11. 1.84 g/cm^3

3-12. 3.45%

3-14. 3.37 g/cm^3

3-16. a. $(\bar{2}12)$

     b. (014)

     c. (101)

3-17. a. $[\bar{1}41]$

     b. [121]

     c. $[1\bar{2}0]$

3-21. b, f, h, and i

3-23. a. (113)

     b. $(3a, 3a, a)$

     c. $[\bar{1}10]$

3-24. a. (632)

     b. $\frac{1}{3}, \frac{2}{3}, 1$

     c. $0.143a$

3-28. 0.2043 nm, 6.07 keV

3-30. $a = {\sim}0.402$ nm, the FCC metal is Al

3-32. 0.018° or 64.8 seconds of arc

3-35. a. 2.98×10^7

     b. 3.01×10^{-5}

     c. 2.5×10^{15}

     d. 3.01×10^{-2}

3-36. a. 4.52

     b. 9.17

Chapter 4

4-1.   −210 kJ/mol

4-4.   1240 mers/molecule

4-5.   a. isoprene, 0.443 mol; butadiene, 0.557 mol

     b. 5.33 g

4-8.   24,660 amu

4-11. $\Delta V_{100}/\Delta V_{200} = 0.091$

4-13. 960 kJ

4-14. −12°C

4-19. a. 74.9 mol% SiO_2, 12.6 mol% Na_2O, 12.5 mol% CaO

     b. $1Na_2O$–$1CaO$–$6SiO_2$

4-22. a. 8.03 wt% Li_2O, 27.4 wt% Al_2O_3, and 64.6 wt% SiO_2

     b. 2.33 wt%

4-24. 7.32 g/cm^3

4-26. a. $N_c = 6$

     b. $N_c = 4$

     c. $N_c = 4$

     d. $N_c = 6$

4-30. 6.38 g/cm^3

4-32. 400 kg cement, 692 kg sand, 998 kg aggregate, 599 kg water

Chapter 5

5-2.   −34.4 kcal/mol, or −144 kJ/mol

5-5. $[P_{H_2O}]/[P_{H_2}] = \{\exp[-\frac{1}{2}(\Delta \hat{G}^{\circ}_{H_2O} - \Delta G^{\circ}_{MO})]\}/RT$

5-7.   10.3 kcal/mol

5-10. ~0.38

5-12. 731 kg

5-18. At 1200°C: L (20 at% As); 100% L

At 1000°C: L (14 at% As), GaAs (50 at% As); 83% L, 17% GaAs

At 200°C: L (0.5 at% As), GaAs (50 at% As); 61% L, 39% GaAs

At 20°C: Ga (0 at% As), GaAs (50 at% As); 60% Ga, 40% GaAs

5-20. At 2800°C: L (50 wt% Re); 100% L

At 2551°C: L (43 wt% Re), β (54 wt% Re); 36% L, 64% β

At 2448°C: α (45 wt% Re), β (54 wt% Re); 44% α, 56% β

At 1000°C: α (38 wt% Re), β (59 wt% Re); 43% α, 57% β

5-26. a. Mullite or γ: 41.8 wt%

     b. Lower

     c. 1840°C

5-28. a. 9.49 g

     b. 8.36 g

     c. 1.64 g

     d. 1.13 g

     e. 0.51 g

5-29. a. γ (2.05 wt% C), Fe_3C (6.7 wt% C); 68.8% γ, 31.2 Fe_3C or 36% γ and 64% Eut

     b. γ (1.25 wt% C), Fe_3C (6.7 wt% C); 59% γ, 41 Fe_3C or 26% γ and 74% Eut

     c. α (0.02 wt% C), Fe_3C (6.7 wt% C); 48% α, 52 Fe_3C

5-32. a. 2.92×10^{-17}

     b. 2×10^{-4}

     c. 0.0002

5-33. a. 120°

     b. 126.9°, 126.9°, and 106.2°

Chapter 6

6-2. 10^{11} atoms/cm²-s

6-4. a. $D_\alpha(C) = 4 \times 10^{-7} \exp(-80.3 \text{ kJ}/RT)$ m²/s
   $D_\gamma(C) = 1.2 \times 10^{-5} \exp(-134 \text{ kJ}/RT)$ m²/s

   b. 1899 K or 1626°C

   c. 0.128

6-6. a. 4.56×10^{-8} cm³ gas (STP)/cm²-s

   b. 149.999 atm

   c. 149.991 atm

6-8. a. 258 kJ/mol

   b. 1.33×10^{-3}

6-10. 1.11×10^{17} cm⁻³

6-12. $D(Mo) = 10^{-3} \exp(-4T_M/T)$ cm²/s

6-14. a. 20%

   b. 1.5%

6-16. a. 189 kJ/mol

   b. 98.9 kJ/mol

   c. Grain boundary diffusion

   d. Surface diffusion

6-18. 8.65×10^{-4} cm

6-20. 45.4 nm

6-23. b. Diffusion limited growth

   c. 2.09×10^4

6-25. a. 2

   b. 9

   c. 4, 3

6-28. a. 6.15×10^{-8} m

   b. 3.17×10^{-15} J/m³

6-30. 3.09×10^{16} nuclei/m³

6-34. 2.52 mm/min

6-36. Volume diffusion

6-38. 68.5 kcal/mol

6-39. 4.46×10^{10} J/mol-m

Chapter 7

7-1. a. 2.00107 m

   b. 3.55 m

7-3. a. 20×10^6 psi

   b. 49,965 psi

   c. 150,000 psi

   d. 62.5 lb-in./in.³

   e. 16,060 lb-in./in.³

   f. 15

7-5. a. 0.30

   b. 0.350

c. 97.8 MPa

   d. ~29.4 ksi

7-8. 3.67 m

7-10. b. 1.33 kJ/kg

   c. 432 kJ/kg

7-12. a. 35.3×10^6 psi

   b. 73.5 ksi

   c. 125 ksi

   d. 46.2

   e. 85

   f. 204 ksi

7-14. 22,900 N

7-19. a. 0.0585 J, 0.0173°C

   b. ~17.3 J/cm³, 5.11°C

7-21. a. 10^{-7}

   b. 4×10^{-4} m/s, 4×10^3 m/s

7-23. 88.3 days

7-25. 1860 seconds

7-28. a. 75 ksi

   b. 0.230 cm

7-30. 75.8 ksi-in.$^{1/2}$

7-34. 0.00179 m

7-36. a. 6%

   b. 12%

   c. Temperature

7-37. $E_c = 79.8$ kJ/mol, $m = 3.91$

7-39. b. 400 MPa

   c. 251,000 cycles

   d. 158,000 cycles

Chapter 8

8-2. 1.91

8-4. 0.621

8-6. diameter = 7.99 cm, height = 16.0 cm

8-9. 180 MPa

8-10. 2.94 MN

8-12. a. 0.08, or 8%

   b. 0.10, or 10%

8-14. 75

8-16. 1.82 MN

8-21. 1090 MPa

8-22. 1.44 MN

8-24. 72°C

8-25. b. 350 kJ/mol

8-30. a. 7.44×10^{10} Pa-s

   b. 1030 K

8-36. $A = 4.40 \times 10^{12}$ (min)$^{-2/5}$, $E = 444$ kJ/mol

Chapter 9

9-7. c. $\sigma_{UTS} = 120$ ksi, $E_{IS} = 50$ ft-lb

9-14. c is reduced from 3.03×10^{-3} to 3.36×10^{-4} m

9-15. $\sigma_0 = 900$ ksi, $K_{1C} = 67.5$ MPa-m$^{1/2}$

9-19. 740 kg

9-23. 575 K

9-25. a. 130 GPa
       b. 165 MPa

9-27. a. 32.0 GPa
       b. 4.40 GPa

9-30. 0.0716

9-31. Fiber, 72.1 GPa; matrix, 3.1 GPa

9-33. a. 002 m
       b. 0.799

9-35. a. 33.9 GPa-m^3/Mg
       b. 0.11

9-36. 5

9-39. a. 0.99 cm
       b. 3.31 cm
       c. −4.55 MPa

9-41. In order of increasing thermal shock resistance: ZrO_2, TiN, Si_3N_4, BeO, SiC

9-44. 87.9 g ZrO_2, 12.1 g Y_2O_3

Chapter 10

10-2. −2.075 V

10-5. 0.504

10-6. 2.48 hours

10-8. a. 8.03×10^{-4} A/cm^2
       b. 6.31×10^{-4} A

10-9. −0.509 V, 8.13×10^{-7} A/cm^2

10-11. a. Zn
       b. Fe
       c. Mg
       d. Cd

10-17. a. 2.36×10^{19} ions
       b. 5.23×10^{-3} cm

10-19. PBR Al_2O_3/Al = 1.28; PBR AlN/Al = 1.26; PBR TiO_2/Ti = 1.76; PBR TiN/Ti = 0.739

10-22. 2.05 kg

10-23. 612 m

10-27. 1301 K or 1028°C

10-29. 82.5 K

10-31. Yes, because the thermal stress of 260 MPa exceeds half the yield stress

10-33. $\Delta\sigma = 22,800$ psi

10-35. 1.83×10^8

Chapter 11

11-3. 58.5 K

11-5. W–Th, $\lambda = 0.48$ μm; W–Ba, $\lambda = 0.79$ μm; W–Cs, $\lambda = 0.91$ μm

11-6. a. 0.487 cm^2
       b. 3.55×10^{-3} cm^2

11-8. 2.61×10^{13}

11-10. a. 0.429 A
       b. 51.5
       c. 994°C
       d. 1508 Ω

11-12. $1/R_{total} = L(V_A/\rho_A + V_B/\rho_B)$
       $R_{total} = (1/L)(V_A\rho_A + V_B\rho_B)$

11-15. Wt (Cu)/Wt (Al) = 2.01

11-17. a. 288
       b. 45.4

11-21. b. 18.2 T
       c. 6.22 K

11-23. a. 3.72×10^{-7} F
       b. 3.72×10^{-8} C/m^2
       c. 2.83×10^{-8} C/m^2

11-27. 5.86

11-28. 1.22 cm

11-30. 0.0312 A

11-32. 62.4 nm/s

11-34. 10.9 N, spark igniter

Chapter 12

12-1. ~326,000 Ω-cm

12-3. 3.33×10^6

12-6. a. 0.763
       b. 0.0449

12-8. 5.51×10^{-12} s; frequency response is ~1.81×10^{11} Hz

12-9. a. 7.18×10^{18}/cm^3
       b. 3.59×10^{-5}

12-12. μ(Si) > μ(Cu)

12-14. -8.59×10^{-6} A

12-18. a. 1.20 at%
       b. 0.462 wt%
       c. 43,200 $(\Omega\text{-cm})^{-1}$

12-20. a. $p = 1.76 \times 10^{15}$/cm^3, $n = 1.19 \times 10^5$/cm^3
       b. $p = 6 \times 10^{16}$/cm^3, $n = 3.5 \times 10^3$/cm^3

c. 7.89 Ω-cm

d. 0.232 Ω-cm

12-22. 0.077 m^2/V-s

12-25. $C_L/C_0 = (1 - f)^{k_e - 1}$

12-27. a. Misfit $= -1.43 \times 10^{-3}$; AlAs in compression, GaAs in tension

b. Misfit $= 6.34 \times 10^{-3}$; InP in compression, CdS in tension

c. Misfit $= -2.82 \times 10^{-3}$; ZnSe in compression, GaAs in tension

Chapter 13

13-2. a. 2.03×10^8 m/s

b. $42.5°$

c. No

13-4. 5.19×10^{-3} cm

13-6. For Cu: at 0.50 μm, $R = 0.625$; at 0.95 μm, $R = 0.987$

For Au: at 0.50 μm, $R = 0.504$; at 0.95 μm, $R = 0.980$

13-9. 0.36

13-11. At 0.45 μm, $R = 0.0193$

At 0.60 μm, $R = 0.00553$

At 0.45 μm, $R = 0.0168$

13-13. a. 4.83×10^{-3} s

b. 176

13-16. a. $12:1$

b. $33.3:1$

13-17. For Si, $\lambda = 1.11$ μm; for GaP, $\lambda = 0.555$ μm; for ZnSe, $\lambda = 0.481$ μm

13-20. 0.904

13-22. Maximum cell power $= \frac{1}{4} I_{sc} V_{sc}$, fill factor $= \frac{1}{4}$

13-24. a. Ga, 56.8 wt%; As, 30.5 wt%; P, 12.6 wt%

b. ~2.2 eV

13-26. a. 6.25×10^{-6} W

b. 5×10^{-3} W

13-27. 0.0786 W

13-31. 0.126

Chapter 14

14-1. a. 3000

b. 16.5 V

14-3. 2.07×10^{-6}

14-9. a. 800 kJ/m^3

b. 200 kJ/m^3

14-10. a. 800 kJ/m^3

b. 50 kJ/m^3

14-14. b. 0.5×10^5 kJ/m^3

c. $H = 10^5$ A/m

14-17. a. 2.64×10^{-7} °C

b. 4.17×10^5 hours

14-19. a. 2.52×10^5 A/m

b. 1.26×10^5 A/m

14-23. 12 A

14-24. ~16.6 T-kA/m

Chapter 15

15-1. 2×10^6 devices

15-5. $h(t) = m/t(t/c)^m$

15-9. a. Weibull

b. 2.52×10^{-3}

c. 2.11×10^{-6}/hour, 2.72×10^{-3}/hour

15-12. a. 24

15-14. 6.22×10^5 hours

15-18. 529 hours

15-19. Reduce n

15-23. 547 nm

15-24. b. 12.1

c. 49.9

d. 11,900

15-27. 69% decrease

15-32. a. 0.65 eV

b. 4.88%

INDEX

DOCUMENTATION FOR *ENGINEERING MATERIALS SCIENCE* COMPUTER MODULES*

Description

The enclosed software was developed to accompany the book, *Engineering Materials Science,* by M. Ohring, Academic Press (1995). It consists of a single diskette (3.5 in.) containing nine separate modules that illustrate important aspects of materials science. The software runs on IBM compatible computers; there is no Macintosh version. Individual modules are entitled:

F1. Galvanic Cells
F2. Free Energy of Oxides
F3. Semiconductor Diode
F4. Tensile Behavior of Materials
F5. Atomic Electron Transitions
F6. Structure
F7. Phase Diagrams
F8. Diffraction
F9. Diffusion

The software is **interactive** and, for the most part, is **computationally** oriented. Using it, many problems dealing with different facets of the subject can be solved with a few keystrokes. Book chapters providing background for particular modules are noted.

Installation and Running the Software

To install *Engineering Materials Science* onto the hard drive

1. Place diskette into drive A
2. Type A: then ↵
3. Type CD\ then ↵
4. Type INSTALL then ↵

To run the program, when C:\MATERIAL⟩ appears, type START.
Note: The software can also be installed in WINDOWS by typing C:\MATERIAL\START on the Command Line of Program Manager.

Software Modules

F1 Galvanic Cells (Chapter 10)

In this module the electrochemical cell potential is displayed for two user-selected half cells. For each half cell the metal and the electrolyte concentration must be entered. A database of standard electrode potentials at 25°C can be accessed for this purpose. The overall cell potential that is displayed upon

---

* Conventions and symbols appearing in the software do not always match those used in the text.

pressing F5 is

$$E_{cell} = E_{cell}^{\circ} + 0.059/n \cdot \log[(A^{n+})/(C^{n+})] \text{ V,} \qquad (10\text{-}9)$$

where

$$E_{cell}^{\circ} = E_{anode}^{\circ} - E_{cathode}^{\circ},$$

and A^{n+}, C^{n+} are the respective anode and cathode electrolyte concentrations.

F2 Free Energy of Oxides (Chapter 5)

The free energy of metal oxides are tabulated in a database as

$$M + O_2 = MO_2; \qquad \Delta G_{MO_2}^{\circ} = \Delta H^{\circ} - T \Delta S^{\circ} \qquad (5\text{-}3)$$

using thermodynamic data. By selecting a single metal, ΔG° vs. T is displayed. When two metals $M(1)$ and $M(2)$ are sequentially selected, ΔG° for the chemical reaction $M(1) + M(2)O = M(1)O + M(2)$ is calculated and displayed as ΔG° vs. T. If ΔG° is negative, the reaction is favored as written, whereas if ΔG° is positive, the reaction cannot occur.

F3 Semiconductor Diode (Chapter 12)

In this module the user creates his or her semiconductor diode and observes its behavior when a bias voltage is applied. On the first screen the doping concentrations for n- and p-type silicon are entered. (Doping levels selected should lie within the specified limits suggested.) The Fermi energy levels are automatically calculated and displayed on the respective energy band diagrams. By pressing F5, the p and n semiconductors are brought into contact and a diode is created. Equilibration of Fermi levels causes band bending in the vicinity of the junction. Both positive and negative bias voltages can be then applied by the user after pressing F2. The resulting current–voltage characteristics of the diode are schematically displayed. Simultaneously, the relative shifting of the conduction and valence band edges is shown. Lastly, the motion of electrons and holes in the diode is simulated through animation by pressing F3. Under forward bias the carriers recombine at the junction. When the junction is reverse biased, carriers are drawn away to create a depleted space charge region. Note: this simulation does not reflect the doping levels initially selected.

F4 Tensile Behavior of Materials (Chapter 7)

The tensile behavior of a material from the elastic range to fracture is approximately determined by its Young's modulus, yield stress, ultimate tensile stress (UTS), % strain to UTS, and % elongation to failure. In this module the user either chooses a material from a database or creates one by entering values for the indicated mechanical properties. Once selected, a possible tensile stress–strain curve is displayed. The extent of work hardening can be varied to change the shape of this curve. Lastly, the stretching and failure of a tensile bar is simulated by animation. By pressing F4 the elastic portion of the stress–strain curve is magnified.

F5 Atomic Electron Transitions (Chapter 2)

When electrons in an atom fall from occupied levels (E_2) to unoccupied levels of lower energy (E_1), the energy difference is often manifested in emitted photons. The photon energy is thus

$$E_2 - E_1 = \Delta E = hc/\lambda \qquad (2\text{-}6)$$

or

$$\Delta E \text{ (eV)} = 1.24/\lambda \text{ (}\mu\text{m)}. \qquad (2\text{-}7)$$

In this module many of the core electron energy levels for a selected group of elements are stored in a database. The user selects one of these elements and the ground state (vacant K, L, ...) level. As indicated by the animation, an impinging electron or photon knocks out an electron creating a vacant level in the process. As a more energetic, outer level electron falls into the vacant level, a photon, usually an X ray, is emitted. The energies and wavelengths of each possible electron transition for the atom in question can then be sequentially accessed.

F6 Structure (Chapter 3)

This module enables directions and planes with arbitrary Miller indices to be visualized in cubic crystal systems. Directions lying in planes and perpendicular to planes are also visually displayed. Interactive exercises are provided to test the user's ability to identify the indices of planes and directions. All that is required is to enter three positive or negative integers.

F7 Phase Diagrams (Chapter 5)

This module displays a number of binary phase diagrams illustrating solid solution, eutectic, peritectic, and eutectoid cooling behavior and enables complete equilibrium phase analyses to be performed by use of tie lines and the lever rule. First, one of the phase diagrams indicated, i.e., Pb–Sn, Cu–Ni, Ge–Si, Pt–Re, As–Ga, or Fe–Fe$_3$C, is selected. Next, the desired composition and temperature state on the phase diagram is pinpointed with either the use of horizontal (composition) and vertical (temperature) arrow keys, or by means of a mouse. The computer automatically displays the initial composition and temperature, as well as the phases present, and the chemical composition and relative amounts of these phases.

F8 Diffraction (Chapter 3)

This module displays schematic X-ray powder diffractometer traces of body-centered and face-centered cubic metals. Diffracting planes can be indexed and interplanar spacings identified. How this is done is shown in the tutorial (also see Example 3-6). A calculation aid allows selection of both X-ray sources and elemental specimens by pressing the vertical arrow keys. Diffraction patterns are then displayed. Exercises are provided to enable users to determine lattice parameters and structures of unknown cubic metals from diffracton patterns.

F9 Diffusion (Chapter 6)

A number of kinetic phenomena are quantitatively modeled in this module.

1. In the case of the **continuous source** the defining equation governing diffusion is

$$[C(x,t) - C_0]/[C_s - C_0] = \text{Erfc} \, [x/(4Dt)^{1/2}]. \tag{6-3}$$

Note that the initial concentration C_0 has been chosen to be zero in this program. The user enters any three of the following quantities: t (s), x (cm), D (cm²/s), and $C(x,t)/C_0$, and the fourth quantity is automatically calculated. Concentration profiles for three different times are calculated and displayed. (Note: In selecting values of variables to enter into the equation, $x = (2Dt)^{1/2}$ should be approximately satisfied. For example, by entering t (s) = 3600, D (cm²/s) = 1E −8, $C(x,t)/C_0$ = 0.5, the diffusion distance x is calculated to be 5.723×10^{-3} cm.)

2. The Gaussian solution describes the case where a **finite surface source,** Q (in units of atoms/cm²) is present. (Note that in the textbook, S rather than Q is used.) The solution to the diffusion equation is

$$C(x,t) = \{Q/(\pi Dt)^{1/2}\} \exp - (x^2/4Dt). \tag{6-5}$$

Through user selection of any four of the following quantities: t (s), x (cm), D (cm²/s), Q, and $C(x,t)$, the fifth quantity is automatically calculated. A Gaussian concentration profile is calculated and displayed. For example, selecting Q = 1E14/cm², t = 3600, D = 1E −11, and x = 1E −3, yields C = 2.866E14/cm³. Again, when selecting values of variables to enter, the equation $x = (2Dt)^{1/2}$ should be approximately satisfied.

As a computational aid, values of Erfc z and $\exp -z^2$ are displayed for any inputted value of the argument z.

3. The rates of chemical reactions are given by the well-known Arrhenius relation, Rate $\sim \exp - Q/RT$. There are several types of problems that can be addressed by this Arrhenius calculator program. By entering two rates at two different temperatures the activation energy Q can be calculated. Similarly, by entering the rate at one temperature, the rate can be calculated at a second temperature.